Advances in Remediation of Contaminated Sites: 2nd Edition

Advances in Remediation of Contaminated Sites: 2nd Edition

Guest Editors

Guining Lu
Zenghui Diao
Yaoyu Zhou
Kaibo Huang

MDPI

Basel • Beijing • Wuhan • Barcelona • Belgrade • Novi Sad • Cluj • Manchester

Guest Editors

Guining Lu
School of Environment
and Energy
South China University
of Technology
Guangzhou
China

Zenghui Diao
College of Resources
and Environment
Zhongkai University of
Agriculture and Engineering
Guangzhou
China

Yaoyu Zhou
College of Resources
and Environment
Hunan Agricultural
University
Changsha
China

Kaibo Huang
School of Environmental
Science and Engineering
Hainan University
Haikou
China

Editorial Office
MDPI AG
Grosspeteranlage 5
4052 Basel, Switzerland

This is a reprint of the Special Issue, published open access by the journal *Processes* (ISSN 2227-9717), freely accessible at: https://www.mdpi.com/journal/processes/special_issues/2NNXMV3017.

For citation purposes, cite each article independently as indicated on the article page online and as indicated below:

Lastname, A.A.; Lastname, B.B. Article Title. *Journal Name* **Year**, *Volume Number*, Page Range.

ISBN 978-3-7258-6163-7 (Hbk)
ISBN 978-3-7258-6164-4 (PDF)
https://doi.org/10.3390/books978-3-7258-6164-4

Contents

About the Editors

Guining Lu

Guining Lu is a Professor at South China University of Technology, Guangzhou. He is a PhD jointly trained by South China University of Technology and Rutgers University. He special focusses on soil/water pollution control and remediation technology, and structure-activity relationship of environmental organic pollutants. He is interested in pollution control, soil remediation, e-waste, mining area, heavy metals, persistent organic pollutants, and polybrominated diphenyl ethers.

Zenghui Diao

Zenghui Diao is a Professor at College of Resources and Environment, Zhongkai University of Agriculture and Engineering, China. He received his Ph.D. degree in environmental engineering from Sun Yat-sen University in 2013. He worked in Chinese Academy of Sciences, China four years from 2014 to 2018, and joined Zhongkai University of Agriculture and Engineering as a professor since June 2018. His research interests mainly focus on environmental remediation and plant protection, water treatment, and soil remediation.

Yaoyu Zhou

Yaoyu Zhou is a Professor and Doctoral Supervisor at Hunan Agricultural University, Changsha, serving as Dean of the College of Environment and Ecology. Head of the Ph.D. program in Ecology. Principal Investigator of the "14th Five-Year Plan" National Key R&D Program—Young Scientists Project; Leading Talent in Scientific and Technological Innovation of Hunan Province. Ph.D. in Environmental Science and Engineering, Hunan University; postdoctoral and visiting research at The Hong Kong Polytechnic University and Tsinghua University. Research focuses on environmental pollution control and ecological restoration. As first or corresponding author, published 90+ papers in international top journals. with an H-index of 60 and over 26,000 citations. Recognized in the World's Top 2% Scientists (2020–2024) and the Lifetime Scientific Impact list (2022–2024).

Kaibo Huang

Kaibo Huang is an Associate professor at Hainan University, Haikou. He is a PhD jointly trained by South China University of Technology and Purdue university. He mainly focusses on environmental fate of contaminants, with a particular interest in tropical offshore. He has expertise in non-targeted screening, soil washing, photo-transformation, surfactant, e-waste, polychlorinated biphenyls, bi-metal, emerging contaminants.

processes

Editorial

Editorial for the Special Issue "Advances in Remediation of Contaminated Sites: 2nd Edition"

Kaibo Huang [1], Xiaodong Du [2], Zenghui Diao [3], Yaoyu Zhou [4] and Guining Lu [2,5,*]

1 Key Laboratory of Agro-Forestry Environmental Processes and Ecological Regulation of Hainan Province, School of Environmental Science and Engineering, Hainan University, Haikou 570228, China; huangkaibo@hainanu.edu.cn
2 School of Environment and Energy, South China University of Technology, Guangzhou 510006, China; dxdon@scut.edu.cn
3 College of Resources and Environment, Zhongkai University of Agriculture and Engineering, Guangzhou 510225, China; zenghuid86@163.com
4 College of Resources and Environment, Hunan Agricultural University, Changsha 410128, China; zhouyy@hunau.edu.cn
5 The Key Lab of Pollution Control and Ecosystem Restoration in Industry Clusters, Ministry of Education, South China University of Technology, Guangzhou 510006, China
* Correspondence: lutao@scut.edu.cn; Tel.: +86-20-3938-0569

With the acceleration of industrialization and urbanization, anthropogenic activities such as mineral exploitation, industrial production, and agricultural cultivation have led to increasingly severe contamination of soil, groundwater, and sediment environments [1–3]. Contaminated sites not only pose threats to ecological system stability—such as disrupting soil microbial communities and inhibiting plant growth—but also endanger human health through pathways such as food chain accumulation and groundwater ingestion [4–6]. As non-renewable land resources become increasingly scarce, developing efficient, eco-friendly, and cost-effective contaminated site remediation technologies has become a core task to achieve global Sustainable Development Goals [7–9].

Remediation of contaminated sites is a complex interdisciplinary field integrating environmental chemistry, soil science, microbiology, and ecological engineering [10–12]. Over the past decade, significant progress has been made in remediation technologies, from traditional physical–chemical methods to emerging biological and intelligent remediation strategies [13,14]. However, challenges remain, including low remediation efficiency for complex combined pollution, high costs of large-scale engineering applications, and unclear long-term ecological risks of remediation technologies [15,16]. Addressing these bottlenecks requires continuous innovation in theoretical research and technological development.

The following Special Issue "Advances in Remediation of Contaminated Sites: 2nd Edition" focuses on the aforementioned core issues and includes 24 high-quality research works, systematically structured around four dimensions: *pollution characterization and risk assessment*, *environmental fate of pollutants*, *remediation technology*, and *review articles and bibliometric analysis*. It forms a complete research chain encompassing "baseline characterization–mechanism analysis–technological innovation–trend guidance". In terms of research proportion, remediation technology (46%) focuses on breakthroughs in materials and processes; environmental fate of pollutants (21%) analyzes the behavioral laws of pollution; pollution characterization and risk assessment (21%) lays the foundation for remediation decision-making; and review articles and bibliometric analysis (12%) guide the development direction of the field.

 https://doi.org/10.3390/pr13123792

1. Pollution Characterization and Risk Assessment

Pollution characterization and risk assessment serve as the prerequisite for remediation decision-making. Through three-dimensional analysis of "concentration distribution–source identification–risk classification", they provide accurate baselines for technology selection [17,18]. This Special Issue covers typical scenarios such as industrial areas, agricultural lands, drinking water sources, and tailings ponds, integrating traditional sampling and emerging detection technologies to construct a multi-dimensional investigation system.

Targeting the pollution characteristics of different sites, research teams have accurately characterized pollutant distribution and sources. In industrial areas, Chen et al. focused on Baotou, a major industrial city in China. By utilizing the Geo-accumulation Index (Igeo) and Potential Ecological Risk Index (PERI), they found that the concentrations of Hg, Pb, Zn, and Cu in the soil of Kundulun District (dominated by iron and steel and power industries) exceeded the standard significantly, with a PERI of 386—2.3 times that of Qingshan District (dominated by machinery manufacturing). This finding confirms the decisive role of industrial structure in pollution distribution. In agricultural areas, Wang et al. focused on the "soil–rice" system in paddy fields of Guangxi province in China. The results of field experiments revealed that single lime treatment could reduce Cd content in rice by 73.15%; however, flooding would increase As accumulation by 28%. In addition, the synergistic treatment of "flooding + foliar silicon fertilizer" could simultaneously control pollution, reducing Cd content in rice to 0.18 mg/kg (below the national standard of 0.2 mg/kg), providing a promotable scheme for safe agricultural production in paddy fields. In drinking water source scenarios, Wei et al. focused on groundwater in Zhanjiang city in China and confirmed through δD–$\delta^{18}O$ isotope analysis that atmospheric precipitation represents the main recharge source. Additionally, the Fe and Mn concentrations in the middle-deep confined aquifer exceeded Class III standards due to the presence of manganese–iron ore deposits, providing a scientific basis for the hierarchical protection of drinking water sources.

For special scenarios such as tailings ponds and riparian environments, research teams have assessed pollution release risks under extreme conditions. Shu et al. focused on low-sulfur lead–zinc tailings in the karst area of northern Guangxi province in China and challenged the traditional notion that "high acid-neutralizing capacity equals low risk" through batch leaching experiments. Although the acid-neutralizing capacity of the tailings reached 166.57 kg H_2SO_4/t, under acid rain conditions (pH = 2–3), the leaching concentrations of Pb, Zn, and Cd were 5.2 times higher than those under neutral conditions. Moreover, due to the well-developed pore structure of coarse-grained tailings, their leaching amount was 18% higher than that of fine-grained tailings, serving as a warning for the prevention and control of acid rain in tailings ponds.

Emerging detection technologies have significantly improved the efficiency of pollution investigation. Zhang et al. focused on a metallurgical slag site in Gejiu, Yunnan province, China, integrating visible–near-infrared (Vis-NIR) and X-ray fluorescence (XRF) spectroscopy and using Competitive Adaptive Reweighted Sampling (CARS) to screen characteristic variables. The constructed quantitative model for soil Cd content achieved a coefficient of determination (R^2) of 0.9505, with calculation time shortened by 68% compared to single XRF technology, enabling on-site rapid and non-destructive monitoring. Additionally, Li et al. provided a supplementary basis for the quantitative analysis of pollution concentration through the study of the adsorption mechanisms of modified biochar on soil Cd.

2. Environmental Fate of Pollutants

The environmental fate of pollutants is crucial for understanding pollution diffusion paths and optimizing remediation strategies [19,20]. This Special Issue explores the behavioral mechanisms of pollutants in the "soil–plant–groundwater" system from three dimensions: heavy metal migration, organic pollutant degradation, and the interaction effects of compound pollution.

The speciation of heavy metals directly determines their bioavailability, and research teams have deeply analyzed the key factors affecting speciation transformation. Qin et al. compared Miscanthus floridulus from the Dabaoshan mining area in Guangdong province in China and the non-mining area in Boluo County, Huizhou city, China, and found that under 100 mg/L Cu stress, the chlorophyll reduction rate of the mining ecotype (18%) was only 43% of that of the non-mining ecotype (42%), and the ascorbic acid content was 2.1 times that of the latter. This finding confirms that plants in mining areas develop Cu tolerance by enhancing their antioxidant systems. The authors further studied the effects of Pb stress on the photosynthetic system and found that under high Pb concentration (240 mg/L), the maximum photochemical efficiency (Fv/Fm) of Miscanthus floridulus from the mining area remained 0.78 (compared to only 0.62 for the non-mining area), revealing the protection mechanism of its photosynthetic apparatus. At the mineral interface level, Li et al. found that the adsorption capacity of rod-like lepidocrocite (exposing the (001) crystal plane) for Cr(VI) (28.6 mg/g) was 1.8 times that of flake-like lepidocrocite (exposing the (010) crystal plane), by using in situ ATR-FTIR and two-dimensional correlation spectroscopy (2D-COS). Additionally, L-aspartic acid exhibited a significantly higher inhibition rate (45%) on adsorption by the (001) crystal plane than by the (010) crystal plane (22%), providing a molecular mechanism for the targeted design of Cr(VI) adsorption materials.

For organic pollutants such as microplastics and persistent organic pollutants, research teams have analyzed their degradation and aging laws. Wang et al. compared the changes in polylactic acid (PLA) and polyethylene (PE) under UV–water coupling. Due to the hydrolysis and cleavage of ester bonds, PLA formed 5–10 μm pores on its surface, and its particle size decreased from 1 mm to 0.3 mm; in comparison, PE only underwent surface oxidation, with the carbonyl index (CI) increasing from 0.02 to 0.06 and no significant change in particle size. Through Py-GC-MS, the characteristic degradation products of the two materials were identified (lactic acid and acetic acid for PLA and 2-hexanone for PE), providing a basis for the environmental risk assessment of microplastics.

For scenarios with coexisting multiple pollutants, research teams have analyzed the impact of interaction effects on pollution fate. Li et al. focused on in situ uranium leaching, addressing the issue whereby natural groundwater inflow dilutes leachate and reduces mining efficiency. They establish a "groundwater flow ratio" quantitative model, taking Inner Mongolia's C1 and C2 uranium mines as study objects. By analyzing how injection volume and filter parameters affect the flow ratio, the authors propose adjusting filter depth (to confined aquifers) and injection mode (intermittent injection) to reduce natural groundwater inflow by 30–40%, providing a quantitative basis for leaching process optimization.

3. Remediation Technology

Remediation technology is the core to solving pollution problems [16,21]. This Special Issue has achieved breakthroughs in four directions: functional materials, bioremediation, chemical and electrochemical remediation, and intelligent optimization. It emphasizes the characteristics of "greenization–accuracy–low cost" and covers the remediation needs of heavy metals, organic pollutants, and radioactive pollution.

Research teams have focused on functional carriers such as biochar, micro-electrolytic materials, and nanocomposites to improve pollution removal efficiency. In the field of heavy metal remediation, Li et al. used tobacco straw as raw material and prepared modified biochar (HAP-TSB550) through co-pyrolysis with hydroxyapatite (HAP). The specific surface area increased from 3.63 m^2/g to 18.36 m^2/g, and the maximum adsorption capacity for Cd(II) reached 14.50 mg/g (9.24 times that of unmodified biochar). Through XPS analysis, the synergistic mechanism of "Ca^{2+} exchange (42%) + P-O complexation (38%) + electrostatic adsorption (20%)" was confirmed. Zhang et al. explored the resource utilization of agricultural waste and found that citric acid-modified spent Agaricus bisporus (SCAB) achieved a Zn(II) removal rate of 75.9% (twice that of raw waste) and maintained 58% selectivity even in the presence of coexisting Cu^{2+} and Pb^{2+}, providing a scheme for low-cost heavy metal wastewater treatment. In the field of radioactive pollution remediation, Gong et al. prepared nanoflower-like MgAl-LDO/konjac gum (LDO-KGM) through freeze-drying and calcination. The maximum adsorption capacity for U(VI) reached 3019.56 mg/g (2.3 times that of pure MgAl-LDO), and the efficiency remained over 84.5% after five cycles, providing a green material for nuclear wastewater treatment. In the field of organic pollution remediation, Dai et al. prepared spherical Fe-C micro-electrolytic materials, achieving a removal capacity of 105.48 mg/g for crystal violet through the synergistic effect of "adsorption (35%) + micro-electrolytic degradation (65%)".

Bioremediation technology has attracted considerable attention due to its environmental friendliness and low cost. This issue focuses on technology optimization and pilot-scale verification. In the field of phytoremediation, Wang et al. targeted polymetallic mining areas in Yunnan province in China and screened native accumulator plants such as *Juncus effusus* (BCF = 1.8) and *Phragmites capitata* (BCF = 1.5) through the Bioaccumulation Factor (BCF), providing a basis for plant selection in mining area remediation. In the field of microbial remediation, Chen et al. proposed zero-valent iron (ZVI)-enhanced microbial technology for the combined ammonia nitrogen (NH_4^+-N) and nitrate nitrogen (NO_3^--N) pollution in acidic groundwater (pH = 3.8). ZVI increased the pH to 6.2 by slowly releasing Fe^{2+}, shortening the denitrification cycle from 15 days to 10 days. Pilot-scale experiments further confirmed that the combined process of alkaline reagent injection–intermittent aeration–ZVI–microbial agent addition achieved over 80% removal rates for both NH_4^+-N and NO_3^--N, breaking the application limitation of microbial remediation in acidic environments. In the field of constructed wetlands, Song et al. monitored the 4–5 year operation data of the Hongze constructed wetland in Jiangsu Province, China, and found that the COD removal rate increased from 7.6% to 15.14%, while the NH_4^+-N removal rate decreased from 78.33% to 46.04%. Additionally, the abundance of ammonia-oxidizing bacteria (AOB) in the aeration tank was 5.3 times that of anaerobic ammonia-oxidizing bacteria, confirming that nitrification dominates NH_4^+-N removal. It is recommended to replace wetland plants every 3 years to maintain efficiency.

For refractory pollutants such as persistent organic pollutants (POPs) and heavy metals, research teams have developed chemical and electrochemical remediation technologies. Liang et al. prepared Pd/metal foam electrodes for the degradation of 2,2′,4,4′-tetrabromodiphenyl ether (BDE-47) and found that the 15.16% Pd/Ni electrode achieved a debromination efficiency of 94.3%. Through in situ Raman spectroscopy, the mechanism of "stepwise debromination dominated by hydrogen atom transfer" was confirmed, providing a direction for the selection of electrochemical remediation materials. Tang et al. provided theoretical support for advanced oxidation technology, using a Support Vector Machine (SVM) combined with Morgan molecular fingerprints to predict the reaction rate constants of water pollutants with sulfate radicals, achieving an R^2 of 0.89 and a low MAE of 0.12 for a test set. Through SHAP analysis, 10 key features were identified, such as "number of hydroxyl groups" and "aromatic ring structure", providing a tool for parameter optimization of advanced oxidation processes (SR-AOPs).

Machine learning and numerical simulation have promoted the transformation of remediation decision-making from "empirical" to "data-driven". Wang et al. constructed a computational framework integrating process models and random forest regression. The authors simulated 120 remediation scenarios in an arsenic-contaminated site, trained the model to establish the correlation between "excavation depth–chemical dosage–cost-removal rate", and finally screened out the optimal strategy of "1.5 m partial excavation + 2% biochar addition + natural attenuation", reducing costs by 35% compared to the traditional full excavation scheme. Zhang et al. further applied this method to an arsenic-polycyclic aromatic hydrocarbon compound-contaminated site of a retired shipyard in Guangzhou, verifying its applicability in complex pollution scenarios and providing technical support for the accurate optimization of remediation schemes.

4. Review Articles and Bibliometric Analysis

Review articles and bibliometric analysis provide trend guidance for the development of the field through systematic sorting and data mining. This Special Issue includes two types of studies: technical reviews and bibliometric analysis.

Alasmary et al. systematically reviewed the remediation technologies for Pb pollution in military shooting ranges, pointing out that bullet weathering is the main pollution source. When soil pH < 5.5, the proportion of exchangeable Pb increases to 15%, significantly enhancing its bioavailability. Among existing technologies, chemical stabilization (phosphate addition) has the lowest cost (200 RMB/mu); however, its long-term stability requires verification. It is recommended to develop "stabilization + *Leymus chinensis-rhizobium* combined bioremediation" composite technology in the future. Liu et al. focused on forest ecosystems and reviewed the mechanisms of microbial degradation of organic pollutants: acidic soil (pH = 4.5–5.5) is conducive to PAH degradation, and fungi (such as white-rot fungi) contribute 60% of the degradation amount. Genetically engineered *Pseudomonas* can improve the degradation efficiency of petroleum hydrocarbons by 40%; however, attention should be paid to the impact of exogenous bacteria on the native ecology, providing a technical pathway for pollution remediation in forest ecosystems.

Meng et al. analyzed the research trends of soil minerals and carbon stability using 1834 articles from the Web of Science (2013–2023) with tools such as CiteSpace (version 6.3.R1) and VOSviewer (version 1.6.20). The number of publications in this field increased at an average annual rate of 12.3%, with China (32%) and the United States (25%) leading the field. However, Sino–US cooperative papers only accounted for 8%. Research hotspots focused on "clay mineral–organic carbon adsorption" and "iron oxide catalytic transformation", with *Nature Communications* as the core journal. It is recommended to strengthen the research on the coupling mechanism of "climate change–minerals–carbon cycle" in the future to promote the synergy between carbon sequestration and pollution remediation.

5. Conclusions and Future Prospects

The Second Issue of "Advances in Remediation of Contaminated Sites" showcases cutting-edge research across pollution characterization, environmental fate, and remediation technology, reflecting the interdisciplinary and practical nature of the field. Key achievements include (1) the application of multi-source spectral fusion and machine learning for rapid, accurate pollution detection; (2) the revelation of adaptive mechanisms of plants and microbes in contaminated environments; (3) the development of high-performance functional materials (e.g., modified biochar and nanoflower-like oxides) for pollution control; and (4) the integration of intelligent optimization into remediation strategy design, improving cost-effectiveness.

However, several challenges remain: (1) remediation of complex combined pollution (e.g., heavy metal–organic pollutant co-contamination) requires more efficient integrated technologies; (2) long-term ecological risks of remediation technologies (e.g., secondary pollution from nanomaterials) require further evaluation; (3) scaling up lab-scale technologies to engineering applications requires addressing cost and stability issues. Future research should focus on (1) developing synergistic remediation technologies for combined pollution; (2) applying omics and synthetic biology to enhance biological remediation efficiency; (3) establishing long-term monitoring systems for remediation sites; and (4) promoting the integration of intelligent technologies (e.g., AI and the IoT) into site management.

We would like to express our sincere gratitude to all authors for their valuable contributions, the reviewers for their rigorous comments, and the editorial team of *Processes* for their professional support. It is our hope that this issue will inspire more innovative research and promote the application of remediation technologies, contributing to the sustainable management of contaminated sites worldwide.

Author Contributions: K.H.: Conceptualization, methodology, formal analysis, data curation, writing—original draft preparation, writing—review and editing, X.D.: writing—original draft preparation, writing—review and editing Z.D.: writing—review and editing, Y.Z.: writing—review and editing, G.L.: Conceptualization, methodology, writing—original draft preparation, writing—review and editing. All authors have read and agreed to the published version of the manuscript.

Funding: This research received no external funding.

Institutional Review Board Statement: Not applicable.

Informed Consent Statement: Not applicable.

Data Availability Statement: Not applicable.

Acknowledgments: As Guest Editors of the Special Issue "Advances in Remediation of Contaminated Sites: 2nd Edition", we would like to express our deep appreciation to all authors whose valuable work was published in this issue and thus contributed to the success of the edition.

Conflicts of Interest: The authors declare no conflicts of interest.

List of Contributions

1. Wang, Q.; Du, W.; Jin, X.; Wang, J.; Lu, Y.; Huang, D.; Nong, J.; Huang, H.; Xie, T.; Han, B. The Effects of Water Management, Foliar Fertilizers, and Lime Application on the Accumulation of Cd and As in Rice Grains Based on a Field Trial. *Processes* **2024**, *12*, 2241. https://doi.org/10.3390/pr12102241.
2. Wei, H.; Zou, Q.; Chen, Z.; Cao, Y.; Wang, S.; Zhu, F.; Liu, X. Analysis of Hydrochemical Characteristics and Causes of Drinking Water Sources in South China: A Case Study in Zhanjiang City. *Processes* **2023**, *11*, 1196. https://doi.org/10.3390/pr11041196.
3. Shu, X.; Zhang, D.; Zhang, Q.; Ai, T.; Zhang, X.; Liu, J. Impact of Acid Rain on Release Characteristics of Heavy Metals in Low-Sulfur Tailings with Strong Acid Neutralization Capacity: A Case Study from Northern Guangxi, China. *Processes* **2024**, *12*, 2492. https://doi.org/10.3390/pr12112492.
4. Zhang, Z.; Wang, Z.; Luo, Y.; Zhang, J.; Feng, X.; Zeng, Q.; Tian, D.; Li, C.; Zhang, Y.; Wang, Y.; et al. Quantitative Analysis of Soil Cd Content Based on the Fusion of Vis-NIR and XRF Spectral Data in the Impacted Area of a Metallurgical Slag Site in Gejiu, Yunnan. *Processes* **2023**, *11*, 2714. https://doi.org/10.3390/pr11092714.

5. Li, T.; Li, X.; Shen, C.; Chen, D.; Li, F.; Xu, W.; Wu, X.; Bao, Y. Nano-Hydroxyapatite Modified Tobacco Stalk-Based Biochar for Immobilizing Cd(II): Interfacial Adsorption Behavior and Mechanisms. *Processes* **2024**, *12*, 1924. https://doi.org/10.3390/pr12091924.
6. Qin, J.; Yan, Z.; Jiang, X.; Zhao, H.; Liu, S.; Dai, M.; Xiong, D.; Chen, X. Differences in Physiological Metabolism and Antioxidant System of Different Ecotypes of Miscanthus floridulus under Cu Stress. *Processes* **2022**, *10*, 2712. https://doi.org/10.3390/pr10122712.
7. Li, X.; Li, T.; Jin, X.; Wei, Y.; Bao, Y.; Yao, Q.; Li, F.; Xu, W.; Wu, X. Effects of L-Aspartic Acid on Cr(VI) Adsorption onto the Lepidocrocite with Different Exposed Facets: Batch Experiments and In Situ ATR-FTIR Analysis. *Processes* **2024**, *12*, 2598. https://doi.org/10.3390/pr12112598.
8. Wang, X.; Chen, J.; Jia, W.; Huang, K.; Ma, Y. Comparing the Aging Processes of PLA and PE: The Impact of UV Irradiation and Water. *Processes* **2024**, *12*, 635. https://doi.org/10.3390/pr12040635.
9. Li, Q.; Li, Z.; Xie, T.; Ding, Y.; Gan, N.; Tan, Y.; Zhang, C. Analysis of Natural Groundwater Flowing into the Flow Field of In Situ Leaching Mining. *Processes* **2023**, *11*, 471. https://doi.org/10.3390/pr11020471.
10. Zhang, X.; Zhao, C.; Xue, F.; Xia, B.; Lu, Y.; Ying, R.; Hu, Z. Adsorption of Zinc(II) Ion by Spent and Raw Agaricus bisporus in Aqueous Solution. *Processes* **2024**, *12*, 717. https://doi.org/10.3390/pr12040717.
11. Gong, P.; Wu, J.; Shang, Z.; Chen, M.; Li, B.; Wang, T.; Sun, L.; Dang, Z.; Zhu, N.; Wu, P. Construction of Novel Nanoflowering MgAl-Double Oxide Konjac Gum for Efficient Enrichment of Uranium (VI) from Wastewater. *Processes* **2025**, *13*, 876. https://doi.org/10.3390/pr13030876.
12. Dai, M.; Wei, X.; Lin, C.; Xie, C.; Lai, Z.; Zhu, W.; Guo, J.; Peng, C. Preparation and Exploration of Spherical Fe-C Micro-Electrolysis Materials for the Removal of Crystal Violet. *Processes* **2024**, *12*, 2538. https://doi.org/10.3390/pr12112538.
13. Wang, B.; Hou, J.; Wu, X.; Niu, X.; Zhou, F. Assessing Phytoremediation Potential: Dominant Plants in Soils Impacted by Polymetal(loid)lic Mining. *Processes* **2024**, *12*, 833. https://doi.org/10.3390/pr12040833.
14. Chen, J.; Luo, Y.; Zhang, J.; Lu, Z.; Han, Y.; Chen, X.; Li, M.; Fan, H.; Wei, X. In Situ Remediation of Combined Ammonia and Nitrate Nitrogen Contamination Using Zero-Valent Iron-Enhanced Microorganisms in Acidic Groundwater: A Laboratory and Pilot-Scale Study. *Processes* **2024**, *12*, 2768. https://doi.org/10.3390/pr12122768.
15. Song, S.; Wang, B.; Yang, T.; Gu, Y.; Sheng, S.; Zhao, D.; An, S.; Li, A. Performance and Bacteria Communities of a Full-Scale Constructed Wetland Treating the Secondary Effluent after Multi-Years' Operation. *Processes* **2023**, *11*, 1469. https://doi.org/10.3390/pr11051469.
16. Liang, C.; Liu, H.; Liang, J.; Du, X.; Tao, X.; Lu, G. Electrochemical Reduction Performance and Mechanism of 2,2′,4,4′-Tetrabromodiphenyl Ether (BDE-47) with Pd/Metal Foam Electrodes. *Processes* **2025**, *13*, 853. https://doi.org/10.3390/pr13030853.
17. Tang, T.; Song, D.; Chen, J.; Chen, Z.; Du, Y.; Dang, Z.; Lu, G. Utilizing Machine Learning Models with Molecular Fingerprints and Chemical Structures to Predict the Sulfate Radical Rate Constants of Water Contaminants. *Processes* **2024**, *12*, 384. https://doi.org/10.3390/pr12020384.
18. Wang, X.; Li, R.; Tian, Y.; Zhang, B.; Zhao, Y.; Zhang, T.; Liu, C. A Computational Framework for Design and Optimization of Risk-Based Soil and Groundwater Remediation Strategies. *Processes* **2022**, *10*, 2572. https://doi.org/10.3390/pr10122572.
19. Zhang, B.; Wang, X.; Liu, C. Screening and Optimization of Soil Remediation Strategies Assisted by Machine Learning. *Processes* **2024**, *12*, 1157. https://doi.org/10.3390/pr12061157.
20. Alasmary, Z. Lead (Pb) Contamination in Soil and Plants at Military Shooting Ranges and Its Mitigation Strategies: A Comprehensive Review. *Processes* **2025**, *13*, 345. https://doi.org/10.3390/pr13020345.
21. Liu, P.; Wen, S.; Zhu, S.; Hu, X.; Wang, Y. Microbial Degradation of Soil Organic Pollutants: Mechanisms, Challenges, and Advances in Forest Ecosystem Management. *Processes* **2025**, *13*, 916. https://doi.org/10.3390/pr13030916.
22. Meng, X.; Xia, B.; Gao, W.; Chen, W.; He, Q.; Qian, J.; Chen, Z.; Chen, H.; Zhang, X.; Ying, R. Trends and Hotspots in Soil Minerals' Impacts on Carbon Stability Research: A Bibliometric Analysis Based on Web of Science. *Processes* **2025**, *13*, 821. https://doi.org/10.3390/pr13030821.

References

1. Liu, Y.-R.; van der Heijden, M.G.A.; Riedo, J.; Sanz-Lazaro, C.; Eldridge, D.J.; Bastida, F.; Moreno-Jiménez, E.; Zhou, X.-Q.; Hu, H.-W.; He, J.-Z.; et al. Soil contamination in nearby natural areas mirrors that in urban greenspaces worldwide. *Nat. Commun.* **2023**, *14*, 1706. [CrossRef] [PubMed]
2. Jin, X.; Guo, C.; Huang, Q.; Tao, X.; Li, X.; Xie, Y.; Dang, Z.; Zhou, J.; Lu, G. Arsenic redistribution associated with Fe (II)-induced jarosite transformation in the presence of polygalacturonic acid. *Sci. Total Environ.* **2024**, *935*, 173444. [CrossRef]
3. Saktrakulkla, P.; Li, X.; Martinez, A.; Lehmler, H.-J.; Hornbuckle, K.C. Hydroxylated polychlorinated biphenyls are emerging legacy pollutants in contaminated sediments. *Environ. Sci. Technol.* **2022**, *56*, 2269–2278. [CrossRef]
4. Gallego, S.; Montemurro, N.; Béguet, J.; Rouard, N.; Philippot, L.; Pérez, S.; Martin-Laurent, F. Ecotoxicological risk assessment of wastewater irrigation on soil microorganisms: Fate and impact of wastewater-borne micropollutants in lettuce-soil system. *Ecotoxicol. Environ. Saf.* **2021**, *223*, 112595. [CrossRef]
5. McGinnis, M.; Sun, C.; Dudley, S.; Gan, J. Effect of low-dose, repeated exposure of contaminants of emerging concern on plant development and hormone homeostasis. *Environ. Pollut.* **2019**, *252*, 706–714. [CrossRef]
6. Tang, B.; Luo, X.J.; Huang, C.C.; Ren, Z.-H.; Zeng, Y.-H.; Mai, B.-X. Characterizing the influence of metabolism on the halogenated organic contaminant biomagnification in two artificial food chains using compound-and enantiomer-specific stable carbon isotope analysis. *Environ. Sci. Technol.* **2018**, *52*, 10359–10368. [CrossRef]
7. Wei, J.; Wang, X.; Tu, C.; Long, T.; Bu, Y.; Wang, H.; Jeyakumar, P.; Jiang, J.; Deng, S. Remediation technologies for neonicotinoids in contaminated environments: Current state and future prospects. *Environ. Int.* **2023**, *178*, 108044. [CrossRef]
8. Naidu, R.; Biswas, B.; Nuruzzaman, M.; Singh, B.K. Bioremediation of heavy metal (loid) s in agricultural soils and crops. *Nat. Rev. Bioeng.* **2025**, 1–14. [CrossRef]
9. Bates, M.E.; Grieger, K.D.; Trump, B.D.; Keisler, J.M.; Plourde, K.J.; Linkov, I. Emerging technologies for environmental remediation: Integrating data and judgment. *Environ. Sci. Technol.* **2016**, *50*, 349–358. [CrossRef] [PubMed]
10. Shah, V.; Daverey, A. Phytoremediation: A multidisciplinary approach to clean up heavy metal contaminated soil. *Environ. Technol. Innov.* **2020**, *18*, 100774. [CrossRef]
11. Ren, W.; Liu, H.; Mao, T.; Teng, Y.; Zhao, R.; Luo, Y. Enhanced remediation of PAHs-contaminated site soil by bioaugmentation with graphene oxide immobilized bacterial pellets. *J. Hazard. Mater.* **2022**, *433*, 128793. [CrossRef] [PubMed]
12. Cao, T.; Liu, Y.; Gao, C.; Yuan, Y.; Chen, W.; Zhang, T. Understanding Nanoscale Interactions between Minerals and Microbes: Opportunities for Green Remediation of Contaminated Sites. *Environ. Sci. Technol.* **2024**, *58*, 14078–14087. [CrossRef]
13. Katubi, K.M.; Amari, A.; Harharah, H.N.; Eldirderi, M.M.; Tahoon, M.A.; Ben Rebah, F. Aloe vera as promising material for water treatment: A review. *Processes* **2021**, *9*, 782. [CrossRef]
14. Finn, M.; Yackel, N.; Giampietro, G.; Mazyck, D. Dissolving Activated Carbon Pellets for Ibuprofen Removal at Point-of-Entry. *Processes* **2023**, *11*, 1470. [CrossRef]
15. Wang, J.; Feng, X.; Anderson, C.W.N.; Xing, Y.; Shang, L. Remediation of mercury contaminated sites—A review. *J. Hazard. Mater.* **2012**, *221*, 1–18. [CrossRef] [PubMed]
16. Azad, M.A.K.; Amin, L.; Sidik, N.M. Genetically engineered organisms for bioremediation of pollutants in contaminated sites. *Chin. Sci. Bull.* **2014**, *59*, 703–714. [CrossRef]
17. Zheng, C.; Zeng, X.; Wang, T.; Yang, J.; Ma, Y.; Chen, X.; Huang, K. Non-target screening of emerging contaminants in tropical island rivers: A case study of the Nandu River. *Environ. Pollut.* **2025**, *374*, 126225. [CrossRef]
18. Chen, X.; Ren, Y.; Li, C.; Shang, Y.; Ji, R.; Yao, D.; He, Y. Pollution Characteristics and Ecological Risk Assessment of Typical Heavy Metals in the Soil of the Heavy Industrial City Baotou. *Processes* **2025**, *13*, 170. [CrossRef]
19. Zhong, Z.; Du, X.; Wang, Z.; Wu, L.; Tao, X.; Dang, Z.; Lu, G. Dissolution release of antimony from E-waste microplastics in aquatic solution: Kinetics, mechanisms and surface-release-layer thickness. *Water Res.* **2025**, *288*, 124619. [CrossRef]
20. Qin, J.; Jiang, X.; Qin, J.; Zhao, H.; Dai, M.; Liu, H.; Chen, X. Effects of lead pollution on photosynthetic characteristics and chlorophyll fluorescence parameters of different populations of *Miscanthus floridulus*. *Processes* **2023**, *11*, 1562. [CrossRef]
21. Wu, P.; Du, X.; Liang, J.; Huang, J.; Tao, X.; Zou, M.; Zhou, J.; Dang, Z.; Lu, G. Verification of positive impact of critical soil mineral on persulfate-based in-situ remediation of polybrominated diphenyl ether (PBDE): Robust role of goethite under multivariant conditions. *J. Environ. Chem. Eng.* **2025**, *13*, 116664. [CrossRef]

processes

MDPI

Article

Pollution Characteristics and Ecological Risk Assessment of Typical Heavy Metals in the Soil of the Heavy Industrial City Baotou

Xiangmei Chen [1,2], Yongqiang Ren [2,*], Chi Li [3], Yan Shang [2], Rui Ji [3], De Yao [1] and Yingchun He [4]

1 College of Chemical Engineering, Inner Mongolia University of Technology, Hohhot 010051, China; geochem_chen19@oit.edu.cn (X.C.); yaode@imut.edu.cn (D.Y.)
2 Department of Civil Engineering, Ordos Institute of Technology, Ordos 017000, China; shangy312@163.com
3 School of Georesources and Environmental Engineering, Inner Mongolia University of Technology, Hohhot 010051, China; nmglichi@imut.edu.cn (C.L.); 20231800753@imut.edu.cn (R.J.)
4 Inner Mongolia Institute of Technology, China University of Geosciences, Ordos 017000, China; 15174910065@163.com
* Correspondence: ryongqiang@oit.edu.cn

Abstract: Urban soil samples were collected from the major heavy industrial city of Baotou in Inner Mongolia, China, to investigate the concentration, spatial distribution, and pollution levels of heavy metals. The study employed the geoaccumulation index (Igeo), ecological risk index, and spatial interpolation methods to comprehensively assess urban soil pollution. The results indicated that apart from arsenic (As) and nickel (Ni), the concentrations of heavy metals such as cadmium (Cd), chromium (Cr), manganese (Mn), mercury (Hg), copper (Cu), zinc (Zn), and lead (Pb) were significantly higher than the corresponding background values in the study areas. According to the geoaccumulation index (Igeo), the overall pollution level in the study area ranged from uncontaminated to low pollution. However, variations existed among different urban districts. Among Baotou's four main urban areas, the soil pollution level in Kundulun District was notably higher compared to the other three urban areas. Mercury (Hg), lead (Pb), zinc (Zn), and copper (Cu) exhibited relatively higher pollution levels across the four district sites. The observed pollution characteristics are closely linked to the distinct industrial attributes of the urban districts: the Kundulun District of Baotou, Inner Mongolia, is renowned for its significant presence of industrial activities such as steel manufacturing, power generation, and coal combustion. These industries play a crucial role in the local economy but also contribute substantially to heavy metal emissions, leading to notable environmental impacts. Similar to the Kundulun District, the Qingshan District of Baotou, Inner Mongolia, is significantly influenced by industrial activities, which have led to elevated concentrations of certain heavy metals and formed higher potential ecological risk index (*PERI*) hotspots. Implications and Recommendations. The disparity in industrial activities across the four urban districts of Baotou is a principal factor contributing to variations in pollution levels and ecological risks. In conclusion, this research highlights the necessity of aligning industrial zoning with effective environmental management strategies to combat heavy metal pollution in urban soils. By implementing these integrated approaches, Baotou can safeguard its environment and public health, paving the way for a sustainable future.

Keywords: urban soil; heavy metals; pollution level; hotspots

1. Introduction

Due to the rapid pace of industrialization and urbanization, the accumulation of heavy metals in urban soil has become a major issue. This environmental issue poses a threat to human health and the environment. This accumulation directly impacts human health and contributes to secondary pollution in water bodies, and atmospheric scientists [1–5] worldwide have focused on heavy metal pollution in soil [6–10], examining various aspects such as the sources of heavy metal pollutants, spatial and temporal distributions, chemical forms, migration and transformation, soil environmental background values, environmental capacity, health and environmental standards, bioavailability, risk assessment, pollution control, and remediation. Previous research indicates that urban soil is influenced not only by the climate and parent material but also by human activities. Factors such as the history of urban construction, land use practices, functional zoning, and proximity to pollution sources contribute to the varying degrees and types for soil pollution [11–13]. These variations are the primary reason for the significant differences observed in the geochemical characteristics of soils in different cities.

The input of anthropogenic external pollutants is the main cause of environmental geochemical problems in urban soil [14–16], which comprise three main urban industrial waste types: coal burning, domestic waste, and motor vehicle exhaust act as the main sources of urban soil pollution [17–19]. The geochemical pollution of some urban soils is related to mining activities in the area, while the heavy metal content of some urban soils is related to the soil-forming parent material [20–22]. Human activities are primarily responsible for the ecological and geochemical issues observed in urban soils. Consequently, these soils have become the principal repositories for the containment and purification of pollutants in urban environments.

Once urban soil becomes polluted, the pollutants often bind with soil organic matter or minerals, rendering them resistant to degradation and artificial removal, complicating recovery efforts. Heavy metals in soil primarily harm the human body through direct exposure, such as skin contact or inhalation [23–30]. Additionally, these metals can enter the human body indirectly through the food chain, thereby posing a significant threat to human health. Consequently, studying the characteristics and risk assessment of heavy metal pollution in soil is of paramount importance.

Baotou's special heavy industry system, covering iron and steel, rare earth and other industries, makes the typical heavy metal pollution in its soil unique. The study of heavy metal pollution characteristics in Baotou soil can help to reveal the contamination and accumulation law of heavy metals in soil, the spatial distribution and migration patterns under the multi-industry composite pollution source. The study of Baotou soil can provide key data and case support for the construction of a more comprehensive and precise theoretical system of soil heavy metal pollution and promote the in-depth development of environmental science, soil science and other related disciplines in the direction of heavy metal pollution research.

Therefore, this study focuses on the soil environment of Baotou, a city of heavy industry, and takes typical heavy metals as the objects of study to investigate their pollution characteristics in depth, and applies scientific and rigorous ecological risk assessment methods to comprehensively analyze the potential risks to the ecosystem. By employing Geographic Information System (GIS) technology in conjunction with field sampling data, we have mapped the spatial distribution of heavy metals to elucidate their migration pathways and impact extents. This study aims to fill the gap in this field of research in Baotou, provide accurate and reliable data support and a scientific and reasonable decision-making basis for soil environmental protection, pollution management and ecological restoration in Baotou. It is also expected to provide valuable references and experiences for other cities with similar industrial background and soil pollution problems around the world, and to help promote the in-depth development and technological progress in the field of environmental science in soil heavy metal pollution.

2. Materials and Methods

2.1. Study Area

The study area (109°16′–111°25′) is located in the western part of the Inner Mongolia Autonomous Region (Figure 1). It is the industrial center of Inner Mongolia and the largest city in the autonomous region with an important steel and rare earth production base and an important transportation hub in China. It is located at the southern end of the Mongolian Plateau, bordering the Yellow River in the south and traversed by the Yinshan Mountains, forming three terrain regions: the northern plateau, central mountain, and southern plain. The land area is 27,768 km^2, the central area is 315 km^2, and the population in 2023 was 2.762 million. The average altitude of the urban area is 1067 m.

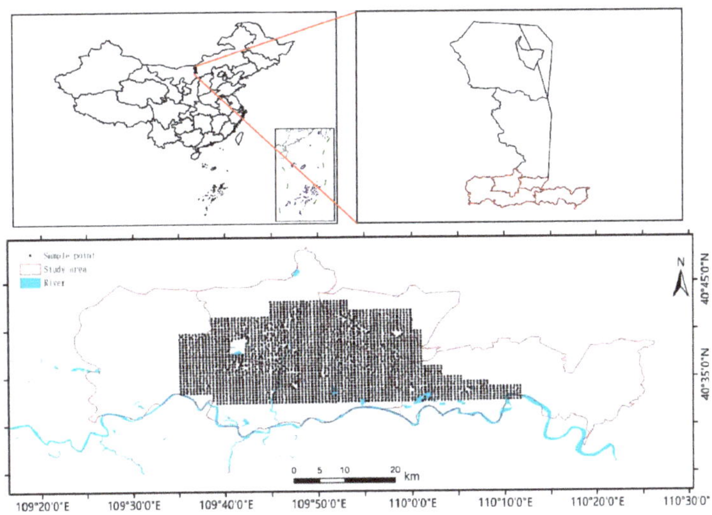

Figure 1. Map of the study area and sampling points.

2.2. Sample Collection

Soil samples were mainly taken from the four main urban districts of Baotou City (Kundulun District, Qingshan District, East River District, and Jiuyuan District). Surface soil samples (0–20 cm) were collected. The sampling locations are shown in Figure 1. Soil sampling was carried out in strict accordance with the relevant regulations in the technical specification for soil environmental monitoring (GBT36197-2018) [31]. The sampling density was 4 points/km^2, 2820 samples were collected, and the sampling area was 705 km^2. A 2–3 point combined sampling method was used to remove debris in the soil when collecting the samples. The sample mass was generally required to be approximately 500 g.

2.3. Sample Processing and Analysis

The soil samples in this study were pretreated [32] in strict accordance with the standards outlined in the "Technical Specifications for Soil Environmental Monitoring" (HJ/T 166-2004) [33]. This research analyzed the concentrations of nine heavy metals (As,

Cd, Cr, Cu, Mn, Ni, Pb, Zn and Hg) in the soil. The soil samples were then analyzed for elements other than mercury (Hg) using inductively coupled plasma mass spectrometry (ICP-MS). The concentration of mercury in the soil was quantified using atomic fluorescence spectrometry (AFS).

2.4. Geoaccumulation Index

The assessment of soil environmental quality typically relies on single pollution indices, where a smaller index indicates lighter pollution and a larger index signifies heavier contamination. When evaluating the overall soil environmental quality of a region for comparison with external regions or historical data, in addition to single pollution indices, composite pollution indices are frequently employed. Given the significant regional background variations in soils, the use of the accumulation index of soil pollution provides a more accurate reflection of anthropogenic contamination levels.

The geoaccumulation index (I_{geo}), first introduced by the German scholar Müller [34], serves as a critical tool for assessing minor alterations in the background factors of soil environments influenced by natural geological processes. Additionally, it functions as an evaluation index to quantify the contribution of anthropogenic activities to heavy metal pollution in soils [35]. The classification of this index into various grades is detailed in Table 1. The calculation process is illustrated by Equation (1).

Table 1. Cumulative index classification.

Classification	Clean	Slightly Polluted	Slightly-Moderate Polluted	Moderately Polluted	Moderately-Strongly Polluted	Extremely-Strongly Polluted	Strongly Polluted
I_{geo}	<0	[0, 1)	[1, 2)	[2, 3)	[3, 4)	[4, 5)	≥5

$$I_{geo} = \log_2(C_i / KB_i) \tag{1}$$

where I_{geo} is the geoaccumulation index for a specific element; C_i is the measured concentration of the heavy metal in the soil sample, expressed in mg/kg; and 1.5 is a coefficient used to account for variations in background values caused by differences in local lithology. This factor adjusts for natural fluctuations and is typically set to 1.5. B_i is the background value of the heavy metal in the soil, which reflects the natural concentration of the element in the region. For this study, the background values of nine typical heavy metals were obtained from the benchmark values of the Hetao region and the Inner Mongolia soil environmental background values. The selected background values for the Hetao region are as follows: As (9.61), Cd (100.75), Cr (54.13), Cu (18.88), Mn (506.01), Ni (24.21), Pb (19.67), Zn (52.80), and Hg (16.84).

2.5. Potential Ecological Risk Index

The Hakanson Potential Ecological Risk Model [36] is a widely recognized method for assessing the ecological risk posed by heavy metal pollution in soils and sediments. Proposed by Swedish scientist Lars Hakanson in 1980, this model has become one of the most commonly applied frameworks for evaluating ecological risks associated with heavy metals. The model integrates multiple factors, including total heavy metal concentration, toxicity, mobility, sensitivity, and background values, providing a comprehensive assessment of potential ecological risks.

The toxicity coefficients of Hg, Cd, As, Cu, Pb, Cr, Zn, Ni, and Mn were 40, 30, 10, 5, 5, 2, 1, 5, and 1, respectively [37]. The potential ecological risk coefficient and index evaluation criteria in Hakanson's original text are shown in Table 2.

Table 2. Potential ecological risk coefficient and index evaluation criteria.

Potential Ecological Risk Factor (E_r^i)	Potential Ecological Risk Index (RI)	Ecological Hazard Level
<40	<150	slight
[40, 80)	[150, 300)	moderate
[80, 160)	[300, 600)	strong
[160, 320)		very strong
≥320	≥600	extremely strong

The calculation process is shown in Equations (2)–(4).

$$E_r^i = T_r^i \cdot C_J^i \tag{2}$$

$$C_J^i = C_{iD}/C_{iB} \tag{3}$$

$$RI = \sum_{i=1}^{n} E_r^i \tag{4}$$

where RI is the index that provides a comprehensive measure of the overall ecological risk posed by multiple heavy metals in the environment; E_r^i quantifies the ecological risk associated with each specific heavy metal, taking into account both its toxicity and the degree of contamination; T_r^i is the unique toxicity response factor for each heavy metal that reflects its potential harm to the ecosystem; C_J^i compares the measured concentration of a heavy metal in the sample to its background value, highlighting any anthropogenic contribution to pollution; C_{iD} is the measured concentration of the ith heavy metal element; and C_{iB} is the background value of heavy metals in the soil.

2.6. Spatial Differentiation of Heavy Metals in Soil

Soil is a variant of the earth's surface with a high degree of heterogeneity and a certain temporal and spatial continuity. The occurrence and development of soil are easily affected by many factors. There are two main types of spatial variation in soil heavy metals: random variation and structural variation. Kriging interpolation is the most widely used interpolation method in geostatistics, which is most suitable for both structural and random geographical elements [38]. In this study, universal Kriging was selected to compare and analyze the interpolation simulation results of soil heavy metal content in the study area in order to obtain the optimal spatial distribution interpolation results for each heavy metal element.

The accuracy of spatial prediction results is verified by an independent verification method. The evaluation criteria of Kriging interpolation accuracy are as follows: mean error (ME) and standardized mean error (MSE) are close to 0, standardized root mean square error is close to 1, root mean square error ($RMSE$) is the smallest, and average standard error is the closest to $RMSE$. In this section, the ME and $RMSE$ are used as accuracy verification indicators for comparative analysis. The index calculation formula is as follows:

$$ME = \frac{1}{n}\sum_{i=1}^{n}(\hat{Z}_i - Z_i) \tag{5}$$

$$RMSE = \sqrt{\frac{1}{n}\sum_{i=1}^{n}(\hat{Z}_i - Z_i)^2} \tag{6}$$

where $\hat{Z}_i$ is the predicted value at the i-th validation point; Z_i is the actual measured value at the i-th validation point; and n is the number of sample points.

Mean error (ME): This metric measures the average bias in the predictions. It indicates whether the model systematically overestimates or underestimates the values. An ME close to 0 suggests that the model's predictions are unbiased, meaning there is no systematic error. A positive ME indicates a tendency to overestimate, while a negative ME indicates a tendency to underestimate. In studying the spatial distribution of typical heavy metal contamination in the soil of the heavy industrial city of Baotou, it is based on the principle of spatial correlation, which means that things closer together are considered to be more similar to each other than things farther away. The closer the value is to 0, the smaller the error and the better the simulation effect.

Root mean square error ($RMSE$): This metric provides a measure of the magnitude of prediction errors. It penalizes larger errors more heavily due to the squaring operation, making it sensitive to outliers. A smaller RMSE value indicates better model performance, as it signifies that the predicted values are closer to the actual measurements, and a smaller value indicates a better simulation effect.

2.7. Quality Control and Data Analysis

The accuracy of the total amount of elements was ensured by repeated analysis of GSS-1–GSS-6 standard substances [39]. The logarithmic standard deviation of the calculated measured values and the monitored values of the GSS-1–GSS-6 reference materials were calculated to ensure the precision of the sample analysis. The relative errors of most elements are less than 20%.

3. Results and Discussion

3.1. Trace Metal Concentrations

The characteristics of the soil heavy metal content in the four main urban areas of Baotou City are shown in Table 3 and Figure 2, which can further identify the distribution pattern of heavy metals in urban soil. The distribution of nine kinds of heavy metal elements in the Baotou urban area is extremely uneven, among which the high-anomaly areas of Hg, Cr, Mn, Zn, and Pb appear in the Kundulun District. High-anomaly areas of Hg, Cr, Mn, Ni, Cu, Zn, and Pb were found in the East River District. High-anomaly areas of Hg, Cr, Mn, Zn, and Pb appeared in the Jiuyuan District, while only high-anomaly areas of Hg, Cr, Mn, and Zn appeared in the Qingshan District. The uneven distribution of heavy metals in the soils of these four urban areas is helpful for investigating the relationship between urbanization and its impact on soil composition [36]. The study's results reveal a significant heterogeneity in the spatial distribution of heavy metal concentrations within the soil of the investigated region. While the mean concentrations of all elements remain below the risk threshold as per GB36600-2018 [40] standards, the average levels of lead (Pb), zinc (Zn), and mercury (Hg) marginally surpass the national mean. Additionally, arsenic (As), mercury (Hg), and zinc (Zn) concentrations exceed the global average.

The comparative results with other typical industrial cities are presented in Table 4. In comparison to Anshan, Baotou exhibits more severe Mn contamination in its surface soil, a phenomenon likely linked to industrial emissions and mining activities in the area. When compared to Tangshan, Baotou's soil shows generally higher average concentrations of As, Hg, Cr, Ni, Cu, and Zn, indicating that Baotou surpasses Tangshan in terms of heavy metal pollution. Additionally, while the average concentrations of Cd, As, Hg, Cr, Zn, and Pb in Baotou's soil are slightly higher than those in Chongqing, the difference is not significant. A comprehensive analysis reveals that Wuhan ranks first in heavy metal pollution, followed by Anshan. Baotou and Chongqing exhibit similar levels of pollution, with urban industrial activities being the primary source of soil heavy metal contamination.

Table 3. Statistical characteristics of soil heavy metal content in the four districts of Baotou City.

Elements (mg/kg)		Cd	As	Hg	Cr	Mn	Ni	Cu	Zn	Pb
Study area	Mean	0.16	8.26	0.077	65.84	615.51	23.87	21.82	80.77	32.28
	SD	0.42	8.11	0.19	19.73	673.56	6.95	14.02	109.08	69.63
Kundulun	Mean	0.17	9.27	0.07	70.33	857.78	26.65	24.04	109.67	48.43
	MAX	1.45	292.00	2.06	336.50	15,623.00	66.90	339.20	2837.00	1235.40
	MIN	0.04	0.70	0.00	2.30	296.00	9.90	10.90	33.20	4.20
	SD	0.16	14.95	0.134	24.56	80.08	8.14	18.74	188.45	106.72
Jiuyuan	Mean	0.14	7.60	0.07	59.83	501.10	21.12	19.01	66.50	25.33
	MAX	3.63	26.20	3.11	175.00	3200.00	51.30	294.80	985.10	639.50
	MIN	0.04	2.30	0.01	34.00	275.00	12.40	10.30	32.70	6.80
	SD	0.17	2.44	0.192	12.60	156.10	5.40	11.47	61.30	31.50
Qingshan	Mean	0.16	7.42	0.07	67.19	555.98	24.10	21.45	68.10	23.48
	MAX	8.00	26.20	3.53	328.00	1661.00	48.00	180.40	517.90	477.70
	MIN	0.04	2.10	0.01	37.60	328.00	15.20	13.00	38.70	15.60
	SD	0.42	1.99	0.260	18.43	115.75	4.26	10.61	32.24	25.33
East River	Mean	0.20	9.25	0.11	75.65	580.27	27.34	27.32	84.69	32.78
	MAX	19.00	22.10	1.98	330.70	330.70	103.80	128.60	362.10	172.20
	MIN	0.06	0.60	0.01	32.80	32.80	16.40	13.00	42.10	9.60
	SD	0.03	2.63	0.012	8.34	93.62	4.69	5.01	12.01	3.60
Background value of the Hetao area		0.10	9.61	0.02	54.13	506.01	24.21	18.88	52.80	19.67
Risk screening value		0.60	25.00	3.40	250.00		190.00	100.00	300.00	170.00

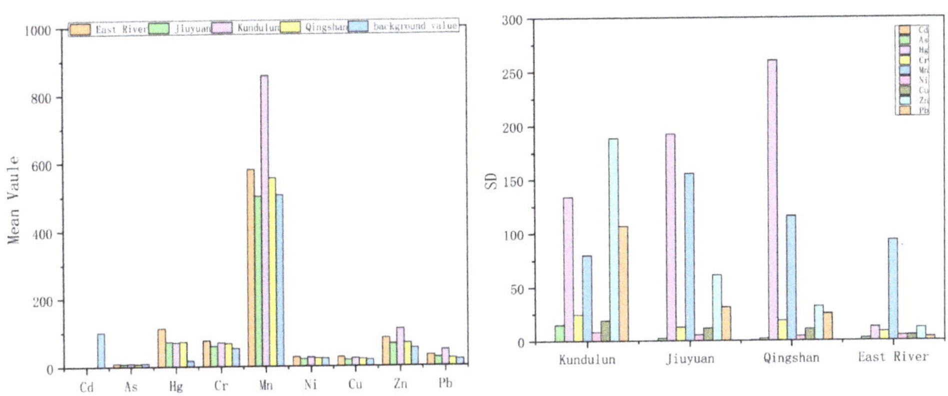

(a) (b)

Figure 2. Average concentration of heavy metals in soil in four main urban areas of Baotou City. (**a**) Mean concentration of heavy metal elements in the soil of the four main urban districts; (**b**) Standard deviation of heavy metal element concentrations in the soil of the four main urban districts.

Table 4. Comparison of soil heavy metal content with other heavy industrial cities.

Elements (mg/kg)	Baotou	Anshan	Tangshan	Wuhan	Chongqing	China	International Standard
Cd	0.158	0.86	0.10	0.86 ± 0.20	0.132	0.68	0.35
As	8.255	3.72	6.79	41.15 ± 24.40	5.80	13.39	6
Hg	0.077		0.065		0.053	0.31	0.06
Cr	65.836	69.93	46.20	58.15 ± 9.05	76.56	63.04	
Mn	615.510	291.5				576.17	1000
Ni	23.873	33.48	17.33	158.64 ± 80.86	30.50	26.18	50
Cu	21.825	52.29	20.97	41.80 ± 12.55	23.99	40.78	30
Zn	80.767	213.49	63.38	349.31 ± 151.77	75.58	137.72	50
Pb	32.280	45.05		155.56 ± 179.99	25.55	47.34	35

3.2. Trace Element Pollution Assessment

Figure 3 visually represents the soil heavy metal pollution levels across the four main urban districts of Baotou, namely Kundulun District, Qingshan District, Jiuyuan District, and another district. The assessment is based on comprehensive sampling and analysis of soil samples from these areas, employing indices such as the geoaccumulation index (I_{geo}), ecological risk index (*ERI*), and spatial interpolation methods. The average of each element decreased in the order of Cd (−10.23) < As (−0.93) < Ni (−0.66) < Cu (−0.49) < Mn (−0.47) < Cr (−0.36) < Pb (−0.32) < Zn (−0.24) < Hg (0.68). The average value of Hg in the soil of Baotou City is slightly polluted, and the average value of other elements is not polluted. From the perspective of single elements, it can be determined that there are strong pollution areas of Hg element in the four main urban areas of Baotou City, and other elements reflect different pollution degree characteristics in the four urban areas. The primary cause of this phenomenon is the variation in industrial zones distributed across the four urban districts. In the Kundulun District, Zn and Pb have strong pollution areas, As and Mn have extremely strong pollution areas, Cu has medium–strong pollution areas, and Cr has moderate pollution areas. In the Qingshan District, Cu, Zn, and Pb exist in moderate to strong pollution areas. In the Jiuyuan District, As, Mn, Zn, and Pb have extremely strong pollution areas, Cu has medium to strong pollution areas, and Cr has moderate pollution areas. In the East River District, there is a strong pollution area for Pb and a moderate pollution area for Cr, Cu, and Zn. Highly and extremely polluted areas require targeted interventions and remedial efforts in these areas to address rising pollution levels and ensure the environmental health of these urban areas. The aforementioned pollution characteristics are associated with the unique industrial features of Baotou City, such as the steel and power generation industries, and coal combustion.

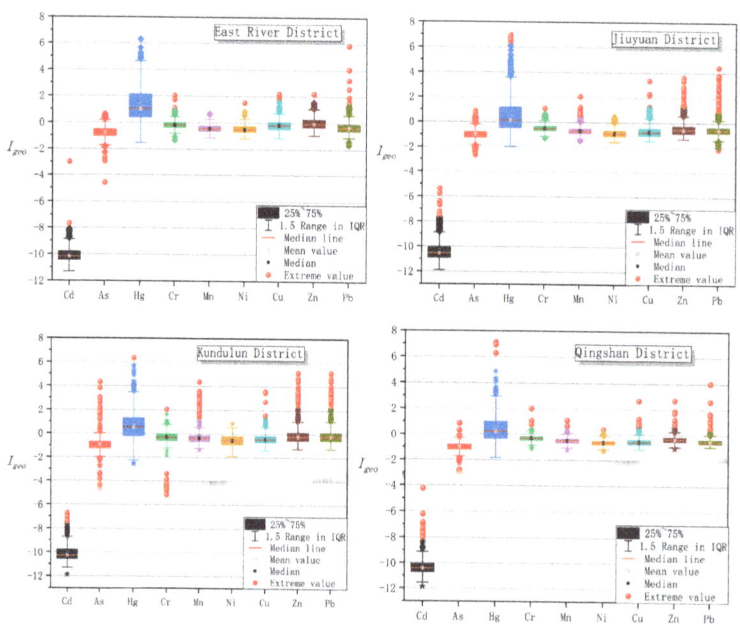

Figure 3. Box plots of the geoaccumulation index of heavy metals in the main urban areas of Baotou City (Kundulun District, Qingshan District, Donghe District, Jiuyuan District). The red line in the box plot represents the mean value.

In the heavy industrial city of Baotou, the operation of aluminum plants and aluminum slag processing facilities significantly influences the typical patterns of soil heavy metal pollution. The production process at these aluminum plants, particularly during the stages of ore extraction and beneficiation, serves as the initial source of heavy metal contamination. During the extraction and beneficiation of bauxite, various heavy metals such as cadmium (Cd) and chromium (Cr) are present in the ores. These metals are brought out in significant quantities along with the mined materials. Some of these heavy metals enter the tailings during the beneficiation process. If tailings management is inadequate, heavy metals can leach into the surrounding soils through rainwater runoff and dust dispersion, thereby increasing the concentration of corresponding heavy metals in the soil. Improper handling or storage of tailings can lead to environmental issues, including contamination of nearby water bodies and agricultural lands. Dust emissions from exposed tailings piles can also contribute to atmospheric deposition, further exacerbating soil pollution.

In the oxidation–aluminum production phase, high-temperature smelting causes the volatilization of heavy metal impurities from the ores into the atmosphere. Volatile metals like lead (Pb) and mercury (Hg) can escape in gaseous form under these conditions. Subsequently, these metals return to the soil surface via dry and wet deposition, leading to a gradual increase in soil contamination around the plant.

The spatial pattern of soil contamination tends to decrease with distance from the plant due to atmospheric dispersion. Areas downwind of the prevailing winds exhibit higher levels of contamination because of the directional transport of pollutants. For instance, volatile metals released during smelting may settle within a certain radius around the factory, creating a gradient of pollution intensity that diminishes with distance. This phenomenon results in a clear directional pattern of heavy metal accumulation in the soil, with concentrations peaking in the downwind direction.

3.3. Potential Ecological Risk

Potential ecological risk (*PERI*) was used to assess the potential ecological risk of heavy metal pollutants in soil. In this study, through the analysis shown in Figure 4, it was found that the Kundulun and Jiuyuan Districts have higher ecological risks than the East River District and Qingshan District. Hg, As, and Pb were classified as high risk, Cu and Zn were classified as medium risk, and the remaining elements were classified as low risk. The *RI* values of the Kundulun District and Jiuyuan District were also higher than those of the East River District and Qingshan District. The *RI* values of the Kundulun District ranged from 37.92 to 127,138.24, with an average value of 17,525.55. The *RI* values of the Jiuyuan District ranged from 52.54 to 219,716.63, with an average value of 28,411.45, which was higher than that of the Qingshan District (8875.98) and East River District (128,347.61).

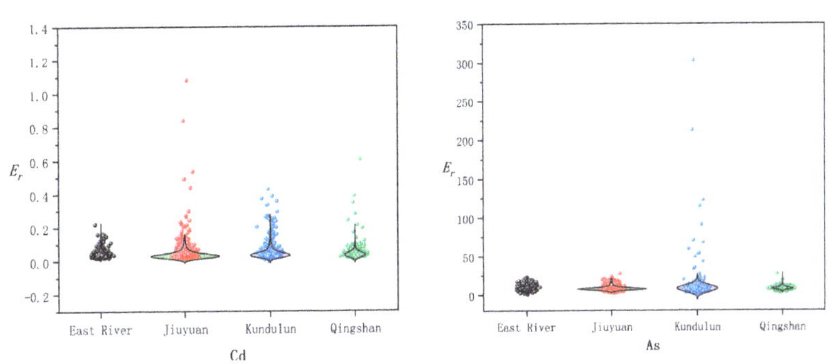

Figure 4. *Cont.*

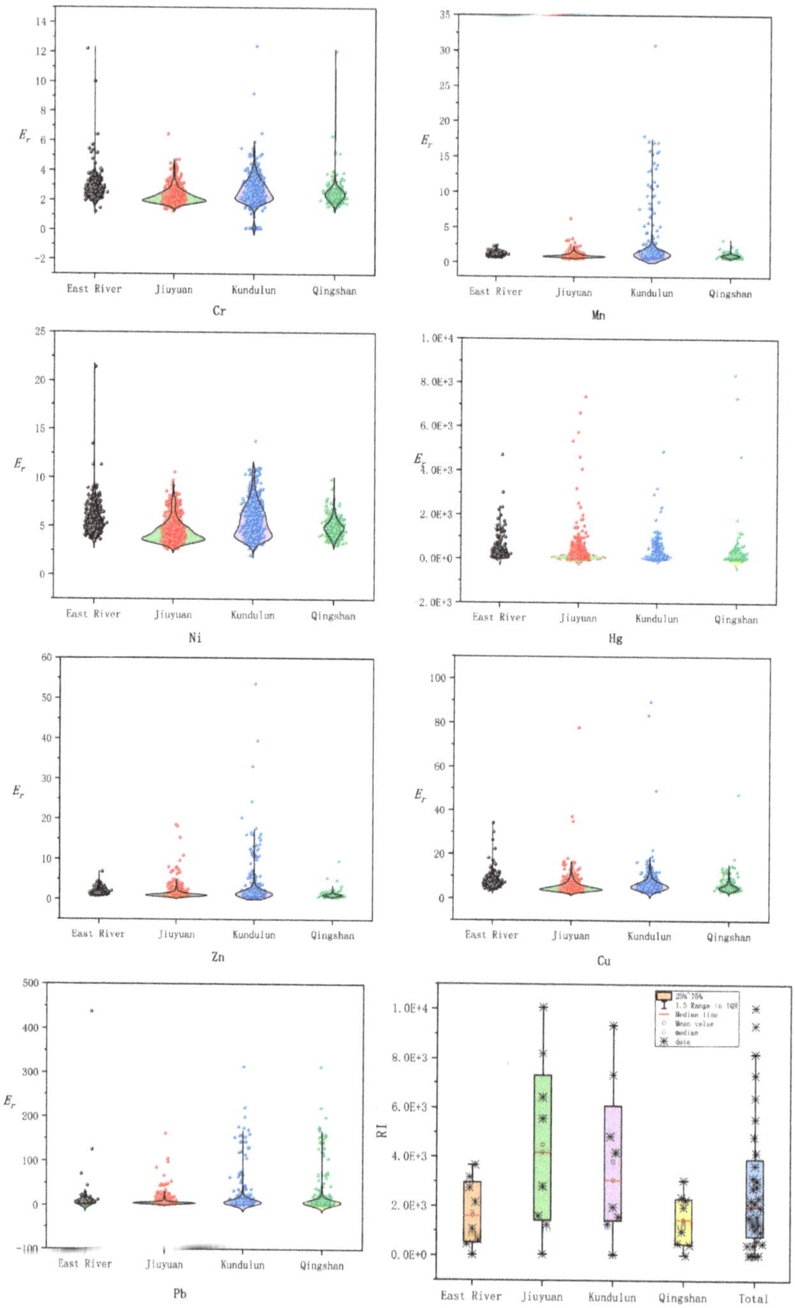

Figure 4. Potential ecological risk index (E_r) of heavy metals in the main urban areas of Baotou City (Kundulun District, Qingshan District, East River District, and Jiuyuan District): violin diagrams and risk index (*RI*) box diagram.

3.4. Spatial Variability Characteristics of Heavy Metals

Analyzing the spatial distribution is a crucial method for pinpointing areas with high concentrations of heavy metal contamination and for tracing potential origins of these metals in soil environments. This approach is particularly useful in urban environments where multiple pollution sources can complicate the identification of specific contributors to soil contamination. Geostatistical analysis, including methods like Kriging interpolation, helps in examining possible pollution sources and provides a detailed understanding of the spatial variability of contaminants. In this research, the Kriging interpolation technique was employed within geostatistical analysis to interpolate the spatial distribution of nine heavy metals [41]. Figure 5 illustrates the spatial distribution map of heavy metal concentrations across nine urban soil samples from Baotou City.

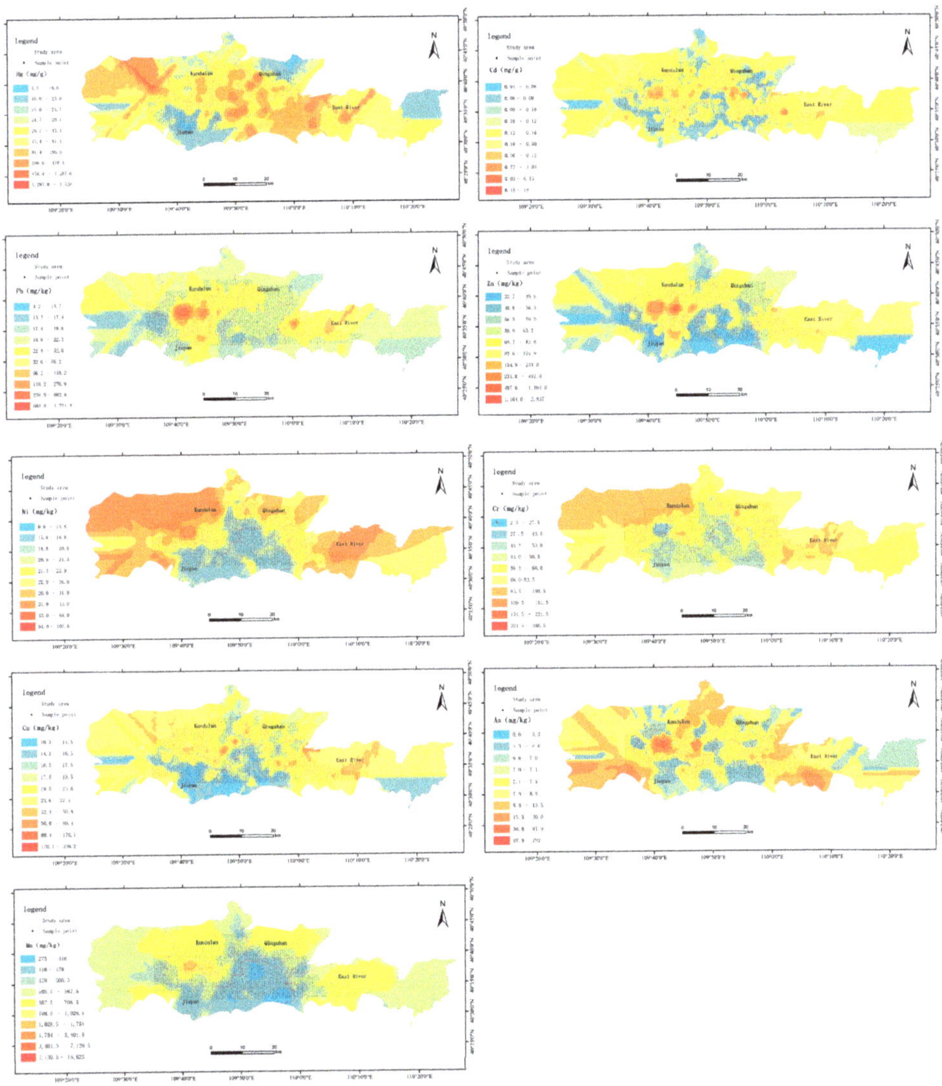

Figure 5. Spatial distribution characteristics of heavy metals in urban soil from Baotou City.

From the perspective of spatial distribution characteristics, the figure shows that there is a large spatial difference in the concentration of trace elements in Baotou urban soil, showing hotspots in the city center [36].

The high-value area of Cd is mainly concentrated in the central and eastern regions of Baotou City, including the Kundulun District, Qingshan District, and East River District. It can be found that the high-value area of Cd is mainly distributed in the industrial waste slag stacking area (Baotou Steel Tailings Dam and Power Plant Ash Storage Pool) and near the Baotou Aluminum Plant in the study area. The distribution is planar, and the accumulation of Cd is closely related to industrial emissions.

The content of As in the soil of Baotou City is higher in the urban area, showing a high background to abnormal distribution. In the middle of the three districts, the suburban area dominated by the Jiuyuan District is distributed from low background to negative anomaly. The abnormal area of As is mainly in the tailings dam of Baotou Steel, and the high content area of the East River District is mainly distributed in the southern area of the Baotou Aluminum Plant. Along the Kundulun River on the south side of the Kundulun District and the Sidao River on the south side of the Qingshan District, two north–east and nearly north–south anomaly zones are formed, respectively. Many factories and enterprises usually select a location near the river.

Hg is banded with high background and abnormal distribution in the urban area of Baotou City from southeast to northwest, and the piedmont alluvial fan in the northwest and northeast of the urban area is the low background area. In the southwestern suburbs of the city is a low background to negative anomaly area.

Cr, Ni, and Cu display similarities in spatial distribution: in the northwest side of the Kundulun District of Baotou City, the piedmont alluvial fan on the northeast side of the Qingshan District, and the suburbs to the east of the East River District show high background and patchy distribution, and local star-shaped scattered mass anomaly areas are formed. Ni and Cu show high background to positive anomalies in the ash storage tank of a power plant. In the tailings dam area of Baotou Steel, Ni, and Cu have a high background, and Cr has a low background. Cr, Ni, and Cu are located in the south of the Qingshan District and the west of the East River District.

One of the core strengths of the Kriging interpolation method lies in its ability to fully account for the spatial autocorrelation of soil heavy metal content data. In the study of Baotou's soil, the distribution of heavy metals is not random but is influenced by various geographical factors such as industrial layout, wind direction, and water flow, exhibiting a certain spatial structure. Around the Baotou steel industrial area, due to the effects of atmospheric deposition and wastewater discharge, soil heavy metal pollution shows a trend of gradually decreasing from the factory center outward, with a more pronounced pollution belt in the prevailing wind direction. The Kriging interpolation method accurately captures this spatial autocorrelation by calculating the semivariogram between sample points. Thus, when predicting the heavy metal content at unknown points, it not only considers the proximity but also the structural changes in the data's spatial distribution, making the interpolation results more consistent with the actual spatial distribution patterns. This allows for more precise mapping of the spatial distribution of heavy metal content, aiding in the identification of pollution hotspots and diffusion pathways. The Kriging interpolation method, in contrast, is based on the assumption of stationarity of regionalized variables, meaning it assumes that the statistical characteristics (such as mean and variance) of soil heavy metal content remain stable within a certain range. However, in a heavy industrial city like Baotou, due to long-term industrial development and urban changes, the situation of soil heavy metal pollution may be complex, with local pollution source variations and changes in land use types, leading to non-stationary statistical characteristics of soil heavy metal content. For instance, in an old industrial area of Baotou that has undergone enterprise upgrades and transformations, some regions may experience reduced pollution emissions, resulting in localized trends of change in soil heavy metal content. This could violate the stationarity assumption, thereby affecting the accuracy of the Kriging interpolation method. In such cases, the interpolation results may exhibit deviations in local areas, failing to accurately reflect the actual changes in soil heavy metal content.

4. Conclusions

In this study, the contamination profiles of potentially toxic elements (PTEs) such as cadmium (Cd), arsenic (As), mercury (Hg), chromium (Cr), manganese (Mn), nickel (Ni), copper (Cu), zinc (Zn), and lead (Pb) in soil samples from the Kundulun, Jiuyuan, Qingshan, and East River Districts of Baotou were evaluated. The average concentrations of Cd, Cr, Mn, Hg, Cu, Zn, and Pb in urban soil samples significantly exceeded their respective background levels. Notably, Hg, Pb, Zn, and Cu emerged as the primary contaminants, with their geoaccumulation index (I_{geo}) values spanning from low to heavily polluted categories. Additionally, the contamination severity across the districts followed the sequence: Kundulun > Jiuyuan > Qingshan > East River. Regarding the assessment of potential ecological risks, mercury (Hg), lead (Pb), and arsenic (As) exhibited a moderate risk, significantly exceeding the risk levels of other potentially toxic elements (PTEs). A notably higher ecological threat was identified in the Kundulun and Jiuyuan Districts, as indicated by the risk index (*RI*) values. The spatial distribution of PTEs demonstrated a higher concentration and distinct hotspots in Kundulun and Qingshan compared to the other districts. Recommendations for the prevention and remediation of the current pollution status in the study area are proposed as follows: (1) To address the common heavy metal contamination, such as lead and cadmium, in Baotou's soil, immobilizing agents can be added. For instance, at a vegetable cultivation site in Baotou, the soil contains 200 mg/kg of lead and 2 mg/kg of cadmium. By incorporating 2% lime and 1% apatite as immobilizing agents, the soil's pH can be increased, causing lead and cadmium to form insoluble compounds, thereby reducing their bioavailability. The phosphate ions in apatite can bind with heavy metal ions to form precipitates. (2) In areas of Baotou with mild to moderate cadmium contamination, planting centipede grass is advisable. Centipede grass has a strong ability to accumulate cadmium, with its above-ground parts containing over 1000 mg/kg of cadmium when the soil concentration is 5 mg/kg. By harvesting the above-ground parts of the centipede grass, heavy metals can be removed from the soil. Typically, 2–3 harvests per year over a period of 3–5 years can significantly reduce soil cadmium levels. (3) Based on the spatial distribution characteristics of heavy metal pollution in Baotou's soil, land use should be planned rationally. Severely polluted areas should be designated for industrial use or green spaces, avoiding agricultural production or residential development. In a heavy metal pollution hotspot in Baotou, where multiple heavy metal concentrations exceed standards, the area can be planned as an urban park or an industrial park expansion site. By planting landscape plants with high heavy metal tolerance or constructing industrial facilities, the opportunity for residents to come into contact with contaminated soil is minimized, thereby reducing health risks.

Author Contributions: Conceptualization, X.C.; methodology, D.Y. and C.L.; software, Y.R. and Y.H.; validation, Y.S. and R.J. All authors have read and agreed to the published version of the manuscript.

Funding: This work was sponsored in part by the scientific research project of Academician and expert workstation of mine geology and environment, Ordos Institute of Technology (2021OITYSZJGZZ-008, 2023OITYSZJGZZ-004); the key project of natural science and technology of Ordos Institute of Technology (KYZD2023004, KYZD2019002); Erdos Mining Area Geohazard Prevention and Geoenvironmental Protection Engineering Research Center (RZ2300001544).

Data Availability Statement: The data that support the findings of this study are available from the corresponding author upon reasonable request, but are not publicly available due to proprietary data.

Conflicts of Interest: The authors declare no conflicts of interest.

References

1. Liu, X. Study on the Behavior and Risk of Major Heavy Metals in Typical Soil-Plant Systems in Arid Areas. Ph.D. Thesis, Lanzhou University, Lanzhou, China, 2009.
2. Li, X.X. Spatial distribution of heavy metal in urban soil of China. *Adv. Mater. Res.* **2014**, *989*, 454–457. [CrossRef]
3. Jahandari, A.; Abbasnejad, A.; Jamasb, R. Concentration, likely sources, and ecological risk assessment of potentially toxic elements in urban soils of Shiraz City, SW Iran: A preliminary assessment. *Arab. J. Geosci.* **2020**, *13*, 936. [CrossRef]
4. Wang, X. Study on the Occurrence Characteristics, Source Apportionment and Health Risk of Heavy Metals in Soil and Surface Dust in Jinan. Ph.D. Thesis, Shandong Normal University, Jinan, China, 2024.
5. Tang, S.; Wang, C.; Song, J.; Ihenetu, S.C.; Li, G. Advances in studies on heavy metals in urban soil: A bibliometric analysis. *Sustainability* **2024**, *16*, 860. [CrossRef]
6. Long, Z.; Zhu, H.; Bing, H.; Tian, X.; Wang, Z.; Wang, X.; Wu, Y. Contamination, sources and health risk of heavy metals in soil and dust from different functional areas in an industrial city of Panzhihua City, Southwest China. *J. Hazard. Mater.* **2021**, *420*, 126638. [CrossRef]
7. Chabukdhara, M.; Nema, A.K. Heavy metals assessment in urban soil around industrial clusters in Ghaziabad, India: Probabilistic health risk approach. *Ecotoxicol. Environ. Saf.* **2013**, *87*, 57–64. [CrossRef]
8. Dartan, G.; Taşpınar, F.; Toröz, İ. Assessment of heavy metals in agricultural soils and their source apportionment: A Turkish district survey. *Environ. Monit. Assess.* **2015**, *187*, 99. [CrossRef]
9. Davis, H.T.; Marjorie Aelion, C.; McDermott, S.; Lawson, A.B. Identifying natural and anthropogenic sources of metals in urban and rural soils using GIS-based data, PCA, and spatial interpolation. *Environ. Pollut.* **2009**, *157*, 2378–2385. [CrossRef]
10. Gope, M.; Masto, R.E.; George, J.; Hoque, R.R.; Balachandran, S. Bioavailability and health risk of some potentially toxic elements (Cd, Cu, Pb and Zn) in street dust of Asansol, India. *Ecotoxicol. Environ. Saf.* **2017**, *138*, 231–241. [CrossRef]
11. Li, Y.; Dong, Z.; Feng, D. Study on the risk of soil heavy metal pollution in typical developed cities in eastern China. *Sci. Rep.* **2022**, *12*, 3855. [CrossRef]
12. Elamin, M.E.; Bradberry, S.; Dear, J.W. Poisoning by metals. *Medicine* **2020**, *48*, 205–207. [CrossRef]
13. Li, Y.; Shen, R.; Wu, H.; Yu, L.; Wang, Z.; Wang, D. Liver changes induced by cadmium poisoning distinguishedby confocal Raman imaging. *Spectrochim. Acta A Mol. Biomol. Spectrosc.* **2020**, *225*, 117483. [CrossRef] [PubMed]
14. Ma, Y.; Huo, R.; Liu, C. Speciation and fractionation of rare earth elements in a Lateritic profile from southern China: Identification of the carriers of Ce anomalies. In Proceedings of the Goldschmidt Conference, Davos, Switzerland, 18–23 August 2002.
15. Guo, G.; Wu, F.; Xie, F.; Zhang, R. Spatial distribution and pollution assessment of heavy metals in urban soils from southwest China. *J. Environ. Sci.* **2012**, *24*, 410–418. [CrossRef] [PubMed]
16. Xue, S.; Ke, W.; Zeng, J.; Tabelin, C.B.; Xie, Y.; Tang, L.; Xiang, C.; Jiang, J. Pollution prediction for heavy metals in soil-groundwater systems at smelting sites. *Chem. Eng. J.* **2023**, *473*, 145499. [CrossRef]
17. He, B.; Yun, Z.; Shi, J.; Jiang, G. Research progress of heavy metal pollution in China: Sources, analytical methods, status, and toxicity. *Chin. Sci. Bull.* **2012**, *58*, 134–140. [CrossRef]
18. Zheng, F.; Guo, X.; Tang, M.; Zhu, D.; Wang, H.; Yang, X.; Chen, B. Variation in pollution status, sources, and risks of soil heavy metals in regions with different levels of urbanization. *Sci. Total Environ.* **2023**, *866*, 161355. [CrossRef]
19. Li, Y.; Wang, Y.; Li, Y.; Li, T.; Mao, H.; Talbot, R.; Nie, X.; Wu, C.; Zhao, Y.; Hou, C.; et al. Characteristics and potential sources of atmospheric particulate mercury in Jinan, China. *Sci. Total Environ.* **2017**, *574*, 1424–1431. [CrossRef]
20. Wang, X.; Liu, E.; Yan, M.; Zheng, S.; Fan, Y.; Sun, Y.; Li, Z.; Xu, J. Contamination and source apportionment of metals in urban road dust (Jinan, China) integrating the enrichment factor, receptor models (FA-NNC and PMF), local Moran's index, Pb isotopes and source-oriented health risk. *Sci. Total Environ.* **2023**, *878*, 163211. [CrossRef]
21. Huang, J.; Guo, S.; Zeng, G.M.; Li, F.; Gu, Y.; Shi, Y.; Shi, L.; Liu, W.; Peng, S. A new exploration of health risk assessment quantification from sources of soil heavy metals under different land use. *Environ. Pollut.* **2018**, *243*, 49–58. [CrossRef]

22. Laidlaw, M.A.S.; Zahran, S.; Pingitore, N.; Clague, J.; Devlin, G.; Taylor, M.P. Identification of lead sources in residential environments: Sydney Australia. *Environ. Pollut.* **2014**, *184*, 238–246. [CrossRef]
23. Liu, Z.; Zhou, J.; Zhang, J.; Mao, Y.; Huang, X.; Qian, G. Evaluation for the heavy metal risk in fine particulate matter from the perspective of urban energy and industrial structure in China: A meta-analysis. *J. Clean. Prod.* **2020**, *244*, 118597. [CrossRef]
24. Jalali, M.; Ghanbari, M.; Jalali, M. Geochemical and environmental health threat evaluation of heavy metals in groundwater of Asad Abad, Hamedan, Iran. *Environ. Earth Sci.* **2019**, *78*, 699. [CrossRef]
25. Chen, H.; Teng, Y.; Lu, S.; Wang, Y.; Wu, J.; Wang, J. Source apportionment and health risk assessment of trace metals in surface soils of Beijing metropolitan, China. *Chemosphere* **2016**, *144*, 1002–1011. [CrossRef] [PubMed]
26. Chen, M.; Chen, X.; Xing, Y.; Liu, Y.; Zhang, S.; Zhang, D.; Zhu, J. Arsenic and cadmium in soils from a typical mining city in Huainan, China: Spatial distribution, ecological risk assessment, and health risk assessment. *Bull. Environ. Contam. Toxicol.* **2021**, *107*, 1080–1086. [CrossRef]
27. Facchinelli, A.; Sacchi, E.; Mallen, L. Multivariate statistical and GIS Based approach to identify heavy metal sources in soils. *Environ. Pollut.* **2001**, *114*, 313–324. [CrossRef]
28. Fang, H.; Gui, H.; Li, J.; Yu, H.; Wang, M.; Jiang, Y.; Wang, C.; Chen, C.; Zhang, Y.; Huang, Y. Risks assessment associated with different sources of metals in abandoned soil of metals in abandoned soil of Zhuxianzhuang coal mine, Huaibei coalfield (Anhui, China). *Bull. Environ. Contam. Toxicol.* **2021**, *106*, 370–376. [CrossRef]
29. Gao, L.; Hu, X.; Ma, C.; Kuang, H.; Qi, H.; He, Z. Geoenvironmental risk evaluation of high-efficiency eco-economic zone in Weifang city, China. *Nat. Hazards Rev.* **2021**, *21*, 05020005. [CrossRef]
30. Jain, A.; Wadhera, S.; Malhotra, P.; Singh, C.; Lad, D.P.; Khadwal, A.R.; Prakash, G.; Varma, N.; Reddy, P. Association between Heavy Metal Exposure and Idiopathic Aplastic Anemia in Adolescents and Adults: A Case Control Study from Northern India. *Blood* **2023**, *142*, 7323. [CrossRef]
31. *GBT 36197-2018*; Soil Quality—Guidance on Sampling Techniques. Standardization Administration of China: Beijing, China, 2018.
32. Chen, X.; Ren, Y.; Li, C.; Shang, Y.; Ji, R.; Yao, D.; He, Y. Studyon Factors Influencing the Migration of Heavy Metals from Soil to Vegetables in a Heavy Industry City. *Sustainability* **2024**, *16*, 11084. [CrossRef]
33. *HJ/T 166-2004*; Environmental Monitoring Technical Requirements—Soil Sampling. Ministry of Environmental Protection of the People's Republic of China: Beijing, China, 2004.
34. Muller, G. Index of geo-accumulation in sediments of the Raine River. *GeoJournal* **1969**, *2*, 108–118.
35. Peng, C.; Wang, M.; Chen, W.; Chang, A.C.; Crittenden, J.C. Mass balance-based regression modeling of Cd and Zn accumulation in urban soils of Beijing. *J. Environ. Sci.* **2017**, *53*, 99–106. [CrossRef]
36. Hakanson, L. An ecological risk index for aquatic pollution control sedimentological approach. *Water Res.* **1980**, *14*, 975–1001. [CrossRef]
37. Zhengqi, X.; Shijun, N.; Xianguo, T.; Chengjiang, Z. Calculation of toxicity coefficient of heavy metals in assessment of potential ecological risk index. *J. Environ. Sci. Technol.* **2008**, *2*, 112–115.
38. Longtao, X.; Jianjun, P.; Haoran, B.; Peiyu, Z.; Xiang, K.; Fei, W.; Yanxiang, Z. GIS-based spatial distribution and pollution assessment of heavy metals in farmland soils: A case study of a township in Jiangning District, Nanjing. *Soil J.* **2020**, *57*, 316–325.
39. *GSB 07-3903-2021*; Heavy Metal Elements in Soil(Loam) Certified Reference Material. China Environmental Press: Beijing, China, 2021.
40. *GB36600-2018*; Soil Environmental Quality—Soil Pollution Risk Control Standard for Construction Land (Trial). China Environmental Press: Beijing, China, 2018.
41. Ayaz, H.; Nawaz, R.; Nasim, I.; Irshad, M.A.; Irfan, A.; Khurshid, I.; Okla, M.K.; Wondmie, G.F.; Ahmed, Z.; Bourhia, M. Comprehensive human health risk assessment of heavy metal contamination in urban soils: Insights from selected metropolitan zones. *Front. Environ. Sci.* **2023**, *11*, 1260317. [CrossRef]

processes MDPI

Article

The Effects of Water Management, Foliar Fertilizers, and Lime Application on the Accumulation of Cd and As in Rice Grains Based on a Field Trial

Qiming Wang [1], Wanchuan Du [2], Xiaodan Jin [1], Jian Wang [2], Yannan Lu [1], Dating Huang [2], Jiaying Nong [1], He Huang [2,*], Tian Xie [2] and Biao Han [1,*]

[1] Scientific Research Academy of Guangxi Environmental Protection, Nanning 530022, China; qmwang_gx@163.com (Q.W.); jinxiaodan0773@163.com (X.J.); l_yannan@163.com (Y.L.); 15878198561@163.com (J.N.)

[2] Guangxi Bohuan Environmental Consulting Services Co., Ltd., Nanning 530007, China; watrix@163.com (W.D.); bskwangjian@163.com (J.W.); 18776741950@163.com (D.H.); tianckstar@163.com (T.X.)

* Correspondence: river_63@163.com (H.H.); hannbiaoh@163.com (B.H.)

Abstract: The aim of this study was to explore the feasible measurement of the control of cadmium (Cd) and arsenic (As) accumulation in rice grains. A field experiment was carried out to research the effect of different treatments, including spraying silicon (Si)/selenium (Se) foliar fertilizers, the application of lime (CaO), water management (continuous flooding), and the co-application of foliar fertilizers and flooding, on Cd and As accumulation in rice grains in Guangxi Province. The results indicate that Cd accumulation in rice grains decreased under different treatments and Cd content in rice grains reached the threshold of 0.2 mg kg^{-1}. In the single technical treatments, CaO application, flooding, spraying foliar Se fertilizer, and spraying foliar Si fertilizer decreased Cd content by 73.15%, 60.44%, 45.76%, and 36.07%, respectively. However, flooding and CaO amendment enhanced As accumulation in rice grains. The co-application of flooding and spraying foliar fertilizer can simultaneously reduce Cd and As in rice grains. In addition, they resulted in lower Cd content than the single technical treatments. Among the treatments, the lowest bioaccumulation factors of Cd and As were found after the co-application of flooding and Si foliar fertilizer, which decreased these factors by 74.02% and 22.72%, respectively. These results suggest that spraying foliar Si fertilizer combined with flooding may be a promising method to synchronously inhibit the accumulation of Cd and As in rice.

Keywords: flooding; foliar spraying; CaO application; rice accumulation; heavy metals

1. Introduction

The heavy metals in soil are mainly derived from anthropogenic and natural sources of pollution. With the rapid development of industry (e.g., mining, smelting, etc.), an increasing amount of heavy metals has been brought into farmland soils [1,2]. In addition, heavy metal-contaminated areas have aroused increasing attention. The natural content of heavy metals mainly depends on the parent material of the soil and the topography [3]. In China, according to the national soil contamination survey in 2014, over 7% of sampling sites' Cd contents exceeded the standard value. The natural background value of Cd ranges from 0.02 to 6.2 mg kg^{-1} around the world. In China, higher contents of natural Cd in surface soils are commonly found in Guangxi Province [4]. In Guangxi, the natural content of Cd in surface soils ranges from 0.001 to 13.430 mg kg^{-1}, which is 1.75 times the average level found in other regions of China [5]. Today, increasing attention has been paid to heavy metal-contaminated farmland, which is caused by natural factors.

Rice (*Oryza sativa* L.) is a staple food for billions of people in the world, and rice has a notable feature in its accumulation of cadmium (Cd) [6]. It has been demonstrated that rice consumption is one of the major exposure pathways to Cd for humans. This toxic metal is deleterious to human health. One toxicologic study reported that Cd can do harm to the nephridium and skeleton, causing many diseases [7]. For example, itai-itai disease has been caused by Cd [8].

Due to Cd's high risks and potential adverse effects on human health, methods for remediating heavy metal-contaminated soil, such as phytoremediation [9], chemical washing [10], and chemical immobilization [11], have been previously developed. Additionally, many agronomic mitigation strategies have been used to minimize the accumulation and translocation of Cd in rice grains from soils, such as planting cultivars with water and nutrient management, the application of foliar fertilizer, and soil amendments [12–15]. As one kind of water management regime, soil flooding can control Cd accumulation in rice grains by regulating the pH, Eh, and Cd fractions in the soil [13]. Foliar fertilizer can strengthen the antioxidant defense system and decrease the heavy metal content in plants [16].

The biogeochemical behavior of Cd is different from As in soil [17], which results in decreasing Cd accumulation while increasing the risk of As accumulation in rice under the same technology. The anaerobic condition in soil decreases Cd accumulation but increases As accumulation in rice [18]. Similar results have been found in field experiments; continuous flooding results in low Cd content but high As content in rice [19]. When the soil environment is in a reduced state, As(V) is reduced to As(III), and the iron and manganese oxides—which have strong binding ability with As—are reduced, thereby increasing As mobility and bioavailability. Therefore, attention should be paid to As accumulation in rice when remediating Cd-contaminated paddy soil. In addition, an effective treatment should be further studied to synchronously regulate Cd and As uptake in rice grains.

In recent studies, the treatment of heavy metal contamination in rice and soil has mostly included potted experiments with consistent conditions [20,21]. In field conditions, heavy metals in rice grains are easily influenced by various environmental factors such as temperature and precipitation. Therefore, field plot experiments are necessary to demonstrate the effect of these treatments on heavy metal accumulation in rice [22].

In this study, a field-scale experiment was carried out in a typical area with a high Cd background to investigate the effects of water management, the application of foliar fertilizer, and lime (CaO) amendments on Cd and As accumulation in rice. The objectives of this study include the following: (i) to evaluate the effect of a single technical treatment on Cd and As accumulation in rice grains; (ii) to find out the effects of combined water management and spraying foliar fertilizer on Cd and As accumulation in rice grains; and (iii) to elucidate the associated factors influencing Cd accumulation in rice. This study could provide a practical basis for field-scale Cd remediation in high-Cd-background areas.

2. Materials and Methods

2.1. Experimental Field and Materials

The studied field is located in the center of the Guangxi Zhuang Autonomous Region of China. This region has a subtropical monsoon climate, with a mean annual precipitation of 1360 mm and a mean annual temperature of 22.6 °C. Rice is the staple crop here, and Wuyouhuazhan is one of the most widely grown cultivars of rice (*Oryza sativa* L.) in the local area. According to a previous investigation, there were no industrial or mining pollution sources around the experimental field, and the irrigation water met the limit values of Grade III of the Environmental Quality Standards for Surface Water (GB 3838-2002) [23]. Additionally, it has been reported that this area has the characteristic of a geological background of high amounts of heavy metals in soils [24]. The basic properties of the soil are listed in Table 1. According to the Environmental Quality Standard for Soils of China (GB 15618-2018) [25], Cd content in the soil exceeded the limitation.

Table 1. Physical–chemical properties of soil in the experiment field.

pH	SOM	CEC	Cd
6.69 ± 0.10	40.82 ± 9.79 g kg^{-1},	9.67 ± 0.66 cmol kg^{-1}	0.72 ± 0.12 mg kg^{-1}

SOM: soil organic matter; CEC: cation exchange capacity. Data are means ± SD (n = 6).

The tested rice cultivar (Wuyouhuazhan) and soil amendments were purchased in markets. The Si foliar fertilizer was purchased from Environmental Protection Bridge (Hunan) Ecological Environment Engineering Co., Ltd. (Changsha, China). The Se foliar fertilizer was purchased from Guangxi Penshibao Co., Ltd. (Beihai, China)

2.2. Field Experiment

A total of seven treatments were set up in this study (Table 2): (1) control (CK)—the plot followed the local conventional operating practices; (2) the foliar spray of silicon (Si) fertilizer (T2)—the foliar spray of Si fertilizer alone was performed at the booting stage and initial filling stage at a dose of 7500 mL ha^{-2}; (3) the foliar spray of selenium (Se) fertilizer (T3)—the foliar spray of Se fertilizer alone was performed at the booting stage and initial filling stage at a dose of 3840 mL ha^{-2}; (4) water management (T4)—the experiment plot was flooded with a water depth of 3–5cm during the entire rice growth season; (5) water management and the foliar spray of Se fertilizer (T5); (6) water management and the foliar spray of Si fertilizer (T6); and (7) the application of soil amendment (T7)—CaO was incorporated into arable layer with basic fertilizers by mechanical plowing at the addition of 1125 kg ha^{-1}.

Table 2. Test treatments and operating procedures.

Abbreviation	Water Management	Foliar Spraying	Application of Soil Amendment
CK	Conventional water management	No	No
T1	Conventional water management	Foliar spray of Si fertilizer	No
T2	Conventional water management	Foliar spray of Se fertilizer	No
T3	Continuous flooding throughout the growth stage	No	No
T4	Continuous flooding throughout the growth stage	Foliar spray of Si fertilizer	No
T5	Continuous flooding throughout the growth stage	Foliar spray of Se fertilizer	No
T6	Conventional water management	No	Application of CaO

CK: control; T1: only foliar spray of Si fertilizer; T2: only foliar spray of Se fertilizer; T3: flooding of soil with water; T4: T1 combined with T3; T5: T1 combined with T4; T6: application of CaO.

Before use, the Si and Se foliar fertilizers were diluted 50 times with water. The Si and Se foliar fertilizer spraying was conducted during the period of the booting stage and the early filling stage of rice. The area of each experimental plot area was 30 m^2 (5 m × 6 m), with a protective row 1.5 m wide around each sub-plot. Each treatment was repeated three times randomly. Before rice transplantation, chemical fertilizers containing 20% N, 10% P, and 10% K were incorporated into the arable layer at doses of 450 kg ha^{-2} as basic fertilizers.

2.3. Sample Collection and Pre-Treatment

Rice samples and the corresponding 0–20 cm arable layer fresh soil samples were collected in October of 2020. Each soil sample was a mixture of 5 sub-samples in each experiment plot. In addition, the rice samples were collected in the corresponding soil sampling sites. A total of 2 kg of soil and 1 kg of rice grains was collected for one experiment plot. The soil was air-dried at room temperature (25 ± 1 °C), and then, one part of the soil was sieved through a 20-mesh nylon screen for the determination of pH and the content of available Cd. The other part was passed through a 100-mesh nylon screen for the determination of total As and Cd content. Rice grain samples were washed with deionized water and oven-dried at 105 °C for 1 h, and then, they were left at 70 °C until obtaining a constant weight. After husking, the grain was ground and passed through a 100-mesh nylon screen for digestion. The prepared soil and rice samples were kept in sealed polyethylene bags and placed in a desiccator.

2.4. Sample Analysis

Soil pH was measured in a 2.5:1 (*v*/*w*) water-to-soil suspension using a digital pH meter (Mettler Toledo FE28, SUI). Total soil Cd was digested with HNO_3-HF-$HClO_4$ [26]. The available Cd in soil was extracted with 0.005 M DTPA (diethylene triamine pentaacetate acid) [27]. Total soil As was digested with HCl and H_2SO_4 at a rate of 1:1 (*v*/*v*). The rice grain samples were digested using the microwave-assisted method described in a previous study [28]. Total As and Cd contents in rice grains were determined following digestion with HCl/HNO_3 (3/1, *v*/*v*) and HCl/HNO_3 (3/1, *v*/*v*). The content of Cd in rice and soil was determined via atomic absorption spectrophotometry (Agilent 280Z, Santa Clara, CA, USA). The content of As was determined via atomic fluorescence spectrometry (Titan AFS-8330, Beijing, China). Replicate samples, blanks, standard reference soil (GBW07454, GSS-25), and rice (GBW10044, GSB-22) from the National Research Center for Standards in China were included in chemical analyses for quality control.

2.5. Data Analysis

The enrichment behavior of heavy metals in the soil–rice system was evaluated using bioaccumulation factor (BCF). The bioaccumulation factor was established according to the following equation [29].

$$BCF_i = C_{i,rice}/C_{i,soil}$$

where $C_{i,rice}$ and $C_{i,soil}$ are the contents of heavy metal i in rice and soil, respectively.

The data were calculated using Microsoft Excel 2016. The relationships between soil properties and the BCF of heavy metals in rice grains were determined using Pearson's correlation analysis. Significant differences were evaluated using an analysis of variance (ANOVA). All these analyses were conducted using SPSS 19.0 for Windows.

3. Results

3.1. The Soil Contents of Cd and As in Different Treatments

The information on heavy metal content in soil is listed in Table 3. The contents of total As, total Cd, and available Cd were 7.48~10.68 mg kg^{-1}, 0.57~0.74 mg kg^{-1}, and 0.18~0.32 mg kg^{-1}, respectively. According to with the soil environmental quality risk control standard for soil contamination of agricultural land (GB 15168-2018), the total As did not exceed the risk screening values, but the total Cd exceeded the relative risk screening values. Soil pH ranged from 6.41 to 6.52. There was no significant difference in soil pH between the CK and other treatments except for the T6 treatment. The application of soil conditioner (T6) significantly improved the value of pH by 0.24 and decreased the available Cd content by 26%.

Table 3. The heavy metals of soil in different treatments.

Treatment	pH	Total As (mg kg^{-1})	Total Cd (mg kg^{-1})	Available Cd (mg kg^{-1})
CK	6.47 ± 0.04 b	8.69 ± 0.64 a	0.74 ± 0.10 a	0.31 ± 0.03 a
T1	6.52 ± 0.07 b	10.68 ± 1.01 a	0.72 ± 0.06 a	0.32 ± 0.02 a
T2	6.47 ± 0.05 b	9.09 ± 1.17 a	0.62 ± 0.17 a	0.27 ± 0.06 a
T3	6.44 ± 0.06 b	8.84 ± 0.42 a	0.63 ± 0.08 a	0.24 ± 0.01 b
T4	6.50 ± 0.10 b	9.94 ± 1.78 a	0.61 ± 0.11 a	0.18 ± 0.01 b
T5	6.41 ± 0.03 b	8.54 ± 1.21 a	0.57 ± 0.09 a	0.22 ± 0.02 b
T6	6.71 ± 0.08 a	7.48 ± 0.27 a	0.59 ± 0.12 a	0.23 ± 0.01 b

Note: The different letters a and b indicate significant differences at the 0.05 probability level.

3.2. The Content of As in Rice Grains

In this study, the content of As in rice grains ranged from 0.191 to 0.258 mg kg^{-1} (Figure 1), which was lower than the national standard limit of 0.2 mg kg^{-1} according to the national food safety standards' contaminant limit in food (GB 2762-2017) [30]. Thus, in this area, the food chain risk of rice As was low. The content of As in rice grains followed the order of T6 > T3 > CK > T1 > T2 > T4 > T5. Flooding (T3) and the application of CaO (T6) increased the content of As in rice grains by 10.85% and 12.01%, respectively. The foliar application of Si and Se decreased As content by 11.02% (T1) and 12.30% (T2), respectively. The lowest content of As was found in the T5 treatment (0.19 mg kg^{-1}), the combination of flooding and the foliar spray of silicon decreased As content by 17.08%.

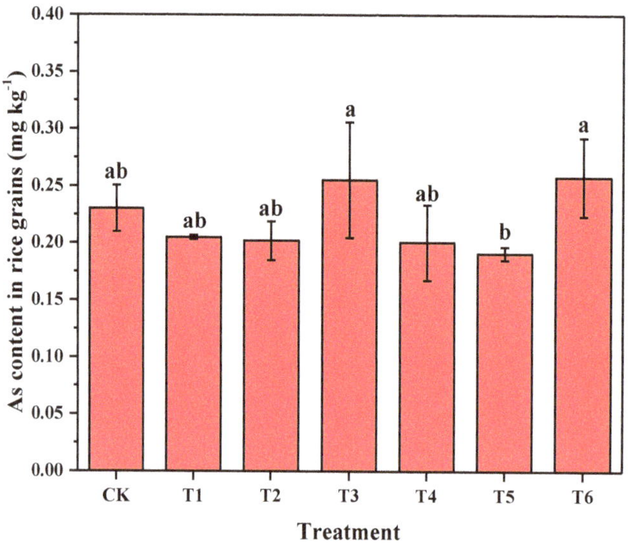

Figure 1. Content of As in rice grains in different treatments. (Note: the different letters a and b indicate significant differences at the 0.05 probability level).

3.3. The Content of Cd in Rice Grains

The content of Cd in rice grains was 0.25 mg kg^{-1} in the CK treatment (Figure 2), which exceeded the standard limit of 0.2 mg·kg^{-1} (GB 2762-2017) [30]. These results indicate that this area might have a rice Cd food chain risk.

After the treatments, the content of Cd in rice grains decreased and was lower than 0.2 mg kg^{-1}. After foliar fertilizer spraying, the Cd content in rice grains decreased by 36.07% and 45.76% in the T1 and T2 treatments, respectively, indicating that the application of Se has a better result than Si in reducing cadmium content in rice. Additionally, both water regulation and the application of soil amendments decreased the content of Cd in rice grains, which decreased by 60.44% and 73.15% in T3 and T6, respectively. The combination of foliar spray fertilizer and water regulation can significantly decrease the Cd content by 79.37% and 76.29% in the T4 and T5 treatments, respectively.

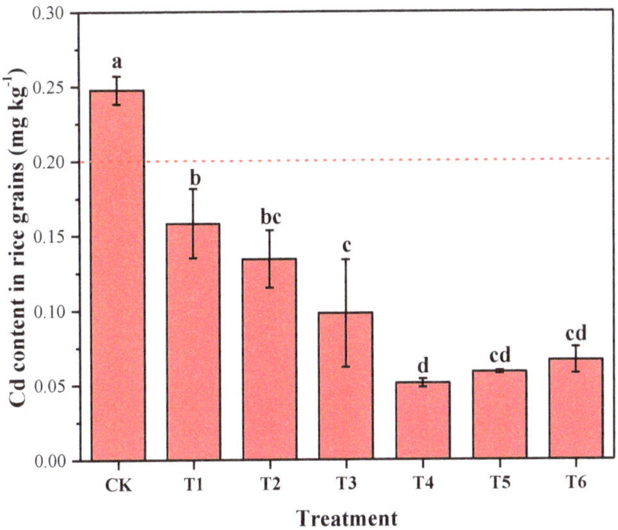

Figure 2. Content of Cd in rice grains in different treatments. (Note: the different letters a, b, c, and d indicate significant differences at the 0.05 probability level. The dashed line represents the threshold of rice grains in GB 2762-2017).

3.4. Bioaccumulation Factor of Heavy Metals in Rice Grains

As shown in Figure 3, the BCF of Cd was higher than that of As, which indicated that Cd was more likely to accumulate in rice grains than As. Additionally, all treatments decreased the BCF of Cd in rice grains. The BCF of Cd followed the ascending order of T4 < T5 < T6 < T3 < T2 < T1 < CK treatments. The co-application of flooding and spraying foliar fertilizer has better results in decreasing Cd accumulation in rice grains than the single technical treatment. In addition, spraying foliar Si fertilizer is better than Se fertilizer. For As, the BCF of As followed the ascending order of T1 < T4 < T2 < T5 < CK < T3 < T6 treatments. T3 and T6 enhanced the As accumulation in rice grains by 8.25% and 30.92%, respectively. The lowest As content was found in the T1 treatment, which decreased the BCF of As by 27.03%.

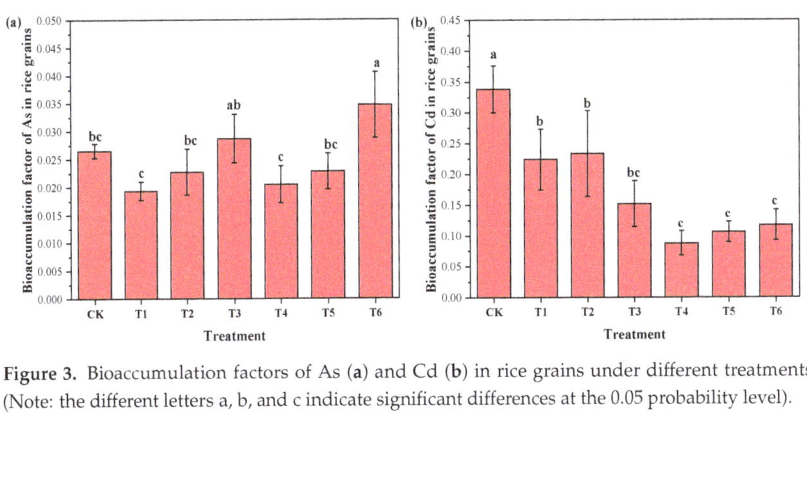

Figure 3. Bioaccumulation factors of As (**a**) and Cd (**b**) in rice grains under different treatments. (Note: the different letters a, b, and c indicate significant differences at the 0.05 probability level).

4. Discussion

This study demonstrated that on a field scale, Cd and As accumulations in rice grains were greatly affected by spraying foliar Si/Se fertilizer, continuous flooding, and the application of CaO. As a beneficial element thereof, Si can enhance plant growth, improve the ability of plants to overcome adverse stresses, and alleviate the toxicity of heavy metals [31,32]. In this study, the application of foliar Si fertilizer decreased As and Cd accumulation in rice grains, which was in accordance with a previous study [16]. According to previous studies, the main reason for Si's decrease in Cd uptake and accumulation in rice grains may be mainly due to the fact that rice can take up monosilicic acid and predominantly deposit this in the leaves of special silica cells, forming a Si cuticle double layer [15,33], which can inhibit Cd transport in rice. Additionally, monosilicic acid in rice leaves can compete with As for the same Si transporters during absorption and xylem unloading [34], which block the transport of As. Moreover, Se foliar fertilizer has a better effect on decreasing Cd and As accumulation in rice grains than Si foliar fertilizer. Although Se is not an essential nutrient for plant growth, many studies have shown that Se and Cd have antagonistic effects on plants [35]. Additionally, Se plays an important role in plant antioxidant systems, which can resist the oxidative stress of Cd and keep Cd in the wall of leaves, thereby helping to stimulate plant growth, alleviate heavy metal toxicity in rice, and, thus, inhibit the migration of Cd to grains [36,37]. For As, Se can promote the synthesis of plant chelates in root cells, increase the vacuolar binding of As to plant chelates, and, thus, reduce the accumulation of As in rice grains [38]. The heading stage and filling stage are the key periods of heavy metal accumulation in rice [39,40]. The application of foliar fertilizer can be effective in decreasing the heavy metal accumulation in rice grains in these periods.

It has been demonstrated that water management influences heavy metal fractions in soils, thereby affecting plant availability. When the soil floods, the anaerobic condition decreases the content of available Cd in soils because Cd is combined with sulfur (S) to form CdS at a low redox potential [41]. The accumulation of heavy metals in rice largely depends on the content of available heavy metals [29]. In this study, available Cd content in soil is significantly positively correlated with the BCF of Cd in rice grains (Figure 4b), which is in accordance with a previous study [42]. Soil pH can affect the adsorption–desorption and precipitation–dissolution of heavy metals in soil, and, thus, it is strongly related to the bioaccumulation of heavy metals in soil–rice systems [29]. The application of CaO significantly decreased Cd accumulation in rice mainly due to the improvement of soil pH, which has been reported before [43]. In this study, soil pH is negatively correlated with the bioaccumulation of Cd (Figure 4). Cd exists in cation forms in soil. The improvement of soil pH increased the soil's negative charge and caused more formations of strongly bound hydroxyl species of Cd in soil [44].

However, there were large differences in bioavailability between Cd and As in soil under the flooding condition [45]. One previous study reported that flooding caused an anaerobic environment in soil and that the redox potential decreased, which caused As(V) to be reduced to As(III), thereby enhancing the toxicity and mobility of As in soil [46]. In addition—different from Cd—As exists predominantly in anionic forms; the improvement of pH would have opposite effects on the mobility and availability of As in soil [17]. Water management and the application of CaO changed the redox environment and pH in soil, and, thus, affected the fractions of Cd and As in soil. A large number of studies have also shown that the accumulation of Cd in rice decreased while that of As in rice increased [17,47]. However, in this study, the increase in As accumulation in rice grains is not significant in the T3 and T6 treatments, which may mainly be due to the low content of As in soil.

Due to the different chemical properties, the responses of Cd and As to soil moisture are opposite. A single flooding treatment increases the risk of As bioavailability. The co-application of flooding and the application of foliar fertilizers can simultaneously decrease Cd and As accumulation in rice grains and significantly reduce Cd uptake in rice. The Cd content in rice grains reaches the threshold of 0.2 mg·kg^{-1} in the co-application treatment. A previous study also indicated that the co-application of multiple technologies has the advantage of remediating heavy metal pollution in soil and rice compared to single technical measures [48]. Zeng et al. [16] found that continuous flooding co-applied with foliar-sprayed Si could effectively decrease Cd and As accumulation more in rice grains.

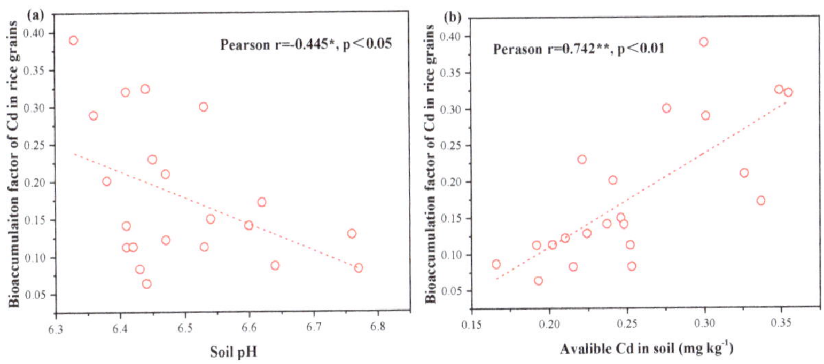

Figure 4. Correlation between soil properties and BCF of Cd in rice grains. (**a**) The relationship between soil pH and BCF of Cd in rice grains, (**b**) The relationship between soil available Cd and BCF of Cd in rice grains. The circles represent the BCF of the sample in different soil pH or available Cd content. The lines represent the linear fit curve. * and ** indicate the significantly correlation between soil properties and BCF at the 0.05 and 0.01 probability level, respectively.

5. Conclusions

All treatments in this study can effectively decrease Cd accumulation in rice grains and reach the Chinese national standard limit. The available content of Cd was positively related to the BCF of Cd in rice grains. Flooding and the application of CaO significantly decreased the bioavailability of Cd content in soil and, thus, decreased Cd content in rice. In addition, the two treatments studied have better results in inhibiting Cd accumulation in rice when compared to foliar applications with Si and Se. However, flooding and CaO application increased As accumulation in rice due to the changes in redox potential and pH in the soil. The co-application of flooding and foliar applications with Si/Se fertilizer can inhibit the accumulation of Cd and As in rice grains simultaneously, and their Cd and As contents were less than those with the single technical treatment. Especially in the T4 treatment, the co-application of flooding and Si foliar fertilizer decreased the Cd and As contents in rice grains by 74.02% and 22.72%. In conclusion, it was feasible to solve the problem of Cd excess in rice in areas with high geological background soil when taking into consideration the low cost, easy availability, and high efficiency of the co-application of flooding and foliar application with fertilizer. However, this mechanism should be further investigated on a field scale to optimize these co-application measures.

Author Contributions: Conceptualization, H.H.; Methodology, Q.W., W.D., X.J. and J.W.; Software, W.D., Y.L. and D.H.; Formal analysis, X.J., J.N. and H.H.; Investigation, Y.L.; Writing—original draft, Q.W.; Writing—review & editing, H.H., T.X. and B.H.; Visualization, J.W.; Funding acquisition, T.X. All authors have read and agreed to the published version of the manuscript.

Funding: This research was supported by the Guangxi Key Research and Development Program (Grant No. Guike AB23075157 and AB21196037) and the Science and Technology Program of Nanning City (RC20220104).

Data Availability Statement: The original contributions presented in this study are included in the article; further inquiries can be directed to the corresponding authors.

Conflicts of Interest: The authors Wanchuan Du, Jian Wang, Dating Huang, He Huang, and Tian Xie are employed by Guangxi Bohuan Environmental Consulting Services Co., Ltd. The remaining authors declare that the research was conducted in the absence of any commercial or financial relationships that could be construed as potential conflicts of interest. Furthermore, the funding sponsors had no role in the design of the study; in the collection, analysis, or interpretation of the data; in the writing of the manuscript; or in the decision to publish the results.

References

1. Zhang, R.; Zhao, Y.; Lin, J.; Hu, Y.; Hanninen, H.; Wu, J. Biochar application alleviates unbalanced nutrient uptake caused by N deposition in Torreya grandis trees and seedlings. *For. Ecol. Manag.* **2019**, *432*, 319–326. [CrossRef]
2. Li, Z.; Ma, Z.; van der Kuijp, T.J.; Yuan, Z.; Huang, L. A review of soil heavy metal pollution from mines in China: Pollution and health risk assessment. *Sci. Total Environ.* **2014**, *468*, 843–853. [CrossRef] [PubMed]
3. Jin, G.; Fang, W.; Shafi, M.; Wu, D.; Li, Y.; Zhong, B.; Ma, J.; Liu, D. Source apportionment of heavy metals in farmland soil with application of APCS-MLR model: A pilot study for restoration of farmland in Shaoxing City Zhejiang, China. *Ecotoxicol. Environ. Saf.* **2019**, *184*, 109495. [CrossRef] [PubMed]
4. China MoEPo. *Chinese Environmental Monitoring Station Background Concentrations of Elements in Chinese Soils*; Chinese Environmental Science Publisher: Beijing, China, 1990. (In Chinese)
5. Li, D.; Li, W.; Qin, L.; Li, Y.; Ning, L.; Xu, H.; Ren, Z.; Zhang, Y.; Wang, J. Cadmium bioavailability well assessed by DGT and factors influencing cadmium accumulation in rice grains from paddy soils of three parent materials. *J. Soil. Sediment.* **2018**, *18*, 2552–2561. [CrossRef]
6. Reeves, P.G.; Chaney, R.L. Bioavailability as an issue in risk assessment and management of food cadmium: A review. *Sci. Total Environ.* **2008**, *398*, 13–19. [CrossRef]
7. Satarug, S.; Vesey, D.A.; Gobe, G.C. Health Risk Assessment of Dietary Cadmium Intake: Do Current Guidelines Indicate How Much is Safe? *Env. Health Persp.* **2017**, *125*, 284–288. [CrossRef] [PubMed]
8. Kobayashi, J. Pollution by cadmium and the itai-itai disease in Japan. In *Toxicity of Heavy Metals in the Environment*; Marcel Dekker, Inc.: New York, NY, USA, 1978; pp. 199–260.
9. Gil-Loaiza, J.; White, S.A.; Root, R.A.; Solís-Dominguez, F.A.; Hammond, C.M.; Chorover, J.; Maier, R.M. Phytostabilization of mine tailings using compost-assisted direct planting: Translating greenhouse results to the field. *Sci. Total Environ.* **2016**, *565*, 451–461. [CrossRef]
10. Makino, T.; Sugahara, K.; Sakurai, Y.; Takano, H.; Kamiya, T.; Sasaki, K.; Itou, T.; Sekiya, N. Remediation of cadmium contamination in paddy soils by washing with chemicals: Selection of washing chemicals. *Env. Pollut.* **2006**, *144*, 2–10. [CrossRef]
11. Li, B.; Zhou, S.; Wei, D.; Long, J.; Peng, L.; Tie, B.; Williams, P.N.; Lei, M. Mitigating arsenic accumulation in rice (*Oryza sativa* L.) from typical arsenic contaminated paddy soil of southern China using nanostructured α-MnO2: Pot experiment and field application. *Sci. Total Environ.* **2019**, *650*, 546–556. [CrossRef]
12. Zhang, Q.; Chen, H.; Huang, D.; Xu, C.; Zhu, H.; Zhu, Q. Water managements limit heavy metal accumulation in rice: Dual effects of iron-plaque formation and microbial communities. *Sci. Total Environ.* **2019**, *687*, 790–799. [CrossRef]
13. Wan, Y.; Huang, Q.; Camara, A.Y.; Wang, Q.; Li, H. Water management impacts on the solubility of Cd, Pb, As, and Cr and their uptake by rice in two contaminated paddy soils. *Chemosphere* **2019**, *228*, 360–369. [CrossRef]
14. Huang, Y.; Sheng, H.; Zhou, P.; Zhang, Y. Remediation of Cd-contaminated acidic paddy fields with four-year consecutive liming. *Ecotoxicol. Environ. Saf.* **2020**, *188*, 109903. [CrossRef] [PubMed]
15. Gao, M.; Zhou, J.; Liu, H.; Zhang, W.; Hu, Y.; Liang, J.; Zhou, J. Foliar spraying with silicon and selenium reduces cadmium uptake and mitigates cadmium toxicity in rice. *Sci. Total Environ.* **2018**, *631–632*, 1100–1108. [CrossRef] [PubMed]
16. Zeng, P.; Wei, B.; Zhou, H.; Gu, J.; Liao, B. Co-application of water management and foliar spraying silicon to reduce cadmium and arsenic uptake in rice: A two-year field experiment. *Sci. Total Environ.* **2022**, *818*, 151801. [CrossRef]
17. Zhao, F.; Wang, P. Arsenic and cadmium accumulation in rice and mitigation strategies. *Plant Soil* **2020**, *446*, 1–21. [CrossRef]
18. Ge, J.W.; Wu, S.; Wu, H.T.; Lin, J.Y.; Cai, Y.J.; Zhou, D.M.; Gu, X.Y. Prediction of As and Cd dissolution in various soils under flooding condition. *Sci. Total Environ.* **2024**, *948*, 174853. [CrossRef] [PubMed]
19. Ma, R.; Shen, J.; Wu, J.; Tang, Z.; Shen, Q.; Zhao, F.J. Impact of agronomic practices on arsenic accumulation and speciation in rice grain. *Env. Pollut.* **2014**, *194*, 217–223. [CrossRef] [PubMed]
20. Fan, H.J.; Tang, S.S.; Long, J.; He, R.J.; Xiao, Z.M.; Hou, H.B.; Peng, P.Q. Effects of exogenous chloride ions on the migration and transformation of Cd in a soil-rice system. *Front. Environ. Sci.* **2024**, *12*, 1403989. [CrossRef]
21. Ullah, A.; Ma, Y.B.; Li, J.M.; Tahir, N.; Hussain, B. Effective Amendments on Cadmium, Arsenic, Chromium and Lead Contaminated Paddy Soil for Rice Safety. *Agronomy* **2020**, *10*, 359. [CrossRef]
22. Feng, Y.J.; Huang, Y.M.; Yu, H.Y.; Zhang, X.Z. Screening of wheat materials with low cadmium accumulation in grains and characteristics of cadmium accumulation. *J. Triticeae Crops.* **2021**, *41*, 842–850.
23. *GB 3838-2002*; Environmental Quality Standards for Surface Water. Ministry of Ecology and Environment, The People's Republic of China: Beijing, China, 2002.
24. Guo, C.; Wen, Y.; Yang, Z.; Li, W.; Guan, D.; Ji, J. Factors controlling the bioavailability of soil cadmium in typical karst areas with high geogenic background. *J. Naningjing Univ. (Nat. Sci.)* **2019**, *55*, 10. (In Chinese)
25. *GB 15618-2018*; Environmental Quality Standard for Soil. China National Environmental Protection Agency: Beijing, China, 2018.
26. Element, C. Method 3051A microwave assisted acid digestion of sediments, sludges, soils, and oils. *Z Für Anal Chem.* **2007**, *111*, 362–366.

27. Baker, D.E.; Amacher, M.C. Nickel, Copper, Zinc, and Cadmium. In *Methods of Soil Analysis, Part 2: Chemical and Microbiological Properties*; American Society of Agronomy, Inc.: Madison, WI, USA, 1982; pp. 323–336.
28. Ma, L.; Wang, L.; Jia, Y.; Yang, Z. Arsenic speciation in locally grown rice grains from Hunan Province, China: Spatial distribution and potential health risk. *Sci. Total Environ.* **2016**, *557–558*, 438–444. [CrossRef]
29. Huang, H.; Mao, J.; Tan, J.; Zhong, K.; Chen, J.X.; Huang, D.; Gu, X.Y.; Zhang, C.L. Heavy metal contamination, accumulation, and risk assessment in a paddy field near Pb-Zn mine, in Guangxi Province, China. *J. Soil. Sediment* **2022**, *23*, 1345–1355. [CrossRef]
30. *GB 2762-2022*; National Food Safety Standard Maximum Levels of Contaminants in Foods. State Administration for Market Regulation, National Health Commission: Beijing, China, 2022.
31. Ma, J.F.; Yamaji, N. Silicon uptake and accumulation in higher plants. *Trends Plant Sci.* **2006**, *11*, 392–397. [CrossRef] [PubMed]
32. Memari-Tabrizi, E.F.; Yousefpour-Dokhanieh, A.; Babashpour-Asl, M. Foliar-applied silicon nanoparticles mitigate cadmium stress through physio-chemical changes to improve growth, antioxidant capacity, and essential oil profile of summer savory (*Satureja hortensis* L.). *Plant Physiol. Biochem.* **2021**, *165*, 71–79. [CrossRef] [PubMed]
33. Namiki, M.; Feng, M.J.; Takashi, I. Identification of the Silicon Form in Xylem Sap of Rice (*Oryza sativa* L.). *Plant Cell Physiol.* **2005**, *46*, 279–283.
34. Liu, C.; Wei, L.; Zhang, S.; Xu, X.; Li, F. Effects of nanoscale silica sol foliar application on arsenic uptake, distribution and oxidative damage defense in rice (*Oryza sativa* L.) under arsenic stress. *Rsc Adv.* **2014**, *4*, 57227–57234. [CrossRef]
35. Wang, Z.; Wang, Y.; Lü, J.L.; Li, T.Q.; Li, S.Q.; Nie, M.; Shi, G.Y.; Zhao, X.H. Silicon and selenium alleviate cadmium toxicity in Artemisia selengensis Turcz by regulating the plant-rhizosphere. *Environ. Res.* **2024**, *252*, 119064. [CrossRef]
36. Gu, X.; Wen, X.; Yi, N.; Liu, Y.; Wu, J.; Li, H.; Liu, G. Effect of foliar application of silicon, selenium and zinc on heavy metal accumulation in wheat grains in field studies. *Environ. Pollut. Bioavailab.* **2022**, *34*, 246–252. [CrossRef]
37. Cui, J.; Liu, T.; Li, F.; Yi, J.; Liu, C.; Yu, H. Silica nanoparticles alleviate cadmium toxicity in rice cells: Mechanisms and size effects. *Environ. Pollut.* **2017**, *228*, 363. [CrossRef]
38. Younoussa, A.; Wan, Y.N.; Yu, Y.; Wang, Q.; Li, H.F. Effect of selenium on uptake and translocation of arsenic in rice seedlings (*Oryza sativa* L.). *Ecotoxicol. Environ. Saf.* **2018**, *148*, 869–875.
39. Cai, Y.X.; Zhang, S.H.; Cai, K.Z.; Huang, F.; Pan, B.G.; Wang, W. Cd accumulation, biomass and yield of rice are varied with silicon application at different growth phases under high concentration cadmium-contaminated soil. *Chemosphere* **2020**, *242*, 125128. [CrossRef] [PubMed]
40. Huang, B.Y.; Zhao, F.J.; Wang, P. The relative contributions of root uptake and remobilization to the loading of Cd and As into rice grains: Implications in simultaneously controlling grain Cd and As accumulation using a segmented water management strategy. *Environ. Pollut.* **2021**, *293*, 118497. [CrossRef] [PubMed]
41. Hu, P.; Huang, J.; Ouyang, Y.; Wu, L.; Wang, S.; Zhu, L.; Han, C.; Zhou, L.; Huang, Y.; Luo, Y.; et al. Water management affects arsenic and cadmium accumulation in different rice cultivars. *Environ. Geochem. Health* **2013**, *35*, 767–778. [CrossRef] [PubMed]
42. Chen, J.; Huang, H.; Zhang, C.; Huang, D.; Zhu, Y.; Chai, X. The mechanism of Cd sorption by silkworm excrement organic fertilizer and its effect on Cd accumulation in rice. *J. Soil. Sediment.* **2022**, *22*, 2184–2195. [CrossRef]
43. Du, Y.Y.; Wang, X.; Ji, X.H.; Zhang, Z.X.; Saha, U.K.; Xie, W.C.; Xie, Y.H.; Wu, J.M.; Peng, B.; Tan, C.Y. Effectiveness and potential risk of CaO application in Cd-contaminated paddy soil. *Chemosphere* **2018**, *204*, 130–139. [CrossRef] [PubMed]
44. Bolan, N.S.; Adriano, D.; Mani, P.; Duraisamy, A. Immobilization and phytoavailability of cadmium in variable charge soils. II. Effect of lime addition. *Plant Soil* **2003**, *251*, 187–198. [CrossRef]
45. Hu, P.; Ouyang, Y.; Wu, L.; Shen, L.; Luo, Y. Effects of water management on arsenic and cadmium speciation and accumulation in an upland rice cultivar. *J. Environ. Sci.* **2015**, *27*, 225–231. [CrossRef] [PubMed]
46. Azam, M.S.; Shafiquzzaman, M.; Haider, H. Arsenic release dynamics of paddy field soil during groundwater irrigation and natural flooding. *J. Environ. Manag.* **2023**, *343*, 118204. [CrossRef]
47. Han, R.X.; Wang, Z.; Wang, S.Q.; Sun, G.X.; Xiao, Z.F.; Hao, Y.L.; Nriagu, J.; Teng, H.H.; Li, G. A combined strategy to mitigate the accumulation of arsenic and cadmium in rice (*Oryza sativa* L.). *Sci. Total Environ.* **2023**, *896*, 16522. [CrossRef] [PubMed]
48. Zeng, P.; Liu, J.W.; Zhou, H.; Wei, B.Y.; Gu, J.F.; Liao, Y.; Liao, B.H.; Luo, X.F. Co-application of combined amendment (limestone and sepiolite) and Si fertilizer reduces rice Cd uptake and transport through Cd immobilization and Si-Cd antagonism. *Chemosphere* **2023**, *316*, 137859. [CrossRef] [PubMed]

processes

Article

Analysis of Hydrochemical Characteristics and Causes of Drinking Water Sources in South China: A Case Study in Zhanjiang City

Hang Wei [1,2], Qi Zou [1,2], Zhiliang Chen [1,2,*], Yingjie Cao [3], Shuang Wang [4], Fen Zhu [4] and Xulong Liu [5,*]

1 State Environmental Protection Key Laboratory of Urban Ecological Environment Simulation and Protection, South China Institute of Environmental Sciences, Ministry of Ecology and Environment of China, Guangzhou 510535, China
2 Guangdong Engineering Technology Research Center of Heavy Metal Pollution Control and Restoration in Farmland Soil, Guangzhou 510535, China
3 School of Environmental Science and Engineering, Guangdong Provincial Key Lab of Environmental Pollution Control and Remediation Technology, Sun Yat-sen University, Guangzhou 510275, China
4 The Fourth Geological Brigade of Guangdong Provincial Geological Bureau, Zhanjiang 524057, China
5 Guangdong Provincial Key Laboratory of Remote Sensing and Geographical Information System, Guangzhou Institute of Geography, Guangdong Academy of Sciences, Guangzhou 510070, China
* Correspondence: chenzhiliang@scies.org (Z.C.); lxlong020@126.com (X.L.); Tel./Fax: +86-020-29119688 (Z.C.); +86-020-87685006 (X.L.)

Abstract: The evaluation of groundwater environmental quality and the identification of recharge sources are very important for groundwater utilization. In this study, hydrochemistry and isotope analysis methods are used to investigate the recharge sources and hydrochemical processes of groundwater in Zhanjiang City. The results show that all samples of groundwater were drawn on the left of the global meteoric water line (GMWL: $\delta D = 8\delta^{18}O + 10$) and local meteorological water lines (LMWL1: $\delta D = 8.17\delta^{18}O + 11.74$ and LMWL2: $\delta D = 7.50\delta^{18}O + 6.18$), indicating that the groundwater was mainly recharged by meteoric precipitation and influenced by the effect of evaporation. In the middle and deep confined aquifers, the isotope data depleted with the depth, indicating that there is a relatively weak hydraulic connection between them. In addition, compared with unconfined groundwater, the isotope data of confined groundwater showed relative depletion, indicating that the confined aquifer may be partially recharged from other confined aquifers. The main chemical types in the groundwater were Na*Ca-HCO_3. There are three major natural hydrochemical processes controlling the source of groundwater ions: silicate weathering, carbonate dissolution, and the cation exchange reaction. In addition, the differences in physical and chemical properties between unconfined groundwater and confined groundwater are significant. Due to the differences in anthropogenic activities and land-use types, the nitrate of the unconfined groundwater exceeds the groundwater standards. Due to the geological background of Zhanjiang City, iron and manganese exceed the third standard of groundwater in confined groundwater. Due to groundwater exploitation, TDS levels in confined groundwater have been increasing. Closed groundwater extraction is not sustainable, and it is depleting ancient water reserves. This study highlights the effectiveness of hydrochemistry and isotope analysis methods for identifying the recharge area and recharge mode of groundwater, andit is significant for fully understanding groundwater hydrochemistry and scientifically managing and protecting groundwater.

Keywords: groundwater; hydrochemical characteristics; stable isotope; factor analysis

1. Introduction

Economic development intensifies the demand for freshwater worldwide [1]. As an important freshwater resource, the quality and quantity of groundwater have received much attention. Due to its wide distribution, a large amount of groundwater is used for irrigation and industrial and domestic use. In China, approximately 400 cities exploit groundwater as a water resource [2]. Reports show that in approximately 220 cities in China, groundwater is used as drinking water resource. Most of this use occurs in the inland northwest part of China, where the environment is harsh and precipitation and surface water resources are scarce [3]. With the overexploitation of groundwater and the increase in anthropogenic activities, groundwater has faced serious environmental problems, including nitrate pollution, saline water mixing, the depletion of groundwater, and land subsidence [1,4]. The hydrochemical characteristics of groundwater are one of the important properties that determine the utility of groundwater resources. Researchers have studied the hydrochemical characteristics and recharge patterns of centralized, deep groundwater as a drinking water resource in parts of Northern China, such as Anhui and Fuyang [5,6]. Centralized deep groundwater in typical cities in Northwest China is affected by "rock weathering" and "evaporation." Between these factors, evaporation is the main mechanism. In order to manage and exploit groundwater resources rationally, it is crucial to study the hydrochemical characteristics of deep groundwater [7–9].

Compared with inland cities, in which the environment is harsh, precipitation is scarce, and surface water resources are scarce, Zhanjiang City is located in the southwest of Guangdong Province, and its rainfall belongs to the rainbelt [10]. As most of the surface water flows directly into the sea and the resources are not evenly distributed in time and space, the surface water of Zhanjiang City is in relatively short supply. Groundwater has always been one of the main water sources of this city's domestic, industrial, and agricultural production water. Since the 1950s, groundwater has become the central water supply source of Zhanjiang City. Due to the demand of economic development, groundwater extraction in Zhanjiang City has gradually increased, and the annual groundwater extraction accounts for about 50.60% of the total groundwater development and utilization in Guangdong Province [10]. There is a long history of groundwater development and utilization in Zhanjiang city, especially the concentrated and large-scale exploitation of middle and deep confined groundwater. Insufficient groundwater recharge and a drop in the water level have led to a series of ecological and environmental geological problems, such as groundwater level drop hopper, land subsidence, seawater intrusion, and groundwater pollution.

Hydrochemical characteristics reflect the environmental quality, the distribution of elements, and their migration/transformation [11,12]. The source of solutes in groundwater may be anthropogenic or natural [13]. Coupled with the input of artificial contaminants, the hydrochemical characteristics may change. Some activities may improve groundwater quality, while others cause groundwater pollution, such as nitrate and heavy metal contamination [5,14]. The major ions are widely used to identify the fundamental processes of water chemistry, such as the weathering of rocks in drainage basins, the evaporation/crystallization of water, atmospheric deposition, and interactions with anthropogenic influences [15,16]. Moreover, the hydrochemical characteristics of groundwater are mostly influenced by recharge sources, water–rock interactions, solute transport, and chemical processes [17,18]. In groundwater management studies, hydrochemistry and isotope-tracing methods are widely used to identify the recharge sources, recharge rates, and flow patterns of groundwater [19–22]. Therefore, with different recharge sources, hydrochemical and isotope indices also significantly differ, making them an effective "fingerprint" for studying groundwater recharge sources [23]. Yuan et al. discussed the runoff conditions of karst groundwater in Xianren Cave, Xichang, by using hydrochemical characteristics, ^{2}H, ^{18}O, and ^{3}H [24]. Jebreen et al. studied the recharge source of karst aquifer in Palestine by using ^{2}H and ^{18}O isotopes [25].

Groundwater can be divided into three types: unconfined groundwater (aquifer buried depth < 30 m), middle confined groundwater (buried depth of 30–200 m), and deep confined groundwater (buried depth of 200–450 m). Unconfined groundwater is usually replenished by rainfall and river runoff, and the water body circulates rapidly, causing it to be easily affected by anthropogenic factors [26]. In deep confined groundwater, storage is limited, the water circulation is slow, and the renewability is relatively weak [27].

Middle confined groundwater is between unconfined groundwater and deep confined groundwater [28]. The quality of unconfined groundwater in Zhanjiang City is affected by anthropogenic activities and environmental changes. The indices of ammonia nitrogen, iron, and manganese exceed the standards. Middle and deep confined groundwaters are buried deep, and the pollution degree is light due to the pollution isolation effect of the top waterproof layer, which mainly manifests as a high iron content.

A scientific understanding of the hydrochemical characteristics of groundwater in Zhanjiang City, analyzing and evaluating the current situation of groundwater quality, identifying the recharge area and recharge mode of groundwater drinking water source, and rationally developing and utilizing groundwater are of great significance for ensuring the sustainable utilization of groundwater resources. The aims of this study are: (1) to analyze and evaluate the current groundwater situation; (2) to reveal the hydrochemical evolution of unconfined, middle, and deep confined groundwater; and (3) to identify the source of the groundwater and the way through which artificial contamination may enter the groundwater.

2. Materials and Methods

2.1. Information of Study Area

Zhanjiang City is located in the southwest of Guangdong Province, China. Its annual average temperature, precipitation, and evaporation are 23.1 °C, 1620.0 mm, and 1189.1 mm, respectively. The groundwater drinking water resource in Potou District of Zhanjiang City, which can be used as the drinking water resource for the surrounding towns, was selected as the study area. The geographical coordinates are 110°25′12″–110°38′29″ E and 21°11′24″–21°25′46″ N. Combined with regional hydrogeological conditions and observation data, the scope of analysis of the hydrological recharge area is the Zhanjiang urban area, including tPotou District, Chikan District, Xiashan District, and Mazhang District, with a total land area of 1460 km^2. The geographical coordinates are 110°06′56″–110°39′23″ E and 20°51′41″–21°27′28″ N.

2.2. Sample Collection and Measurement

Fifteen groundwater sites were sampled in March 2022 to investigate the hydrochemistry of groundwater in Zhanjiang City. Fifteen groundwater samples were collected, comprising five unconfined aquifers (sample numbers starting with Q; used as well water or pumping water, generally for laundry and daily drinking by some residents), three middle confined aquifers (sample numbers starting with Z; used as pumping water) and seven deep confined aquifers (sample numbers starting with S; used as pumping water), as shown in Figure 1.

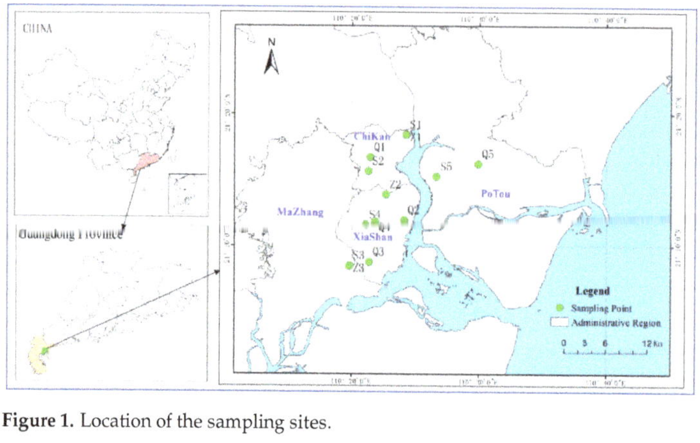

Figure 1. Location of the sampling sites.

For the collection of groundwater hydrochemistry samples, 100 mL brown polyethylene bottles were first cleaned. Samples were then collected three times. The bottles were sealed using a film, refrigerated in a timely manner in a refrigerator at 4 °C, and tested within 7 days. The concentration of major cations in the groundwater samples, including CalCium (Ca^{2+}), Magnesium (Mg^{2+}), Natrium (Na^{+}), and Potassium (K^{+}), was detected by ICP-MS (ICP-MS, 7800 ICP-MS system, Agilent Technologies Co., Ltd., CA, USA). The pH, temperature, and total dissolved solid (TDS) content were measured immediately using a portable multiparameter water analyzer (HQ40d, HACH Co., Ltd., CO, USA). The concentration of HCO_3^- was determined by the titration method. The Cl^-, SO_4^{2-}, and NO_3^- were measured by ion chromatography (ICS 600 system, Thermo Scientific Dionex Co., Ltd., MA, USA).

To collect the groundwater isotope samples, groundwater was filtered through 0.45 μm membranes, stored in 10 mL glass bottles, and immediately sealed with a parafilm sealer to avoid evaporation. The $\delta^{18}O$ and δD of the water samples were measured in the Isotope Analysis Laboratory of the School of Environmental Science and Engineering, Sun Yat-sen University. The $\delta^{18}O$ and δD values were determined using a liquid water isotope analyzer (Delta Advantage V, Thermo Fisher Co., Ltd., MA, USA), with accuracy levels of 0.1‰ and 0.3‰, respectively. The measurements were expressed in terms of the Vienna Standard Mean Ocean Water as the thousandth difference [δ(‰)]. Isotopic ratios ($\delta^{18}O$ and δD) were calculated for standard samples as follows: δsample = (δsample/δstandard) − 1) × 1000(‰), where δsample and δstandard represent the ratio of heavy to light isotopes of the sample and standard, respectively. The results are shown in Figure 2.

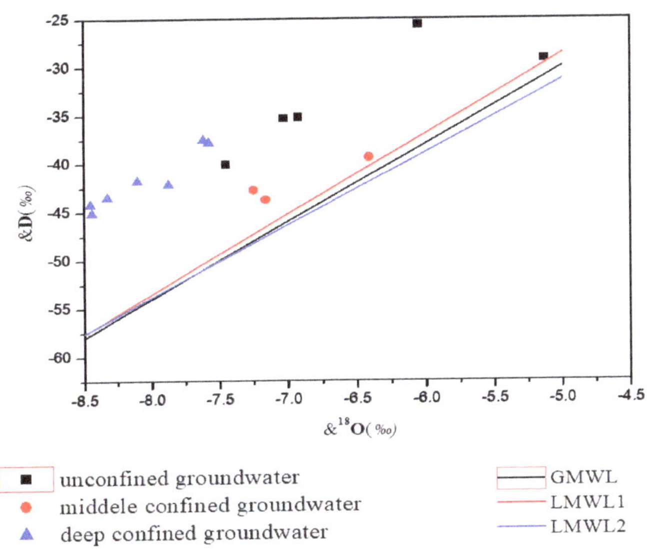

Figure 2. Stable isotopic compositions of groundwater in the study area.

3. Results and Discussion

3.1. Characteristic of Isotopes

The ^{18}O and D isotopes of the samples can reveal information about the water fluxes that are difficult to determine by other methods. Isotope analysis methods have been widely used in many hydrogeological studies [29]. The analytical results of the isotopes are shown in Table 1 and Figure 2. The isotopic compositions of ^{18}O and D were influenced by regional climate and local processes [30]. Based on the monthly rainfall data from the International Atomic Energy Agency's Global Network of Isotopes in Precipitation, the local meteorological water lines (LMWLs) of Hong Kong (LMWL1) and Haikou (LMWL2) stations could be calculated as $\delta D = 8.17\delta^{18}O + 11.74$ and $\delta D = 7.50\delta^{18}O + 6.18$, respectively [31], which were regarded as the LMWL of Zhanjiang City. It can be found in Table 1 that the isotopic compositions of δD in the unconfined, middle, and deep confined groundwater were 25.56–40.00‰ (average of 33.01‰), 39.34–43.70‰ (average of 41.91‰), and 37.51–45.09‰ (average of 41.69‰), respectively. While the isotopic compositions of $\delta^{18}O$ in unconfined, middle and deep confined groundwater were 5.14–7.45‰ (average of 6.52‰), 6.41–7.25‰ (average of 6.94‰), and 7.58–8.45‰ (average of 8.06‰), respectively. The evaporation intensity of the unconfined, middle and deep confined groundwater was investigated by the deuterium excess values (*d-excess* = $\delta D - 8\delta^{18}O$). More negative *d-excess* values implied a stronger evaporation effect [32]. It was found that the *d–excess* values of the unconfined, middle and deep confined groundwater ranged from 11.96–15.33‰ (average of 13.21‰), 19.62–22.88‰ (average of 20.95‰), and 20.91–23.44‰ (average of 13.21‰), respectively. It was indicated that the ion contents of groundwater in Zhanjiang City were controlled by silicate weathering and influenced by carbonate dissolution and the cation exchange reaction.

Table 1. Summary of hydrochemical data in the study area.

	Sample ID	pH	Ca^{2-} (mg/L)	Mg^{2-} (mg/L)	K^+ (mg/L)	Na^+ (mg/L)	Cl^- (mg/L)	SO_4^{2-} (mg/L)	HCO_3^- (mg/L)	NO_3^- (mg/L)	NO_2^- (mg/L)	TDS (mg/L)
Unconfined aquifer	Q1	4.15	24.96	9.28	15.88	43.96	44.10	44.87	2.55	25.00	0.05	175.60
	Q2	7.02	115.19	10.28	27.48	62.44	29.98	3.49	180.97	9.00	7.42	169.41
	Q3	5.91	84.94	9.29	24.65	63.28	38.23	40.52	6.24	50.00	0.01	322.90
	Q4	7.21	128.72	18.43	72.48	19.85	50.25	70.93	281.17	35.00	0.02	239.80
	Q5	4.62	54.39	14.18	8.30	59.25	55.54	89.54	2.99	8.00	ND	216.32
Middle confined aquifer	Z1	6.39	29.31	4.23	22.75	11.76	2.66	19.72	61.74	ND	ND	92.10
	Z2	6.19	24.32	8.84	9.47	11.20	10.43	10.72	52.35	2.00	ND	109.23
	Z3	5.57	17.81	12.92	11.11	28.62	28.53	30.98	10.42	ND	ND	122.12
Deep confined aquifer	S1	8.11	15.59	1.01	6.18	227.90	5.19	7.15	251.52	ND	ND	332.40
	S2	6.61	14.24	5.31	18.65	18.52	16.84	8.34	92.65	0.60	ND	136.02
	S3	6.32	21.69	11.29	12.63	23.93	17.38	14.30	138.29	ND	0.02	212.18
	S4	6.65	22.65	11.25	14.71	17.84	4.43	5.96	95.59	0.60	ND	142.58
	S5	6.43	15.91	5.34	20.17	13.72	24.15	12.98	50.72	ND	ND	199.88
	S6	6.42	24.13	3.87	16.30	29.24	15.07	1.19	91.18	9.00	0.01	226.08
	S7	6.51	23.85	3.24	17.42	33.78	3.49	3.59	73.11	ND	ND	167.58

The different hydrogen and oxygen isotopic compositions of the surface water and groundwater in different layers could indicate the difference in the recharge sources and water cycle process, and the interaction relationship between them to a certain extent [33]. As shown in Figure 2, compared with the confined groundwater, the unconfined groundwater is the most enriched in isotopes. In addition, all samples of unconfined groundwater were on the left of the global meteoric water line (GMWL: $\delta D = 8\delta^{18}O + 10$), LMWL1, and LMWL2, indicating that the groundwater was mainly recharged by meteoric precipitation and influenced by the effect of evaporation. For confined aquifers, the isotope data depleted with the depth, indicating that the hydraulic connection between the middle and deep confined aquifers is relatively weak. The samples were drawn along the GMWL and LMWL, indicating that the confined groundwater originated from precipitation. However, compared with the unconfined groundwater, its isotope data was relatively depleted, indicating that the depletion of the precipitation isotope data may partially recharged from other middle and deep confined aquifers.

3.2. Hydrochemical Characteristics

Understanding the basic characteristics of groundwater chemistry is the basis for investigating groundwater sources and hydrochemical processes. The hydrochemical data of Zhanjiang City are shown in Table 1. In unconfined groundwater, the TDS ranged from 169.41 mg/L to 322.90 mg/L, with an average of 224.81 mg/L. There was a range of 92.10 mg/L to 122.12 mg/L in middle confined groundwater, with an average of 107.82 mg/L, while deep confined groundwater had a TDS of 202.39 mg/L on average, ranging from 142.58 to 332.40 mg/L. The TDS of all groundwater samples was within the desirable limit of 1000 mg/L provided by the Chinese Guidelines, indicating that the groundwater in this aquifer system is predominantly freshwater [34]. The humid climates in Zhanjiang City could replenish groundwater by precipitation. The natural flow of the groundwater level above the mean sea level is the main reason that most unconfined groundwater is not salinized. In addition, the confined groundwater level is higher than the sea level, thereby preventing seawater intrusion into the confined groundwater.

The pH of the unconfined groundwater varied widely, ranging from 4.15 to 7.21 and averaging 5.78. Meanwhile, the middle confined groundwater had a pH ranging from 5.57 to 6.39, with an average of 6.05, while the deep confined groundwater had an average pH of 6.72, ranging from 6.32 to 8.11. With the increase in the depth of the groundwater, the pH slightly increased, which could be ascribed to the influence of anthropogenic activities. Additionally, Chinese guidelines recommend a groundwater pH range of 6.5–8.5 for drinking water sources; however, most groundwater samples were below that range [35]. Therefore, the groundwater is generally acidic, and the pH of the unconfined groundwater was higher than that of middle and deep confined groundwaters.

Among the cations, Ca^{2+} was the dominant ion in unconfined and middle confined aquifers, with concentration ranges of 24.96–128.72 and 17.81–29.31 mg/L, respectively. The Ca^{2+} in groundwater may come from the rock dissolution [36]. Meanwhile, Na^+ was the dominant ion in the deep confined aquifer, with a concentration range of 13.72–227.90 mg/L. Different to Ca^{2+}, rock–water interactions, the mixing of saline water, and ion exchange in similar regions induced Na^+ [37]. Seawater intrusion may be an important reason for the high concentration of Na^+ in the deep confined groundwater. Unconfined and middle confined aquifers had the following mean concentrations of major cations: $Ca^{2+} > Na^+ > K^+ > Mg^{2+}$, while the deep confined groundwater followed the order $Na^+ > Ca^{2+} > K^+ > Mg^{2+}$ for its main cations. For anions, the order of mean concentration of these three major anions was $HCO_3^- > Cl^- > SO_4^{2-}$. Due to the differences in anthropogenic activities and land-use types, the nitrate concentration in the unconfined groundwater exceeded the groundwater standards [38]. Nitrates can come from different sources, such as the excessive use of nitrogen-containing fertilizers and manures, domestic wastewater, and poorly maintained wells and sewers [39]. The changes in the TDS, major ions, and nitrates in groundwater were found to be related to human activities, such as agricultural practices (irrigation and fertilization), wastewater discharge, and groundwater extraction [40,41]. In addition, iron and manganese contents exceeded the third standard of groundwater in confined groundwater, which could be ascribed to the geological background of Zhanjiang City.

3.3. Hydrochemical Pattern of Groundwater

A Piper plot can aid in the understanding of a hydrochemical pattern and the underlying processes that control the hydrochemical characteristics of groundwater. The major ion concentrations shown in Table 1 were plotted, as shown in Figure 3. The distribution of water samples was close to the Na + K line and the Ca line and far from the Mg end. Thus, the main chemical types in groundwater were Na*Ca-HCO_3, which is a typical feature of recharge water sources [42]. The formation and enrichment of Na^+ are mainly related to the water-bearing middle and overlying soil layer. As the aqueous middle in Zhanjiang City contains many clay minerals, Na^+ is subjected to ion exchange, the Ca^{2+} and Mg^{2+} in water are replaced with Na^+, and the same amount of Na^+ is released into the groundwater, leading to the increase in its content and the high occurrence of Na^+. Precipitation had been speculated to carry wastewater into groundwater and to promote the dissolution of rocks by the CO_2 it takes [43]. It had been reported that the concentration of major cations (Ca^{2+}, Mg^{2+}, Na^+, and K^+) and the HCO_3^- (alkalinity) concentration are increased by enhancing the CO_2-mediated silicate mineral weathering [44]. The anion triangle at the bottom right shows a large difference, with Q1, Q3, Q5, and Z4 water samples near the SO_4^{2-} end and the remaining water samples near the $CO_3^{2-} + HCO_3^-$ end. HCO_3^- is mainly derived from the dissolution of carbonate rocks [1] and some compound fertilizers, such as NH_4HCO_3 [45]. In addition, the dissolution of CO_2 in the atmosphere and soil and the biodegradation process could cause a small part of CO_2 to enter the water body and recharge the groundwater. This finding was consistent with previous studies, which found that the hydrochemical composition of the shallow groundwater is affected by the surface water and that the water quality of the deep groundwater is stable [46].

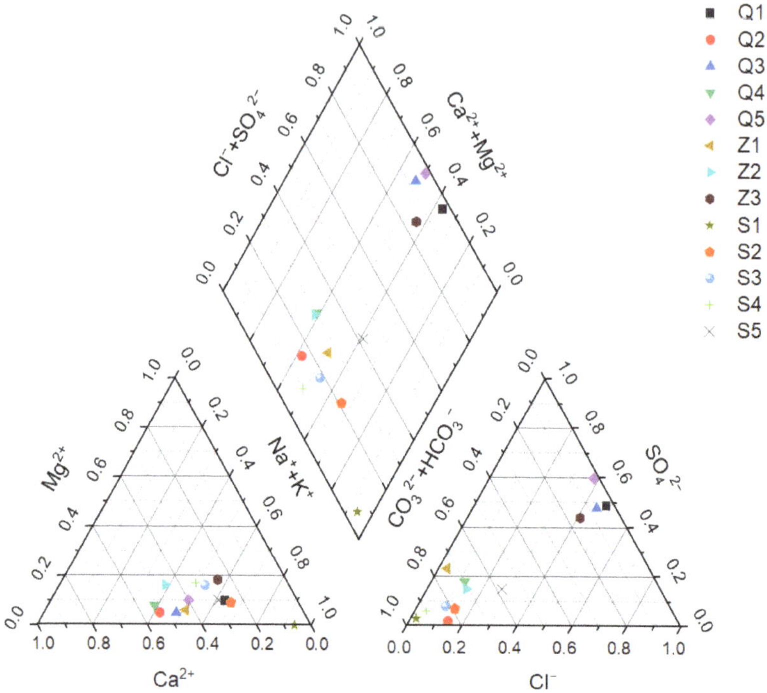

Figure 3. Piper diagram of chemical compositions for groundwater.

By using the Gibbs diagram [47], the relationship between the TDS and cation ratio Na/(Na + Ca), and anion ratio Cl/(Cl + HCO_3) could be assessed to determine the mechanisms controlling the groundwater chemistry [48,49]. As shown in Figure 4, the Gibbs diagram indicated that most of the samples had high Na/(Na + Ca) and low Cl/(Cl + HCO_3) levels and moderate TDS, of which the majority of samples were plotted in the rock-dominant zone, whereas the evaporation-dominant zone and the precipitation-dominant zone had very few samples. The results also revealed that water–rock interactions determine the hydrochemical characteristics of groundwater, followed by evaporation and precipitation.

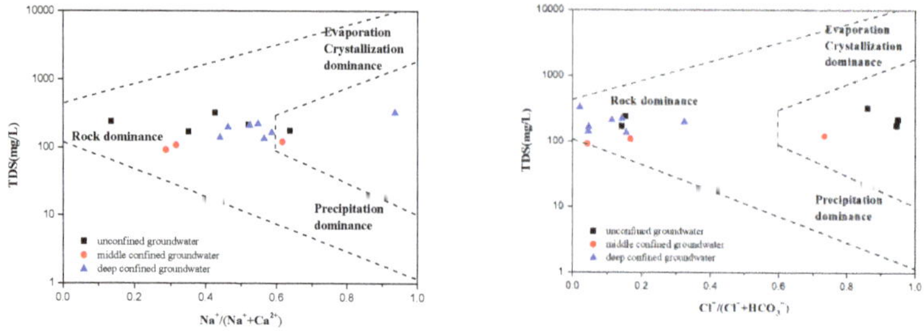

(a) (b)

Figure 4. Gibbs diagram for (**a**) cation ratios and (**b**) anion ratios.

3.4. Correlation Relationship

The relationship between major ions and TDS could be used to investigate the major hydrogeochemical evolution processes occurring in groundwater and to determine the source of the ions in groundwater [50]. Herein, Pearson's correlation coefficient (r) was used to manifest the strength of the linear relationships between hydrochemical parameters (pH, Ca^{2+}, Mg^{2+}, K^+, Na^+, Cl^-, SO_4^{2-}, HCO_3^-, NO_3^-, and Eh) and the TDS in the groundwater samples, as shown in Figure 5. The statistical significance levels (p-values) were lower than 0.05. TDS and major ions, including Na^+ (0.67), Ca^{2+} (0.30), NO_3^- (0.53), and HCO_3^- (0.35), showed good and moderately positive correlations, suggesting significant contributions from these ions in the groundwater, representing the dissolution of salt rock and carbonate rock and indicating the occurrence of a water–rock interaction [51,52]. Meanwhile, it appears that anthropogenic contaminants have contaminated magnesium (e.g., fertilizer applications for magnesium), since there is a good and moderate correlation between Mg^{2+} with Cl^- and SO_4^{2-} (0.9 > r > 0.5).

Figure 5. Correlation between major hydrochemical parameters of groundwater.

4. Conclusions

In this study, hydrochemistry and isotope analysis methods are used to investigate the recharge sources and hydrochemical processes of groundwater in Zhanjiang City. A total of 15 groundwater sites were sampled in March 2022. Hydrochemical plots such as Piper and Gibbs diagrams are used to analyze the controlling types and the hydrochemical processes of groundwater. In addition, Pearson's correlation coefficient (r) was used to manifest the strength of the linear relationships between hydrochemical parameters and TDS in the groundwater samples.

The isotopic composition of ^{18}O and D in the unconfined, middle, and deep confined groundwater indicate that the groundwater was mainly recharged by meteoric precipitation and was influenced by the effect of evaporation. The TDS of all groundwater samples was within the desirable limit of 1000 mg/L, indicating that the groundwater in this aquifer system is predominantly freshwater. The major cation and anion concentrations in unconfined and middle confined groundwater followed the order of $Ca^{2+} > Na^+ > K^+ > Mg^{2+}$ and $HCO_3^- > Cl^- > SO_4^{2-}$, respectively, while the major ions in the deep confined groundwater followed the order of $Na^+ > Ca^{2+} > K^+ > Mg^{2+}$and $HCO_3^- > Cl^- > SO_4^{2-}$. The Piper diagram showed that the main chemical types in groundwater were Na*Ca-HCO3, which is a typical feature of recharge water sources. The results of the Gibbs diagrams revealed that water–rock interactions determine the hydrochemical characteristics of groundwater, followed by evaporation and precipitation. The correlation analysis further demonstrated that the dissolution of salt rock and carbonate rock are the dominant factors affecting the hydrochemical characteristics of groundwater, indicating the occurrence of a water–rock interaction.

This study highlights the effectiveness of hydrochemistry and isotope analysis methods for identifying the recharge area and recharge mode of groundwater, and it is significant for fully understanding groundwater hydrochemistry and scientifically managing and protecting groundwater.

Author Contributions: Conceptualization, H.W. and Z.C.; funding acquisition, H.W., Q.Z. and Z.C.; software, H.W. and F.Z.; project administration, H.W.; writing—original draft, H.W.; methodology, Y.C. and X.L.; investigation, Q.Z.; Writing—review and editing, Z.C., X.L. and S.W.; data curation, F.Z. All authors have read and agreed to the published version of the manuscript.

Funding: This work was supported by the Project of Water Resource Department of Guangdong province (2017–2018); the National Key R&D Program of China (No. 2019YFC1805305); the Key R&D Program of Guangdong Province, China (No. 2020B1111350002); International Science and Technology Cooperation Projects of Guangdong Province, China (No. 2021A0505030045); the Special Basic Research Fund for Central Public Research Institutes of China (No. PM-zx703-202104-088, No. PM-zx703-202002-021).

Data Availability Statement: All data generated or analyzed during this study are included in this published article.

Acknowledgments: The authors wish to thank the anonymous reviewers and the editor.

Conflicts of Interest: The authors declare no conflict of interest.

References

1. Liu, S.; Tang, Z.; Gao, M.; Hou, G. Evolutionary process of saline-water intrusion in Holocene and Late Pleistocene groundwater in southern Laizhou Bay. *Sci. Total. Environ.* **2017**, *607–608*, 586–599. [CrossRef]
2. Wang, H.; Liu, S.; Du, S. The Investigation and Assessment on Groundwater Organic Pollution. *Org. Pollut. Monit. Risk Treat* **2013**, *4*, 87–110. [CrossRef]
3. Wang, S.Y.; Sun, C.J.; Chen, W.; Zhou, S.J.; Zhang, X. Analysis of Water Chemistry Characteristics and Hydraulic Relationships of Different Water Bodies in Typical Mountain-oasis Systems in the Northwest Inland Area. *Environ. Sci.* **2022**, *44*, 1416–1428. (In Chinese)
4. Soltanian, M.; Dargahi, A.; Asadi, F.; Ivani, A.; Setareh, P. Variation of PhysicoChemical Quality of Groundwater Watershed in Gharehsou during 2003–2012. *J. Maz. Univ. Med. Sci.* **2015**, *24*, 275–287.
5. Hu, M.; Wang, Y.; Du, P.; Shui, Y.; Cai, A.; Lv, C.; Bao, Y.; Li, Y.; Li, S.; Zhang, P. Tracing the sources of nitrate in the rivers and lakes of the southern areas of the Tibetan Plateau using dual nitrate isotopes. *Sci. Total Environ.* **2019**, *658*, 132–140. [CrossRef]
6. Kong, L.; Wang, Z.; Bing, W. An Analysis of the Hydro-chemical Characteristics and Causes of Drinking Water Source of Concentrated Deep Groundwater in Fuyang City. *China Rural. Water Hydropower* **2020**, *3*, 78–82. (In Chinese)
7. Al-Barakah, F.N.; Al-jassas, A.M.; Aly, A.A. Water quality assessment and hydrochemical characterization of Zamzam groundwater, Saudi Arabia. *Appl. Water Sci.* **2017**, *7*, 3985–3996. [CrossRef]
8. Zhao, W.; Liang, H.; Gong, S.J. Hydrochemical Characteristics and Oxygen Isotope Tracing Study of Shallow Groundwater Seawater—Intrusion in Tianjin. *J. Cap. Norm. Univ. (Nat. Sci. Ed.)* **2019**, *40*, 66–69.
9. Wang, Y.; Guo, Y.; Zhou, Y.; Li, S.; Wang, Q. Quantifications of spatial and temporal variations in groundwater discharge into a river using hydrochemical and isotopic tracers. *Arid. Land Gegraphy* **2020**, *43*, 290–298. (In Chinese)
10. Zhang, W.; Yang, X.; Huang, L.; Wang, L.; Zheng, N.; Xie, Y. Characteristics and means of comprehensive management of typical groundwater overexploitation areas in southern China—Taking Zhanjiang City of Guangdong Province as an example. *China Water Resour.* **2022**, *7*, 86–88. (In Chinese)
11. Dargahi, A.; Azizi, A.; Karami, A.; Amirian, F.; Mohammadi, M.; Almasi, A. Evaluating the chemical and microbial quality of drinking water in harsin city. *Int. J. Pharm. Technol.* **2016**, *8*, 16709–16719.
12. Sheikhy, N.T.; Sefie, A.; Aris, A.Z. The long-term impacts of anthropogenic and natural processes on groundwater deterioration in a multilayered aquifer. *Sci. Total Environ.* **2018**, *630*, 931–942. [CrossRef]
13. Xiao, J.; Zhang, F.; Jin, Z. Spatial characteristics and controlling factors of chemical weathering of loess in the dry season in the middle Loess Plateau, China. *Hydrol. Process.* **2016**, *30*, 4855–4869. [CrossRef]

14. Su, F.; Wu, J.; Wang, D.; Zhao, H.; Wang, Y.; He, X. Moisture movement, soil salt migration, and nitrogen transformation under different irrigation conditions: Field experimental research. *Chemosphere* **2022**, *300*, 134569. [CrossRef]
15. Alcalá, F.J.; Custodio, E. Using the Cl/Br ratio as a tracer to identify the origin of salinity in aquifers in Spain and Portugal. *J. Hydrol.* **2008**, *359*, 189–207. [CrossRef]
16. Aminiyan, M.M.; Aminiyan, F.M. Comprehensive integrated index-based geochemistry and hydrochemical analyses of groundwater resources for multiple consumptions under coastal conditions. *Environ. Sci. Pollut. Res. Int.* **2020**, *27*, 21386–21406. [CrossRef]
17. Santoni, S.; Huneau, F.; Garel, E.; Aquilina, L.; Vergnaud-Ayraud, V.; Labasque, T.; Celle-Jeanton, H. Strontium isotopes as tracers of water-rocks interactions, mixing processes and residence time indicator of groundwater within the granite-carbonate coastal-aquifer of Bonifacio (Corsica, France). *Sci. Total Environ.* **2016**, *573*, 233–246. [CrossRef]
18. Zabala, M.E.; Manzano, M.; Vives, L. The origin of groundwater composition in the Pampeano Aquifer underlying the Del Azul Creek basin, Argentina. *Sci. Total Environ.* **2015**, *518–519*, 168–188. [CrossRef]
19. Mark, Y.S.; Eric, K. The groundwater recharge regime of some slightly metamorphosed neoproterozoic sedimentary rocks: An application of natural environmental tracers. *Hydrol. Process.* **2014**, *28*, 3104–3117.
20. Tamborski, J.J.; Cochran, J.K.; Bokuniewicz, H.J. Application Of 224ra And 222rn For Evaluating Seawater Residence Times In A Tidal Subterranean Estuary. *Mar. Chem.* **2017**, *189*, 32–45. [CrossRef]
21. Zhu, X.; Zhang, M.; Wu, X.; Pan, M. The relationship between δ18O characteristics of the precipitation (heavy rainfall or rainstorm) and its water vapor sources in Guilin, China. *Carsologica Sin.* **2017**, *36*, 139–161. (In Chinese)
22. Yang, N.; Wang, G. Moisture sources and climate evolution during the last 30 kyr in northeastern Tibetan Plateau: Insights from groundwater isotopes (^{2}H, ^{18}O, ^{3}H and ^{14}C) and water vapour trajectories modeling. *Quat. Sci. Rev.* **2020**, *242*, 106426. [CrossRef]
23. Chowdhury, A.H.; Scanlon, B.R.; Reedy, R.C.; Young, S. Fingerprinting groundwater salinity sources in the Gulf Coast Aquifer System, USA. *Hydrogeol. J.* **2017**, *26*, 197–213. [CrossRef]
24. Yuan, J.; Xu, F.; Liu, H.; Deng, G. Application of hydrochemical and isotopic analysis to research a typical karst groundwater system: A case study at xianrendong, Xichang City. *Sci. Technol. Eng.* **2019**, *19*, 76–83. (In Chinese)
25. Jebreen, H.; Wohnlich, S.; Banning, A.; Wisotzky, F.; Niedermayr, A.; Ghanem, M. Recharge, geochemical processes and water quality in karst aquifers: Central West Bank, Palestine. *Environ. Earth Sci.* **2018**, *77*, 261. [CrossRef]
26. Li, Z.; Xiao, J.; Evaristo, J.; Li, Z. Spatiotemporal variations in the hydrochemical characteristics and controlling factors of streamflow and groundwater in the Wei River of China. *Environ. Pollut.* **2019**, *254*, 113006. [CrossRef] [PubMed]
27. Chen, Z.; Yang, B.; Mengoni, A.; Dong, J.; Peng, X. Distribution of polychlorinated dibenzo-p-dioxins, polychlorinated dibenzofurans in sediments from the Xiangjiang River, China. *Environ. Monit. Assess.* **2012**, *184*, 7083–7092. [CrossRef] [PubMed]
28. Yin, G.; Zhu, H.; Chen, Z.; Su, C.; He, Z.; Chen, X.; Qiu, J.; Wang, T. Spatial Distribution and Source Apportionment of Soil Heavy Metals in Pearl River Delta, China. *Sustainability* **2021**, *13*, 9651. [CrossRef]
29. Yan, J.; Chen, J.; Zhang, W. Study on the groundwater quality and its influencing factor in Songyuan City, Northeast China, using integrated hydrogeochemical method. *Sci. Total Environ.* **2021**, *773*, 144958. [CrossRef]
30. Ako, A.A.; Shimada, J.; Hosono, T.; Kagabu, M.; Richard, A.; Nkeng, G.E.; Tongwa, A.F.; Ono, M.; Eyong, G.E.T.; Tandia, B.K.; et al. Flow dynamics and age of groundwater within a humid equatorial active volcano (Mount Cameroon) deduced by δD, δ18O, 3H and chlorofluorocarbons (CFCs). *J. Hydrol.* **2013**, *502*, 156–176. [CrossRef]
31. IAEA/WMO. Global Network of Isotopes in Precipitation. *The GNIP Database*. 2022. Available online: https://nucleusiaeaorg/wiser (accessed on 31 October 2022).
32. Jin, J.; Wang, Z.; Zhao, Y.; Ding, H.; Zhang, J. Delineation of Hydrochemical Characteristics and Tracing Nitrate Contamination of Groundwater Based on Hydrochemical Methods and Isotope Techniques in the Northern Huangqihai Basin, China. *Water* **2022**, *14*, 3168. [CrossRef]
33. Kumar, S.K.; Rammohan, V.; Sahayam, J.D.; Jeevanandam, M. Assessment of groundwater quality and hydrogeochemistry of Manimuktha River basin, Tamil Nadu, India. *Environ. Model. Assess.* **2009**, *159*, 341–351. [CrossRef]
34. Liu, J.; Gao, Z.; Zhang, Y.; Sun, Z.; Sun, T.; Fan, H.; Wu, B.; Li, M.; Qian, L. Hydrochemical evaluation of groundwater quality and human health risk assessment of nitrate in the largest peninsula of China based on high-density sampling: A case study of Weifang. *J. Clean. Prod.* **2021**, *322*, 129164. [CrossRef]
35. *GB/T 14848–2017*; Standards for Groundwater Quality. General Administration of Quality Supervision: Beijing, China, 2017.
36. Kenniche, S.; Bekkoussa, B.; M'Nassri, S.; Teffahi, M.; Taupin, J.-D.; Patris, N.; Zaagane, M.; Majdoub, R. Hydrochemical characterization, physicochemical and bacteriological quality of groundwater in Sidi Kada Mountains, northwest of Algeria. *Arab. J. Geosci.* **2022**, *15*, 1–18. [CrossRef]
37. Jones, S.; Rosen, M. Controls on the chemical composition of groundwater from alluvial aquifers in the Wanaka and Wakatipu basins, Central Otago, New Zealand. *Hydrogeol. J.* **1998**, *6*, 264–281. [CrossRef]
38. Böhlke, J.-K. Groundwater recharge and agricultural contamination. *Hydrogeol. J.* **2002**, *10*, 153–179. [CrossRef]
39. Busico, G.; Kazakis, N.; Colombani, N.; Mastrocicco, M.; Voudouris, K.; Tedesco, D. A modifed SINTACS method for groundwater vulnerability and pollution risk assessment in highly anthropized regions based on NO_3^- and SO_4^{2-} concentrations. *Sci. Total Environ.* **2017**, *609*, 1512–1523. [CrossRef]

40. Samarghandi, M.R.; Mohammadi, M.; Karami, A.; Tabandeh, L.; Dargahi, A.; Amirian, F. Residue Analysis of Pesticides, Herbicides, and Fungicides in Various Water Sources Using Gas Chromatography-Mass Detection. *Pol. J. Environ. Stud.* **2017**, *26*, 2189–2195. [CrossRef]
41. Liu, F.; Wang, S.; Yeh, T.J.; Zhen, P.; Wang, L.; Shi, L. Using multivariate statistical techniques and geochemical modelling to identify factors controlling the evolution of groundwater chemistry in a typical transitional area between Taihang Mountains and North China Plain. *Hydrol. Process.* **2020**, *34*, 1888–1905. [CrossRef]
42. Zhang, H.; Xu, G.; Zhan, H.; Chen, X.; Liu, M.; Wang, M. Identifcation of hydrogeochemical processes and transport paths of a multiaquifer system in closed mining regions. *J. Hydrol.* **2020**, *589*, 125344. [CrossRef]
43. Ren, X.; Li, P.; He, X.; Su, F.; Elumalai, V. Hydrogeochemical processes afecting groundwater chemistry in the central part of the Guanzhong Basin, China. *Arch. Environ. Contam. Toxicol.* **2021**, *80*, 74–91. [CrossRef]
44. Hosono, T.; Masaki, Y. Post-seismic hydrochemical changes in regional groundwater flow systems in response to the 2016 Mw 7.0 Kumamoto earthquake. *J. Hydrol.* **2020**, *580*, 124340. [CrossRef]
45. Zhang, Y.; Su, C.L.; Ma, Y.H.; Liu, W.J. Indicators of Groundwater Evolution Processes Based on Hydrochemistry and Environmental Isotopes: A Case Study of the Dongyuan Drinking Water Source Area in Ji'nan City. *Environ. Sci.* **2019**, *40*, 2667–2674.
46. Zhao, Y.; Han, J.; Zhang, B.; Gong, J. Impact of transferred water on the hydrochemistry and water quality of surface water and groundwater in Baiyangdian Lake, North China. *Geosci. Front.* **2021**, *12*, 101086. [CrossRef]
47. Gibbs, R.J. Mechanisms Controlling World Water Chemistry. *Science* **1970**, *170*, 1088–1090. [CrossRef] [PubMed]
48. Thakur, T.; Rishi, M.S.; Naik, P.K.; Sharma, P. Elucidating hydrochemical properties of groundwater for drinking and agriculture in parts of Punjab, India. *Environ. Earth Sci.* **2016**, *75*, 1–15. [CrossRef]
49. Liu, F.; Wang, S.; Wang, L.; Shi, L.; Song, X.; Yeh, T.C.J.; Zhen, P. Coupling hydrochemistry and stable isotopes to identify the major factors afecting groundwater geochemical evolution in the Heilongdong Spring Basin, North China. *J. Geochem. Explor.* **2019**, *205*, 106352. [CrossRef]
50. Kim, K.-H.; Yun, S.-T.; Choi, B.-Y.; Chae, G.-T.; Joo, Y.; Kim, K.; Kim, H.-S. Hydrochemical and multivariate statistical interpretations of spatial controls of nitrate concentrations in a shallow alluvial aquifer around oxbow lakes (Osong area, central Korea). *J. Contam. Hydrol.* **2009**, *107*, 114–127. [CrossRef]
51. Su, C.; Wang, Y.; Pan, Y. Hydrogeochemical and isotopic evidences of the groundwater regime in Datong Basin, Northern China. *Environ. Earth Sci.* **2013**, *70*, 877–885. [CrossRef]
52. Wu, J.; Li, P.; Wang, D.; Ren, X.; Wei, M. Statistical and multivariate statistical techniques to trace the sources and afecting factors of groundwater pollution in a rapidly growing city on the Chinese Loess Plateau. *Hum. Ecol. Risk Assess.* **2020**, *26*, 1603–1621. [CrossRef]

processes

MDPI

Article

Impact of Acid Rain on Release Characteristics of Heavy Metals in Low-Sulfur Tailings with Strong Acid Neutralization Capacity: A Case Study from Northern Guangxi, China

Xiaohua Shu [1], Dinghua Zhang [1], Qian Zhang [2,*], Tao Ai [3], Xuehong Zhang [1] and Jie Liu [1]

1 Collaborative Innovation Center for Water Pollution Control and Water Safety in Karst Area, Guilin University of Technology, Guilin 541006, China; sxh9911@glut.edu.cn (X.S.); 13851794939@163.com (D.Z.); zhangxuehong@x263.net (X.Z.); liujie@glut.edu.cn (J.L.)
2 School of Life and Environmental Science, Guilin University of Electronic Technology, Guilin 541000, China
3 School of Environmental Science and Engineering, Guangdong University of Petrochemical Technology, Maoming 525000, China; aitaocqu@163.com
* Correspondence: qzhang0613@guet.edu.cn

Abstract: Tailing ponds are major sources of heavy metal pollution. Previous studies primarily focused on tailings with high sulfur content, with limited attention to low-sulfur tailings. This study explored the release behavior of Pb, Zn, and Cd from low-sulfur tailings under simulated acid rain conditions, considering factors such as pH, particle size, and weathering degree. Samples were collected from a lead–zinc tailing pond in the karst regions of northern Guangxi, China. Batch leaching experiments indicated that even with high acid neutralization capacity (ANC = 166.57–167.45 kg H_2SO_4/t), substantial heavy metal leaching occurred under acidic conditions (pH 2–3), with Pb, Zn, and Cd concentrations increasing 4–6 times compared to neutral conditions. Leachate concentrations were slightly higher in coarser particles than in finer ones, while weathering further enhanced metal release, particularly for Cd. These findings suggest that acid neutralization alone may not be sufficient to prevent heavy metal leaching in low-sulfur tailings exposed to acid rain. However, due to the laboratory scale of this study, further validation through field-scale or mesocosm experiments is necessary to confirm the observed trends and assess their implications for environmental risk management in karst regions.

Keywords: acid neutralization capacity; heavy metal releases; leaching behavior; simulated acid rain

1. Introduction

Tailings are the primary solid waste generated during ore beneficiation and are often stored in tailing ponds. When exposed to the natural environment, tailings interact with air and precipitation, resulting in complex ion dissolution processes [1,2]. The leaching of heavy metal ions poses significant risks to both environmental and human health. Tailings are characterized by large quantities, fine particle sizes, and the presence of various adsorbed heavy metals and contaminants on their surfaces [3]. Open-pit storage of these tailings not only occupies a substantial amount of land but also poses serious environmental risks to the surrounding ecosystem. On one hand, fine tailing particles can easily be dispersed as dust under the influence of wind, and once airborne, these dust particles can be transported over long distances by atmospheric turbulence, contributing to atmospheric pollution. On the other hand, tailings are prone to the leaching and release of heavy metals due to the flushing and infiltration of rainwater, leading to the transportation of contaminants via surface runoff and leachate into nearby surface and groundwater systems [3]. Such processes can result in the contamination of downstream farmlands and water bodies [4–6]. In the lead–zinc tailing pond located in northern Guangxi, China (Latitude: 24.2293° N, Longitude: 109.3641° E), heavy metals in downstream agricultural soils have been reported to exceed national safety standards (Pb: 1000 mg/kg, Zn: 1000 mg/kg, Cd: 20 mg/kg), posing potential risks to local agriculture and water systems [7]. This elevated concentration, in the range of 1150–1350 mg/kg for Pb, 1300–1800 mg/kg for Zn, and 5–15 mg/kg for Cd, combined with the high bioavailability and strong mobility of these heavy metals, results in persistent and recurrent contamination of the local environment, making it a significant environmental concern [8]. Hence, understanding the leaching mechanisms and influencing factors of heavy metals in tailings is essential for developing effective environmental management strategies.

Acid mine drainage (AMD) from sulfide-rich tailings is a widespread environmental hazard that dramatically increases the release of heavy metals, particularly in areas that are vulnerable to acid rain, such as South China [9–11]. Acid rain, characterized by sulfuric and nitric acids, typically exhibits low pH values ranging from 2.0 to 4.0 in this region, creating conditions that enhance the mobility of heavy metals. However, the behavior of heavy metals in low-sulfur tailings with high acid neutralization capacity remains largely unknown. Unlike sulfide-rich tailings, which are prone to generating AMD and releasing toxic metals under acidic conditions, the tailings in northern Guangxi are rich in limestone, imparting a high acid neutralization capacity that theoretically mitigates heavy metal leaching [12,13]. Low-sulfide tailings, like those studied here, account for approximately 15–20% of the total tailings produced in China, primarily found in regions with significant limestone deposits. Nevertheless, even with this buffering capacity, ion exchange reactions between hydrogen ions (H^+) and heavy metal cations (e.g., Pb^{2+}, Zn^{2+}, and Cd^{2+}) can still occur under acidic conditions, as acid rain can temporarily overwhelm the acid neutralization capacity (ANC) of the tailings. This process facilitates the leaching of heavy metals, increasing their environmental risks [14]. As a result, investigating the leaching characteristics of low-sulfur tailings under varying acidic conditions becomes crucial for assessing their potential environmental risks.

To comprehensively evaluate these behaviors, it is necessary to consider multiple factors that influence the release of heavy metals from tailings. The release of toxic metals from tailings is influenced by various factors, including pH, particle size, weathering degree, metal concentration, bioavailability, geochemical composition, bacterial activity, and vegetation growth. Numerous studies have demonstrated that acid rain can enhance the mobility of heavy metals in soil ecosystems due to ion exchange reactions between heavy metals and hydrogen ions, which are dominant under acidic conditions. An acidic environment facilitates the release or dissolution of metal ions, leading to elevated concentrations of these ions in both soil and water bodies. For example, the displacement of cations such as Pb, Cd, and Zn by H^+ reduces the cation exchange capacity (CEC) of the soil and enhances the leaching potential of these metals [15]. Moreover, particle size plays a significant role in leaching behavior, with fine particles exhibiting greater reactivity and leachability due to their larger specific surface area. Likewise, the degree of weathering also influences metal mobility, as prolonged exposure to environmental conditions can alter the fraction distribution and bioavailability of metals, thereby affecting their leaching behavior. Considering these influential factors, further research is needed to elucidate the complex interactions governing heavy metal leaching in tailings.

The ANC of tailings is also a critical factor that influences the release of heavy metals under acidic conditions, primarily driven by the presence of carbonate minerals (e.g., calcite and dolomite), as well as calcium and magnesium oxides and hydroxides, which react with hydrogen ions to neutralize acidity. Tailings with high ANC are expected to buffer acid inputs more effectively, thus reducing the potential for acid generation and metal leaching. Conversely, tailings with low ANC may generate low pH conditions, leading to enhanced metal extraction and mobilization. Therefore, understanding the role of ANC in controlling metal leaching is essential for effective environmental management and pollution control. While static methods like acid-base accounting (ABA) and net acid generation (NAG) tests have greatly advanced the prediction of acid generation potential in mine waste materials, research systematically addressing the correlation between ANC and metal release characteristics in low-sulfur tailings remains limited [16,17].

Despite the high acid neutralization capacity of low-sulfur tailings, the persistent heavy metal contamination observed in downstream farmlands indicates that these tailings can still pose significant environmental risks under acid rain conditions [18]. The lead–zinc tailings in northern Guangxi, China, are located in a typical karst region rich in carbonate minerals, which significantly contribute to their high acid neutralization capacity. Mining activities in this region have been ongoing since the 1950s, with a small flotation plant established near the mining pit in 1958 to process ores. Previous studies have reported that the agricultural soils downstream of this lead–zinc tailing pond exhibit severe heavy metal contamination (e.g., Pb and Zn), with concentrations far exceeding national safety standards, even after 40 years of cultivation. Furthermore, soil organic carbon remains at low levels, indicating a long-term adverse impact on soil quality [19,20]. Low organic carbon levels reduce the soil's capacity to retain heavy metals, thereby increasing the potential for metal discharge and environmental risks. These findings suggest that the presence of carbonate-rich minerals in the tailings may not be sufficient to mitigate heavy metal release under certain environmental conditions, necessitating further research on leaching mechanisms. However, studies focusing on the release behavior of heavy metals in tailings with high acid neutralization capacity are still limited. Therefore, this study aims to investigate the leaching behavior of heavy metals in these tailings under different pH conditions, particle sizes, and vertical profile depths, providing valuable insights for the comprehensive utilization and management of tailing ponds. Thus, exploring the relationship between ANC, acid rain pH, and heavy metal release in these tailings is crucial for effective pollution control and management.

This research is designed to (1) assess the potential for acid formation in Pb/Zn mine tailings from Guangxi through the application of ABA and NAG techniques, (2) analyze the distribution of heavy metal speciation in different layers of the tailing pond, (3) investigate the influence of ANC on the leaching behavior of heavy metals under varying simulated acid rain (SAR) pH, particle size, and weathering degree conditions, and (4) assess the leaching characteristics of heavy metals in the disturbed layer of the tailing pond. The findings of this study are expected to provide valuable insights for scientists, engineers, environmental regulators, and policymakers to better understand the release behavior of heavy metals in low-sulfur tailings and the associated environmental risks, ultimately aiding in the development of effective environmental management strategies for low-sulfur tailings in karst regions.

2. Materials and Methods

2.1. Sample Preparation

To provide a clearer understanding of the sampling locations, a map (Figure 1) illustrating the specific sites within the lead–zinc tailing pond in northern Guangxi is included. Nine surface tailing samples were collected from a lead–zinc tailing pond in northern Guangxi, China, using a stratified sampling strategy based on a "cross" distribution map to cover various locations within the pond area. The sampling depth profile ranged from 0 to 30 cm, which was divided into three distinct layers: surface layer (0–10 cm), middle layer (10–20 cm), and deep layer (20–30 cm). This sampling design was adopted to investigate the variations in weathering degrees and heavy metal concentrations (Pb, Zn, and Cd) across different depths of the tailing deposits.

The tailing samples were collected from the disturbed layer (0–30 cm depth) below the surface of the tailing pond. This layer is frequently exposed to external disturbances, such as wind erosion, surface runoff, and rainfall infiltration, which can alter the physicochemical properties of the tailings and influence the distribution and migration of heavy metals. It also serves as the primary zone where interactions between tailings, plants, and microorganisms occur during remediation and reclamation processes. Understanding the properties and behavior of this layer is therefore crucial for evaluating both the environmental risks associated with tailings and for designing effective remediation strategies.

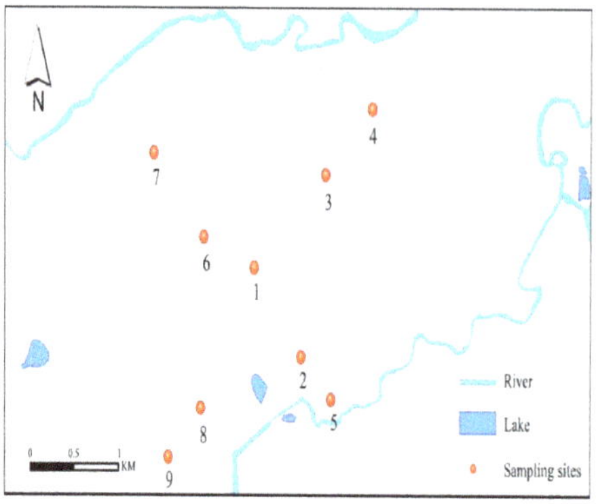

Figure 1. Map of sampling locations (numbers indicate sampling sites) within the lead–zinc tailing pond in Northern Guangxi, China.

The collected samples were first air-dried at ambient temperature in a greenhouse to minimize moisture content without affecting the chemical characteristics of the samples. Once dried, at a low temperature ($\leq$40 °C) to prevent chemical changes such as oxidation of metal sulfides, the samples were sieved sequentially through 40-mesh, 100-mesh, and 200-mesh sieves. The sieving process was performed gently to avoid generating heat or excessive mechanical force that could alter the particle characteristics, ensuring that the nominal chemical properties and grain size distribution were maintained. The processed samples were then stored in sealed plastic bags for subsequent leaching experiments.

Prior to conducting the leaching tests, the basic physicochemical properties of the tailing samples were thoroughly characterized. The analyses included measurements of the original pH and electrical conductivity of the tailings, mineral composition using X-ray diffraction (XRD), and elemental composition using X-ray fluorescence (XRF). In addition, the total heavy metal content was determined, and the chemical partition of heavy metals in the tailings was analyzed to understand their distribution patterns in different fractions. These preliminary analyses provided a comprehensive understanding of the tailing materials, laying a foundation for evaluating the leaching behavior of heavy metals under simulated acid rain conditions.

2.2. Analysis of Basic Physicochemical Properties of the Tailings

The pH of the tailings was measured using a standard soil pH determination method. Specifically, the tailings were mixed with deionized water (purity $\geq$ 18.2 MΩ·cm, initial pH ~6.5, conductivity < 1 μS/cm) at a ratio of 1:2.5, stirred for 2 min, and left to settle for 30 min before measuring the pH with a pH meter. To determine the total heavy metal content, 0.1 g of the tailing sample was digested using a mixed acid solution consisting of 4 mL concentrated nitric acid, 2 mL concentrated hydrochloric acid, and 1 mL hydrofluoric acid. The digestion was conducted over 24 h, followed by microwave digestion. The heavy metal concentrations were then measured using an Inductively Coupled Plasma Optical Emission Spectrometer (ICP-OES, Perkin-Elmer Optima 7000DV, Waltham, MA, USA).

The elemental composition of the tailings was analyzed using X-ray fluorescence spectrometry (XRF), and the mineralogical composition was characterized using X-ray diffraction analysis (XRD, X'Pert PRO, PANalytical, Almelo, The Netherlands). Electrical conductivity (EC) was measured using a conductivity meter (DDS-801, Shanghai INESA Scientific Instrument Co., Ltd., Shanghai, China).

2.3. Acid Production Capacity Analysis of the Tailings

The acid production capacity of the tailings was evaluated using the acid-base accounting (ABA) method in the static leaching experiment, which included indicators such as acid neutralization capacity (ANC), net acid generation (NAG), net acid producing potential (NAPP), and maximum potential acidity (MPA) [21].

The ANC of the tailing samples was determined by mixing 1 g of the bulk tailings with 25 mL of 0.2 M HCl in a 100 mL beaker [21]. After the reaction and cooling, the excess acid was titrated with 0.2 M NaOH to calculate the ANC value based on the corresponding formula. Subsequently, NAPP was calculated using the following equation, where MPA represents the maximum potential acidity:

$$NAPP\ (kgH_2\,SO_4/t) = MPA - ANC \quad (1)$$

MPA was determined by multiplying the pyritic sulfur content (%) of the sample, measured using an elemental analyzer, by a conversion factor of 30.6. This conversion factor is based on stoichiometric calculations, assuming complete oxidation of pyritic sulfur to sulfuric acid, with negligible uncertainty. While pyritic sulfur is the dominant contributor to MPA in these tailings, other sulfur-bearing minerals such as chalcopyrite may contribute locally. The measurement uncertainty for pyritic sulfur and ANC is estimated to be ±0.5% and ±5%, respectively, reflecting the limitations of the analytical methods.

The pyritic sulfur content was measured using an elemental analyzer. NAG was determined by mixing 2.5 g of the pulverized tailing sample with 250 mL of 15% H_2O_2 in a 500 mL conical flask. The mixture was left to react in a fume hood for 24 h and then boiled for 1 h. After cooling to ambient temperature, the final NAG-pH was measured, and the solution acidity was determined by titration with 0.1 M NaOH to reach pH 7. Both ANC and NAG values were expressed in terms of kg H_2SO_4/t [22].

2.4. Heavy Metal Speciation in Tailings

The modified BCR sequential extraction method was used to analyze the speciation of heavy metals in the tailings, aiming to determine their mobility and potential environmental hazards [23]. The detailed experimental procedures are shown in Table 1 below:

Table 1. Four-step sequential extraction of heavy metals.

Step	Speciation	Extraction Procedure
1	Weak Acid Extractable	Add 40 mL of 0.11 mol/L acetic acid solution to 1 g of tailing sample, shake at 22 ± 5 °C for 16 h, and then centrifuge at 3000× g for 20 min.
2	Reducible	Add 40 mL of 0.5 mol/L hydroxylamine hydrochloride solution (pH 2) to the residue from Step 1, shake for 16 h, and then centrifuge at 3000× g for 20 min.
3	Oxidizable	Add 10 mL of 8.8 mol/L H_2O_2 (pH 2–3) to the residue from Step 2, digest at room temperature for 1 h, then place in an 85 °C water bath for 1 h with the lid open and continue heating in the water bath until the volume is reduced to 3 mL. Add another 10 mL of H_2O_2 and continue digesting in the 85 °C water bath until the volume is reduced to about 1 mL. After cooling, add 50 mL of 1 mol/L ammonium acetate solution, shake for 16 h, and then centrifuge at 3000× g for 20 min.
4	Residual	Weigh 0.1 g of the tailing residue from Step 3 and add 4 mL concentrated nitric acid, 2 mL concentrated hydrochloric acid, and 1 mL hydrofluoric acid. Soak the mixture for 24 h, then perform microwave digestion. Measure the total heavy metal content using an inductively coupled plasma optical emission spectrometer (ICP-OES).

2.5. Static Leaching Experiments of Tailings

2.5.1. Effect of Simulated Acid Rain pH on Heavy Metal Leaching from Tailings

Simulated acid rain was prepared based on the typical composition of acid rain observed in Guangxi over the years, using a sulfuric acid/nitric acid molar ratio of 4:1 [24].

The pH of the simulated acid rain was adjusted to 2.0, 3.0, 4.0, and 5.0 using ultrapure water in the laboratory, which resulted in lower ionic strength compared to natural rainwater. This lower ionic strength may enhance metal mobility in the laboratory compared to field conditions, where rainwater typically contains additional dissolved ions. Then, 4 g of tailing sample sieved through a 100-mesh sieve was mixed with 20 mL of the prepared acid rain solutions of different pH values and gently stirred for 2 min to ensure uniform contact, followed by a settling period without further stirring to replicate static leaching conditions. The 100-mesh fraction was chosen to represent common particle sizes in weathered tailings, while the 1:5 solid/liquid ratio ensured adequate interaction and reproducibility in laboratory conditions. The pH and electrical conductivity (EC) of the leachates were measured at 0.5 h, 2 h, 4 h, 8 h, 12 h, 24 h, 48 h, and 96 h intervals. After each measurement, the leachate was filtered through filter paper, diluted to a final volume of 25 mL, and the concentrations of heavy metals (Pb, Zn, and Cd) in the solution were determined using an inductively coupled plasma optical emission spectrometer (ICP-OES).

2.5.2. Effect of Tailing Particle Size on Heavy Metal Ion Leaching

To evaluate the effect of particle size on heavy metal leaching, 4 g of tailing samples with different particle sizes (40 mesh, 100 mesh, 200 mesh) were added to 20 mL of simulated acid rain at pH 3.0. While similar trends may be expected at other low pH values (e.g., pH 2.0), the overall leaching rates and particle size effects could vary at higher pH levels, where metal solubility and mobility are reduced. The pH and EC of the leachates were measured at 0.5 h, 2 h, 4 h, 8 h, 12 h, 24 h, 48 h, and 96 h intervals. After each time point, the leachate was filtered through filter paper and diluted to a final volume of 25 mL, and the concentrations of Pb, Zn, and Cd in the solution were analyzed using ICP-OES.

2.5.3. Effect of Simulated Acid Rain on Heavy Metal Leaching from Tailings in the Disturbed Layer at Different Depths

Tailing samples from different depths (0 cm, 15 cm, 30 cm) were collected and sieved through a 100-mesh sieve. The estimated water content was in the range 10–20% in surface layers and 5–10% in deeper layers, while the local tailings' permeability is generally reported to be low-to-moderate (10^{-6} to 10^{-4} cm/s), based on previous studies of similar tailings. For each depth, 4 g of the sample was mixed with 20 mL of simulated acid rain at pH 3.0. The pH and EC of the leachates were recorded at 0.5 h, 2 h, 4 h, 8 h, 12 h, 24 h, 48 h, and 96 h intervals. After each measurement, the leachate was filtered and diluted to a final volume of 25 mL, and the concentrations of Pb, Zn, and Cd were measured using ICP-OES.

3. Results and Discussion

3.1. Characterization of Tailing Samples

3.1.1. Mineralogy and Chemical Composition

The elemental composition of the tailing samples was determined using X-ray fluorescence (XRF), as presented in Table 2. The samples were homogenized by grinding to a fine powder ($\leq$200 mesh) prior to analysis and correspond to depths of 0 cm, 15 cm, and 30 cm, respectively. The results indicate that the primary constituents of the Pb/Zn tailings are quartz and calcium–aluminum–iron compounds. X-ray diffraction (XRD) analysis further revealed that the major mineral phases include quartz, sericite, pyrite, limonite, and plagioclase (Figure 2). The high calcium content, potentially derived from plagioclase, may contribute to the tailings' acid neutralization capacity [25]. These minerals may correspond with the presence of calcium identified in the XRF analysis, reinforcing their potential role in neutralizing acidity and influencing metal release and retention. The measured pH values of the tailings ranged from 6.54 to 6.95, indicating a slightly acidic nature. This pH range is notably higher than that of typical non-ferrous metal tailings, likely due to the low sulfur content (0.37%), which resulted in a reduced acid generation potential [26].

Table 2. Content of main elements in tailings (wt.%).

Al_2O_3	CaO	Fe_2O_3	K_2O	MgO	MnO	Na_2O	P_2O_5	SiO_2	TiO_2
10.2–10.3	11.6–11.7	5.8–6.4	3.2–3.3	3.4–3.5	0.5–0.53	0.21–0.23	0.1–0.12	61.2–62.1	0.5–0.54

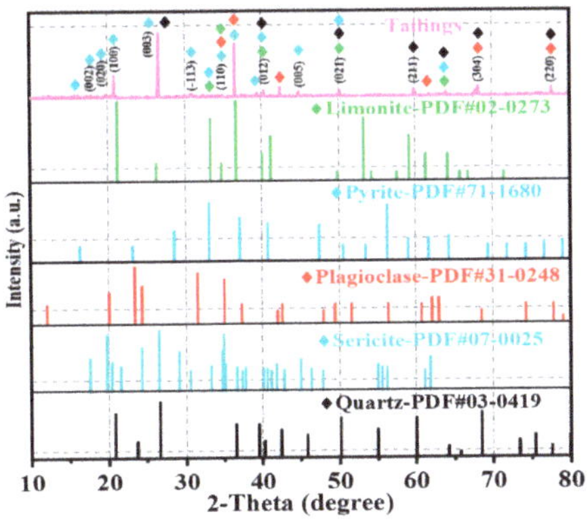

Figure 2. XRD analysis of tailings.

3.1.2. Heavy Metal Concentrations and Environmental Implications

As presented in Table 3, the concentrations of Pb, Zn, and Cd in the tailings exceed the limits established by the Standard for Pollution Control on the Storage and Disposal Site for General Industrial Solid Wastes [27].

Table 3. Content of the main heavy metals in tailings (mg/kg).

Heavy Metal	Pb	Zn	Cd
content	1864.4–2291.0	3143.8–4194.7	23.7–31.3
Limit	1000	1000	20

The XRD analysis revealed an absence of detectable mineral phases associated with Pb-Zn-bearing minerals (Figure 2).

This is likely due to the low content of these metals, making them indistinguishable from the dominant mineral phases. The presence of pyrite, despite its relatively low abundance, suggests a limited but notable acid generation potential. The low sulfur content and predominance of silicate minerals further corroborate the limited acid-generating capacity of the tailings. Additionally, plagioclase, as a source of calcium, can release calcium ions under acidic conditions, contributing to acid neutralization. Limonite, a hydroxide mineral, may also participate in neutralization reactions [28]. These findings enhance our understanding of the roles of various mineral phases in metal release and retention. Furthermore, factors such as particle size, pH, and weathering degree influence the mobility of heavy metals, particularly the leaching behavior of Pb, Zn, and Cd under simulated acid rain conditions.

3.1.3. Particle Morphology and Leaching Behavior

The scanning electron microscopy (SEM) images (Figure 3a) revealed that the tailing particles exhibited irregular shapes and a wide particle size distribution. This variation in particle morphology and size suggested a complex internal structure and surface characteristics, which could influence the leaching behavior of heavy metals [29]. These morphological characteristics could significantly influence the leaching behavior of heavy metals under acidic conditions, as irregular shapes and higher surface areas increase metal release potential. Further analysis through energy-dispersive X-ray spectroscopy (EDS) (as shown in the elemental mapping of Cd, Zn, and Pb in Figure 3b–d) identified elemental concentrations of Zn (56.31%), Cd (5.48%), and Pb (38.21%), confirming their significant presence within the tailings. These results align with the inductively coupled plasma (ICP) measurements, thereby reinforcing the association between these metals and mineral phases.

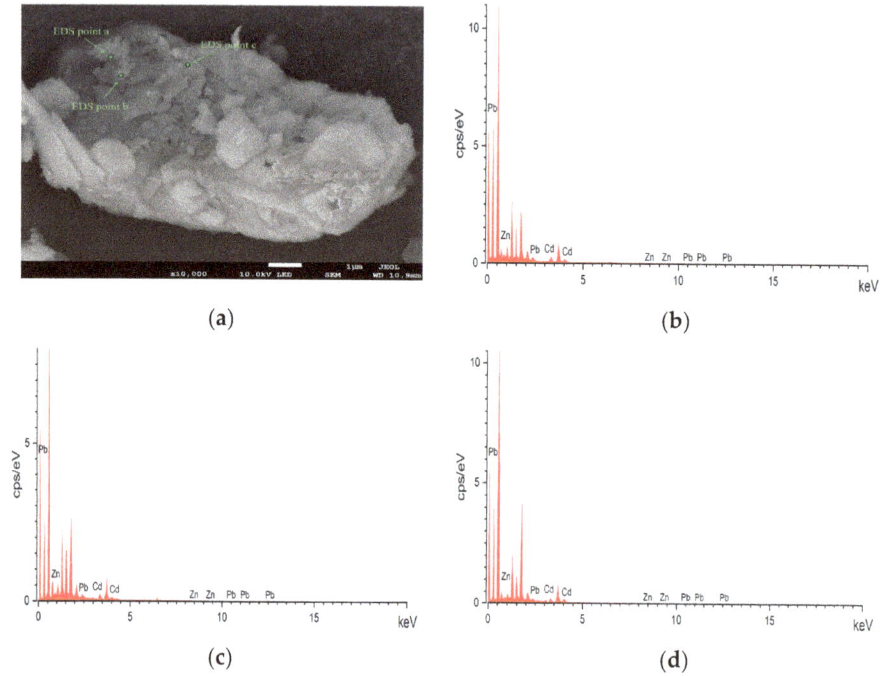

Figure 3. (**a**) Scanning electron microscope image of the tailings; (**b**) elemental analysis of point a, as detected by EDS; (**c**) elemental analysis of point b, as detected by EDS; (**d**) elemental analysis of point c, as detected by EDS.

This comprehensive characterization provides stronger evidence for metal mobilization mechanisms. Existing studies provide evidence of the association between heavy metals and sulfide minerals in similar tailings. For example, Saedi et al. and Liu et al. showed that in tailings from a lead–zinc mining area, Pb and Zn mainly exist in the forms of galena (PbS) and sphalerite (ZnS) [30,31]. In addition, He et al. confirmed through SEM-EDS analysis the coexistence of heavy metals with sulfide minerals [32]. Our findings also demonstrate the coexistence of heavy metals with sulfide minerals, particularly the strong association of Zn, Cd, and Pb with sulfides, as shown by the EDS elemental mapping results. This further strengthens the conclusion about the association between heavy metals and sulfide minerals in our samples.

3.2. Heavy Metal Partitioning in Tailings

Building upon the basic physicochemical properties of tailings, this section delves into the chemical speciation of heavy metals, which directly influences their environmental risk and mobility. The environmental risk posed by heavy metal contamination in tailings is influenced not just by the overall metal content but also by their chemical speciation, since distinct forms of metal species demonstrate varying levels of mobility and bioavailability. Therefore, it is essential to understand the distribution of heavy metal speciation within tailings to accurately predict and assess their environmental impact. Figure 4 illustrates the proportion of various heavy metal species present in the surface, middle, and deep layers of the disturbed layer in the tailing pond.

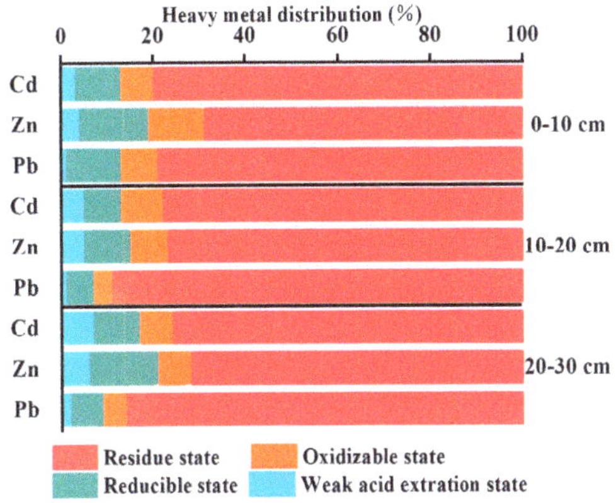

Figure 4. Distribution of heavy metals in tailings at different depths.

In different depths of the disturbed layer, most of the heavy metals were found predominantly in the residual fraction, indicating that a large proportion of heavy metals in the tailings had already undergone oxidation. The maximum proportions of residual fractions for Pb, Zn, and Cd in the upper, middle, and lower layers were 85.25%, 69.3%, and 72.6%, respectively. The high proportion of Pb in the residual fraction (85.25%) suggests that Pb has a relatively low potential for mobility in these tailings. The proportion of reducible fractions for all three metals was significantly lower than that of oxidizable fractions, which was related to the low organic matter content in the tailings [33].

To provide a clearer understanding of the oxidation states and chemical forms of these metals, we analyzed their distribution across different fractions. Lead (Pb) is predominantly present in the residual fraction, suggesting it is incorporated into stable mineral structures, possibly as sulfide minerals like PbS (lead sulfide). In the oxidizable fraction, Pb is associated with organic matter and sulfide minerals, which may oxidize under environmental conditions, potentially releasing Pb into more mobile forms. Zinc (Zn) is similarly found mainly in the residual fraction, possibly as ZnS (zinc sulfide), and in the oxidizable fraction associated with organic matter and sulfides. Cadmium (Cd) is primarily present in the oxidizable fraction, indicating its association with organic matter and sulfides, which, upon oxidation, may release Cd into more mobile and bioavailable forms, contributing to its higher environmental risk.

The non-residual fraction of heavy metals indicates their potential mobility under varying environmental conditions, as heavy metals in oxidizable and reducible forms can transform into exchangeable forms, thereby increasing their likelihood of release. Among the three metals, Cd had the highest proportion of non-residual fractions at 27.4%, suggesting that Cd may have a relatively high mobility and a greater potential for environmental contamination [34]. Based on the content of non-residual fractions, the potential mobility of the heavy metals in these tailings followed the following order: Cd > Zn > Pb.

3.3. *Impact of Acid Generation Potential of Tailings*

The acid generation potential of the tailings was evaluated, with the results summarized in Table 4. The acid neutralization capacity (ANC) values for the two samples were 195.51 kg H_2SO_4/t, respectively, indicating a robust capacity for neutralizing external acidity [35]. This substantial neutralization capacity can be attributed to the high content of calcium and magnesium oxides, hydroxides, and carbonate compounds, as revealed by XRF analysis.

Table 4. The average content of the main heavy metals in tailings (mg/kg).

ANC	MPA	NAPP	NAG-pH	NAG	Sulfur Content
195.51	11.33	−184.19	7.52	−1.59	0.37%

Note: The units for ANC, MPA, NAPP, and NAG are all expressed as kg H_2SO_4/t.

In contrast, the net acid generation (NAG) values were negative, measured at −1.59 kg H_2SO_4/t, indicating a stronger acid neutralization potential that surpasses the net acid production. This suggests that the tailings possess an overall net capacity to neutralize potential acid production, thereby significantly lowering the likelihood of long-term acidification. While NAG values are still related to potential long-term acidification through the slow oxidation of residual sulfides, the negative results indicate that the risk of acid production over prolonged exposure is minimal. Thus, while immediate acidification is unlikely, the overall potential for long-term acidification has been notably reduced due to the enhanced neutralization capacity.

The analysis was conducted on all nine tailings samples to provide a comprehensive view of the acid generation and neutralization characteristics. The study reflects the average values across these nine samples, offering a balanced and representative assessment of the tailings' geochemical behavior in terms of sulfur content, metal concentrations, and their variability. This approach ensures a broader understanding of the acid generation and neutralization potential observed in the collected samples [36–38]. Considering that these tailings have been exposed to weathering for many years and still demonstrate a capacity for acid production, and given the significant volume of material stored in the tailing pond, the risk of acid-related pollution cannot be entirely dismissed. The combined assessment of net acid producing potential (NAPP) and NAG-pH values classifies the tailings as non-acid-generating materials [39]. This classification can be explained by two primary factors: (1) the long-term closure of the tailings pond, leading to the depletion of sulfur in various migratory forms; and (2) the inherently low initial sulfur content in the tailings. Consequently, while the high ANC values suggest a reduced risk of immediate acidification, the potential for long-term acid generation remains a concern that warrants continuous monitoring. The findings suggest that despite the strong acid neutralization capacity (ANC) of low-sulfur tailings, there is still a risk of heavy metal leaching under extreme acidic conditions, necessitating a closer examination of the influence of acid rain pH on metal leaching behavior.

3.4. *Effect of Simulated Acid Rain pH on Leachate Characteristics*

After evaluating the potential for acid generation and its implications on metal release, it is essential to investigate the direct impact of acid rain pH on the leaching characteristics of heavy metals. The static leaching experiments were conducted using the surface layer of tailing samples sieved through a 100-mesh screen to evaluate the impact of external acid rain solutions (pH = 2, 3, 4, and 5) on the pH and electrical conductivity (EC) of the leachates. The results are presented in Figure 5a,b, and the relationship between the leaching concentrations of heavy metals (Pb, Zn, and Cd) and acid rain pH is depicted in Figure 6a–c.

As illustrated in Figure 5a, the pH of the leachates increased rapidly to approximately 6.0 within 12 h of exposure to acid rain. For initial acid rain pH values of 3, 4, and 5, the leachate pH rose to around 6.3 immediately upon contact, indicating that the initial pH of the acid rain had a limited impact on the final leachate pH. This observation suggested that the tailings possessed a strong neutralization capacity against external acidic inputs. The rapid pH rise at t = 0 for pH 3 demonstrates the effective buffering action of the tailings, driven by the presence of acid-neutralizing minerals. However, at pH 4 and 5, the buffering response was slower, likely due to the smaller pH gradient, resulting in a more gradual neutralization process. In the study region, acid rain is relatively frequent, with pH levels typically ranging from 3.5 to 4.5, but occasional extreme events with pH as low as 2.3 have been recorded, making the buffering capacity observed at pH 3 environmentally relevant.

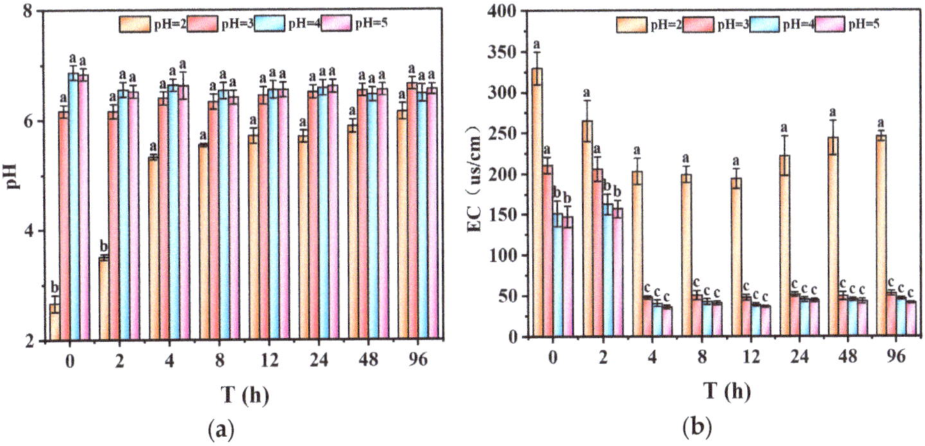

Figure 5. (**a**) Relationship between the pH of leachates and time; (**b**) relationship between the EC of leachates and time. Different lowercase letters indicate significant differences at the $p < 0.05$ level.

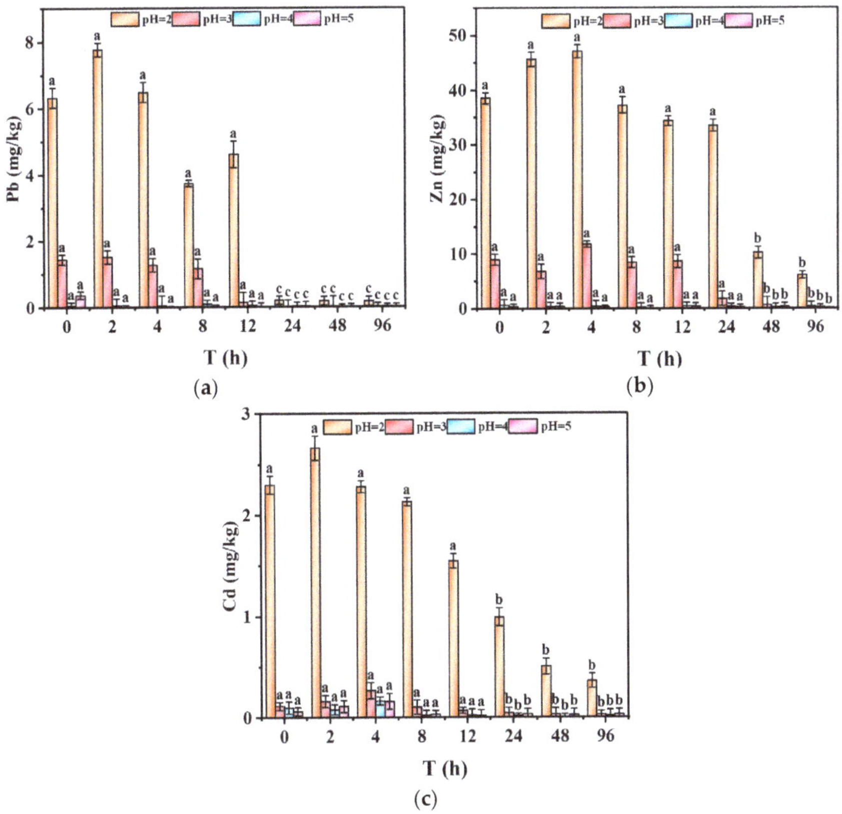

Figure 6. Relationship between leaching concentration of (**a**) Pb, (**b**) Zn, and (**c**) Cd and pH. Different lowercase letters indicate significant differences at the $p < 0.05$ level.

In contrast, at pH 2, the stronger acidity leads to more aggressive disruption of acid-sensitive minerals such as carbonates and oxides, causing a slower initial pH increase and a lower equilibrium pH compared to pH 3. The higher proton concentration at pH 2 enhances mineral dissolution and increases the release of adsorbed heavy metals, which explains the discrepancy between pH 2 and pH 3 leachate curves. This behavior highlights the non-linear nature of acid-tailing interactions, where small pH decreases can result in disproportionately higher metal mobilization due to the increased dissolution and exchange reactions.

Over time, the pH of the leachate remained relatively stable at elevated levels and did not exhibit a decreasing trend, even under conditions of pH 2 acid rain. During the initial 4 h, the tailings rapidly neutralized the acid rain pH from 2 to around 6, after which the pH remained stable or slightly increased. Under pH 3 and 4 conditions, stability was achieved even more rapidly. This observation aligns with the high ANC and low NAG values of the tailings, further confirming their strong neutralization capacity against external acidic inputs. This robust buffering capacity could be attributed to two primary factors: (1) the high calcium and magnesium content in the tailings (10.967%), which suggested a substantial presence of acid-consuming minerals such as oxides, hydroxides, and carbonates [40]; and (2) the inherently low sulfur content, which limited acid generation potential [41].

As illustrated in Figure 5b, the EC of the leachates initially exhibited a decrease before stabilizing. This trend can be explained by the initial release of surface-adsorbed salts and heavy metal ions, which resulted in elevated initial EC values. As the dissolution and adsorption processes approached equilibrium, the rate of ionic release diminished, resulting in the stabilization of the EC [41]. However, due to ongoing chemical reactions, some insoluble components continued to dissolve over prolonged acid rain exposure, leading to minor fluctuations in EC. These fluctuations in electrical conductivity were consistent with variations observed in pH values.

The comparison between Figure 5a,b clearly shows that leachates with lower pH values exhibited higher EC, indicating that a decrease in acid rain pH promoted the leaching of soluble salts and heavy metal ions from the tailings. The correlation between pH and EC suggests that increasing acidity in the leachate leads to the mobilization of more metal ions and soluble salts, which results in elevated EC values.

Figure 6a–c illustrated the leaching patterns of Pb, Zn, and Cd under different acid rain pH levels. It is clear that the pH of acid rain strongly affected the leaching concentrations of these heavy metals, as lower pH values lead to a marked increase in their release [42]. For instance, at pH 5, the leaching concentration of Zn was only 0.076 mg/L, while at pH 2, it surged dramatically to 7.53 mg/L. This significant increase in metal release under pH 2 conditions emphasizes the higher leaching potential associated with extreme acidic inputs, which could pose greater environmental risks if similar conditions arise. Continuous leaching of the three heavy metals was observed under pH 2 and 3 conditions, whereas minimal to no leaching occurred under pH 4 and 5 conditions. This indicated that lower pH values can significantly enhance the leaching of heavy metals from tailings, which poses a greater threat to the surrounding environment. This is consistent with existing studies. These studies indicate that under acidic conditions, especially when the pH is below 3, the solubility of metal ions such as Pb, Zn, and Cd increases, thereby enhancing their mobility. Additionally, increasing pH induces the precipitation of metal hydroxides or carbonates, which reduces metal mobility [43]. Under high pH conditions, these speciation changes significantly decrease the bioavailability of heavy metals [44].

This phenomenon may be associated with the speciation distribution of heavy metals, as Zn and Cd exhibited higher proportions within the weak acid-extractable fraction [45]. Under acidic leaching conditions, oxidizable fractions can transform into weak acid-extractable fractions, thereby increasing metal mobility. While this study focused on absolute leaching concentrations, future work should consider normalizing the results to the mass of metal contributed by the 4 g of sample or its non-refractory fractions. This approach would allow for a more accurate assessment of the relative leaching behavior of each metal and help clarify the mechanisms driving metal mobilization under acidic conditions. Establishing standardized curves and normalization protocols would also enhance the comparability of results across different studies, providing more robust conclusions about the potential environmental impacts of tailings leaching.

The leaching profiles of the three metals displayed a biphasic pattern, featuring a rapid initial release stage succeeded by a more gradual release phase [46]. This pattern can be attributed to the initial rapid dissolution of soluble salts and surface-adsorbed heavy metal ions into the leachate, while the subsequent stage involved the gradual release of exchangeable heavy metal ions. It is important to note that the samples were dried at a low temperature ($\leq$40 °C) to minimize alterations in metal speciation; however, results might differ slightly with undried samples. Additionally, potential metal losses onto the flask walls, particularly at higher pH levels, could contribute to the observed decrease in metal concentrations. Visible precipitates, especially of Zn and Pb, were observed at more neutral pH levels, suggesting the formation of metal hydroxides or carbonates under these conditions. The rapid release phase observed under pH 2 conditions reflects the stronger initial dissolution of minerals and desorption of heavy metals due to higher proton concentrations. As the pH of the leachate increased over time, the oxidative dissolution of the tailings diminished, and the release and desorption of heavy metals reached a state of equilibrium, resulting in a reduced leaching rate. This two-stage release behavior is consistent with the trend observed in the EC data, where EC initially decreased before stabilizing.

The comparison between heavy metal release concentrations and leachate pH (Figure 6a) showed that within the initial 9 h under pH 2 acid rain conditions, the leachate pH increased rapidly, accompanied by a rise in the concentrations of the three heavy metals. This observation suggested that the strong acid neutralization capacity of the tailings was insufficient to completely prevent the release of heavy metals. Notably, a strong acid neutralization capacity does not imply that the tailings do not contribute to environmental contamination. During the initial stage, when the leachate pH was low, heavy metals were rapidly released into the leachate at high concentrations. In contrast, during the later stages, when the leachate pH was elevated, heavy metal concentrations in the leachate were significantly lower or even undetectable. This indicated that acidic conditions facilitate the leaching of heavy metals from the tailings [47].

In summary, one of the key factors influencing the release of heavy metals from tailings is the acidification of the tailing pile. China, particularly the southern regions, is a region prone to acid rain, characterized by both high frequency and intensity. For example, in Guangxi, the lowest recorded pH of acid rain has reached as low as 2.32 [48]. Although acid rain with pH values below 3.0 is relatively rare—occurring in less than 5% of total acid rain events and primarily during extreme pollution episodes—it still poses a potential risk for heavy metal leaching from tailings. Therefore, effective management of tailings acidification is crucial to prevent the release of heavy metals and to mitigate the associated risks to the surrounding environment and human health. Tailings management strategies should prioritize minimizing exposure to extreme acidic conditions, especially those approaching pH 2, to limit heavy metal mobilization and reduce environmental hazards.

3.5. *Effect of Tailings Particle Size on Leaching Behavior*

To further understand the influence of physical properties on metal leaching, this section evaluates the effect of tailing particle size, which determines the exposure and interaction area between tailings and acid rain. Under the condition of simulated acid rain, at pH 3, the variations in leachate pH and EC over time for upper layer tailings of varying particle sizes (40 mesh, 100 mesh, and 200 mesh) are shown in Figure 7a,b. Additionally, the relationship between the leaching concentrations of heavy metals (Pb, Zn, and Cd) and particle size is illustrated in Figure 8a–c.

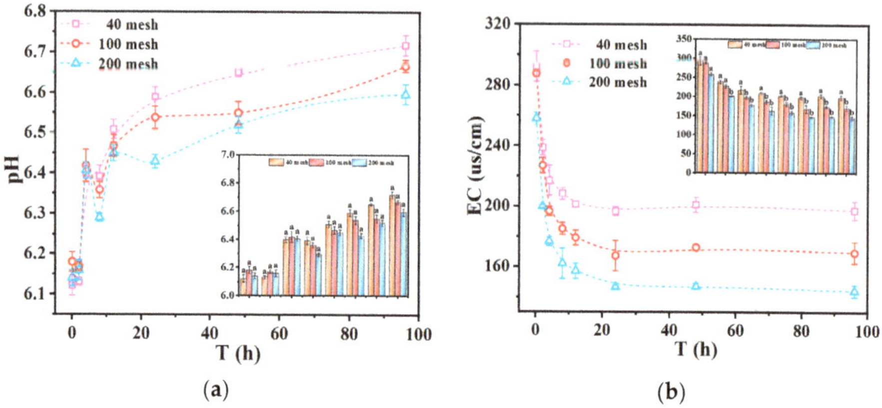

Figure 7. Relationship between the (**a**) pH and (**b**) EC of leachates and particle size. Different lowercase letters indicate significant differences at the $p < 0.05$ level.

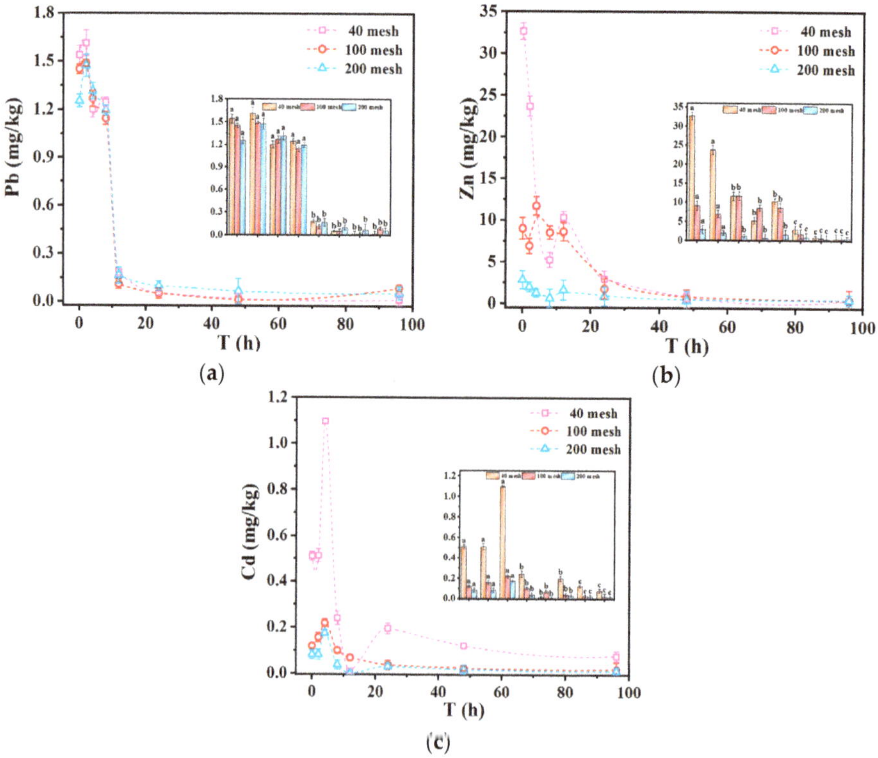

Figure 8. Relationship between leaching concentration of (**a**) Pb, (**b**) Zn, and (**c**) Cd and particle size. Different lowercase letters indicate significant differences at the $p < 0.05$ level.

As shown in Figure 7a, the leachate pH rapidly increased initially and then gradually stabilized, indicating that the tailings have a strong acid neutralization capacity. The neutralization capacity of tailings with different particle sizes followed the following order: 40 mesh > 100 mesh > 200 mesh. Due to the larger contact area between the larger tailing particles and the acid rain, there is an increased ability to dissolve soluble components, resulting in higher conductivity values. This is consistent with the fact that larger tailing particles provide a higher effective contact area [49]. This trend was attributed to the larger effective contact area of coarse tailing particles compared to finer ones [50]. Although larger particles have a smaller specific surface area, they are less susceptible to agglomeration during acid rain leaching, which allows for a greater effective contact area with the leachate.

As depicted in Figure 7b, the EC of leachate increased with particle size, following the following order: 40 mesh > 100 mesh > 200 mesh. This could be attributed to the larger tailing particles providing a greater effective contact area with acid rain, which enhanced the dissolution of soluble components, thereby resulting in elevated EC values. Over time, the EC of the leachate showed an initial decrease followed by stabilization. This trend may be explained by the initial release of ions adsorbed on the surface of the tailings when the leachate pH was lower, leading to higher EC values. As the soluble components on the tailings' surface were gradually depleted and an adsorption–desorption equilibrium was reached, the EC of the leachate decreased and eventually stabilized [51].

The trends in leaching concentration for Pb, Zn, and Cd with varying particle sizes, as shown in Figure 8a–c, demonstrate a consistent pattern; the leaching concentration increased with larger particle size, following the following hierarchy of order: 40 mesh > 100 mesh > 200 mesh. This phenomenon could be attributed to the more effective and extensive interaction between larger tailing particles and simulated acid rain, enabling the heavy metals to be adsorbed on the surfaces of coarser particles. Conversely, finer tailing particles tend to agglomerate significantly upon exposure to acid rain, which can block interstitial spaces between particles and reduce the effective contact area with the leachate. This observation is based on indirect evidence from leaching behavior. Future research should include particle size distribution analysis of the leachate to confirm the extent of agglomeration and its influence on metal mobility. Additionally, analyzing the initial metal content of different tailing fractions would provide further insights into how agglomeration affects metal leaching dynamics. The reduction in effective contact area due to agglomeration likely limited ion-exchange reactions between H^+ in the acid rain and heavy metal ions adsorbed on the tailing surface [52]. Similarly, studies have shown that smaller particles have a larger specific surface area, leading to faster metal release. This is because smaller particles provide more reactive surfaces, thereby accelerating the leaching process of heavy metals [53]. In contrast, larger particles have greater pore space, which not only affects water infiltration but also alters the leaching dynamics [54]. Moreover, due to the increased reactive area of larger particles under acidic conditions, their metal release rate can also be faster. As a result, the concentrations of leached heavy metals from finer tailings were lower than those observed in coarser particles, which had greater effective surface contact with the leachate.

Comparing the effects of acid rain pH and particle size on the leaching concentrations of heavy metals, it can be inferred that particle size had a less significant influence on leaching concentrations than acid rain pH. Specifically, the differences in leaching concentrations observed among tailings of varying particle sizes are relatively minor, while the variations in concentrations under different pH conditions can be substantially greater—sometimes by an order of magnitude [55]. Therefore, when assessing the risk of heavy metal pollution from tailings, priority should be given to the influence of acid rain pH over that of particle size.

3.6. Effect of Disturbed Layer Depth on Heavy Metal Leaching Under Acidic Conditions

Considering that both chemical and physical properties influence heavy metal leaching, this section investigates the effect of depth on metal release to better understand the vertical migration patterns within the tailing pond. This experiment evaluated the impact of pH 3 simulated acid rain on the leaching concentrations of heavy metals (Pb, Zn, and Cd) in tailings from different depths within this disturbed layer (0 cm, 15 cm, and 30 cm). The results are presented in Figure 9a–c. As observed, the leaching concentrations of the three heavy metals showed a consistent trend with depth; as the depth of the disturbed layer increased, the leaching concentrations of heavy metals also increased. This phenomenon may be attributed to the long-term leaching of rainwater, which facilitated the downward migration of heavy metals, especially those adsorbed on the surface of the upper tailings, resulting in higher concentrations in the lower layers. The weathering process plays a significant role here, as it transforms primary sulfide minerals into secondary oxides or hydroxides over time, changing metal speciation and increasing their mobility [56]. For example, some studies have demonstrated that deeper, more weathered layers show higher solubility of metal ions, particularly under acidic conditions [57]. These findings align with the results of this study, where greater weathering was associated with increased mobility of Cd and Zn. Another contributing factor could be the depletion of surface layers through runoff and leaching, which removes heavy metals from the uppermost parts of the tailings over time. Additionally, variations in mineral phases at different depths could affect the association and mobility of metals, leading to distinct distribution patterns along the vertical profile. This trend was most pronounced for Cd, aligning with findings from previous studies [58]. However, the trend was not entirely consistent across all samples, possibly due to the complex physical and chemical changes that have occurred in the disturbed layer over the years, influenced by factors such as wind erosion and hydraulic flushing, especially considering that the tailing pond had remained inactive for an extended period.

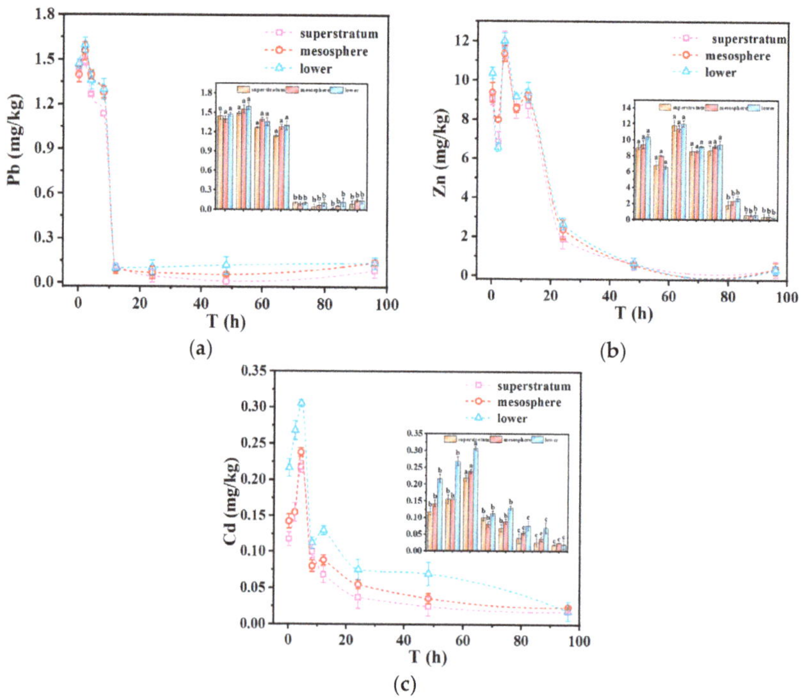

Figure 9. Relationship between leaching concentration of (**a**) Pb, (**b**) Zn, and (**c**) Cd and depth. Different lowercase letters indicate significant differences at the $p < 0.05$ level.

The three heavy metals demonstrated comparable trends concerning both depth and time. Notably, the leaching concentrations of Pb and Zn almost completely overlapped after 12 h of immersion, suggesting that the chemical composition of tailings at different depths within the disturbed layer may be relatively uniform. Furthermore, the variations in the leaching concentrations of the three heavy metals across different depths were not particularly pronounced, indicating a potential lack of strong correlation between the vertical depth of the tailings and heavy metal leaching concentrations. Alternatively, it could be due to the relatively minor differences in sampling depth within the disturbed layer [59].

The leaching behavior of the three heavy metals exhibited a two-stage pattern: an initial rapid release phase occurring within the first 12 h, followed by a slower release phase. This suggested that the heavy metals migrating downward are likely those that were adsorbed on the tailing surfaces. As shown in Figure 9c, even after 12 h of exposure to acid rain, Cd still exhibited noticeable differences in leaching concentrations at various depths, suggesting a notable potential for downward migration. This suggested that Cd possesses the highest downward mobility among the three metals, while Pb and Zn displayed comparatively weaker downward mobility.

This behavior may be associated with the chemical speciation of these metals within the tailings. The non-residual fraction is generally regarded as an indicator of the potential mobility of heavy metals. The non-residual fraction of Cd was 37.8%, significantly higher than that of Pb, which was only 14.7%. This suggests that Cd has a greater potential for migration compared to Pb, thereby making it more susceptible to leaching and contamination under acidic conditions.

4. Conclusions

The findings suggest that the pH of simulated acid rain (SAR) significantly influences the release behavior of heavy metals, particularly for Pb, Zn, and Cd. However, it is important to note that the pronounced mobilization of heavy metals was primarily observed at extremely low pH (pH 2), which is less common in natural settings. Under highly acidic conditions (pH 2–3), the concentrations of these metals in the leachate increased by 4–6 times compared to neutral conditions (pH 7), with Cd showing the highest release potential. This indicates that the ANC of low-sulfur tailings is not sufficient to prevent heavy metal leaching under extreme acidic conditions. It should also be noted that the leaching data have not been normalized to the quantities of metals originally present in the tailings, which could affect the interpretation of the results. Moreover, particle size also plays a role in heavy metal leaching, as coarser particles exhibited higher leaching concentrations due to increased surface area contact; however, the effect of particle size is less pronounced than that of SAR pH. Analysis of the disturbed layer (0–30 cm) further indicated that heavy metal concentrations increased with depth, particularly for Cd, whose non-residual fraction was significantly higher, indicating a strong potential for downward migration. These results highlight the need for effective tailing management strategies that consider not only ANC but also particle size distribution and vertical leaching patterns to mitigate environmental risks. Future studies should aim to refine the experimental protocols, especially in terms of simulating rainwater contact and normalizing metal mobilization data, to provide more representative results for field conditions.

Author Contributions: Conceptualization, funding acquisition, project administration, writing—original draft, supervision, X.S.; data curation, formal analysis, writing—original draft, investigation, D.Z.; conceptualization, supervision, writing—review and editing, Q.Z.; methodology, T.A.; writing—review and editing, X.Z.; writing—review and editing, J.L. All authors have read and agreed to the published version of the manuscript.

Funding: This work was financially supported by the National Natural Science Foundation of China (No. 52230006) and the Guangxi Key R&D Program Project (GuiKeAB21220049).

Data Availability Statement: The data presented in this study are available on request from the corresponding author.

Conflicts of Interest: The authors declare no conflicts of interest.

References

1. Xu, D.-M.; Zhan, C.-L.; Liu, H.-X.; Lin, H.-Z. A critical review on environmental implications, recycling strategies, and ecological remediation for mine tailings. *Environ. Sci. Pollut. Res.* **2019**, *26*, 35657–35669. [CrossRef] [PubMed]

2. Zhang, Q.; Li, Y.; Ma, W.; Bai, X.; Ru, X.; Zhang, L.; Zhong, S.; Shu, X. Three-dimensional recyclable FeS2/reduced graphene oxide aerogel with high porosity reticulated structure for efficient removal of tylosin tartrate. *Sep. Purif. Technol.* **2023**, *324*, 124463. [CrossRef]
3. Wang, G.; Cao, W.; Liang, G.; Xiang, J.; Chen, Y.; Liu, H. Leaching Behavior of Heavy Metals from Pb–Zn Tailings and Remediation by Ca- or Na-Montmorillonite. *Water Air Soil Pollut.* **2023**, *234*, 101. [CrossRef]
4. Lin, K.; Jian, J.; Zhang, Y.; Liu, Y.; Li, S.; Zhao, Y.; Xu, H. Study on Plant-blanket to reduce heavy metal migration caused by precipitation and to improve the soil environment of pyritic tailings. *Sci. Total Environ.* **2024**, *939*, 173376. [CrossRef]
5. Sun, R.; Gao, Y.; Yang, Y. Leaching of heavy metals from lead-zinc mine tailings and the subsequent migration and transformation characteristics in paddy soil. *Chemosphere* **2022**, *291*, 132792. [CrossRef]
6. Luo, D.; Liang, Y.; Wu, H.; Li, S.; He, Y.; Du, J.; Chen, X.; Pu, S. Human-Health and Environmental Risks of Heavy Metal Contamination in Soil and Groundwater at a Riverside Site, China. *Processes* **2022**, *10*, 1994. [CrossRef]
7. Pan, Y.; Chen, M.; Wang, X.; Chen, Y. Ecological risk, source apportionment, and influencing factors of heavy metals in soil in a typical lead-zinc mining watershed, Guangxi, China. *J. Environ. Chem. Eng.* **2024**, *12*, 112731. [CrossRef]
8. Huang, L.-M.; Deng, C.-B.; Huang, N.; Huang, X.-J. Multivariate statistical approach to identify heavy metal sources in agricultural soil around an abandoned Pb–Zn mine in Guangxi Zhuang Autonomous Region, China. *Environ. Earth Sci.* **2013**, *68*, 1331–1348. [CrossRef]
9. Zhang, Q.; Ma, W.; Peng, Q.; Shu, X. Stabilization and Utilization of Pyrite under Light Irradiation: Discussion of Photocorrosion Resistance. *ACS Omega* **2020**, *5*, 28693–28701. [CrossRef]
10. Shu, X.; Qin, Z.; Nie, C.; Zhang, D.; Du, H.; Zhang, Q.; Dang, Z. Inhibition photooxidation of pyrite under illumination via altering photogenerated carrier migration pathways: Role of DTC-TETA surface passivation. *Sci. Total Environ.* **2024**, *926*, 171988. [CrossRef]
11. Shu, X.; Tian, W.; Xiong, S.; Zhang, W.; Zhang, Q. Straw Biochar at Different Pyrolysis Temperatures Passivates Pyrite by Promoting Electron Transfer from Biochar to Pyrite. *Processes* **2022**, *10*, 2148. [CrossRef]
12. Ganapathi, H.; Phukan, M. Environmental Hazards of Limestone Mining and Adaptive Practices for Environment Management Plan. In *Environmental Processes and Management: Tools and Practices*; Singh, R.M., Shukla, P., Singh, P., Eds.; Springer International Publishing: Cham, Switzerland, 2020; pp. 121–134.
13. Zhang, Q.; Li, Y.; Peng, X.; Bai, X.; Zhang, L.; Zhong, S.; Shu, X. Pyrite from acid mine drainage promotes the removal of antibiotic resistance genes and mobile genetic elements in karst watershed with abundant calcium carbonate. *J. Hazard. Mater.* **2024**, *471*, 134344. [CrossRef] [PubMed]
14. Król, A.; Mizerna, K.; Bożym, M. An assessment of pH-dependent release and mobility of heavy metals from metallurgical slag. *J. Hazard. Mater.* **2020**, *384*, 121502. [CrossRef] [PubMed]
15. Paltseva, A. (Ed.) How Do Cations Exchange in The Soil? In *The Urban Soil Guide: A Field and Lab Manual*; Springer Nature: Cham, Switzerland, 2024; pp. 77–81.
16. Balci, N.; Demirel-Floyd, C. Prediction of Acid Mine Drainage (AMD) and Metal Release Sources at the Küre Copper Mine Site, Kastamonu, NW Turkey. *Mine Water Environ.* **2017**, *37*, 56–74. [CrossRef]
17. Matsumoto, S.; Ishimatsu, H.; Shimada, H.; Sasaoka, T.; Kusuma, G.J. Characterization of Mine Waste and Acid Mine Drainage Prediction by Simple Testing Methods in Terms of the Effects of Sulfate-Sulfur and Carbonate Minerals. *Minerals* **2018**, *8*, 403. [CrossRef]
18. Xiao, D.; Zhao, X.; Li, K.; Zhao, X.; Liu, H.; Li, X.; Luo, G. Influence of acid rain on slope instability mechanism—A case study in Sichuan provincial highway, China. *Bull. Eng. Geol. Environ.* **2021**, *80*, 3659–3673. [CrossRef]
19. Bhaduri, A.M.; Fulekar, M.H. Antioxidant enzyme responses of plants to heavy metal stress. *Rev. Environ. Sci. Bio/Technol.* **2012**, *11*, 55–69. [CrossRef]
20. Moncur, M.C.; Ptacek, C.J.; Blowes, D.W.; Jambor, J.L. Release, transport and attenuation of metals from an old tailings impoundment. *Appl. Geochem.* **2005**, *20*, 639–659. [CrossRef]
21. Davis, A.; Whitehead, C.; Lengke, M.; Collord, J. Acid–Base Accounting Tests in Combination with Humidity Cells Help to Predict Waste Rock Behavior. *Mine Water Environ.* **2019**, *38*, 467–487. [CrossRef]
22. Burgot, J.-L. Acid-Base Titration Curves. In *Ionic Equilibria in Analytical Chemistry*; Burgot, J.-L., Ed.; Springer: New York, NY, USA, 2012; pp. 135–156.
23. Rauret, G.; López-Sánchez, J.F.; Sahuquillo, A.; Rubio, R.; Davidson, C.; Ure, A.; Quevauviller, P. Improvement of the BCR three step sequential extraction procedure prior to the certification of new sediment and soil reference materials. *J. Environ. Monit.* **1999**, *1*, 57–61. [CrossRef]
24. Huang, Y.; Cui, X. Dominant cloud microphysical processes of a torrential rainfall event in Sichuan, China. *Adv. Atmos. Sci.* **2015**, *32*, 389–400. [CrossRef]
25. Luo, Y.; Tan, W.; Suib, S.L.; Qiu, G.; Liu, F. Epitaxial growth mechanism of heterogeneous catalytic oxidation of Mn(II) on manganite under oxic conditions. *Chem. Geol.* **2020**, *547*, 119670. [CrossRef]
26. Binnemans, K.; Jones, P.T. Solvometallurgy: An Emerging Branch of Extractive Metallurgy. *J. Sustain. Metall.* **2017**, *3*, 570–600. [CrossRef]
27. *GB 18599-2001*; Standard for Pollution Control on the Storage and Disposal Site for General Industrial Solid Wastes (Modified 2013). Ministry of Ecology and Environment of the People's Republic of China: Beijing, China, 2013.

28. Pokrovsky, O.S.; Shirokova, L.S.; Zabelina, S.A.; Jordan, G.; Bénézeth, P. Weak impact of microorganisms on Ca, Mg-bearing silicate weathering. *Npj Mater. Degrad.* **2021**, *5*, 51. [CrossRef]
29. Akhter, F.; Rao, A.A.; Abbasi, M.N.; Wahocho, S.A.; Mallah, M.A.; Anees-ur-Rehman, H.; Chandio, Z.A. A Comprehensive Review of Synthesis, Applications and Future Prospects for Silica Nanoparticles (SNPs). *Silicon* **2022**, *14*, 8295–8310. [CrossRef]
30. Saedi, A.; Jamshidi-Zanjani, A.; Mohseni, M.; Darban, A.K. Effect of mechanochemical activation on hydration properties of lead-zinc sulfide tailings for concrete construction. *Case Stud. Constr. Mater.* **2023**, *18*, e01996. [CrossRef]
31. Liu, Y.; Molinari, S.; Dalconi, M.; Valentini, L.; Ricci, G.; Carrer, C.; Ferrari, G.; Artioli, G. The leaching behaviors of lead, zinc, and sulfate in pyrite ash contaminated soil: Mineralogical assessments and environmental implications. *J. Environ. Chem. Eng.* **2023**, *11*, 109687. [CrossRef]
32. He, Z.; Xu, Y.; Yang, X.; Shi, J.; Wang, X.; Jin, Z.; Zhang, D.; Pan, X. Passivation of heavy metals in copper–nickel tailings by in-situ bio-mineralization: A pilot trial and mechanistic analysis. *Sci. Total Environ.* **2022**, *838*, 156504. [CrossRef]
33. Benidire, L.; Pereira, S.; Aboudrar, W.; Hafidi, M.; Castro, P.; Boularbah, A. Remediation of metal-contaminated mine tailings by the application of organic and mineral amendments. *J. Soils Sediments* **2022**, *22*, 482–495. [CrossRef]
34. Wang, M.; Chen, Z.; Song, W.; Hong, D.; Huang, L.; Li, Y. A review on Cadmium Exposure in the Population and Intervention Strategies Against Cadmium Toxicity. *Bull. Environ. Contam. Toxicol.* **2021**, *106*, 65–74. [CrossRef]
35. Prakash, J.; Agrawal, S.B.; Agrawal, M. Global Trends of Acidity in Rainfall and Its Impact on Plants and Soil. *J. Soil Sci. Plant Nutr.* **2023**, *23*, 398–419. [CrossRef] [PubMed]
36. Qafoku, N.P. Chapter Two—Climate-Change Effects on Soils: Accelerated Weathering, Soil Carbon, and Elemental Cycling. In *Advances in Agronomy*; Sparks, D.L., Ed.; Academic Press: Cambridge, MA, USA, 2015; Volume 131, pp. 111–172.
37. dos Santos, B.R.; Bacalhau, F.B.; dos Santos Pereira, T.; Souza, C.F.; Faez, R. Chitosan-Montmorillonite microspheres: A sustainable fertilizer delivery system. *Carbohydr. Polym.* **2015**, *127*, 340–346. [CrossRef] [PubMed]
38. Li, X.; Li, Q.; Su, Y.; Yue, Q.; Gao, B.; Su, Y. A novel wheat straw cellulose-based semi-IPNs superabsorbent with integration of water-retaining and controlled-release fertilizers. *J. Taiwan Inst. Chem. Eng.* **2015**, *55*, 170–179. [CrossRef]
39. Moon, Y.; Song, Y.; Moon, H.-S. The potential acid-producing capacity and factors controlling oxidation tailings in the Guryong mine, Korea. *Environ. Geol.* **2008**, *53*, 1787–1797. [CrossRef]
40. Mohapatra, D.P.; Kirpalani, D.M. Process effluents and mine tailings: Sources, effects and management and role of nanotechnology. *Nanotechnol. Environ. Eng.* **2016**, *2*, 1. [CrossRef]
41. Koizumi, S.; Gao, X.; Ueda, S.; Kitamura, S.-Y. Design of a Novel Fertilizer Made of Steelmaking Slag Using the Glassy Phase of the CaO–SiO_2–FeO System. Part I: Optimization on the Slag Chemistry and Microstructure. *J. Sustain. Metall.* **2021**, *7*, 87–98. [CrossRef]
42. Tabelin, C.B.; Silwamba, M.; Paglinawan, F.C.; Mondejar, A.J.S.; Duc, H.G.; Resabal, V.J.; Opiso, E.M.; Igarashi, T.; Tomiyama, S.; Ito, M.; et al. Solid-phase partitioning and release-retention mechanisms of copper, lead, zinc and arsenic in soils impacted by artisanal and small-scale gold mining (ASGM) activities. *Chemosphere* **2020**, *260*, 127574. [CrossRef]
43. Yu, J.; Liu, Y.; Luo, W.; Zhao, J.; Zhou, D.; Wang, X.; Zhu, L. Research on the pollution characteristics and effective control of heavy metals in co-cultivation rice field. *Chin. J. Eco-Agric.* **2024**, *32*, 1–12. [CrossRef]
44. Xu, M.; Ren, M.; Yao, Y.; Liu, Q.; Che, J.; Wang, X.; Xu, Q. Biochar decreases cadmium uptake in indica and japonica rice (*Oryza sativa* L.): Roles of soil properties, iron plaque, cadmium transporter genes and rhizobacteria. *J. Hazard. Mater.* **2024**, *477*, 135402. [CrossRef]
45. Meng, W.; Chen, Z.; Yu, Z.-L.; Wu, P.; Li, X.-X. Reservoir sediment records of polymetallic contamination (Cd, Pb, Zn, As, Hg) in the zinc smelting area of Weining County, Guizhou Province, China. *Environ. Earth Sci.* **2021**, *80*, 378. [CrossRef]
46. Botz, M.; Marsden, J. Heap Leach Production Modeling: A Spreadsheet-Based Technique. *Min. Metall. Explor.* **2019**, *36*, 1041–1052. [CrossRef]
47. Zhang, F.; Li, C.; Shi, Y.; Meng, L.; Zan, F.; Wu, X.; Wang, L.; Sheng, A.; Crittenden, J.C.; Chen, J. Evaluation on leachability of heavy metals from tailings: Risk factor identification and cumulative influence. *Environ. Sci. Pollut. Res.* **2023**, *30*, 64565–64575. [CrossRef] [PubMed]
48. Yadav, D.S.; Jaiswal, B.; Gautam, M.; Agrawal, M. Soil Acidification and its Impact on Plants. In *Plant Responses to Soil Pollution*; Singh, P., Singh, S.K., Prasad, S.M., Eds.; Springer: Singapore, 2020; pp. 1–26.
49. Soro, A.T.; Mufalo, W.; Arima, T.; Tabelin, C.B.; Igarashi, T. Geochemical Characterization of Rock Samples from Selected Fiji Mine Sites to Evaluate On-Site Environmental Vulnerabilities. *Minerals* **2023**, *13*, 661. [CrossRef]
50. Kachhwal, L.K.; Yanful, E.K.; Rennie, C.D. A semi-empirical approach for estimation of bed shear stress in a tailings pond. *Environ. Earth Sci.* **2012**, *66*, 823–834. [CrossRef]
51. Lari, E.; Steinkey, D.; Pyle, G.G. Effects of Seasonal Changes on the Toxic Impacts of Oil Sands Process-Affected Water on Daphnia magna. *Arch. Environ. Contam. Toxicol.* **2018**, *74*, 408–413. [CrossRef]
52. Bashir, A.; Malik, L.A.; Ahad, S.; Manzoor, T.; Bhat, M.A.; Dar, G.N.; Pandith, A.H. Removal of heavy metal ions from aqueous system by ion-exchange and biosorption methods. *Environ. Chem. Lett.* **2019**, *17*, 729–754. [CrossRef]
53. Jiang, C.; Séquaris, J.-M.; Wacha, A.; Bóta, A.; Vereecken, H.; Klumpp, E. Effect of metal oxide on surface area and pore size of water-dispersible colloids from three German silt loam topsoils. *Geoderma* **2014**, *235–236*, 260–270. [CrossRef]
54. Li, J.; Huang, Y.; Li, W.; Guo, Y.; Ouyang, S.; Cao, G. Study on dynamic adsorption characteristics of broken coal gangue to heavy metal ions under leaching condition and its cleaner mechanism to mine water. *J. Clean. Prod.* **2021**, *329*, 129756. [CrossRef]

55. Zhang, Q.; Li, J.; Li, Y.; Zhang, L.; Zhong, S.; Ru, X.; Shu, X. Investigation into the synergistic roles of exposed crystal planes and S vacancies during the photo-Fenton degradation of sulfadiazine via pyrite with three different morphologies. *Sep. Purif. Technol.* **2025**, *353*, 128538. [CrossRef]
56. Courtin-Nomade, A.; Rakotoarisoa, O.; Bril, H.; Grybos, M.; Forestier, L.; Foucher, F.; Kunz, M. Weathering of Sb-rich mining and smelting residues: Insight in solid speciation and soil bacteria toxicity. *Geochemistry* **2012**, *72*, 29–39. [CrossRef]
57. Lawrence, C.; Harden, J.; Maher, K. Modeling the influence of organic acids on soil weathering. *Geochim. Cosmochim. Acta* **2014**, *139*, 487–507. [CrossRef]
58. Bharti, S. Recent advances in heavy metal removal from wastewater using nanomaterials and cloud point extraction: A comprehensive review. *Int. J. Environ. Sci. Technol.* **2024**. [CrossRef]
59. Aneke, F.I.; Mostafa, M.M.H.; El Kamash, W. Pre-compression and capillarity effect of treated expansive subgrade subjected to compressive and tensile loadings. *Case Stud. Constr. Mater.* **2021**, *15*, e00575. [CrossRef]

processes

MDPI

Article

Quantitative Analysis of Soil Cd Content Based on the Fusion of Vis-NIR and XRF Spectral Data in the Impacted Area of a Metallurgical Slag Site in Gejiu, Yunnan

Zhenlong Zhang [1], Zhe Wang [1,*], Ying Luo [1], Jiaqian Zhang [1], Xiyang Feng [1], Qiuping Zeng [1], Duan Tian [1], Chao Li [1], Yongde Zhang [1], Yuping Wang [2], Shu Chen [1] and Li Chen [3]

[1] College of Environment and Resources, Southwest University of Science & Technology, Mianyang 621010, China
[2] Division of International Applied Technology, Yibin University, Yibin 644000, China
[3] College of Natural Resources and Environment, Northwest A&F University, Xianyang 712100, China
* Correspondence: wz2004@126.com

Abstract: Vis-NIR and XRF spectroscopy are widely used in monitoring heavy metals in soil due to their advantages of being fast, non-destructive, cost-effective, and non-polluting. However, when used individually, XRF and vis-NIR may not meet the accuracy requirements for Cd determination. In this study, we focused on the impact area of a non-ferrous metal smelting slag site in Gejiu City, Yunnan Province, fused the pre-selected vis-NIR and XRF spectra using the Pearson correlation coefficient (PCC), and identified the characteristic spectra using the competitive adaptive reweighted sampling (CARS) method. Based on this, a quantitative model for soil Cd concentration was established using partial least squares regression (PLSR). The results showed that among the four fusion spectral quantitative models constructed, the model combining vis-NIR spectral second-order derivative transformation and XRF spectral first-order derivative transformation (D2(vis-NIR) + D1(XRF)) had the highest coefficient of determination (R^2 = 0.9505) and the smallest root mean square error (RMSE = 0.1174). Compared to the estimation models built using vis-NIR and XRF spectra alone, the average computational time of the fusion models was reduced by 68.19% and 63.92%, respectively. This study provides important technical means for real-time and large-scale on-site rapid estimation of Cd content using multi-source spectral fusion.

Keywords: soil Cd pollution; visible-near infrared; X-ray fluorescence; spectral fusion; competitive adaptive reweighted sampling

1. Introduction

Heavy metal pollution in soil has become a global environmental issue [1]. In particular, due to human activities such as mining, industrial waste discharge, and improper use of pesticides, land degradation continues to worsen [2]. According to a national survey, the concentration of Cd in soil was 7.0%, which is significantly higher than that of other heavy metals [3]. Cd could further enter the human body through the food chain, posing a serious threat to human health [4,5]. Indeed, timely determination of Cd concentrations in soil can prevent the spread of Cd and provide a reference for the management of Cd pollution in environmental systems. Traditional methods for determining the concentration of Cd in soil still primarily rely on geochemical methods [6], but their disadvantages, such as long analytical cycles, complicated sample preparation, high analytical cost, and impracticality of on-site monitoring, are not suitable for this study [7]. Therefore, it is necessary to establish an accurate, stable, and rapid method for the determination of soil Cd concentration and to complement it with the integration of multi-source spectral data.

The visible-near infrared (vis-NIR) technology possesses advantages such as time-saving, effectiveness, affordability, and environmental friendliness [8]. Vis-NIR spectroscopy has shown its maturity in predicting cadmium-related soil physicochemical properties, such as iron oxides and organic matter [9]. Previous studies have indicated that vis-NIR spectra can indirectly measure the Cd content in soil [10]. The Chinese standard for risk control of soil pollution on agricultural land stipulates (GB 15618-2018) that the screening limit for Cd in agricultural soil is 0.3 mg/kg [11]. However, the direct measurement of Cd in soil using vis-NIR spectroscopy is limited by the weak signal of low Cd concentrations, affecting the monitoring effect [6,12]. Additionally, the combination of vis-NIR spectroscopy and remote sensing techniques can infer soil heavy metal concentrations at large spatial scales [13]. Portable X-ray fluorescence (pXRF) has been widely applied in heavy metal detection because of its advantages of speed, non-destructiveness, low cost, simultaneous multi-element analysis, and on-site detection [14]. However, the test results of this instrument are highly uncertain due to the influence of soil physicochemical properties, detection limits of elements, and detection principles, making it difficult to predict heavy metal pollution at large spatial scales [15]. Compared to the individual use of vis-NIR or X-ray fluorescence (XRF), the fusion of XRF and NIR spectra can expand the coverage of soil properties [16]. Therefore, the integrated application of proximal soil remote sensing techniques, such as XRF and vis-NIR, for rapid estimation of heavy metal concentrations in soil has been validated [17,18], which can effectively improve prediction accuracy, enhance the monitoring efficiency, and reduce the cost [19]. Among numerous spectral fusion methods, the serial fusion of XRF and vis-NIR spectra has the advantages of simplicity and strong operability [18]. Furthermore, research has shown that combining principal component analysis (PCA) with serial fusion methods can further improve model accuracy [20]. Thus, the utilization of serial XRF and vis-NIR spectra for estimating heavy metal concentrations is a feasible approach. Compared with other fusion methods, feature layer crosstalk is more suitable for rapid estimation of soil Cd content in the field.

The difficulty in predicting Cd concentrations using tandem spectroscopy is the elemental Cd does not have distinct characteristic bands in the vis-NIR spectral region [21]. Moreover, China's low screening limit for Cd in soils further increases the difficulty of identifying characteristic spectral bands in serial spectra. Tan et al. [22] used competitive adaptive reweighted sampling (CARS) for selecting characteristic spectral bands and found that it improved the prediction of arsenic, chromium, lead, and zinc concentrations in soil compared to traditional models. The study suggested that feature spectrum selection plays a key role in enhancing the generalization ability of estimation models. The Pearson correlation coefficient (PCC) has been proven to be statistically based and interpretable for the selected characteristic bands [23]. However, there is limited research on using PCC in conjunction with multisource spectral fusion for estimating Cd content in soil, which can provide technical references for further expanding and optimizing fusion methods. However, it is difficult to address the interrelationships between independent variables by using only PCC to select feature spectra [24]. Estimation models for Cd content built using PCC-selected characteristic spectra may suffer from overfitting, leading to unreliable results. Therefore, it is necessary to combine PCC with other methods to select characteristic spectra [25]. Among these methods, CARS primarily selects feature variables based on the principle of "survival of the fittest", thus removing irrelevant spectral information and retaining variables that effectively enhance the model's adaptability [22,26].

Agricultural land around the impact area of a nonferrous metal processing and smelting slag site in the city of Gejiu, Yunnan Province, has been affected by the nonferrous metal smelting process, and its farmland soil is mainly contaminated by heavy metals such as Cd and Pb [27]. In order to characterize the Cd contamination non-destructively and rapidly in the affected area's soil, a model capable of quantitatively monitoring Cd content is needed to address the limitations of traditional precise analytical monitoring methods in obtaining on-site and rapid measurements of Cd concentration in the study area's soil. Therefore, this study aims to combine XRF and vis-NIR spectra in a serial fusion approach, combined with PCC analysis, to construct a technical roadmap for spectral preprocessing, feature spectrum selection, and quantitative estimation using the PCC_CARS_PLSR (Pearson Correlation

Coefficient–Competitive Adaptive Reweighted Sampling–Partial Least Squares Regression) multi-method approach. This approach enables the rapid assessment of Cd content in the soil, a characteristic pollutant in the agricultural land of the non-ferrous metal selection and smelting slag site in Gejiu City, Yunnan Province. Additionally, Cd^{2+} poses a threat to the natural environment and human health due to its tendency to migrate and diffuse in soil. The detection limit of Cd in XRF spectrometry is lower than that of other heavy metal elements, and the monitoring accuracy is lower than that of other heavy metal elements. In this study, the accuracy of single-spectrum models and spectral fusion models predicting the content of Cd using different transformation methods were compared. This study provides a technical means for rapid estimation of Cd content in the field by multi-source spectral fusion, which lays the foundation for the remediation of Cd-contaminated soil. At the same time, this study also lays a research foundation for future research on dynamic, real-time, and large-scale quantitative monitoring of soil Cd pollution based on hyperspectral remote sensing images.

2. Materials and Methods

2.1. Study Area

The study area is located within a 5–20 km radius of the non-ferrous metal smelting slag repository in Gejiu City, Honghe Hani, and Yi Autonomous Prefecture, Yunnan Province, China. It covers an approximate area of 6.52 km^2. The average elevation is around 1300 m, with significant elevation differences. The area has a distinct vertical climate and belongs to a subtropical highland monsoon climate with abundant rainfall. Within the study area, there are 965,935 m^2 of construction land and 2,609,041 m^2 of farmland. After preliminary processing, it was found that the soil pH ranged from 5.32 to 8.80, and the soil moisture content ranged from 20% to 40%. The wastewater and sludge generated during non-ferrous metal processing have caused severe heavy metal pollution in the surrounding environment. Cd is one of the most heavily polluted heavy metals in the study area, posing a serious threat to crops and human health [28]. The geographical location of the study area and the geological map are shown in Figures 1 and 2. Most of the study area is underlain by Quaternary soil deposits, except for small portions of dolomite, siltstone, siliceous rock, and chert.

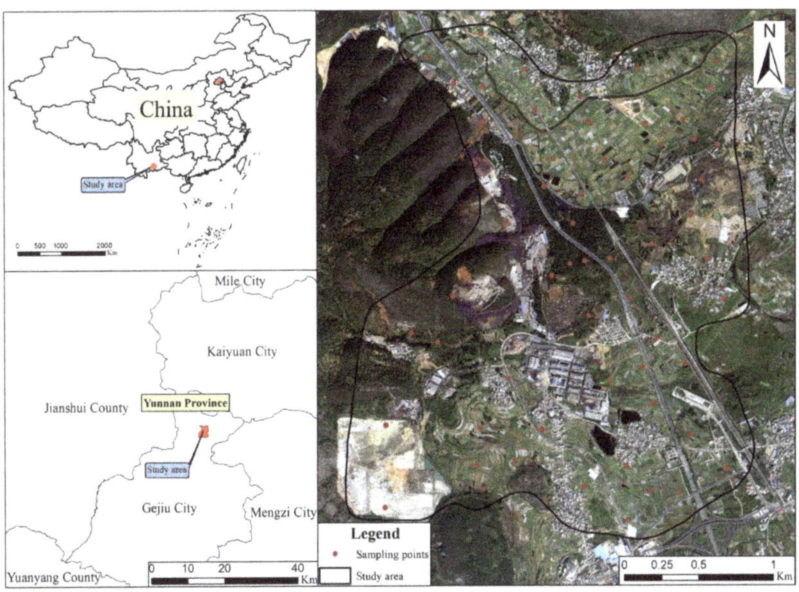

Figure 1. Geographical location of the study area.

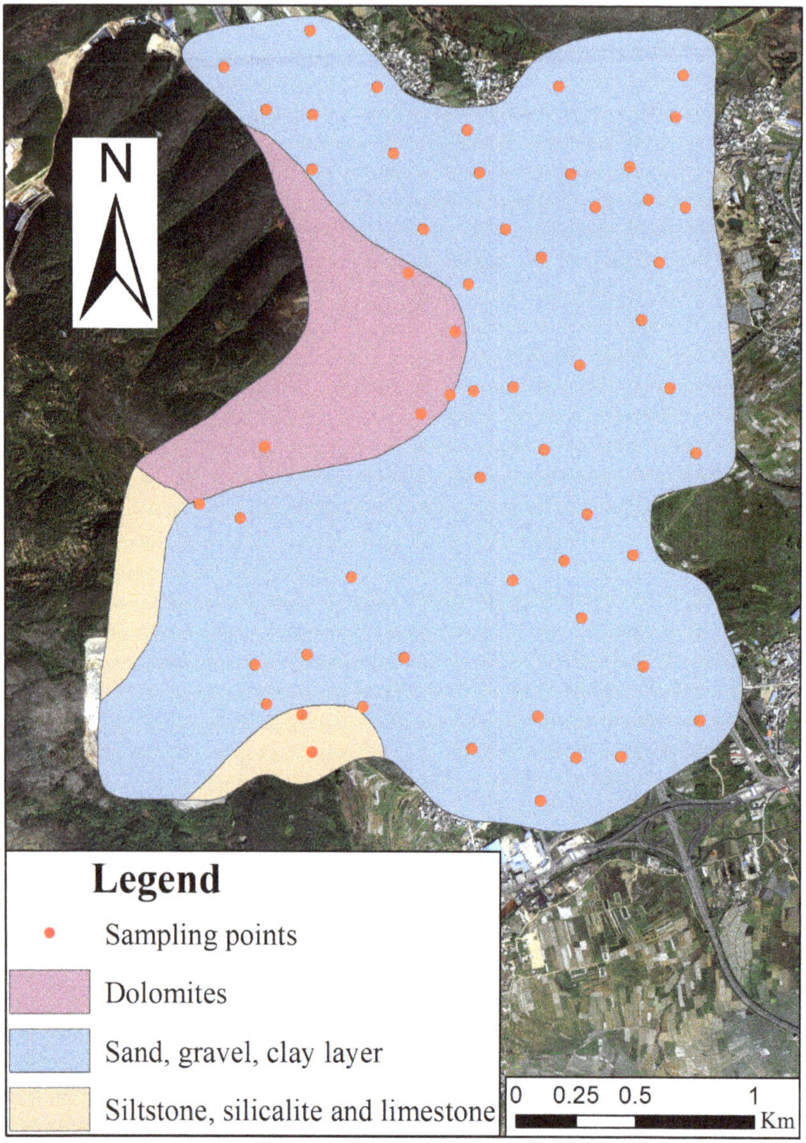

Figure 2. Geological map.

2.2. Data Collection

2.2.1. Soil Sample Collection

To establish a reliable quantitative estimation model for soil Cd concentration in the vicinity of the slag repository (HJ/T 166-2004), 58 soil samples were collected using a random distribution method. The GPS was used to record the locations of the sampling points. According to the "Technical Specifications for Soil Environmental Monitoring" [29], metal tools should be avoided during soil sample collection. First, the soil surface was cleared of mulch, such as tree branches and weeds. Then, five surface samples (0–5 cm) were taken in a 10 × 10 m area using a diagonal sampling method. The samples were mixed together, discarding impurities such as stones and plant roots. Finally, about 1 kg of mixed soil samples were sealed in polyethylene plastic bags [30]. The soil samples were dried in the laboratory, then ground and sieved using a 100-mesh nylon sieve (0.150 mm) to remove larger particles. The soil samples were divided into two parts: one for chemical analysis and the other for spectral analysis.

2.2.2. Chemical Analysis

The soil samples were digested in a graphite digestion apparatus. First, 0.1 g of the sample was weighed and placed in a Teflon digestion tank. Then, 5 mL of nitric acid (ρHNO_3 = 1.42 g/mL) was added to the vessel to soak the sample for 0.5 h to remove organic matter. After that, 2 mL of hydrofluoric acid (ρHF = 1.49 g/mL) and 1 mL of perchloric acid ($\rho HClO_4$ = 1.49 g/mL) were added. Finally, the digestion vessel was placed in a graphite digestion block and digested at 180 °C for 4 h. The total Cd concentration in the solution was determined using an Inductively Coupled Plasma Mass Spectrometer (ICP-MS, 800DV, TMO, Waltham, MA, USA) (argon gas source valve pressure reducing valve at about 550 kPa, circulating water pressure indication between 50 and 310 kPa.). Glassware and polyethylene bags were soaked in nitric acid for 12 h before use, and ultrapure water was used to prepare the solutions for analysis [31]. All experimental samples were digested simultaneously with a Cd concentration reference standard in the soil. Sample replicates (approximately 20%) and standard reference materials (GSS-5/GBW07405) [28] were included in each batch of sample digestion and chemical analysis, maintaining quality assurance, quality control, and blank control. The relative standard deviations were less than 5%, and the recoveries of Cd from the standard reference materials were 90–95%, respectively.

2.2.3. Measurement of vis-NIR Spectra Using a Portable Spectrometer

Visible-NIR spectra of soil samples were measured in a darkroom using a PSR-2500 portable spectrometer manufactured by Spectrum Evolution (operating instructions are available on the Spectral Evolution website (https://spectralevolution.com/products/software/ (accessed on 4 June 2023))). The soil samples were placed in black containers with a diameter of 10 cm and a depth of 1.5 cm. Prior to measurement, the instrument was preheated for at least 30 min with a 100 W halogen lamp set as the sole light source. Calibration with a white reference panel was performed for optimization. The probe viewing angle was kept at 15°, and the light source incidence angle was 30° during acquisition. The distance between the light source and the center of the soil surface was 50 cm, while the probe was positioned 15 cm above the soil surface. The container was then divided into three directions at an angle of 120°, and five spectra were collected in each direction, resulting in a total of 15 spectra per soil sample. The arithmetic mean of the collected spectra was calculated to obtain the final soil reflectance spectra to facilitate subsequent analysis.

2.2.4. Measurement of XRF Spectra Using an X-ray Fluorescence Spectrometer

The XRF spectra of the soil samples were measured using the Niton XL3t 950 XRF spectrometer from Thermo Fisher Scientific, Waltham, MA, USA. After passing through a 200-mesh nylon sieve (0.074 mm), the soil samples were placed in sample cups. The surface of each sample was leveled and covered with a layer of polyester film. Then, the sample cup was placed on the instrument's test stand, and scanning was conducted for 80 seconds per sample. Three scans were performed for each sample, and the average spectrum of the three scans was taken as the result. After removing the low-energy values at the edges, each spectrum retained 3522 data points ranging from 0.39 keV to 53.205 keV. The spectrometer was connected to the computer via a data cable, and the spectral data were exported to the computer using the NDT program, the manual for which is available on the Thermo Fisher Scientific website (https://www.thermofisher.cn/order/catalog/product/10131166?SID=srch-srp-10131166 (accessed on 5 June 2023)).

2.3. Data Analysis

In this study, data preprocessing, feature band selection, construction of single-spectrum models, construction of spectral fusion models, and model accuracy evaluation were implemented using the Python 3.10 programming language in PyCharm Community Edition 2022.1.2 software. The specific processing workflow is illustrated in Figure S1.

2.3.1. Spectral Pre-Processing Methods

Spectral preprocessing aimed to standardize the format of XRF and vis-NIR spectral data and reduce errors caused by system noise. The spectral data were exported to Microsoft Excel in batches and subjected to wavelet transform (WT) and Savitzky–Golay (SG) smoothing preprocessing. WT applied scaling and translation operations to gradually refine the signal (function) at multiple scales, ultimately achieving automatic focus on the details of the spectral profile [32]. SG smoothing was employed for denoising the spectral curve, and used a second-order polynomial with a window size of 12 for smoothing. Spectral transformations enhanced the spectral signal and increased the correlation between spectral wavelengths and Cd content [33,34]. The raw spectra (RS) were subjected to several transformations, including mean centering transformation (CT), standard normal variate transformation (SNV), multiplicative scatter correction (MSC), first-order derivative transformation (D1), second-order derivative transformation (D2), detrending transformation (DT), continuum removal (CR), and reciprocal logarithm transformation (CL). Due to the presence of zero values in XRF spectra, the CL transformation cannot be applied as the denominator. The transformed results of XRF and vis-NIR spectra are shown in Figures S2 and S3, respectively. Finally, the spectral data were subjected to standardization processing.

2.3.2. Feature Spectral Selection

The feature spectral selection process consisted of preliminary screening using the PCC method and subsequent screening using the CARS method.

(1) The correlation analysis was conducted using the PCC method [23]. The correlation coefficients between the Cd concentrations in soil and the spectral bands of both types of spectra were calculated, and the absolute values of the correlation coefficients were obtained. The equation for calculating the correlation coefficient is as follows:

$$r_{(X,Y)} = cov_{(X,Y)} / \sqrt{var(X)}\sqrt{var(Y)} \quad (1)$$

where $r_{(X,Y)}$ is the correlation coefficient between the two variables X and Y; $cov_{(X,Y)}$ is the covariance of the two variables; $var(X)$, $var(Y)$ are the variance of the variables.

After conducting multiple experiments, it was found that selecting 200 highly correlated spectral bands from the vis-NIR spectra and 600 highly correlated spectral bands from the XRF spectra ensured model quality and reduced model computation time.

(2) CARS utilized adaptive sampling to retain spectral bands with relatively large absolute coefficients in the PLSR model [35]. Then, a Monte Carlo cross-validation method was employed to model each subset of wavelength variables, and the optimal subset was selected as the feature spectral band set based on the root mean square error of cross-validation (RMSECV).

2.3.3. Estimation Model

The Kennard–Stone (KS) algorithm [36] was used to calculate the Euclidean distance of the dataset. The samples corresponding to the maximum and minimum distances were selected as the training set. This process was repeated until the training set reached the specified quantity (80% for training, 20% for validation). Among the 58 soil samples, 46 were selected for the training set, and 12 were selected for the validation set. The formula for calculating the Euclidean distance is as follows:

$$d_x(p,q) = \sqrt{\sum_{j=1}^{J} \left[x_p(j) - x_q(j)\right]^2} \quad p,q \in [1,N] \tag{2}$$

where J is the total number of bands; N is the total number of samples; $x_p(j)$ and $x_q(j)$ are date values at p and q on the j band.

A prediction model was developed using the PLSR method. PLSR is a novel multivariate statistical analysis method that combines the advantages of PCA and linear regression modeling and is more effective in distinguishing between spectral information and noise. Compared with traditional linear models, the most important feature of PLSR is its utilization of data dimensionality reduction and comprehensive information selection techniques. PLSR can model independent variables with multiple correlations, which can improve model accuracy when using spectroscopic techniques to build predictive models for soil Cd content [37].

2.3.4. Accuracy and Efficiency Evaluation

The accuracy of the estimation model was evaluated using R^2 (coefficient of determination) and RMSE (root mean square error). A higher R^2 value indicated better model fit, with values closer to 1 indicating higher accuracy. A smaller RMSE value indicated a better predictive ability for the model. However, it should be noted that the CARS method for feature spectrum selection exhibits randomness [38]. This was reflected when the RMSE of the validation set was greater than that of the training set, indicating that the predictive model had obtained a local solution. In such cases, it is necessary to repeat the variable selection and modeling process until the RMSE of the validation set becomes smaller than that of the training set [39]. The formulas for calculating R^2 and RMSE are as follows:

$$R^2 = \sum_{i=1}^{n} (\hat{y}_i - \overline{y_i})^2 / \sum_{i=1}^{n} (y_i - \overline{y_i})^2 \tag{3}$$

$$\text{RMSE} = \sqrt{\frac{1}{n} \sum_{i=1}^{n} (\hat{y}_i - \overline{y_i})^2} \tag{4}$$

where $\overline{y}$ is the mean value of the sample observations; $\hat{y}$ is the predicted value of the sample; n is the number of samples to be verified.

Due to the relatively slow execution speed of CARS, real-time updating of soil heavy metal Cd concentration is required for rapid on-site estimation. Therefore, the algorithm execution time was considered the efficiency evaluation criterion for different spectral transformation methods in the prediction model evaluation, aiming to distinguish the characteristics of different models in the inversion of Cd concentration.

2.3.5. Building a Spectral Fusion Model

First, we conducted a statistical analysis to assess the accuracy and efficiency of the individual spectral models. The models were evaluated based on their R^2, RMSE, and computation time on the validation set. The number of models selected varied depending on the different transformation methods applied to the XRF and vis-NIR spectra. Next, the selected models corresponding to the XRF and vis-NIR spectra were concatenated, resulting in a fused spectrum (FS). The FS spectrum contained a total number of spectral bands equal to the sum of the bands in the vis-NIR and XRF spectra. Subsequently, the FS spectrum underwent another round of spectral preprocessing to enhance its quality and minimize potential noise or artifacts. Following that, the PCC_CARS_PLSR method was employed to establish a predictive model for estimating the soil heavy metal Cd content using the fused spectrum. In this modeling process, 600 variables were selected through PCC screening to ensure the inclusion of relevant spectral information contributing to accurate predictions. Finally, the accuracy and efficiency of the FS model were evaluated.

This evaluation involved assessing the accuracy of the model using R^2 and evaluating its efficiency by measuring the required training and prediction time. Through this comprehensive evaluation, we gained insights into the predictive accuracy and computational efficiency of the FS model in estimating soil heavy metal Cd content.

3. Results

3.1. Descriptive Statistical Analysis

3.1.1. Descriptive Statistical Analysis of Soil Cd Concentration

Descriptive analysis was carried out based on the results of laboratory ICP-MS accurate quantitative analysis tests of Cd content in soil samples from the study area (Table 1). The results of the statistical analysis showed that the mean value of soil Cd was higher than its corresponding median value, indicating a positively skewed mode of normal distribution of Cd in soil. The Cd pollution in the study area was serious, with the average Cd concentration exceeding the background level in Yunnan Province by 30 times, 66 times higher than the risk screening value and 13 times higher than the risk control value in the Chinese Soil Environmental Quality Risk Control Standard for Agricultural Land (GB15618-2018) [11], which poses a serious threat to human health. The standard deviation of soil Cd was 16.3516, indicating significant differences in Cd concentration among different directions in the non-ferrous metal smelting slag depot. The Cd element was influenced by topography, wind direction, and water sources continuously flowing into surrounding areas. The coefficient of variation for Cd concentrations in the study area was 82.25%, suggesting a significant influence of anthropogenic factors on Cd concentrations.

Table 1. Descriptive information of soil samples (unit: mg/kg).

Element	Mean	Median	Standard Deviation	Minimum	Maximum	Coefficient of Variation	Background Value of Soil in Yunnan Province
Cd	19.8806	12.7938	16.3516	2.9005	69.4155	82.25%	0.66

3.1.2. Descriptive Analysis of Spectral Curves

Using the spectral preprocessing method in 2.3.1, eight (RS, CT, SNV, MSC, D1, D2, DT, and CR) and nine (RS, CT, SNV, MSC, D1, D2, DT, CR, and CL) spectral transformations were applied to the XRF and vis-NIR spectra, respectively, to obtain seventeen post-transformation spectra, as shown in Table S1, Figures S2 and S3.

The transformed XRF and vis-NIR spectra exhibited significant differences in value ranges. The vis-NIR spectra had a maximum value of 56.48 and a minimum value of −29.15, whereas the XRF spectra had a maximum value of 859.00 and a minimum value of −137.46. This substantial difference in value ranges posed a challenge in directly concatenating the two spectra, as it would increase the difficulty in identifying fused spectral features. However, this disparity can be mitigated through spectral normalization techniques, which can eliminate such differences and facilitate feature selection using both PCC and CARS. The wavelength range for the XRF spectra was 0.39 to 53.205, while the vis-NIR spectra ranged from 350 to 2500. The number of spectral bands in the XRF and vis-NIR spectra was 3522 and 768, respectively. Hence, during PCC-based feature selection, it was advisable to increase the number of iterations for XRF spectrum selection to retain a greater amount of spectral information and ensure the interpretability of the data. The non-overlapping spectral ranges of the two types of spectra do not impact the uniqueness of the model variables.

3.2. Feature Spectrum Selection

3.2.1. Preliminary Feature Spectrum Selection Based on PCC

The PCC analysis between the spectral data and Cd concentration is depicted in Figure S4. It was observed that the correlation between the transformed vis-NIR spectra and Cd concentration was enhanced. The highest correlation coefficient with Cd concentration was 0.47 for the D2 transformation. For the XRF spectra, most transformations improved their correlation with Cd, with the CR transformation showing the highest correlation coefficient of 0.51. The D1, D2, and CR transformations for both spectra exhibited correlation coefficients above 0.4, providing fundamental data for the establishment of a Cd concentration estimation model. However, relying solely on PCC for feature spectrum selection faces challenges in addressing inter-variable correlations. Building a Cd concentration estimation model based solely on PCC-selected feature spectra is prone to overfitting, resulting in unreliable Cd concentration estimations. Therefore, it was necessary to combine PCC with other methods for feature spectrum selection.

3.2.2. Feature Spectrum Selection Based on CARS Algorithm

The CARS algorithm primarily operates on the principle of "survival of the fittest" to select the feature variables with stronger adaptability, thereby eliminating irrelevant spectral information and retaining variables that can effectively improve the model's adaptability. Building upon the feature spectra obtained from the correlation analysis between soil Cd concentration and spectral values based on PCC, the CARS algorithm was applied to filter these feature spectra further. The number of iterations in the CARS algorithm was a crucial factor influencing the quality of the feature spectra. By setting the Monte Carlo sampling to 100 iterations, the relationship between the number of iterations and RMSECV is depicted in Figure S5. As the number of iterations increased, RMSECV exhibited fluctuating changes, with the curve initially decreasing, reaching a minimum point, and then rising again. The minimum RMSECV was achieved when the number of iterations was 79. Considering the combined impact of iteration count and computational time, setting the iteration count to 100 for the feature spectrum selection based on the CARS algorithm was deemed suitable.

3.3. *Evaluation of Estimation Model Accuracy and Efficiency*

3.3.1. Evaluation of Single-Spectrum Estimation Models

From Table 2, it can be observed that vis-NIR exhibited lower accuracy in estimating Cd concentration. The Cd concentration estimation models established using nine spectral transformation methods had an average R^2 of 0.3529 and an average RMSE of 0.5388. By examining Figures 3 and S6, it is evident that most of the transformation models have large intercepts and small slopes, resulting in predicted values that are lower than the actual measurements. Among them, the trend lines of the D2 transformation model outperformed other models (R^2 = 0.6849, RMSE = 0.2690). Therefore, the D2 transformation model was selected as the optimal estimation model for soil Cd content in the vis-NIR range.

Table 2. Accuracy and efficiency statistics of 17 Cd concentration estimation models.

Spectral Type	Transformation	RMSE	R^2	Time (s)
vis-NIR	RS	0.7459	0.4054	18
	CT	0.4914	0.1123	339
	SNV	0.5288	0.2971	60
	MSC	0.5141	0.4402	30
	D1	0.5065	0.3128	34
	D2	0.2690	0.6849	110
	DT	0.5703	0.1405	51
	CR	0.5334	0.4781	14
	CL	0.6894	0.064	37

Table 2. *Cont.*

Spectral Type	Transformation	RMSE	R^2	Time (s)
XRF	RS	0.3321	0.1082	71
	CT	0.4116	0.2628	59
	SNV	0.3018	0.7495	44
	MSC	0.4531	0.3841	71
	D1	0.1143	0.9079	113
	D2	0.1048	0.8868	164
	DT	0.3088	0.3772	16
	CR	0.1588	0.7442	68

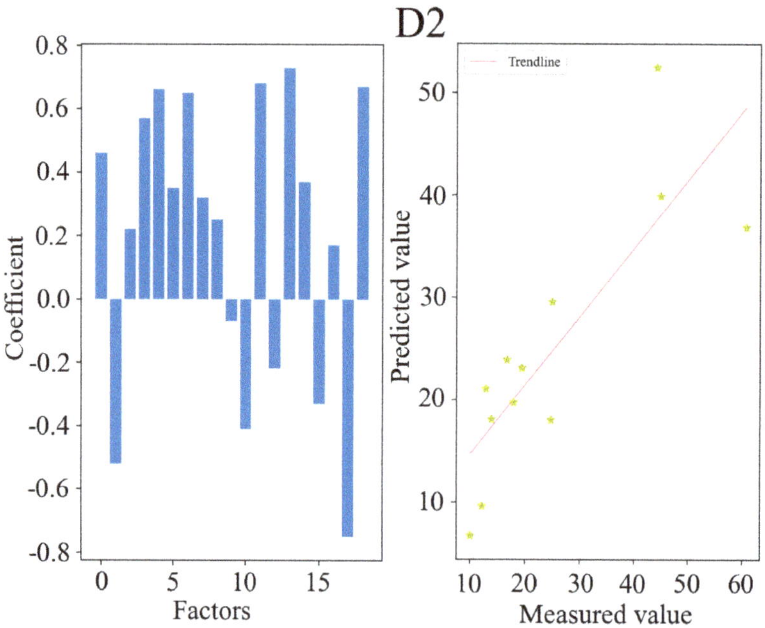

Figure 3. PLSR model with D2 transformation of vis-NIR (the rest of the transformations are in Figure S6).

The additional information in Table 2 indicates that XRF achieved better results in Cd concentration estimation. The excitation spectra of XRF can easily identify heavy metal elements in the soil. By examining Figures 4 and S7, it can be observed that the Cd estimation models based on XRF data had a higher number of spectral bands compared to the vis-NIR estimation models (Figures 3 and S6). The closer the scatter points of the XRF estimation models were to the 1:1 line, the higher the degree of fit of the Cd concentration estimation model. Among them, the scatter plots of the Cd concentration estimation models based on the SNV, D1, D2, and CR transformations exhibited smaller deviations from the 1:1 line compared to the other four methods (CT, MSC, DT, and CL), indicating better fit and higher estimation accuracy, with R^2 values of 0.7495, 0.9079, 0.8868, and 0.7442, and RMSE values of 0.3018, 0.1143, 0.1048, and 0.1588, respectively. Therefore, the SNV, D1, D2, and CR methods can be considered advantageous spectral transformation approaches for XRF spectra, providing a basis for further spectral fusion studies.

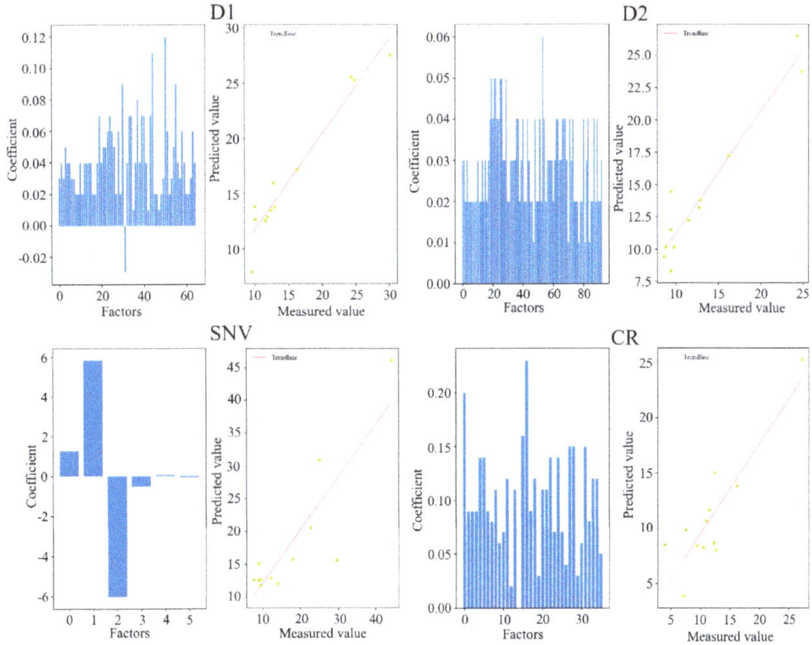

Figure 4. PLSR model with different transformations of XRF (the remaining transformations are in the attached Figure S7).

3.3.2. Evaluation of Spectral Fusion Models

Compared to the single-spectrum models, the fusion models exhibited higher accuracy. Table 3 presents the accuracy and computational time information of four fusion models: the concatenation of the D2 transformation of vis-NIR and the D1 transformation of XRF (D2(vis-NIR) + D1(XRF)); the concatenation of the D2 transformation of vis-NIR and the D2 transformation of XRF (D2(vis-NIR) + D2(XRF)); the concatenation of the D2 transformation of vis-NIR and the SNV transformation of XRF (D2(vis-NIR) + SNV(XRF)); and the concatenation of the D2 transformation of vis-NIR and the CR transformation of XRF (D2(vis-NIR) + CR(XRF)). Their R^2 values were 0.9505, 0.8832, 0.8974, and 0.9096, respectively, while the RMSE was 0.1174, 0.1486, 0.1904, and 0.1309, respectively. It is worth noting that although the fusion models had more spectral bands, their computational time was shorter compared to the single-spectrum models.

Table 3. Accuracy and efficiency statistics of four Cd concentration estimation models.

Spectral Type	Transformation	RMSE	R^2	Time (s)
FS	D2(vis-NIR) + D1(XRF)	0.1174	0.9505	65
	D2(vis-NIR) + D2(XRF)	0.1486	0.8832	34
	D2(vis-NIR) + SNV(XRF)	0.1904	0.8974	18
	D2(vis-NIR) + CR(XRF)	0.1309	0.9096	23

The scatter plot in Figure 5 exhibits a linear distribution, indicating a good level of prediction. While the computational time of the fusion model D2(vis-NIR) + D1(XRF) was slightly longer, this fusion model demonstrates a linear and evenly distributed scatter plot, showcasing better prediction accuracy and stability. Moreover, it exhibited the least deviation from the 1:1 line, the highest degree of fit, and the best estimation model accuracy. Therefore, the D2(vis-NIR) + D1(XRF) fusion model can be regarded as the optimal estimation model for soil Cd content.

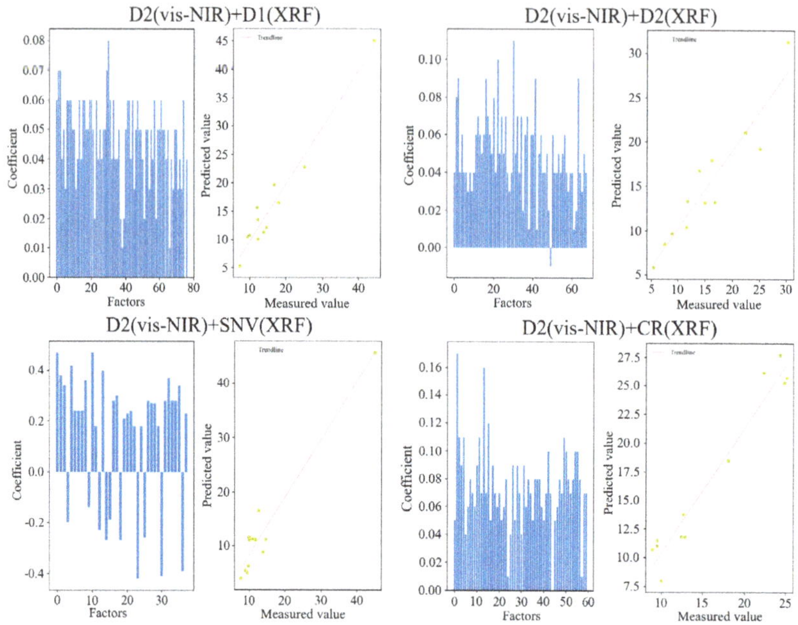

Figure 5. PLSR model with different transformation methods of FS.

Furthermore, Table 4 provides a comprehensive efficiency evaluation of the dominant transformation models for different spectral types, with mean estimates within the confidence interval ($p < 0.05$). Therefore, considering the average RMSE, average R^2, and average computational time, the ranking of estimation accuracy and efficiency for soil Cd concentration is as follows: the FS model was superior to the XRF model, and the XRF model was superior to the vis-NIR model. Taking D2(vis-NIR) as the reference, the average R^2 of XRF improved by 20.03%, while the average R^2 of the FS model improved by 32.90%. The average computational time of XRF decreased by 11.82%, while the average computational time of the FS model decreased by 68.19%. Compared with XRF spectroscopy, the average R^2 of FS was improved by 10.72%; the average operation time of FS was reduced by 63.92%. Overall, the FS model significantly improved the accuracy of the model, reduced algorithm runtime, and exhibited better stability.

Table 4. Evaluation table of the combined efficiency of the three models.

Spectral Type	Transformation	Mean RMSE	Mean R^2	Mean Time (s)
vis-NIR	D2	0.5334	0.6849	110
XRF	D1, D2, SNV, CR	0.1699	0.8221	97
FS	D2(vis-NIR) + D1(XRF), D2(vis-NIR) + D2(XRF), D2(vis-NIR) + SNV(XRF), D2(vis-NIR) + CR(XRF)	0.1468	0.9102	35

4. Discussion

Accurately establishing a model to estimate soil Cd concentrations is highly challenging in practical research. Currently, the primary method for monitoring soil Cd concentrations involves field technicians using specialized instruments to obtain Cd concentrations at certain sampling points, followed by spatial interpolation techniques to estimate Cd concentrations in specific areas. With the development of sensor devices, it is now possible to measure an increasing amount of soil information non-destructively in the field. The development of machine learning has made it possible to analyze complex variable relationships using computers, providing more options for estimating soil Cd concentrations [18,21]. However, there are still numerous issues to be discussed regarding the estimation of Cd concentrations using spectroscopic techniques. Our discussion focuses on the following aspects.

First, as Cd is typically present in the form of compounds in soil, vis-NIR spectroscopy can indirectly estimate the concentration of Cd in soil by utilizing other substances that adsorb Cd, such as organic matter, carbonate minerals, clay minerals, and manganese iron oxides. For instance, Xia et al. [40] analyzed the correlation between eight soil heavy metals and organic matter and established a PLSR model for heavy metals. However, when the soil environment undergoes change, this method still requires on-site sampling to determine the concentrations of intermediate substances for quantitative inversion of heavy metal concentrations [41]. This effect prompted a better understanding of how to obtain information about in situ soil. In this regard, the XRF/NIR spectroscopy coupling technique proposed by Professor Horta expands the coverage of soil properties, improves measurement accuracy, and does not damage the original soil or generate hazardous laboratory wastewater [16]. Li et al. [18] combined XRF, NIR, and MIR spectroscopy and applied PCA to establish a model for estimating soil heavy metals. They found that the prediction model for Cd concentrations achieved an R^2 value as high as 0.98. In our study, a quantitative estimation model for soil Cd concentration was established by combining the serially linked vis-NIR and XRF spectra through the PCC_CARS_PLSR process. Through comprehensive comparison in Figure 6, it was found that the fusion model exhibited higher accuracy and shorter computation time, with the highest R^2 of 0.9896 achieved by the D2(vis-NIR) + D1(XRF) estimation model, indicating its greater potential for practical monitoring of soil Cd content.

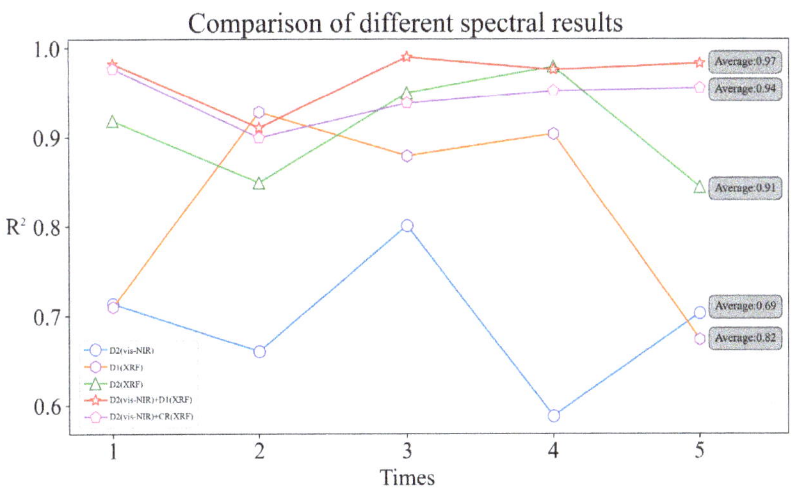

Figure 6. Comparison of R^2 for each of the 5 spectra with 5 operations.

Furthermore, soil spectral preprocessing was an effective approach to enhancing the accuracy of soil heavy metal estimation models. Spectral preprocessing typically involves spectral denoising and spectral transformation [42]. In our study, we utilized WT and spectral smoothing for spectral denoising. WT refined the received spectra from the sensor, reducing system noise. On the other hand, spectral smoothing employed a filtering window to minimize baseline drift in the spectral curves. Spectral transformation techniques enhanced spectral information, improving the correlation between spectra and soil Cd elements and providing additional valuable spectral information for feature spectrum selection [43]. Hou et al. [44] achieved a high-accuracy estimation model by applying SG smoothing in combination with MSC and CL for preprocessing reflectance spectra. Therefore, employing a combination of multiple spectral preprocessing methods can provide more effective feature spectral information for high-accuracy estimation models. In our research, we combined the WT spectral denoising method with nine spectral transformations (CT, SNV, MSC, D1, D2, DT, CR, and CL). After spectral smoothing, a spectral variable matrix was obtained, and the accuracy of estimation models constructed using different preprocessing methods for individual spectra was evaluated. This evaluation served as a basis for selecting the optimal fusion method. Among the nine preprocessing methods applied to the vis-NIR spectra, the WT combined D2 preprocessing method yielded the highest accuracy with an R^2 of 0.6849. Similarly, among the eight preprocessing methods applied to the XRF spectra, the WT combined D1 preprocessing method yielded the highest accuracy with an R^2 of 0.9079. When compared with the optimal preprocessing methods for vis-NIR and XRF spectra in Gholizadeh et al. [45], our selected preprocessing methods demonstrated superior accuracy and stability.

In addition, spectral fusion can be categorized into three stages: data-level fusion; feature-level fusion; and decision-level fusion [16]. Data-level fusion, although straightforward, was limited by the effectiveness of data dimensionality reduction and feature spectral selection on the accuracy of heavy metal estimation models [46]. Feature-level fusion includes feature concatenation and feature operation [21]. Pozza et al. [20] combined PCA to concatenate the vis-NIR and XRF spectra, establishing a soil Pb concentration estimation model with a Lin's Concordance Correlation Coefficient (LCCC) of 0.95. Feature concatenation involves directly concatenating the selected feature spectra into a fused spectrum. Feature operation involved operations such as summation, equal-weight superposition, and outer-product analysis (OPA) on the vis-NIR and XRF feature spectra to obtain the fused spectrum [47]. Feature operation requires the dimensionality reduction in both the vis-NIR and XRF spectra to a unified number of spectral bands. Decision-level fusion typically employs Granger–Ramanathan averaging (GRA) to model the estimation results of individual spectral models. Xu et al. [46] used spectral resampling combined with OPA and GRA to estimate soil Cd concentration, resulting in LCCC values of 0.82 and 0.73, respectively.

We concatenated the vis-NIR and XRF spectra that underwent preliminary screening using PCC. Moreover, we employed CARS to identify the characteristic spectra, leading to the establishment of a quantitative estimation model for soil Cd concentration based on PLSR. This fusion of spectra can be categorized as feature-level fusion. Additionally, based on Table 2, various preprocessing techniques were applied to the individual spectra in this study. We used PCC_CARS integration to select the feature spectra, constructing a single-spectrum estimation model. Furthermore, on this basis, we selected the optimal fusion method for vis-NIR and XRF feature spectra, resulting in the final establishment of the fused spectral estimation model. This spectral fusion can also be categorized as decision-level fusion. Therefore, the fused spectral estimation model in this study is the outcome of multi-level fusion, with an average R^2 of above 0.94 for Cd concentration prediction in the validation set (Figure 6), which is similar to the accuracies of the soil Cd content estimation models developed by Xu et al. [46], Li et al. [18], and Wang et al. [21] The use of other spectral fusion strategies indicated that this multilayer fusion approach was a higher-order fusion strategy with higher accuracy.

In summary, this study fused the PCC-filtered vis-NIR and XRF spectra and employed the CARS method to identify characteristic spectra, resulting in the establishment of a quantitative estimation model for soil Cd concentration based on PLSR. The model demonstrated good prediction accuracy and stability. These findings provide theoretical guidance for non-destructive and rapid monitoring of soil Cd and enrich the methods for spectral techniques in monitoring soil Cd concentration. Moreover, they offer a technical approach for the on-site rapid estimation of soil Cd content based on multisource spectral fusion and lay the foundation for future research on dynamic, real-time, and large-scale quantitative monitoring of soil heavy metal pollution using high-resolution hyperspectral remote sensing images.

5. Conclusions

In conclusion, the feature spectra were screened based on PCC-CARS coupling, and a soil Cd content estimation model with a higher integrated efficiency was developed using PLSR with the strategy of feature layer cascading and decision layer fusion. The comprehensive efficiency (accuracy and computational time) of the estimation models based on different spectral types was ranked as follows: fusion spectral model > XRF spectral model > vis-NIR spectral model. Among them, the D2(vis-NIR) + D1(XRF) fusion model exhibited the highest coefficient of determination and the smallest root mean square error. Importantly, compared with other spectral fusion methods, the concatenated spectral approach was characterized by simplicity and strong operability, which makes it more suitable for dynamic, real-time, and quantitative monitoring of soil heavy metal pollution. Future work should further focus on constructing models that can be used for on-site rapid and accurate estimation of soil Cd content, providing technical references for subsequent research on dynamic, real-time, and large-scale quantitative monitoring of soil heavy metal pollution based on high-resolution hyperspectral remote sensing images. This is of great significance for the dynamic monitoring of soil pollution and agricultural product safety, as well as the safe utilization of arable land.

Supplementary Materials: The following supporting information can be downloaded at: https://www.mdpi.com/article/10.3390/pr11092714/s1, Figure S1: Flow chart of modeling process. Figure S2: 9 transformations of vis-NIR spectra. Figure S3: XRF spectra with 8 transformations. Figure S4: Correlation coefficients between Cd concentration and spectral values. Figure S5: Relationship between Iterations and RMSECV. Figure S6: PLSR model for different transformation methods of vis-NIR. Figure S7: PLSR model with different transformation methods of XRF. Figure S8: Combined spectra of the two spectra. Table S1: Descriptive information of the spectral curves.

Author Contributions: Data curation, Z.Z., Y.L., J.Z. and D.T.; Funding acquisition, S.C.; Investigation, X.F., Q.Z. and C.L.; Methodology, Z.Z.; Project administration, Z.W. and Y.W.; Resources, Z.W. and Y.Z.; Software, Z.Z.; Supervision, Y.L.; Validation, L.C.; Writing—original draft, Z.Z.; Writing—review and editing, Z.W. All authors have read and agreed to the published version of the manuscript.

Funding: This work was supported by grants from the Ministry of Science and Technology of the People's Republic of China [grant numbers 2019YFC1803500, 2019YFC1803504]; the Natural Science Foundation of Sichuan Province [grant numbers 2018SZ0298, 2023YFS0390]; the National Natural Science Foundation of China [grant number No. 41402248]; Biological and Chemical Engineering Laboratory of Panzhihua College [grant numbers JDH-2019-E-01, GR-2020-E-02]; the Bureau of Science and Technology Panzhihua City [grant number 2017CY-N-8]; the Bureau of Science and Technology Aba Qiang Tibetan Autonomous Prefecture [grant number R22YYJSYJ0004, R23YYJSYJ0010]; the Southwest University of Science and Technology [grant numbers 17LZX613, 18LZX638].

Data Availability Statement: The data and materials used and analyzed during the current study are available from the corresponding author upon reasonable request.

Acknowledgments: The authors thank the Analytical Testing Center of Southwest University of Science and Technology for providing total heavy metal analysis services.

Conflicts of Interest: The authors declare no conflict of interest.

References

1. Tian, L.; Liu, X.; Zhang, B.; Liu, M.; Wu, L. Extraction of Rice Heavy Metal Stress Signal Features Based on Long Time Series Leaf Area Index Data Using Ensemble Empirical Mode Decomposition. *Int. J. Environ. Res. Public Health* **2017**, *14*, 1018. [CrossRef]
2. Liu, M.; Liu, X.; Zhang, B.; Ding, C. Regional Heavy Metal Pollution in Crops by Integrating Physiological Function Variability with Spatio-Temporal Stability Using Multi-Temporal Thermal Remote Sensing. *Int. J. Appl. Earth Obs. Geoinf.* **2016**, *51*, 91–102. [CrossRef]

3. Zhao, F.J.; Ma, Y.; Zhu, Y.G.; Tang, Z.; McGrath, S.P. Soil Contamination in China: Current Status and Mitigation Strategies. *Environ. Sci. Technol.* **2015**, *49*, 750–759. [CrossRef]
4. Xu, M.X.; Wu, S.H.; Zhou, S.L.; Liao, F.Q.; Ma, C.-M.; Zhu, C. Hyperspectral reflectance models for retrieving heavy metal content: Application in the archaeological soil. *J. Infrared Millim. Waves* **2012**, *30*, 109–114. [CrossRef]
5. Jia, Z.; Li, S.; Wang, L. Assessment of Soil Heavy Metals for Eco-Environment and Human Health in a Rapidly Urbani-zation Area of the Upper Yangtze Basin. *Sci. Rep.* **2018**, *8*, 3256. [CrossRef]
6. Wang, F.; Gao, J.; Zha, Y. Hyperspectral Sensing of Heavy Metals in Soil and Vegetation: Feasibility and Challenges. *ISPRS J. Photogramm. Remote Sens.* **2018**, *136*, 73–84. [CrossRef]
7. Wu, Y.; Chen, J.; Wu, X.; Tian, Q.; Ji, J.; Qin, Z. Possibilities of Reflectance Spectroscopy for the Assessment of Contaminant Elements in Suburban Soils. *Appl. Geochem.* **2005**, *20*, 1051–1059. [CrossRef]
8. Liu, M.; Liu, X.; Wu, L.; Duan, L.; Zhong, B. Wavelet-Based Detection of Crop Zinc Stress Assessment Using Hyperspectral Reflectance. *Comput. Geosci.* **2011**, *37*, 1254–1263. [CrossRef]
9. Zhang, Y.; Hartemink, A.E. Data Fusion of Vis-NIR and PXRF Spectra to Predict Soil Physical and Chemical Properties. *Eur. J. Soil Sci.* **2020**, *71*, 316–333. [CrossRef]
10. Kemper, T.; Sommer, S. Estimate of Heavy Metal Contamination in Soils after a Mining Accident Using Reflectance Spectroscopy. *Environ. Sci. Technol.* **2002**, *36*, 2742–2747. [CrossRef]
11. *GB15618–2018*; Environmental Quality Standards for Soils. MEPRC (Ministry of Environmental Protection of the People's Republic of China). MEP: Beijing, China, 2018. (In Chinese)
12. Shi, T.; Chen, Y.; Liu, Y.; Wu, G. Visible and Near-Infrared Reflectance Spectroscopy—An Alternative for Monitoring Soil Con-tamination by Heavy Metals. *J. Hazard. Mater.* **2014**, *265*, 166–176. [CrossRef] [PubMed]
13. Yan, Z.; Liu, Y. Extraction of Physical and Chemical Information from Soil Based on Hyperspectral Remote Sensing Based on Plantation of Jerusalem Artichoke. *Arab. J. Geosci.* **2020**, *13*, 908. [CrossRef]
14. Xia, F.; Fan, T.; Chen, Y.; Ding, D.; Wei, J.; Jiang, D.; Deng, S. Prediction of Heavy Metal Concentrations in Contaminated Sites from Portable X-Ray Fluorescence Spectrometer Data Using Machine Learning. *Processes* **2022**, *10*, 536. [CrossRef]
15. Liu, S.; Peng, B.; Li, J. Ecological Risk Evaluation and Source Identification of Heavy Metal Pollution in Urban Village Soil Based on XRF Technique. *Sustainability* **2022**, *14*, 5030. [CrossRef]
16. Horta, A.; Malone, B.; Stockmann, U.; Minasny, B.; Bishop, T.F.A.; McBratney, A.B.; Pallasser, R.; Pozza, L. Potential of Integrated Field Spectroscopy and Spatial Analysis for Enhanced Assessment of Soil Contamination: A Prospective Review. *Geoderma* **2015**, *241–242*, 180–209. [CrossRef]
17. Nawar, S.; Cipullo, S.; Douglas, R.K.; Coulon, F.; Mouazen, A.M. The Applicability of Spectroscopy Methods for Estimating Potentially Toxic Elements in Soils: State-of-the-Art and Future Trends. *Appl. Spectrosc. Rev.* **2020**, *55*, 525–557. [CrossRef]
18. Li, F.; Xu, L.; You, T.; Lu, A. Measurement of Potentially Toxic Elements in the Soil through NIR, MIR, and XRF Spectral Data Fusion. *Comput. Electron. Agric.* **2021**, *187*, 106257. [CrossRef]
19. Ji, W.; Adamchuk, V.I.; Chen, S.; Mat Su, A.S.; Ismail, A.; Gan, Q.; Shi, Z.; Biswas, A. Simultaneous Measurement of Multiple Soil Properties through Proximal Sensor Data Fusion: A Case Study. *Geoderma* **2019**, *341*, 111–128. [CrossRef]
20. Pozza, L.E.; Bishop, T.F.A.; Stockmann, U.; Birch, G.F. Integration of Vis-NIR and PXRF Spectroscopy for Rapid Measurement of Soil Lead Concentrations. *Soil Res.* **2020**, *58*, 247. [CrossRef]
21. Wang, Q.; Li, F.; Jiang, X.; Hao, J.; Zhao, Y.; Wu, S.; Cai, Y.; Huang, W. Quantitative Analysis of Soil Cadmium Content Based on the Fusion of XRF and Vis-NIR Data. *Chemom. Intell. Lab. Syst.* **2022**, *226*, 104578. [CrossRef]
22. Tan, K.; Ma, W.; Chen, L.; Wang, H.; Du, Q.; Du, P.; Yan, B.; Liu, R.; Li, H. Estimating the Distribution Trend of Soil Heavy Metals in Mining Area from HyMap Airborne Hyperspectral Imagery Based on Ensemble Learning. *J. Hazard. Mater.* **2021**, *401*, 123288. [CrossRef]
23. Shah, D.; Zaveri, T. Hyperspectral Endmember Extraction Using Pearson's Correlation Coefficient. *IJFSE* **2021**, *24*, 89. [CrossRef]
24. Lee, K.Y.; Kang, S.; Jeon, E.-I.; Yu, S.; Kwon, O. Exploring Correlations between Hyper-Spectral Signatures Acquired in the Laboratory and in-Situ Observation for Heavy Metal Concentrations in Soil. *Spat. Inf. Res.* **2018**, *26*, 497–505. [CrossRef]
25. Balabin, R.M.; Smirnov, S.V. Variable Selection in Near-Infrared Spectroscopy: Benchmarking of Feature Selection Methods on Biodiesel Data. *Anal. Chim. Acta* **2011**, *692*, 63–72. [CrossRef]
26. Duan, H.W.; Zhu, R.G.; Xu, W.D.; Qiu, Y.Y.; Yao, X.D.; Xu, C.J. Hyperspectral imaging detection of total viable count from vacuum packing cooling mutton based on GA and CARS algorithms. *Spectrosc. Spectr. Anal.* **2017**, *37*, 847–852.
27. Luo, Y.; Wang, Z.; Zhang, Z.-L.; Zhang, J.-Q.; Zeng, Q.-P.; Tian, D.; Li, C.; Huang, F.-Y.; Chen, S.; Chen, L. Contamination Char-acteristics and Source Analysis of Potentially Toxic Elements in Dustfall-Soil-Crop Systems near Non-Ferrous Mining Areas of Yunnan, Southwestern China. *Sci. Total Environ.* **2023**, *882*, 163575. [CrossRef]
28. Luo, Y.; Wang, Z.; Zhang, Z.-L.; Huang, F.-Y.; Jia, W.-J.; Zhang, J.-Q.; Feng, X.-Y. Characteristics and Source Analysis of Potentially Toxic Elements Pollution in Atmospheric Fallout around Non-Ferrous Metal Smelting Slag Sites—Taking Southwest China as an Example. *Environ. Sci. Pollut. Res.* **2023**, *30*, 7813–7824. [CrossRef]
29. *HJ/T166-2004*; Technical Specification for Soil Environmental Monitoring. China National Environmental Protection Agency: Beijing, China, 2004.
30. Chen, M.; Ke, D.; Wang, W.; Zhang, W.; Li, X. Soil Heavy Metal Pb Concentration Quantitative Inversion Method Based on Hyperspectral Remote Sensing. *IOP Conf. Ser. Earth Environ. Sci.* **2022**, *1087*, 012050. [CrossRef]

31. Wu, H.; Yang, F.; Li, H.; Li, Q.; Zhang, F.; Ba, Y.; Cui, L.; Sun, L.; Lv, T.; Wang, N.; et al. Heavy Metal Pollution and Health Risk Assessment of Agricultural Soil near a Smelter in an Industrial City in China. *Int. J. Environ. Health Res.* **2020**, *30*, 174–186. [CrossRef] [PubMed]
32. Wang, G.; Fang, Q.; Teng, Y.; Yu, J. Determination of the Factors Governing Soil Erodibility Using Hyperspectral Visible and Near-Infrared Reflectance Spectroscopy. *Int. J. Appl. Earth Obs. Geoinf.* **2016**, *53*, 48–63. [CrossRef]
33. Mashimbye, Z.E.; Cho, M.A.; Nell, J.P.; De Clercq, W.P.; Van Niekerk, A.; Turner, D.P. Model-Based Integrated Methods for Quantitative Estimation of Soil Salinity from Hyperspectral Remote Sensing Data: A Case Study of Selected South African Soils. *Pedosphere* **2012**, *22*, 640–649. [CrossRef]
34. Asadzadeh, S.; De Souza Filho, C.R. A Review on Spectral Processing Methods for Geological Remote Sensing. *Int. J. Appl. Earth Obs. Geoinf.* **2016**, *47*, 69–90. [CrossRef]
35. Li, H.; Liang, Y.; Xu, Q.; Cao, D. Key Wavelengths Screening Using Competitive Adaptive Reweighted Sampling Method for Multivariate Calibration. *Anal. Chim. Acta* **2009**, *648*, 77–84. [CrossRef]
36. Kennard, R.W.; Stone, L.A. Computer aided design of experiments. *Technometrics* **1969**, *11*, 137–148. [CrossRef]
37. Yoon, S.; Choi, J.; Moon, S.-J.; Choi, J.H. Determination and Quantification of Heavy Metals in Sediments through La-ser-Induced Breakdown Spectroscopy and Partial Least Squares Regression. *Appl. Sci.* **2021**, *11*, 7154. [CrossRef]
38. Vohland, M.; Ludwig, M.; Thiele-Bruhn, S.; Ludwig, B. Determination of Soil Properties with Visible to Near- and Mid-Infrared Spectroscopy: Effects of Spectral Variable Selection. *Geoderma* **2014**, *223–225*, 88–96. [CrossRef]
39. Yu, L.; Hong, Y.; Geng, L.; Zhou, Y.; Nie, Y. Hyperspectral estimation of soil organic matter content based on partial least squares regression. *Nongye Gongcheng Xuebao/Trans. Chin. Soc. Agric. Eng.* **2015**, *31*, 103–109. [CrossRef]
40. Xia, F.; Peng, J.; Wang, Q.L.; Zhou, L.Q.; Shi, Z. Prediction of heavy metal content in soil of cultivated land: Hyperspectral technology at provincial scale. *J. Infrared Millim. Waves* **2015**, *34*, 593–598, 605. [CrossRef]
41. Goodarzi, R.; Mokhtarzade, M.; Zoej, M. A Robust Fuzzy Neural Network Model for Soil Lead Estimation from Spectral Features. *Remote Sens.* **2015**, *7*, 8416–8435. [CrossRef]
42. Liu, Y.; Liu, Y.; Chen, Y.; Zhang, Y.; Shi, T.; Wang, J.; Hong, Y.; Fei, T.; Zhang, Y. The Influence of Spectral Pretreatment on the Selection of Representative Calibration Samples for Soil Organic Matter Estimation Using Vis-NIR Reflectance Spectrosco-py. *Remote Sens.* **2019**, *11*, 450. [CrossRef]
43. Zhou, S.; Yuan, Z.; Cheng, Q.; Zhang, Z.; Yang, J. Rapid in Situ Determination of Heavy Metal Concentrations in Polluted Water via Portable XRF: Using Cu and Pb as Example. *Environ. Pollut.* **2018**, *243*, 1325–1333. [CrossRef] [PubMed]
44. Hou, L.; Li, X.; Li, F. Hyperspectral-Based Inversion of Heavy Metal Content in the Soil of Coal Mining Areas. *J. Environ. Qual.* **2019**, *48*, 57–63. [CrossRef]
45. Gholizadeh, A.; Coblinski, J.A.; Saberioon, M.; Ben-Dor, E.; Drábek, O.; Demattê, J.A.M.; Borůvka, L.; Němeček, K.; Chabrillat, S.; Dajčl, J. Vis–NIR and XRF Data Fusion and Feature Selection to Estimate Potentially Toxic Elements in Soil. *Sensors* **2021**, *21*, 2386. [CrossRef] [PubMed]
46. Xu, D.; Chen, S.; Xu, H.; Wang, N.; Zhou, Y.; Shi, Z. Data Fusion for the Measurement of Potentially Toxic Elements in Soil Using Portable Spectrometers. *Environ. Pollut.* **2020**, *263*, 114649. [CrossRef] [PubMed]
47. Xu, D.; Chen, S.; Viscarra Rossel, R.A.; Biswas, A.; Li, S.; Zhou, Y.; Shi, Z. X-Ray Fluorescence and Visible near Infrared Sensor Fusion for Predicting Soil Chromium Content. *Geoderma* **2019**, *352*, 61–69. [CrossRef]

processes

MDPI

Article

Nano-Hydroxyapatite Modified Tobacco Stalk-Based Biochar for Immobilizing Cd(II): Interfacial Adsorption Behavior and Mechanisms

Tianfu Li, Xiaofei Li *, Chaoran Shen, Dian Chen, Fuhua Li, Weicheng Xu, Xiaolian Wu and Yanping Bao

School of Environmental and Chemical Engineering, Foshan University, Foshan 528000, China
* Correspondence: xfli1103@fosu.edu.cn

Abstract: Biochar, an eco-friendly, porous carbon-rich material, is widely studied for immobilizing heavy metals in contaminated environments. This study prepared tobacco stalks, a typical agricultural waste, into biochar (TSB) modified by hydroxyapatite (HAP) at co-pyrolysis temperatures of 350 °C and 550 °C to explore its Cd(II) adsorption behavior and relevant mechanisms. XRD, SEM–EDS, FTIR, and BET analyses revealed that HAP successfully incorporated onto TSB, enriching the surface oxygen-containing functional groups (P–O and carboxyl), and contributing to the enhancement of the specific surface area from 2.52 (TSB350) and 3.63 m^2/g (TSB550) to 14.07 (HAP–TSB350) and 18.36 m^2/g (HAP–TSB550). The kinetics of Cd(II) adsorption onto TSB and HAP–TSB is well described by the pseudo-second-order model. Isotherm results revealed that the maximum adsorption capacities of Cd(II) on HAP–TSB350 and HAP–TSB550 were approximately 13.17 and 14.50 mg/g, 2.67 and 9.24 times those of TSB350 and TSB550, respectively. The Cd(II) adsorption amounts on TSBs and HAP–TSBs increased significantly with increasing pH, especially in HAP–TSB550. Ionic strength effects and XPS analysis showed that Cd(II) adsorption onto HAP–TSBs occurred mainly via electrostatic interaction, cation exchange with Ca^{2+}, complexation with P–O and –COOH, and surface precipitation. These findings will provide a modification strategy for the reutilization of tobacco agricultural waste in the remediation of heavy metal contaminated areas.

Keywords: tobacco stalk-based biochar; hydroxyapatite; cadmium; adsorption behavior; mechanisms

1. Introduction

As industries and agriculture progress rapidly, anthropogenic activities such as intensive mining, extensive industrial waste emissions, and the liberal application of agrochemicals frequently occur [1]. Poor management practices associated with these activities can inadvertently release heavy metals into the environment, particularly into aquatic and soil systems, thereby aggravating the serious problem of heavy metal pollution [2,3]. Cadmium (Cd), in particular, is acknowledged internationally for its high toxicity among heavy metals. It exhibits extreme environmental mobility and is frequently found in concentrations exceeding the standard limit, presenting a substantial threat to human health by causing kidney damage, bone demineralization, and increasing the risk of cancer [4,5]. Therefore, the control and remediation of Cd contamination has attracted widespread attention.

Biochar, which is a carbon-enriched material obtained from the pyrolysis of biomass under oxygen-restricted conditions, is characterized by its high structural stability, substantial specific surface area, elevated cation exchange capacity, and a plethora of oxygen-containing functional groups [6,7]. These properties collectively confer upon biochar a significant adsorptive capacity for heavy metals, positioning it as a viable and effective agent for the amelioration of heavy metal pollution in soils and aquatic environments [8,9]. The sources of biomass for the production of biochar are diverse, including crop residues (such as corn, rice, wheat, and tobacco stalks), garden waste (such as trimmings and fallen leaves), animal manure, and plant residues [10]. China has the largest tobacco cultivation area in the world, spanning over 1.6 million hectares annually and generating an impressive three to five million tons of tobacco stalk waste [11,12]. This tobacco stalk waste, which poses a risk for the dissemination of diseases and insect pests, is precluded from use for direct soil amendment, and thereby classified as a hazardous agricultural byproduct [13]. Current disposal methods—predominantly, leaving the waste in the fields or incinerating it—are inefficient and contribute to environmental pollution and resource wastage [14]. Consequently, the pyrolysis of tobacco stalks into biochar offers an efficacious strategy not only for the valorization of agricultural residues but also for the remediation of heavy metal pollution, thereby enhancing the sustainable use of agricultural waste.

Currently, the adsorption effect of biochar alone is not ideal, and its application is still limited by its limited surface functional groups and specific surface area [15]. Therefore, the modification or functional regulation of biochar has become a research direction to stimulate the structural advantages and promote the large-scale application of biochar [16]. Modification methods of biochar can be divided into physical methods (ball milling, steaming, microwaving, etc.) [17–19], chemical methods (acid, alkali treatment, element doping, and mineral loading, etc.) [20–23], and biological methods (microbial colonization, anaerobic digestion, etc.) [24,25]. Among the various approaches, the application of cost-effective, high-performance minerals onto biochar is distinguished by its simplicity, affordability, and superior adsorptive properties, typically outperforming other biochar modifications in treatment efficacy [26].

Hydroxyapatite (HAP) is a favored adsorbent for its environmentally benign attributes, including non-toxicity, affordability, and minimal water solubility [27]. It has been extensively documented for its robust heavy metal adsorption capacity, which is ascribed to mechanisms such as ion exchange, surface complex formation, and dissolution–precipitation [26,28]. However, the high surface energy of HAP can result in aggregation, thereby potentially reducing its adsorptive efficiency [29]. Consequently, the integration of hydroxyapatite onto biochar to fabricate enhanced adsorbents has emerged as a focal point of research interest [26–28,30]. For example, Wang et al. (2018) found that HAP-modified rice straw-derived biochar exhibited more efficient adsorption for Pb(II), Cu(II), and Zn(II) compared to the raw biochar and HAP [29]. Chen et al. (2021) observed that HAP formed on sludge-based biochar showed excellent adsorption capacity for Cu(II) and Cd(II), using interfacial processes including ion exchange, surface complexion, and Cd-π or Cu-π surface binding [26]. Luo et al. (2022) also found that when HAP and biomass (wheat straw and dairy cow manure) were mixed and underwent co-pyrolysis under an N_2 atmosphere, they showed more Pb/Cd immobilization compared with the pristine biochar [27]. Although studies report similar results more and more, the type of biomass and the temperature and method of pyrolysis into biochar will also produce obvious differences in the physical and chemical properties and adsorption performance [16]. At present, research on the interfacial adsorption behavior of Cd(II) and its related mechanism in tobacco stalk-derived biochar (TSB) with HAP modification (HAP–TSB) is still lacking. Elucidating the role of HAP–TSB in modulating the mobility and bioavailability of heavy metals is crucial for devising strategies aimed at remediating specific kinds of environmental contamination and enhancing soil or water quality, thereby facilitating ecological restoration.

In our investigation, we selected tobacco stalks as the feedstock to synthesize a novel carbon-based composite material through co-pyrolysis at two different temperatures with hydroxyapatite. The microstructure and surface physicochemical properties of the synthesized materials were measured by employing scanning electron microscopy–energy dispersive spectroscopy (SEM–EDS), X-ray diffraction (XRD), Fourier transform infrared spectroscopy (FTIR), etc. Subsequent batch adsorption experiments were meticulously designed to explore the adsorption performance and behavior of Cd(II) on the TSBs and HAP–TSBs, encompassing studies on adsorption isotherms, kinetics, the influence of pH, and ionic strength. On this basis, X-ray photoelectron spectroscopy (XPS) was then employed to further reveal the interfacial interaction mechanisms of Cd(II) on the TSBs and HAP–TSBs surfaces. This study seeks to provide theoretical foundations and actionable strategies to facilitate the full resource utilization of tobacco stalk waste, specifically targeting the remediation of cadmium pollution in both water and soil environments.

2. Materials and Methods

2.1. Experimental Materials

Tobacco stalk waste was collected from Guizhou Province, China, and subjected to rigorous cleaning using deionized water to eliminate residual soil and dust. Next, the biomass was dried in an oven at a temperature of 60 °C, crushed, and then stored for use in subsequent experiments and measurements. Hydroxyapatite powder (≥97% purity) and cadmium chloride (≥99% purity) were procured from Macklin (Shanghai Macklin Biochemical Co., Ltd., Shanghai, China). In addition, 1 M HCl and 1 M NaOH were employed to regulate the pH of the reaction system, while HNO_3 (5 wt%) was applied for the acidification of solution samples.

2.2. Tobacco Stalk-Derived Biochar and Its Modified Materials Preparation

Preparation of the tobacco stalk-derived biochar (TSB): The pretreated tobacco stalks were enclosed in aluminum foil and positioned inside a tubular furnace under a continuous nitrogen flow of 200 mL/min. The furnace temperature was ramped up to 350 °C and 550 °C at a heating rate of 10 °C /min, respectively, and then sustained for 2 h. After natural cooling, the TSB was pulverized, passed through a 100 mesh sieve, and stored in sealed bags under vacuum conditions for preservation. The two samples were labeled as TSB350 and TSB550, respectively.

Preparation of the hydroxyapatite modified biochar (HAP–TSB): TSB and HAP were homogeneously mixed in an aqueous solution at a weight ratio of 5:1 (*w*:*w*), after which the mixture was dried in an oven at a temperature of 60 °C until the moisture content was reduced to nearly 0% [27]. This specific ratio was chosen based on a review of previous studies and literature [31], balancing considerations for effectiveness and cost-efficiency. Subsequently, the mixture was transferred into a tubular furnace with a nitrogen atmosphere for pyrolysis. The pyrolysis conditions and subsequent treatment of the obtained composite were prepared through the established procedures for tobacco stalk biomass (TSB). The resulting functionalized biochar samples, derived from pyrolysis at specific temperatures, were designated as HAP–TSB350 and HAP–TSB550, respectively.

2.3. Batch Adsorption Experiment

To assess the adsorption performance of the prepared adsorbents for Cd(II) ions, a series of parameters were systematically investigated, including the initial concentration of Cd(II), contact time, pH, and ionic strength. For the adsorption isotherm experiments, 40 mg of each biochar and its modified samples were weighed and introduced into separate 30 mL glass bottles. Then 20 mL aliquots of Cd(II) solutions with various initial concentrations from 15 to 80 mg/L were added into their respective bottles. The adsorption process was carried out at 25 °C with continuous agitation at a rate of 180 r/min for 24 h. Throughout the experiment, the pH of the solutions were meticulously adjusted to a target value of 5.0 using 1 M HCl and NaOH solutions. After the reaction was completed, the supernatant was separated by filtration using a 0.45 μm polyether sulfone membrane filter. The filtered liquid was then acidified with HNO_3 (5 wt%), and the Cd(II) contents in the filtrate were measured with an atomic absorption spectrometer (AAS, model AA-6880, Shimadzu, Japan).

To ascertain the adsorption time effects, 400 mg of TSB and its modified samples were individually introduced into 200 mL of Cd(II) solution at 20 mg/L, with the pH adjusted to 5.0. The samples were then stirred using a magnetic stirrer set at 180 rpm to ensure thorough mixing of the solid–liquid system. At each pre-specified time point (2, 5, 10, 20, 30, 60, 120, 360, 480, and 720 min), 2 mL of the supernatant was extracted and subsequently passed through a 0.45 μm polyether sulfone membrane, and the Cd(II) concentration was then measured according to the above-mentioned methods.

To explore the influence of pH on the adsorption behavior of Cd(II) onto TSBs and HAP–TSBs, the solution pH was systematically adjusted to 2.5, 3.0, 3.5, 4.0, 5.0, and 6.0, and maintained using 1 M HCl and NaOH throughout the reaction. Ionic strength, including 0.01, 0.1, and 0.5 M NaCl, was selected in this study to elucidate the possible interfacial adsorption pathway of Cd(II). Once the reaction reached equilibrium, the mixture was filtered by employing a 0.45 μm polyether sulfone filter membrane, after which the Cd(II) concentration was then measured. To ensure the results were reliable, each experimental condition was replicated three times, and the data analysis was based on the calculated mean of these replicates.

2.4. Analytical Methods

The sorption capacity (Q_e, mg/g) was derived from the calculation that incorporated the initial concentration (C_0, mg/L) and the equilibrium concentration (C_e, mg/L) of the ions. Additionally, the adsorbent dosage (m/V, g/L) within the adsorption system was considered. The calculation was based on the following equation:

$$Qe = \frac{(C_0 - C_e) \times V}{m} \tag{1}$$

To evaluate the adsorption capacity of the TSBs and HAP–TSBs, the experimental adsorption isotherm results were analyzed employing the Langmuir and Freundlich models [32,33]. Their corresponding equations are presented in Text S1 of the Supporting Information. Additionally, the kinetic data obtained from the adsorption time studies are analyzed employing pseudo-first-order and pseudo-second-order models to assess the nature of interfacial interactions and adsorption efficiency [30]. The models are shown in Text S2 of the Supporting Information. These equations facilitate the determination of kinetic parameters that describe the adsorption process.

2.5. Characterization of Experimental Samples

The crystalline phases of the synthesized TSBs and HAP–TSBs were characterized using an X-ray diffractometer (XRD, D8 Advance, Bruker, Germany) with Cu Kα radiation (λ = 1.5418 Å) conducted at 40 kV and 40 mA. The morphologies and microstructures of the TSBs and HAP–TSBs were examined with a scanning electron microscope (SEM, Carl Zeiss Microscopy, München, Germany) that was coupled with an energy dispersive spectrometer (EDS). The specific surface areas of the TSBs and HAP–TSBs were analyzed utilizing an AUTOSORB IQ Quantchrome analyzer (Quantchrome, Boynton Beach, FL, USA). The functional groups on the TSBs' and HAP–TSBs' surfaces were identified using Fourier transform infrared spectroscopy (FTIR, Perkin-Elmer, Shelton, CT, USA). The changes in surface chemical species, including C, O, and Cd elements, on the TSB and HAP–TSB surface after Cd(II) adsorption were ascertained using X-ray photoelectron spectroscopy with a ULVAC-PHI analyzer (XPS, ULVAC-PHI, Incorporated, Tokyo, Japan), which employed a monochromatic Al Kα X-ray (1486.6 eV).

3. Results and Discussions

3.1. Characterization of the Biochar with Modification

The crystal phase components of TSB and HAP–TSB were investigated by examining their XRD patterns. As depicted in Figure 1a,b, the diffraction peaks corresponding to the Whewellite phase are observable in TSB350. However, these peaks are nearly absent in TSB550, with additional peaks corresponding to calcite becoming apparent. This suggests that an increase in hydrolysis temperature facilitated a phase transformation. Furthermore, the XRD patterns of both HAP–TSB350 and HAP–TSB550 exhibit distinct diffraction peaks that can be attributed to the crystalline structure of hydroxyapatite (HAP). This observation confirmed that HAP had been successfully incorporated onto the surfaces of the TSB.

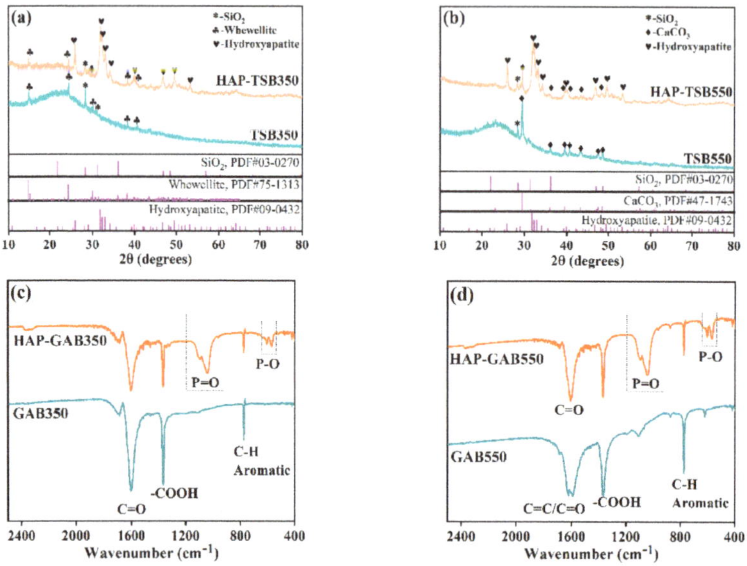

Figure 1. XRD patterns of HAP–TSB350 (**a**) and HAP–TSB550 (**b**), and their corresponding FTIR analyses (**c**,**d**).

The surface functional groups of TSB and HAP–TSB were characterized employing Fourier transform infrared (FTIR) spectroscopy. As shown in Figure 1c,d, the FTIR results of both HAP–TSB350 and HAP–TSB550 reveal characteristic stretching vibration absorption bands at 963, 1042, and 1094 cm^{-1}, and bending vibration absorption bands at 568 and 609 cm^{-1}, which correspond to P–O bonds. These findings indicate that the composite materials were enriched with phosphorus-containing structures, as referenced in the literature [28,34]. For TSB350, a prominent peak is observed at 1604 cm^{-1}, which can be attributed to the vibration of C=O groups [35]. In contrast, TSB550 exhibits a split peak with distinct overlapping bands at 1634 and 1591 cm^{-1}, corresponding to aromatic C=C stretches and the symmetric stretching of carboxylate groups [36]. This observation suggests that elevated temperatures may degrade the oxygen-containing active groups, thereby promoting the formation of aromatic structures [37]. Interestingly, the spectral shape in the corresponding wavenumber region for HAP–TSB550 showed almost no change compared to TSB550. This suggests that the incorporation of HAP helped to preserve the oxygen-containing active groups and reduced the relative proportion of aromatic structures. Additionally, the bands at 1370 and 782 cm^{-1} are indicative of the asymmetric stretching vibrations of the carboxylate group (–COO^{-}) and the aromatic C–H bonds, respectively [35,36,38,39]. Furthermore, the relative intensity of the aromatic C–H peak in TSB550 is considerably higher than that in TSB350, which implies an enhancement in aromaticity with increased hydrolysis temperature [40]. The FTIR analysis demonstrated that TSBs modified with HAP possess carboxylic groups and newly formed P–O functional groups, which might be responsible for Cd(II) adsorption, and exhibit better performance than raw TSBs.

The SEM images revealed that the TSBs exhibited distinctly porous architecture. In contrast, the surface porosity was partially obscured by the presence of hydroxyapatite crystals following the HAP modification (Figure 2a–d). Moreover, the mapping image (Figure 2e–h) and EDS spectra (Figure 2i–l) revealed that the P content of the hydroxyapatite modified biochars HAP–TSB350 and HAP–TSB550 was 5% and 2.5%, respectively. These levels are markedly higher than the P content found in their unmodified counterparts TSB350 (1.2%) and TSB550 (0.2%), indicating the successful integration of a considerable amount of hydroxyapatite (HAP) onto the biochar structure. The specific surface areas of

TSB350 and TSB550 were measured to be 2.52 m^2/g and 3.63 m^2/g, respectively, suggesting that higher hydrolysis temperatures can yield biochar with increased specific surface areas [37]. Upon modification with hydroxyapatite (HAP), the specific surface areas of HAP–TSB350 and HAP–TSB550 notably increased to 14.07 m^2/g and 18.36 m^2/g, respectively, which might be ascribed to the dispersion of HAP by TSB surfaces [41]. These enhancements are considerably higher than those in the raw TSBs, implying that the HAP–TSBs possess an increased availability of surface sites for toxic ion adsorption.

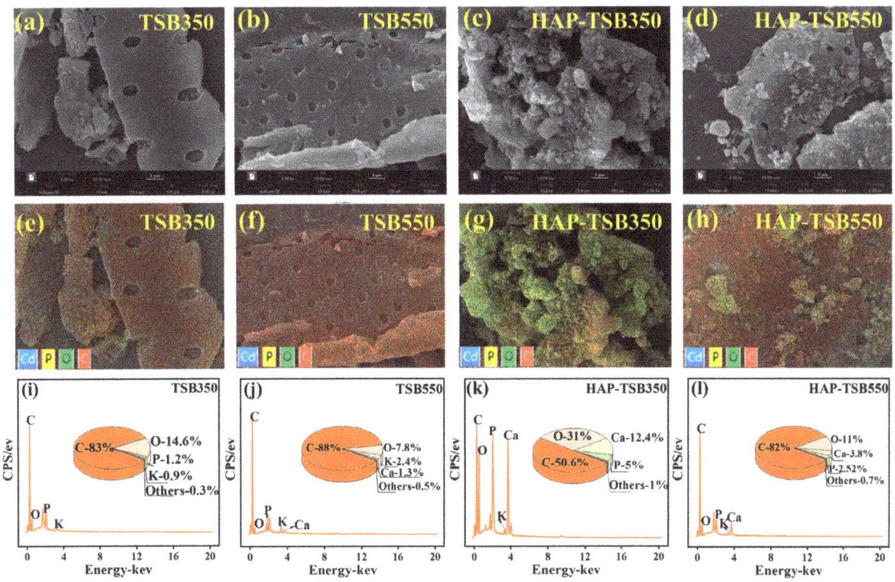

Figure 2. SEM images of TSB350 (**a**), TSB550 (**b**), HAP–TSB350 (**c**), and HAP–TSB550 (**d**) and their elemental mapping (**e**–**h**) and EDS results (**i**–**l**).

3.2. Adsorption Isotherm

Adsorption isotherm studies are pivotal for quantifying the extent to which Cd(II) ions are adsorbed onto the solid phase interfaces of adsorbents, thereby providing insights into their mobility and retention characteristics. Figure 3 presents the sorption isotherms for Cd(II) onto the TSBs and HAP–TSBs at varying concentrations. The isotherms for Cd(II) sorption on these materials exhibit a characteristic L-shaped curve with a plateau, which is indicative of possible homogeneous or heterogeneous precipitation events occurring alongside the interfacial interaction process [17]. Furthermore, the adsorption–isotherm curve for Cd(II) on TSB350 is observed to be elevated relative to that of TSB550 (Figure 3a). Additionally, both HAP–TSB350 and HAP–TSB550 demonstrate significantly enhanced sorption capacities when compared to their non-modified TSB counterparts prepared at equivalent pyrolysis temperatures (Figure 3b).

To comprehensively assess the maximum adsorption capacity and to deduce the adsorption mechanisms, the equilibrium data were interpreted through Langmuir and Freundlich models. The graphical representations of the Langmuir and Freundlich isotherm fits are displayed in Figure 3a and 3b, respectively, with the corresponding parameters of the isotherms listed in Table 1. For TSB350 and TSB550, the experimental data were found to be well-fitted with both the Langmuir and Freundlich models, as evidenced by the determined R-squared (R^2) values. Specifically, TSB350 was more suitably fitted by the Langmuir model, implying the existence of a monolayer adsorption process [28,41,42]. Conversely, TSB550 was better represented by the Freundlich model, which implies a favorable scenario for multilayer adsorption and chemical adsorption [34,43].

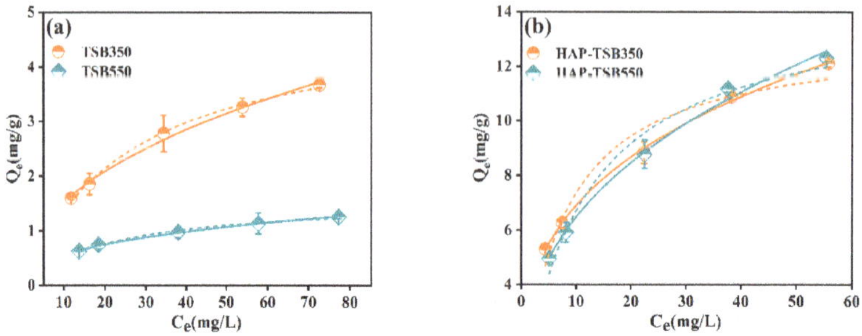

Figure 3. The sorption isotherms of Cd(II) on the TSBs (**a**) and HAP–TSBs (**b**) at a temperature of 298 K. The dashed lines denote Langmuir model fits, while the solid lines correspond to Freundlich model fits.

Table 1. Equilibrium adsorption isotherm fitting parameters.

Parameters	TSB350	TSB550	HAP–TSB350	HAP–TSB550
Langmuir				
Q_{max}	4.92	1.57	13.17	14.50
K_L	0.038	0.048	0.127	0.086
R^2	0.995	0.988	0.939	0.966
Freundlich				
1/n	0.45	0.38	0.33	0.39
K_F	0.54	0.24	3.23	2.66
R^2	0.991	0.995	0.998	0.993

Meanwhile, the experimental data for Cd(II) adsorption onto HAP–TSBs were found to be exclusively well-fitted by the Freundlich model. This phenomenon might be ascribed to the incorporation of HAP, which introduced additional active adsorption sites, resulting in complex interactions between Cd(II) ions and the modified biochar surfaces during the adsorption process. Conclusively, the Freundlich model is deemed appropriate for representing the adsorptive behavior of Cd(II) on the surfaces of TSBs and HAP–TSBs. From this perspective, the n values are found to be less than 1, signifying that the adsorption initially occurs at high energy sites, which are occupied before the subsequent uptake at lower energy sites on the heterogeneous surfaces [44,45]. The Freundlich constant K_F for TSB350 was 0.54, which was higher than the value of 0.24 observed for TSB550, indicating that the lower pyrolysis temperature of TSB350 resulted in a higher availability of high-affinity adsorption sites for Cd(II) [37,46]. Moreover, the K_F constants for the HAP-modified biochars (HAP–TSB350 and HAP–TSB550) were 3.23 and 2.66, respectively, which are substantially higher than those for their corresponding unmodified biochars, suggesting that the incorporation of HAP onto the TSBs introduces additional P–O binding sites, exhibiting strong bonding to Cd(II) [46]. Nevertheless, our tobacco stalk-derived biochars show lower adsorption performance than the results documented by Zhou et al. [47] and Yu et al. [14]. This might be attributed to differences in pyrolysis procedures, including temperature, heating rate, pyrolysis time, and mixing states, affecting the molecular components and reactivity [30]. On the other hand, composites modified with hydroxyapatite display significantly enhanced efficiency. Overall, the substantial improvement in the adsorptive capacity of tobacco stalk-derived biochar offers strategies for advancing the resourceful utilization of waste byproducts.

3.3. Adsorption Kinetics

The kinetic data for Cd(II) sorption onto TSBs and HAP–TSBs are displayed in Figure 4. The interfacial interactions between Cd(II) and the surfaces of the TSBs and HAP–TSBs occurred rapidly in the initial stage, then the rate of interaction decelerated as the system reached sorption equilibrium. The Cd(II) adsorption equilibrium for HAP–TSB350 and HAP–TSB550 was achieved in around 60 min, with corresponding adsorption capacities of 5.7 mg/g and 6.6 mg/g, respectively. In contrast, the equilibrium for TSB350 and TSB550 was reached more quickly—in approximately 30 min and 5 min, respectively—with lower adsorption capacities of 1.9 mg/g and 1.1 mg/g. Meanwhile, the adsorption capacities for the HAP–TSBs substantially exceeded those of the TSBs at equivalent time points, suggesting that the incorporation of HAP significantly enhanced the adsorption kinetics and overall capacity. The different responses in Cd(II) adsorption kinetic behavior between TSBs and HAP–TSBs can likely be attributed to variations in the distribution and concentration of adsorption sites, including the presence of HAP phases as well as oxygen-containing functional groups on the TSBs' surfaces [41].

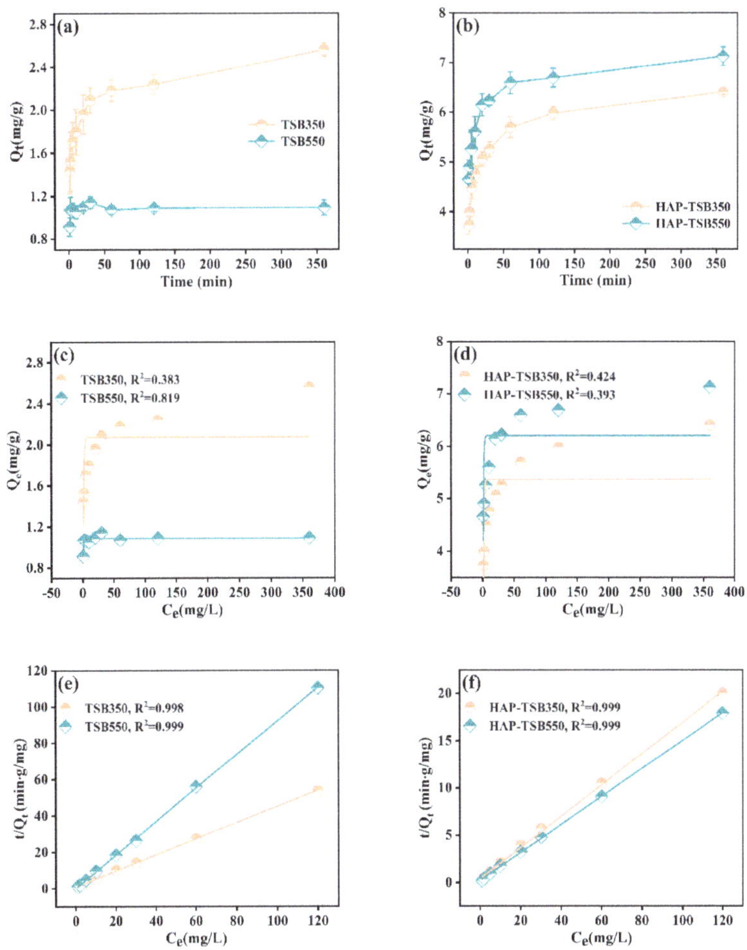

Figure 4. The kinetic plots for Cd(II) adsorption on TSBs' (**a**) and HAP–TSBs' (**b**) surfaces. The corresponding results fitted via the pseudo-first-order and pseudo-second-order model for TSBs (**c**,**e**) and HAP–TSBs (**d**,**f**).

The adsorption data were also subjected to kinetic modeling to shed light on the underlying sorption mechanisms. The pseudo-second-order model (Figure 4e–f) demonstrated a superior fit for the kinetic data compared to the pseudo-first-order model (Figure 4c,d) according to the correlation coefficients (R^2) and the calculated equilibrium adsorption amounts (Q_e) (Table 2). Generally, adherence to a pseudo-second-order model implies that the rate-limiting step in the adsorption process is likely a chemical sorption mechanism [17,26]. Consequently, the uptake of Cd(II) onto the adsorption sites of the TSBs and HAP–TSBs is predominantly governed by chemical interactions such as ion exchange, surface complexation, and/or precipitation. These chemical reactions are the primary drivers, rather than physical adsorption, in the sorption process [17]. Furthermore, the rate of Cd(II) adsorption onto the TSBs, particularly TSB550, was considerably higher than adsorption onto the HAP–TSBs. This increased rate is primarily due to a lower density of active surface sites on TSBs that are available for Cd(II) adsorption, which allows for a more rapid attainment of equilibrium in the interfacial reaction. Conversely, the TSBs modified with HAP exhibited superior adsorption capacity compared to the unmodified TSBs even at the same time intervals. This enhanced performance is attributed to the abundance of functional groups present in these HAP–TSBs, which facilitate more extended interfacial adsorption processes, thereby slowing the rate at which equilibrium is achieved.

Table 2. Kinetic parameters of Cd(II) adsorption onto the TSBs and HAP–TSBs.

	Pseudo-First-Order Model			Pseudo-Second-Order Model		
	Q_e	K_1	R^2	Q_e	K_2	R^2
TSB350	2.08	0.89	0.383	2.26	0.24	0.997
TSB550	1.09	1.85	0.819	1.09	3.94	0.999
HAP–TSB350	5.37	0.91	0.424	6.04	0.07	0.999
HAP–TSB550	1.13	6.21	0.393	6.76	0.11	0.999

3.4. Effects of pH and Ionic Strength

pH is a critical environmental parameter when assessing the utility of a material, as it significantly affects the surface charge distribution of the solid phase, its sorption capabilities, and Cd(II) speciation in aqueous solutions [34]. As documented in previous studies [17], the prevailing species of Cd(II) in aqueous solutions at a pH of 2.5 to 6.0 are the positively charged hydrated ions. The non-occurrence of hydroxide precipitation under these conditions suggests that Cd(II) is attached to the surfaces of the TSBs and HAP–TSBs via direct interactions with the active adsorption sites.

The pH-dependent sorption curves of Cd(II) onto the TSBs and HAP–TSBs are depicted in Figure 5. It is evident that the sorption capacity for Cd(II) augmented with an increase in pH from 2.5 to 6.0. Moreover, the sorption capacities of the HAP–TSBs were significantly greater than those of the unmodified TSBs. Furthermore, at lower pH values, a higher concentration of H^+ or H_3O^+ ions was present, which can compete for and occupy active sites [48], and particularly affected the less oxygen-rich functional groups of TSB550, leading to a reduced adsorption capacity for Cd(II). However, the presence of hydroxyapatite (HAP) on the TSB surface introduced a plethora of adsorption sites characterized by P–O bonds and resulted in a decrease in surface charge [26,49]. This modification enhanced the TSBs' ability to adsorb Cd(II), even at lower pH levels, by providing additional binding sites. With the progression of pH elevation, the deprotonation of active groups on the TSBs and HAP–TSBs surfaces resulted in an increasingly negatively charged interface [49]. Consequently, this promoted a substantial electrostatic attraction to Cd(II) in the aqueous solutions. Broadly speaking, as the pH increased, the degree of deprotonation of the active groups diminished, enabling a greater availability of P–O and oxygen-containing adsorption sites [26]. This availability facilitated the easy occupation of the vacant orbitals of Cd atoms, which in turn significantly enhanced the formation of surface complexes [17]. Therefore, the adsorption of Cd(II) on the HAP–TSBs was significantly more effective than that on the TSBs, even at high pH values.

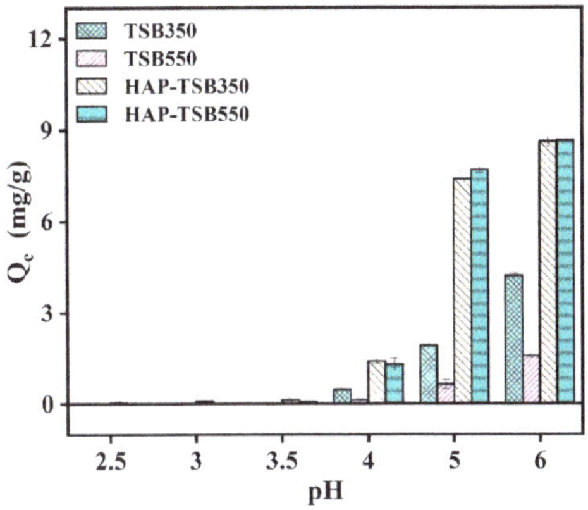

Figure 5. Effects of pH on Cd(II) adsorption onto TSB350, TSB550, HAP–TSB350, and HAP–TSB550 surfaces.

Frequently, observing the interfacial adsorption behavior across various ionic strengths provides insight into the mechanisms responsible for the adsorption of heavy metals by materials in an aqueous system [50]. Figure 6 illustrates the impact of ionic strength, varying from 0.01 to 0.5 mol/L, on the efficacy of the adsorbents. The results show that Cd(II) adsorption onto both the TSBs and HAP–TSBs declined with an increase in the initial NaCl concentration. The evident reliance of sorption on ionic strength suggests that cation exchange and/or outer-sphere complexation, facilitated by electrostatic interactions [51], were the predominant mechanisms responsible for the immobilization of Cd(II) on the surfaces of the TSBs and HAP–TSBs.

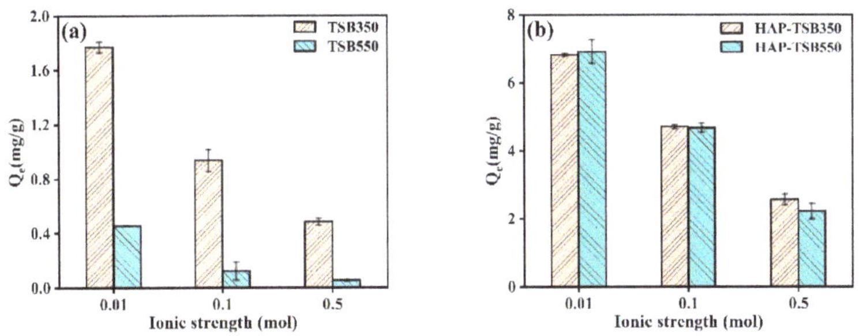

Figure 6. Effects of ionic strength on Cd(II) adsorption onto the TSBs' (**a**) and HAP–TSBs' (**b**) surfaces.

Specifically, the Cd(II) adsorption capacity of TSB350 dropped to 0.48 mg/g under a 0.5 M NaCl solution, a value that significantly exceeded the 0.05 mg/g observed for TSB550. This disparity suggests that TSB350, synthesized at a relatively low pyrolysis temperature, possesses more abundant oxygen-containing active groups that are available for chemically binding Cd(II), thus enhancing its adsorption capacity. Furthermore, the adsorption of Cd(II) on HAP–TSB350 and HAP–TSB550 was reduced to 2.56 and 2.22 mg/g under the condition of 0.5 M NaCl, respectively. Despite this reduction, these values are still significantly higher than the adsorption amounts of the unmodified TSBs. This sustained high adsorption capacity suggests that the HAP phases actively contribute to the interfacial adsorption process and promote the formation of Cd(II) inner-sphere surface complexes.

3.5. X-ray Photoelectron Spectroscopy Analysis

X-ray photoelectron spectroscopy (XPS) was employed to analyze the TSBs and HAP–TSBs before and after Cd(II) adsorption, providing detailed insights into the specific surface chemical species alterations that occurred [34,50]. In contrast to the XPS survey spectra of the TSBs, the presence of characteristic P 2p and Ca 2p peaks in the spectra of the HAP–TSBs confirms the successful deposition of HAP onto the TSB surface (Figure 7a). Concurrently, the Cd 3d peaks detected in the XPS spectra of Cd–HAP–TSB350 and Cd–HAP–TSB550 are more pronounced than those observed in the Cd–TSBs. This suggests that the composite materials have a higher affinity for Cd(II) immobilization. Moreover, the relative intensity of the Ca 2p peak, normalized to that of the C 1s peak, indicated a reduction in the adsorption of Cd(II) onto the HAP–TSBs. This finding suggests that cation exchange is a contributing mechanism for the interaction of Cd(II) with the surfaces of HAP–TSBs [26].

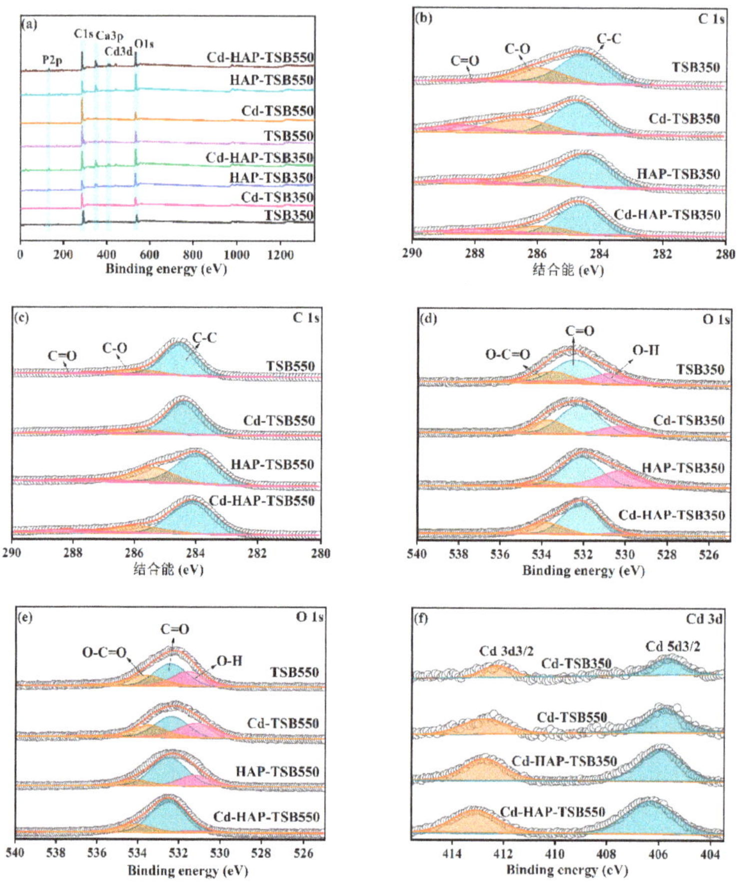

Figure 7. The spectra of the TSBs and HAP–TSBs before and after the adsorption of Cd(II): (**a**) survey spectra, (**b**,**c**) C 1s, (**d**,**e**) O 1s, and (**f**) Cd 3d.

The deconvoluted C 1s spectra are depicted in Figure 7b,c. A comparison of the C 1s spectra with TSB350, HAP–TSB350, TSB550, and HAP–TSB550 reveals that the relative ratio of the peaks corresponding to the C–O group at binding energies of 285.9 to 286.6 eV is notably reduced following Cd adsorption (Table 3), suggesting the formation of Cd-carboxyl complexes on the surfaces of the materials [52]. Simultaneously, there is a significant increase in the peak ratio of the C=O group within the binding energy range of 288.3 to 288.5 eV after Cd(II) adsorption, implying that carbonate precipitation may have occurred during the adsorption process [30].

Table 3. The deconvolution of C and O species for the TSBs', HAP–TSBs', Cd–TSBs', and Cd–HAP–TSBs' reaction systems and the corresponding peak area ratio.

	C Speciation Proportion (%)			**O Speciation Proportion (%)**		
	C–C	C–O	C=O	O–C=O	C=O	O–H
TSB350	65.59	31.57	2.84	21.18	47.28	31.54
Cd–TSB350	50.62	28.25	21.13	23.30	55.87	20.83
HAP–TSB350	68.84	19.33	11.83	8.49	49.55	41.96
Cd–HAP–TSB350	61.52	18.08	20.40	26.26	67.55	6.19
TSB550	85.06	10.96	3.98	27.17	42.60	30.23
Cd–TSB550	77.33	10.76	11.91	28.55	43.63	27.82
HAP–TSB550	63.83	33.02	3.15	12.18	64.93	22.89
Cd–HAP–TSB550	69.79	20.08	10.13	19.59	76.24	4.17

After the sorption of Cd(II), the O 1s XPS spectra of Cd–TSB350, Cd–HAP–TSB350, Cd–TSB550, and Cd–HAP–TSB550 exhibit an elevated peak ratio for the carbonate structures (O–C=O or CO_3^{2-}) (Figure 7d,e). Simultaneously, there is a pronounced reduction in the O–H peak. This suggests that, following sorption, carbonate-related products, including cadmium carbonates, are formed and active oxygen-containing groups on the sorbent surfaces are engaged in complexation with Cd(II) [26,53]. Notably, the O 1s spectra of Cd–HAP–TSBs exhibit a more pronounced alteration in the relative ratio of O–H to O–C=O peaks (Table 3), which might be ascribed to interfacial reactions involving the high oxygen-containing active sites of the HAP phase [30]. Furthermore, the Cd 3d XPS spectra revealed that the peaks corresponding to Cd 3d3/2 and Cd 5d3/2 shifted to higher binding energies on the HAP–TSBs. Additionally, there was a notable increase in the intensity of these peaks on the surface of the Cd–HAP–TSBs. These observations suggest that the HAP modification enhanced Cd(II) adsorption and led to the formation of more stable surface complexes.

4. Conclusions

In this investigation, the TSBs and HAP–TSBs produced through pyrolysis at different temperatures were systematically evaluated by analyzing their microstructures, elemental distribution, specific surface area, functional groups, adsorption performance for Cd(II), and underlying mechanisms governing interfacial interactions. The TSBs modified with HAP resulted in a notable expansion of their surface-specific area and an alteration in the composition of active sites, such as P–O and carboxyl groups. Among these synthesized materials, HAP–TSB550 exhibited superior and effective Cd(II) adsorption, which is primarily attributed to its large specific surface area in conjunction with the presence of incorporated HAP. The adsorption kinetics of Cd(II) onto the HAP–TSBs were found to conform closely to the pseudo-second-order model, whilst the isotherms were well described by the Freundlich model. The dominant mechanisms of Cd(II) adsorption onto the HAP–TSBs include cation exchange, surface inner-sphere complexation, surface precipitation, and outer-sphere complexation (electrostatic attraction), which are facilitated by the co-existence of abundant oxygen-containing and phosphate groups. Overall, the adsorption performance of Cd(II) on HAP–TSB350 is as efficient as on HAP–TSB550, demonstrating that modification with HAP at a low pyrolysis temperature is more economical and has greater applicability. The HAP–TSBs developed in this study provide insight into their adsorption mechanisms and present a feasible approach for the effective reutilization of tobacco stalk waste and the remediation of areas with low-level heavy metal pollution, enhancing overall environmental quality.

Supplementary Materials: The following supporting information can be downloaded at: https://www.mdpi.com/article/10.3390/pr12091924/s1, Text S1: Langmuir and Freundlich isotherm models; Text S2: Adsorption kinetic models.

Author Contributions: T.L.: Investigation, Data curation, Methodology, Writing—original draft; X.L.: Conceptualization, Funding acquisition, Supervision, Writing—review & editing; C.S.: Investigation, Visualization; D.C.: Investigation, software; F.L.: Formal analysis, Writing—review and editing; W.X.: Visualization, Writing—review and editing; X.W.: Software, Writing—review & editing; Y.B.: Validation, Writing—review and editing. All authors have read and agreed to the published version of the manuscript.

Funding: This work was supported by the National Natural Science Foundation of China (Nos. 42307278, 42207302); the Guangdong Basic and Applied Basic Research Foundation, China (Nos. 2022A1515110918, 2022A1515110444); and the National Key Research and Development Program of China (No. 2023YFD17-00802).

Data Availability Statement: All relevant data has been included in the article.

Conflicts of Interest: The authors declare no competing financial interests.

References

1. Oh, S.Y.; Yoon, M.K. Biochar for Treating Acid Mine Drainage. *Environ. Eng. Sci.* **2013**, *30*, 589–593. [CrossRef]
2. Xiang, M.T.; Li, Y.; Yang, J.Y.; Lei, K.G.; Li, Y.; Li, F.; Zheng, D.F.; Fang, X.Q.; Cao, Y. Heavy metal contamination risk assessment and correlation analysis of heavy metal contents in soil and crops. *Environ. Pollut.* **2021**, *278*, 116911. [CrossRef] [PubMed]
3. He, L.Z.; Xu, Y.; Zhang, M.E.; Gul, S.; Zhang, X.K.; Zhong, H.; Tang, Y.X.; Dong, D.B.; Xu, Y.; Liu, D.; et al. Effect of remediation technologies on soil fertility in heavy metal(loid)-contaminated soils: A critical review. *Crit. Rev. Environ. Sci. Technol.* **2024**, *54*, 1417–1435. [CrossRef]
4. Liu, T.; Yuan, X.M.; Luo, K.; Xie, C.; Zhou, L.Y. Molecular engineering of a new method for effective removal of cadmium from water. *Water. Res.* **2024**, *253*, 121326. [CrossRef]
5. Lange, K.; Viklander, M.; Blecken, G.T. Investigation of intra-event variations of total, dissolved and truly dissolved metal concentrations in highway runoff and a gross pollutant trap—Bioretention stormwater treatment train. *Water. Res.* **2022**, *216*, 118284. [CrossRef] [PubMed]
6. Gao, Y.R.; Chen, H.B.; Fang, Z.; Niazi, N.K.; Adusei-Fosu, K.; Li, J.H.; Yang, X.; Liu, Z.Z.; Bolan, N.S.; Gao, B.; et al. Coupled sorptive and oxidative antimony(III) removal by iron-modified biochar: Mechanisms of electron-donating capacity and reactive Fe species. *Environ. Pollut.* **2023**, *337*, 122637. [CrossRef]
7. Inyang, M.I.; Gao, B.; Yao, Y.; Xue, Y.W.; Zimmerman, A.; Mosa, A.; Pullammanappallil, P.; Ok, Y.S.; Cao, X.D. A review of biochar as a low-cost adsorbent for aqueous heavy metal removal. *Crit. Rev. Environ. Sci. Technol.* **2016**, *46*, 406–433. [CrossRef]
8. Tran, H.T.; Bolan, N.S.; Lin, C.; Binh, Q.A.; Nguyen, M.K.; Luu, T.A.; Le, V.G.; Pham, C.Q.; Hoang, H.G.; Vo, D.V.N. Succession of biochar addition for soil amendment and contaminants remediation during co-composting: A state of art review. *J. Environ. Manag.* **2023**, *342*, 118191. [CrossRef]
9. Biswal, B.K.; Balasubramanian, R. Use of biochar as a low-cost adsorbent for removal of heavy metals from water and wastewater: A review. *J. Environ. Chem. Eng.* **2023**, *11*, 110986. [CrossRef]
10. Fakayode, O.A.; Aboagarib, E.A.A.; Zhou, C.S.; Ma, H.L. Co-pyrolysis of lignocellulosic and macroalgae biomasses for the production of biochar—A review. *Bioresour. Technol.* **2020**, *297*, 122408. [CrossRef]
11. Yang, X.; Lu, K.; McGrouther, K.; Che, L.; Hu, G.; Wang, Q.; Liu, X.; Shen, L.; Huang, H.; Ye, Z.; et al. Bioavailability of Cd and Zn in soils treated with biochars derived from tobacco stalk and dead pigs. *J. Soils Sediments* **2017**, *17*, 751–762. [CrossRef]
12. Zhang, W.; Li, C.; Li, G.; Lin, Q.; Zhao, X.; He, Y.; Liu, Y.; Luo, Z. Biochar alters inorganic phosphorus fractions in tobacco-growing soil. *J. Soil Sci. Plant Nut.* **2021**, *21*, 1689–1699. [CrossRef]
13. Zheng, X.; Song, W.; Guan, E.; Wang, Y.; Hu, X.; Liang, H.; Dong, J. Response in physicochemical properties of tobacco-growing soils and N/P/K accumulation in tobacco plant to tobacco straw biochar. *J. Soil Sci. Plant Nutr.* **2020**, *20*, 293–305. [CrossRef]
14. Yu, X.N.; Zhou, H.J.; Ye, X.F.; Wang, H.L. From hazardous agriculture waste to hazardous metal scavenger: Tobacco stalk biochar-mediated sequestration of Cd leads to enhanced tobacco productivity. *J. Hazard. Mater.* **2021**, *413*, 125303. [CrossRef] [PubMed]
15. Awasthi, M.K. Engineered biochar: A multifunctional material for energy and environment. *Environ. Pollut.* **2022**, *298*, 118831. [CrossRef]
16. Chen, H.B.; Gao, Y.R.; Li, J.H.; Fang, Z.; Bolan, N.; Bhatnagar, A.; Gao, B.; Hou, D.Y.; Wang, S.S.; Song, H.; et al. Engineered biochar for environmental decontamination in aquatic and soil systems: A review. *Carbon Res.* **2022**, *1*, 4. [CrossRef]
17. Xiao, J.; Hu, R.; Chen, G.C. Micro-nano-engineered nitrogenous bone biochar developed with a ball-milling technique for high-efficiency removal of aquatic Cd(II), Cu(II) and Pb(II). *J. Hazard. Mater.* **2020**, *387*, 121980. [CrossRef]
18. Duan, W.Y.; Oleszczuk, P.; Pan, B.; Xing, B.S. Environmental behavior of engineered biochars and their aging processes in soil. *Biochar* **2019**, *1*, 339–351. [CrossRef]
19. Medeiros, D.; Nzediegwu, C.; Benally, C.; Messele, S.A.; Kwak, J.H.; Naeth, M.A.; Ok, Y.S.; Chang, S.X.; El-Din, M.G. Pristine and engineered biochar for the removal of contaminants co-existing in several types of industrial wastewaters: A critical review. *Sci. Total Environ.* **2022**, *809*, 151120. [CrossRef]

20. Li, M.P.; Dong, C.; Guo, C.X.; Yu, L.G. Magnetic activated biochar Fe_3O_4-MOS made from moringa seed shells for the adsorption of methylene blue. *Processes* **2022**, *10*, 2720. [CrossRef]
21. Lan, G.X.; Yan, X.X.; Deng, P.Y.; Li, T.Z.; Xia, Y.P.; Zhu, Z.H.; Wu, Y.; Fu, C.A. Preparation of iron salt-modified sludge biochar and its uptake behavior for phosphate. *Processes* **2022**, *10*, 2122. [CrossRef]
22. Chen, H.B.; Gao, Y.R.; Fang, Z.; Li, J.Y.; Pillai, S.C.; Song, H.C.; Sun, C.H.; Bolan, N.; Yang, X.; Vithanage, M.; et al. Investigating the electron-scale adsorption mechanisms using DFT calculations and experimental studies in self-assembly magnetic biochar gel incorporated with graphene nanosheets for enhanced Sb(III) removal. *Chem. Eng. J.* **2024**, *487*, 150740. [CrossRef]
23. Ban, S.E.; Lee, E.J.; Lim, D.J.; Kim, I.S.; Lee, J.W. Evaluation of sulfuric acid-pretreated biomass-derived biochar characteristics and its diazinon adsorption mechanism. *Bioresour. Technol.* **2022**, *348*, 126828. [CrossRef]
24. Bolan, S.; Hou, D.Y.; Wang, L.W.; Hale, L.; Egamberdieva, D.; Tammeorg, P.; Li, R.; Wang, B.; Xu, J.P.; Wang, T.; et al. The potential of biochar as a microbial carrier for agricultural and environmental applications. *Sci. Total Environ.* **2023**, *886*, 163968. [CrossRef] [PubMed]
25. Kumar, A.; Singh, E.; Mishra, R.; Kumar, S. Biochar as environmental armour and its diverse role towards protecting soil, water and air. *Sci. Total Environ.* **2022**, *806*, 150444. [CrossRef]
26. Chen, Y.N.; Li, M.L.; Li, Y.P.; Liu, Y.H.; Chen, Y.R.; Li, H.; Li, L.S.Z.; Xu, F.T.; Jiang, H.J.; Chen, L. Hydroxyapatite modified sludge-based biochar for the adsorption of Cu^{2+} and Cd^{2+}: Adsorption behavior and mechanisms. *Bioresour. Technol.* **2021**, *321*, 124413. [CrossRef]
27. Luo, Y.; Li, Z.P.; Xu, H.C.; Xu, X.Y.; Qiu, H.; Cao, X.D.; Zhao, L. Development of phosphorus composite biochar for simultaneous enhanced carbon sink and heavy metal immobilization in soil. *Sci. Total Environ.* **2022**, *831*, 154845. [CrossRef]
28. Zhu, Z.Q.; Yang, Y.H.; Fan, Y.M.; Zhang, L.H.; Tang, S.; Zhu, Y.N.; Zhou, X.B. Strontium-doped hydroxyapatite as an efficient adsorbent for Cd(II) removal from wastewater: Performance, kinetics, and mechanism. *Environ. Technol. Innov.* **2022**, *28*, 102575. [CrossRef]
29. Wang, Y.Y.; Liu, Y.X.; Lu, H.H.; Yang, R.Q.; Yang, S.M. Competitive adsorption of Pb(II), Cu(II), and Zn(II) ions onto hydroxyapatite-biochar nanocomposite in aqueous solutions. *J. Solid State Chem.* **2018**, *261*, 53–61. [CrossRef]
30. Sun, L.M.; Wu, J.S.; Wang, J.S.; Xu, M.; Zhou, W.Y.; Du, Y.C.; Li, Y.L.; Li, H.Y. Fabricating hydroxyapatite functionalized biochar composite using steel slag and Hami melon peel for Pb(II) and Cd(II) removal. *Colloids Surf. Physicochem. Eng. Aspects.* **2023**, *666*, 131310. [CrossRef]
31. Gao, R.; Fu, Q.; Hu, H.; Wang, Q.; Liu, Y.; Zhu, J. Highly-effective removal of Pb by co-pyrolysis biochar derived from rape straw and orthophosphate. *J. Hazard. Mater.* **2019**, *371*, 191–197. [CrossRef] [PubMed]
32. Langmuir, I. The Adsorption of gases on plane surfaces of glass, mica and platinum. *J. Am. Chem. Soc.* **1918**, *40*, 1361–1403. [CrossRef]
33. Freundlich, H.M.F. Über die adsorption in lösungen Z. *Phys. Chem.* **1906**, *57*, 385–470. [CrossRef]
34. Liao, J.; He, X.S.; Zhang, Y.; Zhang, L.; He, Z.B. The construction of magnetic hydroxyapatite-functionalized pig manure-derived biochar for the efficient uranium separation. *Chem. Eng. J.* **2023**, *457*, 141367. [CrossRef]
35. Li, J.Y.; Gao, Y.R.; Li, C.B.; Wang, F.L.; Chen, H.B.; Yang, X.; Jeyakumar, P.; Sarkar, B.; Luo, Z.B.; Bolan, N.; et al. Pristine and Fe-functionalized biochar for the simultaneous immobilization of arsenic and antimony in a contaminated mining soil. *J. Hazard. Mater.* **2024**, *469*, 133937. [CrossRef]
36. Qian, L.B.; Zhang, W.Y.; Yan, J.C.; Han, L.; Chen, Y.; Ouyang, D.; Chen, M.F. Nanoscale zero-valent iron supported by biochars produced at different temperatures: Synthesis mechanism and effect on Cr(VI) removal. *Environ. Pollut.* **2017**, *223*, 153–160. [CrossRef] [PubMed]
37. Das, S.K.; Ghosh, G.K.; Avasthe, R.K.; Sinha, K. Compositional heterogeneity of different biochar: Effect of pyrolysis temperature and feedstocks. *J. Environ. Manag.* **2021**, *278*, 111501. [CrossRef]
38. Yang, F.; Jiang, Y.T.; Dai, M.; Hou, X.T.; Peng, C.S. Active biochar-supported iron oxides for Cr(VI) removal from groundwater: Kinetics, stability and the key role of FeO in electron-transfer mechanism. *J. Hazard. Mater.* **2022**, *424*, 127542. [CrossRef]
39. Chia, C.H.; Gong, B.; Joseph, S.D.; Marjo, C.E.; Munroe, P.; Rich, A.M. Imaging of mineral-enriched biochar by FTIR, Raman and SEM-EDX. *Vib. Spectrosc.* **2012**, *62*, 248–257. [CrossRef]
40. Sun, Y.Q.; Xiong, X.N.; He, M.J.; Xu, Z.B.; Hou, D.Y.; Zhang, W.H.; Ok, Y.S.; Rinklebe, J.; Wang, L.L.; Tsang, D.C.W. Roles of biochar-derived dissolved organic matter in soil amendment and environmental remediation: A critical review. *Chem. Eng. J.* **2021**, *424*, 130387. [CrossRef]
41. Zhu, S.H.; Irshad, M.K.; Ibrahim, M.; Chen, Q.; Shang, J.Y.; Zhang, Q.R. The distinctive role of nano-hydroxyapatite modified biochar for alleviation of cadmium and arsenic toxicity in aqueous system. *J. Water Process. Eng.* **2022**, *49*, 103054. [CrossRef]
42. Li, X.F.; Guo, C.L.; Jin, X.H.; Yao, Q.; Liu, Q.Q.; Zhang, L.J.; Lu, G.N.; Reinfelder, J.R.; Huang, W.L.; Dang, Z. Molecular-scale study of Cr(vi) adsorption onto lepidocrocite facets by EXAFS, in situ ATR-FTIR, theoretical frequency calculations and DFT+ U techniques. *Environ. Sci. Nano* **2022**, *9*, 568–581. [CrossRef]
43. Zhang, Y.P.; Adi, V.S.K.; Huang, H.L.; Lin, H.P.; Huang, Z.H. Adsorption of metal ions with biochars derived from biomass wastes in a fixed column: Adsorption isotherm and process simulation. *J. Ind. Eng. Chem.* **2019**, *76*, 240–244. [CrossRef]
44. Li, Z.; Li, M.; Che, Q.; Li, Y.D.; Liu, X. Synergistic removal of tylosin/sulfamethoxazole and copper by nanohydroxyapatite modified biochar. *Bioresour. Technol.* **2019**, *294*, 122163. [CrossRef]

45. Liu, N.; Wang, M.X.; Liu, M.M.; Liu, F.; Weng, L.; Koopal, L.K.; Tan, W.F. Sorption of tetracycline on organo-montmorillonites. *J. Hazard. Mater.* **2012**, *225–226*, 28–35. [CrossRef]
46. Liu, C.W.; Ye, J.; Lin, Y.; Wu, J.; Price, G.W.; Burton, D.; Wang, Y.X. Removal of Cadmium (II) using water hyacinth (*Eichhornia crassipes*) biochar alginate beads in aqueous solutions. *Environ. Pollut.* **2020**, *264*, 114785. [CrossRef]
47. Zhou, Z.; Xu, Z.; Feng, Q.; Yao, D.; Yu, J.; Wang, D.; Lv, S.; Liu, Y.; Zhou, N.; Zhong, M.-E. Effect of pyrolysis condition on the adsorption mechanism of lead, cadmium and copper on tobacco stem biochar. *J. Clean. Prod.* **2018**, *187*, 996–1005. [CrossRef]
48. Li, Z.W.; Niu, R.Y.; Yu, J.H.; Yu, L.Y.; Cao, D. Removal of cadmium from aqueous solution by magnetic biochar: Adsorption characteristics and mechanism. *Environ. Sci. Pollut. Res.* **2024**, *31*, 6543–6557. [CrossRef]
49. Wu, W.L.; Liu, Z.H.; Azeem, M.; Guo, Z.Q.; Li, R.H.; Li, Y.G.; Peng, Y.R.; Ali, E.F.; Wang, H.L.; Wang, S.S.; et al. Hydroxyapatite tailored hierarchical porous biochar composite immobilized Cd(II) and Pb(II) and mitigated their hazardous effects in contaminated water and soil. *J. Hazard. Mater.* **2022**, *437*, 129330. [CrossRef]
50. Li, X.F.; Guo, C.L.; Jin, X.H.; He, C.C.; Dang, Z. Mechanisms of Cr(VI) adsorption on schwertmannite under environmental disturbance: Changes in surface complex structures. *J. Hazard. Mater.* **2021**, *416*, 125781. [CrossRef]
51. Wang, H.Y.; Gao, B.; Wang, S.S.; Fang, J.; Xue, Y.W.; Yang, K. Removal of Pb(II), Cu(II), and Cd(II) from aqueous solutions by biochar derived from $KMnO_4$ treated hickory wood. *Bioresour. Technol.* **2015**, *197*, 356–362. [CrossRef] [PubMed]
52. Yu, W.; Hu, J.; Yu, Y.; Ma, D.; Gong, W.; Qiu, H.; Hu, Z.; Gao, H.W. Facile preparation of sulfonated biochar for highly efficient removal of toxic Pb(II) and Cd(II) from wastewater. *Sci. Total Environ.* **2021**, *750*, 141545. [CrossRef] [PubMed]
53. Cai, T.; Liu, X.; Zhang, J.; Tie, B.; Lei, M.; Wei, X.; Peng, O.; Du, H. Silicate-modified oiltea camellia shell-derived biochar: A novel and cost-effective sorbent for cadmium removal. *J. Clean. Prod.* **2021**, *281*, 125390. [CrossRef]

processes

MDPI

Article

Differences in Physiological Metabolism and Antioxidant System of Different Ecotypes of *Miscanthus floridulus* under Cu Stress

Jianqiao Qin [1], Zhiqiang Yan [2], Xueding Jiang [3,*], Huarong Zhao [4,*], Shasha Liu [1], Min Dai [1], Dexin Xiong [1] and Xi Chen [5]

1 Guangdong Provincial Key Laboratory of Environmental Health and Land Resource, School of Environmental and Chemical Engineering, Zhaoqing University, Zhaoqing 526061, China
2 Shunde Polytechnic, Shunde 528300, China
3 School of Environmental and Chemical Engineering, Foshan University, Foshan 528000, China
4 School of Environmental Science and Engineering, Guilin University of Technology, Guilin 541004, China
5 School of Environmental Science and Engineering, Sun Yat-Sen University, Guangzhou 510275, China
* Correspondence: jiangxueding@fosu.edu.cn (X.J.); zhaohuar@mail3.sysu.edu.cn (H.Z.)

Abstract: To reveal the similarities and differences in the resistance mechanisms of different ecotypes to Cu stress, a pot experiment was used to systematically compare the physiological responses of non-mining ecotype *Miscanthus floridulus* (collected from Boluo County, Huizhou City) and mining ecotype *Miscanthus floridulus* (collected from Dabaoshan mining area) under different Cu concentrations. The results showed that chlorophyll a, chlorophyll b and total chlorophyll in the leaves of the two ecotypes of *M. floridulus* were negatively correlated with Cu stress concentration ($p < 0.01$), but the extent of decrease for the ecotypes in the mining area was lower than that for the ecotypes in the non-mining area. The values of chlorophyll a/b for both ecotypes increased with increasing Cu treatment concentration, indicating that Cu is more harmful to chlorophyll b than to chlorophyll a for *M. floridulus*. Cu stress can lead to the accumulation of malondialdehyde (MDA) in the leaves of *M. floridulus* with the amount of MDA accumulation observed being greater in the non-mining ecotype than in the mining ecotype ($p < 0.05$). The content of antioxidant substances (ascorbic acid and reduced glutathione) in the mining ecotype *M. floridulus* was significantly higher than that in the non-mining ecotype. The activity of SOD in the leaves of non-mining ecotypes was inhibited by Cu stress and the activity of POD was increased by Cu stress. However, the increase in POD in the mining ecotypes was greater than that in the non-mining ecotypes and the activities of the two enzymes in the mining ecotypes were significantly higher than those in the non-mining ecotypes at the highest concentration of Cu. Cu had different effects on PPO activity in the leaves of the two ecotypes of *M. floridulus*. The plant leaves of the non-mining ecotype at 400 and 800 mg·kg^{-1} were significantly fewer than those of the control group ($p < 0.05$), which were 87.1% and 65.2% of the control group, respectively. The PPO activity in the plant leaves of the mining ecotype was higher than that in the leaves of the non-mining ecotype and was significantly higher at 400 and 800 mg·kg^{-1} than that of the control group ($p < 0.05$), at 226.5% and 268.1% of the control group, respectively. These results indicate that the mining ecotype *M. floridulus* is more resistant to copper stress, that resistant ecotypes have been formed, and that small-molecule antioxidant substances play an important role in increasing resistance levels.

Keywords: copper; *Miscanthus floridulus*; non-enzymatic substance; antioxidant system

1. Introduction

With increasing growth in Cu-containing pollutants in industry, agriculture, transportation and other fields, especially in recent years, the frequent use of Cu-containing pesticides in agricultural production, the excessive mining of copper ores and the huge discharge of Cu-containing pollutants in industrial production have resulted in serious harm to animals, plants and the environment, and even endangered human health [1–3]. These harms to the ecosystem have attracted the attention of researchers [4,5]. The phytoremediation of mining wasteland with (super) enriched plants and tolerant plants is currently considered to be an economical and effective method [6,7]. Therefore, screening new repair species and exploring their mechanisms of tolerance to heavy metals has become a focus of academic research [8,9].

Previous studies have shown that, due to its long-term survival in a polluted environment, *Miscanthus floridulus* in metal mining areas may have undergone resistance evolution adapted to the polluted environment and formed resistant ecotypes [10–12]. Liao Bin et al. [13] found that there were significant differences in the critical concentration and symptoms of Cu poisoning between the two ecotypes of *Commelina communis*. The differences in the resistance mechanisms of different ecotypes to heavy metal stress may be caused by variation in genetic mechanisms, the functioning of the antioxidant enzyme system or the detoxification of heavy metal chelation, but there is no consensus at present [14–16]. *M. floridulus* is a perennial herb plant of the Miscanthus genus, which is widely distributed in southern China and is highly adaptable [17,18]. Sun Jian et al. [19] investigated heavy metal pollution in the soil and plants in a lead–zinc mining area in Chenzhou, Hunan Province, and found that *M. floridulus* had a large capacity for absorption and transport of lead and copper. Studying the physiological and ecological differences between resistant ecotypes and sensitive ecotypes is a valuable approach for revealing the mechanisms of plant resistance and underpins the widespread use of phytoremediation technology [20–22].

The Dabaoshan mine, located at the junction of Qujiang County and Wengyuan County, Shaoguan City, Guangdong Province, is a large iron polymetallic sulfide-associated deposit [23]. Zhao et al. [24] investigated the soil of Dabaoshan and found that heavy metal Cd pollution in the soil of Dabaoshan was the most severe, with more than 81.19% of the soil contaminated by Cd. Furthermore, more than 72.91% of the soil evidenced Cu pollution. According to a study of plants in the Dabaoshan mining area by Chen et al. [25], the rhizosphere of *M. floridulus* can activate Cu and other heavy metals significantly and may be considered a pioneer plant for local vegetation restoration. Qin et al. [26,27] conducted a number of studies on soil heavy metal content, soil enzyme activity and vegetation restoration in the Dabaoshan lead-zinc mining area. The results showed that *M. floridulus* was beneficial in accelerating the ecological restoration of abandoned mining areas. Previous studies have also shown that *M. floridulus* can grow normally in the heavily polluted tailings ponds in the Dabaoshan mining area and that its roots can absorb and fix heavy metal copper. Moreover, its biomass is large, so it represents good plant repair material [28,29].

To better apply *M. floridulus* to soil heavy metal pollution remediation practice, it is necessary to conduct in-depth research concerning the changes to its physiological and biochemical mechanisms. In this study, through soil culture experimentation, the physiological metabolism and antioxidant enzyme activity response of two ecotypes of *M. floridulus* under Cu stress were compared to provide a theoretical basis for the rational application of *M. floridulus* in the future phytoremediation of Cu-contaminated soil.

2. Materials and Methods

2.1. Materials

The experimental material was *M. floridulus* seedlings. New seedlings of the current year were collected in early March. The mining ecotype *M. floridulus* was collected from Dabaoshan Mining area, Shaoguan, Guangdong (24°33′36.6″ N, 113°43′14.0″ E), while the non-mining ecotype *M. floridulus* was collected from the hills and mountains of Boluo County, Huizhou, Guangdong (23°08′57.8″ N, 114°21′10.0″ E), in the same subtropical monsoon climate zone. The seedlings with soil were collected in the wild and immediately brought back to the school's greenhouse. The roots were rinsed with running water to prepare for planting experiments.

2.2. The Experimental Soil

The soil for this test was taken from the teaching vegetable garden of South China Agricultural University. The air-dried soil, screened to 2 mm, was put into plastic pots, with 1.2 kg per pot. The base fertilizer standard was 100 mg N kg^{-1} dry soil, added as H_2NCONH_2, and 80 mg P kg^{-1} and 100 mg K kg^{-1}, added as KH_2PO_4. This was mixed well and set aside.

The total amounts of Zn, Pb, Cu, and Cd in the soil were determined by digestion using HCl, HF, and perchloric acid, and then by ICP-OES (Optima5300DV, Perkin-Elmer, Sheldon, CT, USA) [30]. The basic chemical properties of the soil were determined using soil agrochemical analysis methods [31].

The basic chemical properties and heavy metal contents of the soil at the plant sample collection site and the test soil are shown in Table 1.

Table 1. The basic chemical properties of the soil at the sampling points of two ecotypes of *M. floridulus* and the soil for the pot experiment [27].

Soil Sample Point	pH	Organic C ($g{\cdot}kg^{-1}$)	Available P ($mg{\cdot}kg^{-1}$)	Available N ($mg{\cdot}kg^{-1}$)	Heavy Metal Content ($mg{\cdot}kg^{-1}$)			
					Zn	Pb	Cu	Cd
Dabaoshan Mining Area	6.2	14.7 ± 0.9 b	32.2 ± 2.0 b	30.2 ± 1.9 b	1768.7 ± 91.1 a	1253.3 ± 71.3 a	1701.3 ± 77.5 a	9.1 ± 0.9 a
Boluo County	6.6	13.8 ± 0.9 b	26.6 ± 1.8 b	28.4 ± 2.7 b	135.2 ± 13.1 b	242.6 ± 44.1 b	48.4 ± 9.5 b	1.1 ± 0.2 b
Soil samples tested	6.5	28.4 ± 1.8 a	61.5 ± 9.9 a	55.5 ± 9.1 a	80.2 ± 8.9 b	8.2 ± 7.2 c	1.5 ± 0.7 c	0.30 ± 0.1 b

Note: Data in the table are means ± SD (n = 3); different letters in the same vertical column indicate significant difference according to SSR test ($p < 0.05$), the same below.

2.3. The Experimental Design

The levels of Cu stress treatment were: CK (control), 50 $mg{\cdot}kg^{-1}$, 100 $mg{\cdot}kg^{-1}$, 200 $mg{\cdot}kg^{-1}$, 400 $mg{\cdot}kg^{-1}$, and 800 $mg{\cdot}kg^{-1}$. Cu was added in the form of $CuSO_4{\cdot}5H_2O$. After the soil was thoroughly mixed, deionized water was added to the soil pre-culture. After the soil was stable for 14 days, the *M. floridulus* seedlings were transplanted. Plants of the same weight and height were selected, separated according to the mining ecotypes and non-mining ecotypes, and assigned to each treatment concentration. Each ecotype was planted with 3 pots per treatment and 3 plants per pot. After transplantation, the soil in the basin was kept at 65% of the field water capacity. The physiological indexes were determined after 60 days of Cu stress.

2.4. Test Index and Method

The photosynthetic pigment of leaves was determined by an extraction method [32] using a 752 N ultraviolet spectrophotometer.

Preparation of supernatant extract: take 0.2 g of the blade to be tested (the last but one blade at the top), grind it quickly in an ice bath to homogenate, add 20 mL of precooled 0.05 $mmol{\cdot}L^{-1}$, pH 7.8 phosphoric acid buffer solution in batches, and centrifuge it at 4 °C for 10 min at 7000 $r{\cdot}min^{-1}$. The supernatant extract was used for the determination of protein content, superoxide dismutase (SOD), peroxidase (POD), polyphenol oxidase (PPO) activity and malondialdehyde (MDA) content.

The activity of SOD, POD and PPO was determined by the method introduced by Li Hesheng et al. [32]; the activity of SOD was determined by the nitrogen blue tetrazole (NBT) method; the activity of POD was determined by the guaiacol method; and the activity of PPO was determined by the colorimetric method.

The method of Zhang and Qu [33] was used for the determination of MDA. An amount of 2 mL of enzyme solution was taken and 2 mL of 0.67% TBA (thiobarbituric acid) was added. After mixing, the solution was boiled in a 100 °C water bath for 30 min, cooled down, and centrifuged again. The absorbance values of the supernatant at 450 nm, 532 nm and 600 nm were measured.

The permeability of the cell membrane was measured using a conductance meter [33], with slight changes in the procedure as follows: take 0.2 g of fresh leaves; wash and cut them with high-purity water; soak them in 20 mL of high-purity water; vacuum them with a vacuum extractor for 20 min; take them out and leave them to stand for 20 min; shake the leaves gently during this process, then measure the conductivity S1 with a conductivity meter (DDS-12, Shanghai, China) at a constant temperature of 20~25 °C; place them in a boiling water bath for 15 min after measurement; after cooling to room temperature, measure the conductivity S2; $S1/S2 \times 100\%$ is the relative injury rate.

For the determination of ascorbic acid (AsA) content, the chemical colorimetric method introduced by Chen and Wang [34] was used. For determination of reduced glutathione (GSH) content, GSH extraction involved the same sample extraction method as for MDA. The method of Chen Jianxun et al. [34] was used for determining the GSH content and the mercapto reagent DTNB was used for determination.

Three strains were selected and tested for all indexes; all tests were repeated more than 3 times.

2.5. Data Processing

The statistical analysis of data was performed using a combination of Microsoft Excel 2007 and IBM SPSS Statistics 20.0 software. The significance of differences between means was analyzed using Duncan's multiple comparisons test (SSR test, $p < 0.05$).

3. Results

3.1. Effect of Cu Stress on Chlorophyll Content in Leaves of *M. floridulus*

Table 2 of the experimental results shows that, with increase in Cu treatment concentration, chlorophyll a, chlorophyll b, carotenoids and the total amount of chlorophyll in the leaves of the two ecotypes of *M. floridulus* decreased.

Table 2. Effect of different Cu treatment concentrations on chlorophyll content of two ecotypes.

Cu Treatment Concentration ($mg \cdot kg^{-1}$)		0	50	100	200	400	800
Chl a ($mg \cdot g^{-1}FW$)	Non-mining ecotype	1.30 ± 0.02	1.32 ± 0.01	1.21 ± 0.07	1.19 ± 0.03	1.01 ± 0.14	1.12 ± 0.11
	Mining ecotype	1.27 ± 0.02	1.27 ± 0.01	1.28 ± 0.01	1.21 ± 0.04	1.17 ± 0.12	1.14 ± 0.02
Chl b ($mg \cdot g^{-1}FW$)	Non-mining ecotype	0.77 ± 0.06	0.91 ± 0.07	0.50 ± 0.13	0.43 ± 0.04	0.30 ± 0.08	0.35 ± 0.10
	Mining ecotype	0.52 ± 0.08	0.47 ± 0.01	0.56 ± 0.01	0.48 ± 0.05	0.41 ± 0.12	0.38 ± 0.01
Carotenoid ($mg \cdot g^{-1}FW$)	Non-mining ecotype	0.42 ± 0.01	0.43 ± 0.01	0.37 ± 0.03	0.38 ± 0.02	0.33 ± 0.02	0.37 ± 0.04
	Mining ecotype	0.38 ± 0.01	0.37 ± 0.01	0.39 ± 0.01	0.37 ± 0.01	0.35 ± 0.05	0.34 ± 0.01
Total chlorophyll ($mg \cdot g^{-1}FW$)	Non-mining ecotype	2.07 ± 0.05	2.23 ± 0.07	1.71 ± 0.11	1.62 ± 0.05	1.31 ± 0.12	1.47 ± 0.20
	Mining ecotype	1.80 ± 0.09	1.74 ± 0.02	1.84 ± 0.02	1.70 ± 0.10	1.59 ± 0.21	1.52 ± 0.02
Chl a/b	Non-mining ecotype	1.70 ± 0.16	1.44 ± 0.09	2.49 ± 0.47	2.74 ± 0.18	3.44 ± 0.52	3.25 ± 0.66
	Mining ecotype	2.44 ± 0.32	2.61 ± 0.06	2.25 ± 0.06	2.50 ± 0.18	2.90 ± 0.60	3.00 ± 0.07

Under low concentration Cu stress (50 $mg \cdot kg^{-1}$), the various chlorophyll components and the total amount of chlorophyll of the non-mining ecotype plants increased slightly, but the difference did not reach a significant level ($p > 0.05$), indicating that the low concentration Cu stress slightly stimulated the chlorophyll synthesis of *M. floridulus*. When the concentration of Cu was greater than 100 $mg \cdot kg^{-1}$, the various components of chlorophyll, and the total amount of chlorophyll, began to decline significantly; the difference between the two ecotypes was significant ($p < 0.05$). The content of chlorophyll a, chlorophyll b, carotenoids and total chlorophyll in the 400 mg·kg^{-1} treatment group were 83.5%, 38.5%, 78.1% and 62.2% of the control group, respectively.

The contents of chlorophyll and the total amount of chlorophyll in the leaves of the mining ecotype *M. floridulus* showed a downward trend with increase in Cu concentration, but the degree of decline was not as large as that for the non-mining ecotype *M. floridulus*. Chlorophyll a and chlorophyll b concentrations were statistically different from the control group for the 400 mg·kg^{-1} and 800 mg·kg^{-1} treatment groups ($p < 0.05$). The total amount of chlorophyll was statistically different from the control for the 800 mg·kg^{-1} treatment group ($p < 0.05$), being 84.4% of the total chlorophyll in the control group.

As shown in Figure 1, with increase in Cu treatment concentration, the chlorophyll a/b value in the leaves of the two ecotypes of *M. floridulus* showed a trend of first decreasing and then increasing. The plant growth of the non-mining ecotype was more obvious. However, while the chlorophyll a/b values were higher than the control for most treatment groups, the chlorophyll a/b value for the 50 mg·kg^{-1} treatment group was slightly lower than that of the control group (85.5% of the control group). For the 400 mg·kg^{-1} treatment, the chlorophyll a/b value was 2.04 times that of the control group. The change in the chlorophyll a/b value of ecotype plants in the mining area was gentle, but also showed a trend of first decreasing and then increasing, being 1.25 times that of the control at 800 mg·kg^{-1}. These results suggest that Cu has a more harmful effect on chlorophyll b than chlorophyll a in the leaves of *M. floridulus*.

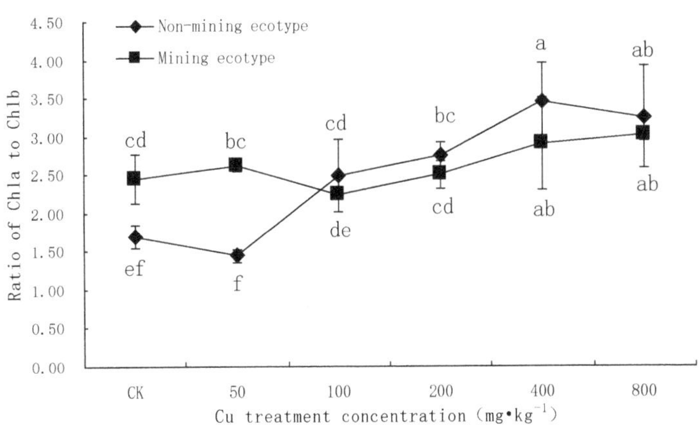

Figure 1. Chlorophyll a/b of two ecotypes of *M. floridulus* under different Cu treatment. Note: Error bars indicate standard deviation; different letters in the same group indicate significant difference at $p < 0.05$ according to Duncan's multiple range tests; the same below.

3.2. The Effect of Cu Stress on the Membrane Protection System of *M. floridulus*

(1) Effects of Cu stress on MDA content in leaves of *M. floridulus*

Malondialdehyde (MDA) is the product of membrane lipid peroxidation in plants, and its content reflects the degree of damage to plant cells caused by membrane lipid peroxidation generated by reactive oxygen species [35]. As can be seen from Figure 2, with increase in Cu treatment concentration, the MDA content in the leaves of the non-mining ecotype *M. floridulus* increased significantly and was significantly increased compared with the control ($p < 0.05$). The MDA levels were 108.5%, 122.1%, 135.1%, 151.3% and 213.2% of the control for each treatment concentration group, respectively. The amount of MDA in the leaves of the mining ecotype *M. floridulus* gradually increased with Cu treatment concentration, but the increase was not significant. The content of MDA in the treatment concentration groups was 123.1%, 135.2%, 152.3%, 159.5% and 189.5% of the control group, respectively. Under each treatment concentration, the MDA content of the non-mining ecotype plants was significantly higher than that of the mining ecotype plants ($p < 0.05$). The gradual increase in MDA content may have been due to the increased Cu treatment concentration, which gradually accumulated free radical content in the plant that could not be removed in time, resulting in severe oxidation of the cell membrane lipids and cell damage. Through regression analysis, for the non-mining ecotype b(MDA) = 0.0084x + 12.61 (x is the concentration of Cu, $R^2 = 0.95$, $p < 0.05$), and for the mining ecotype b(MDA) = 0.0032x + 9.85 (x is the concentration of Cu, $R^2 = 0.83$, $p < 0.05$). The results indicate that Cu had a substantial influence on the content of MDA in the leaves of the non-mining ecotype *M. floridulus*. MDA levels were significantly higher in non-mining ecotype plants than in mining ecotype plants, indicating that non-mining ecotype plants experienced more severe lipid oxidation in their cell membranes.

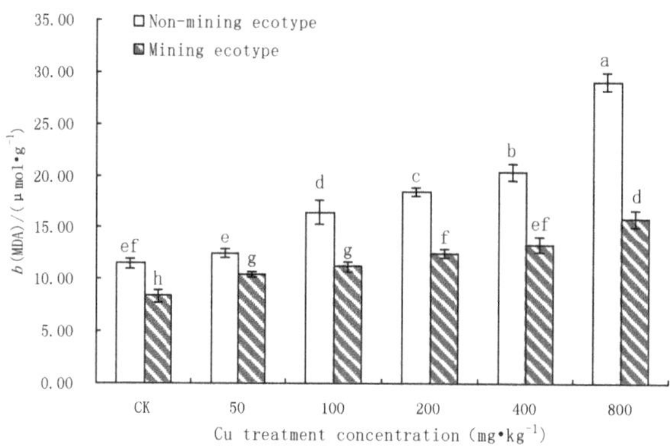

Figure 2. MDA concentrations of two ecotypes of *M. floridulus* under different Cu treatment. Note: Error bars indicate standard deviation; Different letters in the same group indicate significant difference at $p < 0.05$ according to Duncan's multiple range tests; the same below.

(2) Effect of Cu Stress on Cell Membrane Permeability in Leaves of *M. floridulus*

The permeability of the cell membrane is one of the indicators to evaluate the response of plants to pollution; an increase in cell membrane permeability is mainly manifested in an increase in relative electrical conductivity [36]. As can be seen from Figure 3, in the case of CK, there was no significant difference in the membrane permeability of the two ecotypes ($p > 0.05$) and the relative electrical conductivity of the leaf membrane showed an increasing trend with increase in Cu treatment concentration. Under a Cu treatment concentration of 400–800 mg·kg^{-1}, the relative electrical conductivity of plant leaves of the non-mining ecotype was 131.5%, 135.1%, 151.7% and 156.8% of the control, and that of the mining ecotype was 118.9%, 124.4% and 133.1% of the control, respectively. There were significant differences between the two ecotypes and the control ($p < 0.05$). The cell membrane permeability of the non-mining ecotype plants was higher than that of the mining ecotype plants; the difference was significant at the 400–800 mg·kg^{-1} treatment concentrations ($p < 0.05$).

3.3. Effect of Cu Stress on AsA Content of *M. floridulus*

As can be seen in Figure 4, the AsA content of the non-mining ecotype *M. floridulus* gradually increased with Cu concentration, with a similar pattern of change to that of the mining ecotype *M. floridulus*. However, in general, the ascorbic acid content of the mining ecotype *M. floridulus* was significantly higher than that in the non-mining ecotype ($p < 0.05$). At higher Cu concentrations (400 mg·kg^{-1} and 800 mg·kg^{-1}), the AsA content of both ecotypes of *M. floridulus* was significantly different from that of the corresponding control group ($p < 0.05$). The AsA content of the mining ecotype *M. floridulus* was 2.74 times

(400 mg·kg^{-1} Cu treatment) and 1.90 times (800 mg·kg^{-1} Cu treatment) that of the non-mining ecotype *M. floridulus*, respectively.

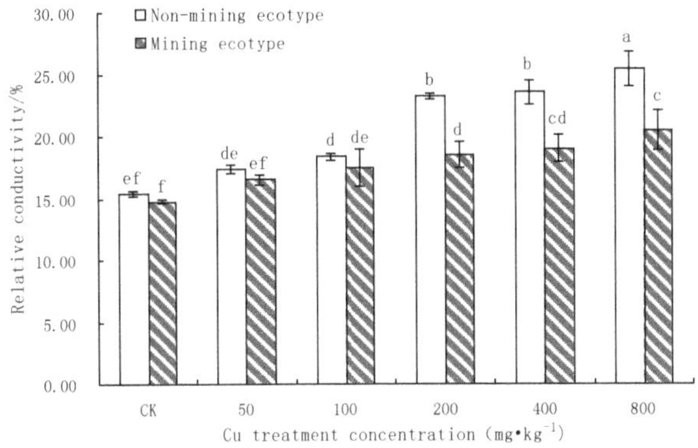

Figure 3. Change in relative conductivity of two ecotypes of *M. floridulus* under different Cu treatment. Note: Error bars indicate standard deviation; Different letters in the same group indicate significant difference at $p < 0.05$ according to Duncan's multiple range tests; the same below.

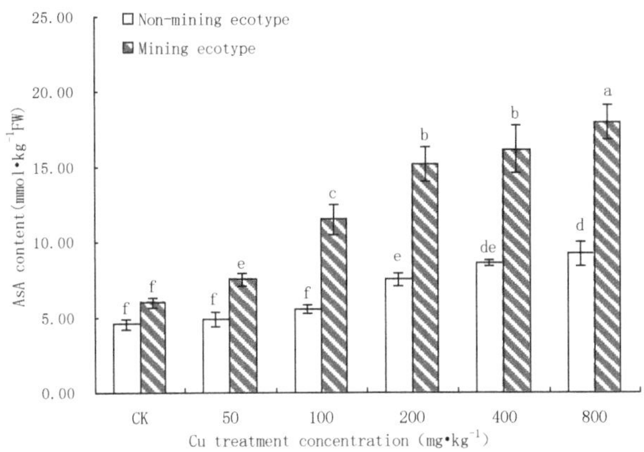

Figure 4. AsA concentrations of two ecotypes of *M. floridulus* under different Cu treatment. Note: Error bars indicate standard deviation; Different letters in the same group indicate significant difference at $p < 0.05$ according to Duncan's multiple range tests; the same below.

3.4. Effect of Cu Stress on the Content of GSH in M. floridulus

As can be seen from Figure 5, when the Cu concentration was 50 mg·kg^{-1}, the GSH content of the two ecotypes of *M. floridulus* increased compared with that of the control group. However, with continuous increase in Cu concentration, the content changes were different: the GSH content of the non-mining ecotype *M. floridulus* decreased rapidly, and the GSH content of the non-mining ecotype *M. floridulus* was significantly different from that of the control group when the Cu concentration was 400 mg·kg^{-1} and 800 mg·kg^{-1} ($p < 0.01$). The GSH content of the ecotype *M. floridulus* also decreased but did not reach a highly significant level compared with the control group ($p > 0.01$). In general, under the higher Cu treatment concentration, the GSH content of ecotype *M. floridulus* in the mining area was significantly higher than that in the non-mining area: The GSH content of the mining ecotype *M. floridulus* was 2.54 times (400 mg·kg^{-1} Cu treatment) and 3.81 times (800 mg·kg^{-1} Cu treatment) that of the non-mining ecotype *M. floridulus*.

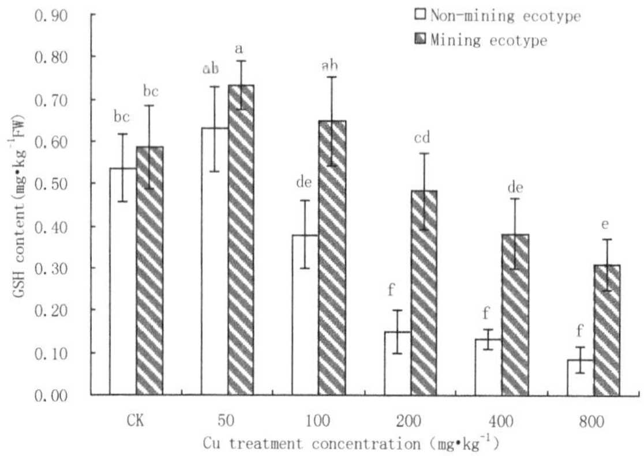

Figure 5. GSH concentrations of two ecotypes of *M. floridulus* under different Cu treatment. Note: Error bars indicate standard deviation; Different letters in the same group indicate significant difference at $p < 0.05$ according to Duncan's multiple range tests; the same below.

3.5. Effect of Cu Stress on SOD Activity of M. floridulus

Many studies have shown that the SOD activity of plant leaves is enhanced under heavy metal stress [37]. Figure 6 shows that Cu had a different effect on the SOD activity in the leaves of the two ecotypes of *M. floridulus*. The SOD activity in the leaves of the non-mining ecotype plants exposed to Cu stress first increased and then decreased and reached the highest level at 100 mg·kg^{-1}, which was 143.5% that of the control. The levels for concentrations of 400 mg·kg^{-1} and 800 mg·kg^{-1} were significantly lower than those of the control group ($p < 0.05$), being 85.2% and 61.5% of the control group, respectively.

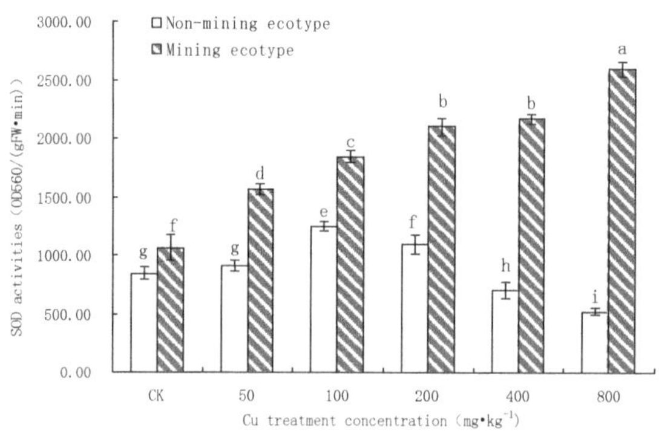

Figure 6. Activity of SOD of two ecotypes of *M. floridulus* under different Cu treatment. Note: Error bars indicate standard deviation; Different letters in the same group indicate significant difference at $p < 0.05$ according to Duncan's multiple range tests; the same below.

The SOD activity in the plant leaves of the mining ecotype was much higher than that of the non-mining ecotype. The SOD activity increased significantly after Cu stress ($p < 0.05$). The SOD activity in the treatment groups was 141.1%, 171.5%, 195.6%, 205.1% and 245.5% that of the control group, respectively.

3.6. Effect of Cu Stress on POD Activity of M. floridulus

Figure 7 shows that Cu has a different effect on the POD activity in the leaves of the two types of *M. floridulus*. The POD activity of non-mining ecotype plants exposed to Cu stress first increased and then decreased and reached its highest level at 200 mg·kg^{-1}, which was 185.2% of that of control. Following this, it decreased significantly, and for the 400 mg·kg^{-1} and 800 mg·kg^{-1} treatment groups was lower than that of the control group (93.5% and 75.5%, respectively). The POD activity in plant leaves of the mining ecotype was much higher than in those of the non-mining ecotype. The POD activity increased significantly after Cu stress ($p < 0.05$). The POD activity in the treatment groups was 155.9%, 245.1%, 315.6%, 351.8% and 430.5% that of the control group, respectively.

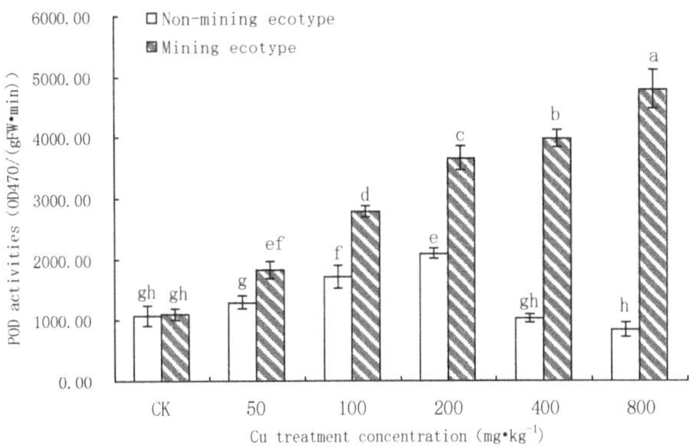

Figure 7. Activity of POD of two ecotypes of *M. floridulus* under different Cu treatment. Note: Error bars indicate standard deviation; Different letters in the same group indicate significant difference at $p < 0.05$ according to Duncan's multiple range tests; the same below.

3.7. Effect of Pb Stress on PPO Activity of *M. floridulus*

Figure 8 shows that Cu had a different effect on the PPO activity in the leaves of the two ecotypes of *M. floridulus*. The PPO activity in the leaves of the non-mining ecotype plants exposed to Cu stress first increased and then decreased, reaching its highest level at 100 mg·kg^{-1}, which was 138.7% of that of the control. Following this, the PPO activity of the 400 and 800 mg·kg^{-1} treatment groups was significantly lower than that of the control group ($p < 0.05$) at 87.1% and 65.2% of the control, respectively. The PPO activity in the plant leaves of the mining ecotype was higher than that in the leaves of the non-mining ecotype. The PPO activity significantly increased after Cu stress ($p < 0.05$). The PPO activity of each treatment concentration group was 151.1%, 190.8%, 207.5%, 226.5% and 268.1% of the control group, respectively.

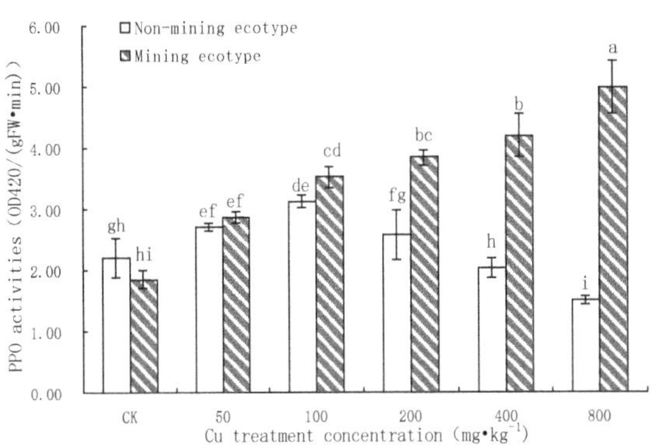

Figure 8. Activity of PPO of two ecotypes of *M. floridulus* under different Cu treatment. Note: Error bars indicate standard deviation; Different letters in the same group indicate significant difference at $p < 0.05$ according to Duncan's multiple range tests; the same below.

3.8. The Correlation between the Physiological Indexes and the Mass Fraction of Cu

As can be seen from Table 3, among the physiological indexes of the non-mining ecotype *M. floridulus*: MDA, RPP and AsA were significantly positively correlated with Cu treatment concentration ($p < 0.05$); GSH, SOD and PPO were negatively correlated with Cu concentration ($p < 0.05$); and there was no significant correlation between POD and Cu concentration. Among the other indexes, SOD, POD and PPO were significantly positively correlated ($p < 0.01$); MDA was negatively correlated with GSH, SOD and PPO ($p < 0.05$); and RPP was negatively correlated with GSH and PPO ($p < 0.05$).

Table 3. Pearson product-moment correlation matrix for MDA, Cu concentrations and *SOD, POD, PPO* of *Miscanthus floridulus*.

Type	Item	Cu	Chl	MDA	RPP	AsA	GSH	SOD	POD	PPO
Non-mining ecotype	Cu	1								
	Chl	−0.671 **	1							
	MDA	0.972 **	−0.717 **	1						
	RPP	0.855 **	−0.757 **	0.902 **	1					
	AsA	0.884 **	−0.810 **	0.891 **	0.936 **	1				
	GSH	−0.763 **	0.842 **	−0.818 **	−0.903 **	−0.883 **	1			
	SOD	−0.713 **	0.310	−0.560 *	−0.437	−0.530 *	0.316	1		
	POD	−0.465	0.086	−0.296	−0.080	−0.203	−0.008	0.852 **	1	
	PPO	−0.747 **	0.354	−0.635 **	−0.548*	−0.649 **	0.478 *	0.853 **	0.650 **	1
Mining ecotype	Cu	1								
	Chl	−0.667 **	1							
	MDA	0.920 **	−0.698 **	1						
	RPP	0.797 **	−0.466	0.909 **	1					
	AsA	0.831 **	−0.622 **	0.932 **	0.852 **	1				
	GSH	−0.794 **	0.424	−0.718 **	−0.738 **	−0.762 **	1			
	SOD	0.868 **	−0.543 *	0.955 **	0.883 **	0.939 **	−0.686 **	1		
	POD	0.875 **	−0.583 *	0.952 **	0.873 **	0.963 **	−0.740 **	0.970 **	1	
	PPO	0.861 **	−0.501 *	0.922 **	0.855 **	0.909 **	−0.702 **	0.976 **	0.959 **	1

Note: Cu: Cu treatment concentration; Chl: total chlorophyll; RPP: cell membrane permeability; MDA: malondialdehyde; AsA: ascorbic acid; GSH: reduced glutathione; POD: peroxidase enzyme; SOD: superoxide dismutase; PPO: polyphenol oxidase; * and ** indicate significance under $p < 0.05$ and $p < 0.01$, respectively.

As can be seen from Table 3, among the physiological indexes of the mining ecotype *M. floridulus*: MDA, RPP, AsA, SOD, POD and PPO were significantly positively correlated with Cu treatment concentration ($p < 0.05$). Among the other indexes, SOD, POD and PPO were significantly positively correlated ($p < 0.01$); MDA was positively correlated with SOD, POD and PPO ($p < 0.05$); and RPP was significantly positively correlated with SOD, POD and PPO ($p < 0.05$).

4. Discussion

4.1. Analysis of the Effect of Cu Stress on Chlorophyll Content of M. floridulus

Cu is a trace mineral element necessary for the normal life activities of plants, and chlorophyll a and chlorophyll b are composed of a metal pigment protein [38]. Therefore, when the concentration of Cu is low, it may participate in and promote the synthesis of plant pigment protein, which is conducive to the various life activities of plants and will not cause obvious damage to them. When the content of Cu is higher than that required by plant life, a significant quantity of oxygen free radicals cannot be quickly removed, which destroys the structure and functioning of mitochondria and chloroplasts [39]. It also causes the oxidation of pigment proteins, inhibits the synthesis of chlorophyll, especially causing obvious damage to chlorophyll b, hindering the photosynthesis of plants and affecting their normal growth and development [40]. Many studies have confirmed that excessive copper can cause a decrease in chlorophyll content in plants [41,42]. Prasad et al. [41] suggested that a reduction in chlorophyll content caused by copper was mainly a result of the decomposition of chlorophyll caused by copper. Some studies have also indicated that copper can replace magnesium in the reaction center of chlorophyll, thus destroying the functioning and structure of chlorophyll [42].

The present study showed that the total amount of chlorophyll a, chlorophyll b, carotenoid, and chlorophyll in the leaves of the two types of *M. floridulus* first increased and then decreased as the Cu stress concentration increased. This may be because low Cu concentration affected the stimulation of chlorophyll synthesis, which would have a greater toxic effect as the Cu concentration increased. The total chlorophyll, chlorophyll a and chlorophyll b of the two ecotypes were negatively correlated with the Cu stress concentration ($p < 0.05$), suggesting that chlorophyll can be used as an indicator to show the degree of Cu stress.

Based on comparison between the two ecotypes, the chlorophyll content in the leaves of the non-mining ecotype decreased more rapidly; Cu stress caused more significant damage to chlorophyll b in the non-mining ecotype. In addition, in terms of appearance, the leaves of the non-mining species became more greenish-yellow than those of the mining species as the Cu concentration increased, presumably because the *M. floridulus* of the mining species has a resistance mechanism that reduces the damage caused by Cu to chlorophyll. This chlorophyll protection mechanism may enable the normal photosynthesis and substance synthesis of plants in polluted conditions.

4.2. Effects of Cd Stress on Physiological Metabolism of M. floridulus

Cell membranes are selective permeable membranes that control and regulate the transport and exchange of substances inside and outside the cell. Their permeability is one of the indicators used to assess the response of plants to pollution. Under adverse conditions, an increase in the permeability of the plant cell plasma membrane is mainly manifested as an obvious increase in relative conductivity [43]. Malondialdehyde (MDA) is a product of membrane lipid peroxidation. The intracellular MDA level indirectly indicates the level of reactive oxygen species and the degree of cellular damage in the plant. When plants are subjected to heavy metal stress, the balance between the production and removal of free radicals in the cell is impaired, leading to an increase in MDA content and the production of a large number of reactive oxygen species [44,45].

The results of this study showed that the membrane permeability of the leaves of the two ecotypes increased with Cu concentration and that the membrane permeability of the two ecotypes had a significant positive correlation with Cu stress concentration. With treatment at 800 mg·kg^{-1}, the number of non-mining ecotype plants was 1.6 times that of the control, while the number of mining ecotype plants was 1.3 times that of the control. This may have been due to the protective mechanism of the mining ecotype *M. floridulus*, which reduced damage to the cell membrane of the plant under Cu stress. The MDA content of the leaves of both ecotypes of *M. floridulus* increased after Cu treatment compared to the control. Under 800 mg·kg^{-1} treatment, the non-mining ecotype plants were 2.2 times as many as the control, while the mining ecotype plants were 1.8 times as many as the control, indicating that the two ecotypes of M. floridulus had different responses to Cu stress. The mining ecotype plants showed significantly stronger resistance and less damage under Cu stress, which was similar to the experimental results of a study by Xie Mingji et al. [46] involving comparison of the MDA content of mining type and non-mining type *Elsholtzia chinensis* under copper stress.

AsA is a very important antioxidant in plants. Mediated by AsA, H_2O_2 accepts NADPH electrons and is reduced to H_2O, thus eliminating H_2O_2 toxicity. AsA can also react directly with superoxide free radicals and hydroxyl free radicals to remove these toxic molecules [47,48]. Studies have shown that AsA can inhibit membrane lipid peroxidation [49]. In this experiment, the content of AsA in the mining area was significantly higher than that in the non-mining area and the content of AsA increased significantly with increase in the treatment concentration; there was an obvious positive correlation between the content of AsA and treatment concentration. This result suggests that AsA can reduce the peroxidation damage caused by copper, which is consistent with the results of other studies [47]. GSH is a specific peptide ubiquitous in plants and is an important free radical scavenger that stabilizes the SH group in proteins and plays an important role in maintaining the structural integrity of membranes and preventing membrane lipid peroxidation. Under higher concentration copper treatment, the GSH content of the mining ecotype *M. floridulus* was significantly higher than that of the non-mining ecotype, indicating that GSH plays a positive role in alleviating the peroxidation damage caused by copper [47,50].

4.3. Effects of Cu Stress on the Antioxidant Enzyme System of *M. floridulus*

Studies have shown that excessive heavy metals produce reactive oxygen species in plants and break the dynamic balance of their scavenging mechanism, resulting in peroxide damage [51]. There are two types of protection mechanisms in plants against reactive oxygen species, namely, enzyme-induced and non-enzyme-induced mechanisms. The enzyme system includes SOD, POD, etc. Non-enzymatic systems mainly consist of minor molecules, such as AsA and reduced GSH [52,53].

Normally, the metabolism of ROS in plants is kept in equilibrium and cells are protected from damage. However, under stress, the production rate of ROS in the plant exceeds the plant's ability to remove ROS and excessive accumulation of ROS leads to peroxidation damage. In order to keep the ROS in plants at a certain equilibrium level, the antioxidant protection enzyme system can be quickly activated in plants under various stresses to reduce the damage of ROS [54]. The SOD can break down O_2^- into H_2O_2 and O_2, reducing the accumulation of O_2^- in the body. POD is an enzyme containing Fe, which can decompose H_2O_2, the product of SOD, into H_2O [55,56]. PPO is a stress-response enzyme system widely present in plants and its activity is closely related to the plant's metabolic strength and environmental stress response [56]. In plant respiration, the end of the respiratory chain oxidase, of which PPO is one, directly transfers electrons released during the oxidation of respiratory substrate intermediate products to O_2. PPO can catalyze the oxidation of phenolic compounds to quinones and decrease in activity of this enzyme affects the respiration of plants [56,57].

As can be seen from the results of this study, the response of antioxidant enzymes to Cu stress is very different for the two ecotypes. The SOD and POD activities of the non-mining ecotypes exposed to Cu first increased and then decreased in each treatment group, while the SOD and POD activities of the mining ecotypes exposed to Cu were significantly increased ($p < 0.05$). In addition, the SOD and POD activities in the leaves of the mining ecotype plants were much higher than those of non-mining ecotype plants and remained at a high level under high concentration Cu (800 $mg \cdot kg^{-1}$) stress. This indicates that the mining ecotype *M. floridulus* is more resistant to Cu stress than the non-mining ecotype.

PPO is barely mentioned in studies of heavy metal stress on plants. In this study, it was found that the response of PPO activity to Cu stress in the leaves of the two ecotypes was very different. The PPO activity in the non-mining ecotype plants first increased and then decreased with increase in Cu treatment concentration, while that in the mining ecotype plants increased significantly ($p < 0.05$). Correlation analysis showed that the activity of PPO in the leaves of non-mining ecotype *M. floridulus* was negatively correlated with the concentration of Cu ($p < 0.05$) and the activity of PPO in the leaves of mining ecotype *M. floridulus* was significantly positively correlated with Cu concentration ($p < 0.05$). The relationship between PPO activity and Cu concentration in the leaves of the two ecotypes of *M. floridulus* was very different and the internal relationships and specific mechanisms underpinning this need further study.

Stress also promotes the production of ROS in plants. Under normal circumstances, the production and removal of ROS in plant cells are in a balanced state, while increase in ROS in plants under stress can, on the one hand, induce an increase in the activities of related protective enzymes, such as SOD, POD and CAT, and, on the other hand, directly destroy biological macromolecules, resulting in the loss of enzyme activity [58]. In general, the antioxidant enzyme activity of mining ecotype plants was higher than that of non-mining ecotype plants. The most significant difference was that the antioxidant enzyme activity of the non-mining ecotype plants was not high under the stress of a high concentration of Cu at 800 $mg \cdot kg^{-1}$, while the enzyme activity of the mining ecotype plants was maintained at a high level under the stress of this concentration of Cu. This behavior of mining ecotype *M. floridulus* implies that the plants possess a tolerance mechanism enabling adaptation to heavy metals [59,60]. This may be because mining ecotype plants have previously been damaged to a certain extent by long-term adverse conditions and that the two ecotypes have become physiologically and ecologically different in different habitats. The results show that the mining ecotype *M. floridulus* is better able to protect itself under the stress of Cu.

5. Conclusions

(1) Chlorophyll a, chlorophyll b and total chlorophyll in the leaves of two ecotypes of *M. floridulus* were negatively correlated with Cu stress concentration ($p < 0.01$), but the extent of decrease for the ecotypes in the mining area was lower than that for ecotypes in the non-mining area. The values of chlorophyll a/b for both ecotypes increased with increasing Cu treatment concentration, indicating that Cu is more harmful to chlorophyll b than to chlorophyll a for *M. floridulus*.

(2) Under Cu stress, the content of antioxidant substances (GSH, AsA) in the mining ecotype was significantly higher than that in the non-mining ecotype. The membrane permeability increased for both ecotypes at high concentrations of copper treatment, and the MDA content of the non-mining ecotype was significantly higher than that of the mining ecotype. The experimental data obtained showed that, under copper stress, the non-mining ecotype *M. floridulus* suffered more severe peroxidation damage than the mining ecotype. The endogenous GSH and AsA of *M. floridulus* play an important role in scavenging free radical accumulation caused by excess copper.

(3) The SOD activity in the leaves of the non-mining ecotype was inhibited by Cu stress, and the POD activity was increased by Cu stress, but the increase for the mining ecotype was larger than that for the non-mining ecotype. At the highest Cu concentrations, both enzyme activities were significantly higher in the mining ecotype plants than in the non-mining ecotype plants. The results suggest that, in the long-term adaptation process, the mining ecotype *M. floridulus* becomes a resistant ecotype, and that the non-enzymatic system plays an important role in raising the level of resistance.

Author Contributions: Project administration and writing—review and editing, J.Q.; conceptualization, X.J. and H.Z.; formal analysis, Z.Y. and M.D.; methodology, S.L. and X.C.; investigation, D.X. All authors have read and agreed to the published version of the manuscript.

Funding: This work was supported by the Guangdong Provincial Key Laboratory of Environmental Health and Land Resource (project number: 2020B121201014); Special Project of Key Areas of Colleges and Universities in Guangdong Province (Science and Technology Promoting Rural Revitalization) "Research and Development of Key Technologies for Resource Utilization of Manure from Large-Scale Livestock and Poultry Breeding in Rural Areas of Western Guangdong" (No.: 2021ZDZX4023), and the Innovation Team Project of Colleges and Universities in Guangdong Province (2021KCXTD055).

Data Availability Statement: The data presented in this study are available on request from the corresponding author.

Acknowledgments: The authors acknowledge the Guangdong Provincial Key Laboratory of Environmental Health and Land Resource; Special Project of Key Areas of Colleges and Universities in Guangdong Province (Science and Technology Promoting Rural Revitalization) "Research and Development of Key Technologies for Resource Utilization of Manure from Large-Scale Livestock and Poultry Breeding in Rural Areas of Western Guangdong" and the Innovation Team Project of Colleges and Universities in Guangdong Province.

Conflicts of Interest: The authors declare no conflict of interest.

References

1. Li, J.; He, J.Z.; Ma, Y.H.; Zhu, Y.G.; Zhang, L. Molecular mechanism of biological copper tolerance and biological joint remediation of copper polluted environment. *Acta Ecol. Sin.* **2007**, *6*, 2615–2626.
2. Li, D.L.; Chen, J.W.; Zhang, H.; Li, J.J. Effects of copper pollution on soil bacterial community structure and heavy metal resistance genes. *J. Environ. Sci.* **2021**, *41*, 1082–1090.
3. Hector, A.; Bagchi, R. Biodiversity and ecosystem multifunctionality. *Nature* **2007**, *448*, 188–190. [CrossRef]
4. Song, J.; Shen, Q.; Wang, L.; Qiu, G.; Shi, J.; Xu, J.; Philip, C.B.; Liu, X. Effects of Cd, Cu, Zn and their combined action on microbial biomass and bacterial community structure. *Environ. Pollut.* **2018**, *243*, 510–518. [CrossRef]
5. Chen, J.; Zhang, H.; Li, J.; Liu, Y.; Shi, W.; Hu, H. The toxic factor of copper should be adjusted during the ecological risk assessment for soil bacterial community. *Ecol. Indic.* **2020**, *111*, 106072. [CrossRef]
6. Zheng, B.J.; Xu, S.C.; Tu, Y.H.; Zou, F.; Liang, D.D.; Wang, H.; Mo, Y.W. Research progress on copper toxicity to crops. *J. Shaoxing Univ. Arts Sci.* **2022**, *42*, 58–64.
7. Lin, M.Z.; Jin, M.F.; Chen, Y.; Zeng, Q.L. Photosynthetic characteristics and mechanism of wedelia chinensis under soil Cu pollution stress. *J. Anhui Agric. Univ.* **2018**, *45*, 657–663.
8. Yang, K.; Huang, C.; Lu, S.; Peng, G.Y.; Huang, C.G. Selection of reference genes in real-time quantitative PCR of root tissue of *Commelina purpurea* under copper stress. *Acta Phytophysiol. Sin.* **2021**, *57*, 195–204.
9. Qin, J.Q.; Zhao, H.R.; Dai, M.; Zhao, P.; Chen, X.; Liu, H.; Lu, B.Z. Speciation Distribution and Influencing Factors of Heavy Metals in Rhizosphere Soil of *Miscanthus Floridulus* in the Tailing Reservoir Area of Dabaoshan Iron Polymetallic Mine in Northern Guangdong. *Processes* **2022**, *10*, 1217. [CrossRef]
10. Wu, B.H.; Luo, S.H.; Luo, H.Y.; Huang, H.Y.; Xu, F.; Feng, S.; Xu, H. Improved phytoremediation of heavy metal contaminated soils by *Miscanthus floridulus* under a varied rhizosphere ecological characteristic. *Sci. Total Environ.* **2021**, *808*, 151995. [CrossRef]
11. Nie, G.; Zhong, M.Y.; Cai, J.B.; Yang, X.Y.; Zhou, J.; Appiah, C.; Tang, M.Y.; Wang, X.; Feng, G.Y.; Huang, L.K.; et al. Transcriptome characterization of candidate genes related to chromium uptake, transport and accumulation in *Miscanthus sinensis*. *Ecotoxicol. Environ. Saf.* **2021**, *221*, 112445. [CrossRef] [PubMed]
12. Li, Q.F.; Du, W.B.; Li, Z.A.; Wang, Z.F.; Peng, S.L. Heavy metals accumulation in mining area's *Miscanthus sinensis* populations and its relationship with soil characters. *Chin. J. Ecol.* **2006**, *25*, 255–258.
13. Liao, B.; Deng, D.M.; Yang, B.; Shu, W.S.; Lin, L.; Lan, C.Y. Study on copper tolerance and accumulation of *Commelina communis*. *J. Environ. Sci.* **2003**, *6*, 797–801.
14. Wen, C.H.; Duan, C.Q.; Chang, X.X. Differentiation in *Datura stramonium* L. populations exposed to heavy-metal pollution at different durations: RAPD analysis. *Acta Ecol. Sin.* **2001**, *21*, 1239–1245.
15. Peng, S.L.; Du, W.B.; Li, Z.A. A review of heavy metal accumulation and tolerance by plants of different ecotype. *J. Jishou Univ.* **2004**, *25*, 19–26.
16. Yu, H.; Zheng, X.; Weng, W.; Yan, X.; Chen, P.; Liu, X.; Peng, T.; Zhong, Q.; Xu, K.; Wang, C.; et al. Synergistic effects of antimony and arsenic contaminations on bacterial, archaeal and fungal communities in the rhizosphere of *Miscanthus sinensis*: Insights for nitrification and carbon mineralization. *J. Hazard. Mater.* **2021**, *411*, 125094. [CrossRef]
17. Barbosa, B.; Boléo, S.; Sidella, S.; Costa, J.; Duarte, M.P.; Mendes, B.; Cosentino, S.L.; Fernando, A.L. Phytoremediation of heavy metal-contaminated soils using the perennial energy crops *Miscanthus* spp. and *Arundo donax* L. *Bioenerg. Res.* **2015**, *8*, 1500–1511. [CrossRef]
18. Zadel, U.; Nesme, J.; Michalke, B.; Vestergaard, G.; Plaza, G.A.; Schroder, P.; Radl, V.; Schloter, M. Changes induced by heavy metals in the plant-associated microbiome of *Miscanthus × Giganteus*. *Sci. Total Environ.* **2020**, *711*, 134433. [CrossRef]
19. Sun, J.; Tie, B.Q.; Qin, P.F. Investigation of contaminated soil and plants by heavy metals in Pb-Zn mining area. *J. Plant Resour. Environ.* **2006**, *15*, 63–67.
20. Zhu, Y.G. Microinterface processes in soil-plant systems and their eco-environmental effects. *J. Environ. Sci.* **2003**, *23*, 205–210.
21. Wang, H.B.; Shu, W.S.; Lan, C. Ecology for heavy metal pollution: Recent advances and future prospects. *Acta Ecol. Sin.* **2005**, *25*, 596–605.
22. Zhou, S.; Deng, R.; Hursthouse, A. Risk Assessment of Potentially Toxic Elements Pollution from Mineral Processing Steps at Xikuangshan Antimony Plant, Hunan, China. *Processes* **2020**, *8*, 29. [CrossRef]
23. Fu, S.; Zhou, Y.; Zhao, Y.; Gao, Q.Z.; Peng, X.Z.; Dang, Z.; Zhang, C.B.; Yang, X.Q.; Yang, Z.J.; Dou, L.; et al. Study on heavy metals in soils contaminated by acid mine drainage from Dabaoshan Mine, Guangdong. *Environ. Sci.* **2007**, *28*, 805–812.
24. Zhao, H.R.; Wang, D.; Zhu, B.T.; Jin, X. Evaluation of heavy metal pollution of soil in the Dabaoshan mining area based on the information diffusion theory. *J. Agro-Environ.Sci.* **2019**, *38*, 79–86.
25. Chen, J.Y.; Liu, G.B.; Cui, J.L.; Xiao, T.F. Migration process and risk assessment of heavy metals in soil plant system of Dabaoshan mining area in Guangdong Province. *Environ. Sci.* **2019**, *40*, 5629–5639.
26. Qin, J.Q.; Xia, B.C.; Hu, M.; Zhao, P.; Zhao, H.R.; Lin, X.F. Analysis of the vegetation succession of tailing wasteland of Dabaoshan Mine, Guangdong Province. *J. Agro-Environ. Sci.* **2009**, *28*, 2085–2091.
27. Qin, J.Q.; Zhao, H.R.; Liu, H.; Dai, M.; Zhao, P.; Chen, X.; Wu, X.G. The Difference of Lead Accumulation and Transport in Different Ecotypes of *Miscanthus floridulus*. *Processes* **2022**, *10*, 2219. [CrossRef]
28. Zhao, H.R.; Xia, B.C.; Qin, J.Q.; Zhang, J. Hydrogeochemical and mineralogical characteristics related to heavy metal attenuation in a stream polluted by acid mine drainage: A case study in Dabaoshan mine, China. *J. Environ. Sci.* **2012**, *24*, 979–989. [CrossRef]

29. Zhao, H.R.; Xia, B.C.; Fan, C.; Zhao, P.; Shen, S. Human health risk from soil heavy metal contamination under different land uses near Dabaoshan mine, southern China. *Sci. Total Environ.* **2012**, *417*, 45–54. [CrossRef]
30. Cui, J.L.; Luo, C.L.; Tang, C.W.; Chang, T.; Li, X. Speciation and leaching of trace metal contaminants from e-waste contaminated soils. *J. Hazard. Mater.* **2017**, *329*, 150–158. [CrossRef]
31. Lu, R. *Methods of Soil and Agricultural Chemistry*; Beijing Science and Technology Press: Beijing, China, 1999; pp. 235–285.
32. Li, H.S.; Sui, Q.; Zhao, S.J.; Zhang, W.H. *Experimental Principle and Technology of Plant Physiology and Biochemistry*; Higher Education Press: Beijing, China, 2000; pp. 134–138.
33. Zhang, Z.L.; Qu, W.J. *A Guide Book of Plant Physiology Experiment*; Higher Education Press: Beijing, China, 2004.
34. Chen, J.X.; Wang, X.F. *Experimental Instruction in Plant Physiology*; South China University of Technology Press: Guangzhou, China, 2002; pp. 124–125.
35. Brun, L.A.; Maillet, J.; Hinsinge, P.; Pepin, M. Evalution of copper availability to plants in copper contaminated vineyard soils. *Environ. Pollut.* **2001**, *111*, 293–302. [CrossRef] [PubMed]
36. Tang, C.F.; Liu, Y.G.; Zeng, G.M. Effects of cadrnium stress on active oxygen generation, lipid peroxidation and antioxidant enzyme actives in radish seedlings. *J. Plant Physiol. Mol. Biol.* **2004**, *30*, 469–474.
37. Zou, C.F.; Wu, G.R.; Shi, G.X. The role of antioxidant systems in Cu^{2+} stress resistance in *Alternanthera philoxeroides*. *Acta Bot. Sin.* **2001**, *43*, 389–494.
38. Nagalakshmi, N.; Prasad, M.N.V. Copper-induced oxidative stress in *Scenedesmus bijugatus*: Protective role of free radical scavengers. *Bull. Environ. Contam. Toxicol.* **1998**, *61*, 623–628. [CrossRef]
39. Ouzounidou, G. Copper-induced changes on growth metal conent and phyotosynthetic function of *Alyssummon tanum* plants. *Environ. Exp. Bot.* **1994**, *34*, 165–172. [CrossRef]
40. Huang, Y.S.; Qiu, G.H. Physiological differences of *Festuca rubra* copper tolerant merlin and sensitive S59 to copper treatment at their early developmental stages. *Chin. J. Appl. Environ. Biol.* **1998**, *4*, 126–131. (In Chinese)
41. Prasad, M.N.V.; Malec, P. Physiological Responses of *Lemna trisulca* L. (duckweed) to Cadmium and Copper Bioaccumulation. *Plant Sci.* **2001**, *161*, 881–889. [CrossRef]
42. Kupper, H.; Kupper, F.; Spiller, M. Environmental Relevance of Heavy Metal Substituted Chlorophylls Using the Example of Water Plants. *J. Exp. Bot.* **1996**, *47*, 259–266. [CrossRef]
43. Ren, Q.Q.; Huang, Z.R.; Huang, W.L.; Huang, W.T.; Chen, H.H.; Yang, L.T.; Ye, X.; Chen, L.S. Physiological and molecular adaptations of Citrus grandis roots to long-term copper excess revealed by physiology, metabolome and transcriptome. *Environ. Exp. Bot.* **2022**, *203*, 105049. [CrossRef]
44. Huang, H.; Li, S.; Guo, J.L. Effects of cadmium stress on antioxidant system and photosynthesis of maize seedlings. *J. Agro-Environ. Sci.* **2010**, *29*, 211–215.
45. Sujkowska-Rybkowska, M.; Muszyñska, E.; Labudda, M. Structural Adaptation and Physiological Mechanisms in the Leaves of *Anthyllis vulneraria* L. from Metallicolous and Non-Metallicolous Populations. *Plants* **2020**, *9*, 662. [PubMed]
46. Xie, M.J.; Ke, W.S.; Wang, W.X. MDA accumulation and antioxidation capacity of two *Elsholtzia splendens* populations under copper stress. *Chin. J. Ecol.* **2005**, *24*, 935–938.
47. Li, X.Y.; Wang, X.; Lu, L.-f; Yin, B.; Zhang, M.; Cui, X.-m. Effects of exogenous NO on ascorbic acid glutathione cycle in tomato seedling roots under copper stress. *Chin. J. Appl. Ecol.* **2013**, *24*, 1023–1030.
48. Du, Y.L.; Zhao, Q.; Chen, L.W.; Yao, X.D.; Zhang, W.; Zhang, B.; Xie, F.T. Effect of drought stress on sugar metabolism in leaves and roots of soybean seedlings. *Plant Physiol. Biochem.* **2020**, *146*, 1–12. [CrossRef]
49. Han, Y.Y.; Zhou, S.; Chen, Y.H.; Kong, X.; Xu, Y.; Wang, W. The involvement of expansins in responses to phosphorus availability in wheat, and its potentials in improving phosphorus efficiency of plants. *Plant Physiol. Biochem.* **2014**, *78*, 53–62. [CrossRef]
50. Li, J.Y.; Zheng, B.H.; He, Y.Z.; Zhou, Y.Y.; Chen, X.; Ruan, S.; Yang, Y.; Dai, C.H.; Tang, L. Antimony contamination, consequences and removal techniques: A review. *Ecotoxicol. Environ. Saf.* **2018**, *156*, 125–134. [CrossRef]
51. Smirnoff, N. The role of active oxygen in the response of plants to water deficit and desiccation. *New Phytol.* **1993**, *125*, 27–58. [CrossRef]
52. Ambrosini, V.G.; Rosa, D.J.; de Melo, G.W.B.; Zalamena, J.; Cella, C.; Simão, D.G.; da Silva, L.S.; dos Santos, H.P.; Toselli, M.; Tiecher, T.L.; et al. High copper content in vineyard soils promotes modifications in photosynthetic parameters and morphological changes in the root system of 'Red Niagara' plantlets. *Plant Physiol. Biochem.* **2018**, *128*, 89–98. [CrossRef]
53. Pietrini, F.; Di Baccio, D.; Iori, V.; Veliksar, S.; Lemanova, N.; Juškaitė, L.; Maruška, A.; Zacchini, M. Investigation on metal tolerance and phytoremoval activity in the poplar hybrid clone "Monviso" under Cu-spiked water: Potential use for wastewater treatment. *Sci. Total Environ.* **2017**, *592*, 412–418. [CrossRef]
54. Oustriere, N.; Marchand, L.; Lottier, N.; Motelica, M.; Mench, M. Long-term Cu stabilization and biomass yields of Giant reed and poplar after adding a biochar, alone or with iron grit, into a contaminated soil from a wood preservation site. *Sci. Total Environ.* **2017**, *579*, 620–627. [CrossRef]
55. Li, C.; Xiao, B.; Wang, Q.H.; Yao, S.H.; Wu, J.Y. Phytoremediation of Zn- and Cr-contaminated soil using two promising energy grasses. *Water Air Soil Pollut.* **2014**, *225*, 2027. [CrossRef]
56. Feigl, G.; Kumar, D.; Lehotai, N.; Tugyi, N.; Molnár, Á.; Ördög, A.; Szepesi, Á.; Gémes, K.; Laskay, G.; Erdei, L.; et al. Physiological and morphological responses of the root system of Indian mustard (*Brassica juncea* L. Czern.) and rapeseed (*Brassica napus* L.) to copper stress. *Ecotoxicol. Environ. Saf.* **2013**, *94*, 179–189. [CrossRef]

57. Marzilli, M.; Di Santo, P.; Palumbo, G.; Maiuro, L.; Paura, B.; Tognetti, R.; Cocozza, C. Cd and Cu accumulation, translocation and tolerance in *Populus alba* clone (Villafranca) in autotrophic in vitro screening. *Environ. Sci. Pollut. Res.* **2018**, *25*, 10058–10068. [CrossRef]
58. He, J.; Li, H.; Luo, J.; Ma, C.; Li, S.; Qu, L.; Gai, Y.; Jiang, X.; Janz, D.; Polle, A. A transcriptomic network underlies microstructural and physiological responses to cadmium in *Populus canescens*. *Plant Physiol.* **2013**, *162*, 424–439. [CrossRef]
59. Zhu, H.; Teng, Y.; Wang, X.; Zhao, L.; Ren, W.; Luo, Y.; Christie, P. Changes in clover rhizosphere microbial community and diazotrophs in mercury-contaminated soils. *Sci. Total Environ.* **2021**, *767*, 145473. [CrossRef]
60. Rehman, M.; Maqbool, Z.; Peng, D.; Liu, L. Morpho-physiological traits, antioxidant capacity and phytoextraction of copper by ramie (*Boehmeria nivea* L.) grown as fodder in copper-contaminated soil. *Environ. Sci. Pollut. Res.* **2019**, *26*, 5851–5861. [CrossRef]

processes

Article

Effects of Lead Pollution on Photosynthetic Characteristics and Chlorophyll Fluorescence Parameters of Different Populations of *Miscanthus floridulus*

Jianqiao Qin [1], Xueding Jiang [2,*], Jianhua Qin [3], Huarong Zhao [4,*], Min Dai [1], Hao Liu [1] and Xi Chen [5]

1 Guangdong Provincial Key Laboratory of Environmental Health and Land Resource, School of Environmental and Chemical Engineering, Zhaoqing University, Zhaoqing 526061, China; qinjianqiaosci@126.com (J.Q.); daimin1007@163.com (M.D.); liuhao13189340196@126.com (H.L.)
2 School of Environmental and Chemical Engineering, Foshan University, Foshan 528000, China
3 School of Foreign Studies, South China Agricultural University, Guangzhou 510640, China; jackyqin04@scau.edu.cn
4 College of Environmental Science and Engineering, Guilin University of Technology, Guilin 541004, China
5 School of Environmental Science and Engineering, Sun Yat-sen University, Guangzhou 510275, China; chenxi90511@163.com
* Correspondence: jiangxueding@fosu.edu.cn (X.J.); zhaohuar@mail3.sysu.edu.cn (H.Z.)

Abstract: This study was conducted in order to study the effect of different concentrations of lead pollution on the photosynthetic characteristics and growth of *Miscanthus floridulus*, and to reveal its photosynthetic adaptability to lead stress. The differences of gas exchange parameters, chlorophyll fluorescence characteristics and photosynthetic pigment of two *Miscanthus floridulus* populations, one population from Boluo an uncontaminated site, and another population from Dabaoshan, a mine site, were compared and studied through nutrient solution culture experiments treated with heavy metal lead (Pb) in green house. The results showed that (1) under Pb stress, the net photosynthetic rate (Pn), transpiration rate (Tr), stomatal conductance (Gs), intercellular carbon dioxide concentration (Ci), and chlorophyll content (Chl) of the leaves of the two populations decreased in different amplitude. Under moderate and severe Pb stress (80 $mg \cdot L^{-1}$, 120 $mg \cdot L^{-1}$, 240 $mg \cdot L^{-1}$), the plant biomass of non-mining population and mining population plants were 54.5%, 39.7%, 29.4% and 70.4%, 54.7%, 50.9% of the control, respectively. (2) Stomatal restriction was the main factor for the Pn's decrease in the leaves of the non-mine population under light Pb stress, while the non-stomatal restriction was the main factor for Pn's decrease in the leaves of the non-mine population under middle and high Pb stresses. (3) Under Pb stress, the maximum photochemical efficiency (Fv/Fm) and potential activity (Fv/Fo) of PS II reaction centers in the two populations of *M. floridulus* decreased. However, Fv/Fm and Fv/Fo showed a smaller decrease, but the capability to utilize light and the potential to activate PSII of the mine population remained higher than that of the non-mine population. The changes of photochemical quenching coefficient (qP) and non-photochemical quenching coefficient (NPQ) of PSII showed that qP value decreased and NPQ value increased in the two populations under Pb stress. On the whole, the resistance mining area population had a low qP reduction and a large increase in NPQ. Electronic transfer rate (ETR) and PSII actual fluorescence efficiency (ΦPSII) of the mine population changed slightly under Pb stress. These results indicated that when under Pb stress, the electron transport activity and photosynthetic apparatus were damaged less in the mine population with high resistance than in the non-mine population with low resistance. Conclusion: the mining population of *M. floridulus* has strong tolerance to Pb, which is suitable for the pioneer species of gramineae in vegetation restoration construction in metal mining area.

Keywords: lead pollution; *Miscanthus floridulus*; gas exchange parameter; chlorophyll fluorescence

1. Introduction

Pb is one of the most serious environmental heavy metal pollutants [1]. Excessive amounts of heavy metals in soil on plant is a kind of coercion or adversity condition; if the stress on the plant itself is limited, it will damage the plant [2]. The toxicology of lead stress is the nucleus, mitochondria, and chloroplast ultrastructure destruction of plants, which reduces chlorophyll and ascorbic acid content, reduces dehydrogenase activity of nitrate reductase, hinders the plant respiration metabolism, photosynthesis, hydrogen reduction, cell division normal physiological function and, ultimately, affects the quality of biomass plants [3,4]. According to statistics, at present, the cultivated land area polluted by heavy metals in China is almost 2.0×10^7 hm^2, accounting for about 1/5 of the cultivated land area. The direct economic loss of grain caused by heavy metal pollution in China is worth tens of billions of yuan every year [5]. These harms to the ecosystem attracted the attention of researchers [6]. Phytoremediation can remove heavy metals using plants and offer the benefits of low cost, as well as being an environmentally sustainable technique [7]. To date, about 700 species of plants were reported to be hyperaccumulators of different contaminants [8]. Metal hyperaccumulators are found in a large number of plant families, but most are found in the Brassicaceae family [9]. Moreover, hyperaccumulators are associated with slow plant growth and low biomass yields, and so, there is an urgent need for identification of other plant species having fast growth and greater biomass production [10]. Phytoremediation would be ideal for the recovery of various types of contaminated soil, since it is effective and environmentally friendly [11].

The *Miscanthus floridulus* is a C4 perennial grass that originates from East Asia, which is widely used as forage and ornamental grass, and it is cultivated as bioenergy crop for lignocellulosic ethanol production, owing to high biomass productivity potential, high efficient photosynthesis, water use efficiency, and great resistance to drought, salt, and heavy metal stress [12,13]. *M. floridulus* is commonly planted in vast marginal land, with limited water supply and heavy metal pollution [14]. It is an ideal plant for phytoremediation due to its fast growth, large biomass, and well-developed roots system, which can accumulate heavy metals in the roots and reduce the mobility and availability [15,16]. Because of its long-term survival in the polluted environment, *M. floridulus* from the metal mining area may have undergone resistance evolution, forming a resistant ecotype [17]. Facultative metal-type plants growing in heavy metal contaminated areas are generally genetically different from plants of the same species in non-contaminated areas [18]. Metalliferous plants are of particular interest for the study of plant anti-pollution mechanisms and phytoremediation of heavy metal contaminated environments. Therefore, identifying their growth limiting factors is of great significance for the practical application of phytoremediation technology [19].

The Dabaoshan mine, located at the junction of Qujiang County and Wengyuan County, Shaoguan City, Guangdong Province, is a large iron polymetallic sulfide-associated deposit [20]. Zhao et al. [21] investigated the soil of Dabaoshan and found that more than 57.98% and 55.09% of the soils in the study area showed Zn and Pb pollution. According to a study of plants in the Dabaoshan mining area by Chen et al. [22], the rhizosphere of *M. floridulus* can activate Pb and other heavy metals significantly and may be considered a pioneer plant for local vegetation restoration. Previous studies also showed that *M. floridulus* can grow normally in the heavily polluted tailings ponds in the Dabaoshan mining area and that its roots can absorb and fix heavy metal Pb [23,24]. Moreover, its biomass is large, and so, it represents good plant repair material [25,26].

Due to the long-term survival in the polluted environment, the resistance evolution in the metal mining area may have occurred in accordance with the polluted environment [16], forming the resistance ecotype. Studying the physiological and ecological differences between resistant ecotypes and sensitive ecotypes is an important way to reveal the mechanism of plant resistance and make phytoremediation technology of heavy metal pollution widely applied [27,28]. At present, relevant studies focus on heavy metal resistance and its accumulation performance [29], while there are few studies on leaf photosynthesis and source-pool allocation of photosynthetic products, which are closely related to plant growth [30].

In this paper, the photosynthetic characteristics and chlorophyll fluorescence parameters of two populations of *M. floridulus* from Dabaoshan (mining area) in Shaoguan City, Guangdong Province, and Boluo County (non-mining area) in Huizhou City, Guangdong Province, were compared through the nutrient solution culture experiment, so as to explore the differences in the physiological and ecological responses of *M. floridulus* populations to Pb pollution in polluted and non-polluted areas. Our aim was to further explore the mechanism of ecological adaptation of Pb resistance in plants. At the same time, the vegetation restoration of tailings and abandoned land is not only faced with heavy metal pollution, but also faced with environmental pressure caused by poor soil water retention and drought. The physiological and ecological characteristics of net photosynthetic rate, transpiration rate, and water use efficiency of *M. floridulus* were studied in order to provide theoretical basis for photosynthetic physiological and ecological aspects of vegetation restoration in abandoned heavy metal mining area.

2. Materials and Methods

2.1. Test Materials

The seeds of the mining population were collected from Dabaoshan mining area in Shaoguan, Guangdong (latitude and longitude: 24°33′36.6″ N, 113°43′14.0″ E), and the seeds of the non-mining population were collected from Boluo County, Huizhou, Guangdong (latitude and longitude: 24°33′36.6″ N) (23°08′57.8″ N, 114°21′10.0″ E). All belonged to the same subtropical monsoon climate region. The soil properties of the seed sources are shown in Table 1.

Table 1. The basic chemical properties of the local soils supporting the two natural populations of *M. floridulus*.

Soil Sample Point	pH	Organic C ($g{\cdot}kg^{-1}$)	Available P ($mg{\cdot}kg^{-1}$)	Available N ($mg{\cdot}kg^{-1}$)	Heavy Metal Contents ($mg{\cdot}kg^{-1}$)			
					Zn	Pb	Cu	Cd
Dabaoshan Mining Area	6.2	14.7 ± 0.9 b	32.2 ± 2.0 b	30.2 ± 1.9 b	1768.7 ± 91.1 a	1253.3 ± 71.3 a	1701.3 ± 77.5 a	9.1 ± 0.9 a
Boluo County	6.6	13.8 ± 0.9 b	26.6 ± 1.8 b	28.4 ± 2.7 b	135.2 ± 13.1 b	242.6 ± 44.1 b	48.4 ± 9.5 b	1.1 ± 0.2 b

Note: Data in the table are means ± SD (n = 3), different letters in the same vertical column indicate significant difference according to SSR test ($p < 0.05$), the same below.

2.2. Experimental Design

Seeds were germinated in nutrient-rich soil, and after 90 days of growth, plants with favorable and consistent growth were selected, washed with tap water, and moved to Hogland nutrient solution for culture. The nutrient solution formula was as follows ($mmol{\cdot}L^{-1}$): $Ca(NO_3)_2{\cdot}4H_2O$ 0.625 + KH_2PO_4 0.025 + $MgSO_4$ 0.25 + KNO_3 0.625 + ($\mu mol{\cdot}L^{-1}$) H_3BO_3 2.50 + $MnCl_2{\cdot}4H_2O$ 0.5 + $ZnSO_4{\cdot}7H_2O$ 0.5 + $CuSO_4{\cdot}5H_2O$ 0.05 + $(NH_4)\ 6Mo_7O_{24}$ 0.005 + Fe-EDTA 0.125 [31]. After seven days of pre-culture in a greenhouse, the treatment was performed. Single Pb setting involved 6 treatments ($mg{\cdot}L^{-1}$): CK(No Pb added), 20, 40, 80, 120, and 240; Pb was added to the nutrient solution with $Pb(NO_3)_2$. Reduce the concentration of KH_2PO_4 in all nutrient solutions to 0.005 $mmol{\cdot}L^{-1}$ to avoid precipitation. There were three replicas in each treatment and three saplings in each replica. The nutrient solution was changed every 4 days, and the pH was adjusted to about 5.8 with 0.1 $mol{\cdot}L^{-1}$ HCl or NaOH. The physiological index was determined after 40 days of culture.

2.3. Determination of Gas Exchange Parameters

Net photosynthetic rate (Pn), transpiration rate (Tr), stomatal conductance (Gs), and intercellular CO_2 concentration (Ci) were measured by LI-6800 portable photosynthesis measuring instrument (Li-6800, LI-COR Bioscience, Lincoln, NE, USA). The measurement conditions were: CO_2 concentration at 380 $\mu mol{\cdot}mol^{-1}$, temperature at 27 °C, relative humidity at 60%, and photosynthetically active radiation (PAR) at 600 $\mu mol\ m^{-2}{\cdot}s^{-1}$. Select healthy and fully unfolded leaves from the upper part of each treated plant for measurement. For each treatment, measure 3 plants and take the average value. Stomatal limit value (Ls) and instantaneous leaf water use efficiency (WUE) were calculated by a formula, namely Ls = 1 − Ci/Ca [32] (Ca is atmospheric CO_2 concentration), WUE = Pn/Tr [33]

2.4. Determination of Chlorophyll Fluorescence Parameters

The kinetic parameters of chlorophyll fluorescence induction were determined by a portable pulse-modulated chlorophyll fluorescence analyzer (PAM-2000, Walz, Effeltrich, Germany). According to Ralph's [34] method, the leaves were dark adapted for 30 min, and initial fluorescence (F_0), maximum fluorescence (F_m), steady-state fluorescence (F_s), maximum fluorescence (F_m'), and minimum fluorescence (F_0') were measured. The photochemical fluorescence quenching coefficient (qP) and non-photochemical quenching coefficient (NPQ) were measured. Based on the above data: variable fluorescence Fv (Fv = $F_m - F_0$), maximum photochemical efficiency of PSII (Fv/Fm), potential activity of PSII (Fv/F_0), energy capture efficiency of PSII (Fv′/Fm′), PSII effective photochemical quantum yield (ϕPSII = (Fm′ − Fs)/Fm′), and the electron transfer rate ETR (ETR = ϕPSII × PAR × 0.5 × 0.84) [35,36].

2.5. Determination of Plant Morphology

Select plants with consistent growth, and measure their height with a scale (0.1 cm); the electronic balance (0.0001 g) was used to measure the biomass of the dried underground and overground parts, and calculate the root shoot ratio; in addition, select the third leaf from top to bottom on the top of the plant, measure the leaf area with the leaf area meter LI-3000A, and then measure the dry weight of the leaf with the electronic balance (0.0001 g). Calculate the specific leaf area: specific leaf area (SLA) = leaf area/dry weight of the leaf [36].

2.6. Determination of Photosynthetic Pigment Content

Using a 752 N type ultraviolet spectrophotometer, take leaves at the same position and maturity as measured for photosynthesis, and use the extraction method to determine photosynthetic pigments (chlorophyll a, chlorophyll b, and carotenoids) [36]. Measurements were performed on 3 plants for each treatment and repeated 3 times.

2.7. Data Processing

The statistical analysis of data was performed using a combination of Microsoft Excel 2007 and IBM SPSS Statistics 22.0 software. Two-way statistical analysis (two-way ANOVA) was used to examine the significant difference between plant and lead treatment. The significance of differences between means was analyzed using Duncan's multiple comparisons test (SSR test, $p < 0.05$).

3. Results

3.1. Growth Morphological Changes of *M. floridulus* under Pb Stress

After 15 days of Pb treatment, both populations grew strongly, with fresh roots and leaves. After 30 days, the population in the mining area still grew well, and new leaves continued to grow, while the low concentration treatment group of the non-mining area population also grew well, but the plants in the 120 mg·L^{-1} and 240 mg·L^{-1} treatment groups began to have the symptoms of leaf chlorosis and wilting.

The changes in plant height and biomass of *M. floridulus* are shown in Table 2. As the degree of Pb stress increased, both populations showed a downward trend. Under 80 mg·L^{-1}, 120 mg·L^{-1}, and 240 mg·L^{-1} Pb stress, the plant height of non-mining population and mining population plants were 48.7%, 39.4%, 31.6% and 71.1%, 56.6%, 51.1% of the control, respectively. Evidently, the plant height of mining population plants decreased less under higher Pb treatment, while that of non-mining population plants decreased significantly. Under moderate and severe Pb stress (80 mg·L^{-1}, 120 mg·L^{-1}, 240 mg·L^{-1}), the plant biomass of non-mining population and mining population plants were 54.5%, 39.7%, 29.4% and 70.4%, 54.7%, 50.9% of the control, respectively. Evidently, the biomass of mining population plants decreased less under higher Pb treatment, while that of non-mining population plants decreased significantly. In particular, the plant biomass of mining population plants was 1.72 times higher than that of non-mining population plants treated with 240 mg·L^{-1}.

Table 2. Growth of two *M. floridulus* populations under different Pb treatment.

Population	Pb Treatment (mg·L^{-1})	Plant Height (cm)	Plant Biomass (g)	Specific Leaf Area (cm^2·mg^{-1})	Root-Shoot Ratio
Non-mining population	0	41.1 ± 3.15 ab	4.25 ± 0.20 a	193.3 ± 3.1 a	0.40 ± 0.07 ef
	20	32.1 ± 1.00 c	3.28 ± 0.02 bc	181.8 ± 8.1 ab	0.41 ± 0.04 def
	40	26.0 ± 1.45 de	2.62 ± 0.34 de	172.3 ± 9.1 bc	0.45 ± 0.03 cde
	80	20.1 ± 2.15 f	2.32 ± 0.04 e	152.5 ± 9.5 d	0.49 ± 0.08 bgd
	120	16.2 ± 1.76 g	1.69 ± 0.42 fg	143.4 ± 20.2 d	0.55 ± 0.03 ab
	240	13.0 ± 1.95 g	1.25 ± 0.19 g	113.4 ± 10.6 e	0.61 ± 0.03 a
Mining population	0	40.8 ± 1.85 ab	4.22 ± 0.31 a	191.8 ± 8.2 a	0.39 ± 0.04 ef
	20	42.0 ± 1.95 a	4.32 ± 0.22 a	196.7 ± 5.0 a	0.37 ± 0.02 f
	40	38.1 ± 1.45 b	3.65 ± 0.59 b	183.7 ± 9.1 ab	0.42 ± 0.01 def
	80	29.0 ± 1.95 cd	2.97 ± 0.27 cd	172.1 ± 9.9 bc	0.45 ± 0.04 def
	120	23.1 ± 1.00 ef	2.31 ± 0.20 e	162.3 ± 9.6 cd	0.48 ± 0.08 bcd
	240	21.0 ± 2.55 f	2.15 ± 0.07 ef	145.5 ± 14.2 d	0.51 ± 0.02 bc

Note: Data in the table are means ± SD (n = 3); different letters in the same vertical column indicate significant difference according to SSR (New Complex Range Method) test (p < 0.05), the same below.

With the increase in Pb treatment concentration, the specific leaf area of different populations of *M. floridulus* showed an overall trend of decline (see Table 2). Under 80 mg·L^{-1}, 120 mg·L^{-1} and 240 mg·L^{-1} Pb stress, the specific leaf area of non-mining population and mining population plants were 78.7%, 74.4%, 58.5% and 89.6%, 84.5%, 75.9% of the control, respectively. Evidently, the specific leaf area of mining populations decreased relatively slight under higher Pb treatment, while that of non-mining populations decreased significantly (p < 0.05). As can be seen from Table 2, the root/shoot ratio of *M. floridulus* of different populations showed an overall upward trend with the increase in Pb stress concentration, but there were significant differences in the root/shoot ratio under high concentrations (120 mg·L^{-1} and 240 mg·L^{-1}) stress: the root/shoot ratios of mining and non-mining populations were 123.1%, 131.2% and 137.5%, 152.5%, respectively, which were significantly different from the control (p < 0.05).

3.2. The Changes of Photosynthesis Indexes

3.2.1. Effects of Pb on Net Photosynthetic Rate (Pn) and Stomatal Conductance (Gs) in Leaves of *M. floridulus*

Photosynthetic rate is the most important indicator to reflect the strength of photosynthesis. A high photosynthetic rate indicates a high level of plant photosynthesis, and vice versa [37]. As can be seen from Figure 1A, there were obvious differences in the net photosynthetic rate of the leaves of mining population and non-mining population: the populations from the mining area (Dabaoshan) had a minor effect on the net photosynthetic rate of leaves under the stress of different concentrations of Pb, while the populations from non-mining area (Boluo) significantly decreased with the increase in treatment concentration. Specifically, with the increase in Pb stress concentration, the leaf Pn value of both populations decreased, and under 40 mg·L^{-1}, 80 mg·L^{-1}, and 160 mg·L^{-1} Pb stress, the Pn values of the mining population and non-mining population were 80.8%, 73.3%, 62.1% and 51.9%, 43.2%, 27.5% of the control population, respectively. Evidently, the decrease in the net photosynthetic rate of the mining population was relatively small under the higher Pb treatment, while the decrease in the non-mining population was significant, especially under the 240 mg·L^{-1} treatment. The net photosynthetic rate of the mining population was 2.17 times that of the non-mining population. Through regression analysis, for non-mining population Pn = −0.035x + 7.56 (x is the concentration of Pb, R^2 = 0.68, p < 0.05), mining population Pn = −0.021x + 9.15 (x is the concentration of Pb, $R^2 = 0.82$, $p < 0.05$), The results indicated that heavy metal Pb had a great effect on leaf Pn of the *M. floridulus* population in the non-mining area.

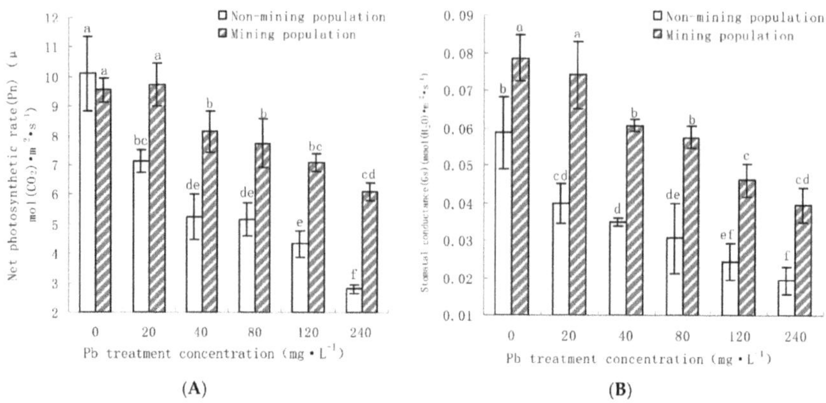

Figure 1. Net photosynthetic rate (Pn) and Stomatal conductance (Gs) of two *Miscanthus floridulus* populations under different Pb treatment((**A**) is Pn, (**B**) is Gs). Note: Error bars indicate standard deviation; Different letters in the same group indicate significant difference at $p < 0.05$ according to Duncan's multiple range tests; the same below.

The effect of Pb on the stomatal conductance of the leaves of the two populations of *M. floridulus* is shown in Figure 2A. Both populations showed a decreasing trend with the increase in Pb concentration, and the non-mining populations under moderate and severe (120 mg·L^{-1} and 240 mg·L^{-1}) Pb stress were 40.7% and 31.8% of the control populations, respectively. The mining population was 59.3% and 50.5% of the control, respectively, indicating that the decline of the non-mining population was greater than that of the mining population.

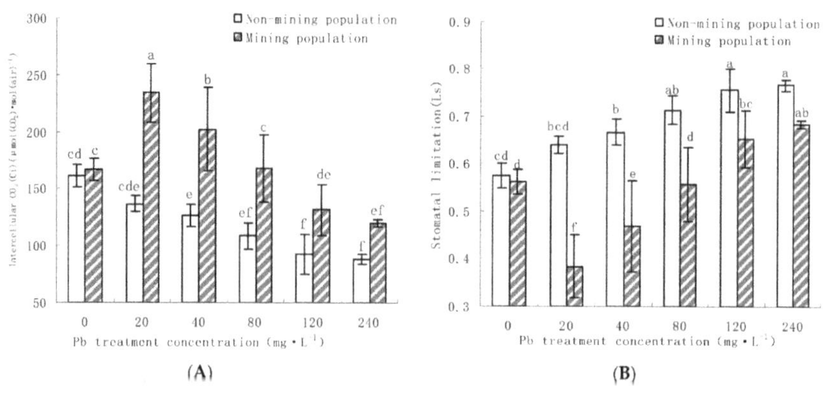

Figure 2. Intercellular CO_2 concentration (Ci) and stomatal limitation (Ls) of two *Miscanthus floridulus* populations under different Pb treatment ((**A**) is Ci, (**B**) is Ls). Note: Error bars indicate standard deviation; Different letters in the same group indicate significant difference at $p < 0.05$ according to Duncan's multiple range tests; the same below.

3.2.2. Effects of Pb on Intercellular Carbon Dioxide Concentration (Ci) and Stomatal Limit (Ls) in Leaves of *M. floridulus*

As can be seen from Figure 2B, the intercellular carbon dioxide concentration (Ci) of mining population and non-mining population showed different trends under the treatment of different concentrations of Pb: the non-mining population decreased with the increase in treatment concentration, and the Ci value decreased significantly under 120 mg·L^{-1} and 240 mg·L^{-1} Pb stress, which was 57.1% and 53.5% of the control, respectively. The Ci value of mining population increased significantly at the low concentration (40 mg·L^{-1} and 80 mg·L^{-1}) treatment ($p < 0.05$), and decreased to a certain extent at the intermediate concentration and elevated concentration treatment. When the concentration of Pb was 240 mg·L^{-1}, the Ci value was 73.2% of the control. The results show that the effect of Pb stress on Ci was much larger in the non-mining population than in the mining population.

Stomatal limitations (Ls) were varied by intercellular CO_2 and atmospheric CO_2 concentration. The stomatal limit of the non-mining population increased significantly with the increase in Pb concentration (Figure 3B), while the mining population decreased at low concentration stress and increased significantly at extreme concentration (120 mg·L^{-1} and 240 mg·L^{-1}). Clearly, under Pb stress, the intracellular limit value of the non-mining population was significantly larger than that of the mining population. The analysis of variance showed that there was a significant difference between Ls and control under high Pb stress ($p < 0.05$).

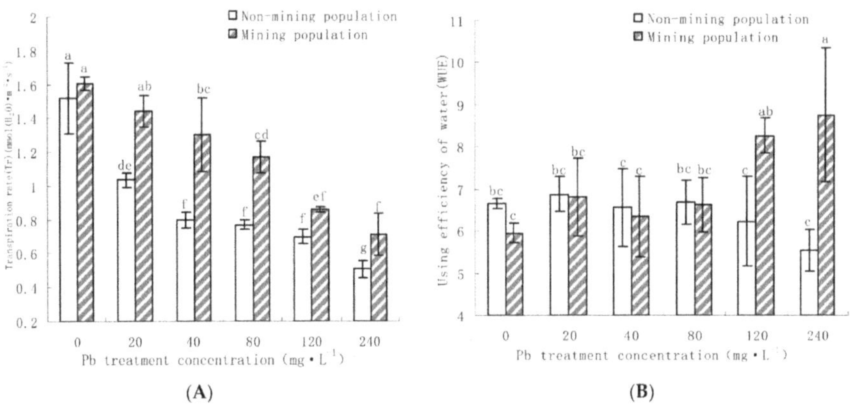

Figure 3. Transpiration rate (Tr) and using efficiency (mmolm^{-2}·s^{-1}) of water (WUE) in two *Miscanthus floridulus* populations under different Pb treatment ((**A**) is Tr, (**B**) is WUE). Note: Error bars indicate standard deviation; Different letters in the same group indicate significant difference at $p < 0.05$ according to Duncan's multiple range tests; the same below.

3.2.3. Effects of Pb on Transpiration Rate (Tr) and Water Use Efficiency (WUE) in Leaves of *M. floridulus*

Transpiration is the main driving force for plants to absorb and transport water. It can maintain the water saturation of various parts of plants, maintain the morphology of cell tissues, promote the distribution of inorganic salts in plants, and dissipate the excess heat generated by plants during photosynthesis and oxidative metabolism [38]. The variation of the transpiration rate is shown in Figure 3A. With increasing Pb stress, the Tr value decreased significantly in both populations, but the TR value decreased in different treatments. Under mild Pb stress (20 mg·L^{-1}, 40 mg·L^{-1}), the Tr value of the non-mining population decreased significantly compared with the control population, but the Tr value of the mining population did not decrease significantly. Under moderate and severe Pb stress (120 mg·L^{-1}, 240 mg·L^{-1}), the Tr value of non-mining population decreased significantly to 46.6% and 33.3%, the Tr value of the mining area population decreased to 52.9% and 48.6%. Overall, the transpiration rate of the mining population was significantly higher than that of the non-mining population at different Pb concentration stresses.

The value of water use efficiency (WUE) is directly determined by the transpiration rate and net photosynthetic rate [39]. As can be seen from Figure 3B, there was no significant difference in water use efficiency between non-mining population and mining population under control and low concentration Pb (20 mg·L^{-1} and 40 mg·L^{-1}) treatment ($p > 0.05$), while high concentration Pb (120 mg·L^{-1} and 240 mg·L^{-1}) treatment, the population in mining area was significantly higher than that in non-mining area ($p < 0.05$). There were also different trends in the variation of the population in mining and non-mining areas. The water use efficiency of non-mining area population increased at low concentration treatment, but decreased sharply with the increase in treatment concentration, and it reached the lowest at 240 mg·L^{-1} treatment. The population water use efficiency increased with the increase in treatment level, and reached the highest level at 240 mg·L^{-1}, which was significantly different from that of the control ($p < 0.05$), and was 148.8% of that of the control.

3.3. Changes of Chlorophyll Fluorescence Parameters

3.3.1. Effects of Pb on Fv/Fm and Fv/F$_0$ of the Leaves of *M. floridulus*

Fv/Fm reflects the maximum photochemical efficiency of PSII, and Fv/F$_0$ reflects the potential activity of PSII, both of which are essential parameters to indicate the status of photochemical reactions [40].

The results showed that with the increase in Pb stress concentration, both Fv/Fm and Fv/F$_0$ decreased (Figure 4), but the decrease amplitude of the two populations was different. Compared with the control, the decrease in Fv/Fm and Fv/F$_0$ in the two populations under severe Pb stress (240 mg·L^{-1}) was 3.7% and 10.1%, respectively. The non-mining population was 7.6% and 22.9%, respectively. The decrease in population in the mining region was significantly smaller than in the non-mining region, indicating that chlorophyll fluorescence was differently affected by Pb stress. Under Pb stress, the net photosynthetic rate of the mined and non-milled populations was significantly positively correlated with Fv/Fm, reaching a correlation coefficient of 0.951 and 0.965, respectively. Under Pb stress, the primary photoconversion efficiency of PSII decreased greatly, and the potential active center of PSII was seriously damaged.

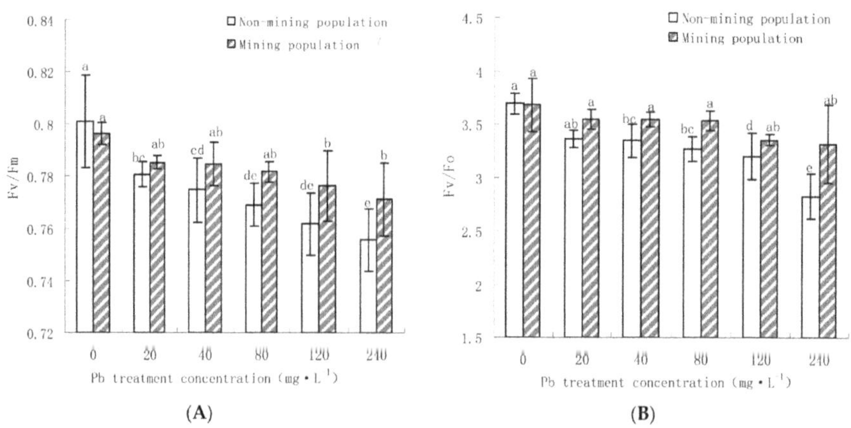

Figure 4. Effects of different treatments on the PSII maximal photochemical efficiency Fv/Fm and the ratio of variable fluorescence to initial fluorescence Fv/F$_0$ of two *Miscanthus floridulus* populations under different Pb treatment ((**A**) is Fv/Fm, (**B**) is Fv/F$_0$). Note: Error bars indicate standard deviation; Different letters in the same group indicate significant difference at $p < 0.05$ according to Duncan's multiple range tests; the same below.

3.3.2. Effects of Pb on the qP and NPQ of the Leaves of *M. floridulus*

The value of photochemical quenching coefficient qP reflects the redox state of the primary electron acceptor QA of PSII and the opening degree of the reaction center of PSII. The higher the value, the higher the electron transfer activity of PSII [41]. As can be seen in Figure 5A, the qP values of the two populations of *M. floridulus* showed a downward trend under Pb stress. Under mild stress (20 mg·L^{-1} and 40 mg·L^{-1}), qP of the two populations decreased slightly compared with the control ($p > 0.05$). As the Pb stress concentration increased, the qP values of the two populations decreased significantly. When Pb concentration was 240 mg·L^{-1}, qP values of the two populations decreased by: the population in the mining area (13.8%) and the population in non-mining area (32.2%) showed a relatively large decrease in the non-mining area, indicating that the electron transport activity of PSII reaction center in leaves of the non-mining population of *M. floridulus* was greatly affected by Pb stress.

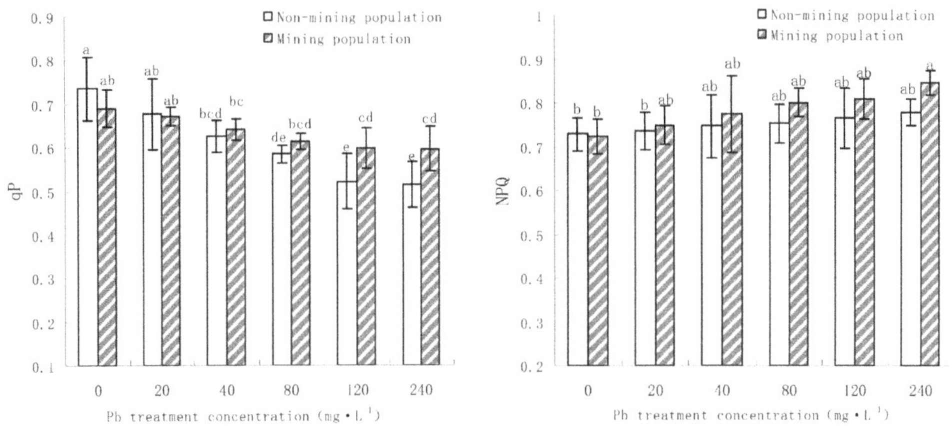

Figure 5. The photochemical quench (qP) and non-photochemical quench (NPQ) of two *Miscanthus floridulus* populations under different Pb treatment ((**A**) is qP, (B) is NPQ). Note: Error bars indicate standard deviation; Different letters in the same group indicate significant difference at $p < 0.05$ according to Duncan's multiple range tests; the same below.

The non-photochemical quenching coefficient NPQ reflects the degree of loss of light energy absorbed by blades in the form of heat dissipation. Heat dissipation is an important mechanism to protect PSII from damage caused by photoinhibition [42]. The results show that the NPQ values of the two populations increased under Pb stress. Compared with the control, the NPQ values of the mining population increased under Pb stress (80 mg·L^{-1}, 120 mg·L^{-1}, and 240 mg·L^{-1}) by 11.2%, 12.5%, and 17.8%, respectively. The non-mining population was 2.8%, 4.2%, and 6.7%, respectively. The increase in NPQ value in mining population was higher than that in non-mining population (Figure 5B), indicating that compared with non-mining population with weak resistance, the highly resistant mining population had higher openness of PSII reaction center and stronger electron transfer and heat dissipation ability under Pb stress.

3.3.3. Effects of Pb on the Leaf of *M. floridulus* on ϕPSII and ETR

With the diameter of PSII and ETR reflecting the actual photochemical reaction efficiency and electron transfer rate with partial closure of PSII reaction center under environmental stress [43]. Figure 6A,B show that: with the increase in Pb stress concentration, the diameter of PSII and ETR of the two populations of *M. floridulus* decreased, but the decreasing amplitude of the two populations was different. Compared with the control, under the Pb concentration (80 mg·L^{-1}, 120 mg·L^{-1}, and 240 mg·L^{-1}) stress, mining population ϕPSII and ETR decreased 17.8%, 21.9%, 24.1% and 15.5%, 18.1%, 20.1%, respectively, The non-mining population ϕPSII and ETR decreased by 31.1%, 40.5%, 52.2% and 22.2%, 31.1%, 48.3%, respectively. It shows that the magnitude of the reduction was much smaller for the mining population than for the non-mining population, and the ETR was much smaller for the mining population. The variance analysis showed that under higher concentration Pb stress (120 mg·L^{-1} and 240 mg·L^{-1}), the difference between the mining population and the other treatments was not significant ($p > 0.05$), whereas the difference between the non-mining population and the different treatments was significant ($p < 0.05$).

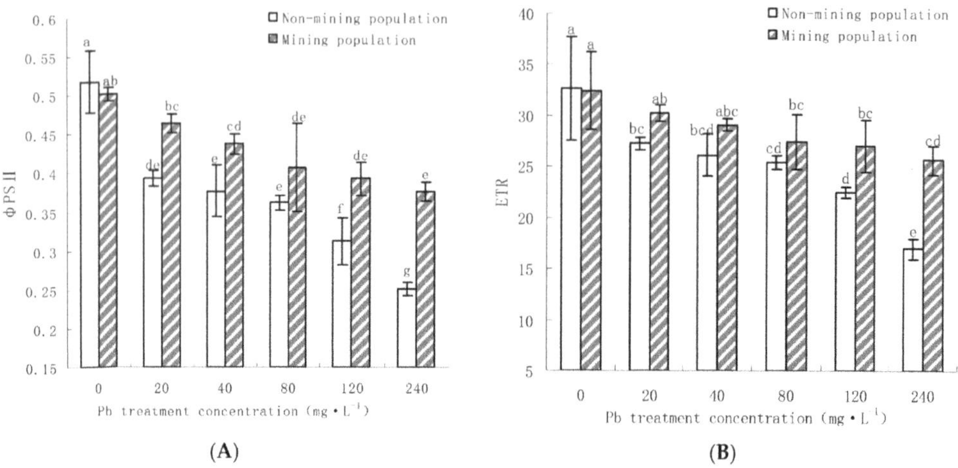

Figure 6. The PSII actual photochemical efficiency (ΦPSII) and the acyclic electron transfer rate (ETR) of two *Miscanthus floridulus* populations under different Pb treatment ((**A**) is ΦPSII, (**B**) is ETR). Note: Error bars indicate standard deviation; Different letters in the same group indicate significant difference at $p < 0.05$ according to Duncan's multiple range tests; the same below.

3.4. *Effects of Pb on Photosynthetic Pigment Content in Leaves of M. floridulus*

The content of photosynthetic pigments will affect the absorption and transfer of light energy by chlorophyll, as well as the amount of ATP and NADPH synthesized by the allocation and conversion between PSII and PSI [44], thus affecting photosynthesis. In stressed environments, plants usually show decreased photosynthesis and photosynthetic pigment content after sensing stress [45,46]. As shown in Table 3, the effect of Pb on the photosynthetic pigment content in the leaves of *M. floridulus* varied with different concentrations and populations of Pb. The contents of Chl-a, Chl-b, and carotenoid were increased compared with the control under low Pb concentration (20 mg·L^{-1} and 40 mg·L^{-1}) stress. However, as the concentration of Pb increased in the non-mining areas, Chl-a, Chl-b, and Carotenoid rapidly decreased, and the trend was essentially the same for each treatment. Chl a/b increased with Pb concentration for both populations. Under 120 mg·L^{-1} and 240 mg·L^{-1} Pb stress, the content of Chl-a and Chl-b in the mining population was significantly higher than that in the non-mining population ($p < 0.05$).

Table 3. Chlorophyll contents (mean ± SD) in *M. floridulus* leaves under different Pb treatment.

Population	Pb Treatment ($mg \cdot L^{-1}$)	Chl-a Content ($mg \cdot g^{-1}$ FW)	Chl-b Content ($mg \cdot g^{-1}$ FW)	Chl a/b Value ($mg \cdot g^{-1}$)	Car Content ($mg \cdot g^{-1}$ FW)
Non-mining population	0	1.31 ± 0.010 a	0.75 ± 0.048 a	1.74 ± 0.155 d	0.43 ± 0.005 a
	20	1.25 ± 0.013 ab	0.68 ± 0.055 a	1.84 ± 0.091 cd	0.41 ± 0.006 a
	40	1.23 ± 0.032 b	0.52 ± 0.102 b	2.36 ± 0.472 bcd	0.39 ± 0.029 b
	80	1.11 ± 0.015 c	0.43 ± 0.041 b	2.58 ± 0.185 bc	0.39 ± 0.025 b
	120	1.01 ± 0.071 d	0.29 ± 0.081 c	3.48 ± 0.523 a	0.34 ± 0.026 c
	240	0.76 ± 0.017 e	0.25 ± 0.006 c	3.04 ± 0.663 ab	0.31 ± 0.033 c
Mining population	0	1.26 ± 0.026 a	0.69 ± 0.009 a	1.82 ± 0.319 c	0.39 ± 0.015 ab
	20	1.29 ± 0.011 a	0.72 ± 0.003 a	1.78 ± 0.061 c	0.41 ± 0.006 a
	40	1.28 ± 0.010 a	0.75 ± 0.018 a	1.71 ± 0.063 c	0.40 ± 0.006 ab
	80	1.20 ± 0.020 b	0.57 ± 0.044 b	2.11 ± 0.179 bc	0.38 ± 0.011 ab
	120	1.18 ± 0.011 b	0.51 ± 0.081 bc	2.31 ± 0.603 ab	0.36 ± 0.044 ab
	240	1.15 ± 0.008 c	0.45 ± 0.005 c	2.56 ± 0.075 a	0.36 ± 0.006 b

Note: Data in the table are means ± SD ($n = 3$), different letters in the same vertical column indicate significant difference according to SSR (New Complex Range Method) test ($p < 0.05$), the same below.

3.5. Correlation between Photosynthetic Physiological Indexes and Pb Concentration

It can be seen from Table 4 that Pn, Gs, Ci, Tr, Fv/Fm, Fv/F_0, qP were significantly negatively correlated with Pb treatment concentration among the photosynthetic physiological indexes of the non-mining population of *M. floridulus* ($p < 0.01$). There was no significant correlation between the NPQ and Pb concentration. Among the other indicators, Pn and Gs, Ci, Tr, Fv/Fm, and ETR were significantly positively correlated ($p < 0.01$). Chl-a was positively correlated with Pn, Gs, Ci, Tr, Fv/Fm, Fv/F_0, and qP ($p < 0.01$).

Table 4. Correlation matrix between photosynthetic physiological factors and Pb concentration of *M. floridulus*.

Type	Item	Pb	Biomass	Pn	Gs	Ci	Tr	Fv/Fm	Fv/F_0	qP	NPQ	ETR	Chla
Non-mining population	Pb	1											
	Biomass	−0.868 **	1										
	Pn	−0.818 **	0.936 **	1									
	Gs	−0.777 **	0.916 **	0.897 **	1								
	Ci	−0.820 **	0.921 **	0.900 **	0.839 **	1							
	Tr	−0.779 **	0.917 **	0.980 **	0.877 **	0.865 **	1						
	Fv/Fm	−0.711 **	0.890 **	0.778 **	0.812 **	0.750 **	0.774 **	1					
	Fv/F_0	−0.846 **	0.841 **	0.873 **	0.788 **	0.802 **	0.811 **	0.701 **	1				
	qP	−0.750 **	0.862 **	0.773 **	0.786 **	0.863 **	0.702 **	0.728 **	0.736 **	1			
	NPQ	0.352	−0.397	−0.439	−0.333	−0.439	−0.356	−0.264	−0.449	−0.450	1		
	ETR	−0.885 **	0.893 **	0.813 **	0.839 **	0.801 **	0.770 **	0.816 **	0.819 **	0.857 **	−0.290	1	
	Chla	−0.986 **	0.877 **	0.819 **	0.773 **	0.832 **	0.775 **	0.750 **	0.847 **	0.759 **	−0.375	0.878 **	1
Mining population	Pb	1											
	Biomass	−0.860 **	1										
	Pn	−0.866 **	0.856 **	1									
	Gs	−0.881 **	0.891 **	0.861 **	1								
	Ci	−0.686 **	0.745 **	0.699 **	0.685 **	1							
	Tr	−0.902 **	0.944 **	0.873 **	0.935 **	0.716 **	1						
	Fv/Fm	−0.456	0.665 **	0.471 *	0.716 **	0.278	0.699 **	1					
	Fv/F_0	−0.565 *	0.545 *	0.405	0.519 *	0.288	0.548 *	0.680 **	1				
	qP	−0.651 **	0.705 **	0.771 **	0.744 **	0.491 *	0.733 **	0.320	0.103	1			
	NPQ	0.653 **	−0.664 **	−0.460	−0.679 **	−0.261	−0.582 *	−0.796 **	−0.604 **	−0.292	1		
	ETR	−0.684 **	0.716 **	0.695 **	0.616 **	0.432	0.710 **	0.615 **	0.617 **	0.376	−0.538 *	1	
	Chla	−0.861 **	0.904 **	0.833 **	0.795 **	0.771 **	0.829 **	0.524 *	0.412	0.669 **	−0.588 *	0.637 **	1

* Correlation is significant at the 0.05 level (2-tailed); ** Correlation is significant at the 0.01 level (2-tailed).

It can be seen from Table 4 that Pn, Gs, Ci, Tr, qP, ETR, Chl-a, and Pb treatment concentration of the mining population of *M. floridulus* were significantly negatively correlated ($p < 0.01$). There was a significant positive correlation between NPQ and Pb concentration ($p < 0.01$). Among the remaining indicators, Pn and Gs, Ci, Tr, and ETR were positively correlated ($p < 0.01$). Chl-a was positively correlated with Pn, Gs, Ci, Tr, and qP ($p < 0.01$).

4. Discussion

(1) Specific leaf area (SLA) is one of the important traits of plant leaves. Since SLA is closely related to the growth and survival strategies of plants, it can reflect the adaptation characteristics of plants to different habitats, and is the preferred index of plants in comparative ecological studies [47,48]. The larger the specific leaf area, the larger the leaf area per unit dry weight, and the thinner the leaf. A relatively large leaf area is conducive to capturing more light energy, which increases the possibility of faster growth of plants [49,50]. Studies by Penning et al. [51] also showed that there is a proportional relationship between the maximum photosynthetic rate of plant leaves and the specific leaf area when the nutrient supply is sufficient. In this experiment, with the increase in Pb stress concentration, the specific leaf area (SLA) of both populations decreased, but the magnitude of the decline of the two populations was different. Compared with the control, the specific leaf area (SLA) of mining and non-mining populations decreased to 84.8% and 75.9% and 74.1% and 58.5%, respectively, under the Pb concentration (120 $mg{\cdot}L^{-1}$ and 240 $mg{\cdot}L^{-1}$) stress. The results showed that the specific leaf area (SLA) of mining population decreased greatly less than that of non-mining population.

Root/shoot ratio is the final comprehensive effect of many basic change processes and self-adaptation and self-regulation in plants, and so, it can be regarded as the basis of plant structure and function [52]. Compared to plants in the non-mining population, plants in the mining population have a further distribution of above-ground biomass, that is, seedlings in weak light conditions use more energy for the growth of various organs in the above-ground part of the plant, thus obtaining additional light energy for photosynthesis. In conclusion, from the external morphology, the plant growth status of the mining area population was better, especially under the stress of high concentration of Pb (120 $mg{\cdot}L^{-1}$ and 240 $mg{\cdot}L^{-1}$).

(2) Chlorophyll (chlorophyll a, chlorophyll b) and carotenoid (carotene and lutein) are the pigments that absorb light energy in the process of plant photosynthesis, among which chlorophyll is the main material that absorbs light energy and directly affects the light energy utilization in vegetation photosynthesis [53,54]. Pb pollution decreased chlorophyll content, Chl a/b ratio and antioxidant enzyme activity, and increased MDA content in tobacco leaves [55]. Chlorophyll fluorescence is commonly used to evaluate the efficiency of photosynthesis and the impact of environmental stress on them [56,57]; the results of this experiment showed that Pb pollution in soil could significantly reduce Fv/F_0, Fv/Fm, ETR, and ΦPSII, while qP and NPQ also showed an obvious downward trend (Figures 4–6). In this experiment, Pb significantly reduced the leaf chlorophyll content of the non-mining population of *M. floridulus*, and the chlorophyll content of the mining population increased under low Pb concentration (20 $mg{\cdot}L^{-1}$ and 40 $mg{\cdot}L^{-1}$) treatment, but decreased significantly under high Pb (120 $mg{\cdot}L^{-1}$ and 240 $mg{\cdot}L^{-1}$) treatment. However, chlorophyll levels were still significantly higher than in the non-mining population. This was likely because some resistance mechanisms were developed in the *M. floridulus* population in mining areas to reduce the damage caused by Pb to chlorophyll.

(3) Farquhar et al. [58] suggested that the reduction in Pn is caused by both gastric and non-gastric restriction. When Pn and Ci vary in the same direction, both decrease at the same time, Ls increase, and Pn's decline is mainly caused by Gs; otherwise, Pn decline could be attributed to the decreased carboxylation ability of mesocytes. In addition, chloroplast structure is closely related to photosynthesis, and the more the number of chloroplast grana and grana lamellae, the denser the arrangement of grana lamellae and the stronger the photosynthetic capacity [59]. In the present study, the decrease in Pn in the non-mining population of *M. floridulus* under mild Pb stress was accompanied by a significant decrease in Ci and a large increase in Ls, indicating that gastric restriction was the main factor responsible for the decrease in Pn in the non-mining population of *M. floridulus* under mild Pb stress. However, with further reduction in Pn under moderate and severe Pb stress, Ci and Ls showed minor changes compared to the control case. These results suggest that the reduction in Pn in the non-mining population under moderate and severe Pb stress was

mainly caused by the reduced photosynthetic activity of mesophyll cells, with non-stomatal restriction playing a major role. In the mining population, Pn decreased only at moderate and severe Pb stresses, while Ci decreased and Ls increased, indicating that the decrease in Pn under Pb stresses was mainly due to the gastral restriction.

(4) Photosynthesis is an extremely important metabolic process in plants, and leaf chlorophyll fluorescence is closely related to various reaction processes in photosynthesis. The influence of any environmental factors on photosynthesis can be reflected through leaf chlorophyll fluorescence dynamics; thus, chlorophyll fluorescence is an ideal internal probe to detect the dynamic changes of plant photosynthesis [60]. In photosynthesis, the value of Fv/Fm is commonly used to indicate the environmental stress on the plant. Fv/Fm measured after dark adaptation reflects the quantum yield when all PSII reaction centers are in the open state, namely the potential maximum quantum yield of PSII [61]. In normal physiological state, Fv/Fm is a relatively stable value, and the value of Fv/Fm is generally around 0.8 [62]. In drought stress [63], salt stress [64], and light stress [65], it was found that the value of Fv/Fm decreased with the increase in stress degree. The photosynthetic inhibition indicated by Fv/Fm to be lower than 0.8 may have been due to photoprotection or damage of the PSII reaction center [66]. In this experiment, the Fv/Fm of the mining population of *M. floridulus* was around 0.8 and the correlation with Pb was low. However, at high Pb stress, the Fv/Fm of the non-mining population of *M. floridulus* dropped to about 0.7, indicating that the effect of Pb stress on photosynthesis was much greater in the non-mining population than in the mining population.

5. Conclusions

(1) Under Pb stress, the net photosynthetic rate (Pn), transpiration rate €, stomatal conductance (Gs), intercellular carbon dioxide concentration (Ci), and chlorophyll content (Chl) of the leaves of the two populations of *M. floridulus* decreased in different amplitude. With the increase in Pb stress, the net photosynthetic rate (Pn) of the mining population decreased slowly, and the plant growth and biomass were less affected.

(2) Stomatal restriction was the main factor in the reduction in Pn in the non-mining population of *M. floridulus* under mild Pb stress and non-stomatal restriction was the main factor in the reduction in Pn under moderate and severe Pb stress.

(3) The maximum photochemical efficiency (Fv/Fm), potential activity (Fv/Fo), and effective photochemical efficiency (Fv′/Fm′) of PSII reaction center decreased under Pb stress. On the whole, the decrease in population in the mining area was small, and the utilization of light energy and the potential activity of PSII were strong. The changes of photochemical quenching coefficient (qP) and non-photochemical quenching coefficient (NPQ) of PSII showed that qP value decreased and NPQ value increased in the two populations under Pb stress. Overall, the population in the resistant mining region had a low qP reduction and a large NPQ increase. With the increase in Pb stress concentration, the actual photochemical reaction efficiency (PSII) and electron transfer rate (ETR) on the mining population did not change much, and the non-mining population decreased significantly. This indicates that the electron transfer activity of PSII reaction center was less affected and the damage degree of photosynthetic mechanism was low. The results show that the mining population of *M. floridulus* had a strong tolerance to Pb, which is suitable for the pioneer species of Gramineae in the vegetation restoration construction in the metal mining area.

Author Contributions: Project administration and writing—review and editing, J.Q. (Jianqiao Qin); conceptualization, X.J. and H.Z.; formal analysis, J.Q. (Jianhua Qin); methodology, M.D. and X.C.; investigation, H.L. All authors have read and agreed to the published version of the manuscript.

Funding: This work was supported by the Guangdong Provincial Key Laboratory of Environmental Health and Land Resource (project number: 2020B121201014), the Special Project of Key Areas of Colleges and Universities in Guangdong Province (Science and Technology Promoting Rural Revitalization) "Research and Development of Key Technologies for Resource Utilization of Manure from Large-Scale

Livestock and Poultry Breeding in Rural Areas of Western Guangdong" (No.:2021ZDZX4023), the Innovation Team Project of Colleges and Universities in Guangdong Province (2021KCXTD055), and 2021 Guangdong Provincial Department of Education "Quality Engineering" College Student Social Practice Teaching Base Construction Project (Yue Jiao Gao Han [2021] No. 29).

Institutional Review Board Statement: Not applicable.

Informed Consent Statement: Not applicable.

Data Availability Statement: The data presented in this study are available on request from the corresponding author.

Acknowledgments: We acknowledge the Guangdong Provincial Key Laboratory of Environmental Health and Land Resource; Special Project of Key Areas of Colleges and Universities in Guangdong Province (Science and Technology Promoting Rural Revitalization) "Research and Development of Key Technologies for Resource Utilization of Manure from Large-Scale Livestock and Poultry Breeding in Rural Areas of Western Guangdong" and the Innovation Team Project of Colleges and Universities in Guangdong Province.

Conflicts of Interest: The authors declare no conflict of interest.

References

1. He, B.H.; Zhao, X.L.; Li, P.; Liang, J.J.; Fan, Q.H.; Ma, X.X.; Zheng, G.D.; Qiu, J. Lead isotopic fingerprinting as a tracer to identify the pollution sources of heavy metals in the southeastern zone of Baiyin, China. *Sci. Total Environ.* **2019**, *660*, 348–357. [CrossRef] [PubMed]
2. Chen, Z.Q.; Liu, X.; Ai, Y.W.; Chen, J.; Luo, X.M.; Chen, J.Y.; Shihong Zhong, S.H. Effects and mechanisms of revegetation modes on cadmium and lead pollution in artificial soil on railway rock-cut slopes. *Sci. Total Environ.* **2018**, *644*, 1602–1611. [CrossRef] [PubMed]
3. Wu, B.H.; Luo, S.H.; Luo, H.Y.; Huang, H.Y.; Xu, F.; Feng, S.; Xu, H. Improved phytoremediation of heavy metal contaminated soils by Miscanthus floridulus under a varied rhizosphere ecological characteristic. *Sci. Total Environ.* **2021**, *808*, 151995. [CrossRef]
4. Zhang, Y.M.; Li, S.; Chen, Z.; Wang, F.; Chen, J.; Wang, L.Q. A systemic ecological risk assessment based on spatial distribution and source apportionment in the abandoned lead acid battery plant zone, China. *J. Hazard. Mater.* **2018**, *354*, 170–179. [CrossRef] [PubMed]
5. Dai, Q.L.; Yuan, J.G.; Fang, W.; Yang, Z.Y. Differences of Pb accumulation among plant tissues of 25 *Zay mays* varieties. *Acta Phytoeco. Sin.* **2005**, *29*, 992–999.
6. Song, J.; Shen, Q.; Wang, L.; Qiu, G.; Shi, J.; Xu, J.; Philip, C.B.; Liu, X. Effects of Cd, Cu, Zn and their combined action on microbial biomass and bacterial community structure. *Environ. Pollut.* **2018**, *243*, 510–518. [CrossRef]
7. Chen, J.; Zhang, H.; Li, J.; Liu, Y.; Shi, W.; Hu, H. The toxic factor of copper should be adjusted during the ecological risk assessment for soil bacterial community. *Ecol. Indic.* **2020**, *111*, 106072. [CrossRef]
8. Zheng, B.J.; Xu, S.C.; Tu, Y.H.; Zou, F.; Liang, D.D.; Wang, H.; Mo, Y.W. Research progress on copper toxicity to crops. *J. Shaoxing Univ. Arts Sci.* **2022**, *42*, 58–64.
9. Lin, M.Z.; Jin, M.F.; Chen, Y.; Zeng, Q.L. Photosynthetic characteristics and mechanism of wedelia chinensis under soil Cu pollution stress. *J. Anhui Agric. Univ.* **2018**, *45*, 657–663.
10. Yang, K.; Huang, C.; Lu, S.; Peng, G.Y.; Huang, C.G. Selection of reference genes in real-time quantitative PCR of root tissue of *Commelina purpurea* under copper stress. *Acta Phytophysiol. Sin.* **2021**, *57*, 195–204.
11. Qin, J.Q.; Zhao, H.R.; Dai, M.; Zhao, P.; Chen, X.; Liu, H.; Lu, B.Z. Speciation Distribution and Influencing Factors of Heavy Metals in Rhizosphere Soil of *Miscanthus Floridulus* in the Tailing Reservoir Area of Dabaoshan Iron Polymetallic Mine in Northern Guangdong. *Processes* **2022**, *10*, 1217. [CrossRef]
12. Barbosa, B.; Boléo, S.; Sidella, S.; Costa, J.; Duarte, M.P.; Mendes, B.; Cosentino, S.L.; Fernando, A.L. Phytoremediation of heavy metal-contaminated soils using the perennial energy crops *miscanthus* spp. And *Arundo donax* L. *Bioenerg. Res.* **2015**, *8*, 1500–1511. [CrossRef]
13. Zadel, U.; Nesme, J.; Michalke, B.; Vestergaard, G.; Plaza, G.A.; Schroder, P.; Radl, V.; Schloter, M. Changes induced by heavy metals in the plant-associated microbiome of miscanthus x giganteus. *Sci. Total Environ.* **2020**, *711*, 134433. [CrossRef]
14. Ren, L.M.; Liu, P.; Zheng, Q.E. A survey of heavy metal content of plants growing on the soil polluted by manganese mine in Daxin County, Guangxi. Subtrop. *Plant Sci.* **2006**, *35*, 5–8.
15. Sun, J.; Tie, B.Q.; Qin, P.F. Investigation of contaminated soil and plants by heavy metals in Pb-Zn mining area. *J. Plant Resour. Environ.* **2006**, *15*, 63–67.
16. Li, Q.F.; Du, W.B.; Li, Z.A.; Wang, Z.F.; Peng, S.L. Heavy metals accumulation in mining area's Miscanthus sinensis populations and its relationship with soil characters. *Chin. J. Ecol.* **2006**, *25*, 255–258.

17. Chen, Z.J.; Tian, W.; Li, Y.J.; Sun, L.N.; Chen, Y.; Zhang, H.; Li, Y.Y.; Han, H. Responses of rhizosphere bacterial communities, their functions and their network interactions to Cd stress under phytostabilization by *Miscanthus* spp. *Environ. Pollut.* **2021**, *287*, 117663. [CrossRef]
18. Ye, Z.H.; Baker, A.J.M.; Wong, M.H.; Willis, A.J. Copper tolerance, uptake and accumulation by Phragmites australis. *Chemosphere* **2003**, *50*, 795–800. [CrossRef]
19. Rascio, N.; Navari-Izzo, F. Heavy metal hyperaccumulating plants: How and why do they do it? And what makes them so interesting. *Plant Sci.* **2011**, *189*, 169–181. [CrossRef]
20. Fu, S.; Zhou, Y.; Zhao, Y.; Gao, Q.Z.; Peng, X.Z.; Dang, Z.; Zhang, C.B.; Yang, X.Q.; Yang, Z.J.; Dou, L.; et al. Study on heavy metals in soils contaminated by acid mine drainage from Dabaoshan Mine, Guangdong. *Environ. Sci.* **2007**, *28*, 805–812.
21. Zhao, H.R.; Wang, D.; Zhu, B.T.; Jin, X. Evaluation of heavy metal pollution of soil in the Dabaoshan mining area based on the information diffusion theory. *J. Agro-Environ.Sci.* **2019**, *38*, 79–86.
22. Chen, J.Y.; Liu, G.B.; Cui, J.L.; Xiao, T.F. Migration process and risk assessment of heavy metals in soil plant system of Dabaoshan mining area in Guangdong Province. *Environ. Sci.* **2019**, *40*, 5629–5639.
23. Qin, J.Q.; Xia, B.C.; Hu, M.; Zhao, P.; Zhao, H.R.; Lin, X.F. Analysis of the vegetation succession of tailing wasteland of Dabaoshan Mine, Guangdong Province. *J. Agro-Environ. Sci.* **2009**, *28*, 2085–2091.
24. Qin, J.Q.; Zhao, H.R.; Liu, H.; Dai, M.; Zhao, P.; Chen, X.; Wu, X.G. The Difference of Lead Accumulation and Transport in Different Ecotypes of *Miscanthus floridulus*. *Processes* **2022**, *10*, 2219. [CrossRef]
25. Zhao, H.R.; Xia, B.C.; Qin, J.Q.; Zhang, J. Hydrogeochemical and mineralogical characteristics related to heavy metal attenuation in a stream polluted by acid mine drainage: A case study in Dabaoshan mine, China. *J. Environ. Sci.* **2012**, *24*, 979–989. [CrossRef]
26. Zhao, H.R.; Xia, B.C.; Fan, C.; Zhao, P.; Shen, S. Human health risk from soil heavy metal contamination under different land uses near Dabaoshan mine, southern China. *Sci. Total Environ.* **2012**, *417*, 45–54. [CrossRef]
27. Qin, J.; Zhao, H.; Liu, H.; Dai, M.; Zhao, P.; Chen, X.; Wu, X. Analysis on characteristics of phytogeochemistry of Elsholtzia haichowensisin Daye Tonglushan copper mine. *Acta Ecol. Sin.* **2001**, *21*, 907–912.
28. Wen, C.H.; Deng, C.Q.; Chang, X.X.; Wang, H.B.; Wang, H.X. Differentiation in Datura stramonium L.populations exposed to heavy-metal pollution at different durations: RAPD analysis. *Acta Ecol. Sin.* **2001**, *21*, 1239–1245.
29. Tian, S.K.; Lu, L.L.; Yang, X.E. The impact of EDTA on lead distribution and speciation in the accumulator Sedum alfredII by synchrotron X-ray investigation. *Environ. Pollut.* **2011**, *159*, 782–789. [CrossRef]
30. Farooqui, A.; Kulshreshtha, K.; Srivastavaa, K.; Singh, S.N.; Farooqui, S.A.; Pandey, V.; Ahmad, P.J. Photosynthesis, stomata1 response and metal accumulation in *Cineraria maritima* L. and *Centauria moschata* L. grown in metal-rich soil. *Sci. Total Environ.* **1995**, *164*, 203–207. [CrossRef]
31. He, B.; Yang, X.E. Seudm alrfedII: A New Lead Accumulatnig Ecotype. *Acta Bot. Sin.* **2002**, *44*, 1356–1370.
32. Berry, J.A.; Downton, W.J.S. Environmental regulation of photosynthesis. In *Photosynthesis: Development, Carbon Metabolism and Plant Productivity, Vol II*; Academic Press: NewYork, NY, USA, 1982; pp. 263–343.
33. Fischer, R.A.; Turner, N.C. Plant productivity in the arid and semiarid zones. *Annual Rev. Plant Physi.* **1978**, *29*, 227–317. [CrossRef]
34. Ralph, P.J.; Schreiber, U.; Gademann, R. Coral photobiology studied with a new imaging pulse amplitude modulated fluorometer. *Phycol. Res.* **2005**, *41*, 335–342. [CrossRef]
35. Demmig-Adams, B.; Adams, W.W., III. Xanthophyll cycle and light stress in nature: Uniform response to excess direct sunlight among higher plant species. *Planta* **1996**, *198*, 460–470. [CrossRef]
36. Li, H.S.; Sui, Q.; Zhao, S.J.; Zhang, W.H. *Experimental Principle and Technology of Plant Physiology and Biochemistry*; Higher Education Press: Beijing, China, 2000; pp. 134–138.
37. Ma, J.; Sun, M.; Qiu, L.; Xie, Y.; Ma, Y.; Liang, W. The 5-Aminolevulinic Acid (5-ALA) Supplement Enhances PSII Photochemical Activity and Antioxidant Activity in the Late Growth Promotion of Pseudostellaria heterophylla. *Plants* **2022**, *11*, 3035. [CrossRef]
38. Zhou, Y.M.; Han, S.J.; Hu, Y.L.; Zhang, H.-S.; Zheng, J. Effects of elevated CO2 concentration on photosynthetic parameters of Pinus koraiensis needles. *Acta Ecol. Sin.* **2008**, *28*, 423–429.
39. Huang, H.; Li, S.; Guo, J.L. Effects of cadmium stress on antioxidant system and photosynthesis of maize seedlings. *J. Agro-Environ. Sci.* **2010**, *29*, 211–215.
40. Li, X.W.; Chen, S.L. Effect of shading on photosynthetic characteristics and chlorophyll fluorescence parameters in leaves of Fritillaria cirrhosa. *Acta Ecol. Sin.* **2008**, *28*, 3438–3445.
41. Fen, D.L.; Liu, Y. Photosynthesis and chlorophyIlfluorescence parameters of the reed (Phragmites communis)grown in the hydro-fluctuation belt of Three Gorges Reservoir Area. *Acta Ecol. Sin.* **2008**, *28*, 2013–2021.
42. Ayyaz, A.; Amir, M.; Umer, S.; Iqbal, M.; Bano, H.; Gul, H.S.; Farooq, M.A. Melatonin induced changes in photosynthetic efficiency as probed by OJIP associated with improved chromium stress tolerance in canola (*Brassica napus* L.). *Heliyon* **2020**, *6*, e04364. [CrossRef]
43. Hoque, M.N.; Tahjib-Ul-Arif, M.; Hannan, A.; Sultana, N.; Akhter, S.; Hasanuzzaman, M.; Akter, F.; Hossain, M.S.; Sayed, M.A.; Hasan, M.T.; et al. Melatonin Modulates Plant Tolerance to Heavy Metal Stress: Morphological Responses to Molecular Mechanisms. *Int. J. Mol. Sci.* **2021**, *22*, 11445. [CrossRef] [PubMed]
44. Yang, N.; Li, Y.M.; Li, J.W.; Teng, R.M.; Chen, Y.; Wang, Y.H.; Zhuang, J. Effects of Exogenous 5-Aminolevulinic acid on the Chlorophyll Synthesis and Fluorescence Characteristics and Gene Expression of Key Enzymes in Tea Plants under Drought Stress. *Tea Sci.* **2022**, *42*, 187–199. (In Chinese)

45. Bussotti, F.; Gerosa, G.; Digrado, A.; Pollastrini, M. Selection of chlorophyll fluorescence parameters as indicators of photosynthetic efficiency in large scale plant ecological studies. *Ecol. Indic.* **2020**, *108*, 105686. [CrossRef]
46. Li, Y.T.; Xu, W.W.; Ren, B.Z.; Zhao, B.; Wang, J.W.; Liu, P.; Zhang, Z.S. High temperature reduces photosynthesis in maize leaves by damaging chloroplast ultrastructure and photosystem II. *J. Agron. Crop Sci.* **2020**, *206*, 548–564. [CrossRef]
47. Yin, C.Y.; Berninger, F.; Li, C.Y. Photosynthetic responses of Populus przewalski subjected to drought stress. *Photosynthetica* **2006**, *44*, 62–68. [CrossRef]
48. Garnier, E.; Shipley, B.; Roumet, C.; Laurent, G. Standardized protocol for the determination of specific leaf area and leaf dry matter content. *Funct. Ecol.* **2001**, *15*, 688–695. [CrossRef]
49. Cornelissen, J.H.C.; Diez, P.C.; Hunt, R. Seedling growth, allocation and leaf attributes in a wide range of woody plant species and types. *J. Ecol.* **1996**, *84*, 755–765. [CrossRef]
50. Wright, I.J.; Wearoby, M. Cross-species relationship between seedling relative growth rate, nitrogen productivity and root vs. leaf function in 28 Australian woody species. *Funct. Ecol.* **2000**, *14*, 97–107. [CrossRef]
51. Penning, F.W.T.; Zhu, D.F.; Cheng, S.H. *Simulation of Ecophysiological Processes of Growth in Several Annual Crops*; Chinese Agricultural Science and Technology Press: Beijing, China, 1991; pp. 34–36.
52. Zhao, C.G. The relationship between root/shoot ratio and quality in tobacco. *Sci. Tech. Infor.* **2006**, *18*, 133.
53. Maxwell, K.; Johnson, G.N. Chlorohpyll fluorescence: A practical guide. *J. Exp. Bot.* **2000**, *51*, 659–668. [CrossRef]
54. Appenroth, K.J.; Stöckel, J.; Srivastava, A.; Strasser, R.J. Multiple effects of chromate on the photosynthetic apparatus of Spirodela polyrhiza as probed by OJIP chlorophyII a fluorescence measurement. *Environ. Pollut.* **2001**, *115*, 49–64. [CrossRef]
55. Yuan, Z.L.; Li, C.M.; Xiong, S.P.; Li, H.; Zhang, G.L.; Ma, X.M. Effect of Cd and Pb pollution on chlorophyll content, activity of protectiase and cell membrance lipid peroxidation change in tobacco leaves. *J. Henan Agric.Univ.* **2005**, *39*, 15–19.
56. Hura, T.; Hura, K.; Ostrowska, A.; Dziurka, K. Rapid plant rehydration initiates permanent and adverse changes in the photosynthetic apparatus of triticale. *Plant Soil* **2015**, *397*, 127–145. [CrossRef]
57. Hu, W.H.; Yan, X.H.; Li, X.H.; Cao, Z.G. Effect of 24-Epibrassionolide on the ChlorophyII Fluorescence Triansient in leaves of Pepper under drought stress. *Bull. Bot. Res.* **2021**, *41*, 53–59.
58. Fghire, R.; Anaya, F.; Ali, O.I.; Benlhabib, O.; Ragab, R.; Wahbi, S. Physiological and photosynthetic response of quinoa to drought stress. *Chil. J. Agric. Res.* **2015**, *75*, 174–183. [CrossRef]
59. Zhang, J.; Li, T.L. Effects of daytime sub-high temperature on photosynthesis and chloroplast ultrastructure of tomato leaves in green house. *Acta Hortic. Sin.* **2005**, *32*, 614–619.
60. Ralph, P.J.; Gademann, R. Rapid light curves: A powerful tool to assess photosynthetic activity. *Aquat. Bot.* **2005**, *82*, 222–237. [CrossRef]
61. Da-Matta, F.M.; Maestri, M.; Mosquim, P.R.; Barros, R.S. Photosynthesis in coffee (*Coffea arabica* and *C. canephora*) as affected by winter and summer conditions. *Plant Sci.* **1997**, *128*, 43–50. [CrossRef]
62. Zhu, H.; Teng, Y.; Wang, X.; Zhao, L.; Ren, W.; Luo, Y.; Christie, P. Changes in clover rhizosphere microbial community and diazotrophs in mercury-contaminated soils. *Sci. Total Environ.* **2021**, *767*, 145473. [CrossRef]
63. Karavatas, S.; Manetas, Y. Seasonal patterns of photosystemIIphotochemical efficiency in evergreen sclerophylls and drought semi-deciduous shrubs under Mediterranean field conditions. *Photosynthetica* **1999**, *36*, 41–49. [CrossRef]
64. Lima, J.D.; Mosquim, P.R.; Damattta, F.M. Leaf gas exchange and chlorophyll fluorescence parameters in Phaseolus vulgaris as affected by nitrogen and phosphorus deficiency. *Photosynthetica* **1999**, *37*, 113–121. [CrossRef]
65. Du, Y.L.; Zhao, Q.; Chen, L.W.; Yao, X.D.; Zhang, W.; Zhang, B.; Xie, F.T. Effect of drought stress on sugar metabolism in leaves and roots of soybean seedlings. *Plant Physiol. Biochem.* **2020**, *146*, 1–12. [CrossRef] [PubMed]
66. Ren, Q.Q.; Huang, Z.R.; Huang, W.L.; Huang, W.T.; Chen, H.H.; Yang, L.T.; Ye, X.; Chen, L.S. Physiological and molecular adaptations of Citrus grandis roots to long-term copper excess revealed by physiology, metabolome and transcriptome. *Environ. Exp. Bot.* **2022**, *203*, 105049. [CrossRef]

processes

MDPI

Article

Effects of L-Aspartic Acid on Cr(VI) Adsorption onto the Lepidocrocite with Different Exposed Facets: Batch Experiments and In Situ ATR-FTIR Analysis

Xiaofei Li [1,*], Tianfu Li [1], Xiaohu Jin [2], Yanfu Wei [3], Yanping Bao [1], Qian Yao [4], Fuhua Li [1], Weicheng Xu [1] and Xiaolian Wu [1]

1 School of Environmental and Chemical Engineering, Foshan University, Foshan 528000, China; 18437612672@163.com (T.L.)
2 National Demonstration Center for Environment and Planning, College of Environment and Planning, Henan University, Kaifeng 475004, China
3 National Observation and Research Station of Coastal Ecological Environments in Macao, Macao Environmental Research Institute, Macau University of Science and Technology, Taipa, Macao 999078, China
4 Guangdong Key Laboratory of Contaminated Environmental Management and Remediation, Guangdong Provincial Academy of Environmental Science, Guangzhou 510045, China
* Correspondence: xfli1103@fosu.edu.cn

Abstract: The adsorption of toxic metals onto iron oxides is a prevalent geochemical process in natural environments. Organic acids are known to modify the adsorption features of toxic ions through either competitive or cooperative effects. Nowadays, the toxic metal adsorption influenced by organic acids on iron oxides with varying facet exposures is not fully understood. This study explored how L-Aspartic acid (LA) influences Cr(VI) adsorption on two different exposure facets of lepidocrocite through batch adsorption experiments, in situ ATR-FTIR spectroscopy, and 2D-COS analysis. The results reveal that LA competes for available binding sites on lepidocrocite, consequently inhibiting the adsorption of Cr(VI). Rod-shaped lepidocrocite (R-LEP) owns more (001) facets and shows stronger Cr(VI) adsorption and LA competition than plate-like lepidocrocite (P-LEP), which mainly has (010) facets. The data for Cr(VI) uptake on both P-LEP and R-LEP within the effect of LA are well-fitted by the pseudo-second-order kinetics and the Freundlich isotherm model, suggesting chemical interaction as the dominant process for Cr(VI) coordination on lepidocrocite. Cr(VI) ions favor interaction with R-LEP over P-LEP, forming inner-sphere complexes on (001) facets. Concurrently, LA's carboxyl groups can compete for the active sites on the lepidocrocite surfaces, engaging in anion exchange with hydroxyl groups, and forming outer-sphere and inner-sphere structures. This competitive effect is particularly pronounced in the R-LEP system. The current findings are expected to broaden insights into how the exposed facets of lepidocrocite influence the fate of Cr(VI) in the organic acid coexistence environment.

Keywords: Cr(VI); L-Aspartic acid; adsorption; lepidocrocite; facet

1. Introduction

Hexavalent chromium (Cr(VI)) is the primary oxidation state of chromium in oxic environments [1]. Known for its toxicity and carcinogenicity, Cr(VI) poses a substantial threat to the health of both wildlife and humans [2,3]. Its environmental presence is largely linked to industrial activities, such as various industrial processes and the weathering of sulfidic ores and pyrite [4,5]. Cr(VI) is more mobile and bioavailable than trivalent chromium (Cr(III)) in biological systems, thereby amplifying its environmental risks [6]. Iron oxides, which are abundant in natural soils and sediments, are crucial in controlling the migration of Cr(VI) through interfacial interactions, such as adsorption, redox, and precipitation [7]. Consequently, elucidating how Cr(VI) interacts with iron oxide is imperative for mitigating environmental risks.

Lepidocrocite, with the chemical formula γ-FeO(OH), is a relatively unstable Fe-containing mineral frequently found in redoximorphic environments like sediments and paddy soils, playing a crucial role in regulating the fate of heavy metals (HMs) [8–10]. Moreover, lepidocrocite is recognized as a typical intermediate in the crystallographic transformation of precursor Fe oxides (ferrihydrite, Schwertmannite, etc.) [11–13]. This positioning enables lepidocrocite to serve as a bridging agent in controlling Fe cycling and trace element partitioning [14]. Therefore, investigating the surface reactivity of lepidocrocite through its interfacial interactions with heavy metals (HMs) holds significant environmental implications [8,15]. As well known, naturally formed lepidocrocite mainly consists of two typical morphologies, including plate- and rod-like shapes, presenting varying ratios of exposed facets [16–18]. These different morphologies correlate with differences in surface reactivity, such as adsorption performance, coordination structures, and interfacial electron transfer capability [8,19–21]. The plate-shaped lepidocrocite is primarily characterized by its (010) facet termination, which features a high density of bidentate hydroxyl groups (μ-OH), thereby supporting oxyanion adsorption [7,22–24]. Molybdenum(VI), for example, generates protonated binuclear monodentate complexes on the (010) facet of lepidocrocite [23]. Similarly, Otte et al. [24] identified doubly protonated MB complexes as the predominant species for As(V) on the same lepidocrocite facet. Rod-shaped lepidocrocite features a notably high percentage of exposed (001) facets, which have rich singly-coordinated hydroxyl sites that can form distinct coordination modes compared to those on the (010) lepidocrocite facet. Li et al. [8] observed that Cr(VI) engages in protonated monodentate binuclear (MB) coordination on the (010) facet and bidentate binuclear (BB) coordination on the (001) facet. Moreover, the increased presence of the (001) facet leads to improved adsorptive capacity for ions. Therefore, the adsorption behavior of trace elements on lepidocrocite is significantly determined by the exposed facets.

Contrary to simplified systems with a single toxic metal and mineral, real-world environments are more complex and contain multiple minerals, inorganic ions, dissolved organic matter, etc. [25,26]. Understanding competitive and cooperative interfacial processes with these coexisting substances is essential for accurately predicting adsorption behavior [7,27,28]. It is well established that organic acids significantly affect the mobility of HMs in natural environments, as they can compete with HMs for binding sites on iron oxide surfaces or form ternary complexes [15,29,30]. For example, under acidic conditions, oxalic acid outcompetes Cr(VI) for the exchangeable groups on lepidocrocite [31]. Furthermore, Li et al. [7] discovered that oxalic acid more readily reduced Cr(VI) adsorption on rod-shaped lepidocrocite compared to plate-like forms, a difference primarily due to the types of exposed facets. Therefore, the competitive adsorption behavior of HMs on iron oxide surfaces is also determined by the exposed facets. As well known, the amino acid-mineral surface interactions are of great interest to a wide range of geochemical topics and processes [32,33]. Numerous studies have mainly focused on the polymerization of amino acids (including glutamate, lysine, and aspartate) on mineral surfaces and their adsorption behavior in the presence of Ca^{2+}, Mg^{2+}, and PO_4^{2-} [25,34,35]. For example, Yang et al. [25] observed that phosphate competition substantially diminished aspartate adsorption capacity on the goethite surface. Conversely, aspartate had a negligible effect on phosphate adsorption, which was predominantly governed by nonprotonated bidentate, monoprotonated bidentate, and diprotonated monodentate complexes. However, limited knowledge is available regarding the interfacial interaction mechanism of Cr(VI) on lepidocrocite in competition with amino acids, especially concerning the in situ adsorption process.

This study primarily investigates the impacts of L-Aspartic acid (LA) on Cr(VI) adsorption onto two morphologically distinct forms of lepidocrocite, which expose different ratios of (001) to (010) facets. The objectives are: (1) to demonstrate Cr(VI) adsorption performance (including kinetics, isotherm, and pH effects) on these two different lepidocrocite when coexisting with L-Aspartic acid, (2) to reveal the dynamic interfacial interaction of Cr(VI) and L-Aspartic acid on the lepidocrocite surface at the molecular level, and (3) to further disclose the facet-dependent Cr(VI) and L-Aspartic acid co-adsorption mechanism on lepidocrocite.

The relevant findings are anticipated to shed new light on the roles of the lepidocrocite facet in controlling the fate of Cr(VI) within the organic acid coexistence environment.

2. Materials and Methods

2.1. Synthesis of Two Morphological Lepidocrocite

Two distinguishable morphologies of lepidocrocite (LEP), plate-shaped (P-LEP) and rod-shaped (R-LEP), were synthesized following the procedures reported by [8,36]. For P-LEP synthesis, 3.98 g of unoxidized $FeCl_2 \cdot 4H_2O$ (Strem Chemicals, Inc., Newburyport, MA, USA) crystals (20 mM Fe) were dissolved in 300 mL of deionized water with 0.2 M NaCl (Macklin Biochemical Co., Ltd., Shanghai, China) and stirred. The solution's pH was adjusted to 6.3 with 1 M NaOH (Macklin Biochemical Co., Ltd., Shanghai, China), and the air was supplied at 200 $mL \cdot min^{-1}$. The oxidation and hydrolysis of Fe ions proceeded until pH stabilization was achieved without further NaOH supplementation. For R-LEP, the synthesis procedure was similar but used 60 mM Fe(II) and 0.75 M NaCl. The mixture's pH was maintained at 6.7, and air was bubbled through at approximately 100 $mL \cdot min^{-1}$. Both end-products underwent centrifugation, multiple washes with pure water, and 24 hours of freeze-drying. The obtained solids were then ground and passed through 200 meshes.

2.2. Macroscopic Adsorption Experiments

The simultaneous adsorption of L-Aspartic acid (LA, Macklin Biochemical Co., Ltd., Shanghai, China) and Cr(VI) (K_2CrO_4, Aladdin Biochemical Technology Co., Ltd., Shanghai, China) was investigated through macroscopic adsorption experiments. In these experiments, 0.02 g of lepidocrocite was suspended in 20 mL of a 0.01 mM NaCl solution containing Cr(VI) and LA at varying concentrations. The reactions were carried out in 30 mL brown bottles under oxic conditions, which were sealed and shaken vigorously at 180 rpm and 25 °C in an orbital shaker to ensure thorough mixing.

Experiments for adsorption isotherms were performed with Cr(VI) concentrations varying between 20 to 200 μM, and LA at levels of 10 and 50 μM, while keeping the system pH constant at 5.0. To determine how pH influences adsorption, we investigated pH-dependent adsorption curves at pH values ranging from 4.0 to 8.0, each with a fixed 100 μM Cr(VI) and two different LA concentrations (10 and 50 μM). Additionally, the kinetics of adsorption were explored with initial Cr(VI) concentrations set at 100 μM and LA concentrations at 10 μM and 50 μM, under a pH of 5.0. Sampling intervals were predetermined, covering a duration from 0 to 2 h. Upon reaching equilibrium or at designated time intervals. The suspension (2 mL) was passed through a 0.22 μm filter to isolate the aqueous phase for subsequent measurement. The remaining suspension was poured into 50 mL brown bottles and spun at 8000 rpm for 10 min. The leftover solid was then cleaned with deionized water and vacuum freeze-dried. After drying, the solid particles were sieved through a 200-mesh screen and kept for subsequent measurements.

2.3. In Situ ATR-FTIR Experiments

The real-time ATR-FTIR studies were conducted on a Nicolet iS10 spectrometer (Thermo Fisher Scientific, Waltham, MA, USA), featuring a multi-bounce horizontal ATR unit and a flow cell system. The OMNIC software package, version 8.2, was used for managing data collection and analysis [37]. A ZnSe ATR crystal, oriented horizontally, was hidden beneath a lepidocrocite layer by evaporating a suspension containing 2.4 mg of lepidocrocite at a concentration of 2.0 $g \cdot L^{-1}$. The pH level of the solution was adjusted by introducing 0.1 M HCl or 0.1 M NaOH with a peristaltic pump set at a flow rate of 500 $\mu L \cdot min^{-1}$. ATR-FTIR spectra were captured at 25°C, scanning from 4000 to 600 cm^{-1} at a 4 cm^{-1} resolution with 125 scans.

Before conducting the adsorption studies, the lepidocrocite sample was equilibrated with a 0.01 M NaCl solution, and its spectrum was recorded to serve as the reference background. For the ATR-FTIR experiments, concentrations of 50 μM LA and 100 μM Cr(VI) were used, with pH values set at 8.0, 7.0, 6.0, 5.0, and 4.0. Spectra collection continued until successive scans showed no notable differences. Furthermore, it was verified that the lepidocrocite film remained intact throughout the adsorption experiments [38]. The ATR-FTIR spectra data were baseline-corrected before analysis [37].

2.4. Analytical Methods

X-ray powder diffraction (XRD, Bruker D8 Advance, Bremen, Germany) was used to confirm the synthesized lepidocrocite and the components of reacted solid samples. The measurements spanned from 10 to 80° in 2θ, with steps of 0.02°. SEM images (Merlin, ZEISS, Oberkochen, Germany) were applied to assess the morphologies of the two lepidocrocites. The specific surface areas of the two synthesized lepidocrocites were obtained from our previous study [8], which were determined using an AUTOSORB IQ analyzer from Quantchrome (Boynton Beach, FL, USA). The Cr(VI) concentrations were assessed by the 1,5-diphenylcarbazide (DPC) technique [39].

After baseline correction and smoothing, the ATR-FTIR data were further analyzed using Two-Dimensional Correlation Spectroscopy (2D-COS) with the 2D Shige software (version 1.3) [40]. For the synchronous map, the auto-peak is indicative of the variations in the intensity of a peak over time, whereas the cross-peak signifies the correlated response to temporal perturbations observed at two distinct bands. The positive cross-peak indicates a simultaneous increase or decrease in the intensities of the two bands, while a negative cross-peak shows that the intensities of the bands change in opposite directions. In the asynchronous spectra, the absence of auto-peaks indicates that there are no self-correlated signals. The asynchronous cross-peak indicates the presence of two uncorrelated bands, suggesting they stem from distinct surface complexes or varied molecular settings within a single complex [41].

3. Results and Discussions

3.1. Characterization of the Solid Samples

The XRD patterns' diffraction peaks were all successfully matched to the lepidocrocite (AMCSD 0020687) (Figure 1a,b) [7,8,42], confirming that the prepared solids corresponded to the crystalline phases of lepidocrocite and were free from any additional impurity phases. In addition, the co-adsorption of LA and Cr(VI) on the two lepidocrocite samples was further analyzed using XRD. As displayed in Figure S1, the diffraction peaks of the reacted samples were identical to those of pure lepidocrocite, suggesting that the presence of adsorbed LA and Cr(VI) did not cause any phase changes under the specified conditions. Therefore, the surface adsorption reaction of LA and Cr(VI) on LEP was the predominant interfacial process. Our previous investigation also showed that oxalic acid can inhibit Cr(VI) adsorption on lepidocrocite by competing for binding sites, with almost no effect on driving dissolution and recrystallization reactions [7].

The two morphological lepidocrocite samples were subsequently monitored through SEM images. As depicted in Figure 1c,d, the samples exhibited two distinct morphologies: plate- and rod-shaped particles denoted as P-LEP and R-LEP. As noted in previous studies, the R-LEP samples exhibited a notably higher abundance of the (001) facet in comparison to the P-LEP samples [8]. In contrast, P-LEP is chiefly distinguished by its (010) facet, with a minimal presence of the (001) facet [43]. Subsequently, the increased thickness of R-LEP suggests a higher proportion of (001) facets relative to (010) facets exposed on the lepidocrocite surface [16]. Furthermore, the reactivity on the lepidocrocite surface is affected by the distribution and density of the reactive sites it contains. On the perfect (010) facet of lepidocrocite, only surface bidentate hydroxyl groups are found, which are not effective for substitution with oxalic acid [9]. In contrast, the (001) facet of lepidocrocite consists of monodentate and bidentate hydroxyl groups [8,18], and the former groups serve as active sites and are favorable for the adsorption of organic acids [7,9]. Consequently, lepidocrocite variants with distinct exposed facets may display unique co-adsorption characteristics at the interface. Our previous study has shown that Cr(VI) adsorption on lepidocrocite (LEP) is highly facet-dependent [8]. Consequently, we deduce that the adsorption of Cr(VI) on R-LEP and P-LEP exhibits different patterns when LA is also present.

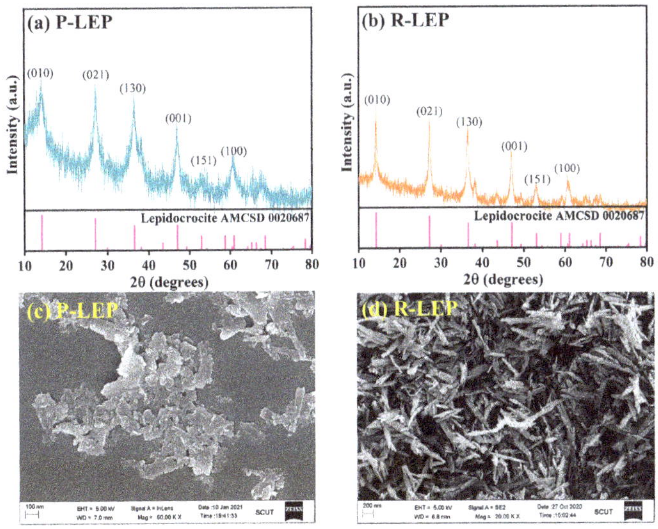

Figure 1. XRD patterns of P-LEP (**a**) and R-LEP (**b**), and their corresponding SEM images (**c**,**d**).

3.2. Cr(VI) Adsorption Kinetics in the Presence of LA

To elucidate the disparities in adsorption dynamics, the temporal progression of Cr(VI) uptake by R-LEP and P-LEP in the presence of LA was studied. The Cr(VI) uptake amounts were determined by normalizing the values concerning the specific surface area (P-LEP: 177 $m^2 \cdot g^{-1}$, R-LEP: 61 $m^2 \cdot g^{-1}$) [8]. As depicted in Figure 2a, at an initial LA concentration of 10 μM, the inhibitory effect on Cr(VI) adsorption was almost complete for both P-LEP and R-LEP. For the P-LEP system, the amount of Cr(VI) adsorbed declined from 0.029 $mg \cdot m^{-2}$ (5.15 $mg \cdot g^{-1}$) to 0.028 $mg \cdot m^{-2}$ (4.90 $mg \cdot g^{-1}$) with an increase in LA concentration from 10 to 50 μM, once equilibrium was reached. These values were significantly lower than the control group's value (0.065 $mg \cdot m^{-2}$) reported in our previous study [7]. This finding suggests that a competitive process takes place among these two anions on the P-LEP surface.

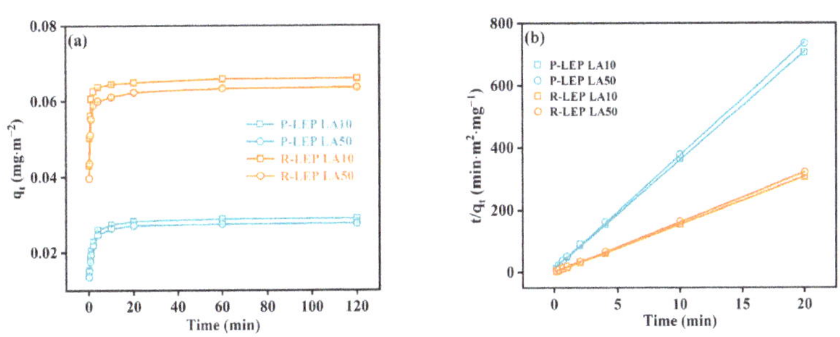

Figure 2. Time-dependent adsorption curves (**a**) and their kinetics plots (**b**) for Cr(VI) on P-LEP and R-LEP under the effects of LA.

In addition, R-LEP demonstrated superior adsorptive efficiency compared to P-LEP. As shown in Figure 2a, the saturation adsorption amount of Cr(VI) reached 0.065 $mg \cdot m^{-2}$ (3.97 $mg \cdot g^{-1}$) and 0.062 $mg \cdot m^{-2}$ (3.78 $mg \cdot g^{-1}$) under the influence of 10 μM and 50 μM LA, respectively. These values were considerably higher than those observed for P-LEP.

However, they were significantly lower than the control group's value (0.137 mg·m^{-2}) reported in our previous study [7,8], also confirming that a strong competition occurred between LA and Cr(VI) on the R-LEP surface. The decreased amplitude of Cr(VI) adsorption on R-LEP under the impact of 10 μM LA compared to the CK was lower than that on P-LEP; this could be attributed to the greater exposure of the (001) facet on R-LEP, which provided rich binding sites for LA adsorption and reduced the inhibition effects on Cr(VI) adsorption. Interestingly, as the LA concentration continued to increase, the decrease in Cr(VI) adsorption amount on R-LEP was more pronounced compared to P-LEP. This enhanced sensitivity could be due to the greater involvement of LA in competing with Cr(VI) for the binding groups on the (001) facet. According to prior research [9], oxalate acid interacts with the (001) facet through an inner-sphere complex, in contrast to the outer-sphere complex it forms with the (010) facet. Consequently, it more readily competes with Cr(VI) for the binding groups on the (001) facet. Moreover, the adsorption data were effectively described by the pseudo-second-order kinetic model (Text S1) [44], which is illustrated in Figure 2b and Table S1. This suggests that the adsorption of Cr(VI) onto the two LEP surfaces under the effects of LA was primarily driven by chemical bonding [45].

3.3. Cr(VI) Adsorption Isotherms

The inherent Cr(VI) adsorption capacity on the two different morphological lepidocrocites at varying LA concentrations was explored using isotherm equations (Supplementary Section S2). Figure 3 indicates that the experimental data fit the Freundlich isotherm equations more closely than the Langmuir isotherm, suggesting a favorable scenario for multi-layer adsorption and chemical adsorption processes [46]. Regarding Cr(VI) maximum adsorption amount on R-LEP, it was 0.068 mg·m^{-2} (4.15 mg·g^{-1}) and 0.065 mg·m^{-2} (3.97 mg·g^{-1}) with the addition of 10 and 50 μM LA, respectively (Table S2). These values were significantly lower than those in the LA-free system (0.137 mg·m^{-2}) [7], while being 2.24 and 2.25 times higher than those of P-LEP. The K_F values for the 10LA/P-LEP and 10LA/R-LEP systems were higher than those for 50LA/P-LEP and 50LA/R-LEP, confirming that Cr(VI) could be more readily adsorbed onto the LEP under the low concentration of LA. In our previous study, we estimated the proportions of the (010) and (001) facets on LEP to be 89.5% and 8.1% for P-LEP, and 67.6% and 30% for R-LEP, respectively [8]. This result also confirmed that competitive adsorption is facet-dependent. Furthermore, the adsorption of organic acids on the (010) facet predominantly occurs through outer-sphere complexes [7,9]. This process is primarily linked to the competitive adsorption that takes place on R-LEP, which has a higher concentration of (001) facet, as opposed to P-LEP, which is characterized by a predominance of (010) facet [8].

3.4. pH-Dependent Cr(VI) Adsorption onto LEP in the Presence of LA

Fluctuations in pH levels within the adsorption system can modify the adsorbent's surface charge and the forms of toxic metals and organic acids, resulting in varied adsorption behaviors for metals [47]. As depicted in Figure 4, there was a marked reduction in Cr(VI) adsorption capacity with increasing pH and LA levels, which implies that Cr(VI) adsorption is influenced by pH and is compromised by LA competition. Furthermore, the inhibitory effect for Cr(VI) uptake was more pronounced on R-LEP compared to P-LEP. The consequences are likely due to differences in the distribution of surface binding sites on the lepidocrocite. As noted in earlier studies, the bidentate μ-OH sites present on the (010) facet have a lower adsorption affinity for oxalate acid [9]. Nevertheless, such a facet is more conducive to the adsorption of oxyanions like chromate, molybdenum, and arsenate, forming monodentate structures [8,23,24]. In this study, it can be found that the inhibitory effects of LA on Cr(VI) adsorption exhibited a trend that first increased and then decreased as the pH continued to rise. These phenomena could be ascribed to the following reasons: (1) the speciation of LA around pH 4.0 primarily exists as a zwitterion with protonated NH_3^+ and γ-COOH [25]. At this condition, the Cr(VI) ions mainly existed as $HCrO_4^-$, which showed a high affinity to the Fe-OH active sites [34], behaving strongly competing with the protonated speciation of the LA for the surface sites. (2) Concurrently, the partial deprotonation of γ-COOH of LA occurred as pH increases and may benefit from its interfacial interactions with the Fe-OH active sites, then exhibited a more significant effect in inhibiting the Cr(VI) adsorption. (3) As the pH continued to increase, the deprotonation of Fe-OH active sites at alkaline conditions could also diminish the chemical adsorption of LA onto the surface [7]. Meanwhile, the majority of the γ-COOH groups occurred deprotonation, leading to a more negative charge and resulting in the interfacial structures transformation from chemical bonding structures (inner-sphere γ-COOH complexation) to the physical adsorption speciation (outer-sphere NH_2 complexation) [25]. Under the pH range from 6.0 to 8.0, partial Cr(VI) ions could also interact with the iron oxide surface via the inner-sphere complexes at the pH range from 6.0 to 8.0 [1,44]. Therefore, the variations in the speciation of aqueous LA with pH increasing also determined the adsorption of Cr(VI) on LEP.

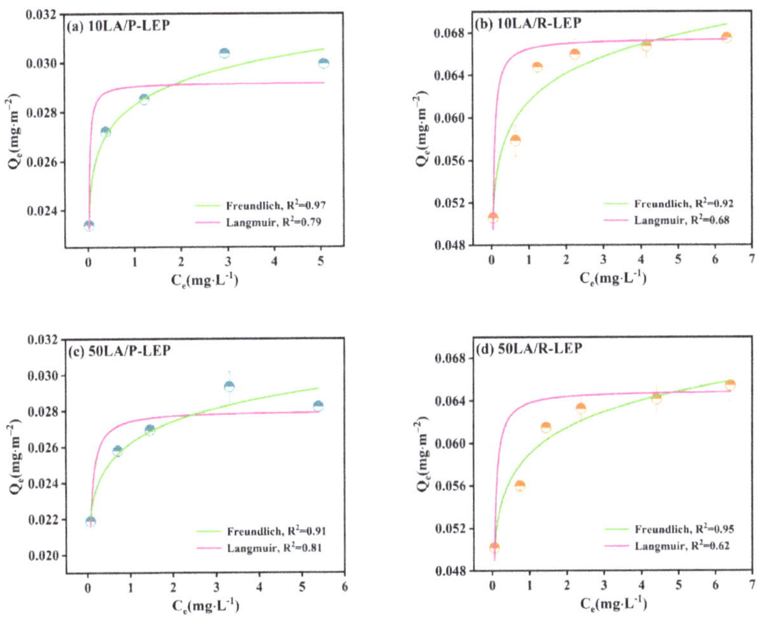

Figure 3. Isotherm analysis for Cr(VI) adsorption on P-LEP and R-LEP with Langmuir and Freundlich models at 10 μM (**a**,**c**) and 50 μM LA concentrations (**b**,**d**).

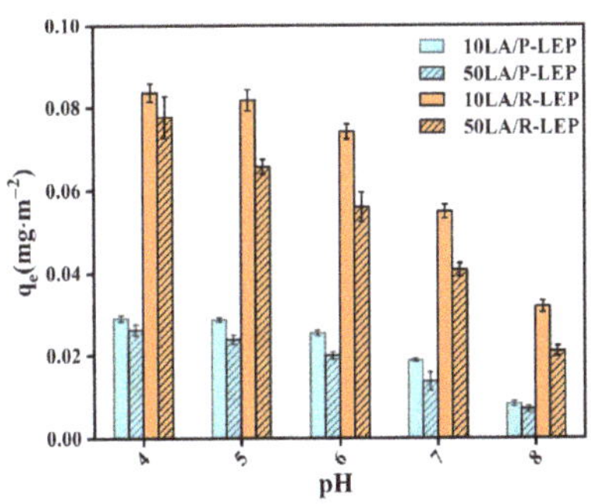

Figure 4. The influence of pH on Cr(VI) adsorption by P-LEP and R-LEP with varying LA concentrations.

3.5. Interfacial Dynamic Adsorption Process of Cr(VI) Under the Effects of LA

To gain insights into the dynamic simultaneous adsorption of Cr(VI) and LA on P-LEP and R-LEP, we employed in situ ATR-FTIR to probe their interfacial vibrational signals. It is crucial to first ascertain the speciation and vibrational characteristics of dissolved Cr(VI) and aqueous LA across a range of pH values. relevant information can be found in studies conducted by Johnston et al. [48] and Yang et al. [25]. Briefly, for dissolved Cr(VI), the peaks observed at 882 and 770 cm^{-1} correspond to the stretching modes of Cr-O and the asymmetric vibrations of the Cr-O-Cr bridge in $Cr_2O_7^{2-}$, respectively [49]. Meanwhile, the 901 cm^{-1} peak is indicative of the symmetric Cr-O stretching in $HCrO_4^-$ [37]. At pH 8.0, CrO_4^{2-} emerges as the predominant species in solution, which is evident from a distinct peak at 880 cm^{-1} [48]. For dissolved LA, there are three typical species: zwitterionic, monoanionic, and dianionic, which correspond to pH levels of 3.0, 6.0, and 11.0, respectively [25]. These species are primarily due to the (de)protonation of γ-COOH functional groups. ATR-FTIR analysis revealed that the $\delta_{as}(NH_3^+)$, $\delta_s(NH_3^+)$, and $\rho r(NH_3^+)$ modes are found at 1637, 1511, and 1272 cm^{-1}, respectively. The two bands at 1601 and 1396 cm^{-1} are attributed to the asymmetric and symmetric $\nu(\alpha\text{-}COO^-)$ vibrations, respectively [25]. As documented, the aspartate NH_2 group seldom engages in surface complexation on iron oxides, particularly at pH levels below 8.0 [50], whereas the -COOH group plays a significant role in the interaction [25]. Consequently, the observed spectral changes in this study are primarily attributed to the distinct interaction processes involving the -COOH group and the LEP surface.

Figure 5a,b display the ATR-FTIR spectroscopic analysis of the impact of varying pH on the simultaneous Cr(VI) and LA adsorption onto P-LEP and R-LEP. The spectra are categorized into three distinct regions: 1700–1100 cm^{-1} for the stretching vibrations of organic functional groups, 1100–980 cm^{-1} for Fe-OH vibrations [51], and 980–830 cm^{-1} for Cr-O vibrations [52]. The absorbance in the spectral region of 1700–1100 cm^{-1} increased significantly as the pH decreased from 8.0 to 4.0, suggesting substantial LA adsorption onto the LEP surface. The high spectral intensity, broader peaks, and well-defined shapes observed for LA adsorbed on R-LEP contrasted with those on P-LEP, suggesting the formation of outer-sphere and inner-sphere LA structures on R-LEP [25], which has a higher proportion of (001) facets. Concurrently, the region between 980–830 cm^{-1} exhibited a positive overlapping peak, with increasing absorbance as the pH decreased. This suggests that Cr(VI) can also be adsorbed onto both P-LEP and R-LEP surfaces under the influence of LA. Furthermore, the spectral position and shape of Cr-adsorbed R-LEP and P-LEP differed from those of dissolved Cr(VI) ions, indicating the formation of inner-sphere Cr(VI) complexes [1,37,48]. Meanwhile, the spectral region for Cr-adsorbed R-LEP exhibited more well-distinguishable than that for Cr-adsorbed P-LEP, proving a different ratio of surface inner-sphere Cr(VI) complexes formed on R-LEP and P-LEP [8]. As previously documented, Cr(VI) can coordinate with P-LEP and R-LEP by forming both binuclear monodentate and binuclear bidentate complexes. The binuclear monodentate Cr(VI) complex is the predominant structure on P-LEP, whereas the binuclear bidentate Cr(VI) complex is more abundant on R-LEP [8]. Additionally, the maximum peak position moved to higher wavenumbers as the pH decreased, indicating that Cr(VI) surface complexes might be transforming into more stable configurations. Interestingly, the region between 1100–980 cm^{-1} showed negative peaks, including a distinct peak at 1026 cm^{-1}, which represents the Fe-OH stretching vibration from the LEP surface. The absorbance of this peak decreased as the pH decreased, suggesting the release of hydroxyl groups from the P-LEP and R-LEP surfaces. This could be due to competition between LA and Cr(VI) for -OH groups on the LEP surface, leading to ligand exchange processes. Moreover, the amplitude decrease of the 1026 cm^{-1} peak was significantly greater in the R-LEP system than in P-LEP. This suggests that the active hydroxyl sites on R-LEP are more readily involved in ligand exchange with LA and Cr(VI) compared to those on P-LEP. This phenomenon can be attributed to the exposed (001) facet, which contains a higher concentration of singly-coordinated -OH [18]. Figure 5c,d display the kinetic curves for the three spectral regions, indicating a strong correlation between the adsorption of LA and Cr(VI) and the release of surface-OH groups. Furthermore, R-LEP, with its increased availability of active sites on (001) facets, facilitates more competitive adsorption of Cr(VI) in the presence of LA, leading to a higher release of -OH groups.

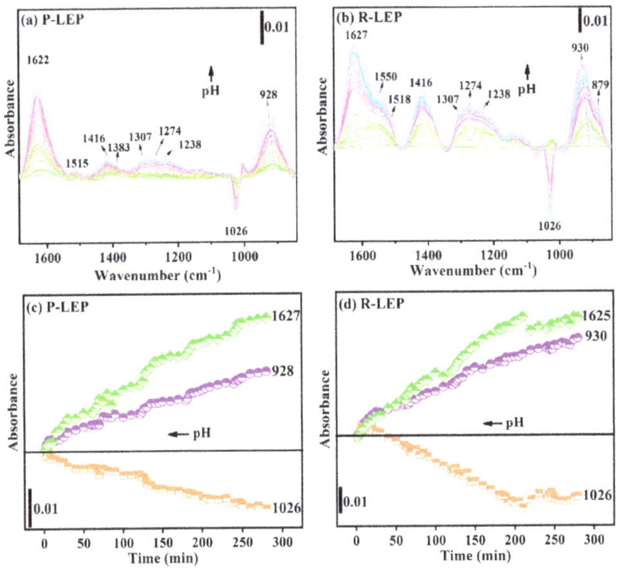

Figure 5. ATR-FTIR data collected as a function of time with flowing pH 8.0, 7.0, 6.0, 5.0, and 4.0 of LA and Cr(VI) on P-LEP (**a**) and R-LEP (**b**). Kinetic curves for the representative peaks of adsorbed LA and Cr(VI) and desorbed -OH groups for P-LEP (**c**) and R-LEP (**d**).

The 2D correlation contour plots were utilized to resolve the overlapping ATR-FTIR peaks [38], thereby improving our understanding of the interaction mechanisms between LA and Cr(VI) on the LEP surfaces. Figure 6a and Table S3 illustrate the synchronous correlation map for the LA-Cr/P-LEP systems, which features five diagonal peaks at 1622, 1416, 1274, 1026, and 928 cm^{-1}. These peaks indicate a significant change in spectral intensity associated with increasing reaction time and decreasing pH. Notably, the peaks at 1416 and 1274 cm^{-1} are less intense compared to the others. The peak at 1416 cm^{-1} was assigned to CH_2 scissoring [22]. Concurrently, the cross peaks at (1622, 1416), (1622, 1274), (1622, 928), (1416, 1274), (1416, 928), and (1274, 928) showed positive signals, indicating a similar trend in spectral intensity changes. In contrast, the negative cross peaks at (1622, 1026), (1416, 1026), (1274, 1026), and (1026, 928) suggest that the peak at 1026 cm^{-1} behaves differently, confirming distinct interfacial reactions, such as adsorption and desorption processes. In the asynchronous correlation map (Figure 6c and Table S4), the peak pairs at (1622, 1026), (1416, 1026), (1274, 1026), and (1026, 928) exhibited negative signals, mirroring the features observed in the synchronous correlation map. This consistency indicates that the peak at 1026 cm^{-1} is distinct and does not overlap with other peaks. As mentioned above, this peak is confirmed to be associated with the vibration of hydroxyl groups on the LEP surface. Concurrently, the peak pair at (1274, 928) exhibited a positive signal, which could be ascribed to the different vibrational sources among these two peaks. The former is assigned to the symmetric vibration of C-O, while the latter is attributed to the Cr-O vibration [9,37]. Additionally, the peak pairs, including (1612, 1274), (1622, 1378), and (1622, 1515), were also found in the synchronous correlation map. The peaks at 1515 and 1274 cm^{-1} are indicative of inner-sphere complexes involving carboxyl groups. The peak at 1612 cm^{-1}, corresponding to the asymmetric stretch of the COO^- group, and the peak at 1378 cm^{-1}, associated with the symmetric stretch, are likely attributed to outer-sphere surface coordination on P-LEP. These observations are consistent with the spectral characteristics of carboxyl groups in various coordination environments as reported in the literature [25,53]. These frequencies fall within the range of asymmetric stretching frequencies for aqueous carboxylates, which are typically found between 1540 and 1650 cm^{-1}, and symmetric stretching frequencies, which are found between 1300 and 1420 cm^{-1} [54]. In addition, it is observed that the maximum peak position at high pH is relatively lower than that at acidic conditions (Figure 5a,b), suggesting the formation of different complexes. Previous studies illustrated that the carboxyl group coordination modes change from outer-sphere to inner-sphere structures with pH decreasing, causing the corresponding peak to shift to the higher wavenumber [9,25]. Additionally, the 928 cm^{-1} peak is associated with the vibrational stretching of Cr-O bonds, indicating the formation of inner-sphere complexes as reported previously [1].

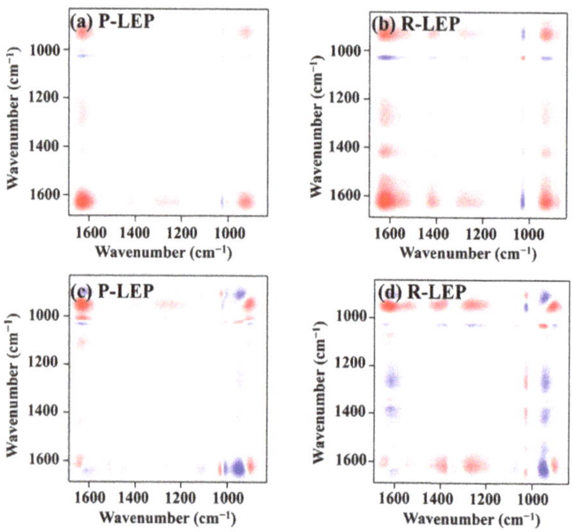

Figure 6. Synchronous contour plots for LA and Cr(VI) adsorbed onto P-LEP (**a**) and R-LEP (**b**) across varying pH levels, with corresponding asynchronous plots (**c**,**d**).

For the R-LEP, the synchronous correlation map (Figure 6b and Table S5) revealed six distinct diagonal peaks at 1627, 1550, 1416, 1274, 1026, and 930 cm^{-1}. Notably, the peak at 1550 cm^{-1}, corresponding to the asymmetric stretching vibration of COO^- ($v_{as}(COO^-)$), was more pronounced than in the LA-Cr/P-LEP system's map, suggesting a greater interaction between LA and R-LEP via the carboxyl groups. This phenomenon is attributed to the higher exposure ratio of the (001) facet on R-LEP [7]. The signal change trends for the other peaks were similar to those observed in the LA-Cr/P-LEP systems. As seen in the asynchronous correlation map (Figure 6d) and Table S6, the negative peaks observed at (1602, 1550) and (1602, 1385) suggest that these three peaks might be associated with the stretching vibrations of the carboxyl groups, and are identified as part of outer-sphere complexes [25]. Meanwhile, the peak at 1602 cm^{-1} shifted significantly to a higher wavenumber at 1627 cm^{-1} with pH decreasing, which might be attributed to the strong interaction between LA and the R-LEP surface [51]. In addition, the peak pair at (1602, 1518) presented a negative signal, which cannot be observed in the synchronous correlation map. We concluded that these two peaks originated from different surface complexes, and the peak at 1518 cm^{-1} was assigned to the bidentate binding from the carboxyl groups [25,51]. The peak pair at (1274, 1026) showed a negative signal, confirming that the 1026 cm^{-1} peak originates from the vibration of surface Fe-OH complex structures. In summary, we conclude that the interfacial reactions involve three main processes: (1) LA interacts with LEP via carboxyl group coordination, forming both outer-sphere and inner-sphere structures; (2) Cr(VI) adsorption on LEP, primarily through the formation of inner-sphere structures, including binuclear bidentate and monodentate structures; (3) competition between LA and Cr(VI) for surface hydroxyl groups, leading to ligand exchange processes.

4. Conclusions

This research illuminated the influence of LA on the adsorption of Cr(VI) by two types of lepidocrocite featuring distinct exposed facets. LA was identified as a competitor for the lepidocrocite surface active sites, thereby inhibiting the adsorption of Cr(VI). The dominant conclusions were summarized as follows:

- The competitive adsorption processes were found to be facet-dependent. Specifically, the rod-shaped lepidocrocite (R-LEP), which possesses a higher ratio of (001) facet, exhibits greater efficacy in Cr(VI) adsorption and resistance to LA competition compared to plate-shaped lepidocrocite (P-LEP), which is primarily composed of (010) facet.
- Cr(VI) interacts with the LEP surface under the influence of LA through a chemical reaction. The inner-sphere Cr(VI) complexation is more favored with R-LEP compared to P-LEP, demonstrating that more (001) facets provide more monodentate binding sites (-OH) for the competitive adsorption between Cr(VI) and LA.
- In addition, LA adsorbs onto LEP through the coordination of its carboxyl groups, forming inner-sphere and outer-sphere structures. The presence of inner-sphere LA complexes, which are primarily associated with the exposed (001) facet. The ligand exchange reaction with the hydroxyl groups on LEP appears to be the predominant interaction mechanism.

The findings shed light on the impact of lepidocrocite's exposed facets and coexisting substances on Cr(VI) migration in complex environments.

Supplementary Materials: The following supporting information can be downloaded at: https://www.mdpi.com/article/10.3390/pr12112598/s1, Figure S1: XRD patterns of P-LEP (a) and R-LEP (b) under the effects of 50 μM LA; Table S1: Kinetics parameters of Cr(VI) adsorption onto P-LEP and R-LEP; Table S2: Equilibrium adsorption isotherm fitting parameters; Table S3: The synchronous 2D correlation contour plot of LA and Cr(VI) co-adsorbed P-LEP; Table S4: The asynchronous 2D correlation contour plot of LA and Cr(VI) co-adsorbed P-LEP; Table S5: Results from the synchronous 2D correlation contour plot of LA and Cr(VI) co-adsorbed R-LEP; Table S6: Results from the asynchronous 2D correlation contour plot of LA and Cr(VI) co-adsorbed R-LEP. References cited in Supplementary Materials [8,55,56].

Author Contributions: X.L.: writing—original draft, conceptualization, funding acquisition, supervision; T.L.: investigation, data curation, methodology, writing—review and editing; X.J.: investigation, visualization, writing—review and editing; Y.W.: formal analysis, writing—review and editing; Y.B.: visualization, writing—review and editing; Q.Y.: software, writing—review and editing; F.L.: writing—review and editing; W.X.: writing–review and editing; X.W.: writing—review and editing. All authors have read and agreed to the published version of the manuscript.

Funding: This work was supported by the Guangdong Basic and Applied Basic Research Foundation, China (Nos. 2022A1515110918, 2022A1515110444, 2023A1515110359), Guangdong Natural Science Foundation (2023A1515012128), the National Natural Science Foundation of China (No. 42307278), the Science and Technology Development Fund of Macau SAR (Grant No. 0094/2022/A2).

Data Availability Statement: All relevant data has been included in the article.

Conflicts of Interest: The authors declare no conflict of interest.

References

1. Johnston, C.P.; Chrysochoou, M. Investigation of chromate coordination on ferrihydrite by in situ ATR-FTIR spectroscopy and theoretical frequency calculations. *Environ. Sci. Technol.* **2012**, *46*, 5851–5858. [CrossRef]
2. Qi, T.Y.; Zhang, S.; Zhang, J.Z.; Li, T.; Xing, L.; Fang, Z.M.; An, S.L.; Xu, Z.F.; Xiao, H.N.; Wang, L.D. In situ reconstruction of active catalysis sites triggered by chromium immobilization for sulfite oxidation. *Environ. Sci. Technol.* **2023**, *57*, 3905–3916. [CrossRef]

3. Kebir, M.; Tahraoui, H.; Chabani, M.; Trari, M.; Noureddine, N.; Assadi, A.A.; Amrane, A.; Ben Hamadi, N.; Khezami, L. Water cleaning by a continuous fixed-bed column for Cr(VI) eco-adsorption with green adsorbent-based biomass: An experimental modeling study. *Processes* **2023**, *11*, 363. [CrossRef]
4. Xie, Y.Y.; Ye, H.; Wen, Z.; Dang, Z.; Lu, G.N. Sulfide-induced repartition of chromium associated with schwertmannite in acid mine drainage: Impacts and mechanisms. *Sci. Total. Environ.* **2022**, *848*, 157863. [CrossRef] [PubMed]
5. Zhao, Q.; Yang, B.B.; Ren, H.W.; Chen, S.H.; Luo, C.H.; Li, Q.D.; Yang, W.; Yan, K.P. Removal of Cr(VI) from aqueous systems using Fe-P slag as a reducing agent. *Hydrometallurgy* **2022**, *211*, 105875. [CrossRef]
6. Chattopadhyay, B.; Utpal, S.; Mukhopadhyay, S.K. Mobility and bioavailability of chromium in the environment: Physico-chemical and microbial oxidation of Cr (III) to Cr (VI). *J. Appl. Sci. Environ.* **2010**, *14*, 97–101. [CrossRef]
7. Li, X.F.; Guo, C.L.; Pillai, S.C.; Jin, X.H.; Yao, Q.; Bao, Y.P.; Jiang, X.D.; Lu, G.N.; Wang, H.L.; Dang, Z. Facet-dependent competitive adsorption mechanisms of chromate and oxalic acid on γ-FeO(OH) nanocrystals. *Langmuir* **2023**, *39*, 14539–14549. [CrossRef]
8. Li, X.F.; Guo, C.L.; Jin, X.H.; Yao, Q.; Liu, Q.Q.; Zhang, L.J.; Lu, G.N.; Reinfelder, J.R.; Huang, W.L.; Dang, Z. Molecular-scale study of Cr(VI) adsorption onto lepidocrocite facets by EXAFS, in situ ATR-FTIR, theoretical frequency calculations and DFT+U techniques. *Environ. Sci. Nano* **2022**, *9*, 568–581. [CrossRef]
9. Borowski, S.C.; Biswakarma, J.; Kang, K.; Schenkeveld, W.D.C.; Hering, J.G.; Kubicki, J.D.; Kraemer, S.M.; Hug, S.J. Structure and reactivity of oxalate surface complexes on lepidocrocite derived from infrared spectroscopy, DFT-calculations, adsorption, dissolution and photochemical experiments. *Geochim. Cosmochim. Acta* **2018**, *226*, 244–262. [CrossRef]
10. Liao, S.; Wang, X.M.; Yin, H.; Post, J.E.; Feng, X.H. Effects of Al substitution on local structure and morphology of lepidocrocite and its phosphate adsorption kinetics. *Geochim. Cosmochim. Acta* **2020**, *276*, 109–121. [CrossRef]
11. Chen, C.M.; Kukkadapu, R.; Sparks, D.L. Influence of coprecipitated organic matter on $Fe^{2+}_{(aq)}$-catalyzed transformation of ferrihydrite: Implications for carbon dynamics. *Environ. Sci. Technol.* **2015**, *49*, 10927–10936. [CrossRef] [PubMed]
12. Schoepfer, V.A.; Burton, E.D.; Johnston, S.G. Contrasting effects of phosphate on the rapid transformation of schwertmannite to Fe(III) (oxy)hydroxides at near-neutral pH. *Geoderma* **2019**, *340*, 115–123. [CrossRef]
13. Jin, X.H.; Li, X.F.; Guo, C.L.; Jiang, M.G.; Yao, Q.; Lu, G.N.; Dang, Z. Fate of oxalic-acid-intervened arsenic during Fe(II)-induced transformation of As(V)-bearing jarosite. *Sci. Total. Environ.* **2020**, *719*, 137311. [CrossRef] [PubMed]
14. Jin, X.H.; Guo, C.L.; Tao, X.Q.; Li, X.F.; Xie, Y.Y.; Dang, Z.; Lu, G.N. Divergent redistribution behavior of divalent metal cations associated with Fe(II)-mediated jarosite phase transformation. *Environ. Pollut.* **2024**, *350*, 124004. [CrossRef]
15. Cao, Q.Q.; Guo, C.L.; Ren, M.H.; Li, X.F.; Xu, Z.R.; Wang, C.P.; Lu, G.N.; Dang, Z. Influence of tartaric acid on the electron transfer between oxyanions and lepidocrocite. *J. Hazard. Mater.* **2024**, *476*, 135082. [CrossRef]
16. Kozin, P.A.; Salazar-Alvarez, G.; Boily, J.F. Oriented aggregation of lepidocrocite and impact on surface charge development. *Langmuir* **2014**, *30*, 9017–9021. [CrossRef]
17. Paikaray, S.; Schröder, C.; Peiffer, S. Schwertmannite stability in anoxic Fe(II)-rich aqueous solution. *Geochim. Cosmochim. Acta* **2017**, *217*, 292–305. [CrossRef]
18. Ding, X.; Song, X.; Boily, J.F. Identification of fluoride and phosphate binding sites at FeOOH surfaces. *J. Phys. Chem. C* **2012**, *116*, 21939–21947. [CrossRef]
19. Hunge, Y.M.; Yadav, A.A.; Kang, S.W.; Kim, H.; Fujishima, A.; Terashima, C. Nanoflakes-like nickel cobaltite as active electrode material for 4-nitrophenol reduction and supercapacitor applications. *J. Hazard. Mater.* **2021**, *419*, 126453. [CrossRef]
20. Hunge, Y.M.; Yadav, A.A.; Kang, S.-W.; Kim, H. Facile synthesis of multitasking composite of Silver nanoparticle with Zinc oxide for 4-nitrophenol reduction, photocatalytic hydrogen production, and 4-chlorophenol degradation. *J. Alloys Compd.* **2022**, *928*, 167133. [CrossRef]
21. Yadav, A.A.; Hunge, Y.M.; Majumder, S.; Mourad, A.H.I.; Islam, M.M.; Sakurai, T.; Kang, S.W. Multiplicative rGO/Cu-BDC MOF for 4-nitrophenol reduction and supercapacitor applications. *J. Colloid Interface Sci.* **2025**, *677*, 161–170. [CrossRef]
22. Das, S.; Hendry, M.J.; Essilfie-Dughan, J.A. Adsorption of selenate onto ferrihydrite, goethite, and lepidocrocite under neutral pH conditions. *Appl. Geochem.* **2013**, *77*, 185–193. [CrossRef]
23. Davantès, A.; Costa, D.; Lefèvre, G. Molybdenum(VI) adsorption onto lepidocrocite (γ-FeOOH): In situ vibrational spectroscopy and DFT+U theoretical study. *J. Phys. Chem. C* **2016**, *120*, 11871–11881. [CrossRef]
24. Otte, K.; Schmahl, W.W.; Pentcheva, R. DFT+U study of arsenate adsorption on FeOOH surfaces: Evidence for competing binding mechanisms. *J. Phys. Chem. C* **2013**, *117*, 15571–15582. [CrossRef]
25. Yang, Y.L.; Wang, S.R.; Xu, Y.S.; Zheng, B.H.; Liu, J.Y. Molecular-scale study of aspartate adsorption on goethite and competition with phosphate. *Environ. Sci. Technol.* **2016**, *50*, 2938–2945. [CrossRef] [PubMed]
26. Meng, X.Q.; Wang, X.M.; Zhang, C.M.; Yan, S.; Zheng, G.Y.; Zhou, L.X. Co-adsorption of As(III) and phenanthrene by schwertmannite and Fenton-like regeneration of spent schwertmannite to realize phenanthrene degradation and As(III) oxidation. *Environ. Res.* **2021**, *195*, 110855. [CrossRef]
27. Flynn, E.D.; Catalano, J.G. Competitive and cooperative effects during nickel adsorption to iron oxides in the presence of oxalate. *Environ. Sci. Technol.* **2017**, *51*, 9792–9799. [CrossRef] [PubMed]
28. Situm, A.; Rahman, M.A.; Allen, N.; Kabengi, N.; Al-Abadleh, H.A. ATR-FTIR and flow microcalorimetry studies on the initial binding kinetics of arsenicals at the organic–hematite interface. *J. Phys. Chem. A* **2017**, *121*, 5569–5579. [CrossRef]
29. Mesuere, K.; Fish, W. Chromate and oxalate adsorption on goethite. 2. Surface complexation modeling of competitive adsorption. *Environ. Sci. Technol.* **1992**, *26*, 2365–2370. [CrossRef]

30. Yao, W.B.; Huang, L.; Yang, Z.H.; Zhao, F.P. Effects of organic acids on heavy metal release or immobilization in contaminated soil. *Trans. Nonferrous Met. Soc. China* **2022**, *32*, 1277–1289. [CrossRef]
31. Hu, S.W.; Li, H.; Wang, P.; Liu, C.X.; Shi, Z.Q.; Li, F.B.; Liu, T.X. Interfacial photoreactions of Cr(VI) and oxalate on lepidocrocite surface under oxic and acidic conditions: Reaction mechanism and potential implications for contaminant degradation in surface waters. *Chem. Geol.* **2021**, *583*, 120481. [CrossRef]
32. Kitadai, N.; Nishiuchi, K. Thermodynamic impact of mineral surfaces on amino acid polymerization: Aspartate dimerization on goethite. *Astrobiology* **2019**, *19*, 1363–1376. [CrossRef] [PubMed]
33. Lambert, J.F. Adsorption and polymerization of amino acids on mineral surfaces: A review. *Orig. Life Evol. Biosph.* **2008**, *38*, 211–242. [CrossRef] [PubMed]
34. Lee, N.; Sverjensky, D.A.; Hazen, R.M. Cooperative and competitive adsorption of amino acids with Ca^{2+} on rutile (α-TiO_2). *Environ. Sci. Technol.* **2014**, *48*, 9358–9365. [CrossRef] [PubMed]
35. Yeasmin, S.; Singh, B.; Kookana, R.S.; Farrell, M.; Sparks, D.L.; Johnston, C.T. Influence of mineral characteristics on the retention of low molecular weight organic compounds: A batch sorption-desorption and ATR-FTIR study. *J. Colloid. Interf. Sci.* **2014**, *432*, 246–257. [CrossRef]
36. Lewis, D.G.; Farmer, V.C. Infrared absorption of surface hydroxyl groups and lattice vibrations in lepidocrocite (γ-FeOOH) and boehmite (γ-AlOOH). *Clay. Miner.* **1986**, *21*, 93–100. [CrossRef]
37. Li, X.F.; Guo, C.L.; Jin, X.H.; He, C.C.; Dang, Z. Mechanisms of Cr(VI) adsorption on schwertmannite under environmental disturbance: Changes in surface complex structures. *J. Hazard. Mater.* **2021**, *416*, 125781. [CrossRef]
38. Jing, L.; Zhu, R.; Liang, X.; Ma, L.; Molinari, M. Synergistic adsorption of Cd(II) with sulfate/phosphate on ferrihydrite: An in situ ATR-FTIR/2D-COS study. *Chem. Geol.* **2018**, *477*, 12–21.
39. Bartlett, R.; James, B. Behavior of Chromium in Soils: III. *Oxidation1. J. Environ. Qual.* **1979**, *8*, 31–35. [CrossRef]
40. Yan, W.; Jing, C.Y. Molecular insights into glyphosate adsorption to goethite gained from ATR-FTIR, two-dimensional correlation spectroscopy, and DFT study. *Environ. Sci. Technol.* **2018**, *52*, 1946. [CrossRef]
41. Noda, I.; Dowrey, A.E.; Ozaki, Y. *Two-Dimensional Correlation Spectroscopy. Applications in Vibrational and Optical Spectroscopy*; John Wiley & Sons, Inc.: Hoboken, NJ, USA, 2004.
42. Christensen, A.N.; Lehmann, M.S.; Convert, P.; Beyer, L.; Cyvin, S.J. Deuteration of crystalline hydroxides. Hydrogen bonds of gamma-AlOO(H,D) and gamma-FeOO(H,D). *Acta Chem. Scand* **1982**, *36*, 303–308. [CrossRef]
43. Song, X.; Boily, J.F. Structural controls on OH site availability and reactivity at iron oxyhydroxide particle surfaces. *Phys. Chem. Chem. Phys.* **2012**, *14*, 2579–2586. [CrossRef]
44. Li, T.F.; Li, X.F.; Shen, C.R.; Chen, D.; Li, F.H.; Xu, W.C.; Wu, X.L.; Bao, Y.P. Nano-hydroxyapatite modified tobacco stalk-based biochar for immobilizing Cd(II): Interfacial adsorption behavior and mechanisms. *Processes* **2024**, *12*, 1924. [CrossRef]
45. Gan, M.; Sun, S.; Zheng, Z.; Tang, H.; Sheng, J.; Zhu, J.; Liu, X. Adsorption of Cr(VI) and Cu(II) by $AlPO_4$ modified biosynthetic Schwertmannite. *Appl. Surf. Sci.* **2015**, *356*, 986–997. [CrossRef]
46. Wang, X.H.; Liu, F.F.; Lu, L.; Yang, S.; Zhao, Y.; Sun, L.B.; Wang, S.G. Individual and competitive adsorption of Cr(VI) and phosphate onto synthetic Fe-Al hydroxides. *Colloid. Surf. A* **2013**, *423*, 42–49. [CrossRef]
47. Liu, B.X.; Chen, C.; Li, W.; Liu, H.Y.; Liu, L.; Deng, S.S.; Li, Y. Effective removal of Cr(VI) from aqueous solution through adsorption and reduction by magnetic S-doped Fe-Cu-La trimetallic oxides. *J. Environ. Chem. Eng.* **2022**, *10*, 107433. [CrossRef]
48. Johnston, C.P.; Chrysochoou, M. Mechanisms of chromate adsorption on hematite. *Geochim. Cosmochim. Acta* **2014**, *138*, 146–157. [CrossRef]
49. Hoffmann, M.M.; Darab, J.G.; Fulton, J.L. An infrared and X-ray absorption study of the structure and equilibrium of chromate, bichromate, and dichromate in high-temperature aqueous solutions. *J. Phys. Chem. A* **2001**, *105*, 6876–6885. [CrossRef]
50. Parikh, S.J.; Kubicki, J.D.; Jonsson, C.M.; Jonsson, C.L.; Hazen, R.M.; Sverjensky, D.A.; Sparks, D.L. Evaluating glutamate and aspartate binding mechanisms to rutile (α-TiO_2) via ATR-FTIR spectroscopy and quantum chemical calculations. *Langmuir* **2011**, *27*, 1778–1787. [CrossRef] [PubMed]
51. Hug, S.J.; Bahnemann, D. Infrared spectra of oxalate, malonate and succinate adsorbed on the aqueous surface of rutile, anatase and lepidocrocite measured with in situ ATR-FTIR. *J. Electron. Spectrosc. Relat. Phenom.* **2006**, *150*, 208–219. [CrossRef]
52. Li, X.F.; Li, T.F.; Jeyakumar, P.; Li, J.Y.; Jin, X.H.; Zhang, J.; Guo, C.L.; Jiang, X.D.; Lu, G.N.; Dang, Z.; et al. Effect of biochar-derived DOM on contrasting redistribution of chromate during Schwertmannite dissolution and recrystallization. *J. Hazard. Mater.* **2024**, *476*, 134988. [CrossRef] [PubMed]
53. Simanova, A.A.; Loring, J.S.; Persson, P. Formation of ternary metal-oxalate surface complexes on α-FeOOH particles. *J. Phys. Chem. C* **2011**, *115*, 21191–21198. [CrossRef]
54. Hay, M.B.; Myneni, S.C.B. Structural environments of carboxyl groups in natural organic molecules from terrestrial systems. Part 1: Infrared spectroscopy. *Geochim. Cosmochim. Acta* **2007**, *71*, 3518–3532. [CrossRef]
55. Langmuir, I. The Adsorption of gases on plane surfaces of glass, mica and platinum. *J. Am. Chem. Soc.* **1918**, *40*, 1361–1403. [CrossRef]
56. Freundlich, H.M.F. Über die adsorption in lösungen Z. *Phys. Chem.* **1906**, *57A*, 385–470. [CrossRef]

processes

MDPI

Article

Comparing the Aging Processes of PLA and PE: The Impact of UV Irradiation and Water

Xucheng Wang, Jinxin Chen, Wenhao Jia, Kaibo Huang and Yini Ma *

Key Laboratory of Agro-Forestry Environmental Processes and Ecological Regulation of Hainan Province, School of Ecology and Environment, Hainan University, Haikou 570228, China
* Correspondence: mayini@hainanu.edu.cn

Abstract: The aging processes of microplastics (MPs) are prevalent in natural environments. Understanding the aging mechanisms of MPs is crucial for assessing their environmental behavior and potential risks. In this study, we selected polylactic acid (PLA) and polyethylene (PE) as representatives of biodegradable and conventional plastics, respectively, to examine changes in their physicochemical properties induced by water and UV light exposure. Laboratory aging resulted in significant fragmentation, characterized by cracks and pores on the surfaces, for both types of MPs, with PLA MPs exhibiting more severe changes, particularly under combined UV and water exposure. Notably, PLA MPs tended to become progressively smaller after aging, whereas PE MPs did not show significant size changes. Chemical analyses of aged MPs using micro-Fourier-transform infrared spectroscopy (FTIR) and X-ray photoelectron spectroscopy (XPS) revealed a substantial increase in the carbonyl index (CI) and oxygen content for PE, suggesting surface oxidation during photo-oxidation. Conversely, PLA MPs displayed a CI decrease, along with an oxygen content increase, indicating the breakdown of ester linkages in PLA and the formation of other oxidation products. Furthermore, we developed and optimized pyrolysis–gas chromatography–mass spectrometry (Py–GC–MS) methods to identify potential chemical degradation products of PE and PLA, considering their differing thermal stabilities. We observed a distinct trend regarding the peaks in the chromatogram of aged MPs and identified the typical oxidation and crosslinking products for PLA. Additionally, after the aging process, both PE and PLA exhibited a significant increase in organic carbon content, with the eluate containing submicron/nano-sized particles. This study provides a scientific foundation for a deeper understanding of the environmental aging mechanisms of various MPs, particularly in regards to the effects of UV irradiation and water exposure.

Keywords: aging; particle size; PLA; PE; Py–GC–MS

1. Introduction

Plastics are widely used and produced around the world, and plastic pollution is a global problem. Because of their stable physical and chemical properties, plastics can be stored for long periods in marine [1], terrestrial [2], atmospheric [3], riverine [4], and sewage [5] environments. Plastic waste is affected by environmental factors (light, temperature, pH, salinity, etc.). It can undergo a series of degradation processes, forming MPs with a particle size of less than 5 mm which are then released into the environment and further transformed into smaller nano plastic particles [6,7]. The degradation and aging of MPs are usually characterized by changes in the specific surface area, porosity, functional groups, chemical structure, and hydrophilic and hydrophobic properties of the MPs [8]. Further potential risks might result from the aging and degradation process, e.g., plastics are more likely to be ingested by plants and animals and retained in them as they change from larger to smaller sizes [9]. At the same time, after a series of degradations and transformations, MPs are more likely to coexist with pollutants in the environment [10], acting as carriers for the accumulation of heavy metals and hydrophobic organic pollutants from the environment, leading to bioconcentration, indirect toxicity, and thus higher biological risks [11,12].

UV irradiation is one of the main drivers of plastic aging, through the energy generated by light aging, attacking the C-H bond in the structure of MPs, generating alkyl radicals, which further react with hydroxyl and oxygen radicals to generate hydroxyl, carbonyl, and carboxyl structures and other intermediate and end products [13,14]. The generation of free radicals and the mechanism of photoreaction might vary depending on the type of plastic involved. UV irradiation leads to the formation of tertiary alkoxy and peroxyl radicals in PS, whereas PE tends to form alkoxy and carbonyl radicals [15]. PS contains a large number of aromatic ring structures, and its high sensitivity to light might produce more aromatic photodegradation products, while PE degradation produces more alkane substances [16]. These aging effects on plastics, produced by photodegradation radical reactions, are usually manifested by surface fragmentation of the polymer, leading to the structural formation of cracks, increasing the specific surface area of the MPs, which is further manifested by the generation of surface lobes and fine particles [17].

Unlike traditional petroleum-based plastic polymers, biodegradable plastics are typically prepared from biopolymers (polylactic acid, thermoplastic starch cellulose esters, and polyhydroxyalkanoate [18]. Biodegradable plastics are generally less stable and durable than conventional plastics, so they are more susceptible to physical, chemical, and biological degradation when entering the natural environment. In addition to UV irradiation, other environmental conditions, particularly the presence of water, significantly affect the degradation of different MPs [19], especially biodegradable plastics such as polylactic acid, which can be hydrolyzed in the presence of water, leading to a decrease in molecular weight and the release of low molecular weight soluble oligomers and monomers [20]. Unlike the environmental photodegradation mechanism of petroleum-based plastics, the hydrolysis process of degradable plastics usually does not start gradually from the surface (surface erosion) [19], but rather proceeds relatively homogeneously from within their structures as water molecules penetrate the plastic (bulk erosion) [21]. There are two phases, the first of which occurs within the polymer with an increase in the number of autocatalyzed carboxyl termini; the second phase is marked by the escape of soluble oligomers from the matrix [20]. Due to the different degradation mechanisms, the physical structure changes during the degradation of PLA, and the mechanical crushing process due to internal degradation will also be different from that of petroleum-based plastics. However, most of the current research on degradable plastics focuses on the improvement of the properties of the material products and composting degradation [22–25]. In addition, these degradation products usually possess more polar groups (hydroxyl and carboxyl groups) that induce and increase the affinity for water, further promoting the degradation reaction [26]. However, there are fewer studies regarding whether or not these products are further released into the water column or react with other structural substances.

In response to the different properties of petroleum-based conventional plastics and biodegradable plastics, the mechanisms by which different types of plastic particles undergo degradation and aging in different environments to produce further fragmentation, thus releasing MPs particles of smaller particle sizes or soluble organics, are not known. In addition, there are fewer studies discussing the mechanisms related to the combined effects of water and light as important influences on aging. Therefore, this study mainly focused on the combination of water and light as an important factor affecting the degradation of plastic particles and explored the degradation process and the mechanism of water and light on plastic particles. PLA and PE were selected as representatives of bio-based and petroleum-based MPs to study their environmental degradation mechanisms and the formation of small particle-size micro–nano plastics, reflecting the differences and commonalities between them after aging. This study focused on the characterization of the changes in the physicochemical properties of the particulate matter itself after aging and the analysis of the components of the eluate using scanning electron microscopy (SEM), FTIR, a total organic carbon analyzer (TOC-L), a Zetasizer Nano device, and Py–GC–MS, respectively. The purpose of this study was to explore: 1. The photodegradation and hydrolysis mechanisms of PE and PLA in aqueous and air environments; 2. The changes in particle size during the degradation of PE and PLA plastics under UV irradiation and the link between the process of generating smaller-sized micro–nano-particle plastic particles and dissolved organic matter. The results of the study will provide a scientific basis for the degradation patterns of two typical plastic polymers and their resulting environmental behaviors.

2. Materials and Methods

2.1. Accelerated Laboratory Aging of Plastic Particles

The materials required for the experiment, PE (CAS: 9002-88-4) and PLA (CAS: 26100-51-6), were all purchased from Shanghai Yangli Electromechanical Technology Co., Ltd., (Shanghai, China), and the detailed information regarding these products is given in Table S1. A UV lamp with a wavelength of 254 nm and a power of 20 W was employed, and the sample was maintained at a distance of about 28 cm from the UV lamp. In order to ensure the uniformity of light exposure among the samples, before the experiment, we pre-positioned a photometer in the experimental box to measure the radiation intensity received by the sample at a position of 0.4 mW/cm^2; after adjusting, we found that after three days of sequential adjustment of the test paper placement, the difference between the samples was controlled to be within the range of 1%, so we followed this protocol in the placement of the actual samples. In addition, we adjusted the height, angle, and direction of the UV lamps to ensure that their position and angle could cover the entire surface of the samples and maintain as uniform an irradiation as possible. The UV aging process was carried out in a house-made sealed experimental box, which was closed with an airtight lid, assisted by aluminum foil, to avoid interference from other factors in the laboratory (Figure 1).

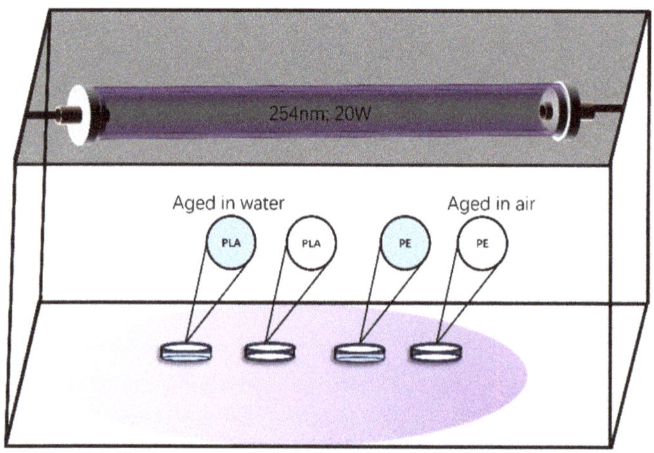

Figure 1. Diagram of UV aging setup for MPs.

A sample of partially aged particles and a sample of partially virgin particles were obtained and eluted separately with pure water. The specific operations were as follows: About 0.5 g of the above dried plastic particles were added to a beaker, to which was added 40 mL of pure water, and the sample was subjected to ultrasound at 40 °C for 30 min, after which 20 mL of the solution was aspirated, and then the pure water was replenished to 40 mL, and the sample was again subjected to ultrasound for 30 min. The eluate and plastic particles were filtered through a 0.22 μm glass fiber, and the eluate was collected for subsequent analysis.

2.2. Characterization of PE and PLA MPs, before and after Aging

Plastic particles were analyzed, before and after aging, using SEM (Thermoscientific; Verios G4 UC, Waltham, MA, USA). The SEM current voltage was 10 Kv, 0.8 nA, and the samples were subjected to critical point drying and gold metallization using a sputter coater. The plastic surfaces were photographed using scanning electron microscopes of different magnifications.

FTIR, with a measuring range of 397–4000 cm^{-1} and a resolution of 4 cm^{-1}, was performed, with 32 scans/sample (Nicolet is 50+ Nicolet continuum, Thermo Fisher Scientific, Waltham, MA, USA). For different areas of each sample, FTIR was performed at different locations. The sample compartment of the device was: R%T. The aged sample particles were placed on a gold mirror for observation, and the lens was adjusted and brought into focus. The energy of the optical bench was first determined to be 4–11, after which the background was scanned, followed by aiming the light source at the samples for detection, with two different areas of each sample being detected. The baselines of the atlas were corrected using the manual mode in the Omnic (Version 9.2), and the maps were smoothed (13 smoothing points). The CI was calculated as the ratio of the integrated absorbance area of the carbonyl (C=O) peak between 1850 and 1650 cm^{-1} to the integrated absorbance area of the methylene (-CH_2) peak between 1500 and 1420 cm^{-1}, I added theaccording to the methods of Maryam Hoseini et al. [27].

$$\text{carbonyl Index (CI)} = \frac{\textit{area under band}\ \mathrm{C=O}}{\textit{area under band}\ \text{–}\mathrm{CH_2}}$$

In this study, XPS (Axis Supra Axis Supra, Japan) was used to qualitatively analyze and quantify the changes in the chemical states of oxygen (O) and carbon (C) on the surface of the MPs after UV aging, and the O/C ratio was used to further analyze the degree of UV aging of the MPs. XPS scanning spectra and high-resolution spectra of virgin and UV-aged MPs were collected at an Al Kα monochromatized X-ray source under a vacuum of 10^{-8} Pa. The O/C ratios of the samples were derived from the percentage of O atoms compared to the percentage of C atoms using the identification of the full spectrum method.

A total of 0.3 mg of the plastic particles collected before and after aging were weighed and placed in a pyrolysis cup for analysis. A comparison of the degradation products, including the time to peak and the relative abundance between peaks, was performed. Test condition: The relevant parameters of the Py–GC–MS (PY3030D-8890/5977B, Frontier, Japan) were as follows: (1) the chromatographic columns are Rtx-5MS, 30 m × 0.25 mm × 0.25 μm; (2) the temperature was increased from 40 °C to 320 °C at a rate of 20 °C/min and held at 320 °C for 14 min; (3) the carrier gas was helium.

The filtered solution was injected into the total organic carbon analyzer for analysis (TOC-L, Shimadzu, Kyoto, Japan). A laser particle size analyzer (Zetasizer Nano S90, Malvern Instruments, Malvern, UK), was used to characterize changes in the average particle size of the soluble compounds. A total of 10 mL of the eluate was removed and added to a four-sided translucent cuvette, which was then placed in the sample cell to set the corresponding refractive index and absorbance of PE and PLA, as well as the corresponding parameters of the solution.

2.3. Data Analysis

All original data were analyzed using Excel (Microsoft Office 2021). SPSS (IBM SPSS Statistics 26) was used for statistical analysis. ANOVA was applied for the test of significant differences in data, and statistical significance was determined as statistically (* $p < 0.05$) and highly (** $p < 0.01$) significant, respectively [28].

3. Results and Discussion

3.1. UV Aging Altered the Morphological Characteristics of the PE and PLA MPs Particles

The surface morphology of both PE and PLA plastic particles changed significantly after 30 days of UV irradiation. As shown in Figure S1, PLA plastic particles showed significant yellowing after undergoing UV aging, while PE MPs did not show significant color changes after UV aging. In addition, we noted differences in the rate of yellowing at the same time points, and the occurrence of such differences might be related to the structure of the plastic or the presence of stabilizers [29]. Carbonyls, vinyl, and quinones, as well as cyclic structures, could be conjugated via p-π and exhibit absorbance at 280 nm, ultimately resulting in the yellowing of the plastic [30,31]. The yellowing effect could be generated by conjugated double bonds generated under the excitation of UV energy with reactive oxygen species, i.e., C-H → C-OH → C=O → O-C=O [32]. For PLA, which contains rich ester bond structure, are more easily to form O-C=O structure during aging. Moreover, in the chain breaking process, it was easy to cross-link PLA to form a ring structure, which PE did not display. Thus, the yellowing phenomenon of PLA was more obvious. The rate of yellowing was slowed down in the presence of water, which might be related to the hydrolysis of PLA, leading to the breakage of double bonds or the eluate of additives into the aqueous solution.

The yellowing and weathering process of plastics usually occurs in conjunction with surface cracking [29]. Observation of the surface microstructure of the plastics using SEM showed that the texture of both initial MPs was relatively uniform and dense, with smooth and regular surfaces, no obvious faults or holes, and no other particles found to be attached (Figure 2). After 30 days of UV irradiation, fine cracks began to appear on the PE surface, while the surface appeared unchanged before and after pure water eluate (Figures 2 and 3). Compared to PE, PLA formed wider intertwining, and connecting cracks after aging, creating holes, as well as signs of fracture and breaking (Figure 2). It also formed fragments of fine particles that adhered to the surface of the aging plastic. The holes of the non-eluted plastic particles were mostly distributed in the sub-surface layer. After being eluted with pure water, we found that dense holes appeared on the surface of the PLA pure water photoaging exposure group, as pointed out by the arrows in Figure 3, whereas this phenomenon did not appear in the air-exposed UV aging group, which indicated that, after the combined effects of hydrolysis and photoaging, the surface of PLA MPs could produce a large number of fine particles, which could be easily dislodged in an aqueous environment to enter into the environment (Figure 3). Dislodged microplastics could exhibit surface adsorption interactions with hydrophobic organic contaminants (e.g., cyclic hydrocarbons, as reported in the literature), and in conjunction with our findings, the increase in pore size due to UV aging could provide ideal conditions for the adsorption of cyclic hydrocarbons, along with the formation of smaller plastic particles, which could result in more potential ecotoxicity [33].

Changes in the particle size distribution of the PLA and PE MPs after different aging conditions were assessed by optical microscopy (Figure 4). The statistical methods for determining particle size distribution are detailed in Text S1 and Table S5. The PE particle size distribution before aging was mainly dominated by particles measuring 201–300 μm (35.2%), followed by those measuring 101–200 μm (24.01%), 301–400 μm (19.72%), and 0–100 μm (8.41%), respectively; After aging in both exposure modes, the particle size distribution pattern did not change significantly, with the exception that the particle size of the 201–300 μm group of PE decreased after anhydrous photoaging, while the particle size of the 101–200 μm group increased accordingly. The specific particle size distributions were: 8.47% (0–100 μm), 39.27% (101–200 μm), 27.98% (201–300 μm), 15.7% (301–400 μm), 2.82% (401–500 μm), and 5.76% (501–600 μm), respectively. This change might be due to surface degradation and the gradual disappearance of the main components of 201–300 μm during the aging process achieved by UV irradiation, resulting in a slow change in particle size to the next level. For PLA, the trend was quite different from that exhibited by PE. The particle size distribution of PLA before aging was mainly 101–200 μm (49.37%), and the percentage content gradually decreased with increasing particle size: 201–300 μm (23.1%), 301–400 μm (12.69%), 401–500 μm (2.33%), and 501–600 μm (1.17%), respectively. After aging, a large number of MPs particles with small particle sizes (0–100 μm) appeared under both exposure conditions, which was more obvious in the pure water UV aging group, with particles in the 201–600 μm particle size range decreasing, and the large-size particles (401–500 μm and 501–600 μm) actually disappearing completely. The specific distribution was the particle size distribution of PLA after anhydrous photoaging, i.e., 45.08% (0–100 μm), 32.72% (101–200 μm), 9.02% (201–300 μm), 9.02% (301–400 μm), 4.17% (401–500 μm), and 0% (501–600 μm), respectively. The particle size distribution of PLA after aging with hydrophobic light was 69.8% (0–100 μm), 23.99% (101–200 μm), 5.42% (201–300 μm), 3.12% (301–400 μm), 0% (401–500 μm), and 0% (501–600μm), respectively. This pattern of change indicated that the PLA aging degradation process was accompanied by a large number of internal fragmentation processes, in addition to surface degradation. In the presence of water, due to the hydrolysis processes, in addition to surface aging, PLA also exhibited inward erosion, accelerating its rate of fragmentation from the inside; thus, the particle size became smaller. Under air exposure conditions, PLA only absorbed moisture in the air, the rate of hydrolysis was greatly reduced, the plastic was less fragmented, and the relative content of small-sized particles was lower than that observed in the pure water exposure group. The formation of small-sized particles due to crumbling caused by internal fragmentation simultaneously exposed a larger specific surface area, further accelerating their degradation and creating a positive feedback effect along with the formation of additional small-sized fragments. According to Isadounene et al., the appearance of significant pores and the fragmentation of PLA was related to its water absorption, which led to the disruption of the chain structure, resulting in higher porosity [34]. This result was consistent with SEM observations that revealed a large number of holes and deep grooves. Smaller-sized particles might have been detached from the plastic particles, thus forming the porous areas. Hydrolysis and photolysis produced a large number of small-sized particles, which on the one hand, entered the environment to be ingested by organisms, increasing the probability of retention in the body [35], and on the other hand, might act as "carriers" for other sorbed organic pollutants, ultimately leading to more serious ecotoxicity effects [36,37].

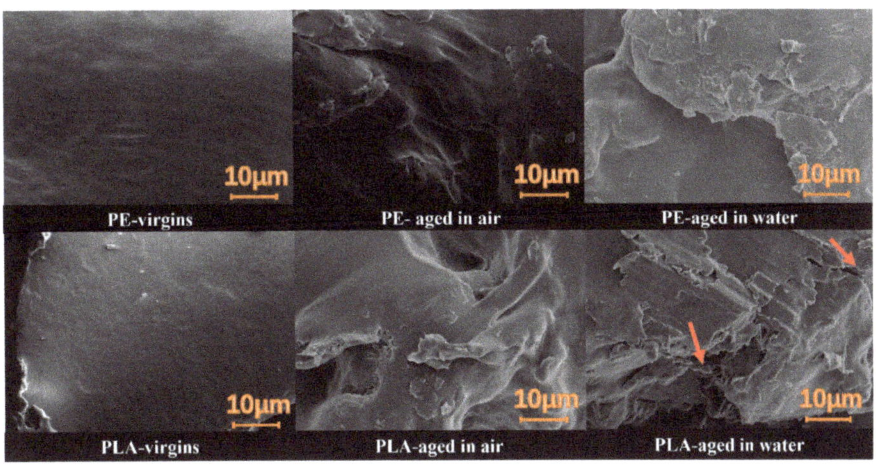

Figure 2. Surface morphology of MPs before and after aging, where the red arrows indicate the area of the MPs broke.

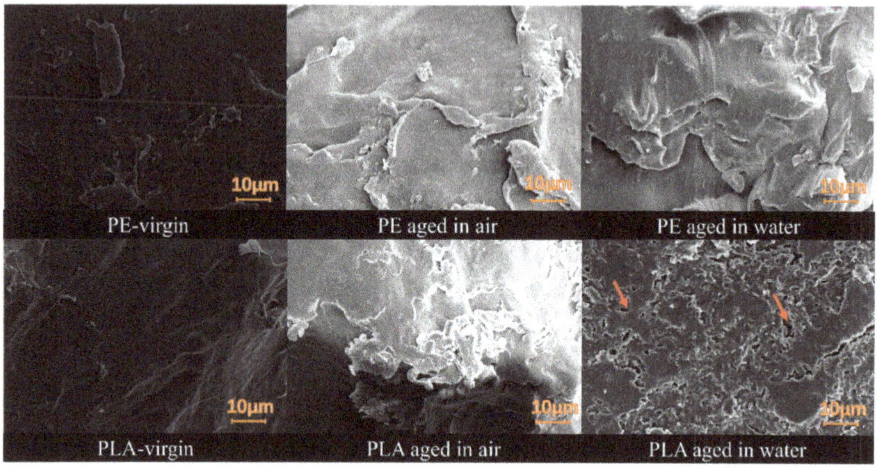

Figure 3. Surface morphology of MPs before and after aging under pure water eluate, where the red arrows indicate the area of the MPs holes.

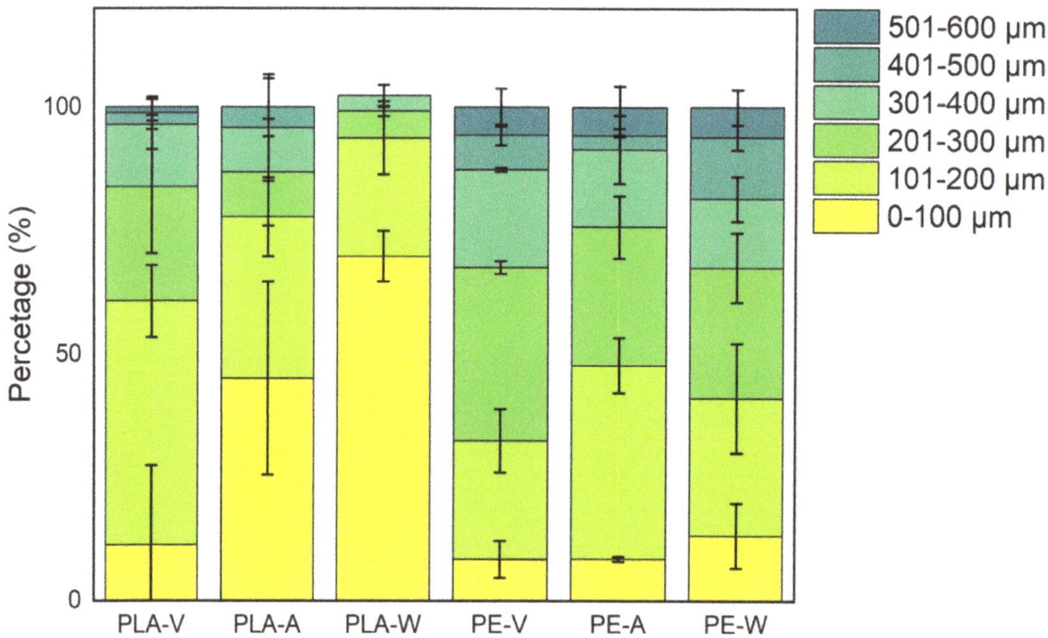

Figure 4. Plastic particle size distribution after aging. V represents virgin plastic particles, A represents photoaging plastic particles under anhydrous conditions, and W represents photoaging plastic particles under aqueous conditions.

3.2. *UV Aging Altered the Chemical Functional Groups of PE and PLA MPs Particles*

FTIR characterization showed that after UV aging under air and water exposure, the PE MPs suffered a significant surface oxidation process. This was mainly characterized by an increase in the vibration of the carbonyl peak and band broadening at 1750 cm^{-1}, and the rate of carbonyl formation was related to the exposure environment—carbonyl formation was faster under pure water exposure conditions (Figure 5a). Correspondingly, after UV aging for 30 days, the CI of PE MPs significantly increased ($p < 0.01$), and the degree of increase was greater in the condition with water, 0.31 ± 0.05, than without water, 0.09 ± 0.05 (Table 1). Under UV irradiation, alkyl radicals underwent cleavage to combine with hydroxyl radicals to form carbon–oxygen bonds, which then formed carbonyl groups with hydroxyl radicals [14]. At the same time, liquid water in this process was exposed to UV irradiation to produce volatile hydroxyl radicals, which contained a higher content of hydroxyl radicals than that of air, and the reaction of the radicals with water molecules could be used as a source of hydroxyl radicals [38]. Therefore, in our experiments, we found that the CI of the photoreaction in combination with water was higher than that of the photo-oxidation without the participation of water, and the scanning electron microscope results showed that the degree of change in the surface morphology of the former was slightly larger than that of the latter, i.e., more peeling and cracking phenomena were produced, which was similar to the results of previous research, and the higher the CI of PE, the greater the degree of aging to which it was subjected [39]. The carbonyl peaks were associated with lactones, esters, alcohols, aldehydes, and ketones [40], and the degradation products were subsequently further resolved based on the Py–GC–MS results. This trend was also reflected in the increase in oxygen content and oxygen-to-carbon ratio in the XPS analysis—the oxygen-to-carbon ratio of PE plastic particles before and after aging showed an sightly increasing trend from 0.13 ± 0.03 before aging to 0.15 ± 0.03 (photoaging under aqueous conditions) and 0.17 ± 0.03 (photoaging under anhydrous conditions) (Table 1). This trend was consistent with the findings of Brandon et al. [41].

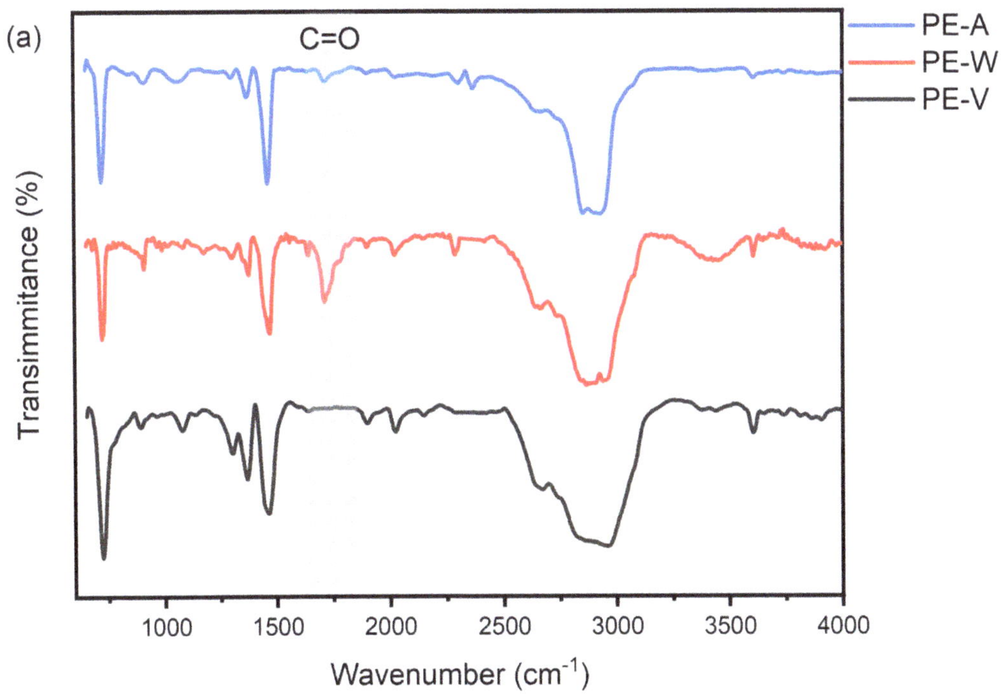

Figure 5. *Cont.*

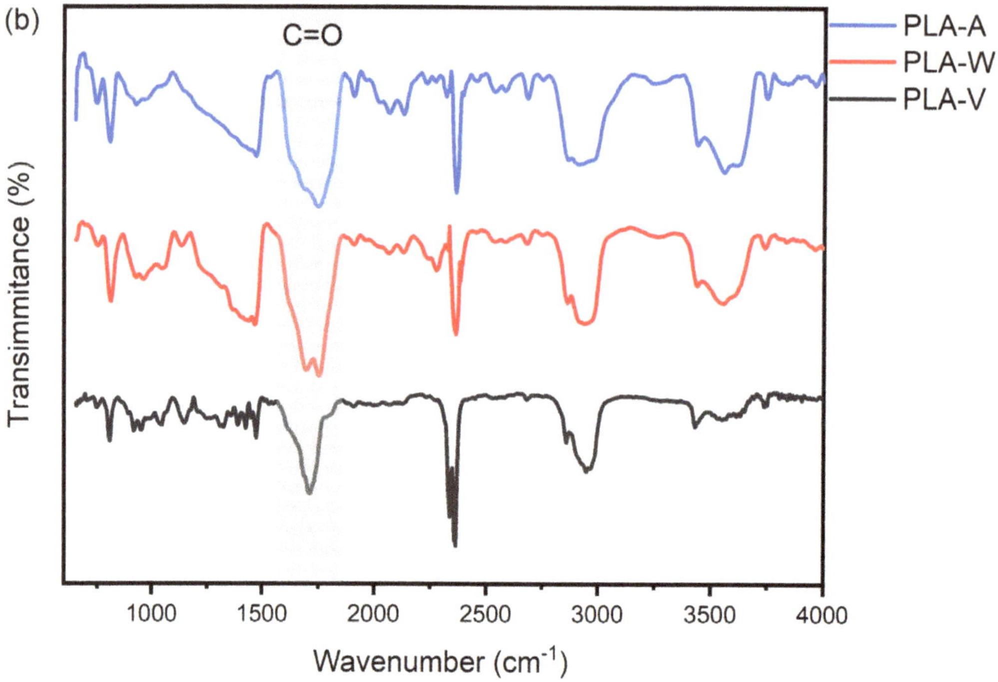

Figure 5. (**a**) μ-FTIR plot of PE after photoaging. (**b**) μ-FTIR plot of PLA after photoaging. W represents photoaging in water, A represents photoaging under anhydrous conditions, and V represents virgin status.

Table 1. Chemical properties of MPs before and after aging.

MPs	C Atomic %	O Atomic %	O/C	CI
PE-V	92.14 ± 2.39	7.86 ± 2.39	0.13 ± 0.03	0.01 ± 0.00
PE-W	86.98 ± 2.06	13.23 ± 2.34	0.15 ± 0.03	0.31 ± 0.05
PE-A	85.31 ± 1.97	14.71 ± 1.95	0.17 ± 0.03	0.09 ± 0.05
PLA-V	72.95 ± 1.34	27.06 ± 1.34	0.37 ± 0.03	19.66 ± 9.62
PLA-W	67.62 ± 1.16	32.38 ± 1.16	0.46 ± 0.01	14.90 ± 8.16
PLA-A	71.22 ± 0.29	28.79 ± 0.29	0.42 ± 0.02	12.13 ± 5.67

Note: W represents photoaging in water, A represents photoaging under anhydrous conditions, and V represents virgin status.

Unlike PE, the structure of PLA is mainly connected by ester bonds, and its structure contains a large number of C=O structures (Figure 5b); therefore, the CI of the original unaged PLA was 19.66 ± 9.62, which was much higher than that of the PE MPs. The changes in the oxidized functional groups on the surface of the PLA MPs before and after UV aging showed an opposite trend to that of the PE MPs—the intensity of the C=O peaks and the CI decreased significantly after photoaging, and the decrease was more pronounced under the anhydrous condition. Based on their opposite trends, we analyzed the mechanisms behind this differing behavior. The hydrolysis mechanism of PLA is shown in Figure 6. During hydrolysis, the polymer backbone breaks, resulting in a portion of the carbonyls in the backbone breaking away from the backbone, and due to the presence of an aqueous medium, this portion of the shedding carbonyls may be released into the water, as shown by a decrease in the carbonyl index, as we have observed. Specifically speaking, the photodegradation of PLA first occurs by chain breakage through photolysis and hydrolysis, followed by conversion through Norrish type I and type II reactions, in which cross-linking or oxidation reactions may also occur. The photodegradation of PE occurs by chain breakage caused by UV light and oxygen, with the generation of ester bonds [42,43]. Some studies found an increase in the CI of PLA in sludge treatment, which was contrary to our results [40]. On the one hand, the presence of sludge and microbial films might have interfered with the spectroscopic analysis. On the other hand, the large pieces of plastic that were used were difficult to degrade further because they were quite thick, so the inside of their structures could not receive UV irradiation. However, in another study, it was found that the CI of PLA in flake form also showed a decreasing trend after undergoing UV irradiation in pure water, which was consistent with the results in this study [44]. In our study, hydrolysis and photolysis occurred simultaneously on the surface and inside of the small-sized plastics, and the hydrolysis of PLA led to the breakage of ester bonds and the formation of lactic acid and carboxylic acid structures [45]; these terminal carboxylic acid structures were also easily further decomposed by UV irradiation to produce the final product CO_2 [19]. More importantly, the PLA MPs used in this study possessed smaller particle sizes, and the degradation process fragmented to form a large number of smaller-sized micro- and nanoparticles, a phenomenon that was more pronounced in the presence of water (Figure 3). The creation of these fine particles made it easier for the oxygenated ends produced by hydrolysis, as well as the carbonyl groups in the main structure of the PLA, to receive the energy of the UV irradiation and to be completely degraded to CO_2. This was consistent with our surface observations of the production of holes and cracks in the presence of water, as well as the particle size distribution maps, where it was observed that the involvement of water contributed to the fracture of PLA into fragments, and the absence of water resulted in the presence of relatively large-sized PLA particulate matter. In the natural environment, the effects of water and light usually co-exist, and previous studies have focused on the photolysis, hydrolysis, and biodegradation of the single-factor and large-scale degradation of thin film plastics [44]; thus, we took the lead in exploring the degradation of micron-sized PLA plastic particles under the combined effects of light and water. Due to the simultaneous presence of hydrolysis and photolysis to accelerate the complete degradation of PLA, a more pronounced decrease in the carbonyl peak area and the carbonyl index occurred after the aging of PLA in the presence of water. Interestingly, the XPS results showed that the oxygen content of PLA plastic particles before and after aging, along with the oxygen to carbon ratio, showed a small increase, with the oxygen to carbon ratios increasing from 0.37 ± 0.03 before aging to 0.46 ± 0.00 (under aqueous conditions) and 0.42 ± 0.02 (under anhydrous conditions) after photoaging. From the previous results, it can be seen that the CI decreased, but the oxygen–carbon ratio did not decrease, suggesting that the carbonyl group might be converted to other oxygen-containing functional groups during the aging process. The O/C ratio of MPs exposed to water increased faster than the ratio under photoaging without water, which might be related to the creation of pores on the surface of PLA, leading to its susceptibility to the water absorption processes [46].

To further clarify the degradation products and their chemical structures produced by different types of plastics during degradation and aging, this study examined high-temperature pyrolysis (600 °C) and low-temperature cracking (300 °C), while combining GC–MS to analyze the cracking products. For PLA, the two pyrolysis temperatures showed different trends before and after aging. At 600 °C, the pyrolysis products were the same before and after aging; at 300 °C, there were significant differences before and after aging, and these differences were concentrated in a series of degradation products found at around 14 min, 15 min, 19 min, and 20 min, which were not characteristic of the virgin PLA (Figure 7 at the red dashed box). This result suggested that low-temperature cracking better reflected the environmental processes of plastic aging in terms of chemical structure, which might be related to the fact that substances produced by aging were more susceptible to low-temperature cracking. At 300 °C, we were able to observe the appearance of oxidation products and additives in the aged PLA at around 14 min, 15 min, 19 min, and 20 min, etc. These oxidation products and additives were compared using the NIST database, respectively: 1,3-dioxepane, 2-heptyl-; Butanedioic acid, dipropyl ester; Ethyl hydrogen succinate; 1,3-dioxepane, 2-pentadecyl-; Butanamide, n-formyl-2-hydroxy-n,3-dimethyl-2-(1-methylethyl)-; 2-t-butyl-5-propyl-[1,3]dioxolan-4-one; and 1,2,4-butanetricarboxylic acid, trimethyl ester (Table S2). These products were esters and non-carbonyl oxygenated compounds with cyclic and chain structures, respectively. The mechanisms of the thermal degradation of PLA included both free radical and non-free radical reactions. We found that succinate (Figure S3, Text S4) can be a source of plastic additives and showed that succinate is prepared from succinic acid by tandem-parallel reactions that proceed sequentially, with the intermediate product being ethyl hydrogen succinate and the final product being diethyl succinate. In our experiments, we found that the intermediate monomer could be derived from succinic acid and could potentially be converted to diethyl succinate [43]. In addition, in another study, we found that PBS (poly butylenes succinate) exhibits a structural similarity to ethyl hydrogen succinate; however, this connection requires further investigation [47]. This similarity was found in studies related to PLA pyrolysis products, which may be related to the preparation process of PLA and UV radiation [48]. PLA formed a series of oxidation, fracture, and double bond products after photoaging, and these products occurred under low-temperature thermal degradation conditions, including molecular chain fracture, depolymerization, and inter- and intramolecular ester exchange reactions [49], forming a series of small molecule lipid compounds [50]. Further normalization of the mapping data revealed all the differences in the relative intensities of the major peaks, with no significant differences in the number of peaks in the aging reactions involving water. The intensity of the three main peaks was analyzed, i.e., Succinic acid, di(2-methylallyl) ester, Ethyl hydrogen succinate, and Succinic acid, 2-methylallyl undecyl ester (Figure S2). In the presence of water, the strength of the main pyrolysis products increased by 1.23, 1.86, and 1.29 times, respectively. This result may be related to the hydrolysis of PLA to form small molecule fragments, and the pyrolysis efficiency of small-size fragments was higher than that of the large-size plastic pieces [50]. Under the traditional 600 °C pyrolysing conditions, the aged PLA appeared to generate similar pyrolysis products before aging, and only individual lactones and additives, such as Butyrolactone, Butanedioic acid, and dipropyl ester, were newly generated after aging (Table S3). This phenomenon might occur because the oxidation and cracking products formed after aging were destroyed at high temperatures, and therefore, the information could not be retained.

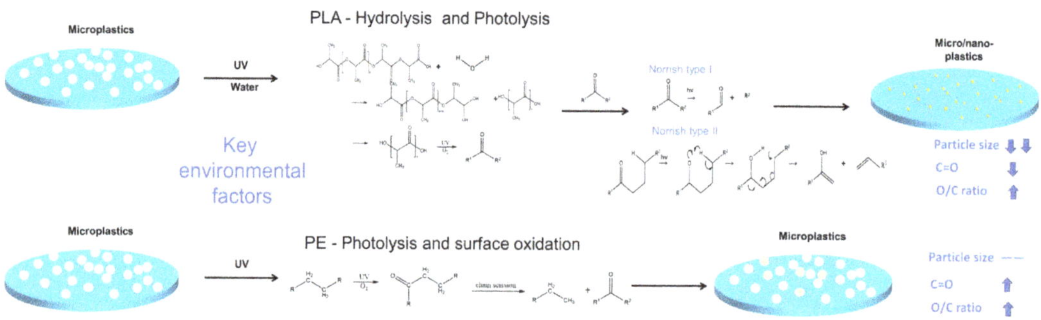

Figure 6. Possible reaction mechanisms of PLA and PE after UV aging. Yellow dots indicate that the MPs have been yellowing.

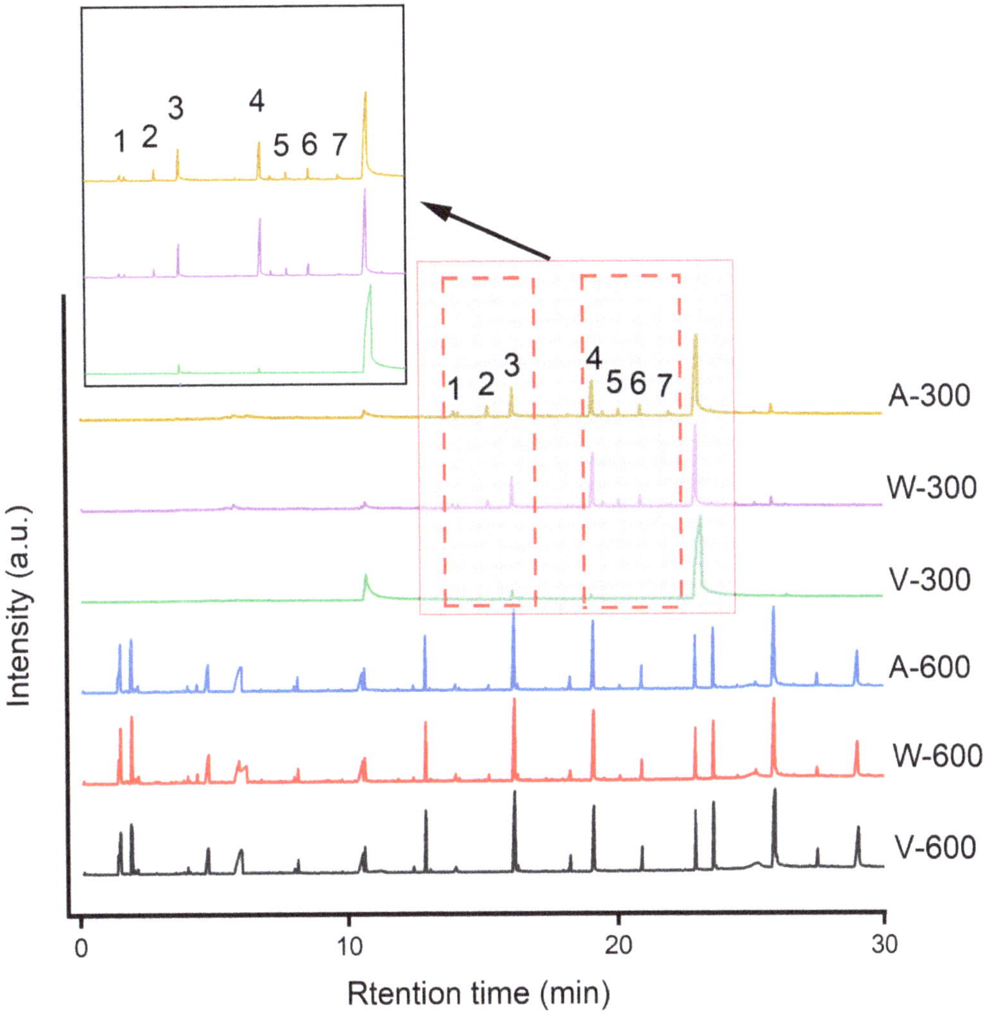

Figure 7. Low and high-temperature pyrolysis chromatograms of PLA before and after aging. A-300 represents chromatograms under anhydrous conditions with a pyrolysis temperature of 300 °C, W-300 represents chromatograms under aqueous conditions with a pyrolysis temperature of 300 °C, V-300 represents virgin PLA particles with a pyrolysis temperature of 300 °C, and 600 represents 600 °C. The red box marks contain the differences of PLA products before and after aging, a total of seven, respectively 1, 2, 3, 4, 5, 6, 7.

For PE, using the same assay, the gas chromatogram at 600 °C showed cleavage information mainly for mono-olefins, dienes, and alkanes, which was similar to the results obtained in the previous study [51]. It was noteworthy that the relative intensities of the peaks changed before and after aging at 600 °C, with an increase in the relative intensities of the small molecules after aging (0–10 min) (yellow portion of Figure 8). We speculated that the increase in the intensity of this portion of the peaks might be related to the formation of relatively active functional groups, such as oxygen-containing functional groups and double bonds, during the aging process of MPs, and the sites where these functional groups were located during the pyrolysis process were more prone to cleavage, forming small-molecule cleavage products. This result was in agreement with the FTIR and XPS results

(Figure 5, Table 1), as well as with the thermal cracking mass spectrometry characterization of PE plastics with different particle sizes in the natural environment [52]. In addition, the photoaging process could also occur as a result of main chain breakage, generating small molecular fragments, which was corroborated by the SEM observation of exfoliation on the surface of the PE plastics (Figure 2). Compared with PLA, PE MPs were cracked at 300 °C under cracking conditions, and the cracking product peaks were too low, which might be related to its structural stability, although the overall peak height of the aged product was still higher than that of the virgin plastic (Figure 8). It was difficult to reach the detection limit in order to identify the structural information of the product; thus, the low-temperature pyrolysis could not adequately characterize the information regarding the aging process of PE (Table S4). In this study, we pioneered the use of Py–GC–MS to characterize the degradation products and aging process of different types of MPs under UV light and aqueous environmental conditions and optimized the analytical conditions based on their thermal stability for better investigating the degradation mechanisms. The results showed that for the PLA MPs with lower thermal stability, low-temperature cracking conditions were more suitable for studying their environmental degradation behavior. For PE, high-temperature pyrolysis could produce more characteristic products, and although no obvious new degradation products were found, the aging process could be reflected in the relative peak heights of the pyrolysis spectra.

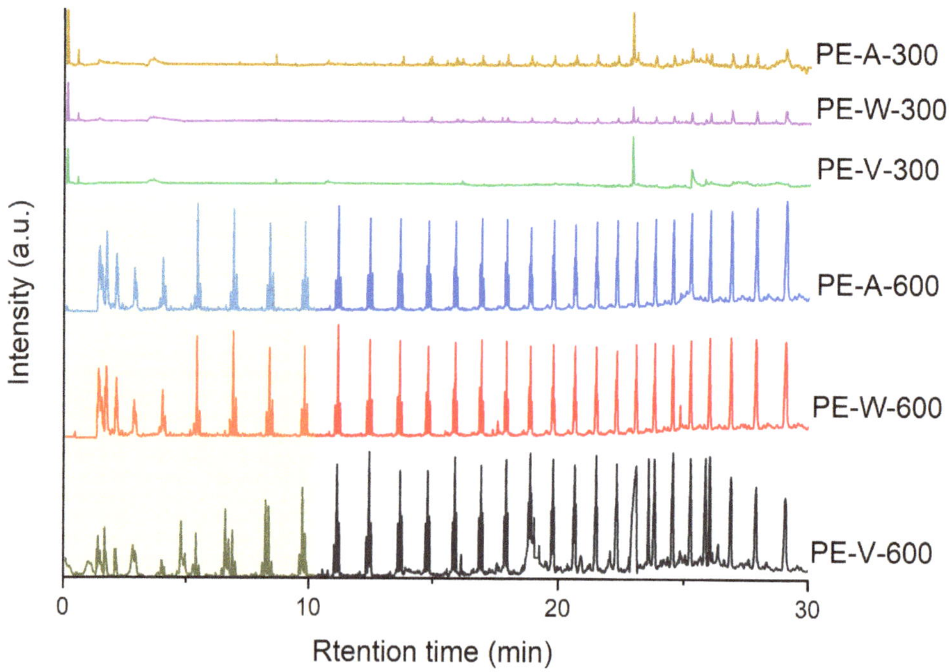

Figure 8. Low and high-temperature pyrolysis chromatograms of PE before and after aging. A-300 represents chromatograms under anhydrous conditions with a pyrolysis temperature of 300 °C, W-300 represents chromatograms under aqueous conditions with a pyrolysis temperature of 300 °C, V-300 represents virgin PLA particles with a pyrolysis temperature of 300 °C, and 600 represents 600 °C. The yellow area indicates the change in the strength of PE pyrolysis products within 10 min before and after aging at the pyrolysis temperature of 600 °C.

3.3. Characterization of MP Eluate after Aging

By eluting PE and PLA MP particles before and after aging and analyzing the total oxygen carbon (TOC) of the eluate (Figure 9), we found that compared to PE, the TOC eluted from PLA, with or without aging, was close to 100 times higher than that of PE, indicating that PLA MPs were more likely to release organic matter into the water column in the environment. It had been demonstrated by others that dissolved TOC could evaporate in a short period, and at the same time, MPs could adsorb TOC present in the natural environment, disrupting low trophic level processes and thus affecting microbial activity and carbon cycling in the oceans [53]. Both PE and PLA MPs particles that underwent UV aging released significantly higher organic carbon by eluate than the samples that did not undergo the aging process ($p < 0.01$). Specifically, the TOC of PE after aging (3.62–3.69 mg/L) was 5.84–5.95 times higher than that before aging (0.62 mg/L), and the TOC of PLA after aging (284.55–330.9 mg/L) was 4.25–4.94 times higher than that before aging (66.97 mg/L), whereas the aging process with the involvement of water produced slightly higher TOC than that of the no-water-involved conditions. Thus, our results indicated an increase in the generation and release of organic matter from the surface of MP particles after undergoing photoaging, and this was more pronounced in PLA. Combined SEM and FTIR analyses revealed that the oxidation process during photoaging and the creation of surface holes might be driving the release of more organic matter from the plastic. Our experimental results were consistent with previous findings that MPs could elute TOC under UV irradiation [54], and that the quality of organics eluted from different types of MPs varied considerably, e.g., the TOC eluted from PS and PVC was higher than that of PE and lower than that of PLA [55]. The eluted TOC might be derived from two sources: the degradation of plastic polymers and the release of additives. Combined with Py–GC–MS analysis, the plastic was irradiated by UV light to produce small molecular fragments and oligomers on the surface of the MPs, which were then eluted and released into pure water. Besides, the additives inside their structures could also release into surrounding water. In the study of Zhong et al., the related phenomenon of additive eluates released into water bodies was also found [56]. In addition, when halogens were present in the water, the eluted organic matter could react with them to further produce DBP precursors and chloroform [54].

As mentioned previously, low molecular polymers, such as plastic degradation products and additives, were released into the water column, with the potential for the release of smaller-sized submicrons and nanoparticles in addition to the dissolved material. Therefore, we used a laser particle size analyzer to further analyze the changes in the particle size of dissolved organic matter in the aqueous environment after UV aging. The results showed that 100–1000 nm sized particulate matter appeared in both aged and unaged PLA eluate, and the particle size of the aged eluted was significantly larger than that of the unaged sample, with average particle sizes of 608.3 nm, 926.6 nm, and 964.13 nm, respectively (Figure 10a). This result might be related to the fragmentation of PLA during aging degradation to form submicron particles, which was also confirmed by SEM analysis—the surface of the aged PLA plastic particles appeared to have more pronounced holes after eluting. Using ultrapure water to elute PE, it was found that the particle size of the particles in the eluate did not differ significantly in each treatment group, and the overall trend was a gradual decrease in the virgin plastic, the water photoaging, and the air photoaging groups (Figure 10b). This result indicated that there was a difference in the particle size of the substances dissolved from degradable and polyolefin plastics, and that such a difference could be linked to the difference in the carbonyl index and the surface defects, as described previously. Numerous studies have shown that small molecules at the nanoscale could be toxic to microbial communities and marine organisms [57,58]. The results of this study showed that aging in the environment produced large amounts of nanoscale polymer fragments and other dissolved small-molecule organic matter. Microplastics that have been aged may have a short-term potential to increase their hydrophilicity, which allows hydrophilic organics to adsorb to their surfaces [59]. In the long term, microplastics may interact with organic pollution in the oceans, increasing its aggregation and sedimentation, carrying it to the bottom of the oceans.

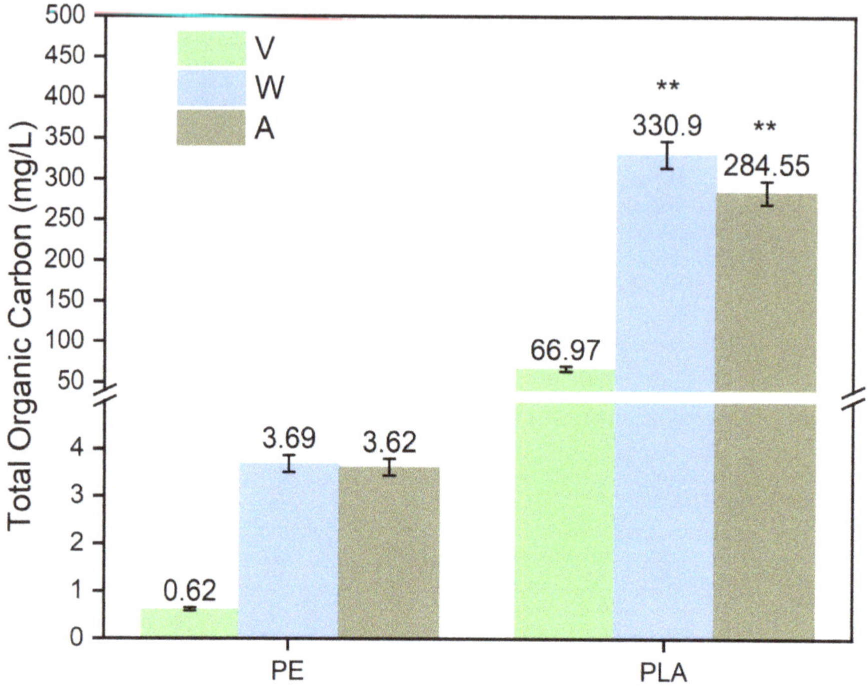

Figure 9. Dissolved organic carbon in eluate before and after aging of plastics. V represents virgin plastic, W represents the photoaging eluate under aqueous conditions, and A represents the photoaging eluate under anhydrous conditions. **: p-value < 0.01.

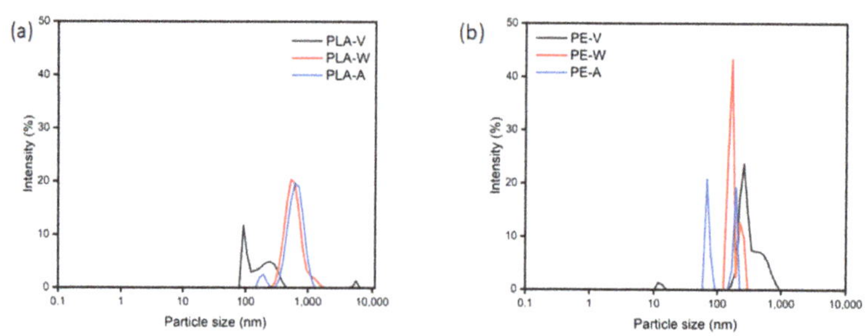

Figure 10. Particle size distribution of dissolved particles in elutes of PLA (**a**) and PE (**b**) plastics before and after aging. V represents virgin plastic, A represents the photoaging of plastic particles under anhydrous conditions, and W represents the photoaging of plastic particles under aqueous conditions.

4. Conclusions

In this study, the aging process of PE and PLA particles, which were representatives of biobased and petroleum-based MPs, were investigated by laboratory UV aging simulations driven by the combined factors of water and light. It was found that under the simultaneous presence of water and light aging, the degree of fragmentation on the surface of the two MPs varied greatly, with a series of holes and a large number of fine particles appearing on the surface of the PLA MPs. Based on this, we analyzed the particle size of the crushed plastics and found that there was a tendency for the particle size of MPs to become smaller step by step after crushing, and this phenomenon was more significant with the participation of water, indicating that hydrolysis played an important role in the degradation process of PLA. Meanwhile, through FTIR characterization, we found that the carbonyl strength and index of PLA MPs decreased after aging, which might be related to the UV degradation of the carbonyl structure of the ester bonds. After XPS analysis, we found that the carbonyl group of PLA decreased, while the oxygen content increased slightly, indicating that some of the ester-bonded compounds might be converted to non-carbonyl oxygenated compounds during the aging process, and this result was also confirmed by Py–GC–MS analysis. For PE, after the same aging experiments, although the degree of aging was not as drastic as that for PLA, PE also showed the phenomena of cracking and peeling, but the particle size distribution did not change significantly. The results of FTIR showed that, contrary to PLA, the CI of PE showed an increasing trend after aging, while XPS also showed that the oxygen content on the surface of PE increased significantly after aging, and the results were similar for aging in both air and water.

In this study, Py–GC–MS, combined with different pyrolysis temperatures, was innovatively applied to investigate the environmental aging degradation of different types of MPs. For PLA MPs with lower thermal stability, low-temperature cracking conditions were more suitable for studying their environmental degradation behavior. PE, on the other hand, required high-temperature cracking due to its high thermal stability. Using Py–GC–MS analysis of PLA and PE MPs before and after aging, we found that degradation products, such as esters and non-carbonyl oxygenated compounds, were formed after the aging of PLA, and aging in water was more significant, whereas more small molecule cleavage products were formed under the high-temperature cleavage conditions of aging PE, and the trend of the change was relatively similar in both air and water. This result further illustrated that the photoaging of PLA in water was significantly affected by hydrolysis. Further analyzing the plastic eluate, we found that the MPs generated finer nano-sized particulate matter after aging, and using the combined SEM and FTIR results, we speculated that the oxidation process and the creation of surface holes during photoaging might be driving the release of additional dissolved organic matter from the plastics. Overall, this study preliminarily investigated the UV chemical aging and hydrolysis processes of polyolefin MPs (PE) and polyester MPs (PLA) in aqueous environments and analyzed and clarified their degradation trends and the characteristics of the degradation products, providing a scientific basis for further identification of their environmental processes and ecological risks. We believe that the aging effect of UV radiation is the key factor affecting the aging of microplastics in terms of strategies to mitigate UV aging. According to our experiments, it is believed that the presence of water may accelerate this phenomenon; therefore, in terms of controlling the aging of microplastics, it is necessary to store the plastics under dry conditions, and the plastics need to be stored in a way that separates the dry from the wet to mitigate the aging behavior of microplastics. On the other hand, the storage of plastics should provide protection from light, and direct exposure to UV radiation must to be minimized. In the future, we hope to continue to identify these degradation products and their derivatives in the environment and to further analyze the mechanisms by which they may undergo further transformations and to assess the potential risks they pose to the environment. It is worthwhile to further investigate whether or not these aging degradation products, as well as their nanoparticles, will further impact the aqueous ecosystem.

Supplementary Materials: The following supporting information can be downloaded at: https://www.mdpi.com/article/10.3390/pr12040635/s1, Text S1. Statistical method of particle size distribution; Text S2. PY-GC/MS data analysis methods; Text S3. Set up information; Text S4. The details of Ethyl hydrogen succinate. Table S1. Information on plastic products; Table S2. Pyrolysis products produced after PLA aging at 300 °C; Table S3. Pyrolysis products produced after PLA aging at 600 °C; Table S4. Pyrolysis products produced after PE aging at 300 °C; Table S5. particle size distributions. Figure S1. Plastic pellets before and after aging pictures. W represents photoaging in water, A represents photoaging under anhydrous conditions; Figure S2. Chromatograms of the main pyrolysis products involved in photoaging at 300 °C with and without water; Figure S3. Mass spectra of Ethyl hydrogen succinate.

Author Contributions: Conceptualization, X.W. and Y.M.; Methodology, X.W., J.C. and Y.M.; Software, X.W. and W.J.; Validation, X.W.; Formal analysis, X.W. and Y.M.; Resources, K.H and Y.M.; Data curation, X.W. and W.J.; Writing—original draft, X.W. and J.C.; Writing—review & editing, X.W., J.C. and K.H.; Supervision, Y.M.; Project administration, Y.M.; Funding acquisition, Y.M. All authors have read and agreed to the published version of the manuscript.

Funding: The support of the National Natural Science Foundation of China (No. 42277377) and the Natural Science Foundation of Hainan Province (422MS035) is acknowledged.

Data Availability Statement: Data are contained within the article and Supplementary Materials.

Conflicts of Interest: The authors declare no conflict of interest.

References

1. Chen, R.; Zhao, X.; Wu, X.; Wang, J.; Wang, X.; Liang, W. Research progress on occurrence characteristics and source analysis of microfibers in the marine environment. *Mar. Pollut. Bull.* **2024**, *198*, 115834. [CrossRef]
2. Rillig, M.C.; Lehmann, A. Microplastic in terrestrial ecosystems. *Science* **2020**, *368*, 1430–1431. [CrossRef] [PubMed]
3. Zhao, C.; Liang, J.; Zhu, M.; Zheng, S.; Zhao, Y.; Sun, X. Occurrence, characteristics, and factors influencing the atmospheric microplastics around Jiaozhou Bay, the Yellow Sea. *Mar. Pollut. Bull.* **2023**, *196*, 115568. [CrossRef]
4. Lin, L.; Zuo, L.Z.; Peng, J.P.; Cai, L.Q.; Fok, L.; Yan, Y.; Li, H.X.; Xu, X.R. Occurrence and distribution of microplastics in an urban river: A case study in the Pearl River along Guangzhou City, China. *Sci. Total Environ.* **2018**, *644*, 375–381. [CrossRef] [PubMed]
5. Park, H.J.; Oh, M.J.; Kim, P.G.; Kim, G.; Jeong, D.H.; Ju, B.K.; Lee, W.S.; Chung, H.M.; Kang, H.J.; Kwon, J.H. National Reconnaissance Survey of Microplastics in Municipal Wastewater Treatment Plants in Korea. *Environ. Sci. Technol.* **2020**, *54*, 1503–1512. [CrossRef] [PubMed]
6. Liu, L.; Xu, M.; Ye, Y.; Zhang, B. On the degradation of (micro)plastics: Degradation methods, influencing factors, environmental impacts. *Sci. Total Environ.* **2022**, *806 Pt 3*, 151312. [CrossRef] [PubMed]
7. Wen, Q.; Liu, N.; Qu, R.; Ge, F. High salinity promotes the photoaging of polystyrene microplastics with humic acid in seawater. *Sci. Total Environ.* **2023**, *901*, 165741. [CrossRef] [PubMed]
8. Xiao, C.; Zhang, M.; Ding, L.; Qiu, X.; Guo, X. New sight of microplastics aging: Reducing agents promote rapid aging of microplastics under anoxic conditions. *J. Hazard. Mater.* **2023**, *451*, 131123. [CrossRef] [PubMed]
9. Fueser, H.; Mueller, M.T.; Weiss, L.; Hoss, S.; Traunspurger, W. Ingestion of microplastics by nematodes depends on feeding strategy and buccal cavity size. *Environ. Pollut.* **2019**, *255 Pt 2*, 113227. [CrossRef]
10. Peng, L.; Fu, D.; Qi, H.; Lan, C.Q.; Yu, H.; Ge, C. Micro- and nano-plastics in marine environment: Source, distribution and threats—A review. *Sci. Total Environ.* **2020**, *698*, 134254. [CrossRef]
11. Hahladakis, J.N.; Velis, C.A.; Weber, R.; Iacovidou, E.; Purnell, P. An overview of chemical additives present in plastics: Migration, release, fate and environmental impact during their use, disposal and recycling. *J. Hazard. Mater.* **2018**, *344*, 179–199. [CrossRef]
12. Dehghanian, Z.; Asgari Lajayer, B.; Biglari Quchan Atigh, Z.; Nayeri, S.; Ahmadabadi, M.; Taghipour, L.; Senapathi, V.; Astatkie, T.; Price, G.W. Micro (nano) plastics uptake, toxicity and detoxification in plants: Challenges and prospects. *Ecotoxicol. Environ. Saf.* **2023**, *268*, 115676. [CrossRef]
13. Gardette, M.; Perthue, A.; Gardette, J.-L.; Janecska, T.; Földes, E.; Pukánszky, B.; Therias, S. Photo- and thermal-oxidation of polyethylene: Comparison of mechanisms and influence of unsaturation content. *Polym. Degrad. Stab.* **2013**, *98*, 2383–2390. [CrossRef]
14. Lu, Q.; Zhou, Y.; Sui, Q.; Zhou, Y. Mechanism and characterization of microplastic aging process: A review. *Front. Environ. Sci. Eng.* **2023**, *17*, 100. [CrossRef]
15. Liu, S.; Huang, W.; Yang, J.; Xiong, Y.; Huang, Z.; Wang, J.; Cai, T.; Dang, Z.; Yang, C. Formation of environmentally persistent free radicals on microplastics under UV irradiations. *J. Hazard. Mater.* **2023**, *453*, 131277. [CrossRef]
16. Jiang, X.; Gallager, S.; Pedrosa Pàmies, R.; Ruff, S.E.; Liu, Z. Laboratory-Simulated Photoirradiation Reveals Strong Resistance of Primary Macroplastics to Weathering. *Work. Pap.* 2023; *ahead of print*. [CrossRef]
17. Zhong, Y.; Ding, Q.; Huang, Z.; Xiao, X.; Han, X.; Su, Y.; Wang, D.; You, J. Influence of ultraviolet-aging and adsorbed pollutants on toxicological effects of polyvinyl chloride microplastics to zebrafish. *Environ. Pollut.* **2023**, *316 Pt 2*, 120617. [CrossRef] [PubMed]
18. Sanchez-Hernandez, J.C.; Capowiez, Y.; Ro, K.S. Potential Use of Earthworms to Enhance Decaying of Biodegradable Plastics. *ACS Sustain. Chem. Eng.* **2020**, *8*, 4292–4316. [CrossRef]
19. Tian, L.; Chen, Q.; Jiang, W.; Wang, L.; Xie, H.; Kalogerakis, N.; Ma, Y.; Ji, R. A carbon-14 radiotracer-based study on the phototransformation of polystyrene nanoplastics in water versus in air. *Environ. Sci. Nano* **2019**, *6*, 2907–2917. [CrossRef]
20. De Jong, S.; Arias, E.R.; Rijkers, D.; Van Nostrum, C.; Kettenes-Van den Bosch, J.; Hennink, W. New insights into the hydrolytic degradation of poly (lactic acid): Participation of the alcohol terminus. *Polymer* **2001**, *42*, 2795–2802. [CrossRef]

21. Elsawy, M.A.; Kim, K.-H.; Park, J.-W.; Deep, A. Hydrolytic degradation of polylactic acid (PLA) and its composites. *Renew. Sustain. Energy Rev.* **2017**, *79*, 1346–1352. [CrossRef]
22. Ohkita, T.; Lee, S.-H. Thermal degradation and biodegradability of poly (lactic acid)/corn starch biocomposites. *J. Appl. Polym. Sci.* **2006**, *100*, 3009–3017. [CrossRef]
23. Gong, W.; Jiang, M.; Han, P.; Liang, G.; Zhang, T.; Liu, G. Comparative analysis on the sorption kinetics and isotherms of fipronil on nondegradable and biodegradable microplastics. *Environ. Pollut.* **2019**, *254*, 112927. [CrossRef]
24. Zimmermann, L.; Dombrowski, A.; Volker, C.; Wagner, M. Are bioplastics and plant-based materials safer than conventional plastics? In vitro toxicity and chemical composition. *Environ. Int.* **2020**, *145*, 106066. [CrossRef]
25. Chang, Q.; Zhu, D.; Hu, L.; Kim, H.; Liu, Y.; Cai, L. Rapid photo aging of commercial conventional and biodegradable plastic bags. *Sci. Total Environ.* **2022**, *822*, 153235. [CrossRef]
26. Porfyris, A.; Vasilakos, S.; Zotiadis, C.; Papaspyrides, C.; Moser, K.; Van der Schueren, L.; Buyle, G.; Pavlidou, S.; Vouyiouka, S. Accelerated ageing and hydrolytic stabilization of poly(lactic acid) (PLA) under humidity and temperature conditioning. *Polym. Test.* **2018**, *68*, 315–332. [CrossRef]
27. Hoseini, M.; Stead, J.; Bond, T. Ranking the accelerated weathering of plastic polymers. *Environ. Sci. Process Impacts* **2023**, *25*, 2081–2091. [CrossRef]
28. Venkataravanappa, R.Y.; Lakshmikanthan, A.; Kapilan, N.; Chandrashekarappa, M.P.G.; Der, O.; Ercetin, A. Physico-Mechanical Property Evaluation and Morphology Study of Moisture-Treated Hemp–Banana Natural-Fiber-Reinforced Green Composites. *J. Compos. Sci.* **2023**, *7*, 266. [CrossRef]
29. Abaroa-Perez, B.; Ortiz-Montosa, S.; Hernandez-Brito, J.J.; Vega-Moreno, D. Yellowing, Weathering and Degradation of Marine Pellets and Their Influence on the Adsorption of Chemical Pollutants. *Polymers* **2022**, *14*, 1305. [CrossRef] [PubMed]
30. Yang, J.; Xia, Z.; Kong, F.; Ma, X. The effect of metal catalyst on the discoloration of poly(ethylene terephthalate) in thermo-oxidative degradation. *Polym. Degrad. Stab.* **2010**, *95*, 53–58. [CrossRef]
31. Biale, G.; La Nasa, J.; Mattonai, M.; Corti, A.; Castelvetro, V.; Modugno, F. Seeping plastics: Potentially harmful molecular fragments leaching out from microplastics during accelerated ageing in seawater. *Water Res.* **2022**, *219*, 118521. [CrossRef]
32. Su, J.; Ruan, J.; Luo, D.; Wang, J.; Huang, Z.; Yang, X.; Zhang, Y.; Zeng, Q.; Li, Y.; Huang, W.; et al. Differential Photoaging Effects on Colored Nanoplastics in Aquatic Environments: Physicochemical Properties and Aggregation Kinetics. *Environ. Sci. Technol.* **2023**, *57*, 15656–15666. [CrossRef]
33. Wang, W.; Wang, J. Comparative evaluation of sorption kinetics and isotherms of pyrene onto microplastics. *Chemosphere* **2018**, *193*, 567–573. [CrossRef]
34. Isadounene, S.; Hammiche, D.; Boukerrou, A.; Rodrigue, D.; Djidjelli, H. Accelerated Ageing of Alkali Treated Olive Husk Flour Reinforced Polylactic Acid (PLA) Biocomposites Physico-Mechanical Properties. *Polym. Polym. Compos.* **2018**, *26*, 223–232. [CrossRef]
35. Rist, S.; Baun, A.; Hartmann, N.B. Ingestion of micro- and nanoplastics in Daphnia magna—Quantification of body burdens and assessment of feeding rates and reproduction. *Environ. Pollut.* **2017**, *228*, 398–407. [CrossRef]
36. Qiu, Y.; Zhang, T.; Zhang, P. Fate and environmental behaviors of microplastics through the lens of free radical. *J. Hazard. Mater.* **2023**, *453*, 131401. [CrossRef]
37. Joppien, M.; Westphal, H.; Stuhr, M.; Doo, S.S. Microplastics alter feeding strategies of a coral reef organism. *Limnol. Oceanogr. Lett.* **2022**, *7*, 131–139. [CrossRef]
38. Jin, F.; Wei, M.; Liu, C.; Ma, Y. The mechanism for the formation of OH radicals in condensed-phase water under ultraviolet irradiation. *Phys. Chem. Chem. Phys.* **2017**, *19*, 21453–21460. [CrossRef] [PubMed]
39. Prata, J.C.; Reis, V.; Paco, A.; Martins, P.; Cruz, A.; da Costa, J.P.; Duarte, A.C.; Rocha-Santos, T. Effects of spatial and seasonal factors on the characteristics and carbonyl index of (micro)plastics in a sandy beach in Aveiro, Portugal. *Sci. Total Environ.* **2020**, *709*, 135892. [CrossRef] [PubMed]
40. Campanaro, A.L.; Simcik, M.F.; Maurer-Jones, M.A.; Penn, R.L. Sewage sludge induces changes in the surface chemistry and crystallinity of polylactic acid and polyethylene films. *Sci. Total Environ.* **2023**, *890*, 164313. [CrossRef] [PubMed]
41. Brandon, J.; Goldstein, M.; Ohman, M.D. Long-term aging and degradation of microplastic particles: Comparing in situ oceanic and experimental weathering patterns. *Mar. Pollut. Bull.* **2016**, *110*, 299–308. [CrossRef] [PubMed]
42. Zaaba, N.F.; Jaafar, M. A review on degradation mechanisms of polylactic acid: Hydrolytic, photodegradative, microbial, and enzymatic degradation. *Polym. Eng. Sci.* **2020**, *60*, 2061–2075. [CrossRef]
43. Orjuela, A.; Yanez, A.J.; Evans, J.; Hassan, A.M.; Miller, D.J.; Lira, C.T. Phase equilibria in binary mixtures with monoethyl succinate. *Fluid. Phase Equilibria* **2011**, *309*, 121–127. [CrossRef]
44. Wang, Z.; Ding, J.; Song, X.; Zheng, L.; Huang, J.; Zou, H.; Wang, Z. Aging of poly (lactic acid)/poly (butylene adipate-co-terephthalate) blends under different conditions: Environmental concerns on biodegradable plastic. *Sci. Total Environ.* **2023**, *855*, 158921. [CrossRef]
45. Beltran-Sanahuja, A.; Casado-Coy, N.; Simo-Cabrera, L.; Sanz-Lazaro, C. Monitoring polymer degradation under different conditions in the marine environment. *Environ. Pollut.* **2020**, *259*, 113836. [CrossRef]
46. Rahman, M.M.; Islam, M.S.; Li, G.S. Development of PLA/CS/ZnO nanocomposites and optimization its mechanical, thermal and water absorption properties. *Polym. Test.* **2018**, *68*, 302–308. [CrossRef]

47. Mochane, M.J.; Magagula, S.I.; Sefadi, J.S.; Mokhena, T.C. A Review on Green Composites Based on Natural Fiber-Reinforced Polybutylene Succinate (PBS). *Polymers* **2021**, *13*, 1200. [CrossRef]
48. Sun, C.; Wei, S.; Tan, H.; Huang, Y.; Zhang, Y. Progress in upcycling polylactic acid waste as an alternative carbon source: A review. *Chem. Eng. J.* **2022**, *446*, 136881. [CrossRef]
49. Usachev, S.V.; Lomakin, S.M.; Koverzanova, E.V.; Shilkina, N.G.; Levina, I.I.; Prut, E.V.; Rogovina, S.Z.; Berlin, A.A. Thermal degradation of various types of polylactides research. The effect of reduced graphite oxide on the composition of the PLA4042D pyrolysis products. *Thermochim. Acta* **2022**, *712*, 179227. [CrossRef]
50. Arrieta, M.P.; Parres, F.; López, J.; Jiménez, A. Development of a novel pyrolysis-gas chromatography/mass spectrometry method for the analysis of poly(lactic acid) thermal degradation products. *J. Anal. Appl. Pyrolysis* **2013**, *101*, 150–155. [CrossRef]
51. Ainali, N.M.; Bikiaris, D.N.; Lambropoulou, D.A. Aging effects on low- and high-density polyethylene, polypropylene and polystyrene under UV irradiation: An insight into decomposition mechanism by Py-GC/MS for microplastic analysis. *J. Anal. Appl. Pyrolysis* **2021**, *158*, 105207. [CrossRef]
52. Picó, Y.; Barceló, D. Pyrolysis gas chromatography-mass spectrometry in environmental analysis: Focus on organic matter and microplastics. *TrAC Trends Anal. Chem.* **2020**, *130*, 115964. [CrossRef]
53. Romera-Castillo, C.; Pinto, M.; Langer, T.M.; Alvarez-Salgado, X.A.; Herndl, G.J. Dissolved organic carbon leaching from plastics stimulates microbial activity in the ocean. *Nat. Commun.* **2018**, *9*, 1430. [CrossRef]
54. Koh, K.Y.; Chen, Z.; Lin, S.; Chandra Mohan, K.; Luo, X.; Chen, J.P. Leaching of organic matters and formation of disinfection by-product as a result of presence of microplastics in natural freshwaters. *Chemosphere* **2022**, *299*, 134300. [CrossRef] [PubMed]
55. Lee, Y.K.; Murphy, K.R.; Hur, J. Fluorescence Signatures of Dissolved Organic Matter Leached from Microplastics: Polymers and Additives. *Environ. Sci. Technol.* **2020**, *54*, 11905–11914. [CrossRef]
56. Zhong, X.; Yi, X.; Cheng, F.; Tong, H.; Xu, W.; Yang, X. Leaching of di-2-ethylhexyl phthalate from biodegradable and conventional microplastics and the potential risks. *Chemosphere* **2023**, *311*, 137208. [CrossRef]
57. Sun, Y.; Zhao, X.; Sui, Q.; Sun, X.; Zhu, L.; Booth, A.M.; Chen, B.; Qu, K.; Xia, B. Polystyrene nanoplastics affected the nutritional quality of Chlamys farreri through disturbing the function of gills and physiological metabolism: Comparison with microplastics. *Sci. Total Environ.* **2024**, *910*, 168457. [CrossRef]
58. Sharma, V.K.; Ma, X.; Lichtfouse, E.; Robert, D. Nanoplastics are potentially more dangerous than microplastics. *Environ. Chem. Lett.* **2022**, *21*, 1933–1936. [CrossRef]
59. O'Connor, I.A.; Golsteijn, L.; Hendriks, A.J. Review of the partitioning of chemicals into different plastics: Consequences for the risk assessment of marine plastic debris. *Mar. Pollut. Bull.* **2016**, *113*, 17–24. [CrossRef] [PubMed]

processes

MDPI

Article

Analysis of Natural Groundwater Flowing into the Flow Field of In Situ Leaching Mining

Qinci Li [1,2], Zhaokun Li [1,2], Tingting Xie [1,2], Ye Ding [1,2], Nan Gan [1,2], Yahui Tan [1,2] and Chong Zhang [1,2,*]

1 Beijing Research Institute of Chemical Engineering and Metallurgy, China National Nuclear Corporation, Beijing 101149, China
2 China Nuclear Mining Science and Technology Corporation, China National Nuclear Corporation, Beijing 101149, China
* Correspondence: zhangchong@bricem.com.cn

Abstract: This paper aims to quantitatively analyze the influence of natural groundwater flowing into the flow field of in situ leaching mining. The computational method was built to evaluate the effect of natural groundwater on the production efficiency of pumping wells for the in situ leaching of uranium, and the "flow ratio of groundwater" and related formulas were defined. C1 and C2 mining areas of an in situ leaching uranium mine in Inner Mongolia were taken as an example, and the effect on the "flow ratio of groundwater" when changing the flow quantity of injection wells and the position and length of the filter in the pumping and injection wells were compared. The results show that the variation in the "flow ratio of groundwater" of a whole mining area or a single pumping well in different production stages can be obtained by the neutral solution concentration value from the mining area's numerical simulation. Regulating the position, length of the filter, and mode of fluid injection in an in situ leaching mine can control the quantity of natural groundwater flowing into the mining area and reduce the fluid exchange between the flow field of in situ leaching uranium mining and natural groundwater.

Keywords: flow ratio of groundwater; natural groundwater; in situ uranium leaching; open flow field; numerical simulation

1. Introduction

In situ leaching uranium mining is a technology that depends on drilling holes to inject some chemical solution into the ore-bearing strata. The solution moves along the ore bed by controlling the hydraulic gradient of the flow field. In this process, the solution reacts with the ore to generate a uranium-bearing solution. Then, the uranium-bearing solution is extracted to the surface through the pumping well for separation and purification [1,2]. The flow field of in situ leaching mining is a comprehensive field with the solution alongside convective motions, solute dispersions, and chemical reactions. The three fields interact to form the solution's flow, diffusion, and reaction. Hydrodynamic convection is the basis of the flow field, which determines the solution diffusion and reaction range. Solute dispersions rely on the range of action from the hydrodynamic convection.

With the development of computer technology, numerical simulation has become an essential quantitative method to study groundwater flow conditions [3–9] and has been widely utilized in the in situ leaching of uranium. The numerical simulation can be used to quantitatively calculate water flow conditions in the flow field of in situ leaching mining and analyze related characteristics and regularities [10–16]. Tan et al. [17] researched the multi-process coupling and the movement of the reaction front during in situ leaching mining and formed relevant theories. Chen et al. [18] studied the leaching kinetics at a uranium deposit in Xinjiang, China, and simulated the flow characteristics of the flow field of in situ leaching mining. Simon et al. [19] established the coupled simulation of hydrodynamic transport rate and geochemical reactions in the in situ leaching of uranium and determined the factors of mineral precipitation that might affect the solution propagation of in situ leaching. Zuo et al. [20] discussed the migration of groundwater contamination in the aquifer at a decommissioned uranium mine and described the characteristics and laws of solution migration after in situ leaching mining. Wang et al. [21], Zhou et al. [22], Liu et al. [23], and Xie et al. [24] used the tracer technique to study the migration characteristics, solution migration range, and control mode of the fluid and related ions during the in situ leaching of uranium. Zhang et al. [16] first proposed the flow microbalance of pumping and injection units in the in situ leaching of uranium and analyzed all pumping and injection units at an in situ leaching mining area in Inner Mongolia.

The flow field of in situ leaching mining is an open flow field [25–27]. It exchanges with natural groundwater sustainably, which leads to natural groundwater flowing into the flow field of in situ leaching mining and fluid exchange with the solution in the flow field. The fluid exchange results in the decline of uranium concentration and the production efficiency of in situ leaching, which adversely affects natural groundwater. If the fluid exchange between the flow field of in situ leaching mining and natural groundwater could be quantitatively studied, the processes of in situ leaching uranium mining would be optimized to improve the production efficiency [28] and control the solution diffusion range effectively to avoid pollution accidents [29]. Therefore, it is significant to establish a method of calculating the quantity of natural groundwater flowing into the flow field of in situ leaching mining and measurably make a related analysis. It is not only of great significance in studying the control of the flow field of in situ leaching and improving the production capacity of uranium mines. It also provides enlightenment on the research of natural groundwater pollution treatment.

Nowadays, many researchers have conducted various studies on natural groundwater flowing into the flow field of in situ leaching mining. Wang et al. [25] studied the relationship between different well patterns and the amount of natural groundwater flowing into the flow field. They analyzed the difference in the influence of natural groundwater on the solution of the flow field between the narrow strip orebody and the square orebody. Zhou et al. [24] set different flow quantities for the pumping and injection wells of a single unit in an in situ leaching mining area, which gained the influence of the balanced relationship between the flow quantities of pumping and injection wells on the quantity of groundwater flowing into the mining area. Xie et al. [27,30] conducted numerical simulations on the hydrodynamic and dispersion characteristics of an in situ leaching mine in Inner Mongolia. They divided the periods of the in situ leaching flow field by the groundwater influence ratio of each pumping well. Xu et al. [31], Ji et al. [32], Cao et al. [33], Li et al. [34], and Zhang et al. [35] put forward a method for regulating the flow rate ratio of pumping and injection wells to control the solution outflow in the flow field of in situ leaching mining and carried out related calculations in different mining areas, respectively.

This paper aims to analyze the influence of natural groundwater on the flow field of in situ leaching uranium mining under complex mining conditions. A quantitative calculation and analysis method of natural groundwater flowing into in situ leaching mining areas based on previous studies was set up, and the "flow ratio of groundwater" and relevant equations were defined. Then, the method was utilized to calculate the corresponding values of C1 and C2 mining areas of an in situ leaching uranium mine in Inner Mongolia and the influence of the flow quantity of injection wells and the position and length of the filter in the pumping and injection wells on the "flow ratio of groundwater" were analyzed. The analysis can help us realize the purpose of reducing the fluid exchange between the flow field of in situ leaching uranium mining and natural groundwater.

2. Materials and Methods

Currently, the analysis of the fluid exchange between the flow field of in situ leaching mining and natural groundwater cannot evaluate the impact of natural groundwater on the production efficiency of the in situ leaching of uranium. Zhou et al. [26] obtained the direction of groundwater velocity by the numerical simulation of the flow field of in situ leaching mining, which was utilized to divide the flow balance blocks and boundaries. Then, they made a flow balance analysis to calculate the quantities of influent natural groundwater and outflow flow field of in situ leaching mining in every boundary. However, this method could only calculate the flow balance between the mean flow field with a single pumping and injection unit and natural groundwater, which could not reflect the effect of groundwater on the production efficiency of the pumping wells for the in situ leaching of uranium. Xu et al. [31], Ji et al. [32], Cao et al. [33], and Li et al. [34] obtained the area results of solute transport from the numerical simulation of the flow field of in situ leaching mining, which could be used to judge the flow quantity of the solution in the flow field of in situ leaching mining that merged into natural groundwater. This method could only qualitatively reflect the fluid exchange between the flow field of in situ leaching mining and natural groundwater. In comparison, the method proposed in this paper could be utilized to carry out quantitative multi-factor analysis on the heterogeneous deposit with multiple mining areas and evaluate the effect of groundwater on the production efficiency of the pumping wells for the in situ leaching of uranium quantitatively.

If injecting a neutral solution (whose concentration is C_0) into all injection wells in a mining area at the same time, this may result in the concentration of the solution extracted from the pumping well not being equal to the injected, which is related to the fluid exchange between natural groundwater and the flow field of in situ leaching uranium mining. Thus, the fluid extracted from the pumping well can be regarded as two "solutions": the neutral solution and natural groundwater. The quantity of natural groundwater flowing into the flow field of in situ leaching uranium mining can be obtained by the variation in the concentration of the solute in the fluid extracted from the pumping well.

As shown in Figure 1, solution A (its concentration is C_0) is injected into the injection well w_{i1}, w_{i2}, w_{i3}, w_{i4} and w_{i5}, which belong to the pumping-injection unit of the pumping well w_{p1} and w_{p2}. The concentration of solute A in the fluid extracted from the pumping well w_{p1} and w_{p2} is C_{p1}^h and C_{p2}^h at the t_h moment, respectively. According to the change in the concentration between C_{p1}^h and C_0, and between C_{p2}^h and C_0, the quantity of groundwater entering the solution extracted from the pumping well w_{p1} and w_{p2} can be calculated.

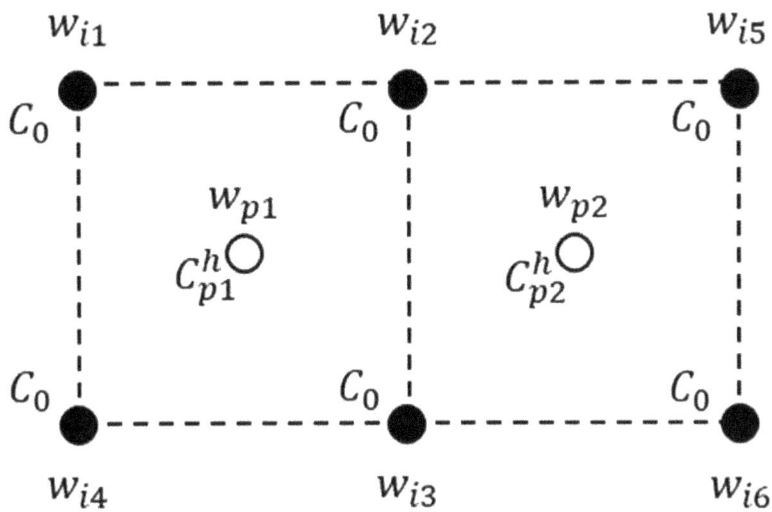

Figure 1. The schematic diagram of the calculation principle for the "flow ratio of groundwater" in two pumping-injection units.

For a mining area including j pumping wells and k injection wells, the indexes of the pumping and injection wells are $w_{pj}(j = 1, 2, 3, 4, \ldots)$ and w_{ik} $(k = 1, 2, 3, 4, \ldots)$, respectively. If the solute A (its concentration is C_0) is injected into different injection wells at the same time, and the total injection time is T, the total injection quality of the solute A in the mining area will be:

$$M_0 = T \cdot q_{ik} \cdot C_0 \tag{1}$$

where q_{ik} is the flow quantity of the injection well w_{ik} in a time unit, $k = 1, 2, 3, 4, \ldots, T$ is the total injection time of solute A, and C_0 is the injected concentration of solute A.

The total injection time T is dispersed to h points of time, the flow quantity of the pumping well w_{pj} is q_{pj} at the t_h moment, and the concentration of the solute A extracted by the pumping well w_{pj} at the t_h moment is C_{pj}^h, so the quality of the solute A extracted by the pumping well w_{pj} in the total injection time T is:

$$M_{pj} = \sum\nolimits_0^T \left[q_{pj} \cdot C_{pj}^h \cdot (t_h - t_{h-1}) \right] \tag{2}$$

where q_j^h is the flow quantity of the pumping well w_{pj} in a time unit at the t_h moment, the concentration of the solute A extracted by the pumping well w_{pj} at the t_h moment is C_{pj}^h, $j = 1, 2, 3, 4, \ldots$.

Therefore, the total accumulative quality of the solute A extracted by all pumping wells of the whole mining area in the total injection time T is:

$$M_p = \sum M_{pj} \tag{3}$$

If $\frac{M_p}{C_0}$ is defined as "the flow quantity of non-groundwater" when the solute A extracted from all pumping wells in the whole mining area, the difference between "the flow quantity of non-groundwater" and the total flow quantity from all the pumping wells in the mining area is the quantity of natural groundwater flowing into the flow field of this mining area, which can be defined as:

$$r_w = \frac{\sum \sum_0^T \left[q_{pj} \cdot (t_h - t_{h-1}) \right] - \frac{M_p}{C_0}}{\sum \sum_0^T \left[q_{pj} \cdot (t_h - t_{h-1}) \right]} = 1 - \frac{\frac{M_p}{C_0}}{\sum \sum_0^T \left[q_{pj} \cdot (t_h - t_{h-1}) \right]} \tag{4}$$

where r_w is the "flow ratio of groundwater" of this mining area, $\sum_0^T \left[q_{pj} \cdot (t_h - t_{h-1}) \right]$ is the total flow quantity of the pumping well w_{pj} in the total injection time T, and $\sum \sum_0^T \left[q_{pj} \cdot (t_h - t_{h-1}) \right]$ is the total flow quantity from all pumping wells in the mining area in the total injection time T.

If set $\sum_0^T \left[q_{pj} \cdot (t_h - t_{h-1}) \right] = Q_{pj}$, and $\sum \sum_0^T \left[q_{pj} \cdot (t_h - t_{h-1}) \right] = \sum Q_{pj} = Q_p$, Equation (4) can be simplified as:

$$r_w = 1 - \frac{\frac{M_p}{C_0}}{Q_p} = 1 - \frac{M_p}{C_0 \cdot Q_p} \tag{5}$$

If the "flow ratio of groundwater" for a single pumping well in the mining area is calculated, "the flow quantity of non-groundwater" of the pumping well w_{pj} in a time unit will be $\frac{q_{pj} \cdot C_{pj}^h}{C_0}$, and the "flow ratio of groundwater" of this pumping well in the total injection time T will be:

$$r_j = \frac{\sum_0^T \left[q_{pj} \cdot (t_h - t_{h-1}) \right] - \frac{\sum_0^T \left[q_{pj} \cdot C_{pj}^h \cdot (t_h - t_{h-1}) \right]}{C_0}}{\sum_0^T \left[q_{pj} \cdot (t_h - t_{h-1}) \right]} = \frac{Q_{pj} - \frac{M_{pj}}{C_0}}{Q_{pj}} = 1 - \frac{\frac{M_{pj}}{C_0}}{Q_{pj}} = 1 - \frac{M_{pj}}{C_0 \cdot Q_{pj}} \tag{6}$$

To sum up, it can be seen that when a neutral solution is injected into different injection wells in the mining area, the "flow ratio of groundwater" for a whole mining area or of a single pumping well at each time point can be calculated by the dispersion of the solute in the flow field. It can realize the accurate quantification of the influence of natural groundwater on the flow field of in situ leaching mining.

3. Case Study

3.1. The Background of the Study Area

The study area is in the Erlian Basin in middle and northern Inner Mongolia. According to the geological data of this area [36], the layers from the top to bottom are the Yierdingmanha group of Paleogene (E_2y), the upper member of Saihan group (K_1s^2), and the lower member of Saihan group (K_1s^1) in Lower Cretaceous.

The targeted aquifer is one layer at the C1 and C2 mining areas of an in situ leaching uranium mine in the study area (Figure 2). The buried depth of the water table of this aquifer is 16.23–19.67 m, and its pressure head is 60.83–92.07 m. This aquifer is stratiform, with an average thickness of ~35.71 m. The distribution of the aquifer is continuous and stable. The water-isolated top of the aquifer is mainly composed of mudstone in the upper member of the Saihan group, and its thickness is ~10–50 m [36]. Moreover, the water-isolated bottom of the aquifer is compounded from the mudstone in the lower member of the Saihan group, and its depth of the top surface is ~80–150 m, with an average value of ~115.43 m [36]. The top and bottom of the targeted aquifer are continuous and have good water-isolated performance.

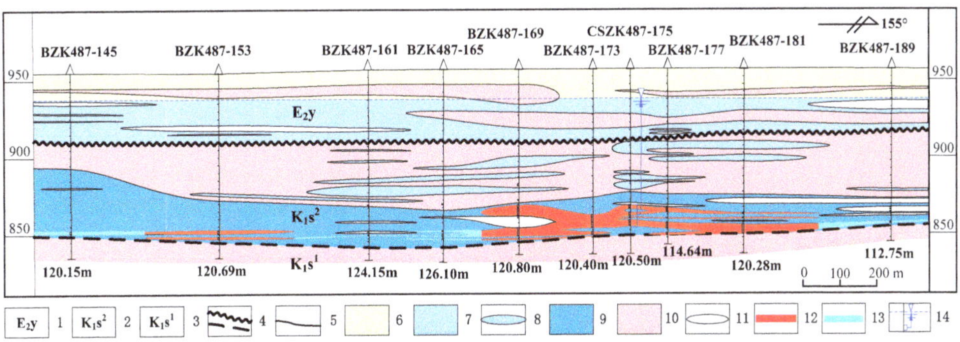

Figure 2. The transversal hydrogeologic profile at an in situ leaching uranium mine in Inner Mongolia. Note: 1—Yierdingmanha group; 2—The upper member of Saihan group; 3—The lower member of Saihan group; 4—Angular/parallel unconformity boundary of the layers; 5—Lithological boundary; 6—The aquifer in Quaternary; 7—The aquifer in Yierdingmanha group; 8—Permeable lens; 9—The targeted aquifer; 10—The water-isolated layer; 11—Water-isolated lens; 12—The uranium ore body; 13—The uranium mineralization body; 14—The location of the pumping test and the water level of the pressure head of the targeted aquifer.

There are 62 pumping-injection units at the C1 and C2 mining areas with production in progress, and the pumping and injection wells are arranged in a five-point pattern. In total, there are 62 pumping wells and 80 injection wells. The spacing between the pumping and injection wells in a single unit is 27 m, and the area of the mining areas is about 98,000 m^2 (Figure 3). The flow quantity of a single pumping well in the mining areas is stable at ~120 m^3/day(m^3/d), and the flow quantities of the pumping and injection wells in the mining areas are balanced, which is called "the pumping and injection balanced principle".

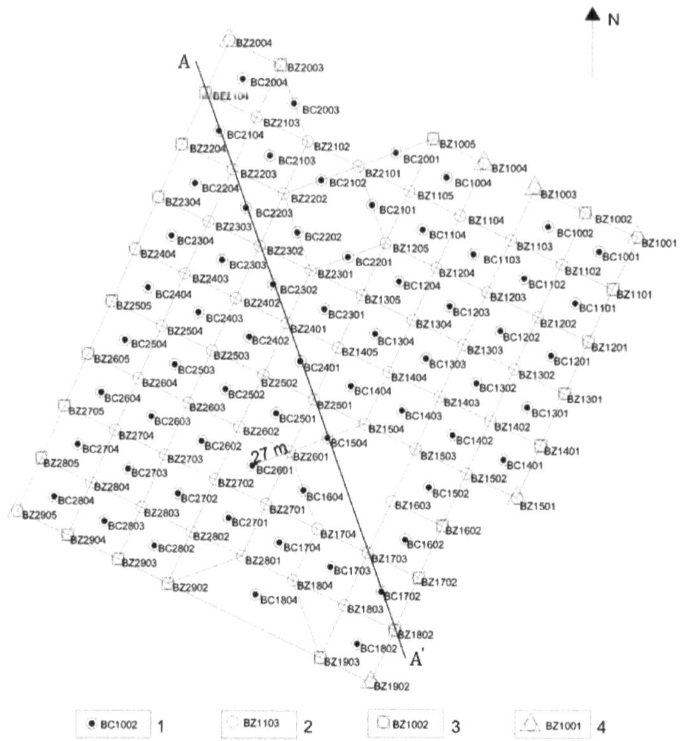

Figure 3. Distribution of the pumping and injection wells at the C1 and C2 mining areas of an in situ leaching uranium mine in Inner Mongolia. Note: 1—Pumping wells; 2—Internal injection wells; 3—Edge injection wells; 4—Top corner injection wells.

3.2. *Numerical Simulation of the Mining Areas*

3.2.1. Model Generalizability

Aquifer System Generalizability

The targeted aquifer in the simulation was the aquifer in the upper member of the Saihan group (K_1s^2). Its water-isolated top and bottom were continuous and stable (Figure 2), which led to no hydraulic relationship between the targeted and other aquifers. In order to simplify the model, the flow field simulation could only aim at the targeted aquifer in the mining areas, ignoring other aquifers.

Boundary Condition Generalizability of Groundwater System

The top and bottom boundary of the model was water-isolated. The perimeter boundary of the model was the first boundary condition with known water levels.

Source and Sink Generalizability

The source and sink of the model were the production operation of the pumping and injection wells. The distribution of the pumping and injection wells was set in the model according to the actual situation. As stated earlier, the flow quantity of a single pumping well in the mining areas is stable at ~120 m^3/d, and the flow quantity of an injection well should follow "the pumping and injection balanced principle". Setting different flow quantities in the same injection well under "the pumping and injection balanced principle" might result in different fluid exchanges between the flow field of in situ leaching mining and natural groundwater. Thus, the influence of the flow quantity of an injection well on the fluid exchange between the flow field of in situ leaching mining and natural groundwater under the same total injected quantity could be studied by setting different flow quantities in the same injection well. According to the distribution of the injection wells, they could be divided into internal injection wells, edge injection wells, and top-corner injection wells. As shown in Figure 3, the internal injection well BZ2203 was related to the pumping wells BC2104, BC2103, BC2204, and BC2203 directly, the edge injection well BZ2104 was related to the pumping wells BC2004 and BC2104 directly, and the top corner injection well BZ2004 was related to the pumping well BC2004 directly. According to the corresponding relationship between the injection well, and the number of its adjacent pumping wells, different flow quantities for different injection wells with different locations could be set.

Solute Generalizability

For each injection well in the mining area, the neutral solute injected point was taken. The injected concentration was 10,000 mg/L, whose background concentration value in natural groundwater was 0 mg/L. The total injection and simulation time was 1370 days. The neutral solute was set to have similar physical and chemical properties to carbinol. Its density was close to the water, and it could easily dissolve in water. This solute was not involved in any chemical reactions, and its adsorption constant was zero. In the simulation of solute dispersion, only mechanical dispersion and the concentration diffusion of the fluid could be considered because the neutral solute did not join in the liquid-solid reaction and adsorption dissolution.

3.2.2. Solution and Identification of the Numerical Simulation

Hydrodynamic Flow and Solute Transport Equation

The hydrogeological conceptual model generalized by the above conditions was calculated by the three-dimensional mathematical model of saturated flow to show its conditions for groundwater flow and solute transport. The governing equations and definite conditions were utilized to express the three-dimensional mathematical model.

The definite conditions usually include initial and boundary conditions [37,38].

The governing hydrodynamic equation is:

$$\frac{\partial}{\partial x}\left(K_x\frac{\partial H}{\partial x}\right)+\frac{\partial}{\partial y}\left(K_y\frac{\partial H}{\partial y}\right)+\frac{\partial}{\partial z}\left(K_z\frac{\partial H}{\partial z}\right)+q=\mu_s\frac{\partial H}{\partial t} \tag{7}$$

The initial condition is:

$$H(x,y,z,t)|_{t=0}=H_0(x,y,z) \tag{8}$$

The boundary condition is:

The top and bottom are:

$$\frac{\partial H}{\partial n}=0 \tag{9}$$

The perimeter is:

$$K\frac{\partial H}{\partial n}\bigg|_{t=0}=v(x,y,z,t) \tag{10}$$

where H is waterhead and the unit is m; n is the normal direction pointing the outside of the boundary; t is time, the unit is the day (d); K is the tensor of permeability coefficient; q is the source and sink; μ_s is storage coefficient; v is the fluid velocity, and the unit is m/day(m/d).

The solute transport equation is:

$$-\frac{\partial}{\partial x}(q_xC)-\frac{\partial}{\partial y}(q_yC)-\frac{\partial}{\partial z}(q_zC)+q_sC_s=\theta\frac{\partial C}{\partial t} \tag{11}$$

The initial condition is:

$$C(x, y, z, t)|_{t=0} = 0 \tag{12}$$

The boundary condition is:

The perimeter is:

$$C = 0 \tag{13}$$

where C is the concentration of the solute, and the unit is mg/L; C_s is the concentration of the solute in sources and sinks, and the unit is mg/L; q_s is the flow quantity of sources and sinks, and the unit is m^3/d; θ is porosity; t is time, the unit is d; q_x, q_y and q_z are the flow quantities in x, y, and z directions, respectively, and the unit is m^3/d.

Identification of the Model

The common stochastic geological modeling method can be divided into stochastic modeling based on conditional simulation and unconditional simulation through how to utilize the conditional dataset [39,40]. Furthermore, the method used in this paper to construct a heterogeneous lithologic model was a geostatistical program called T-PROG, which is based on the Markov chain. This method could reflect geological asymmetry and anisotropy and build the spatial structure of the aquifer more in accordance with reality [41,42].

The groundwater modeling system (GMS) software is utilized to make the simulation. The whole flow field was composed of the lithologic model, groundwater flow model, and solute movement model after the assignment of hydrogeological parameters for the C1 and C2 mining areas. The T-PROG, MODFLOW, MODPATH, and MT3D modules were used for the construction [43].

First, the dataset of 144 boreholes in the mining areas was sorted out and imported into the software. Then, the heterogeneous lithologic model of the targeted aquifer (including the water-isolated top and bottom) in the mining areas was established, with a thickness of ~50–110 m (Figure 4). Then, the top and bottom were taken as the upper and lower boundaries of the model, and the elevation of the top and bottom was ~910–830 m. The T-PROG module was used to assign and interpolate the hydrogeological parameters of the five kinds of lithology (mudstone, silty mudstone, fine sandstone, medium sandstone, and coarse sandstone) in the targeted aquifer. The distribution of heterogeneous hydrogeological parameters in the aquifer was obtained, which could be used to simulate the targeted aquifer in the mining areas.

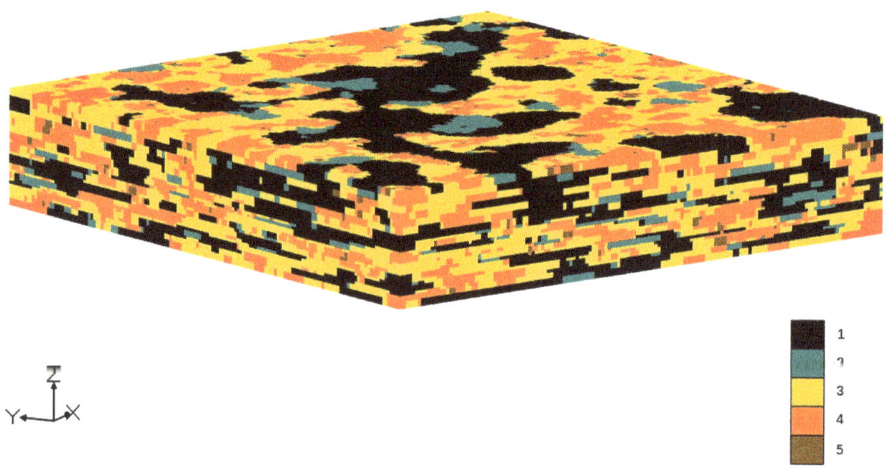

Figure 4. Simulated results of heterogeneous lithological distribution in the study area. Note: 1—Mudstone; 2—Silty mudstone; 3—Fine sandstone; 4—Medium sandstone; 5—Coarse sandstone.

3.3. *Analysis of the "Flow Ratio of Groundwater" in the Mining Areas*

According to the previous generalizations regarding geology, hydrogeology, and manufacturing processes, the water levels and solute transport in the mining areas under different production plans were calculated (Figure 5). The concentration of the solute A extracted by each pumping well at different time points (C_{pj}^{h}) could be obtained from the numerical simulation of the mining areas. Then, the varieties of the "flow ratio of groundwater" in the mining areas or a single pumping well at different production stages could be obtained to make a quantitative multi-factor analysis on the fluid exchange between the flow field of in situ leaching uranium mining and natural groundwater.

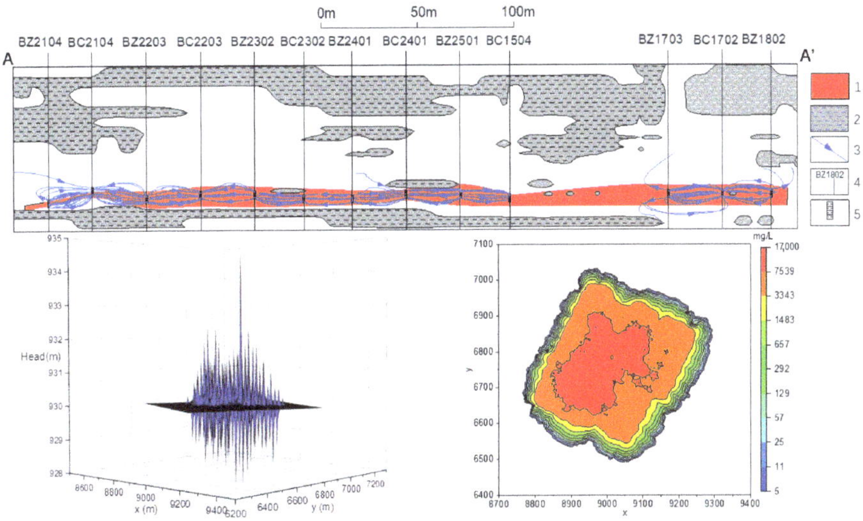

Figure 5. Simulated results of the flow field of in situ leaching uranium mining (Plan A). Note: The above–the profile (A–A′) of the flow field, and the location of the profile is shown in Figure 3 (1—The uranium ore body; 2—Mudstone; 3—Streamline and its direction; 4—Pumping and injection wells; 5—Filter); lower left—water level contour map of the mining areas; lower right—solute transport diagram of the mining areas.

3.3.1. The Flow Quantity of Injection Wells

As previously mentioned, if different flow quantities are set in the same injection well, the fluid exchange between the flow field of in situ leaching mining and natural groundwater will be different. The flow quantity of all pumping wells in plans A and B was 120 m^3/d. Their locations set the flow quantity of the injection wells: in plan A, the flow quantity of top corner injection wells was 30 m^3/d, the flow quantity of edge injection wells was 60 m^3/d, and the flow quantity of internal injection wells was 120 m^3/d (the positions for those injection wells as displayed in Figure 3). The flow quantity for all the injection wells in plan B was 90 m^3/d.

The filters' position in two plans in the middle of the uranium ore bed was set, and the length of the filters was 6 m. The flow quantity of all pumping wells was set to be the same in this case, so the difference between "the flow quantity of non-groundwater" corresponding to the average concentration of all pumping wells at each time point and the flow quantity of a single pumping well could be calculated. The proportion of the difference in the flow quantity of a single pumping well can be used to obtain the "flow ratio of groundwater" of the flow field in the whole C1 and C2 mining areas at each time point.

Figure 6 and Table 1 reveal that the "flow rate of groundwater" for plan A with "small flow injection at the periphery of the well-site" is always lower than that of plan B with the same flow quantity for all injection wells. When the "flow ratio of groundwater" in plan A reaches 70%, 50%, 30%, and 10%, respectively, the arrival time is about the 29th, 66th, 147th, and 561st day of the simulation. In contrast, the "flow ratio of groundwater" for plan B reaches the same values. The arrival time is about the 37th, 89th, 230th, and 1318th day of the simulation, which shows that plan A can control the inflow of natural groundwater faster, so plan A can reduce the impact of the solution in the flow field at the mining area on natural groundwater faster. This is consistent with the viewpoint of Zhang et al. [35], who proposed that "small flow injection at the periphery of the well-site" can effectively control the leaching range of the flow field in in situ leaching mining. This results from the fact that the pumping or injection capacity of each borehole is different when the strata in the study area are heterogeneous; the drilling technology is different. On the premise of "the pumping and injection balanced principle" in the mining areas, different flow quantities are adopted for the injection wells. Even if the total injected quantity in the mining areas remains unchanged, the fluid exchange between the flow field of in situ leaching mining and natural groundwater is different [16]. Figures 7–9 further confirm that the peripheral pumping well (Figure 7), the internal pumping well (Figure 8), and the pumping well with only three injection wells (Figure 9) in plan A with "small flow injection at the periphery of the well-site" also have smaller "flow ratio of groundwater".

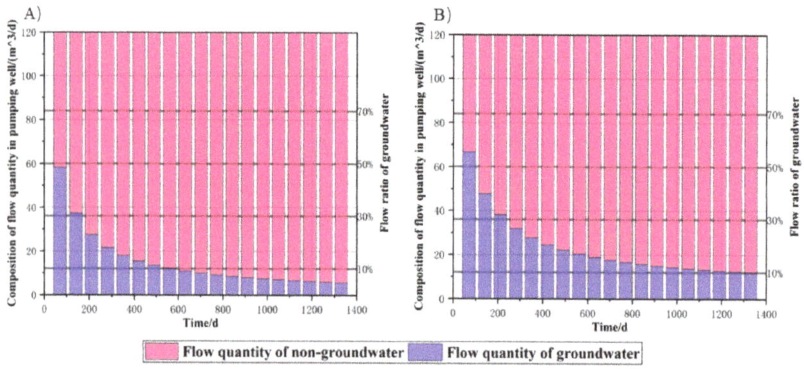

Figure 6. The bar graph of the composition of flow quantity in pumping wells at the whole C1 and C2 mining areas with the change in time (based on the average concentration from all pumping wells). Note: (**A**)—Plan A; (**B**)—Plan B.

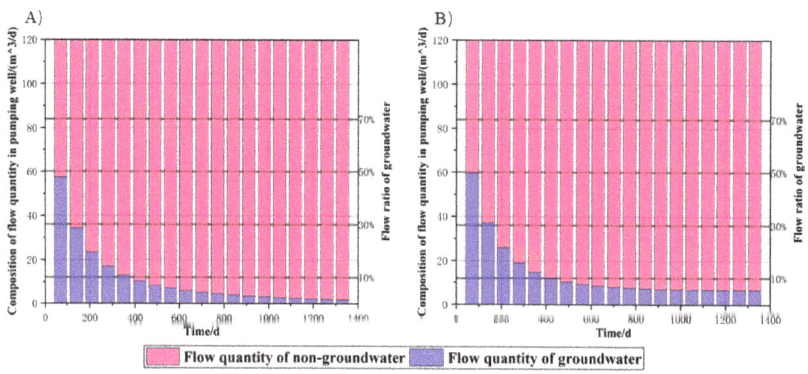

Figure 7. The bar graph of the composition of flow quantity in pumping well BC2204 with the change in time. Note: (**A**)—Plan A; (**B**)—Plan B.

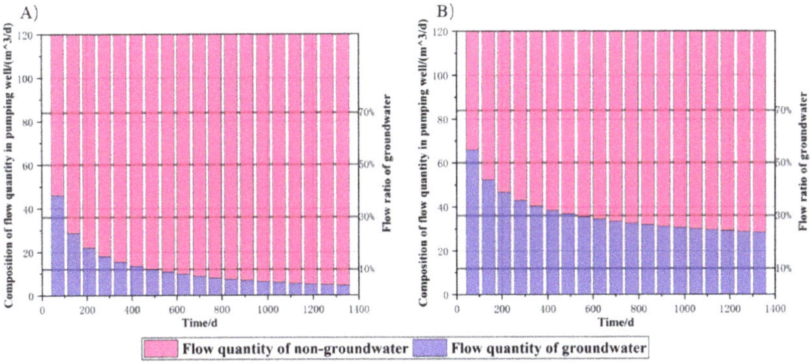

Figure 8. The bar graph of composition of flow quantity in pumping well BC2401 with the change in time. Note: (**A**)—Plan A; (**B**)—Plan B.

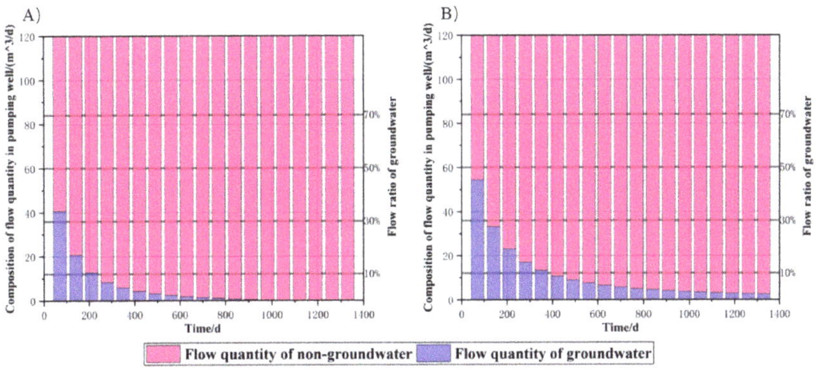

Figure 9. The bar graph of composition of flow quantity in pumping well BC1604 with the change in time. Note: (**A**)—Plan A; (**B**)—Plan B.

Table 1. Comparison of the day for plan A and plan B to reach the corresponding "flow ratio of groundwater".

Arrival Time (Day) / Flow Ratio of Groundwater / Plan	70%	50%	30%	10%
Plan A	29	66	147	561
Plan B	37	89	230	1318

3.3.2. The Position of the Filter

To discuss the impact of the filter's position, the filter in plan A was set in the middle of the uranium ore bed. The filter in plan C was staggered with the uranium ore bed (the bottom of the filter in the pumping wells was 2 m lower than the bottom of the ore bed, and the top of the filter in injection wells was 2 m higher than the top of the ore bed). The flow quantity of all pumping wells in the two plans was 120 m^3/d, the length of all filters was 6m, and the flow quantity of injection wells was set similarly to plan A in Section 3.3.1.

Figure 10 and Table 2 present that the "flow ratio of groundwater" of plan C is always lower than that of plan A. Moreover, the time to reach 70%, 30%, 50%, and 10% in plan C is always earlier than in plan A. It shows that when the filter is staggered with the ore bed, the inflow of natural groundwater can be reduced more effectively, and the fluid exchange between the solution in the flow field and natural groundwater can be reduced.

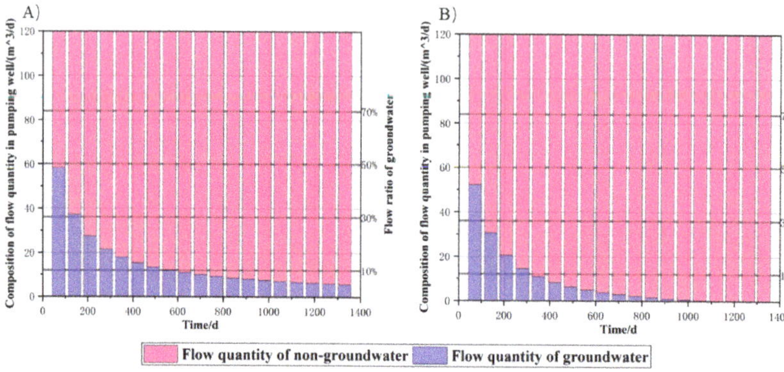

Figure 10. The bar graph of the composition of flow quantity in the pumping well at the whole C1 and C2 mining areas with the change in time (based on the average concentration from all pumping wells). Note: (**A**)—Plan A; (**B**)—Plan C.

Table 2. Comparison of the day for plan A and plan C to reach the corresponding "flow ratio of groundwater".

Arrival Time (Day) / Flow Ratio of Groundwater / Plan	70%	50%	30%	10%
Plan A	29	66	147	561
Plan C	25	55	117	324

This is mainly because when the solution flows from the injection well to the pumping well, its flow velocity decreases and then increases. Therefore, the flow velocity of the solution in the middle area between the pumping and injection well is very slow, and some of the solutions descend by gravity. Therefore, the position of the filter in the pumping well is deeper than that in the injection well and is staggered with the ore bed, which causes more falling solutions to gather around the pumping well and reduces the impact on the natural groundwater.

3.3.3. The Length of the Filter

To understand the influence of the length of the filter on the "flow ratio of groundwater", the length of filters in plans D and E were set as 6 m and 4 m, respectively. Furthermore, all filters were set at the same elevation of 844 m. All wells had the same flow quantity as plan A in Section 3.3.1.

Figure 11 and Table 3 indicate that the time difference between plan D and plan E when the "flow ratio of groundwater" reaches 70%, 50%, 30%, and 10% is about 5, 13, 38, and 289 days, which indicates that the time difference between the two plans when the "flow ratio of groundwater" reaches the same values (such as 70%, 50%, 30%, and 10%) will increase with the decrease in the values of the flow ratio.

Table 3. Comparison of the day for plan D and plan E to reach the corresponding "flow ratio of groundwater".

Arrival Time (Day) / Flow Ratio of Groundwater / Plan	70%	50%	30%	10%
Plan D	30	66	154	758
Plan E	25	53	116	469

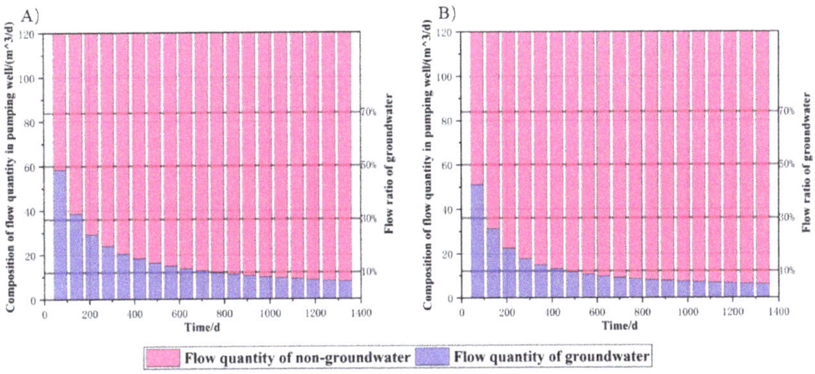

Figure 11. The bar graph of the composition of flow quantity in the pumping well at the whole C1 and C2 mining areas with the change in time (based on the average concentration from all pumping wells). Note: (**A**)—Plan D; (**B**)—Plan E.

The length difference between 6 m and 4 m is a little slight. When the flow quantity of the groundwater flowing into the flow field is large, there is no apparent difference in the arrival time of the two plans. However, as the flow quantity of the groundwater flowing into the flow field decreases, the difference in arrival time between the two plans gradually increases.

4. Conclusions

In this paper, computational and analytical methods were built to calculate the flow quantity of groundwater flowing into the flow field of in situ leaching uranium mines quantitatively and defined the "flow ratio of groundwater" and related formulas. Then, the calculation of an in situ leaching uranium mine in Inner Mongolia was taken as an example, and the position, length of the filter, and the mode of fluid injection at the C1 and C2 mining areas of this in situ leaching mine were regulated, and the following conclusions have been drawn:

(1) In this paper, a method was proposed to evaluate the effect of natural groundwater on the production efficiency of the pumping wells for the in situ leaching of uranium quantitively. The concentration of the neutral solute extracted by each pumping well at different time points could be obtained from the mining areas' numerical simulation. Then, the results of the simulation were able to obtain the varieties of the "flow ratio of groundwater" in the mining areas or a single pumping well at different production stages.

(2) The plan with a "small flow injection at the periphery of the well-site" can control the fluid exchange between the solution in the flow field and natural groundwater faster and realize less groundwater inflow faster.

(3) The plan of having the filters staggered with the ore bed is beneficial to gather more solutions in the flow field around the pumping well and reduce groundwater inflow effectively.

(4) When the flow quantity of the groundwater flowing into the flow field is large, there is no noticeable difference in the arrival time between the plans, with the length of filters between 6 m and 4 m. With the decrease in the inflow of natural groundwater, the difference in the arrival time between those plans gradually increases.

Author Contributions: Conceptualization, Q.L., T.X. and C.Z.; methodology, Q.L., T.X. and Y.T.; software, T.X., Y.D. and N.G.; validation, Z.L.; formal analysis, Y.T.; investigation, Z.L.; writing—review and editing, Q.L., T.X. and C.Z. All authors have read and agreed to the published version of the manuscript.

Funding: This research was funded by the National Natural Science Foundation of CNNC Joint Fund number U2067204, Young-Talent Project of CNNC (The Research on Intelligent Monitoring Technology of Optical Fiber for Critical Hydraulic Parameters of In situ Uranium Leaching), and Young-Talent Project of CNNC (The Research about Intelligent Extraction of Multisource Information in Mining Area based on In situ Leaching Borehole Data).

Data Availability Statement: Data are contained within the article.

Conflicts of Interest: The authors declare no conflict of interest.

References

1. Zhu, Z.Q.; Luo, X.F.; Wang, X.W. *In-Situ Leaching of Uranium Handbook*; The Sixth Research Institute of Nuclear Industry: Hunan, China, 2000.
2. Que, W.M.; Tan, Y.H.; Zeng, Y.J.; Wang, S.D. *Geochemical Kinetics and Mass Transport of In-Situ Uranium Leaching*; Atomic Energy Press: Beijing, China, 2002.
3. Ding, J.H.; Zhou, D.L.; Ma, S.Z. The State-of-Art Trends of Development of Groundwater Modeling Software Abroad. *Site Investig. Sci. Technol.* **2002**, *1*, 37–42.
4. Guo, X.D.; Tian, H.; Zhang, M.G.; Zhu, W.; Cui, J. Visual MODFLOW; GMS; FEFLOW Application Development of Groundwater Value Simulation Software in Our Country. *Ground Water* **2010**, *32*, 5–7.
5. Mategaonkar, M.; Eldho, T.I. Simulation of groundwater flow in unconfined aquifer using meshfree point collocation method. *Eng. Anal. Bound. Elem.* **2011**, *35*, 700–707.
6. Li, F.; Li, J.K.; Ma, Y.; Li, H.E.; Ji, G.Q.; Yuan, M. The research and application progress of numerical simulation on groundwater. *J. Water Resour. Water Eng.* **2018**, *29*, 99–104.
7. Wang, S.; He, S.L.; Yuan, H.Y.; Ji, D.L.; Qi, Z.B. Research progress of groundwater simulation in mining area based on GMS application. *Ground Water* **2022**, *44*, 14–17.
8. Lu, H.; Peng, H.; Xu, Z.D.; Matthews, J.C.; Wang, N.; Iseley, T. A feature selection–based intelligent framework for predicting maximum depth of corroded pipeline defects. *J. Perform. Constr. Facil.* **2022**, *36*, 04022044. [CrossRef]
9. Qin, G.; Xia, A.; Lu, H.; Wang, Y.; Li, R.; Wang, C. A hybrid machine learning model for predicting crater width formed by explosions of natural gas pipelines. *J. Loss Prev. Process Ind.* **2023**, *82*, 104994. [CrossRef]
10. Yabusaki, S.B.; Fang, Y.L.; Waichler, S.R. Building conceptual models of field-scale uranium reactive transport in a dynamic vadose zone-aquifer-river system. *Water Resour. Res.* **2008**, *44*, 853–863. [CrossRef]
11. Li, G.J.; Wang, H.F.; Zhang, Y.; Zhang, B.; Xiao, Z.X. Application of Visual MODFLOW numerical simulation of groundwater to in-situ leaching of uranium. *Uranium Min. Metall.* **2011**, *30*, 1–5.
12. Cao, F.B.; Lian, G.X. Simulation for radical transportation of sulfate during in-situ leaching uranium in a sandstone-type uranium deposit. In *Progress Report on China Nuclear Science & Technology (Vol.4)*; Atomic Energy Press: Beijing, China, 2015; pp. 275–279.
13. Regnault, O.; Lagneau, V.; Fiet, N. *3D Reactive Transport Simulations of Uranium in situ Leaching: Forecast and Process Optimization*; Merkel, B.J., Arab, A., Eds.; Uranium Past and Future Challenges; Springer International Publishing: Cham, Switzerland, 2015; pp. 725–730.
14. Chen, X.X. Numerical Simulation of In-Situ Leaching of Uranium by Acid Method Based on Reactive Solute Transport Theory. Ph.D. Thesis, East China University of Technology, Jiangxi, China, 2020.
15. Zhang, T.; Yuan, L.; Yang, K.; Liu, Y.; Yu, X. Modeling of Multiphysical–Chemical Coupling for Coordinated Mining of Coal and Uranium in a Complex Hydrogeological Environment. *Nat. Resour. Res.* **2021**, *30*, 571–589. [CrossRef]
16. Zhang, C.; Tan, K.X.; Xie, T.T.; Tan, Y.; Fu, L.; Gan, N.; Kong, L. Flow Microbalance Simulation of Pumping and Injection Unit in In Situ Leaching Uranium Mining Area. *Processes* **2021**, *9*, 1288. [CrossRef]
17. Tan, K.X.; Wang, Q.L.; Hu, E.M.; Hu, K.G.; Zhou, Q. Multiple Processes Coupling and Reaction Front Propagation During In-Situ Leach Mining: 1. Theoretical Analyses. *Uranium Min. Metall.* **2005**, *24*, 14–18.
18. Chen, L.; Tan, K.X.; Liu, J.; Xie, Y.S.; Huang, W.; Ma, Q. Characteristics of Leaching Kinetics of a Sandstone Uranium Deposit in Yili Basin. *Met. Mine* **2013**, *3*, 18–20.
19. Simon, R.B.; Thiry, M.; Schmitt, J.M.; Lagneau, V.; Langlais, V.; Bélières, M. Kinetic reactive transport modelling of column tests for uranium In Situ Recovery (ISR) mining. *Appl. Geochem.* **2014**, *51*, 116–129. [CrossRef]
20. Zuo, W.; Tan, K.X. The Characteristics of Groundwater Contamination of a Decommissioned Wellfield in Some in Situ Leaching Uranium Mine, Xinjiang, China. *J. Univ. S. China* **2014**, *28*, 28–31.
21. Wang, X.D.; Wu, L.W.; Duan, B.S.; Xie, T.T.; Yang, L.Z.; Du, Z.M. Interwell tracer test in in-situ leaching uranium in one place of Xinjiang. *Uranium Min. Metall.* **2014**, *33*, 130–133.
22. Zhou, Y.P.; Shen, Z.L.; Sun, Z.X.; Liu, J.H.; He, J.T. Application study on particle tracer technique to calculate the in-situ leaching area of uranium mining. *China Min. Mag.* **2015**, *24*, 117–120.
23. Liu, Z.B.; Wang, H.F.; Wen, Z.Q.; Ding, Y.; Jiang, Y.; Xie, T.T.; Xu, G.L.; Wang, G. Study on the Characteristics of Solution Migration and Drawing and Injection Control at Well Field In-Situ Leaching Uranium. *Uranium Min. Metall.* **2017**, *36*, 23–26.

24. Xie, T.T.; Ding, Y.; Zhou, G.M.; Xu, G.L.; Li, H.X.; Deng, J.X.; Zhang, C.; Xu, Z.H. Numerical Simulation Analysis of Homologous Tracing Data Before and After Acidification in the Conditional Experiment of an In-situ Leaching Uranium Mine. *Uranium Min. Metall.* **2018**, *37*, 1–8.
25. Wang, H.F.; Guo, N.; Xie, T.T.; Li, G.J. The influence of groundwater dilution on pregnant solution of in-situ leaching of uranium. *Uranium Min. Metall.* **2012**, *31*, 9–13.
26. Zhou, Y.P.; Shen, Z.L.; Sun, Z.X.; He, J.T.; Xing, Y.G. Study on the influence of pumping-injection equilibrium on outflow of leaching solution and inflow of groundwater during in-situ leaching of uranium. *Nonferrous Met.* **2013**, *65*, 1–4.
27. Xie, T.T.; Yao, Y.X.; Gan, N.; Zhang, C.; Xu, G.L.; Wen, Z.Q. Analysis and application of seepage-dispersion field characteristics in-situ leaching uranium (to be continued). *Uranium Min. Metall.* **2016**, *35*, 149–158.
28. Panfilov, M.; Uralbekov, B.; Burkitbayev, M. Reactive transport in the underground leaching of uranium: Asymptotic analytical solution for multi-reaction model. *Hydrometallurgy* **2016**, *160*, 60–72. [CrossRef]
29. Greskowiak, J.; Prommer, H.; Liu, C.; Post, V.E.A.; Ma, R.; Zheng, C.; Zachara, J.M. Comparison of parameter sensitivities between a laboratory and field scale model of uranium transport in a dual domain, distributed rate reactive system. *Water Resour. Res.* **2010**, *46*, 4921–4933. [CrossRef]
30. Xie, T.T.; Yao, Y.X.; Gan, N.; Zhang, C.; Xu, G.L.; Wen, Z.Q.; Peng, Y. Analysis and application of seepage-dispersion field characteristics in-situ leaching uranium (Complete). *Uranium Min. Metall.* **2016**, *35*, 229–239.
31. Xu, Q. Study on Influence of the Ratio of Drawing-injection to In-situ Leaching Solution Diffusion Range. *Uranium Min. Metall.* **2017**, *36*, 93–97.
32. Ji, H.B.; Huang, Q.Y.; Zhou, Y.P.; Zhang, Q.L. Influence of In-situ Leaching Solution Diffusion with Drawing Injection flux Distribution and Drawing Injection proportion. *Uranium Min. Metall.* **2017**, *36*, 172–181.
33. Cao, Y.X.; Zuo, W.; Bao, Z.X.; Gu, P.; Ye, T. Practice and Exploring of Different Proportional Control of Drawing and Injection of Leaching Solution for the In-Situ Leaching Uranium and Its Impact on the Groundwater Environment. *Uranium Min. Metall.* **2017**, *36*, 134–143.
34. Li, D.; Duan, B.S. Microscopic Equilibrium Optimization of Leaching Range in In-Situ Leaching Mining Field. *Uranium Min. Metall.* **2017**, *36*, 41–45.
35. Zhang, C.; Xie, T.T.; Tan, K.X.; Yao, Y.X.; Wang, Y.A.; Li, C.G.; Li, Y.M.; Zhang, Y.; Wang, H. Hydrodynamic Simulation of the Influence of Injection Flowrate Regulation on In-Situ Leaching Range. *Minerals* **2022**, *12*, 787. [CrossRef]
36. Huang, Q.F. *Uranium Metallogenic Characteristics and Prospecting Direction in the BARUN Area, The Left County Sunite, Inner Mongolia*; Jilin University: Jilin, China, 2018.
37. Chen, C.X.; Tang, Z.H. *The Numerical Methods of Ground Water Flow*; China University of Geosciences Press: Hubei, China, 2009.
38. Zheng, C.M.; Bennett, D.G. *Applied Contaminant Transport Modeling, Second Edition*; Wiley Interscience: Hoboken, NJ, USA, 2002.
39. Wang, Y.H.; Xia, A.L.; Qin, G.J. Probabilistic modeling for reliability analysis of buried pipelines subjected to spatiotemporal earthquakes. *Probabilistic Eng. Mech.* **2022**, *69*, 103315. [CrossRef]
40. Wang, Y.H.; Xia, A.L.; Zhang, P.; Qin, G.J. Probabilistic physical modeling of randomly corroded surface and its use in reliability analysis of corroded pipelines under spatiotemporal vibration. *Ocean Eng.* **2022**, *262*, 112219. [CrossRef]
41. Dell, A.D.; Bersezio, R.; Felletti, F.; Giudici, M.; Comunian, A.; Renard, P. Comparison of three geostatistical methods for hydrofacies simulation: A test on alluvial sediments. *Hydrogeol. J.* **2012**, *20*, 299–311.
42. Sun, Q. *Improvement and Application of Groundwater Flow Stochastic Simulation Method based on T-PROGS*; China University of Geosciences (Beijing): Beijing, China, 2019.
43. He, Z. *Numerical Simulation of Uranium Mine Based on GMS*; University of South China: Hunan, China, 2016.

processes

Article

Adsorption of Zinc(II) Ion by Spent and Raw *Agaricus bisporus* in Aqueous Solution

Xiaoyu Zhang [1,†], Caiyi Zhao [1,†], Feng Xue [2], Beicheng Xia [3], Yuanyuan Lu [1], Rongrong Ying [1,*] and Zhewei Hu [1]

1 Nanjing Institute of Environmental Sciences, Ministry of Ecology and Environment, Nanjing 210042, China; zhangxiaoyu@nies.org (X.Z.); zhaocaiyi@nies.org (C.Z.); yylu@nies.org (Y.L.); huzhewei@nies.org (Z.H.)
2 Shenzhen Degree of Freedom Eco Technology Co., Ltd., Shenzhen 518116, China; 13770759516@163.com
3 School of Environment Science and Engineering, Sun Yat-sen University, Guangzhou 510275, China; xiabch@mail.sys.edu.cn
* Correspondence: yrr@nies.org
† These authors contributed equally to this work.

Abstract: A significant environmental concern globally is the pollution of water bodies as a result of heavy metals. To date, various attempts have been made to effectively remove heavy metals, such as those that use synthetic and biogenic materials to abate and control water pollution. The biological removal of pollutants from water is an efficient and environmentally friendly technique. In this study, we evaluated the biosorption characteristics of Zn^{2+} ions from aqueous solution by spent composed of raw *Agaricus biosporium* (RAB) and *A. biosporium* (SCAB). We added different biosorption doses, metal ions, and initial concentrations of pollutants to explore the adsorption of Zn^{2+} by RAB and SCAB. We applied pseudo-first- and second-order models to investigate the biosorption kinetics. According to our results, the rate of Zn^{2+} removal from the aqueous solution using raw biomass was significantly lower than that using sodium citrate-treated biomass of SCAB. When the Zn^{2+} concentration increased from 10 mg L^{-1} to 200 mg L^{-1}, the rate of removal of RAB decreased from 73.9% to 38.4%, and that of Zn^{2+} by SCAB decreased from 99.9% to 75.9%. As we increased the biosorbent dose, the rate of Zn^{2+} removal by SCAB increased. Interestingly, Zn^2 biosorption was inhibited by heavy co ions (Cu^{2+} and Pb^{2+}) and light metals (e.g., Na^+, Mg^{2+}, K^+, and Ca^{2+}). When treating wastewater polluted with Zn^{2+}, Pb^{2+}, and Cu^{2+}, SCAB showed good potential. The results of this study provide a scientific basis for an environmentally sound approach to controlling water pollution.

Keywords: biosorbent; *Agaricus bisporus*; Zn^{2+}; kinetics; thermodynamic

1. Introduction

Technological advancements, growing population, mining and other industrial activities, and increasing urbanization have contributed to the rapid increase in the number of water bodies polluted with heavy metals, creating a major environmental problem. Developmental advances in industrial activities, such as mining operations, smelting, printing, metal plating facilities, explosive manufacturing, and dying, as well as consumer products, including paints, plastics, pesticides, alloys, batteries, tanneries, and ceramic glass, are responsible for this alarming extent of heavy metal pollution. Wastewater containing heavy metals generated by industrial units is being directly and indirectly discharged into the environment. The toxic heavy metals discharged into the water bodies influence the flora and fauna of aquatic ecosystem systems and negatively affect human health when these metals are transferred and enriched through the food chain [1]. Chemical precipitation, electrochemical treatment, filtration, and membrane technologies are commonly used to remove toxic metals from the wastewater [2–4]. When the concentration of heavy metals is low (e.g., 1–100 mg L^{-1}) and the volume of wastewater is high, conventional methods are extremely expensive and may even be ineffective [5,6]. Another major problem associated with these conventional treatment processes is the generation of toxic chemical sludge, and its disposal becomes environmentally costly [7]. Thus, learning how to remove heavy metals from wastewater to an environmentally safe level in an ecofriendly and cost-effective manner is essential.

A significant amount of research has been conducted over the past 20 years to examine ways to control heavy metal pollution. In recent years, the removal of heavy metals involving biological methods has attained significant attention because of the cost-effective and environmentally friendly aspects [8]. Fungus [9–11], citrus peel [12], marine algae [13,14], bacteria [15], and other materials [16–18] are cost-effective and ecofriendly. These readily available biomaterials can effectively remove heavy metals in various forms from aqueous solution. However, the low adsorbing capability of raw biomass has limited the practical application of biosorption to heavy metals. Therefore, to enhance the biosorbent's adsorption capacity, chemical pretreatments are often applied, including washing the raw biomaterials with HNO_3 [12], calcium chloride [19], formaldehyde [20], HCl and acetone [21], and NaOH [22]. By masking or removing the surface groups or exposing metal-binding sites, pretreatment may change the functional groups and surface properties [23].

Recent studies have used Fourier transform infrared spectroscopy (FT-IR) or X-ray photoelectron spectroscopy (XPS) [24,25] to examine changes in the functional groups and study the mechanism of biosorption. The biosorption mechanism usually consists of adsorption, ion exchange, complexation, and microprecipitation [16]. The cell wall of biosorbent materials is usually composed of a large number of polysaccharides, proteins, lignin, and other polycomplexes, which can provide sufficient functional groups such as carboxyl, hydroxyl, sulfate, phosphate, amide, and amino for the binding of metal ions [26,27]. However, ion exchange, which is another important mechanism for chemical adsorption, has not attracted sufficient attention in the field of biosorption. The pH change after biosorption has not been reported, although it can help to evaluate the chemical changes in the biosorption process. Several studies have examined the leaching of Na^+, Mg^{2+}, K^+, and Ca^{2+} from biosorbents; however, these changes have not been compared with simultaneous changes in pH [9,28,29]. Furthermore, the combination of a simultaneous pH change and light metal ion release might explain the ion exchange involved in the process of biosorption [16].

One of the most common pollutants—and also among the most harmful—found in wastewater for both irrigational and industrial applications is zinc [30]. Zinc is a trace element that is essential for human beings and for the growth of flora and fauna, but large quantities of zinc can have an adverse impact on plants, animals, and humans [31]. According to the U.S. Environmental Protection Agency, the Zn^{2+} concentration in drinking water must be less than 3 mg L^{-1} [32].

Agaricus bisporus is a kind of large fungus, widely cultivated in China. Its subsidiary entity has the advantages of large size, easy access, low price, and good tolerance to harsh environments. It has the potential to prepare heavy metal bioadsorbent materials, which provides a good basis for the development of bioadsorbent materials and related mechanism research. Although most of the raw biomass materials used for biosorption are cheap and readily available, the cost involved in the preparation of biosorbents must also be considered. Hence, regeneration of the biosorbent can lower processing costs and also recover the adsorbed heavy metals. Thus, nondestructive regeneration of biosorbents by ecofriendly, economical, and effective desorbing eluants is a valuable option for eliminating heavy metals from industrial wastewater. Several studies have investigated the recycling of raw biosorbents [24,33,34]. The number of effective sorption–desorption cycles of many biosorbents, however, do not meet ideal applications. Thus, in this study, we (1) examined the role of the initial biosorbent dosage, the Zn^{2+} concentration, and aqueous metal ions; (2) tested the Zn^{2+} adsorption capacity of sodium citrate-treated *Agaricus bisporus* (SCAB); (3) studied the Zn^{2+} adsorption mechanism from multiple angles; (4) explored the effect of SCAB on the removal of Zn^{2+} from aqueous solution and determined the number of possible sorption–desorption cycles. To reduce and remove water pollution caused by toxic heavy metals, the results of this study offer a scientific basis and an environmentally sound approach.

2. Materials and Methods

2.1. Biosorbent Preparation

The raw fungal biomass of *A. bisporus* (RAB) was obtained from a commercial company. The fruit bodies were washed thoroughly with deionized water to remove impurities and dirt, then dried for 24 h in an 80 °C oven. To obtain a uniform particle size, the dried biomass was crushed and sieved through 250–380 μm American Society for Testing Material–standard sieves. The sieved biomass was treated per gram with between 0.1 L and 10 mM of sodium citrate in a rotary shaker for 0.5 h at 150 rounds per minute. Before use, the treated biomass was dried for 24 h at 80 °C and then stored in a desiccator.

2.2. Zn^{2+} Solution Preparation

Zinc nitrate ($Zn(NO_3)_2{\cdot}6H_2O$) was used to prepare Zn^{2+} solutions. The required quantity of $Zn(NO_3)_2{\cdot}6H_2O$ was dissolved in deionized water to prepare the stock solution of Zn^{2+} (1 g L^{-1}). This solution was used to prepare different working concentrations of the dilute solutions. In all of the experiments, 0.1 M HCl and 0.1 M NaOH solutions were used to adjust the pH of the Zn^{2+} solutions to 4.0. All of the analytical-grade chemicals and reagents were used without further purification.

2.3. Batch Experiments

2.3.1. Effect of Biosorbent on Zn^{2+} Biosorption by RAB and SCAB

Then, 3 to 20 g L^{-1} of SCAB was added to 250 mL Erlenmeyer flasks containing a concentration of 100 mL Zn^{2+} solution. The temperature was maintained at 25 °C and the pH at 4.0. The experiments were performed by using equivalent amounts of raw *A. bisporus* (RAB) biomass under the same conditions to study the efficiency of the SCAB treated with sodium citrate.

2.3.2. Effect of Zn^{2+} Concentration on Zn^{2+} Biosorption by RAB and SCAB

We investigated how SCAB biosorption of Zn^{2+} was affected by the initial Zn^{2+} concentration, which ranged from 20 to 200 mg L^{-1}. The same experimental conditions were used to add RAB, further examining the effect of sodium citrate on SCAB.

2.3.3. Effect of Adsorption Time on Zn^{2+} Biosorption by SCAB

The 13 g L^{-1} SCAB was placed in 22 mg L^{-1}, 65 mg L^{-1}, 85 mg L^{-1}, and 135 mg L^{-1} zinc solution at 25 °C and a pH of 4.0, and the remaining zinc ion concentration in the solution was sampled at 5, 10, 20, 30, 40, and 60 min after the start of the reaction.

2.3.4. Effect of Co-Ions on Zn^{2+} Biosorption by SCAB

The competition for binding sites on the biomass was investigated by adding other metal ions to Zn^{2+} solutions.

Then, 0 mM of Na^+, 1 mM of Mg^{2+}, 5 mM of K^+, and 10 mM of Ca^{2+} were added to the Zn^{2+} solutions to examine how light metal ions affected the biosorption properties of Zn^{2+} by SCAB. The quantity of light metal ions relative to their inmolar equivalents was calculated because the concentration was quite high in practical wastewater.

Next, 50 mg L^{-1}, 75 mg L^{-1}, and 100 mg L^{-1} Cu^{2+} and Pb^{2+} were added to the Zn^{2+} solutions to examine how heavy metal ions affected the biosorption properties of Zn^{2+} by SCAB. The ratios of Cu^{2+}/Pb^{2+} to Zn^{2+} in solution were set to 1:1, 1.5:1, and 2:1 because a 50 mg L^{-1} concentration of Zn^{2+} was maintained.

After the biosorption experiments were completed, the biosorbent was immediately removed by filtering the samples with a 0.45 μm filter membrane. An inductively coupled plasma-optical emission spectrometer (ICP-OES) (Perkin-Elmer Instruments, Optima

5300 DV, Waltham, MA, USA) was used to determine the concentration of Zn^{2+} in the filtered solutions.

2.3.5. Specific Surface Area Detection of Bioadsorbent Materials

The RAB and SCAB were pretreated with vacuum pump at 348 K for 10 h. Then, an automatic nitrogen adsorption instrument (ASAP 2010, Micromeritics, Norcross, GA, USA) was used to detect the specific surface area of multi-point BET.

2.4. Light Metal Ion Release

To determine the relationship between the quantity of light metal ions released and the quantity of Zn^{2+} ions adsorbed, we sampled the solutions following the biosorption of 20–200 mg L^{-1} Zn^{2+} solutions by SCAB. We employed the charge balance method (Sum = $Na^+/2 + K^+/2 + Mg^{2+} + Ca^{2+}$) to calculate the quantity of light metal ions released.

2.5. Results of Fourier Transform Infrared Spectroscopy

Fourier transform infrared (FT-IR) spectroscopy (Thermo Nicolet IS10, Waltham, MA, USA) was used to study the chemical characteristics of RAB and SCAB before and after biosorption of Zn^{2+}. A ratio of 1:200 of the dried and sieved biomass samples was mixed with potassium bromide (KBr) (SP). The samples were compacted under high pressure into pellets. The FT-IR spectra (500–4000 cm^{-1}) were collected, and the pure KBr background from the sample spectra were subtracted.

2.6. Results of Sorption–Desorption Cycles

To study the possibility of regeneration and reuse of SCAB, 10 mM sodium citrate was used as both eluent and the regenerant. Then, 13 g L^{-1} SCAB was used to treat the 100 mg L^{-1} Zn^{2+} solution at 25 °C and a pH of 4.0; 100 mL of 10 mM sodium citrate was used to desorb the Zn^{2+} onto the biosorbent. Both the sorption duration and desorption duration were 30 min. Excessive distilled water was used to wash the biosorbent and remove the sodium citrate. Then, vacuum filtration was used to dry the biosorbent and prepare it for next process. The sorption–desorption cycle was repeated 10 times. After each sorption or desorption process, the contents of Zn^{2+} in the remaining solution were analyzed by using ICP-OES.

2.7. Experimental Data Analysis

2.7.1. Capacity of Biosorption

To evaluate the biosorption capacity, we calculated the amount of Zn^{2+} adsorbed by the biomass according to the following equation:

$$E\% = (C_0 - C_e)/C_0 \times 100, \quad (1)$$

$$q_e = (C_0 - C_e)V/W, \quad (2)$$

where $E\%$ represents biosorption efficiency; q_e represents the amount of Zn^{2+} adsorbed by SCAB (mg g^{-1}); C_0 represents the initial Zn^{2+} concentration (mg L^{-1}); C_e represents the Zn^{2+} concentration at equilibrium (mg L^{-1}); V represents the solution volume (L); and W represents the biosorbent weight (g). The experiments were completed at 25 °C (room temperature).

2.7.2. Equilibrium Isotherms

Langmuir, Freundlich, Dubinin–Radushkevich (D-R), and Temkin isotherms were used to calculate the biosorption equilibrium data.

Sorption occurs on only one ion per binding site, according to the Langmuir model, and the sorbent surface has just one type of binding site [9]. The linear form of this model is expressed as follows:

$$C_e/q_e = 1/Q_0 b + C_e/Q_0, \tag{3}$$

where C_e represents the equilibrium Zn^{2+} concentration in solution (mg L^{-1}); q_e represents the amount of equilibrium absorption (mg g^{-1}); the Langmuir constant Q_0 (mg g^{-1}) represents the capacity of the monolayer adsorption; and b (L mg^{-1}) represents adsorption heat.

The Freundlich model can be used to calculate surface heterogeneity and multilayer adsorption to binding sites on the sorbent's surface [35]. The linear form of this empirical expression follows:

$$\log q_e = \log K_f + \log C_e/n, \tag{4}$$

where C_e represents the Zn^{2+} equilibrium concentration in solution (mg L^{-1}); q_e represents the amount of equilibrium absorption (mg g^{-1}); K_f (mg $g^{1-(1/n)}$ $L^{1/n}$ g^{-1}) represents the adsorbent's adsorption; and $1/n$ represents the heterogeneity of the surface [36].

Because the D-R isotherm does not assume constant sorption potential or a homogeneous surface, it is less specific than the Langmuir isotherm [37]. D-R can be calculated as follows:

$$\ln q_e = \ln q_m - \beta \varepsilon^2, \tag{5}$$

where, β represents a constant related to the mean free energy of adsorption per mole of adsorbate (mol^2 kJ^{-2}); q_m represents the capacity for theoretical saturation; and ε represents the Polanyi potential, which is equal to $RT \ln(1 + (1/C_e))$, where R represents the universal gas constant and T represents the absolute temperature.

A decrease in sorption heat is linear in the Temkin isotherm rather than logarithmic, as in the Freundlich equation. This isotherm can be calculated as follows:

$$q_e = RT \ln(A_T C_e)/b_T, \tag{6}$$

where A_T (L mg^{-1}) and b_T represent the Temkin isotherm constants; R represents the universal gas constant; and T represents the absolute temperature.

2.7.3. Biosorption Kinetics

Several studies have described heavy metal experimental data for biosorbent removal using two kinetic models [38,39].

The first is the Lagergren model, which generally expresses a pseudo-first-order rate model as follows:

$$\log(q_e - q_t) = \log q_e - k_1/2.303, \tag{7}$$

The corresponding pseudo-second-order model is as follows:

$$t/q_t = 1/k_2 q_e^2 + 1/q_e, \tag{8}$$

where q_e and q_t represent the quantities of Zn^{2+} ions adsorbed on the biosorbent at equilibrium and at time t (min.), respectively; k_1 represents the first-order adsorption rate constant (min^{-1}); and k_2 represents the second-order adsorption rate constant (g mg^{-1} min^{-1}).

2.7.4. Biosorption Thermodynamics

Changes in Gibbs free energy (ΔG°), entropy (ΔS°), and enthalpy (ΔH°), which are the most common thermodynamic parameters, were calculated as follows:

$$K_c = C_{e,ad}/C_e, \tag{9}$$

$$\Delta G^\circ = -RT \ln Kc, \tag{10}$$

$$\Delta G^{\circ} = \Delta H^{\circ} - T\Delta S^{\circ}, \quad (11)$$

where $C_{e,ad}$ and C_e represent the Zn^{2+} concentration on the biosorbent and the residual Zn^{2+} concentration at equilibrium (mg L^{-1}); K_c represents the equilibrium constant; R represents the gas constant (8.3145 J mol K^{-1}); and T represents the solution temperature (K).

3. Discussion of Results

3.1. Effect of *A. bisporus* Pretreatment

In this study, when the Zn^{2+} concentration increased from 10 to 200 mg L^{-1}, the rate of Zn^{2+} removal by SCAB lowered from 99.9% to 75.9%, and the rate of RAB removal lowered from 73.9% to 38.4% (Figure 1a). This result indicated that the sodium citrate-treating method improved the biosorption capability of *A. bisporus* biomass and made the decreasing trend less significant. Because the trend in saturation when using a higher initial metal concentration for the biosorbent created a finite number of surface binding sites, the removal rate with high Zn^{2+} concentration decreased [40]. Our findings aligned with a previous study, which found that *Chlorella coloniales* was effective for Cr, Co, Fe, and As bioaccumulation; however, bioaccumulation efficiency decreased when the concentration of heavy metal was high [41].

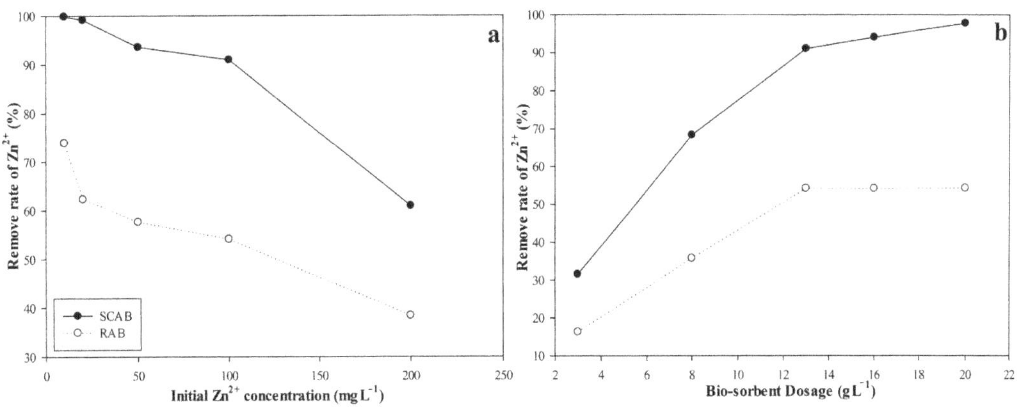

Figure 1. Removal rate of Zn^{2+} by the sodium citrate-treated *Agaricus bisporus* (SCAB) and the raw *Agaricus biosporium* (RAB): (**a**) with different initial Zn^{2+} concentrations; and (**b**) with different biosorbent dosages.

When the biosorbent dosage was changed from 3 to 20 g L^{-1}, the rate of Zn^{2+} removal by SCAB rose from 31.6% to 97.7%. Under matching experimental conditions, the rate of RAB removal increased only from 16.3% to 54.2% (Figure 1b). We attributed this increased rate of Zn^{2+} removal with a high dose of the biosorbent to an increase in the surface area as well as in the number of binding sites. The exchangeable ions that were available for adsorption increased, which was tied to the amount of biosorbent needed in the process of biosorption [42]—that is, when the biomass amount increased, the metal ion biosorption per gram of biomass decreased [41,43–46]. Therefore, when the biomass dose increased, the final amount of heavy metal removal was also enhanced; however, the biosorption capacity of the heavy metal ions decreased [41,47–49]. The removal rate of both SCAB and RAB did not rise significantly when the biosorbent dose was more than 13 g L^{-1}. The reduced effective sorption area that results from biosorbent aggregation likely was responsible [50].

The 13 g L^{-1} SCAB was placed in 22 mg L^{-1}, 65 mg L^{-1}, 85 mg L^{-1}, and 135 mg L^{-1} zinc solution at 25 °C and a pH of 4.0, and the remaining zinc ion concentration in the solution was sampled at 5, 10, 20, 30, 40, and 60 min after the start of the reaction. The kinetic curve for zinc ion showed that the amount of biosorption sharply increases with increasing contact time in the initial stage (0–20 min) and then gradually increases to reach an equilibrium value. The higher reaction rate originates from the large number of convenient adsorption sites on the surface or inside of the adsorbent. As the reaction proceeds, the available sites are gradually saturated, and the adsorption rate cannot be further improved.

3.2. Effect of Co-Ions on Zn^{2+} Biosorption by SCAB

Heavy metal ions in industrial effluents are always accompanied by light metal ions (e.g., Na^+, Mg^{2+}, K^+, and Ca^{2+}). The concentration of light metal ions normally is higher than that of heavy metal ions. The sorption capacity of commercial ion exchange resins is affected by light metal ions [19]. Even at a concentration of 10 mM, the effect of Mg^{2+} on Zn^{2+} uptake was somewhat insignificant. The rate of Zn^{2+} removal by SCAB decreased by about 39.6% when the Na^+ concentration was 10 mM. The removal rate decreased by 16.3% and 12.5%, however, with the same Ca^{2+} and K^+ concentrations (Figure 2), which indicated SCAB's relatively higher affinity for Zn^{2+} than for the light metal ions.

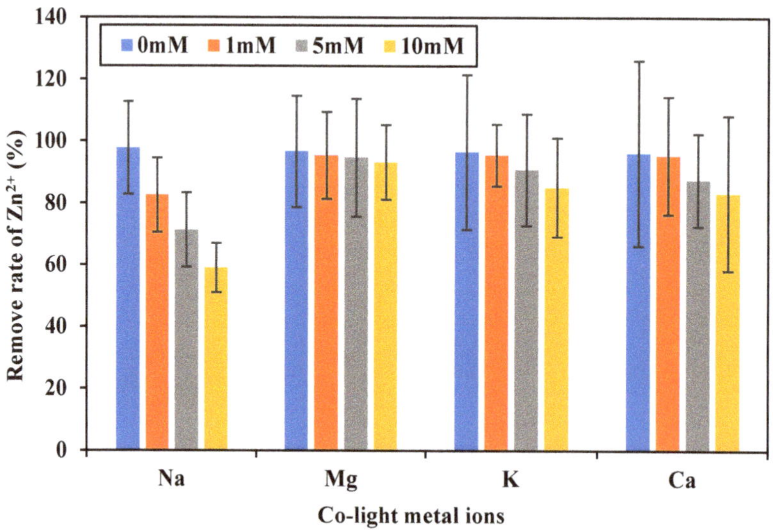

Figure 2. Light metal ions have an impact on the Zn^{2+} biosorption by the sodium citrate-treated *Agaricus bisporus* (SCAB) at an initial concentration of 1 mM.

When multiple metal ions are present in natural and polluted water, it often has an interactive effect that does not occur in the presence of single toxic metallic species. Therefore, how various heavy metal ions interact must be considered in the biosorption process [51]. Using SCAB, we effectively removed Pb^{2+} (99.7–98.6%), but the effect of interfering ions was higher, resulting in a decrease from 90.2 to 44.0% for Zn^{2+} in the presence of Pb^{2+}. This result showed that Zn^{2+} biosorption was hindered by Pb^{2+}. In contrast, when the Cu^{2+} concentration increased from 50 to 100 mg L^{-1}, the rate of Cu^{2+} removal decreased from 96.6% to 80.5% and the rate of Zn^{2+} removal decreased from 39.3% to 20.2%. As the ratio of Pb^{2+} or Cu^{2+} to Zn^{2+} increased, the removal rate of Zn^{2+} in the Cu-Zn binary metal system decreased more than that in the corresponding Pb-Zn binary metal system, which suggested that Cu^{2+} shared more binding sites with Zn^{2+} on the surface of SCAB than Pb^{2+}. In addition, both lead and copper ions could be effectively removed from the solution by SCAB, which could also be applied to these heavy metals as a biosorbent.

3.3. Effect of Biosorption on Solution pH

Research has shown that heavy metal uptake is affected by pH [52,53]. Past studies have primarily examined the pH value in initial biosorption when considering the capacity of biosorbents for heavy metal uptake [53], but they have rarely considered pH value changes following biosorption. In this study, we set the initial solution pH at 4.0. Figure 3 shows the value of the solution pH after Zn^{2+} biosorption by SCAB. As the initial Zn^{2+} concentration increased, the solution pH value also increased (Figure 3a). This result showed that H^+ could be removed from the solution onto the biosorbent surface as a result of the involvement of an ion exchange mechanism. Additionally, the change in the pH solution was limited at a higher initial Zn^{2+} concentration. As shown in Figure 3b, with an increase in the dose of the biosorbent in solution, the solution pH also increased. This result may have been caused by more exchangeable and releasable components, such as Na^+, K^+, Mg^{2+}, and Ca^{2+}, as well as additional binding sites when the dose of the ion exchange mechanism was higher. The involvement of the biosorption process relative to the ion exchange mechanism will be addressed in a follow-up study.

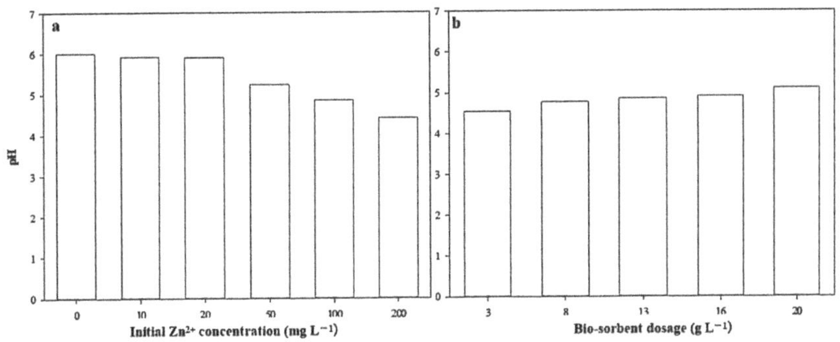

Figure 3. Solution pH value following Zn^{2+} biosorption by the sodium citrate-treated *Agaricus bisporus* (SCAB): (**a**) with various initial Zn^{2+} concentrations; and (**b**) with various biosorbent dosages.

3.4. Ion Exchange during Biosorption

According to Chen et al. [54], a generalized type of functional group such as carboxyl, hydroxyl, sulfate, phosphate, and amine groups (represented as R^{2-}) is part of the metal ion adsorption process. Most complex functional groups have light metal ions such as Ca^{2+} (CaR), which can be stripped away through the ion exchange process. In aqueous solution, H^+ ions could replace Ca^{2+} when the biosorbents were applied according to the following reaction:

$$2H^+ + [CaR] = Ca^{2+} + [H_2R]. \quad (12)$$

(1) During biosorption, a similar exchange occurs between Ca^{2+} ions and heavy metal ions (represented as M^{2+}), as shown in Equation (13):

$$M^{2+} + [CaR] = Ca^{2+} + [MR]. \quad (13)$$

(2) The driving force behind the equilibrium on the right side of the equation was the stronger affinity of M^{2+} for R^{2-}. As demonstrated in Equation (12), when the Ca^{2+} in the solution exchanged the H^+ in the biosorbent, the solution's pH increased.

Because Na^+ was the most common light metal released and contributed the most to the ion exchange of Zn^{2+} in the biosorption process, Equations (12) and (13) can be rewritten as follows:

$$2H^+ + [Na_2R] = 2Na^+ + [H_2R], \quad (14)$$

$$M^{2+} + [Na_2R] = 2Na^+ + [MR]. \quad (15)$$

(3) During the uptake of Zn^{2+}, ion exchange was essential; however, it appeared that the nonstoichiometric exchange of ions meant that the ion exchange mechanism was not solely responsible for the biosorption of Zn^{2+} by SCAB.

Light metal ions (e.g., calcium ions) can saturate many biosorbents and then can exchange them with heavy metal ions from the solution. As a result, the release of the light metal ions and the adsorption of heavy metal ions can occur simultaneously [54]. We observed that light metal ions, including Na^+, Mg^{2+}, K^+, and Ca^{2+}, were released simultaneously with Zn^{2+} adsorption (Figure 4a). As the initial Zn^{2+} concentration increased, the four light metal ions in solution also increased, which demonstrated that light metal ions might be released as a result of an ion exchange with Zn^{2+}. After treating the raw biomass with sodium citrate, the main exchangeable light metal ion was Na^+, in contrast to other studies, which found Ca^{2+} on the surface of SCAB [20,55]. The quantity of light metal ions released was proportional to Zn^{2+} adsorption (Figure 4b). The involvement of an ion exchange mechanism was demonstrated by the simultaneous release of light metal ions and Zn^{2+} adsorption. After Zn^{2+} biosorption by SCAB, the solution pH increased, which showed that light metal ions replaced H^+ from solution on the surface of SCAB (Figure 3). Conversely, when the initial Zn^{2+} concentration increased, pH changes were limited, which showed that Zn^{2+} rather than H^+ replaced the light metal ions on the biosorbent. This was most notable when the initial Zn^{2+} concentrations were high.

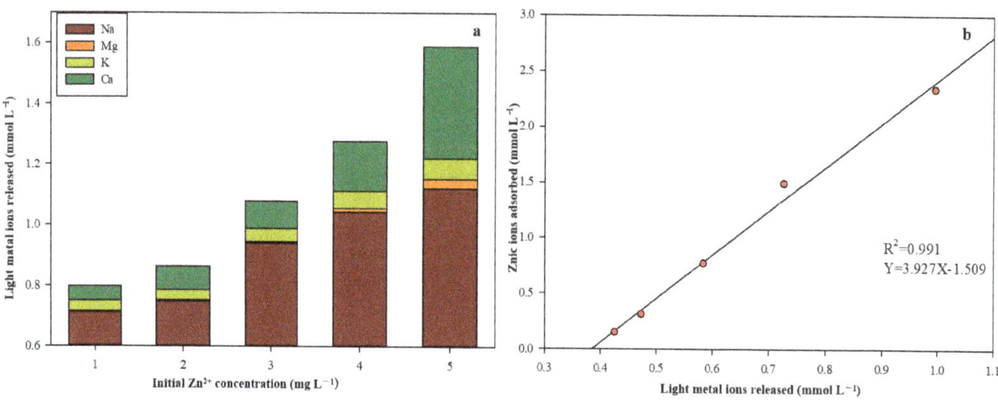

Figure 4. Ions released during Zn^{2+} biosorption by the sodium citrate-treated *Agaricus bisporus* (SCAB): (**a**) four light metal ions released with various initial Zn^{2+} concentrations; and (**b**) the relationship between the quantity of light metal ions released and the quantity of Zn^{2+} ions adsorbed.

3.5. Functional Groups in Zn^{2+} Biosorption

Specific surface area refers to the total area of the unit mass of the material, including the external area and the internal surface area (micropore area). By studying the specific surface area, the contact degree between adsorbed material can be fully understood. Specific surface area is an important index to evaluate the efficacy of adsorbent. It was determined that the specific surface area, micropore area, and surface area of the SCAB were 9.68 m^2/g, 1.05 m^2/g, and 8.62 m^2/g, respectively, which were 3.57 times, 4.56 times, and 4.36 times higher than that of the RAB. The increase in specific surface area will help increase the contact surface between SCAB and zinc ions in solution. More adsorption sites will also be exposed, which is conducive to the binding of zinc ions in the solution with effective adsorption sites.

Four bands showed decreased intensity and band shifts (Figure 5). The absorbance peak of the hydroxyl group stretch decreased from 3423 to 3398 cm^{-1} after biosorption. The absorbance peaks of the aromatic nitro groups' absorbance peaks reached a symmetric stretch of 1380 cm^{-1} and an asymmetric stretch of 1551 cm^{-1}, and absorbance peaked at 617 cm^{-1}. These results likely were caused by the shift in the phosphate groups. We also observed shifts in the FT-IR spectra of the aromatic nitro, phosphate, and hydroxyl group peaks. These results indicated that these groups had an impact on SCAB biosorption of Zn^{2+}.

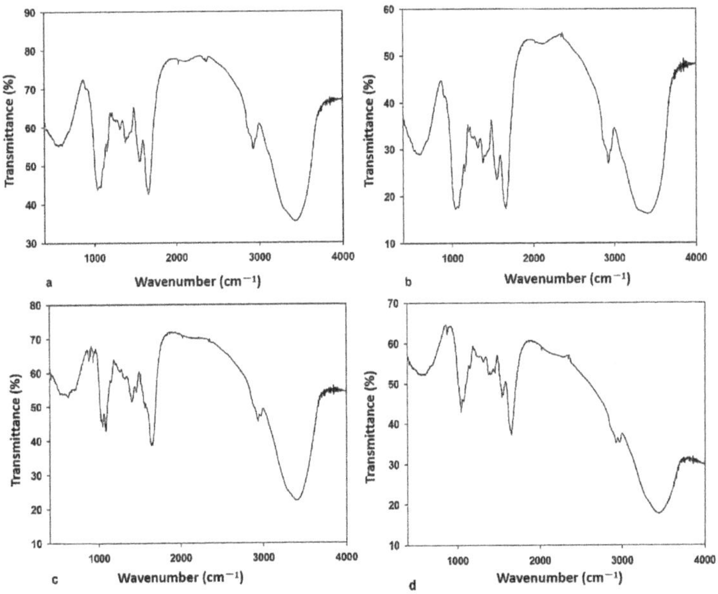

Figure 5. SCAB and RAB FT-IR spectra: (**a**) SCAB before biosorption; (**b**) SCAB after biosorption; (**c**) RAB before biosorption; and (**d**) RAB after biosorption.

The stretching vibration of the absorbance peak of the hydroxyl group increased from 3405 to 3448 cm^{-1}. We attributed the peak at 1637 cm^{-1} to the C=O bond in the carboxyl group, which was evident for Zn-loaded RAB but not for RAB. We attributed the peaks at 2973 and 1383 cm^{-1} to the alky and aliphatic nitro groups, which were not evident for RAB but were for Zn-loaded RAB. The phosphate groups may have been responsible for changes in absorbance peaks in the fingerprint zone (<1000 cm^{-1}). The FT-IR spectra showed that changes in aliphatic nitro, hydroxyl, carboxyl, and phosphate group peaks were responsible for RAB biosorption of Zn^{2+}. Therefore, the primary functional groups that were involved in Zn^{2+} biosorption were affected by the sodium citrate-treating method.

3.6. SCAB Regeneration and Reuse

The regeneration efficiency of biosorbent after metal desorption relies on the exploitation of biosorption technology to remove heavy metals [56]. The rate of Zn^{2+} removal of 100 mg L^{-1} lowered from 96.81% to only 94.11% after 10 biosorption–desorption cycles (Figure 6). The slight SCAB loss from the open-reaction system may have been the cause. Hence, the Zn^{2+} removal efficiency by SCAB did not decrease significantly after 10 biosorption–desorption cycles.

The desorption results showed that most of the adsorbed Zn^{2+} could be effectively eluted from SCAB by the sodium citrate solution after each cycle. During desorption cycle processing, the desorption rate slightly increased, which could be attributed to the incomplete desorption of Zn^{2+} on some binding sites in the previous cycle and to the activation of these Zn^{2+}-binding sites in the subsequent cycle.

In summary, the adsorbent seems to be performing well by reaching 10 cycles with removal efficiency above 90%. Sodium citrate solution could be used as both regenerant and eluant and did not affect the biosorption properties of SCAB.

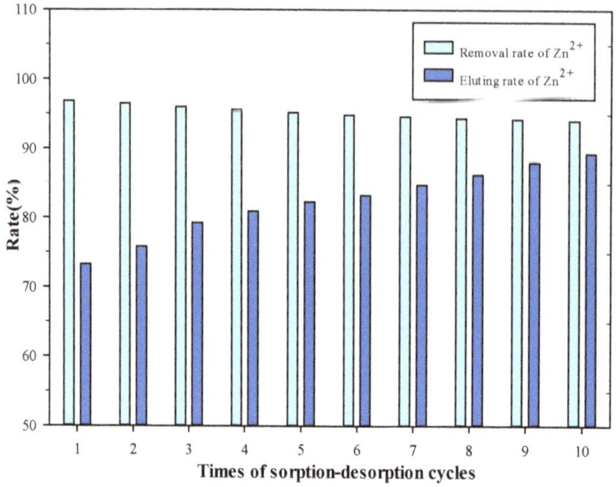

Figure 6. Removal and elution rate for successive cycles of the sodium citrate-treated *Agaricus bisporus* (SCAB) biosorption of Zn^{2+}.

3.7. Equilibrium Isotherms

Certain constants, including values that express biosorbent affinity and surface properties, can be reflected in an adsorption isotherm [55]. To design an optimal biosorption system to remove metal ions from effluents, we established an appropriate relationship for these equilibrium isotherms to characterize these constants. We analyzed batch experimental data using four equilibrium isotherms (i.e., the Langmuir, Freundlich, D-R, and Temkin isotherms). We calculated the corresponding constants for the Zn^{2+} biosorption by SCAB (Table 1). We used the Langmuir isotherm (Equation (3)) to calculate the Zn^{2+} monolayer adsorption capacity (Q_0) of SCAB, which was 19.607 mg g^{-1}. We used the Freundlich isotherm (Equation (4)) to calculate the constants K_f and 1/n, which were 4.518 mg $g^{1-(1/n)}$ $L^{1/n}$ g^{-1} and 0.692, respectively. The Zn^{2+} biosorption of on SCAB was favorable, as shown by the values of 1/n between 0 and 1 [57].

Table 1. Model results for the correlation coefficients and constants.

Isotherm Equations	Parameters		R^2
Langmuir	Q_0 (mg g^{-1}) 19.607	b (L mg^{-1}) 0.593	0.950
Freundlich	K_f (mg $g^{1-(1/n)}$ $L^{1/n}$) g^{-1}) 4.518	1/n 0.268	0.952
Dubinin–Radushkevich	q_m (mg g^{-1}) 50.816	β (mol^2 kJ^{-2}) 7.0×10^{-9}	0.664
Temkin	A_T (L g^{-1}) 1.467	b_T 2370.882	0.878

We used the D-R constants (Equation (5)) to calculate the theoretical saturation capacity (q_m) and the free energy of adsorption per mole of the adsorbate (β), which were 50.816 mg g^{-1} and 7.0×10^{-9} mol^2 kJ^{-2}, respectively. We used the Temkin isotherm to determine the constants of b_T (2370.882) and A_T (1.467 L g^{-1}), which showed that the adsorption of heat for all the molecules in the layer decreased in a linear fashion along with coverage as a result of the adsorbent–adsorbate interactions. We also found that a uniform distribution (up to the maximum) of binding energies characterized this adsorption [58].

The biosorption of Zn^{2+} on SCAB correlated well with the Freundlich (0.952) and Langmuir-type (0.950) isotherms. Compared with the D-R (0.664) and Temkin (0.878) isotherms, the Freundlich (0.952) and Langmuir (0.950) isotherms correlated well with the Zn^{2+} biosorption of SCAB (Table 1). The Langmuir isotherm revealed the adsorbent's surface homogeneity, and the Freundlich adsorption isotherm revealed of the adsorbent's surface heterogeneity. Because the regression coefficients for these two isotherms were high, we found that the SCAB surface had small heterogeneous adsorption patches, which featured adsorption capabilities that were identical [59].

In order to know whether SCAB has an advantage in zinc adsorption, maximum capacity, K_f for biosorption of Zn^{2+} by various biosorbents are listed in Table 2. As can be seen from the table, the K_f value of SCAB was significantly higher than that of the reported zinc-adsorbed biomaterials, which reflected certain advantages in terms of the theoretical maximum adsorption capacity.

Table 2. Maximum capacity, K_f for biosorption of Zn^{2+} by various biosorbents.

Biosorbents	K_f	References
Sargassum sp.	1.44	[60]
Botrytis cinerea	1.13	[61]
Sargassum	0.41	[62]
S. intermedia	1.03	[63]
L. minor	1.06	[64]
P. stratiots	0.25	[61]
Botrytis cinerea	1.13	[58]
Plain Ca-alginate bead	0.51	[65]
SCAB	4.52	This study

3.8. Kinetic Studies

To select the optimum operating conditions for a full-scale batch metal removal process, biosorption kinetics are necessary. These kinetics offer promising biosorption efficiency characteristics [65]. To predict the biosorption kinetics, we applied the pseudo-first-order (Equation (7)) and pseudo-second-order (Equation (8)) models to evaluate the mechanism controlling biosorption.

Table 3 lists the biosorption capacity (q_e), biosorption rate constants ($k1$ and $k2$), and initial sorption rate (h) for Zn^{2+} biosorption by SCAB. We compared the calculated biosorption capacity ($q_{e,cal}$) with the experimental biosorption capacity ($q_{e,exp}$) and found that these values were similar to the pseudo-second-order kinetic values. We observed, however, significant differences between the experimental values and the pseudo-first-order kinetic model values. We also found that the correlation coefficients were much lower for the pseudo-first-order kinetic model than they were for the pseudo-second-order kinetic model (0.999) at all working concentrations. This result indicated that the pseudo-second-order kinetic model was better able to describe Zn^{2+} biosorption by SCAB.

Table 3. Pseudo-first-order model comparison with pseudo-second-order kinetic model; experimental and calculated q_e values for various concentrations of initial Zn^{2+}.

C_0 (mg L^{-1})	$q_{e,exp}$ (mg g^{-1})	First-Order Kinetic Model			Second-Order Kinetic Model			
		$k1$ (1 min^{-1})	$q_{e,cal}$ (mg g^{-1})	R^2	$k2$ (g mg^{-1} min^{-1})	$q_{e,cal}$ (mg g^{-1})	R^2	H (mg g^{-1} min^{-1})
22	1.746	0.099	0.004	0.824	65.437	1.748	0.999	200.000
65	4.978	0.108	0.061	0.928	4.444	5.000	0.999	111.111
75	6.449	0.122	0.242	0.969	1.395	6.494	0.999	58.824
135	9.567	0.219	2.193	0.915	0.379	9.709	0.999	35.714

3.9. Thermodynamic Parameters

Table 4 lists the equilibrium constant (K_c) as well as changes in Gibbs free energy ($\Delta G°$), entropy ($\Delta S°$), and enthalpy ($\Delta H°$). According to these thermodynamic parameters, when the temperature changed from 25 °C to 40 °C, Kc rose from 29.348 to 32.837, which demonstrated SCAB's increased biosorption capacity in keeping with an increase in temperature. As adsorption capacity increased with temperature, the kinetic energy of SCAB particles also rose at higher temperature, which caused the collision frequency between the sorbate and biosorbent to increase [66]. Temperature increases also caused the pore size of the adsorbent to change, thus improving the pore's intraparticle diffusion and increasing the adsorbent's adsorption capacity. A bond rupture in the functional groups on the biosorbent surface, however, could cause the number of active biosorption sites to increase [67]. In this study, we did not consider higher temperatures because desorption started at ≥50 °C.

Table 4. Thermodynamic the sodium citrate-treated *Agaricus bisporus* (SCAB) parameters at 100 mg L^{-1} concentration.

Temperature (K)	K_C	ΔG° (kJ mol^{-1})	ΔH° (kJ mol^{-1})	ΔS° (J mol^{-1} K^{-1})
298	29.348	−8.376		
303	30.696	−8.630	5.739 [a]	0.047 [a]
313	32.837	−9.090		

[a] All of the temperature values are valid.

When the temperature rose from 25 °C to 40 °C, Gibbs free energy (ΔG°) changed from −8.376 to −9.090 kJ mol^{-1}. This higher negative value suggested the spontaneous nature of biosorption [68]. Because chemical adsorption generally is characterized by free energy ranging from 80 to 400 kJ mol^{-1}, the Zn^{2+} biosorption by SCAB was considered to be physical adsorption [56]. The endothermic nature of biosorption was confirmed by the positive ΔH° value. Increased randomness at the interface of the biosorbent and the Zn^{2+} solution during biosorption was demonstrated by the positive ΔS° value.

4. Conclusions

Sodium citrate-treated SCAB showed a significantly higher biosorption capability for Zn^{2+} in solution than in raw biomass. SCAB had a much higher affinity for Zn^{2+} than for light metal ions, and it preferred to adsorb Pb^{2+} and Cu^{2+} when the two ions were also present. We determined that a release of light metal ions and a change in pH following biosorption verified the involvement of the ion exchange mechanism. We confirmed changes in functional groups following pretreatment in biosorption through FT-IR. We completed 10 cycles of biosorption–desorption and verified that the removal rate of Zn^{2+} did not noticeably decrease and most of the absorbed Zn^{2+} could be eluted. The change in Gibbs free energy (ΔG°) from −8.376 to −9.090 kJ mol^{-1} with an increase in temperature showed that the physical adsorption and spontaneous biosorption process followed pseudo-second-order kinetics. The endothermic nature of biosorption was confirmed by the positive ΔH° value. This study demonstrated that SCAB treated with sodium citrate is a better alternative for the treatment of wastewaters contaminated with Zn^{2+}, Pb^{2+}, and Cu^{2+} ions. The world's most widely consumed mushroom species (*A. bisporus*) has high removal efficiency and high accumulation potential for heavy metal ions from mono- and polycomponent aqueous solutions. The use of sodium citrate-treated SCAB offers an efficient, cost-effective, and environmentally sound solution to treat wastewater, particularly at a time when global political views are shifting toward sustainable green technologies.

This study verified the removal efficiency of sodium citrate-treated SCAB under controlled laboratory conditions. Additional field studies are needed to better determine how to use SCAB to treat wastewater polluted by heavy metals. This limitation reveals the necessity of future work in this direction.

Author Contributions: X.Z., R.Y. and C.Z. wrote the paper; F.X. conceived and designed the paper; B.X., Z.H. and Y.L. collected and analyzed the data. All authors have read and agreed to the published version of the manuscript.

Funding: This research was supported by the National Key Research and Development Program of China (2018YFF0213402) and the National Natural Science Foundation of China (NO. 21607049).

Data Availability Statement: The data that support the findings of this study are available upon request from the authors.

Conflicts of Interest: Author Feng Xue was employed by the Shenzhen Degree of Freedom Eco Technology Co., Ltd. The remaining authors declare that the research was conducted in the absence of any commercial or financial relationships that could be construed as a potential conflict of interest.

References

1. Elahi, A.; Arooj, I.; Bukhari, D.A.; Rehman, A. Successive use of microorganisms to remove chromium from wastewater. *Appl. Microbiol. Biotechnol.* **2020**, *104*, 3729–3743. [CrossRef] [PubMed]
2. Jayakumar, V.; Govindaradjane, S.; Rajamohan, N.; Rajasimman, M. Biosorption potential of brown algae, *Sargassum polycystum*, for the removal of toxic metals, cadmium and zinc. Environ. *Sci. Pollut. Res.* **2022**, *29*, 41909–41922. [CrossRef] [PubMed]
3. Berkani, M.; Smaali, A.; Kadmi, Y.; Almomani, F.; Vasseghian, Y.; Lakhdari, N.; Alyane, M. Photocatalytic degradation of Penicillin G in aqueous solutions: Kinetic, degradation pathway, and microbioassays assessment. *Hazard. Mater.* **2022**, *421*, 126719. [CrossRef] [PubMed]
4. Xu, Z.; Gu, S.; Rana, D.; Matsuura, T.; Lan, C.Q. Chemical precipitation enabled UF and MF filtration for lead removal. *J. Water Process Eng.* **2021**, *41*, 101987. [CrossRef]
5. Ayangbenro, A.S.; Babalola, O.O. A new strategy for heavy metal polluted environments: A review of microbial biosorbents. *Int. J. Environ. Res. Public Health* **2017**, *14*, 94. [CrossRef] [PubMed]
6. Nokman, W.; Benluvankar, V.; Packiam, S.M.; Vincent, S. Screening and molecular identification of heavy metal resistant *Pseudomonas putida* S4 in tannery effluent wastewater. *Biocatal. Agric. Biotechnol.* **2019**, *18*, 101052. [CrossRef]
7. Jiang, D.; Yang, Y.; Huang, C.; Huang, M.; Chen, J.; Rao, T.; Ran, X. Removal of the heavy metal ion nickel (II) via an adsorption method using flower globular magnesium hydroxide. *Hazard. Mater.* **2019**, *373*, 131–140. [CrossRef]
8. Şentürk, İ.; Yıldız, M.R. Highly efficient removal from aqueous solution by adsorption of Maxilon Red GRL dye using activated pine sawdust. *Korean J. Chem. Eng.* **2020**, *37*, 985–999. [CrossRef]
9. Bano, A.; Hussain, J.; Akbar, A.; Mehmood, K.; Anwar, M.; Hasni, M.S.; Ullah, S.; Sajid, S.; Ali, I. Biosorption of heavy metals by obligate halophilic fungi. *Chemosphere* **2018**, *199*, 218–222. [CrossRef]
10. Hameed, A.H.A.E.; Eweda, W.E.; Abou-Taleb, K.A.A.; Mira, H.I. Biosorption of uranium and heavy metals using some local fungi isolated from phosphatic fertilizers. *Ann. Agr. Sci.* **2015**, *60*, 345–351. [CrossRef]
11. Choma, A.; Nowak, K.; Komaniecka, I.; Waśko, A.; Pleszczyńska, M.; Siwulski, M.; Wiater, A. Chemical characterization of alkali-soluble polysaccharides isolated from a *Boletus edulis* (Bull.) fruiting body and their potential for heavy metal biosorption. *Food Chem.* **2018**, *266*, 329–334. [CrossRef] [PubMed]
12. Schiewer, S.; Balaria, A. Biosorption of Pb^{2+} by original and protonated citrus peels: Equilibrium, kinetics, and mechanism. *Chem. Eng. J.* **2009**, *146*, 211–219. [CrossRef]
13. Alharbi, N.K.; Al-Zaban, M.I.; Albarakaty, F.M.; Abdelwahab, S.F.; Hassan, S.H.; Fawzy, M.A. Kinetic, isotherm and thermodynamic aspects of Zn^{2+} biosorption by *Spirulina platensis*: Optimization of process variables by response surface methodology. *Life* **2022**, *12*, 585. [CrossRef] [PubMed]
14. Cepoi, L.; Zinicovscaia, I. *Spirulina platensis* as a model object for the environment bioremediation studies. In *Handbook of Algal Science, Technology and Medicine*; Academic Press: Cambridge, MA, USA, 2020; pp. 629–640.
15. Saravanan, A.; Jeevanantham, S.; Kumar, P.S.; Varjani, S.; Yaashikaa, P.R.; Karishma, S. Enhanced Zn (II) ion adsorption on surface modified mixed biomass-*Borassus flabellifer* and *Aspergillus tamarii*: Equilibrium, kinetics and thermodynamics study. *Ind. Crop. Prod.* **2020**, *153*, 112613. [CrossRef]
16. Priya, A.K.; Gnanasekaran, L.; Dutta, K.; Rajendran, S.; Balakrishnan, D.; Soto-Moscoso, M. Biosorption of heavy metals by microorganisms: Evaluation of different underlying mechanisms. *Chemosphere* **2022**, *307*, 135957. [CrossRef]
17. Kalaimurugan, D.; Balamuralikrishnan, B.; Durairaj, K.; Vasudhevan, P.; Shivakumar, M.S.; Kaul, T.; Chang, S.W.; Ravindran, B.; Venkatesan, S. Isolation and characterization of heavy-metal-resistant bacteria and their applications in environmental bioremediation. *Int. J. Environ. Sci. Technol.* **2020**, *17*, 1455–1462. [CrossRef]
18. Bhatt, P.; Bhandari, G.; Turco, R.F.; Aminikhoei, Z.; Bhatt, K.; Simsek, H. Algae in wastewater treatment, mechanism, and application of biomass for production of value-added product. *Environ. Pollut.* **2022**, *309*, 119688. [CrossRef] [PubMed]
19. Matheickal, J.T.; Yu, Q. Biosorption of lead (II) and copper (II) from aqueous solutions by pre-treated biomass of Australian marine algae. *Bioresour. Technol.* **1999**, *69*, 223–229. [CrossRef]
20. Chen, J.P.; Yang, L. Study of a heavy metal biosorption onto raw and chemically modified *Sargassum* sp. via spectroscopic and modeling analysis. *Langmuir* **2006**, *22*, 8906–8914. [CrossRef]

21. Sudha, B.R.; Abraham, T.E. Studies on enhancement of Cr(VI) biosorption by chemically modified biomass of *Rhizopus nigricans*. *Water Res.* **2002**, *36*, 1224–1236.
22. Šoštarić, T.D.; Petrović, M.S.; Pastor, F.T.; Lončarević, D.R.; Petrović, J.T.; Milojković, J.V.; Stojanović, M.D. Study of heavy metals biosorption on native and alkali-treated apricot shells and its application in wastewater treatment. *J. Mol. Liq.* **2018**, *259*, 340–349. [CrossRef]
23. Regine, H.S.; Volesky, F.B. Biosorption: A solution to pollution. *Int. Microbiol.* **2000**, *3*, 17–24.
24. Sarangi, N.V.; Rajkumar, R. Biosorption potential of *Stoechospermum marginatum* for removal of heavy metals from aqueous solution: Equilibrium, kinetic and thermodynamic study. *Chem. Eng. Res. Des.* **2024**, *203*, 207–218. [CrossRef]
25. Gu, S.; Lan, C.Q. Mechanism of heavy metal ion biosorption by microalgal cells: A mathematic approach. *J. Hazard. Mater.* **2024**, *463*, 132875. [CrossRef] [PubMed]
26. Cheng, S.Y.; Show, P.-L.; Lau, B.F.; Chang, J.-S.; Ling, T.C. New prospects for modified algae in heavy metal adsorption. *Trends Biotechnol.* **2019**, *37*, 1255–1268. [CrossRef] [PubMed]
27. Sathya, A.B.; Sivashankar, R.; Kanimozhi, J.; Devika, R.; Balaji, R. Biosorption and different native sources for preparation of biosorbents. *Emerg. Contam.* **2022**, *67*, 23–41.
28. Xu, Y.; Chen, Y. Leaching heavy metals in municipal solid waste incinerator fly ash with chelator/biosurfactant mixed solution. *Waste Manag. Res.* **2015**, *33*, 652–661. [CrossRef] [PubMed]
29. Zhang, S.; Zhang, X.; Xiong, Y.; Wang, G.; Zheng, N. Effective solidification/stabilisation of mercury-contaminated wastes using zeolites and chemically bonded phosphate ceramics. *Waste Manag. Res.* **2015**, *33*, 183–190. [CrossRef] [PubMed]
30. Seid, L.; Lakhdari, D.; Berkani, M.; Belgherbi, O.; Chouder, D.; Vasseghian, Y.; Lakhdari, N. High-efficiency electrochemical degradation of phenol in aqueous solutions using Ni-PPy and Cu-PPy composite materials. *Hazard. Mater.* **2022**, *423*, 126986. [CrossRef]
31. Joshi, H.K.; Vishwakarma, M.C.; Kumar, R.; Sharma, H.; Bhandari, N.S.; Joshi, S.K. The biosorption of Zn^{2+} by various biomasses from wastewater: A review. *J. Water Process Eng.* **2023**, *56*, 104389. [CrossRef]
32. Matouq, M.; Jildeh, N.; Qtaishat, M.; Hindiyeh, M.; Al Syouf, M.Q. The adsorption kinetics and modeling for heavy metals removal from wastewater by Moringa pods. *J. Environ. Chem. Eng.* **2015**, *3*, 775–784. [CrossRef]
33. Shroff, K.A.; Vaidya, V.K. Kinetics and equilibrium studies on biosorption of nickel from aqueous solution by dead fungal biomass of *Mucor hiemalis*. *Chem. Eng. J.* **2011**, *171*, 1234–1245. [CrossRef]
34. Rao, R.A.K.; Khan, M.A.; Rehman, F. Utilization of Fennel biomass (*Foeniculum vulgari*) a medicinal herb for the biosorption of Cd(II) from aqueous phase. *Chem. Eng. J.* **2010**, *156*, 106–113. [CrossRef]
35. Rudy, C.; Milhome, M.A.L.; Cavalcante, R.M.; Silveira, E.R.; Denis, D.K.; Nascimento, R.F. Removal of some polycyclic aromatic hydrocarbons from petrochemical wastewater using low-cost adsorbents of natural origin. *Bioresour. Technol.* **2008**, *99*, 4515–4519.
36. Ofomaja, A.E.; Naidoo, E.B.; Modise, S.J. Kinetic and pseudo-second-order modeling of lead biosorption onto pine cone powder. *Ind. Eng. Chem Res.* **2010**, *49*, 2562–2572. [CrossRef]
37. Ho, Y.S.; Ofomaja, A.E. Kinetics and thermodynamics of lead ion sorption on palm kernel fibre from aqueous solution. *Process Biochem.* **2005**, *40*, 3455–3461. [CrossRef]
38. Ho, Y.S.; Wase, D.A.J.; Forster, C.F. Kinetic studies of competitive heavy metal adsorption by sphagnum moss peat. *Environ. Technol.* **1996**, *17*, 71–77. [CrossRef]
39. Dinesh, M.; Kunwar, P.S. Single- and multi-component adsorption of cadmium and zinc using activated carbon derived from bagasse-an agricultural waste. *Water Res.* **2002**, *36*, 2304–2318.
40. Jaafari, J.; Yaghmaeian, K. Optimization of heavy metal biosorption onto freshwater algae (*Chlorella coloniales*) using response surface methodology (RSM). *Chemosphere* **2019**, *217*, 447–455. [CrossRef]
41. Vimala, R.; Das, N. Biosorption of cadmium (II) and lead (II) from aqueous solutions using mushrooms: A comparative study. *J. Hazard. Mater.* **2009**, *168*, 376–382. [CrossRef]
42. Abia, D.; Nzali, S.; Acayanka, E.; Kamgang, G.Y.; Laminsi, S.; Ghogomu, P.M. Synergetic effect of gliding arc discharge treatment and biosorption for removal of nitrophene and glycine from aqueous solution. *J. Ind. Eng. Chem.* **2015**, *29*, 156–162. [CrossRef]
43. Moyo, M.; Guyo, U.; Mawenyiyo, G.; Zinyama, N.P.; Nyamunda, B.C. Marula seed husk (*Sclerocarya birrea*) biomass as a low cost biosorbent for removal of Pb(II) and Cu(II) from aqueous solution. *J. Ind. Eng. Chem.* **2015**, *27*, 126–132. [CrossRef]
44. Munoz, A.J.; Espínola, F.; Ruiz, E. Removal of Pb(II) in a packed-bed column by a *Klebsiella* sp. 3S1 biofilm supported on porous ceramic Raschig rings. *J. Ind. Eng. Chem.* **2016**, *40*, 118–127. [CrossRef]
45. Dil, E.A.; Ghaedi, M.; Ghezelbash, G.R.; Asfaram, A.; Purkait, M.K. Highly efficient simultaneous biosorption of Hg^{2+}, Pb^{2+} and Cu^{2+} by live yeast *Yarrowia lipolytica* 70562 following response surface methodology optimization: Kinetic and isotherm study. *J. Ind. Eng. Chem.* **2017**, *48*, 162–172.
46. Nadeem, R.; Manzoor, Q.; Iqbal, M.; Nisar, J. Biosorption of Pb(II) onto immobilized and native *Mangifera indica* waste biomass. *J. Ind. Eng. Chem.* **2016**, *35*, 185–194. [CrossRef]
47. Gunasundari, E.; Kumar, P.S. Adsorption isotherm, kinetics and thermodynamic analysis of Cu (II) ions onto the dried algal biomass (*Spirulina platensis*). *J. Ind. Eng. Chem.* **2017**, *56*, 129–144.
48. Pradhan, D.; Sukla, L.B.; Sawyer, M.; Rahman, P.K. Recent bioreduction of hexavalent chromium in wastewater treatment: A review. *J. Ind. Eng. Chem.* **2017**, *55*, 1–20. [CrossRef]

49. Esposito, A.; Pagnanelli, F.; Lodi, A.; Solisio, C.; Vegliò, F. Biosorption of heavy metals by *Sphaerotilus natans*: An equilibrium study at different pH biomass concentrations. *Hydrometallurgy* **2001**, *60*, 129–141. [CrossRef]
50. Sheng, P.X.; Ting, Y.P.; Chen, J.P. Biosorption of heavy metal ions (Pb, Cu, and Cd) from aqueous solutions by the marine alga *Sargassum* sp. in single- and multiple-metal systems. *Ind. Eng. Chem. Res.* **2007**, *46*, 2438–2444. [CrossRef]
51. Zhou, K.; Yang, Z.G.; Liu, Y.C.H.; Kong, X. Kinetics and equilibrium studies on biosorption of Pb (II) from aqueous solution by a novel biosorbent: *Cyclosorus interruptus*. *J. Ind. Eng. Chem.* **2015**, *3*, 2219–2228. [CrossRef]
52. Masoumi, F.; Khadivinia, E.; Alidoust, L.; Mánourinejad, Z.; Shahryari, S.; Safaei, M.; Mousavi, A.; Salmanian, A.H.; Zahiri, H.S.; Vali, H.; et al. Nickel and lead biosorption by *Curtobacterium* sp. FM01, an indigenous bacterium isolated from farmland soils of northeast Iran. *J. Ind. Eng. Chem.* **2016**, *4*, 950–957. [CrossRef]
53. Mohapatra, M.; Rout, K.; Mohapatra, B.K.; Anand, S. Sorption behavior of Pb(II) and Cd(II) on iron ore slime and characterization of metal ion loaded sorbent. *Ind. Eng. Chem. Res.* **2009**, *166*, 1506–1513. [CrossRef] [PubMed]
54. Chen, J.P.; Wang, L.K.; Yang, L.; Lim, S.F. Emerging biosorption, adsorption, ion exchange, and membrane technologies. In *Advanced Physicochemical Treatment Technologies*; Wang, L.K., Hung, Y.T., Shammas, N.K., Eds.; Humana Press: Clifton, NJ, USA, 2007; pp. 367–390.
55. Tran, H.T.; Vu, N.D.; Matsukawa, M.; Okajima, M.; Kaneko, T.; Ohki, K.; Yoshikawa, S. Heavy metal biosorption from aqueous solutions by algae inhabiting rice paddies in Vietnam. *J. Environ. Chem. Eng.* **2016**, *4*, 2529–2535. [CrossRef]
56. Wang, J.; Chen, C. Biosorbents for heavy metals removal and their future. *Biotechnol. Adv.* **2009**, *27*, 195–226. [CrossRef]
57. Kim, Y.; Kim, C.; Choi, I.; Rengaraj, S.; Yi, J. Arsenic removal using mesoporous alumina prepared via a templating method. *Environ. Sci. Technol.* **2004**, *38*, 924–931. [CrossRef]
58. Ayhan, Ş.; Mahmut, Ö. Competitive biosorption of Pb^{2+}, Cu^{2+} and Zn^{2+} ions from aqueous solutions onto *Valonia tannin* resin. *J. Hazard. Mater.* **2009**, *166*, 1488–1494.
59. Ertugay, N.; Bayhan, Y.K. The removal of copper (II) ion by using mushroom biomass (*Agaricus bisporus*) and kinetic modelling. *Desalination* **2010**, *255*, 137–142. [CrossRef]
60. Mahmood, Z.; Zahra, S.; Iqbal, M.; Raza, M.A.; Nasir, S. Comparative study of natural and modified biomass of *Sargassum* sp. for removal of Cd^{2+} and Zn^{2+} from wastewater. *Appl. Water Sci.* **2017**, *7*, 3469–3481. [CrossRef]
61. Tunali, S.; Akar, T. Zn(II) biosorption properties of *Botrytis cinerea* biomass. *J. Hazard. Mater.* **2006**, *131*, 137–145. [CrossRef]
62. Luna, A.S.; Costa, A.L.H.; Costa, A.C.A.; Henriques, C.A. Competitive biosorption of cadmium(II) and zinc(II) ions from binary systems by *Sargassum filipendula*. *Bioresour. Technol.* **2010**, *101*, 5104–5111. [CrossRef]
63. Miretzky, P.; Saralegui, A.; Fernandezcirelli, A. Simultaneous heavy metal removal mechanism by dead macrophytes. *Chemosphere* **2006**, *62*, 247–254. [CrossRef] [PubMed]
64. Sağ, Y.; Kaya, A.; Kutsal, T. The simultaneous biosorption of Cu and Zn on *Rhizopus arrhizus*: Application of the adsorption models. *Hydrometallurgy* **1998**, *50*, 297–314. [CrossRef]
65. Malkoc, E.; Nuhoglu, Y.; Dundar, M. Adsorption of chromium(VI) on pomace—An olive oil industry waste: Batch and column studies. *J. Hazard. Mater.* **2006**, *138*, 142–151. [CrossRef] [PubMed]
66. Tewari, N.; Vasudevan, P.; Guha, B. Study on biosorption of Cr(VI) by *Mucor hiemalis*. *Biochem. Eng. J.* **2005**, *23*, 185–192. [CrossRef]
67. Tunali, S.; Akar, T.; Özcan, A.S.; Kiran, I. Equilibrium and kinetics of biosorption of lead(II) from aqueous solutions by *Cephalosporium aphidicola*. *Sep. Purif. Technol.* **2006**, *47*, 105–112. [CrossRef]
68. Sprynskyy, M. Solid-liquid-solid extraction of heavy metals (Cr, Cu, Cd, Ni and Pb) in aqueous systems of zeolite–sewage sludge. *J. Hazard. Mater.* **2009**, *161*, 1377–1383. [CrossRef]

processes

MDPI

Article

Construction of Novel Nanoflowering MgAl-Double Oxide Konjac Gum for Efficient Enrichment of Uranium (VI) from Wastewater

Ping Gong [1], Jiayan Wu [1], Zhongbo Shang [2], Meiqing Chen [3], Bo Li [1], Tianming Wang [1], Leiye Sun [1], Zhi Dang [1,4,5], Nengwu Zhu [1,4] and Pingxiao Wu [1,4,5,6,*]

1 School of Environment and Energy, South China University of Technology, Guangzhou Higher Education Mega Center, Guangzhou 510006, China; esgongping@mail.scut.edu.cn (P.G.); wwujy@foxmail.com (J.W.); bondboli@163.com (B.L.); tianmingwang@scut.edu.cn (T.W.); essunly131415@mail.scut.edu.cn (L.S.); chzdang@scut.edu.cn (Z.D.); nwzhu@scut.edu.cn (N.Z.)
2 School of Resources and Environment, Henan Institute of Science and Technology, Xinxiang 453003, China; shangzhongbo@126.com
3 School of Environmental Science and Engineering, Guangdong University of Technology, Guangzhou 510006, China; cmq20221@scut.edu.cn
4 The Key Laboratory of Pollution Control and Ecosystem Restoration in Industry Clusters, Ministry of Education, Guangzhou 510006, China
5 Guangdong Provincial Key Laboratory of Solid Wastes Pollution Control and Recycling, Guangzhou 510006, China
6 Guangdong Engineering and Technology Research Center for Environmental Nanomaterials, Guangzhou 510006, China
* Correspondence: pppxwu@scut.edu.cn

Abstract: Given the pressing demand for efficient uranium (U(VI)) enrichment and its elimination from wastewater to curtail the risks of radioactive contamination inherent in nuclear energy applications, it is crucial to design materials with high removal efficiency and straightforward separation processes. In the current study, we incorporated konjac gum (KGM) into MgAl-double oxide (MgAl-LDO) and synthesized an innovative, economical, and environmentally friendly LDO-KGM material by using the freeze-drying-calcination (FDC) method, which provided a solution for U(VI) concentration from aqueous solutions. The nanoflower structures LDO-KGM with abundant pore structure and high specific surface area exhibited an optimal U(VI) adsorptive capacity (3019.56 mg·g^{-1}) at pH = 6.0 and 293 K, which was 2.3 times greater than that of MgAl-LDO (1296.39 mg·g^{-1}). LDO-KGM also showed great adaptability for the immobilization of U(VI) over a broad pH range (4.0 to 9.0) and coexisting ions. U(VI) adsorption onto LDO-KGM adhered to the pseudo-second-order kinetic model ($R^2 \geq 0.99$) and the Langmuir isotherm model ($R^2 \geq 0.99$). The analysis of thermodynamic parameters derived from isotherms at varying temperatures revealed that U(VI) adsorption onto LDO-KGM was an endothermic and spontaneous process. The mechanism underlying U(VI) adsorption by LDO-KGM was mainly complexation, carbonate co-precipitation, and electrostatic adsorption. Furthermore, the adsorption efficiency of LDO-KGM for U(VI) could still retain more than 84.5% after five cycles. The findings indicate that the synthesized LDO-KGM exhibits potential as an exceptionally potent adsorbent for the purification of wastewater contaminated with U(VI).

Keywords: MgAl-double oxide; uranium; enrichment; freeze-drying calcination; immobilization

1. Introduction

The escalating need for energy resources and the extensive extraction of minerals have caused radionuclides to spread and concentrate within the environment, thereby increasing the threat of radioactive contamination [1]. Uranium-containing wastewater amplifies these dangers, as it is released into the natural environment through nuclear fuel cycles, ore mining, processing, and manufacturing activities at uranium-contaminated sites [2]. When uranium goes (U(VI)) into aquatic ecosystems without proper treatment, it can lead to its accumulation in the food chain [3,4]. U(VI) toxicity poses a significant risk to the proper functioning of vital organs, such as the lungs and kidneys, with severe cases possibly resulting in life-threatening outcomes [5]. Thus, the development of materials that can extract U(VI) from nuclear effluent is critical to reduce environmental pollution and address the energy crisis [6]. Various methodologies have been implemented to eliminate U(VI) from wastewater and nuclear effluents, encompassing precipitation [7], membrane separation [8], solvent extraction [9], phytoremediation [10], and adsorption [11,12]. Among them, adsorption stands out as an exceptionally effective technique for removing U(VI) from wastewater [13,14] and is considered especially attractive due to its cost-effectiveness, robust efficiency, straightforward implementation, and environmentally benign nature [15].

Over the past few years, substantial research efforts have been directed towards exploring diverse sorbent materials for U(VI) elimination from wastewater, including konjac gum (KGM) [16], clays [17], metal–organic frameworks (MOFs) [18], layered double hydroxides (LDHs) [19], activated carbons [20], and several other potential candidates. As a quintessential 2D anionic clay, LDHs possess a distinctive layered architecture, a plethora of reactive sites, and superior stability, making them an ideal candidate for the sequestration of U(VI) [21,22]. Notably, layered double oxides (LDO) exhibit a greater number of active sites, larger specific surface area, and superior chemical and thermal stability compared to the pristine LDHs [23] and, thus, have better U(VI) removal capability. Momin et al. used graphitic carbon nitride (GCN) in LDO to prepare GCN-LDO for U(VI) capture, which showed a substantial adsorption capacity, reaching 158.25 $mg \cdot g^{-1}$ [24]. Zhu et al. demonstrated that the novel γ-Fe_2O_3 cross-linked LDO composite (γ-Fe_2O_3/LDO) exhibited an adsorptive capacity for U(VI) approximately 526.32 $mg \cdot g^{-1}$ in aqueous media [25]. However, due to the dense surface and low functional group structure of most LDOs, their adsorption performance is limited to a certain degree [26]. Therefore, the development of Mg-Al LDO with a larger specific surface area and a greater number of surface functional groups, such as Mg-O-H, Al-O, and -OH, is highly sought after for the efficient removal of U(VI).

The presence of numerous hydroxyl and carbonyl groups on the saccharide units of KGM enables it to readily form gels with exhibit high porosity, tunable densities, and advantageous porous structures through the action of hydrogen bonding, and its combination with LDO can effectively solve the structure problem of LDO [27,28]. For example, Xiong et al. synthesized hydroxyapatite (HAP) aerogels with high selectivity and capacity by employing a combination of freeze-drying and templating techniques, utilizing KGM as a template [16]. Wang et al. synthesized a 3D porous aerogel, utilizing KGM as a structural template, aimed at U(VI) adsorption [29]. The freeze-drying calcination (FDC) technique has been recognized for its efficacy in preparing ultra low-density porous materials, which have a wide range of applications, such as catalysis and adsorption [30,31].

In the current study, LDO-KGM with a nanoflower structure was prepared by the FDC method using KGM as a template. This material has a larger specific surface, larger pores, and richer functional groups than other composites. By combining KGM with LDO, the structural limitations of LDO can be effectively addressed. This synergy leverages KGM's advantageous properties to enhance LDO's performance in terms of capturing

U(VI), thereby positioning the composite as a promising material for the selective extraction of U(VI) in wastewater treatment. The morphology and chemical constituents of LDO-KGM were examined using a suite of analytical methods, such as scanning electron microscopy (SEM), X-ray diffraction (XRD), Fourier transform infrared spectroscopy (FTIR), N_2 adsorption–desorption analysis, and X-ray photoelectron spectroscopy (XPS). Meanwhile, the impact of several parameters on U(VI) adsorption was explored, including adsorbent concentration, contact time, adsorption isotherms, different ions, pH, and temperature. In addition, a synergistic approach involving XPS and zeta potential analysis was employed to elucidate the adsorption mechanism. The results offer significant understanding for the development and preparation of eco-friendly, high-performance adsorbents tailored for the purification of wastewater containing U(VI).

2. Materials and Methods

2.1. Materials

The laboratory experiments utilized chemical reagents of analytical grade. Supplied included KGM, $UO_2(NO_3)_2$, $Mg(NO_3)_2{\cdot}6H_2O$, $Al(NO_3)_3{\cdot}9H_2O$, NaOH, HNO_3, NaCl, KCl, $MgCl_2$, $CaCl_2$, $NaNO_3$, Na_2CO_3, and Na_2SO_4 were provided by Macklin Biochemical Technology Co., Ltd. (Shanghai, China). The laboratory experiments were conducted using deionized water (DIW), which had a resistivity of 18.2 MΩ·cm.

2.2. Synthesis of LDO-KGM

LDH was achieved through a simple one-pot hydrothermal process, utilizing urea as the alkaline agent, and then, LDO-KGM was synthesized according to the FDC method (Figure 1). Research showed that FDC was a widely used way to synthesize products with a porous structure, which could provide more active sites [32]. In detail, 0.008 mol $Mg(NO_3)_2{\cdot}6H_2O$, 0.004 mol $Al(NO_3)_3{\cdot}9H_2O$, 0.125 mol urea and x g KGM (x = 0.5, 1, 1.5, 2, 2.5) were added to a beaker containing 50 mL DIW, whilst stirring for at least 30 min until completely mixed. Subsequently, the homogeneous solution was poured into a 100 mL Teflon-lined autoclave for hydrothermal processing at a temperature of 100 °C over a period of 48 h. Following this, the resulting material was permitted to gradually cool to ambient temperature. The resulting precipitates were then centrifuged and washed with DIW to eliminate any residual free metals. The washing process was continued until the supernatant became colorless, and the pH of the solution reached approximately neutral, around pH 7. Finally, the precipitates were lyophilized at −60 °C in a freeze-dried machine and named LDH-KGM-x (X was the amount of KGM added), among which LDH-KGM-2 had the best adsorption capacity. The obtained LDH-KGM-2 was calcined at 250, 350, 450, 550, 650, and 750 °C, respectively, and named LDO-KGM-2-Y (Y was the calcination temperature). The calcination process was conducted in a muffle furnace, with the temperature ramped up at a rate of 5 °C/min. The material was then subjected to calcination at the designated target temperature for a period of 5 h. Following the calcination, the furnace was permitted to cool down naturally to ambient temperature. The final obtained material with the best adsorption properties was LDO-KGM-2-650, denoted as LDO-KGM. The MgAl-LDH and MgAl-LDO used for comparison were synthesized without adding KGM, where MgAl-LDO was synthesized through the calcination of MgAl-LDH at 650 °C.

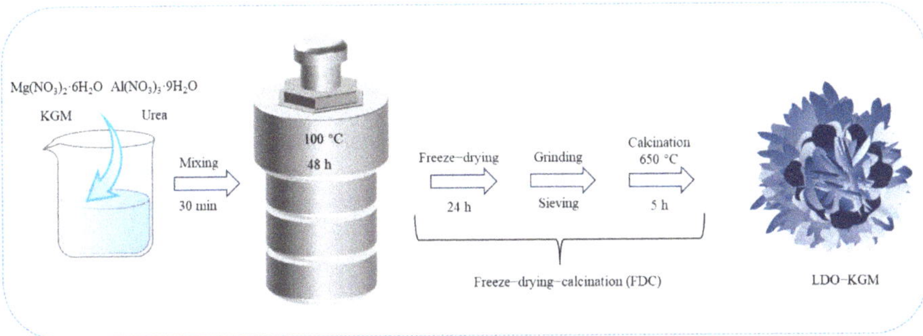

Figure 1. The flowchart of preparation of LDO-KGM.

2.3. U(VI) Immobilization Experiment

A series of batch experiments were carried out to determine the adsorptive immobilization characteristics and the intrinsic mechanisms of U(VI) by the synthesized materials. The kinetic studies on U(VI) immobilization were conducted with the following procedure: a 100 mL conical flask contained 100 mL of a 100 mg·L^{-1} U(VI) solution into which 5.0 mg of the material was introduced, with an initial pH carefully controlled at 6.0 ± 0.1; the mixed solution was then stirred continuously throughout the experiment, and the samples were collected periodically (0, 20, 40, 60, 90, 150, 240, 360, 540, 720, 1080, and 1440 min). In the isotherm experiments, 5.0 mg of material was dispersed to 100 mL of U(VI) solution with concentrations varying between 20 and 200 mg·L^{-1} at pH of 6.0, followed by agitation for a period of 24 h.

After the reaction, the supernatant liquid was passed through a membrane filter with a pore size of 0.22 μm to isolate the solid phase. The filtrate's U(VI) concentration was then measured employing a UV–vis spectrophotometer (UV-2450, Shimadzu, Kyoto, Japan). The absorbance was conducted at a wavelength of 651 nm, utilizing arsenazo (III) as the chromogenic reagent at a concentration of 0.05% [33]. Based on these measurements, the U(VI) adsorption capacity (Q_e, mg·g^{-1}) (Equation (1)) and the adsorption rate (R, %) (Equation (2)) were determined by employing the subsequent equations [33,34]:

$$Q_e = (C_0 - C_e) \times V/M, \tag{1}$$

$$R = (C_0 - C_e)/C_0 \times 100\%, \tag{2}$$

where Q_e denotes the adsorption capacity for U(VI) (mg·g^{-1}), C_0 denotes the initial concentration of U(VI) solution, C_e denotes the concentration of U(VI) at equilibrium (mg·L^{-1}), V denotes the volume of U(VI) solution (L), M denotes the mass of LDO-KGM (g), and R denotes adsorption rate (%).

Moreover, the influence of adsorbent concentration, different ions, pH, and temperature were thoroughly investigated. Each data point was obtained from three replicate experiments to ascertain the dependability of the findings, and error bars were supplied.

2.4. Characterization Methods

The specific surface areas and pore size distributions of the materials were evaluated using a Chemisorption Surface Area Analyzer (Micromeritics ASAP 2460, Shanghai, China), employing the Brunauer–Emmett–Teller (BET) and Barrett–Joyner–Halenda (BJH) methodologies, respectively. The morphologies of the materials were analyzed using the SEM (Hitachi SU8020, Tokyo, Japan). The physical phases of the materials were analyzed using the XRD (D8 advance Sox-I, Bruker Co., Billerica, MA, USA). The characteristic functional groups of the materials within the wavenumber range of 400–4000 cm^{-1} were observed using the FTIR (Nicolet 6700, Thermo Fisher Scientific, Waltham, MA, USA). The chemical states of the materials were studied using XPS (Thermo Escalab 250XI, Thermo Fisher Scientific, Waltham, MA, USA). Additionally, the zeta potential of the materials, which indicates their surface charge, was measured before the reaction using a Zeta Potential Analyzer (Zetasizer Nano ZS90, Malvern Panalytical, Malvern, UK).

3. Results and Discussion

3.1. *Effect of Calcination Temperature and Doping Ratio*

Figure 2a illustrates the differences in adsorptive capacities among different materials. Notably, the enhanced adsorption efficiency for U(VI) by the modified materials was evident, showcasing a significant improvement in their performance compared to their unmodified counterparts. In order to synthesize the materials with the best performance, the impact of the calcination temperature and doping ratio on the adsorption properties were initially examined. This preliminary investigation aimed to determine the optimal conditions for material synthesis, which would result in the best adsorptive behavior. It was observed that the LDO-KGM obtained by calcination at 650 °C with the incorporation of 20 $g \cdot L^{-1}$ KGM demonstrated optimal U(VI) adsorption capacity; hence, it was used for further investigation. An insufficient amount of KGM made the structure of MgAl-LDH loose, and an excess amount caused the collapse of the MgAl-LDH layered structure [35]. Meanwhile, the KGM in LDH-KGM may occupy the adsorption sites; whereas, during the calcination process, the KGM gradually disappeared and only the fluffy and porous KGM structure remained in the calcined LDO-KGM, and therefore, the adsorption performance was enhanced. The KGM in LDH-KGM could be removed to produce pure LDO at 650 °C [16]. With the ongoing rise in temperature, the metal oxides in LDH sintered and formed stable spinel compounds with the decreasing adsorption performance. The increased temperature affected the structural integrity and adsorptive sites of the LDH, thereby impacting its ability to effectively capture U(VI) [36].

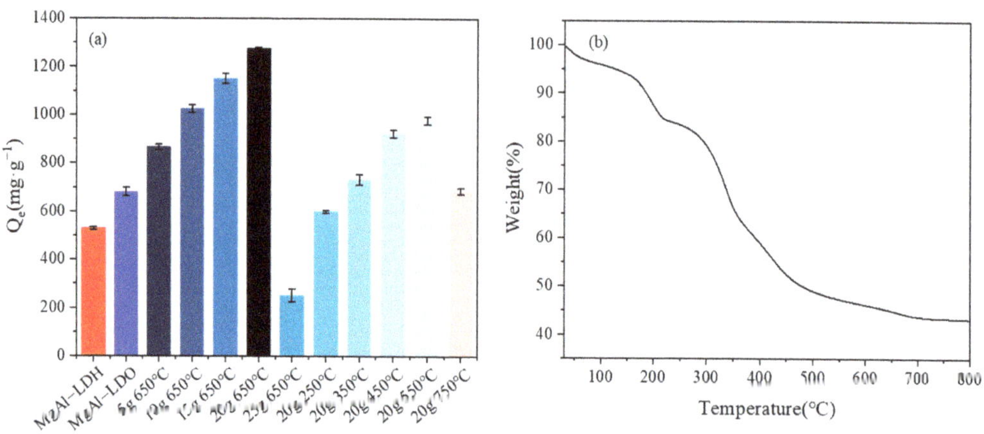

Figure 2. (**a**) Adsorption capacity of U(VI) by materials synthesized under different conditions (C_0 = 100 $mg \cdot L^{-1}$, pH = 6.0, T = 293 K); (**b**) TGA curves of LDO-KGM.

Figure 2b presented the thermogravimetric analysis (TGA) curve of LDO-KGM, revealing a two-stage weight loss process. The initial stage (below 200 °C) corresponded to the evaporation of physically adsorbed water, while the second stage (200–650 °C) was attributed to the thermal decomposition of the KGM template. Above 650 °C, the KGM template was entirely removed, leaving pure LDO, confirming the successful synthesis of LDO-KGM and its robust thermal stability. These TGA results aligned closely with XRD and FTIR data, collectively demonstrating that KGM served as an effective template for LDO-KGM synthesis due to its compatibility with the FDC method and enhanced thermal stability.

3.2. Characterization

The morphology and porosity of MgAl-LDH, MgAl-LDO, and LDO-KGM were examined through SEM images (Figure 3a–f). MgAl-LDH presented a laminated hexahedron-like morphology, which was a characteristic morphology often observed in LDHs [37]. MgAl-LDO exhibited a structure analogous to that of MgAl-LDH, and both were in a laminated hexahedron-like morphology. Nevertheless, the MgAl-LDO exhibited a rougher surface texture compared to MgAl-LDH, attributed to the removal of interlayer water molecules and the breakdown of carbonate ions during calcination. The as-obtained LDO-KGM consisted primarily of nanoflower structures. Due to the addition of KGM, the layered structure of MgAl-LDO underwent a transformation, adopting a flower-like morphology accompanied by an increased specific surface area.

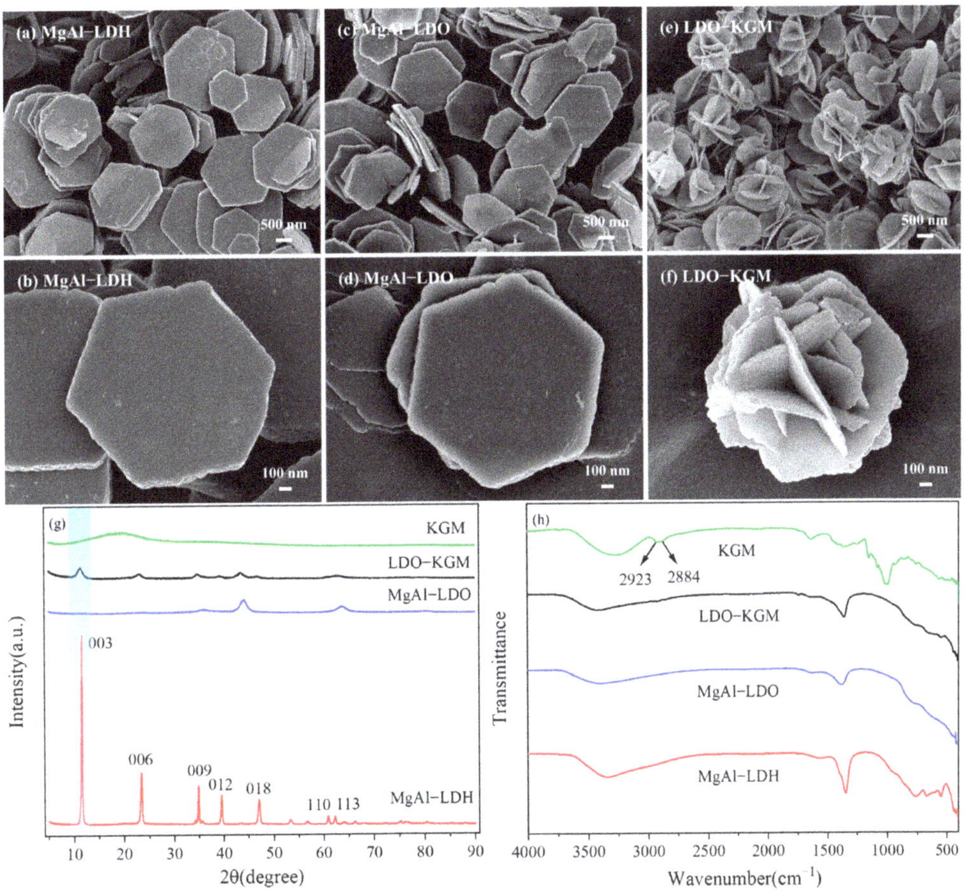

Figure 3. SEM images of (**a**,**b**) MgAl-LDH; (**c**,**d**) MgAl-LDO; (**e**,**f**) LDO-KGM; (**g**) XRD patterns; and (**h**) FTIR spectra of MgAl-LDH, MgAl-LDO, LDO-KGM, and KGM.

The XRD analysis (Figure 3g) provided additional clarity on the crystallographic and chemical configurations of MgAl-LDH, MgAl-LDO, and LDO-KGM. The distinctive peaks indicative of (003), (006), (009), (012), (018), (110), and (113) of MgAl-LDH were identified at 11.55°, 23.38°, 34.80°, 39.41°, 46.97°, 60.78°, and 62.19°, respectively [38], which was consistent with its typical patterns (JCPDS No. 70-2151) [39]. However, in comparison to MgAl-LDH, the MgAl-LDO peaks became weaker and broader, indicating the lower crystallinity and lattice defects of the MgAl-LDO. Interestingly, for MgAl-LDO, carbonate decomposition during calcination led to the disappearance of the 003 reflection [36], which reappeared in LDO-KGM after the addition of KGM (Figure 3g). Besides, many peaks that disappeared in MgAl-LDO reappeared in LDO-KGM. In contrast to pure KGM, the LDO-KGM showed no characteristic peak associated with KGM at $2\theta = 20.68°$. This observation further confirmed the KGM in LDH-KGM, which was removed to yield pure LDO-KGM at 650 °C. The functional groups of three materials and KGM were illustrated in the FTIR spectrum (Figure 3h). As for KGM, adsorption bands at 2923 cm^{-1} and 2884 cm^{-1} were ascribed to the stretching vibration of C-H bonds that were inherent to KGM [16]. Notably, the absence of peaks at 2923 cm^{-1} and 2884 cm^{-1} in LDO-KGM suggested that the KGM had become completely decomposed following the heat treatment process at 650 °C. Drawing from the findings, it can be confidently stated that the pure LDO-KGM has been successfully synthesized.

It had been demonstrated that the specific surface area, electrostatic forces, and structural characteristics of materials have a profound influence on their adsorptive capacity [40]. In an effort to pinpoint the key determinant governing the adsorptive interaction of MgAl-LDH, MgAl-LDO, and LDO-KGM with respect to U(VI), the impact of the specific surface area, as calculated using the BET method, was evaluated. As shown in Table 1, the specific surface area of LDO-KGM (231.22 $m^2 \cdot g^{-1}$) was 30.26 times greater than that of MgAl-LDH (7.64 $m^2 \cdot g^{-1}$) and 1.38 times higher than that of MgAl-LDO (167.17 $m^2 \cdot g^{-1}$), respectively. This marked increase in surface area implied a higher probability of effective collisions between LDO-KGM and U(VI), thereby enhancing the adsorption capacity of LDO-KGM for U(VI) [30]. As shown in Figure S1, the adsorption isotherm of MgAl-LDO revealed a type H2 hysteresis loop, demonstrating a relatively uniform mesoporous structure. In contrast, MgAl-LDH and LDO-KGM exhibited type H3 hysteresis loops, indicative of lamellar or layered materials with non-uniform pore size distributions. While all three materials possessed mesoporous characteristics, their pore size distributions exhibited notable differences. Upon calcination, the transformation of MgAl-LDH into MgAl-LDO resulted in a reduction in the average pore size, likely due to structural reorganization and densification. Notably, the incorporation of a KGM matrix into LDO-KGM led to an increase in the average pore size, attributed to the fluffy and porous structure of KGM, which effectively altered the material's porosity and pore architecture.

Table 1. Structural information of adsorbents.

Adsorbents	$S_{BET}/(m^2 \cdot g^{-1})$	D_a/nm	$V_t/(cm^3 \cdot g^{-1})$
MgAl-LDH	7.95	10.73	0.02
MgAl-LDO	167.17	2.74	0.11
LDO-KGM	231.22	5.62	0.32

3.3. Excellent Performance and Adaptability for U(VI) Removal

3.3.1. Kinetic Studies and Isotherm Studies

The excellent physicochemical properties of the successfully modified material with the greatly improved adsorption capacity for U(VI) were further demonstrated. Next, we continued to investigate the influence of reaction conditions on the immobilization capacity for U(VI). The efficiency of U(VI) immobilization was positively correlated with the dosage of the materials (Figure S2). Among the various materials, the dosage of 0.05 g·L^{-1} was more effective in immobilizing U(VI), and thus, this concentration was chosen for further study. As depicted in Figure 4a, each material exhibited a swift adsorption rate for U(VI) during the initial 180 min. The ranking of U(VI) immobilization capacities was as follows: LDO-KGM > MgAl-LDO > MgAl-LDH. To clarify the mechanism behind the immobilization of U(VI), we fitted the experimental data to both pseudo-first-order and pseudo-second-order kinetic models. The parameters obtained from the fitting are detailed in Table S1. The analysis revealed that for all materials, the pseudo-second-order model demonstrated superior correlation coefficients (R^2 = 0.99) when compared to the pseudo-first-order model (R^2 = 0.92), implying that the process of U(VI) immobilization was largely governed by chemical reactions [41,42]. Consequently, LDO-KGM demonstrates the capability to serve as a highly efficient adsorbent for the mitigation of U(VI) contamination events.

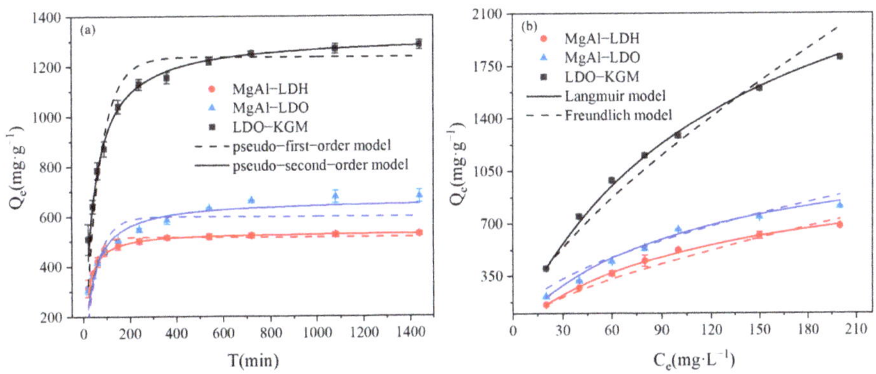

Figure 4. (**a**) Adsorption kinetic of U(VI) on materials (C_0 = 100 mg·L^{-1}, pH = 6.0, T = 293 K); (**b**) adsorption isotherm of U(VI) on materials (C_0 = 20–200 mg·L^{-1}, pH = 6.0, T = 293 K).

In the current study, the Langmuir and Freundlich adsorption isotherm models were utilized to assess the maximum immobilization capacity (Q_m) of the synthesized materials. As shown in Figure 4b, the Q_e of the three materials exhibited a steep rise with an increase in the initial U(VI) concentration until all accessible reactive sites were saturated. The relevant parameters for the model are presented in Table S2. An examination of the correlation coefficient (R^2) indicated that the Langmuir model offered a greater description of the U(VI) immobilization than the Freundlich model, suggesting that U(VI) molecules were homogeneously adsorbed as a monolayer on the reactive sites [43]. The calculated Q_m for LDO-KGM (3019.56 mg·g^{-1}) significantly exceeded those of both MgAl-LDH (1107.74 mg·g^{-1}) and MgAl-LDO (1296.39 mg·g^{-1}). Furthermore, in comparison to the adsorption performance of various studied materials, LDO-KGM is a highly effective adsorbent for U(VI) with good application prospects (Table 2) [30,36,44–48].

Table 2. Comparison of the maximum adsorption amounts of the LDO-KGM and those of other adsorbents towards U(VI).

Adsorbent	Experimental Conditions	Q_m (mg·g^{-1})	Equilibrium Time (h)	References
CMK-3@MgAl LDH	pH = 5.0, T = 298 K	171.00	0.5	[44]
CaMgAl-LDH/nZVI	pH = 5.0, T = 298 K	216.10	4	[45]
rGO/LDH	pH = 4.0, T = 298 K	277.80	2	[46]
Fe-NCNF-LDH	pH = 6.0, T = 298 K	532.25	4	[47]
Al_2O_3 aerogel	pH = 7.0, T = 298 K	746.30	24	[30]
MgO-N aerogel	pH = 4.0, T = 298 K	1061.80	1.67	[48]
MgAl-LDO-500	pH = 6.0, T = 303 K	1098.90	0.25	[36]
LDO-KGM	pH = 6.0, T = 293 K	3019.56	24	This work

3.3.2. Effect of Coexisting Ions and pH

Considering the omnipresent nature of coexisting ions in natural water systems, the impact of diverse coexisting cations and anions on U(VI) adsorption by the prepared samples was explored (Figure 5a,b). As illustrated in Figure 5a, K^+, Na^+, and Mg^{2+} exhibited little influence on the adsorptive capacity of the MgAl-LDH, MgAl-LDO, and LDO-KGM. However, the adsorptive capacity of MgAl-LDH, MgAl-LDO, and LDO-KGM for U(VI) decreased by more than 40% in the presence of Ca^{2+}. The decrease was likely because Ca^{2+} with the high valence was more prone to occupy the adsorption sites on the surfaces of materials [6], and the formation of $CaUO_2(CO_3)_3{}^{2-}$ complexes impeded the movement of positively charged uranyl ions ($UO_2{}^{2+}$) to the adsorbent surface [49].

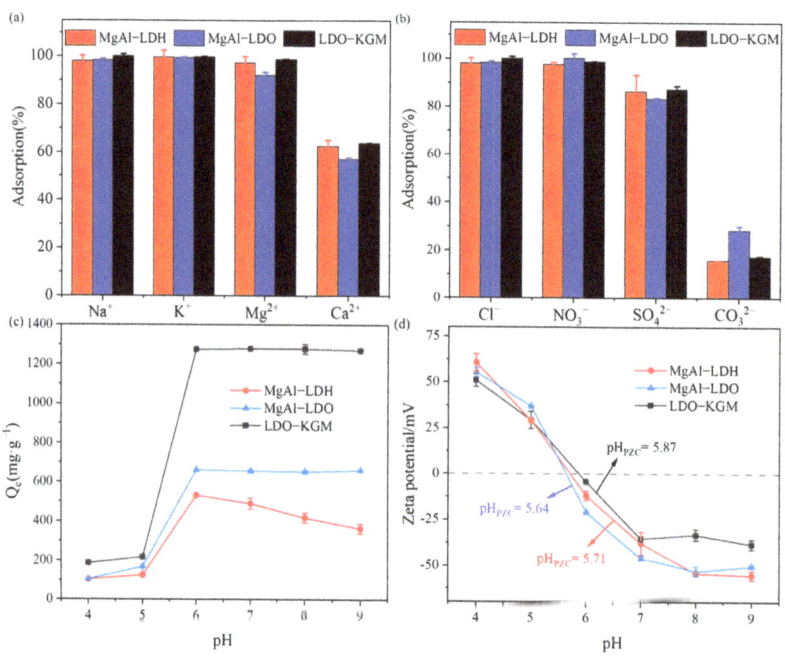

Figure 5. Effect of coexisting cation (**a**) and anion (**b**) on U(VI) immobilization by materials (C_0 = 100 mg·L^{-1} for (**a**,**b**), pH = 6.0, T = 293 K, ionic strength = 0.01 mol·L^{-1}); (**c**) effect of pH on U(VI) immobilization (C_0 = 100 mg·L^{-1}, pH = 4–9, T = 293 K); (**d**) zeta potential of materials before reaction with U(VI).

As shown in Figure 5b, the inhibitory influence of coexisting anions on adsorption followed the order: CO_3^{2-} > SO_4^{2-} >NO_3^- ≈ Cl^-. In the presence of Cl^- and NO_3^-, the U(VI) immobilization capacity decreased negligibly. Nevertheless, the adsorptive capacity of MgAl-LDH, MgAl-LDO, and LDO-KGM for U(VI) was markedly reduced in the presence of CO_3^{2-}, which can likely be ascribed to complex formation (such as UO_2CO_3 and $UO_2(CO_3)_2^{2-}$) between UO_2^{2+} and CO_3^{2-} [50,51]. Similarly, $UO_2(SO_4)_2^{2-}$ complexes were formed between SO_4^{2-} and UO_2^{2+}, which resulted in a reduction in U(VI) adsorption capacities by MgAl-LDH, MgAl-LDO, and LDO-KGM [6]. Therefore, implementing an appropriate desalination step prior to treatment can enhance the removal of U(VI) from saline wastewater containing this ion.

The pH of the solution is recognized as a pivotal parameter that significantly impacts the adsorption of U(VI) onto materials, as it significantly impacts the chemical form of U(VI), the charge on the surface, and the reactivity of the adsorbent's active sites in aqueous media. As depicted in Figure 5c, the adsorptive capacity of MgAl-LDH, MgAl-LDO, and LDO-KGM for U(VI) was considerably enhanced within the range of 4.0–6.0. For MgAl-LDH, a significant decline in the rate of U(VI) removal was detected at a pH greater than 6.0. Therefore, MgAl-LDO and LDO-KGM demonstrated applicability across a broad spectrum of pH values.

3.3.3. Effect of Temperatures

As demonstrated in Figure 6, temperature fluctuations significantly affected the adsorption performance of the adsorbent towards U(VI). The quantity of U(VI) adsorbed escalated with rising temperatures. To assess the thermodynamic nature of U(VI) adsorption, the Gibbs free energy change (Equation (3)) and Van't Hoff equation (Equation (4)), which included entropy change (ΔS), enthalpy change (ΔH), and Gibbs energy (ΔG), were introduced. The thermodynamic fitting outcomes were presented in Figure 6 and Table S3 [52,53]:

$$\ln K_L = \frac{-\Delta H^0}{RT} + \frac{-\Delta S^0}{R} \tag{3}$$

$$\Delta G^0 = \Delta H^0 - T\Delta S^0, \tag{4}$$

where R represents the universal gas constant (8.314 J·mol^{-1}·K^{-1}), T represents the absolute temperature (K), and K_L represents the thermodynamic equilibrium coefficient (L·g^{-1}), which is calculated from a plot of lnKd, and the intercept is lnK_L.

As shown in Table S3, the computed values for both ΔH and ΔG were found to be positive, and the decrease in ΔG values with rising temperature, indicating that the adsorption of U(VI) onto MgAl-LDH, MgAl-LDO, and LDO-KGM were endothermic, with increasing entropy and spontaneous process. Furthermore, the sorption of U(VI) was found to be favored at higher temperatures [54]. The ΔS demonstrated a significantly positive value, indicating that the adsorption of U(VI) by MgAl-LDH, MgAl-LDO, and LDO-KGM was characterized by an irreversible and spontaneous reaction process [55].

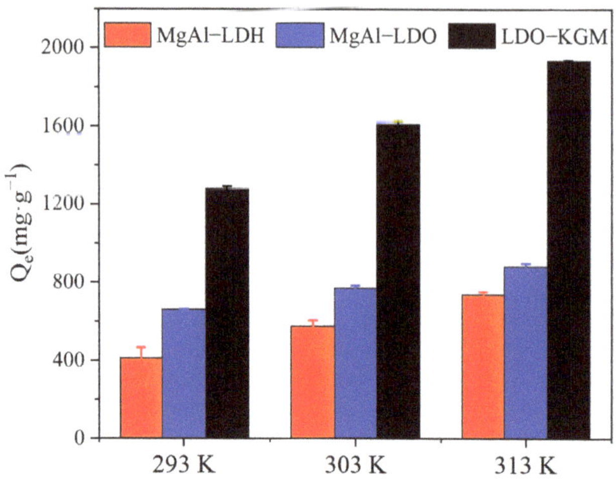

Figure 6. Effect of temperatures on U(VI) immobilization (C_0 = 100 mg·L^{-1}, pH = 6, T = 293–313 K).

3.4. Adsorption Mechanism

Figure 7a–f presented SEM images along with the corresponding elemental distribution maps of MgAl-LDH, MgAl-LDO, and LDO-KGM following the adsorption process, demonstrating a homogeneous distribution of U(VI) on the surfaces of materials. The interaction between MgAl-LDH, MgAl-LDO, and LDO-KGM with U(VI) resulted in the generation of numerous small surface fragments, implying that U(VI) was strongly complexed and immobilized on the surfaces [36]. Furthermore, the uniform distribution of U(VI) on the adsorbent surfaces was closely correlated with the presence of magnesium (Mg) and aluminum (Al), indicating an immobilization process that was not only homogeneous but also intimately linked to the oxygen-containing functional groups present on the materials (Figure 7d–f) [56].

To confirm the mechanism, the influence of adsorbent structural features for U(VI) immobilization was explored by XRD (Figure 7g). The XRD patterns of MgAl-LDH, MgAl-LDO, and LDO-KGM following U(VI) adsorption exhibited a little change, indicating that the crystal structures of these materials were largely preserved without significant destruction [57]. However, the intensities of some MgAl-LDH peaks were reduced gradually, indicating that the adsorption of U(VI) affects the original layered structure of MgAl-LDH. Following the adsorption of U(VI), the XRD pattern of MgAl-LDH, MgAl-LDO, and LDO-KGM exhibited similar additional diffraction peaks within the 2θ range of 15° to 48° (Figure 7g). It was observed that the clear and sharp new peaks appeared at 15.26° and 26.45° in MgAl-LDH, MgAl-LDO, and LDO-KGM. This indicated that U(VI) formed complexes with the hydroxyl groups present on the metal layer of MgAl-LDH, MgAl-LDO, and LDO-KGM. After the immobilization of U(VI), a noticeable weakening of the (003) plane was observed, implying that U(VI) had bound to the carbonate present on the material surfaces [36]. Moreover, the observed diffraction peaks at 24.68°, 25.88°, and 44.39° may be contributed to the uranyl carbonate hydrate species (PDF 34-0578), indicating that the uranyl carbonate hydrate complexes were entrapped in the reconstructed structure [36]. In summary, the sorption mechanism of U(VI) onto the materials may be surface complexation and precipitation with carbonate.

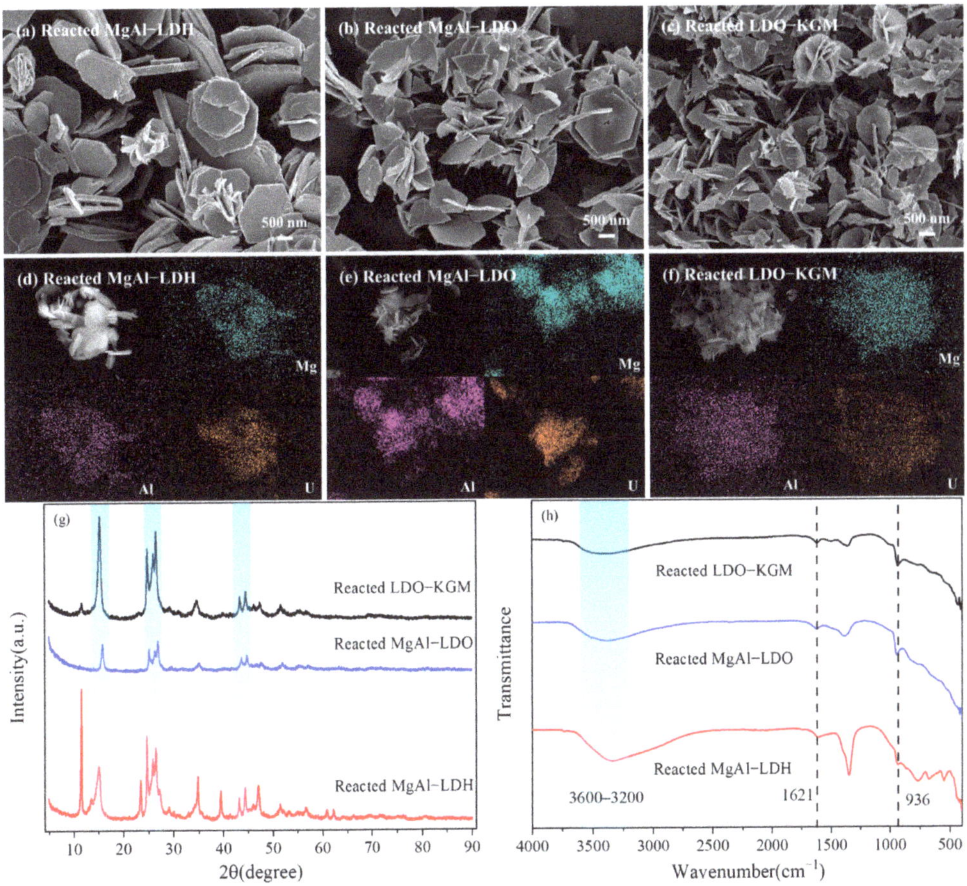

Figure 7. SEM and the corresponding EDS mapping of MgAl-LDH (**a**,**d**), MgAl-LDO (**b**,**e**), and LDO-KGM (**c**,**f**) after reacting with U(VI), Mg (cyan zone), Al (rose zone), U(VI) (orange zone); (**g**) XRD patterns and (**h**) FTIR spectra of MgAl-LDH, MgAl-LDO, and LDO-KGM after reacting with U(VI).

The FTIR spectrum provided functional groups present in the three materials following their interaction with U(VI) (Figure 7h). The prominent and broad adsorption band observed in the range of 3600–3200 cm^{-1} on MgAl-LDH, MgAl-LDO, and LDO-KGM was attributed to the -OH stretching vibrations of the physically adsorbed water, while the distinctive peak at 1621 cm^{-1} was attributed to the bending vibration of the interlayer water molecules [58]. Subsequent to U(VI) adsorption, the peaks corresponding to -OH in MgAl-LDH, MgAl-LDO, and LDO-KGM were red-shifted, which was a stretching vibration of -OH, implying that the H of -OH on the materials was more reactive during the reaction and could be easily replaced by UO_2^{2+}. A novel peak emerged at 936 cm^{-1} on reacted materials, which can be associated with the asymmetric stretching vibration of UO_2^{2+}, further indicating that U(VI) has been effectively incorporated into the structures of MgAl-LDH, MgAl-LDO, and LDO-KGM through the adsorption process [59]. The results were in agreement with the XRD analysis.

The XPS was utilized to analyze the surface chemistry of the materials, both prior to and following the reaction, as depicted in Figure 8. Analysis revealed that the intense peaks of U 4f were introduced by the adsorption process and could be resolved into the following two distinct peaks: U $4f_{7/2}$ at a binding energy of 381.74 eV and U $4f_{5/2}$ at 392.64 eV. This result indicated that no redox reaction between U(VI) and other materials occurs [57]. Instead, U(VI) formed chemical bonds with the reactive surface sites of the material [60]. In comparison with MgAl-LDH and MgAl-LDO, the reacted LDO-KGM displayed enhanced peak intensity at the corresponding binding energies, indicating a more robust binding affinity between U(VI) and LDO-KGM. This stronger interaction aligns with LDO-KGM's superior capacity for immobilizing U(VI).

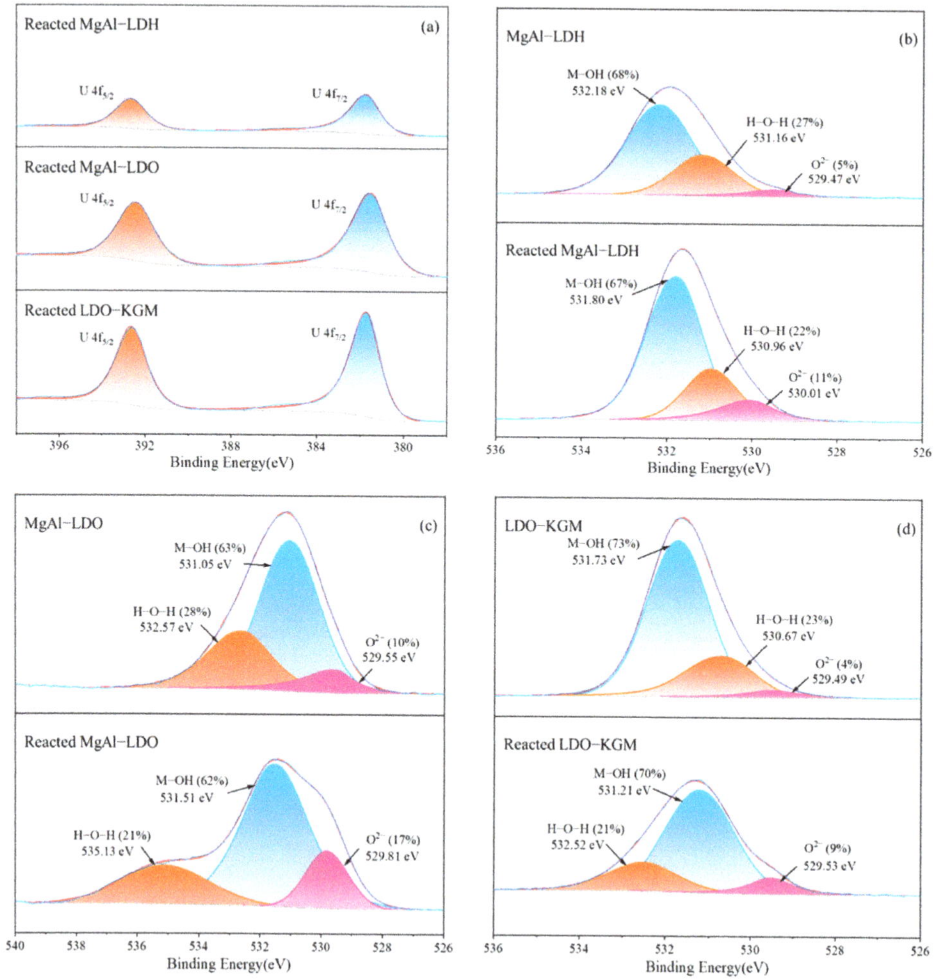

Figure 8. High-resolution XPS spectrum of (**a**) U 4f; (**b–d**) O 1s before and after U(VI) adsorption.

Figure 8b–d presented the O 1s spectra for three materials, both prior to and following the U(VI) immobilization. As illustrated in Figure 8d, LDO-KGM had three O 1s overlapping peaks at 529.49, 530.67, and 531.73 eV, which were ascribed to the vibration of O^{2-}, H-O-H, and M-OH (where M represented Mg or Al), respectively [61,62]. During the reaction, M-O on the LDO-KGM surface was hydrolyzed to yield M-OH, which subsequently participated in surface complexation with U(VI), leading to the formation of M-O-U bonds. Therefore, after the reaction, the proportion of M-OH in the LDO-KGM decreased from 73% to 70%, and the relative abundance of O^{2-} escalated from 4% to 9%.

Following the adsorption of U(VI), alterations in the relative concentrations and positions of these O-donor functional groups were observed, with the binding energy of M-OH bonds shifting to lower values and that of H-O-H and O^{2-} shifting to higher values. The reaction processes of MgAl-LDH and MgAl-LDO were similar to those of LDO-KGM. The findings substantiated that the oxygen-containing functional groups on the materials' surface played a crucial role as enrichment and active sites for coordination with U(VI) during adsorption.

The presence of coexisting cations exerted a significant competitive influence on the adsorption of U(VI) onto the materials, in which the stretching vibration of -OH, and the formation of uranyl carbonate hydrate complexes suggested that the uptake of U(VI) by materials was predominantly influenced by surface complexation and carbonate co-precipitation. To obtain a more comprehensive understanding of the fundamental mechanisms underlying the adsorption of U(VI) by three materials, the point of zero charge (pH_{pzc}) of the materials was investigated (Figure 5c,d).

At pH values below 6.0, the removal efficiency for U(VI) was relatively low, as both materials and UO_2^{2+} were positively charged below pH_{pzc} (Figure 5d). This led to significant electrostatic repulsion between UO_2^{2+} and adsorbents, which inhibit the U(VI) removal behavior of the materials. Furthermore, the solution contained a high concentration of H^+ ($pH < 6.0$), which were more likely to be adsorbed onto the adsorbent surfaces compared to UO_2^{2+} [63]. This preferential adsorption of H^+ contributed to the reduced efficiency in removing U(VI) from the solution. However, the adsorbent still had some adsorption capacity, indicating the occurrence of surface complexes between U(VI) species and the oxygen-containing functional groups present on the adsorbent [64]. At a pH value of approximately 6.0, the removal efficiency for U(VI) was at its peak, attributed to the strong electrostatic attraction between negatively charged materials and positively charged UO_2^{2+}. When the pH surpassed 6.0, the three materials exhibited disparate trends. In alkaline environments, the solution's OH^- concentration increased, and U(VI) could form stable U(VI)-containing complexes (Figure S3), such as $(UO_2)_3(OH)_7^-$ [65], which in turn, underwent electrostatic repulsion with negatively charged materials or complexed with OH^- on the surface of materials. For MgAl-LDH, the adsorption capacity decreased, because the complexation on the surface was weaker than electrostatic repulsion. In contrast, MgAl-LDO and LDO-KGM demonstrated exceptional efficiency in removing U(VI), attributed to their robust surface complexation capabilities.

3.5. *Reusability and Regeneration*

The reusability of adsorbents is an important factor for their practical application. Simultaneously, recycling adsorbents can save cost and reduce the waste of resources; therefore, it was also necessary to study the recycling performance of the adsorbents. On this basis, the reusability of the LDO-KGM was studied in five cycles of regeneration (Figure 9). During this process, the LDO-KGM were filtered out in the process of collecting the supernatants for the recycling experiments. Then, the collected LDO-KGM were eluted with hydrochloric acid of pH 3 and stirred for 24 h to desorb. After desorption, the LDO-KGM, which were restored to the state before adsorption, were repeatedly washed with DI water and then dried for the next adsorption experiment. After five cycles, the removal rates of LDO-KGM were 84.5%, which showed that LDO-KGM had good regeneration characteristics for the removal of U(VI).

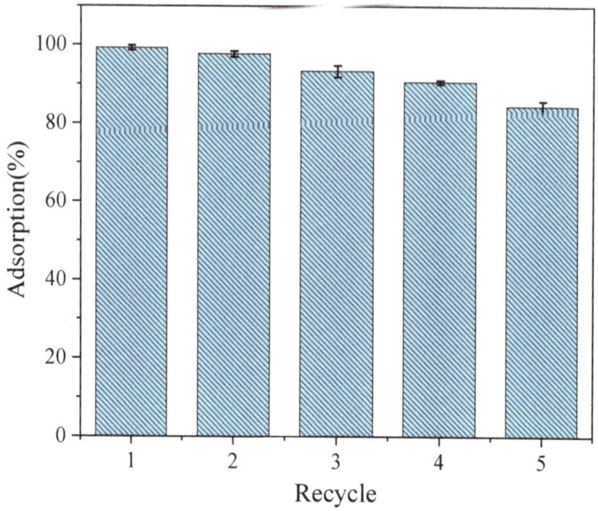

Figure 9. Recycling of LDO-KGM in the adsorption of U(VI) (C_0 = 100 mg·L^{-1}, pH = 6.0, T = 293 K).

4. Conclusions

LDO-KGM, synthesized by the FDC method using KGM as a template, has demonstrated its effectiveness as an adsorbent for removing U(VI) from wastewater. The high specific surface area and plentiful active adsorption sites of LDO-KGM stem from its porous architecture and oxygen-containing functional groups. The U(VI) adsorption behavior followed the pseudo-second-order kinetic model ($R^2 \geq 0.99$) and the Langmuir isotherm model ($R^2 \geq 0.99$), which indicated that the primary adsorption mechanism at the molecular level was chemisorption. The Langmuir isotherm model determined a maximum adsorption capacity of 3019.56 mg·g^{-1}, outperforming other adsorbents derived from LDHs or KGM, as reported in the literature. Moreover, LDO-KGM retained over 84.5% U(VI) adsorption efficiency after five consecutive cycles, underscoring its excellent recyclability. Complexation, carbonate co-precipitation, and electrostatic adsorption were the predominant mechanisms for U(VI) immobilization. Consequently, our findings not only elucidate the adsorption mechanism of U(VI) by LDO-KGM but also advance the design of high-performance LDO-based materials for environmental applications. Importantly, the synthesis method, combined with the material's high capacity and recyclability, suggests strong potential for cost-effective, large-scale implementation in water purification and related fields, such as heavy-metal remediation. This work bridges fundamental research with practical engineering solutions, offering a sustainable pathway for next-generation adsorbent development.

5. Insufficient and Prospect

The current investigation was predominantly confined to idealized laboratory conditions involving single-component U(VI) solutions, while critical challenges inherent to authentic radioactive effluents—particularly synergistic interference from dissolved organic colloids, hypersaline matrices, and coexisting actinides—remained insufficiently characterized. This methodological simplification risks overestimating the material's ability in nuclear wastewater treatment scenarios. To bridge this technology readiness gap, the following two strategic directions are proposed: (1) the systematic evaluation of chemical stability under authentic multi-pollutant systems containing competitive radionuclides (such as Th^{4+}, Pu^{4+}); and (2) optimization through different templating agents (such as cellulose, chitosan) or doping strategies (the introduction of magnetic components, multi-metal synergism) to enhance the adsorption selectivity, cyclic stability, and functional versatility of the materials.

Supplementary Materials: The following supporting information can be downloaded at: https://www.mdpi.com/article/10.3390/pr13030876/s1, Figure S1: (a) N_2 adsorption–desorption isotherms and (b) the average pore size distributions of materials; Figure S2: Effect of materials dosages on U(VI) immobilization (C_0 = 100 mg·L^{-1}, pH = 6.0, T = 293 K); Figure S3: Species distribution of U(VI) as a function of pH (C_0 = 100 mg·L^{-1}); Figure S4: XPS spectra of materials before (a) and after (b) reacted with U(VI); Table S1: The kinetic model parameters for the immobilization of U(VI) on materials at pH 6.0; Table S2: Langmuir and Freundlich isotherm constants for U(VI) immobilization on adsorbents; Table S3: Thermodynamic parameters of adsorbents.

Author Contributions: Conceptualization, N.Z. and Z.D.; methodology, P.G., J.W., Z.S., M.C. and B.L.; software, P.G.; validation, P.G.; formal analysis, P.G.; investigation, P.G.; resources, P.G.; data curation, P.G.; writing—original draft preparation, P.G.; writing—review and editing, J.W., Z.S., M.C., B.L., T.W. and L.S.; visualization, P.G.; supervision, P.W.; project administration, P.W., Z.D. and N.Z.; funding acquisition, P.W. All authors have read and agreed to the published version of the manuscript.

Funding: This research was funded by National Natural Science Foundation of China, grant numbers 42207254 and 42172042.

Data Availability Statement: The original contributions presented in this study are included in the article/Supplementary Materials. Further inquiries can be directed to the corresponding author.

Conflicts of Interest: The authors declare no conflicts of interest.

References

1. Zhang, H.; Li, C.; Chen, X.; Fu, H.; Chen, Y.; Ning, S.; Fujita, T.; Wei, Y.; Wang, X. Layered ammonium vanadate nanobelt as efficient adsorbents for removal of Sr^{2+} and Cs^{+} from contaminated water. *J. Colloid Interface Sci.* **2022**, *615*, 110–123. [CrossRef] [PubMed]
2. Akash, S.; Sivaprakash, B.; Raja, V.C.V.; Rajamohan, N.; Muthusamy, G. Remediation techniques for uranium removal from polluted environment—Review on methods, mechanism and toxicology. *Environ. Pollut.* **2022**, *302*, 119068. [CrossRef] [PubMed]
3. Liu, R.; Zhang, W.; Chen, Y.; Wang, Y. Uranium (VI) adsorption by copper and copper/iron bimetallic central mofs. *Colloids Surf. A Physicochem. Eng. Asp.* **2020**, *587*, 124334. [CrossRef]
4. Yuan, D.; Chen, L.; Xiong, X.; Yuan, L.; Liao, S.; Wang, Y. Removal of uranium (VI) from aqueous solution by amidoxime functionalized superparamagnetic polymer microspheres prepared by a controlled radical polymerization in the presence of dpe. *Chem. Eng. J.* **2016**, *285*, 358–367. [CrossRef]
5. Zheng, N.; Yin, L.; Su, M.; Liu, Z.; Tsang, D.C.W.; Chen, D. Synthesis of shape and structure-dependent hydroxyapatite nanostructures as a superior adsorbent for removal of U(VI). *Chem. Eng. J.* **2020**, *384*, 123262. [CrossRef]
6. Ma, M.; Ye, Z.; Zhang, J.; Wang, Y.; Ning, S.; Yin, X.; Fujita, T.; Chen, Y.; Wu, H.; Wang, X. Synthesis and fabrication of segregative and durable MnO_2@chitosan composite aerogel beads for uranium(VI) removal from wastewater. *Water Res.* **2023**, *247*, 120819. [CrossRef]
7. Mellah, A.; Chegrouche, S.; Barkat, M. The precipitation of ammonium uranyl carbonate (AUC): Thermodynamic and kinetic investigations. *Hydrometallurgy* **2007**, *85*, 163–171. [CrossRef]
8. Kulkarni, P.S.; Mukhopadhyay, S.; Bellary, M.P.; Ghosh, S.K. Studies on membrane stability and recovery of uranium (VI) from aqueous solutions using a liquid emulsion membrane process. *Hydrometallurgy* **2002**, *64*, 49–58. [CrossRef]
9. Singh, H.; Mishra, S.L.; Vijayalakshmi, R. Uranium recovery from phosphoric acid by solvent extraction using a synergistic mixture of di-nonyl phenyl phosphoric acid and tri-n-butyl phosphate. *Hydrometallurgy* **2004**, *73*, 63–70. [CrossRef]
10. Krawczyk-Bärsch, E.; Gerber, U.; Müller, K.; Moll, H.; Rossberg, A.; Steudtner, R.; Merroun, M.L. Multidisciplinary characterization of U(VI) sequestration by acidovorax facilis for bioremediation purposes. *J. Hazard. Mater.* **2018**, *347*, 233–241. [CrossRef]
11. Li, F.; Li, D.; Li, X.; Liao, J.; Li, S.; Yang, J.; Yang, Y.; Tang, J.; Liu, N. Microorganism-derived carbon microspheres for uranium removal from aqueous solution. *Chem. Eng. J.* **2016**, *284*, 630–639. [CrossRef]

12. Li, Y.; Wang, L.; Li, B.; Zhang, M.; Wen, R.; Guo, X.; Li, X.; Zhang, J.; Li, S.; Ma, L. Pore-free matrix with cooperative chelating of hyperbranched ligands for high-performance separation of uranium. *ACS Appl. Mater. Interfaces* **2016**, *8*, 28853–28861. [CrossRef]
13. Zhang, Z.; Dong, Z.; Wang, X.; Dai, Y.; Cao, X.; Wang, Y.; Hua, R.; Feng, H.; Chen, J.; Liu, Y.; et al. Synthesis of ultralight phosphorylated carbon aerogel for efficient removal of U(VI): Batch and fixed-bed column studies. *Chem. Eng. J.* **2019**, *370*, 1376–1387. [CrossRef]
14. Zhang, H.; Ruan, Y.; Liang, A.; Shih, K.; Diao, Z.; Su, M.; Hou, L.; Chen, D.; Lu, H.; Kong, L. Carbothermal reduction for preparing nZVI/BC to extract uranium: Insight into the iron species dependent uranium adsorption behavior. *J. Clean. Prod.* **2019**, *239*, 117873. [CrossRef]
15. Ahmed, W.; Mehmood, S.; Núñez-Delgado, A.; Ali, S.; Qaswar, M.; Khan, Z.H.; Ying, H.; Chen, D. Utilization of citrullus lanatus l. Seeds to synthesize a novel $MnFe_2O_4$-biochar adsorbent for the removal of U(VI) from wastewater: Insights and comparison between modified and raw biochar. *Sci. Total Environ.* **2021**, *771*, 144955. [CrossRef] [PubMed]
16. Xiong, T.; Li, Q.; Liao, J.; Zhang, Y.; Zhu, W. Highly enhanced adsorption performance to uranium(VI) by facile synthesized hydroxyapatite aerogel. *J. Hazard. Mater.* **2022**, *423*, 127184. [CrossRef]
17. Bachmaf, S.; Merkel, B.J. Sorption of uranium(VI) at the clay mineral–water interface. *Environ. Earth Sci.* **2011**, *63*, 925–934. [CrossRef]
18. Liu, S.; Luo, M.; Li, J.; Luo, F.; Ke, L.; Ma, J. Adsorption equilibrium and kinetics of uranium onto porous azo-metal–organic frameworks. *J. Radioanal. Nucl. Chem.* **2016**, *310*, 353–362. [CrossRef]
19. Yang, P.; Li, S.; Liu, C.; Liu, X. Interface-constrained layered double hydroxides for stable uranium capture in highly acidic industrial wastewater. *ACS Appl. Mater. Interfaces* **2021**, *13*, 17988–17997. [CrossRef]
20. Chong, M.N.; Jin, B.; Chow, C.W.K.; Saint, C. Recent developments in photocatalytic water treatment technology: A review. *Water Res.* **2010**, *44*, 2997–3027. [CrossRef]
21. Xie, L.; Zhong, Y.; Xiang, R.; Fu, G.; Xu, Y.; Cheng, Y.; Liu, Z.; Wen, T.; Zhao, Y.; Liu, X. Sono-assisted preparation of Fe(II)-Al(III) layered double hydroxides and their application for removing uranium (VI). *Chem. Eng. J.* **2017**, *328*, 574–584. [CrossRef]
22. Zou, Y.; Liu, Y.; Wang, X.; Sheng, G.; Wang, S.; Ai, Y.; Ji, Y.; Liu, Y.; Hayat, T.; Wang, X. Glycerol-modified binary layered double hydroxide nanocomposites for uranium immobilization via extended x-ray absorption fine structure technique and density functional theory calculation. *ACS Sustain. Chem. Eng.* **2017**, *5*, 3583–3595. [CrossRef]
23. Yao, W.; Yu, S.; Wang, J.; Zou, Y.; Lu, S.; Ai, Y.; Alharbi, N.S.; Alsaedi, A.; Hayat, T.; Wang, X. Enhanced removal of methyl orange on calcined glycerol-modified nanocrystallined Mg/Al layered double hydroxides. *Chem. Eng. J.* **2017**, *307*, 476–486. [CrossRef]
24. Momin, Z.H.; Lingamdinne, L.P.; Kulkarni, R.; Pal, C.A.; Choi, Y.; Chang, Y.; Koduru, J.R. Exploring recyclable alginate-enhanced gcn-ldo sponge for U(VI) and Cd(II) removal: Insights from batch and column studies. *J. Hazard. Mater.* **2024**, *469*, 134015. [CrossRef]
25. Zhu, K.; Chen, C.; Wang, H.; Xie, Y.; Wakeel, M.; Wahid, A.; Zhang, X. Gamma-ferric oxide nanoparticles decoration onto porous layered double oxide belts for efficient removal of uranyl. *J. Colloid Interface Sci.* **2019**, *535*, 265–275. [CrossRef] [PubMed]
26. Yao, W.; Wang, X.; Liang, Y.; Yu, S.; Gu, P.; Sun, Y.; Xu, C.; Chen, J.; Hayat, T.; Alsaedi, A.; et al. Synthesis of novel flower-like layered double oxides/carbon dots nanocomposites for U(VI) and 241Am(III) efficient removal: Batch and exafs studies. *Chem. Eng. J.* **2018**, *332*, 775–786. [CrossRef]
27. Chen, T.; Shi, P.; Zhang, J.; Li, Y.; Duan, T.; Dai, L.; Wang, L.; Yu, X.; Zhu, W. Natural polymer konjac glucomannan mediated assembly of graphene oxide as versatile sponges for water pollution control. *Carbohydr. Polym.* **2018**, *202*, 425–433. [CrossRef]
28. Zhu, J.; Hu, J.; Jiang, C.; Liu, S.; Li, Y. Ultralight, hydrophobic, monolithic konjac glucomannan-silica composite aerogel with thermal insulation and mechanical properties. *Carbohydr. Polym.* **2019**, *207*, 246–255. [CrossRef]
29. Wang, R.; Dong, H.; Li, X.; Zhou, L.; Zhu, W.; Chen, T. Biomass encapsulated zif-8-derived ZnO carbon aerogels for efficient uranium extraction by synergistic adsorption-photoreduction. *Chem. Eng. J.* **2023**, *478*, 147331. [CrossRef]
30. Liao, J.; Zhang, Y.; Zhang, L. Improved catalytic activity on the thermal decomposition of ammonium perchlorate and efficient adsorption of uranium using a novel ultra-low density Al_2O_3-based aerogels. *J. Hazard. Mater.* **2020**, *387*, 122015. [CrossRef]
31. Hu, W.; Sun, T.; Liu, C.; Yu, L.; Ahamad, T.; Ma, Z. Refined microstructure and enhanced mechanical properties in mo-Y_2O_3 alloys prepared by freeze-drying method and subsequent low temperature sintering. *J. Mater. Sci. Technol.* **2021**, *88*, 36–44. [CrossRef]
32. Hu, W.; Kong, X.; Du, Z.; Khan, A.; Ma, Z. Synthesis and characterization of nano tic dispersed strengthening W alloys via freeze-drying. *J. Alloy. Compd.* **2021**, *859*, 157774. [CrossRef]
33. Tang, M.; Chen, J.; Wang, P.; Wang, C.; Ao, Y. Highly efficient adsorption of uranium(VI) from aqueous solution by a novel adsorbent: Titanium phosphate nanotubes. *Environ. Sci. Nano* **2018**, *5*, 2304–2314. [CrossRef]
34. Xiong, J.; Fan, Y.; Luo, F. Grafting functional groups in metal–organic frameworks for U(VI) sorption from aqueous solutions. *Dalton Trans.* **2020**, *49*, 12536–12545. [CrossRef] [PubMed]
35. Sun, Y.; Chen, J.; Kang, Y.; Zhang, H.; Ge, Y.; Yuan, N.; Xing, Y.; Ma, W.; Yang, Z.; Zou, L.; et al. Ultralight carbon aerogel composites derived from MOFs and cross-linked chitosan: Synthesis, characterization, and U(VI) adsorption. *Chem. Eng. J.* **2023**, *455*, 140749. [CrossRef]

36. Chen, M.; Li, S.; Li, L.; Jiang, L.; Ahmed, Z.; Dang, Z.; Wu, P. Memory effect induced the enhancement of uranium (VI) immobilization on low-cost MgAl-double oxide: Mechanism insight and resources recovery. *J. Hazard. Mater.* **2021**, *401*, 123447. [CrossRef]
37. Huang, Z.; Wu, P.; Gong, B.; Fang, Y.; Zhu, N. Fabrication and photocatalytic properties of a visible-light responsive nanohybrid based on self-assembly of carboxyl graphene and ZnAl layered double hydroxides. *J. Mater. Chem. A* **2014**, *2*, 5534–5540. [CrossRef]
38. Chen, J.; Wang, C.; Zhang, Y.; Guo, Z.; Luo, Y.; Mao, C. Engineering ultrafine nis cocatalysts as active sites to boost photocatalytic hydrogen production of MgAl layered double hydroxide. *Appl. Surf. Sci.* **2020**, *506*, 144999. [CrossRef]
39. Lv, X.; Qin, X.; Wang, K.; Peng, Y.; Wang, P.; Jiang, G. Nanoscale zero valent iron supported on MgAl-LDH-decorated reduced graphene oxide: Enhanced performance in Cr(VI) removal, mechanism and regeneration. *J. Hazard. Mater.* **2019**, *373*, 176–186. [CrossRef]
40. Liu, C.; Wu, P.; Zhu, Y.; Tran, L. Simultaneous adsorption of Cd^{2+} and bpa on amphoteric surfactant activated montmorillonite. *Chemosphere* **2016**, *144*, 1026–1032. [CrossRef]
41. Zhu, K.; Chen, C.; Xu, M.; Chen, K.; Tan, X.; Wakeel, M.; Alharbi, N.S. In situ carbothermal reduction synthesis of Fe nanocrystals embedded into n-doped carbon nanospheres for highly efficient U(VI) adsorption and reduction. *Chem. Eng. J.* **2018**, *331*, 395–405. [CrossRef]
42. Wang, G.; Liu, J.; Wang, X.; Xie, Z.; Deng, N. Adsorption of uranium (VI) from aqueous solution onto cross-linked chitosan. *J. Hazard. Mater.* **2009**, *168*, 1053–1058. [CrossRef]
43. Wu, P.; Dai, Y.; Long, H.; Zhu, N.; Li, P.; Wu, J.; Dang, Z. Characterization of organo-montmorillonites and comparison for Sr(II) removal: Equilibrium and kinetic studies. *Chem. Eng. J.* **2012**, *191*, 288–296. [CrossRef]
44. Yin, L.; Hu, Y.; Ma, R.; Wen, T.; Wang, X.; Hu, B.; Yu, Z.; Hayat, T.; Alsaedi, A.; Wang, X. Smart construction of mesoporous carbon templated hierarchical Mg-Al and Ni-Al layered double hydroxides for remarkably enhanced U(VI) management. *Chem. Eng. J.* **2019**, *359*, 1550–1562. [CrossRef]
45. Pang, H.; Wu, Y.; Huang, S.; Ding, C.; Li, S.; Wang, X.; Yu, S.; Chen, Z.; Song, G.; Wang, X. Macroscopic and microscopic investigation of uranium elimination by Ca–Mg–Al-layered double hydroxide supported nanoscale zero valent iron. *Inorg. Chem. Front.* **2018**, *5*, 2657–2665. [CrossRef]
46. Tan, L.; Wang, Y.; Liu, Q.; Wang, J.; Jing, X.; Liu, L.; Liu, J.; Song, D. Enhanced adsorption of uranium (VI) using a three-dimensional layered double hydroxide/graphene hybrid material. *Chem. Eng. J.* **2015**, *259*, 752–760. [CrossRef]
47. Zhu, J.; Liu, Q.; Liu, J.; Chen, R.; Zhang, H.; Li, R.; Wang, J. Ni–Mn LDH-decorated 3d Fe-inserted and n-doped carbon framework composites for efficient uranium(VI) removal. *Environ. Sci. Nano* **2018**, *5*, 467–475. [CrossRef]
48. Xiong, T.; Li, Q.; Li, K.; Zhang, Y.; Zhu, W. Construction of novel magnesium oxide aerogel for highly efficient separation of uranium(VI) from wastewater. *Sep. Purif. Technol.* **2022**, *295*, 121296. [CrossRef]
49. Wang, P.; Yin, L.; Wang, J.; Xu, C.; Liang, Y.; Yao, W.; Wang, X.; Yu, S.; Chen, J.; Sun, Y.; et al. Superior immobilization of U(VI) and 243Am(III) on polyethyleneimine modified lamellar carbon nitride composite from water environment. *Chem. Eng. J.* **2017**, *326*, 863–874. [CrossRef]
50. Zhuang, Z.; Ou, X.; Li, J.; Zhou, Y.; Zhang, Z.; Dong, S.; Lin, Z. Interfacial engineering improved the selective extraction of uranyl from saline water by nano-$Mg(OH)_2$ and the underlying mechanism. *ACS Sustain. Chem. Eng.* **2016**, *4*, 801–809. [CrossRef]
51. Liao, J.; Zhang, Y. Effective removal of uranium from aqueous solution by using novel sustainable porous Al_2O_3 materials derived from different precursors of aluminum. *Inorg. Chem. Front.* **2020**, *7*, 765–776. [CrossRef]
52. Liu, J.; Shi, S.; Li, C.; Hong, X.; Gu, Z.; Li, F.; Zhai, J.; Zhang, Q.; Liao, J.; Liu, N.; et al. U(VI) adsorption by one-step hydrothermally synthesized cetyltrimethylammonium bromide modified hydroxyapatite-bentonite composites from phosphate-carbonate coexisted solution. *Appl. Clay Sci.* **2021**, *203*, 106027. [CrossRef]
53. Liu, X.; Pang, H.; Liu, X.; Li, Q.; Zhang, N.; Mao, L.; Qiu, M.; Hu, B.; Yang, H.; Wang, X. Orderly porous covalent organic frameworks-based materials: Superior adsorbents for pollutants removal from aqueous solutions. *Innovation* **2021**, *2*, 100076. [CrossRef] [PubMed]
54. Huang, S.; Pang, H.; Li, L.; Jiang, S.; Wen, T.; Zhuang, L.; Hu, B.; Wang, X. Unexpected ultrafast and high adsorption of U(VI) and Eu(III) from solution using porous Al_2O_3 microspheres derived from mil-53. *Chem. Eng. J.* **2018**, *353*, 157–166. [CrossRef]
55. Hu, W.; Lu, S.; Song, W.; Chen, T.; Hayat, T.; Alsaedi, N.S.; Chen, C.; Liu, H. Competitive adsorption of U(VI) and Co(II) on montmorillonite: A batch and spectroscopic approach. *Appl. Clay Sci.* **2018**, *157*, 121–129. [CrossRef]
56. Gray, M.L.; Soong, Y.; Champagne, K.J.; Pennline, H.; Baltrus, J.P.; Stevens, R.W.; Khatri, R.; Chuang, S.S.C.; Filburn, T. Improved immobilized carbon dioxide capture sorbents. *Fuel Process. Technol.* **2005**, *86*, 1449–1455. [CrossRef]
57. Tu, J.; Peng, X.; Wang, S.; Tian, C.; Deng, H.; Dang, Z.; Lu, G.; Shi, Z.; Lin, Z. Effective capture of aqueous uranium from saline lake with magnesium-based binary and ternary layered double hydroxides. *Sci. Total Environ.* **2019**, *677*, 556–563. [CrossRef]
58. Chernobay, G.B.; Chesalov, Y.A.; Baltakhinov, V.P.; Popova, G.Y.; Andrushkevich, T.V. In situ ftir study of β-picoline transformations on V–Ti–O catalysts. *Catal. Today* **2011**, *164*, 58–61. [CrossRef]

59. Lefèvre, G.; Noinville, S.; Fédoroff, M. Study of uranyl sorption onto hematite by in situ attenuated total reflection–infrared spectroscopy. *J. Colloid Interface Sci.* **2006**, *296*, 608–613. [CrossRef]
60. Lei, Y.; Li, K.; Liao, J.; Zhang, Y.; Zhang, L.; Zhu, W. Design of 3d alumina-doped magnesium oxide aerogels with a high efficiency removal of uranium(VI) from wastewater. *Inorg. Chem. Front.* **2021**, *8*, 2561–2574. [CrossRef]
61. Mei, H.; Tan, X.; Tan, L.; Meng, Y.; Chen, C.; Fang, M.; Wang, X. Retention of U(VI) by the formation of Fe precipitates from oxidation of Fe(II). *ACS Earth Space Chem.* **2018**, *2*, 968–976. [CrossRef]
62. Wu, C.; Tu, J.; Tian, C.; Geng, J.; Lin, Z.; Dang, Z. Defective magnesium ferrite nano-platelets for the adsorption of As(V): The role of surface hydroxyl groups. *Environ. Pollut.* **2018**, *235*, 11–19. [CrossRef] [PubMed]
63. Sun, Y.; Li, Y. Application of surface complexation modeling on adsorption of uranium at water-solid interface: A review. *Environ. Pollut.* **2021**, *278*, 116861. [CrossRef] [PubMed]
64. Zhang, R.; Chen, C.; Li, J.; Wang, X. Preparation of montmorillonite@carbon composite and its application for U(VI) removal from aqueous solution. *Appl. Surf. Sci.* **2015**, *349*, 129–137. [CrossRef]
65. Foroutan, R.; Peighambardoust, S.J.; Hosseini, S.S.; Akbari, A.; Ramavandi, B. Hydroxyapatite biomaterial production from chicken (femur and beak) and fishbone waste through a chemical less method for Cd^{2+} removal from shipbuilding wastewater. *J. Hazard. Mater.* **2021**, *413*, 125428. [CrossRef]

processes

Article

Preparation and Exploration of Spherical Fe-C Micro-Electrolysis Materials for the Removal of Crystal Violet

Min Dai [1], Xixi Wei [2], Chihpeng Lin [1,*], Chunsheng Xie [1], Zimin Lai [1], Wencan Zhu [1], Junhao Guo [1] and Changsheng Peng [3,*]

[1] Guangdong Provincial Key Laboratory of Eco-Environmental Studies and Low-Carbon Agriculture in Peri-Urban Areas, Zhaoqing University, Zhaoqing 526061, China
[2] Zaozhuang Standard Measurement Research Center, Zaozhuang 277800, China
[3] School of Chemical and Environmental Engineering, Anhui Polytechnic University, Wuhu 241000, China
* Correspondence: ftirlin@foxmail.com (C.L.); pcs005@sohu.com (C.P.)

Abstract: In this paper, the spherical Fe-C micro-electrolysis materials (Fe-C MEM) were prepared using iron powder, activated carbon powder, corn straw, and bentonite as the raw materials. The preparation conditions optimized by single factor test showed Fe-C MEM had a high crystal violet removal and strength under 1:1 of Fe/C ratio, 2% corn straw content, 25% bentonite content, and 900 °C sintering temperature. The porous Fe-C MEM had a high specific surface area of 108.069 m^2/g with an even distribution of zero-valent iron and carbon. The maximum removal capacity of CV by Fe-C MEM was 105.48 mg/g at 25 °C. The CV removal was a spontaneous endothermic process. The mechanism of CV removal by Fe-C MEM was adsorption combined with degradation. Fe-C MEM has a good performance in dye wastewater treatment.

Keywords: Fe-C micro-electrolytic material; optimization; crystal violet; removal mechanism

1. Introduction

Dyes are widely used in textile, leather, plastic, and other fields because of their bright colors and strong adhesion properties. However, most of the dyes are toxic, non-biodegradable, and harmful to the ecosystem and human beings [1]. Crystal violet (CV), a commercial dye, bears a triphenyl methane group in its structure and can persist in the natural environment [2]. Considering the recalcitrance and carcinogenesis, it is urgent to develop an efficient and low-cost method for the removal of CV.

Many technologies, such as precipitation, coagulation, adsorption, catalytic degradation, electrolytic oxidation, etc., can be used in dye wastewater treatment [1]. The technologies are often limited by incomplete removal, secondary pollution, or high operating costs. Adsorption is an effective method of dye removal. However, the high cost of the efficient adsorbents limits the commercial application. In recent years, the Fe-C micro-electrolysis materials (Fe-C MEM) have shown potential application in the field because of their high effectiveness and low cost.

The Fe-C MEM contains carbon and iron, which can combine adsorption (carbon) and ozonation (zero-valent iron (ZVI)) in the process. In addition, the Fe-C MEM uses iron as the cathode, carbon as the anode, and wastewater as the electrolyte to form the galvanic cell reaction. The formation of galvanic cells between Fe and C, a kind of micro-electrolysis reaction, could promote the degradation and removal of the dye wastewater [3,4]. Accordingly, iron filings using the internal carbon to form galvanic cells could be used directly. However, the lack of carbon and the formation of iron compounds covering the surface make the treatment of low efficiency [5]. Then, fly ash, carbon, slag, sponge iron, etc., are mixed [6–8]. The addition of carbon can improve the dispersion of iron and make the material more effective. Then, granular Fe-C MEM are produced for easy separation in the treatment [8]. Zhang et al. used sintering ferric-carbon ceramic coupled with a modified biological aerated filter in wastewater treatment. The ferric-carbon ceramic enhanced phosphorus and ciprofloxacin removal in the process [9]. To further improve the efficiency, porous Fe-C MEM are produced by adding pore former. Ammonium salt works by generating ammonia during sintering. However, the emission of irritant gas and the cost of salts are not conducive to large-scale production and utilization. The addition of biomass also works because the pyrolysis of biomass generates H_2O, CO_2, CO, H_2, etc., and the reducing gases contribute to the reduction of ferric oxides to ZVI [10]. Some researchers focus on the formation of ZVI in the process because ZVI is more reactive than ferric oxides. The addition of red mud, iron ore tailings, ferric iron, and reduction to ZVI is studied [4,9]. In summary, the granular porous Fe-C MEM are promising in dye wastewater treatment. Most research focuses on the reduction of iron and high efficiency of degradation (Fenton treatment, ozone oxidation) in wastewater treatment [3,9,11].

The properties, easy production, low cost, and high efficiency are the keys to Fe-C MEM in commercial applications. Therefore, iron powder, activated carbon powder, corn straw, and bentonite were chosen to explore the production of porous Fe-C MEM by biomass pyrolysis. The effects of Fe-C ratio, sintering temperature, and the addition of corn straw and bentonite were optimized for high efficiency in CV removal. Then, a detailed study of CV removal by Fe-C MEM was conducted, and the mechanism of CV removal was explored.

2. Materials and Methods

2.1. Materials

Bentonite was obtained from Mingxi mineral product plant (Chuzhou, China). Corn straw was obtained from the farmland in the suburb of Qingdao, Shandong Province. CV was obtained from Dengke Chemical Reagent Co., Ltd. (Tianjin, China). Iron powder and activated carbon powder were obtained from Sinopharm Chemical Reagent Co., Ltd. (Shanghai, China). All chemicals were of analytical grade. Deionized water was prepared and used in this work.

2.2. Preparation of Fe-C MEM

Iron powder, activated carbon powder, corn straw, and bentonite were used to prepare the Fe-C MEM. The iron powder was passed through a 200-mesh sieve. The corn straw was ground into a size of 100 mesh. The raw materials were added to the blender and mixed at different ratios with deionized water. The mixture was pressed into spherical particles with a diameter of 8 mm. Then the particles were dried at 105 °C in a vacuum drying oven and sintered in an atmosphere sintering furnace with a heating rate of 10 °C for 1 h. The particles were covered with activated carbon powder to isolate oxygen during sintering. Finally, the prepared Fe-C MEM was stored in the sealing pocket. The preparation process is shown in Figure 1.

2.3. Optimization of Fe-C MEM

The impacts, such as Fe/C ratio (2:1–14:1), corn straw content (0–5%), bentonite content (15–35%), and the sintering temperature (550 °C–1000 °C) were explored to optimize the properties of Fe-C MEM. The properties included CV removal efficiency and strength of Fe-C MEM.

The experiment of CV removal was conducted in a shaker at 120 rpm, 25 °C, and 0.5 g Fe-C MEM was added in 250 mL conical flasks containing 100 mL CV solution with a concentration of 1000 mg/L. The concentration of CV was examined at λ_{max} = 585 nm using a UV–Vis spectrophotometer (L8 Plus, INESA). CV removal was calculated as follows.

$$CV\ removal = \frac{(C_0 - C_t)V}{m} \times 100\% \quad (1)$$

where C_0 and C_t were the concentration of CV solution at beginning and time t in the adsorption. V was the volume of CV solution, and m was the weight of Fe-C MEM.

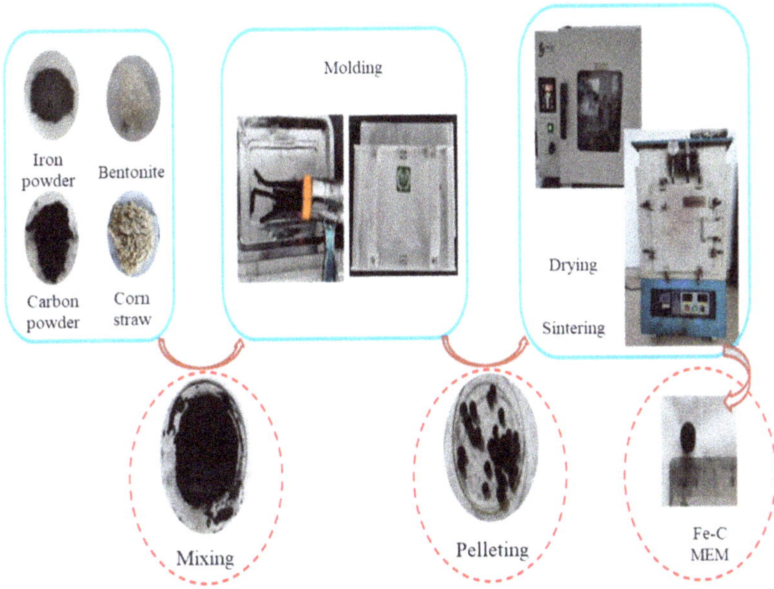

Figure 1. Schematic diagram of Fe-C MEM preparation.

The strength of Fe-C MEM was evaluated by weight loss. Fe-C MEM was shaken in 50 mL deionized water at 200 rpm and 25 °C for 60 min. Then Fe-C MEM was washed and heated at 105 °C to constant quantity in a vacuum oven. Weight loss was calculated as follows.

$$Weight\ loss = \frac{(m - m_2)}{m_2} \times 100\% \quad (2)$$

where m_2 was the weight of Fe-C MEM after being heated.

2.4. CV Removal by Fe-C MEM

To evaluate the CV removal efficiency of Fe-C MEM, batch experiments were carried out. A 0.2 g amount of Fe-C MEM was added in 100 mL CV solution in a shaker at 120 rpm for 72 h. The initial concentration of CV, pH of the solution, removal time, and temperature were studied. The experiment was triplicated.

Regeneration of Fe-C MEM after CV adsorption was conducted by sintering Fe-C MEM at 900 °C again in the atmosphere sintering furnace for 1 h. Then, the sintered Fe-C MEM were used for CV adsorption.

2.5. Characterization

Scanning electron microscopy (SEM, JSM-6610 LV, JEOL, Japan) was used to analyze the surface morphology. X-ray diffraction (XRD, XRD-6100, Shimadzu, Japan) was used to analyze iron in the material. Fourier infrared spectrometer (FTIR, IRAffinity-1S, Shimadzu, Japan) was used to analyze the chemical functional groups on the surface of the materials. The pore size distribution and specific surface area of the iron–carbon micro-electrolytic materials were analyzed by BET (ASAP 2020, Micromeritics, USA).

3. Results

3.1. Optimization of Fe-C MEM Production

The impacts, such as Fe/C ratio, corn straw content, bentonite content, and the sintering temperature, were explored to optimize the properties of Fe-C MEM. The results are shown in Figure 2. In addition to bentonite content, Fe/C ratio, corn straw content, and the sintering temperature show similar variations in CV removal and weight loss of Fe-C MEM. The removal capacity of CV increases first and then decreases, and the weight loss of Fe-C MEM increases. According to the reports, dyes could be adsorbed or degraded by Fe-C MEM [7,12,13]. Therefore, the carbon content and galvanic cells formed in the solution might contribute to CV removal. The removal capacity of CV is the highest at Fe/C 1:1. Then it decreases with the reduction of activated carbon in Fe-C MEM, indicating the synergistic effect of carbon and iron reaches the highest at Fe/C 1:1.

Corn straw underwent rapid thermal cracking and released a large amount of reducing gases, such as CO and CH_4, when the temperature reached 200 °C [10]. The escape of gases formed the connected pores in Fe-C MEM, increasing the specific surface area of the material, thus improving the removal capacity of CV. Excessive addition of corn straw results in an increase in carbon content, leading to a decrease in CV removal. The results are in agreement with those shown in Figure 2a.

The effects of sintering temperature on CV removal by Fe-C MEM are shown in Figure 2c. Fe-C MEM have the highest removal capacity of CV at 900 °C. As reported, the pyrolysis of corn straw finished below 800 °C, and the reduction of ferric oxides usually happened above 900 °C [10]. The synergistic effect promotes the formation of porous material with reactive ZVI. When the sintering temperature is 1000 °C, ferric oxides will react with SiO_2 in the clay, resulting in the reduction of reactive ZVI and the blockage of the pores [14].

Bentonite works as a binder Fe-C MEM. Therefore, CV removal decreases with the increasing bentonite content while the strength of Fe-C MEM increases. Except for bentonite, the addition of iron powder and corn straw leads to a decrease in the strength of Fe-C MEM.

In summary, Fe-C MEM was produced at a 1:1 Fe/C ratio, 2% of corn straw content, 25% of bentonite content, and 900 °C.

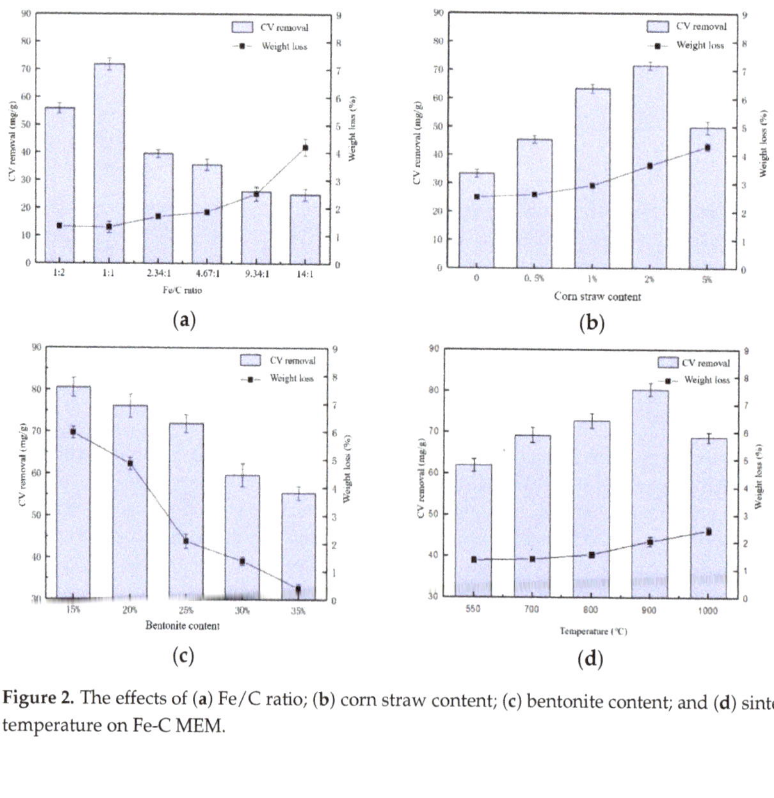

Figure 2. The effects of (**a**) Fe/C ratio; (**b**) corn straw content; (**c**) bentonite content; and (**d**) sintering temperature on Fe-C MEM.

3.2. Characterization of Fe-C MEM

The properties of Fe-C MEM under optimal conditions were analyzed by SEM, XRD, FTIR, and BET.

The XRD patterns show the characteristic peaks at 2θ of 44.6° and 65°, which are in good agreement with the standard data of ZVI (PDF #06-0696) (Figure 3a). The strong diffraction peaks of Fe indicate a good reduction and protection of ZVI in the process of preparation and sintering. There is also a strong diffraction peak of SiO_2 (PDF #27-0605), which might come from the calcined bentonite in the XRD pattern. The FTIR spectra are shown in Figure 3b. The peak at 3400 cm^{-1} is assigned to OH stretching vibration. The two peaks at 1420 cm^{-1} and 1640 cm^{-1} are assigned to COOH [15,16]. The spectra indicate that the carbon in Fe-C MEM have a large amount of carboxyl and hydroxyl groups, which might be beneficial for CV removal [17].

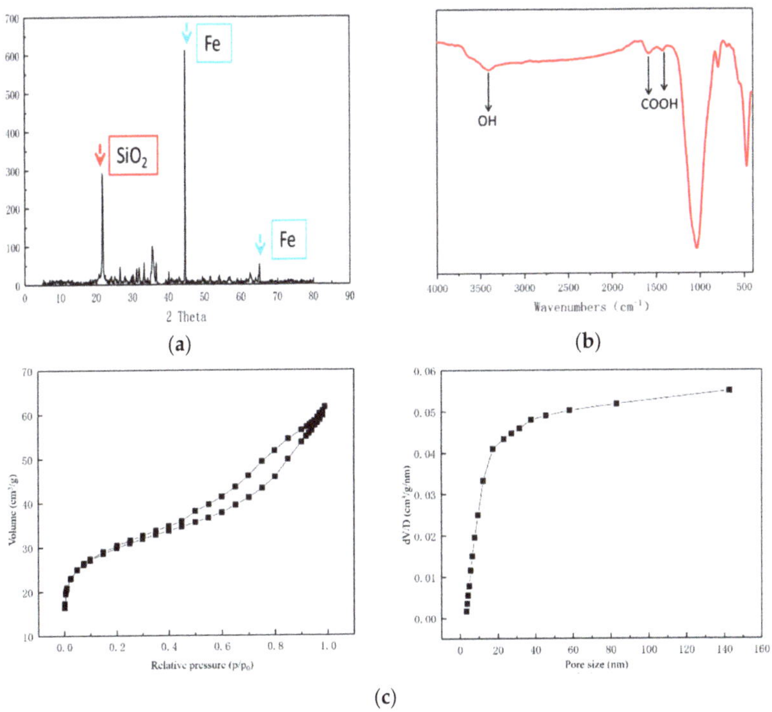

Figure 3. (**a**) XRD patterns of Fe-C MEM; (**b**) FTIR spectra of Fe-C MEM; and (**c**) BET analysis of Fe-C MEM.

The specific surface area of Fe-C MEM is 108.069 m^2/g, and the average pore size is 5.630 nm. According to the International Union of Pure and Applied Chemistry (IUPAC) classification, the nitrogen isothermal adsorption–desorption curve of the material belongs to type IV, with a lag loop in the middle, indicating that the pores in the Fe-C MEM are mainly mesoporous. The result is consistent with that in pore size distribution (Figure 3c). The results indicate the Fe-C MEM have a large specific surface area and well-developed pore structure.

Figure 4 shows the morphology and Fe, C distribution of Fe-C MEM. The pores are obvious in Figure 4a, which might be attributed to the evaporation of water, pyrolysis of corn straw, and the addition of activated carbon. According to the reports, pyrolysis of corn straw generated CO, H_2, CH_4, and biochar, making Fe-C MEM porous [10]. The release of reducing gas could efficiently reduce ferric oxide to obtain ZVI. EDX-mapping (Figure 4b) shows Fe and C distributed evenly in the material. The existence of ZVI and carbon could form a galvanic cell, facilitating CV removal.

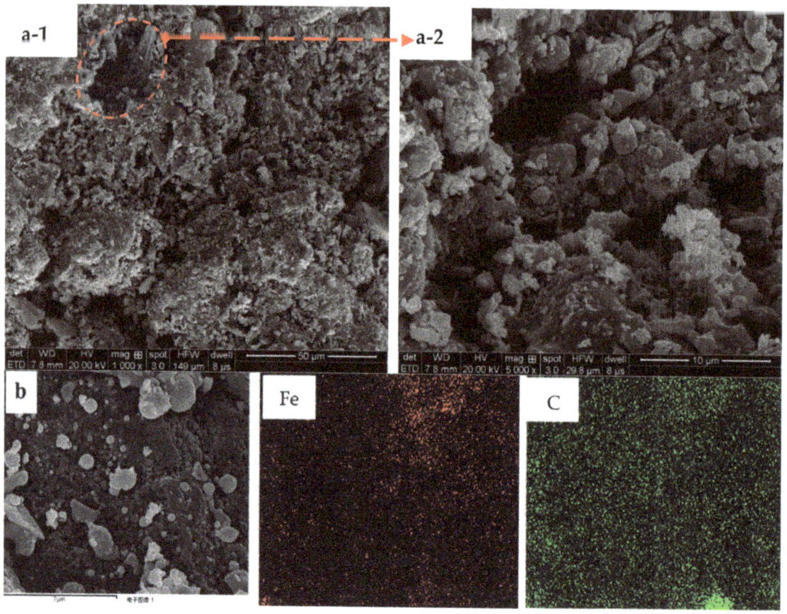

Figure 4. (**a-1**,**a-2**) SEM image of Fe-C MEM; (**b**) EDX-mapping of Fe-C MEM.

3.3. CV Removal by Fe-C MEM

The effects of the initial concentration of CV, time, temperature, and pH on CV removal by Fe-C MEM are shown in Figure 5. At the initial stage, the CV removal is fast. Then, the removal rate gradually decreases until reaching the final equilibrium. CV removal increases with the increasing initial concentration. At an initial concentration of 300 mg/L, the removal capacity of CV is 45.57 mg/g, while it changes to 105.48 mg/g when the initial concentration is 1000 mg/L. As the initial concentration increases, the collision between CV molecules and the active site on the surface of Fe-C MEM increases. Meanwhile, the increase in the initial concentration also increases the driving force for CV molecules to move toward the surface of the material, overcoming certain mass transfer resistance [18,19].

The effect of temperature on CV removal by Fe-C MEM is shown in Figure 5b. CV removal increases when the temperature increases from 25 °C to 45 °C, indicating that the increase in temperature facilitates CV removal. Because the thermal motion speed of CV molecules becomes faster when the temperature increases, the probability of collision between CV and the surface of Fe-C MEM increases, promoting CV removal [20]. According to the reports, the internal pore size and specific surface area of the porous materials could increase under high-temperature conditions while the hydrated radius of CV decreases. Therefore, CV removal could be promoted in high temperatures [21].

The removal capacity of CV increases in the pH ranged from 3 to 11, shown in Figure 5c. CV is positive in an acidic solution. The abundant H^+ would compete for the adsorption site and hinder the removal of CV. And the basic condition could promote CV removal. Meanwhile, the chromogenic group of CV is destroyed under strong alkaline conditions, suggesting a higher CV removal in the study.

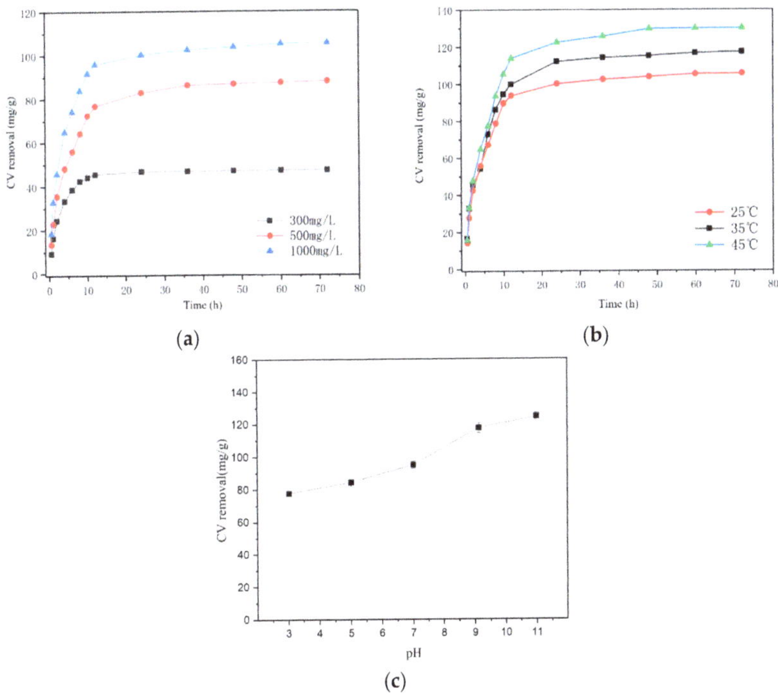

Figure 5. The effects of (**a**) initial concentration of CV, time; (**b**) temperature; and (**c**) pH on CV removal by Fe-C MEM.

Since Fe-C MEM contains activated carbon and biochar, which have a large amount of hydroxyl and carboxyl functional groups, CV could be adsorbed on the porous material [4,17]. Therefore, the results were fitted with the pseudo-first-order and pseudo-second-order adsorption kinetics models (Equations (3) and (4) [22,23], the Langmuir and Freundlich isotherm (Equations (5) and (7) [24,25] to study the CV adsorption mechanism on Fe-C MEM. The equations are as follows:

$$log(q_e - q_t) = logq_e - \frac{k_1 t}{2.303} \tag{3}$$

$$\frac{t}{q_t} = \frac{1}{k_2 q_e^2} + \frac{t}{q_e} \tag{4}$$

$$\frac{C_e}{q_e} = \frac{1}{Q_0 b} + \frac{C_e}{Q_0} \tag{5}$$

$$R_L = \frac{1}{1 + bC_0} \tag{6}$$

$$lnq_e = lnK_F + \frac{1}{n} lnC_e \tag{7}$$

where q_e (mg/g) and C_e (mg/L) are the equilibrium adsorption capacity and the equilibrium concentrations of the solution, respectively. q_t (mg/g) and C_t (mg/L) are the adsorption capacity and concentration of solution at time t, respectively. k_1 and k_2 are the pseudo-first-order and pseudo-second-order adsorption rate constant, respectively. Q_0 (mg/g) is the maximum adsorption capacity, and b is the equilibrium adsorption constant. K_F and n are the Freundlich constants.

The results are shown in Tables 1 and 2. The correlation coefficient R^2 of the pseudo-second-order model is higher than that of the pseudo-first-order model, indicating that the CV adsorption process on Fe-C MEM is consistent with the pseudo-second-order model. The equilibrium adsorption isotherm determines the adsorption affinity of CV on Fe-C MEM. The correlation coefficient R^2 of the Langmuir model is higher than that of the Freundlich model. In addition, the maximum adsorption capacity q_e calculated by the Langmuir model is 105.47 mg/g, which is very close to the practical adsorption capacity of 105.48 mg/g. The calculated R_L in the Langmuir model is between 0 and 1, indicating that Fe-C MEM have a good effect on CV removal [26]. In summary, the fitted results suggest that CV adsorption on Fe-C MEM is monolayer absorption.

Meanwhile, the thermodynamic parameters, including the enthalpy change (ΔH, kJ/mol), Gibbs free energy change (ΔG, kJ/mol), and entropy change (ΔS, J/(mol·K)) for CV removal were calculated. The equations are as follows [27]:

$$\Delta G = -RTln\ K_L \tag{8}$$

$$lnK_L = -\frac{\Delta G}{RT} = -\frac{\Delta H}{RT} + \frac{\Delta S}{R} \tag{9}$$

where K_L is the equilibrium constant. R is the universal gas constant (8.314 J/(mol·K)). The values of ΔH and ΔS were calculated from the plot of ln K_L versus $1/T$.

Table 1. Corresponding parameters of kinetics models.

C_0 (mg/L)	q_e (mg/g)	Pseudo-First-Order Model		Pseudo-Second-Order Model	
		k_1 (h^{-1})	R^2	k_2 (h^{-1})	R^2
300	46.99	0.127	0.938	0.0206	0.995
500	88.03	0.667	0.942	0.108	0.996
1000	105.48	0.846	0.9453	0.0091	0.997

Table 2. Corresponding parameters of Langmuir and Freundlich models.

T (°C)	Langmuir				Freundlich		
	R^2	q_m (mg/g)	b (L/mg)	R_L	R^2	K_f	1/n
25 °C	0.9978	105.47	0.008	0.05–0.27	0.988	23.68	0.4934
35 °C	0.9892	117.03	0.007	0.07–0.21	0.9833	29.74	0.3206
45 °C	0.9945	129.98	0.004	0.05–0.17	0.982	35.82	0.4428

The thermodynamic parameters are shown in Table 3. The ΔH is positive, indicating that the CV adsorption is endothermic [27]. According to the reports, when $\Delta H < 20$ kJ/mol, it is physical adsorption; when $80 < \Delta H < 200$ kJ/mol, it is chemical adsorption [28]. In this study, ΔH is 24.12 kJ/mol, indicating the chemisorption of CV on Fe-C MEM. Meanwhile, the positive ΔS suggests that the disorder of the solid–liquid system increases during the removal process, and the randomness at the solution interface increases. The negative ΔG decreases with the increasing temperature, indicating that CV removal is a spontaneous process. It is more favorable for the process at higher temperatures [29].

Table 3. Thermodynamic parameters for CV removal on Fe-C MEM.

T (K)	ΔH (KJ/mol)	ΔS (J/mol·K)	ΔG (KJ/mol)
25 °C	24.12	124.13	−15.69
35 °C	24.12	124.13	−17.24
45 °C	24.12	124.13	−19.26

In addition to the adsorption, many microcurrent cells could be formed spontaneously in solution between carbon and ZVI and might contribute to CV removal on Fe-C MEM [13]. Therefore, the UV–vis spectra of CV during the decolorization process were analyzed. The results are illustrated in Figure 6. The characteristic bands of CV are located at 580 nm, 300 nm, and 246 nm, respectively. Then, the intensity of the bands decreases during decolorization. Two new bands at 340 nm and 373 nm appear 12 h later, indicating CV is degraded and new species are generated. In the process, ZVI acted as an anode and was oxidized to Fe^{2+}. Then, Fenton-like chain reactions produced hydroxyl radicals (·OH) to destroy CV molecules. Carbon acted as a cathode and accepted and transferred the electrons [3,7,30].

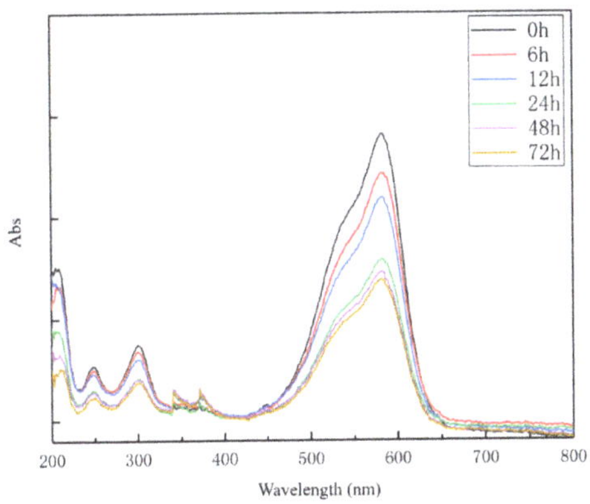

Figure 6. Variation of UV–vis spectra of CV during decolorization process.

Therefore, the removal mechanism of CV by Fe-C MEM was adsorption combined with degradation, as shown in Figure 7. CV molecules diffused from the solution to the surface of Fe-C MEM and were adsorbed. Part of the CV was degraded in the process.

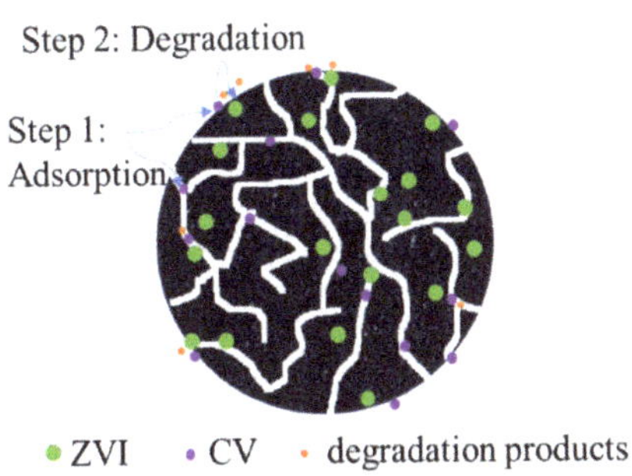

Figure 7. Mechanism of CV removal by Fe-C MEM.

Table 4 presents the Fe-C MEM with other adsorbents for CV removal. The adsorption capacity of CV varied from 24.45 to 303.03 mg/g by different kinds of adsorbents. The Fe/C-based adsorbents showed similar adsorption capacity of CV, which was relatively high in the table. Considering the low cost, Fe-C MEM showed a great potential for CV removal in practice.

Table 4. Comparison of Fe-C MEM with other adsorbents for CV removal.

Adsorbents	Q_m (mg/g)	Reference
Palm Kernel Shell-Derived Biochar	24.45	[2]
Magnetic HKUST-1 MOF	70.42	[31]
Cu-chitosan nano-biocomposite particles	84.75	[32]
Aromatic polyamides (APIs)-1	285.71	[33]
APIs-2	303.03	
Barium encapsulated alginate/carbon composites	50	[34]
Fe/sponge structure peanut shell carbon composite	143.25 (addition of H_2O_2)	[35]
Activated carbon embedded zero-valent iron	95	[12]
Fe-MEM	105.48	This study

3.4. *Reuse of Fe-C MEM*

Fe-C MEM were reused three times, and the results are shown in Figure 8. The removal capacity of CV decreases a little bit after sintering. This might be due to the loss of reactive site on the surface of Fe-C MEM. The formation of ferric minerals covered the surface of ZVI and blocked the pores of Fe-C MEM, leading to a decrease in CV removal capacity. The CV removal efficiency is above 70% of the original after three times regeneration. Therefore, Fe-C MEM could be an effective and low-cost material for dye wastewater treatment.

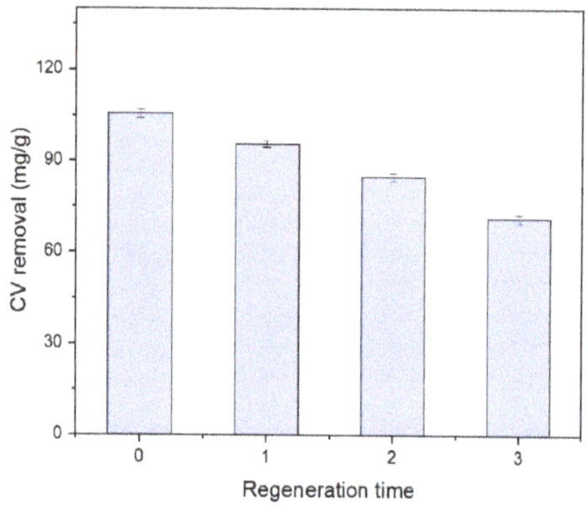

Figure 8. Reuse of Fe-C MEM.

4. Conclusions

The spherical Fe-C MEM was produced with iron powder, activated carbon powder, corn straw, and bentonite by sintering. In the process, corn straw worked as the pore former, and bentonite worked as the binder. The optimal condition for the production was a 1:1 Fe/C ratio, 2% of corn straw content, 25% of bentonite content, and 900 °C. The even distribution of Fe and C, abundant functional groups, and high specific surface area of 108.069 m^2/g contributed to the high CV removal efficiency. CV was removed by adsorption and degradation by Fe-C MEM. The adsorption was endothermic and well fitted the pseudo-second-order kinetics model and Langmuir isotherm model. Fe-C MEM had a good reusability. Therefore, Fe-C MEM could be an effective and low-cost material for dye wastewater treatment.

Author Contributions: Conceptualization, M.D.; data curation, C.L.; formal analysis, X.W.; funding acquisition, M.D., W.Z. and C.P.; investigation, X.W., Z.L., W.Z. and J.G.; methodology, M.D. and C.X.; validation, C.X.; writing—original draft, M.D.; writing—review and editing, C.L. and C.P. All authors have read and agreed to the published version of the manuscript.

Funding: This research was funded by the Guangdong University Innovation Team (2021KCXTD055) and the Guangdong College Students' Innovative Project (S202310580064, S202310580065, X202410580141).

Data Availability Statement: Dataset available on request from the authors.

Conflicts of Interest: The authors declare no conflicts of interest.

References

1. Shabir, M.; Yasin, M.; Hussain, M.; Shafiq, I.; Akhter, P.; Nizami, A.S.; Jeon, B.H.; Park, Y.K. A review on recent advances in the treatment of dye-polluted wastewater. *J. Ind. Eng. Chem.* **2022**, *112*, 1–19. [CrossRef]
2. Kyi, P.P.; Quansah, J.O.; Lee, C.G.; Moon, J.K.; Park, S.J. The removal of crystal violet from textile wastewater using palm kernel shell-derived biochar. *Appl. Sci.* **2020**, *10*, 2251. [CrossRef]
3. Huang, C.; Peng, F.; Guo, H.J.; Wang, C.; Luo, M.T.; Zhao, C.; Xiong, L.; Chen, X.F.; Chen, X.D. Efficient COD degradation of turpentine processing wastewater by combination of Fe-C micro-electrolysis and Fenton treatment: Long-term study and scale up. *Chem. Eng. J.* **2018**, *351*, 697–707. [CrossRef]
4. Wang, S.; Zhao, M.; Zhou, M.; Li, Y.C.; Wang, J.; Gao, B.; Sato, S.; Feng, K.; Yin, W.; Igalavithana, A.D.; et al. Biochar-supported nZVI (nZVI/BC) for contaminant removal from soil and water: A critical review. *J. Hazard. Mater.* **2019**, *373*, 820–834. [CrossRef] [PubMed]
5. Ma, W.; Han, Y.; Xu, C.; Han, H.; Zhong, D.; Zhu, H.; Li, K. The mechanism of synergistic effect between iron-carbon microelectrolysis and biodegradation for strengthening phenols removal in coal gasification wastewater treatment. *Bioresour. Technol.* **2019**, *271*, 84–90. [CrossRef]
6. Cui, X.; Zhang, M.; Ding, Y.; Sun, S.; He, S.; Yan, P. Enhanced nitrogen removal via iron carbon micro-electrolysis in surface flow constructed wetlands: Selecting activated carbon or biochar? *Sci. Total Environ.* **2022**, *815*, 152800. [CrossRef]
7. Yang, Z.; Ma, Y.; Liu, Y.; Li, Q.; Zhou, Z.; Ren, Z. Degradation of organic pollutants in near-neutral pH solution by Fe-C micro-electrolysis system. *Chem. Eng. J.* **2017**, *315*, 403–414. [CrossRef]
8. Lai, B.; Zhou, Y.; Yang, P.; Yang, J.; Wang, J. Degradation of 3, 3′-iminobis-propanenitrile in aqueous solution by Fe0/GAC micro-electrolysis system. *Chemosphere* **2013**, *90*, 1470–1477. [CrossRef]
9. Zhang, L.; Yue, Q.; Yang, K.; Zhao, P.; Gao, B. Enhanced phosphorus and ciprofloxacin removal in a modified BAF system by configuring Fe-C micro electrolysis: Investigation on pollutants removal and degradation mechanisms. *J. Hazard. Mater.* **2018**, *342*, 705–714. [CrossRef]
10. Liu, X.; Rong, R.; Dai, M.; Bian, H.; Peng, C. Preparation of red mud-based zero-valent iron materials by biomass pyrolysis reduction: Reduction mechanism and application study. *Sci. Total Environ.* **2023**, *864*, 160907. [CrossRef]
11. Zhang, P. Effective reinforcement ozone oxidation degradation of N, N-dimethylformamide with cobalt doping micro electrolysis composite. *Sep. Purif. Technol.* **2023**, *314*, 123639. [CrossRef]
12. Wang, Y.; Chen, T.; Zhang, X.; Mwamulima, T. Removal study of crystal violet and methylene blue from aqueous solution by activated carbon embedded zero-valent iron: Effect of reduction methods. *Front. Environ. Sci.* **2021**, *9*, 799264. [CrossRef]
13. Li, X.; Zhang, J.; Qin, Y.; Zhang, X.; Zou, W.; Ding, L.; Zhou, M. Enhanced removal of organic contaminants by novel iron–carbon and premagnetization: Performance and enhancement mechanism. *Chemosphere* **2022**, *303*, 135060. [CrossRef] [PubMed]
14. Takeuchi, N.; Takahashi, H.; Ishida, S.; Horiie, F.; Wakamatsu, M. Mechanistic study of solid-state reaction between kaolinite and ferrous oxide at high temperatures. *J. Ceram. Soc. Jpn.* **2000**, *108*, 876–881. [CrossRef]
15. Keiluweit, M.; Nico, P.S.; Johnson, M.G.; Kleber, M. Dynamic molecular structure of plant biomass-derived black carbon (biochar). *Environ. Sci. Technol.* **2010**, *44*, 1247–1253. [CrossRef]
16. Gusmão, K.A.; Gurgel, L.V.; Melo, T.M.S.; Gil, L.F. Adsorption studies of methylene blue and gentian violet on sugarcane bagasse modified with EDTA dianhydride (EDTAD) in aqueous solutions: Kinetic and equilibrium aspects. *J. Environ. Manag.* **2013**, *118*, 135–143. [CrossRef]
17. Jawad, A.H.; Salleh, N.; ALOthman, Z.A.; Selvasembian, R. Mesoporous Activated Carbon from Sunflower (*Helianthus annuus*) Seed Pericarp for Crystal Violet Dye Removal: Numerical Desirability Optimization and Mechanism Study. *Water Air Soil Pollut.* **2024**, *235*, 1–17. [CrossRef]
18. Hwang, Y.; Lee, Y.C.; Mines, P.D.; Huh, Y.S.; Andersen, H.R. Nanoscale zero-valent iron (nZVI) synthesis in a Mg-aminoclay solution exhibits increased stability and reactivity for reductive decontamination. *Appl. Catal. B Environ.* **2014**, *147*, 748–755. [CrossRef]
19. Qian, L.; Zhang, W.; Yan, J.; Han, L.; Chen, Y.; Ouyang, D.; Chen, M. Nanoscale zero-valent iron supported by biochars produced at different temperatures: Synthesis mechanism and effect on Cr(VI) removal. *Environ. Pollution.* **2017**, *223*, 153–160. [CrossRef]
20. Jyoti Saini, J.S.; Garg, V.K.; Gupta, R.K.; Navish Kataria, N.K. Removal of Orange G and Rhodamine B dyes from aqueous system using hydrothermally synthesized zinc oxide loaded activated carbon (ZnO-AC). *J. Environ. Chem. Eng.* **2017**, *5*, 884–892. [CrossRef]
21. Dong, H.; Ning, Q.; Li, L.; Wang, Y.; Wang, B.; Zhang, L.; Tian, R.; Li, R.; Chen, J.; Xie, Q. A comparative study on the activation of persulfate by bare and surface-stabilized nanoscale zero-valent iron for the removal of sulfamethazine. *Sep. Purif. Technol.* **2020**, *230*, 115869. [CrossRef]

22. Lagergren, S. Zur theorie der sogenannten adsorption geloster stoffe. In *Zeitschrift für Chemie und Industrie der Kolloide*; Springer: Berlin/Heidelberg, Germany, 1898.
23. Ho, Y.S.; McKay, G. Kinetic models for the sorption of dye from aqueous solution by wood. *Process Saf. Environ. Prot.* **1998**, *76*, 183–191. [CrossRef]
24. Freundlich, H. Über die adsorption in lösungen. *Z. Fur Phys. Chem.* **1907**, *57*, 385–470. [CrossRef]
25. Langmuir, I. The adsorption of gases on plane surfaces of glass, mica and platinum. *J. Am. Chem. Soc.* **1918**, *40*, 1361–1403. [CrossRef]
26. Muthukumaran, C.; Sivakumar, V.M.; Thirumarimurugan, M. Adsorption isotherms and kinetic studies of crystal violet dye removal from aqueous solution using surfactant modified magnetic nanoadsorbent. *J. Taiwan Inst. Chem. Eng.* **2016**, *63*, 354–362. [CrossRef]
27. Boukoussa, B.; Hakiki, A.; Moulai, S.; Chikh, K.; Kherroub, D.E.; Bouhadjar, L.; Guedal, D.; Messaoudi, K.; Mokhtar, F.; Hamacha, R. Adsorption behaviors of cationic and anionic dyes from aqueous solution on nanocomposite polypyrrole/SBA-15. *J. Mater. Sci.* **2018**, *53*, 7372–7386. [CrossRef]
28. Ponder, S.M.; Darab, J.G.; Mallouk, T.E. Remediation of Cr(VI) and Pb(II) Aqueous Solutions Using Supported, Nanoscale Zero-valent Iron. *Environ. Sci. Technol.* **2000**, *34*, 2564–2569. [CrossRef]
29. Wu, J.; Gao, H.; Yao, S.; Chen, L.; Gao, Y.; Zhang, H. Degradation of Crystal Violet by catalytic ozonation using Fe/activated carbon catalyst. *Sep. Purif. Technol.* **2015**, *147*, 179–185. [CrossRef]
30. Nasiri, J.; Motamedi, E.; Naghavi, M.R.; Ghafoori, M. Removal of crystal violet from water using β-cyclodextrin functionalized biogenic zero-valent iron nanoadsorbents synthesized via aqueous root extracts of Ferula persica. *J. Hazard. Mater.* **2019**, *367*, 325–338. [CrossRef]
31. Mohammadnejad, M.; Nekoo, N.M.; Alizadeh, S.; Sadeghi, S.; Geranmayeh, S. Enhanced removal of organic dyes from aqueous solutions by new magnetic HKUST-1, Facile strategy for synthesis. *Sci. Rep.* **2023**, *13*, 17981. [CrossRef]
32. Shukla, S.K.; Pandey, S.; Saha, S.; Singh, H.R.; Mishra, P.K.; Kumar, S.; Jha, S.K. Removal of crystal violet by Cu-chitosan nano-biocomposite particles using Box–Behnken design. *J. Environ. Chem. Eng.* **2021**, *9*, 105847. [CrossRef]
33. Loganathan, M.; Raj, A.S.; Murugesan, A.; Kumar, P.S. Effective adsorption of crystal violet onto aromatic polyimides: Kinetics and isotherm studies. *Chemosphere* **2022**, *304*, 135332. [CrossRef] [PubMed]
34. Venkataraman, S.; Viswanathan, V.; Thangaiah, S.G.; Omine, K.; Mylsamy, P. Adsorptive exclusion of crystal violet dye using barium encapsulated alginate/carbon composites: Characterization and adsorption modeling studies. *Environ. Sci. Pollut. Res.* **2023**, *30*, 106718–106735. [CrossRef] [PubMed]
35. Wu, M.; Li, S.; Zhou, S.; Li, F.; Li, T.; Li, H. Fe/sponge structure peanut shell carbon composite preparation for efficient Fenton oxidation crystal violet. *Environ. Sci. Pollut. Res.* **2023**, *30*, 105457–105473. [CrossRef]

processes

MDPI

Article

Assessing Phytoremediation Potential: Dominant Plants in Soils Impacted by Polymetal(loid)lic Mining

Boxin Wang [1,2,3], Juan Hou [4], Xueyong Wu [4], Xuekui Niu [4] and Fengping Zhou [1,*]

1 National-Regional Joint Engineering Research Center for Soil Pollution Control and Remediation in South China, Guangdong Key Laboratory of Integrated Agro-Environmental Pollution Control and Management, Institute of Eco-Environmental and Soil Sciences, Guangdong Academy of Sciences, Guangzhou 510650, China; boxinwang2025@163.com
2 Guangdong Experimental High School, Guangzhou 510650, China
3 Shenzhen Shentai Environmental Technology Co., Ltd., Shenzhen 518045, China
4 Yunnan Research Academy of Eco-Environment Science, Kunming 650034, China; hkhoujuan2007@163.com (J.H.); nywxy1234@163.com (X.W.); ynhknxk@163.com (X.N.)
* Correspondence: fpzhou@soil.gd.cn

Abstract: Phytoremediation, an ecological approach aimed at addressing polymetal(loid)lic-contaminated mining soils, has encountered adaptability challenges. Dominant plant species, well-suited to the local conditions, have emerged as promising candidates for this purpose. This study focused on assessing the phytoremediation potential of ten plant species that thrived in heavy metal(loid)-contaminated mining soils. This investigation covered nine heavy metal(loid)s (As, Cu, Cd, Cr, Hg, Ni, Pb, Sn, and Zn) in both plants and rhizosphere soils. The results revealed a significant impact of mining activities, with heavy metal(loid) concentrations surpassing the Yunnan Province's background levels by 1.06 to 362 times, highlighting a significant concern for remediation. The average levels of the heavy metal(loid)s followed the order of As (3.98×10^3 mg kg^{-1}) > Cu (2.83×10^3 mg kg^{-1}) > Zn (815 mg kg^{-1}) > Sn (176 mg kg^{-1}) > Pb (169 mg kg^{-1}) > Cr (68.1 mg kg^{-1}) > Ni (36.2 mg kg^{-1}) > Cd (0.120 mg kg^{-1}) > Hg (0.0390 mg kg^{-1}). The bioconcentration factors (BCFs), bioaccumulation factors (BAFs), and translocation factors (TFs) varied among the native plants, indicating diverse adaptation strategies. Low BCFs and BAFs (ranging from 0.0183 to 0.418 and 0.0114 to 0.556, respectively) suggested a low bioavailability of heavy metal(loid)s. Among the species, both *J. effusus* and *P. capitata* showed remarkable abilities for As accumulation, while *A. adenophora* demonstrated a notable accumulation ability for various heavy metal(loid)s, especially Cd, with relatively high BCFs (1.88) and BAFs (3.11), and the TF at 1.66 further underscored the crucial role of translocation in preventing root toxicity. These findings emphasized the potential of these plant species in mine ecological restoration and phytoremediation, guiding targeted environmental rehabilitation strategies.

Keywords: phytoremediation; polymetal(loid)lic contamination; native dominant plants; adaption strategies; ecological restoration

1. Introduction

Mining operations are integral to the provision of materials and energy for human society; however, they often lead to extensive soil contamination with heavy metal(loid)s, primarily attributed to the disposal of mine tailings and wastewater discharge. This prevalent issue, notably pronounced in China, poses substantial environmental threats characterized by the enduring, concealed, and cumulative nature of heavy metal(loid)s pollution [1]. The consequences extend to adverse effects on soil quality, water resources, crop productivity, and, ultimately, human health through the food chain. Addressing this issue necessitates the blocking and remediation of heavy metal(loid)s-contaminated areas to mitigate their impact on the ecosystem.

Various soil remediation methodologies, spanning physical, chemical, and biological approaches, have been employed [2]. Nevertheless, physical and chemical methods often present drawbacks such as high costs, irreversible alterations to soil properties, and potential secondary pollution. In contrast, phytoremediation emerges as a promising strategy due to its cost-effectiveness, environmental friendliness, minimal secondary pollution, and operational simplicity, making it suitable for large-scale applications [3]. Phytoremediation essentially refers to the utilization of plants and associated soil microbes to lower the concentrations of contaminants in soil or mitigate their adverse effects [4]. This approach can not only address heavy metal(loid)s but also organic pollutants. Phytoremediation also boasts low installation and maintenance costs compared to alternative remediation options. In terms of cost-effectiveness, phytoremediation can be as economical as 5% of the cost of alternative cleanup methods [5]. Notably, plants typically address contaminants without disturbing the topsoil, thereby conserving its fertility and utility, and may even enhance soil fertility through organic matter inputs [6]. It is particularly suitable for implementation on large field sites even where other remediation technologies may not be cost-effective or practical. Phytoremediation encompasses technologies (phytostabilization, phytoextraction, and phytodegradation) utilizing higher plants to remediate contaminated sites and has garnered significant research attention in the past two decades [7]. In phytostabilization, the heavy metal(loid)s in soil are immobilized, minimizing their transport and bioavailability [8]. Phytoextraction involves plants removing contaminants from the soil and storing them in harvestable tissues [9], and a successful phytoextraction strategy typically seeks high translocation factor (TF) values, facilitating the translocation of metal(loid)s into above-ground plant parts for harvest. The choice of plant species is pivotal, with adaptability to local environments, substantial biomass, and the ability to either transfer contaminants to above-ground biomass (for phytoextraction) or exhibit a high tolerance for the contaminants (for phytostabilization) being critical factors. The bioconcentration factor (BCF) and bioaccumulation factor (BAF), as well as the TF of heavy metal(loid)s in plants, are commonly determined to estimate the phytoremediation potential [10,11].

Plants, especially hyperaccumulators, with notable capacities for accumulating contaminants serve as the foundation for phytoremediation, offering practical solutions for decontaminating polluted soils [12]. These plants typically exhibit three key characteristics [10]. Firstly, they possess the ability to accumulate contaminants in their stems and leaves at concentrations notably higher (ranging from 10 to 100 times) than those observed in common plants thriving in uncontaminated soils. Secondly, they demonstrate a TF exceeding 1, indicating the efficient biotranslocation and bioaccumulation of contaminants (usually heavy metal(loid)s) from the roots to the above-ground biomass. Thirdly, these species exhibit robust growth even in environments with elevated concentrations of contaminants. Consequently, dominant plant species have been extensively screened and utilized in phytoremediation, particularly in the restoration of mine tailings [13–15].

However, challenges persist in mining areas, ranging from nutrient deficiencies to extreme pH levels, loose and erodible soil, and water retention issues [16]. These factors collectively pose obstacles to the survival and optimal functioning of phytoremediation plants. For instance, while *Sedum alfredii* exhibits strong tolerance, absorption, and accumulation abilities for Zn, with a bioaccumulation factor ranging from 1.25 to 1.94, its growth may be hindered in cold, high-altitude regions [17]. Similarly, *Pteris vittata* can effectively absorb high concentrations of As, with a bioaccumulation factor of 4.31–5.23, but its growth is restricted in northern China due to the relatively low temperatures [18,19]. Consequently, it is essential to screen for suitable plants capable of thriving under local contaminated and climatic conditions to ensure the success of phytoremediation efforts. To overcome these challenges, considerable efforts have been directed towards identifying dominant plant species that are well-adapted to local climate conditions while also exhibiting a high accumulating capacity of heavy metal(lodi)s. Native dominant plants, having evolved over millennia in naturally elevated heavy metal environments, are believed to possess inherent traits that confer resistance to heavy metal toxicity. As a result, they play a crucial role in soil restoration endeavors, particularly in heavy metal-contaminated soils in mining areas. Numerous previous studies have underscored the significance of native dominant plants in phytoremediation, highlighting their effectiveness in mitigating heavy metal pollution in mining environments [20–22]. For example, Bothe (2011) has demonstrated that metallophytes, plants thriving in metal-rich environments, have a long history of exposure to heavy metal(loid)s [23]. Singh et al. (2022) have identified dominant plant species suitable for the phytoremediation of arsenic-contaminated soil, noting that ten of these plants exhibited a BCF higher than 1, indicating their potential role in As remediation and site restoration [24]. Moreover, previous research has consistently emphasized that traits such as high biomass production, rapid growth, strong resistance, and suitability for remediating various soil types are key criteria in plant selection that also represent challenges in phytoremediation efforts [25–27].

In the specific case of Gejiu city, in the Yunnan Province, China, where severe contamination has ensued due to the accumulation of multiple tailings, the utilization of native plants for phytoremediation remains underexplored. This study aims to address this knowledge gap by investigating the spatial distribution of heavy metal(loid)s, exploring vegetation abundance and distribution in proximity to the tailings, measuring metal(loid) contents in native dominant plants, and evaluating their enrichment and migration potentials. The research outcomes are anticipated to guide future phytoremediation projects in similar mine tailing areas, providing insights into the remediation of heavy metal(loid)s-contaminated soil under comparable environmental conditions.

2. Experimental

2.1. Sites and Sampling

The investigated tailings area is situated in Gejiu city, Yunnan Province, China (approximately 1911 m altitude, 103°14′ E, 23°31′ N), predominantly composed of non-ferrous metal(loid)s, with As, Cu, Pb, and Zn as the predominant constituents. The tailings' storage facility adopts a valley-type configuration. It has been decommissioned since 2018, with a total volume of 31.5 million m^3. Currently, the site exhibits the spontaneous growth of numerous herbaceous plants. These species, characterized by a substantial biomass and widespread distribution, are also prevalent in other tailings facilities in Gejiu and are recognized as dominant species in the local ecosystem. The specific sampling locations are presented in Figure 1.

The selection of the plant species was based on three key criteria. Firstly, all the chosen species demonstrate rapid growth and possess a substantial biomass in the tailing dumpsites, indicating their potential for effective phytoremediation. Secondly, these species exhibit a preference for growing abundantly on slag heaps or similar contaminated sites, with limited occurrence in uncontaminated areas (e.g., development land or agriculture land), suggesting their adaptive advantage in contaminated environments and their potential for successful colonization and remediation. Lastly, prior research studies and references have reported on the phytoremediation potential of these selected species, affirming their suitability for this study. Accordingly, samples from ten high-biomass plant species were collected within the tailings area, and Table 1 provides a comprehensive overview of the sampling location and information for each plant species. It should be noted that both below-ground and above-ground parts of the plant species were collected separately. For every plant species, a selection was made of three individual plants, with their samples meticulously combined into one composite plant sample. A minimum of two composite samples were gathered for each plant species. Concurrently, about 2000 g of the corresponding rhizosphere soil of each plant species, located approximately 30 cm deep and around a 2/3 area from the center of vertical crown projection, was excavated using a narrow-bladed trowel. All the plant and rhizosphere soil samples were hermetically sealed in polyethylene bags for their subsequent transport to the laboratory. Upon arrival at the laboratory, the plant samples underwent a meticulous cleansing process. Initial rinsing with water was performed to eliminate surface soil particles, followed by three rinses with ultrapure water. The samples were then air-dried. Subsequently, the dried plant samples were placed in an oven at 105 °C for 30 min to arrest the metabolic processes. Further drying at 80 °C continued until a constant weight was attained. The resulting dried plant materials were finely ground using a stainless-steel grinder, deposited into polyethylene bags with assigned labels for subsequent analysis.

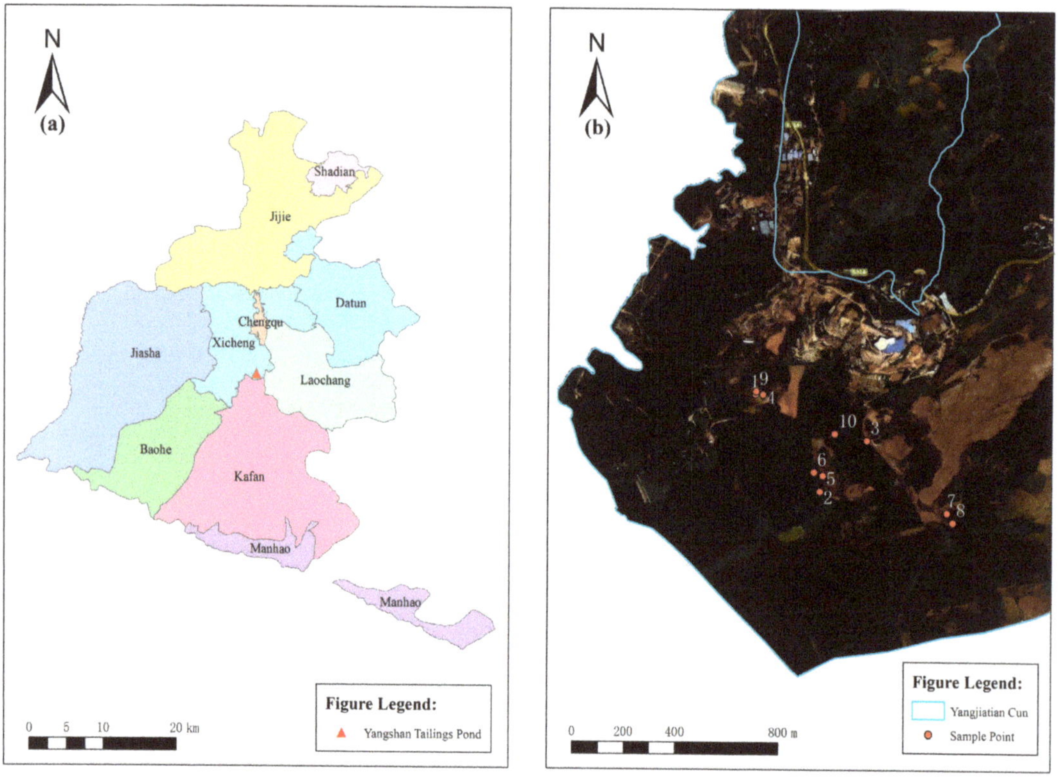

Figure 1. The map of (**a**) Gejiu city, Yunnan Province, China, and (**b**) the sampling location.

Table 1. Plant species investigated in this study.

Pictures of Plants	Sites	Plant Name	Family	Location (Longitude, Latitude, and Altitude)	Average Size of Plants (cm)	
					Above-Ground	Underground
	1	*Ageratina adenophora* (Spreng.) R. M. King & H. Rob.	*Asteraceae*	103°9′ E, 23°18′ N 1.90×10^3 m	78.5	31.5

Table 1. *Cont.*

Pictures of Plants	Sites	Plant Name	Family	Location (Longitude, Latitude, and Altitude)	Average Size of Plants (cm)	
					Above-Ground	Underground
	2	*Buddleja lindleyana* Fortune	*Scrophulariaceae*	103°8′ E, 23°18′ N 1.91×10^3 m	87.0	45.0
	3	*Coriaria nepalensis* Wall. (Morus calva Coriaria kweichowensis Coriaria sinica)	*Coriariaceae*	103°8′ E, 23°18′ N 1.92×10^3 m	40.5	22.0
	4	*Juncus effusus* L.	*Juncaceae*	103°9′ E, 23°18′ N 1.91×10^3 m	60.0	27.0
	5	*Leycesteria formosa* Wall.	*Caprifoliaceae*	103°8′ E, 23°18′ N 1.90×10^3 m	101	30.0
	6	*Puhuaea sequax* (Wall.) H. Ohashi & K. Ohashi	*Fabaceae*	103°8′ E, 23°18′ N 1.92×10^3 m	65.0	80.5
	7	*Pseudognaphalium affine* (D. Don) Anderb.	*Asteraceae*	103°8′ E, 23°18′ N 1.92×10^3 m	35.0	13.5

Table 1. *Cont.*

Pictures of Plants	Sites	Plant Name	Family	Location (Longitude, Latitude, and Altitude)	Average Size of Plants (cm)	
					Above-Ground	Underground
	8	*Pyracantha fortuneana* (Maxim.) H. L. Li	*Rosaceae*	103°8′ E, 23°18′ N 1.91×10^3 m	45.0	24.0
	9	*Persicaria capitata* (Buch.-Ham. ex D. Don) H. Gross	*Polygonaceae*	103° 8′ E, 23°18′ N 1.91×10^3 m	34.5	51.0
	10	*Swertia bimaculata* (Siebold & Zucc.) Hook. f. & Thomson ex C. B. Clarke	*Gentianaceae*	103°8′ E, 23°18′ N 1.92×10^3 m	36.0	12.0

At the same time, the soil samples underwent an initial air-drying phase to remove extraneous materials. They were then subjected to thorough grinding in a mortar and pestle, sieved through a 0.75 μm nylon sieve, and placed into sealed bags.

2.2. Analysis Methods

Both the soil and plant samples underwent acidic digestion before the analyses. In detail, 0.1000 g of the samples was mixed with 8.0 mL of aqua regia solution and subjected to microwave digestion using a MARS6 system (CEM, Charlotte, NC, USA) at a temperature of 190 °C for a duration of 2 h. The resulting solution was diluted to 20 mL by adding 2% HNO_3 solution before being analyzed. The concentrations of As, Cu, Cd, Cr, Hg, Ni, Pb, Sn, and Zn were determined using an Inductively Coupled Plasma Optical Emission Spectrometer (ICP-OES, Avio 500, Perkin Elmer, Hopkinton, MA, USA). Moreover, soil pH measurements were conducted using a pH meter (Five-Easy FE20, Mettler Toledo, Greifensee, Switzerland) with a soil-to-solution ratio of 1: 5. The pH measurement process employed ultrapure water (Milli-Q® IQ Element, Merck Life, Darmstadt, Germany). All the measurement methods were detailed in Table 2.

2.3. Data Analyses

In this study, TF, BCF, and BAF were employed to reflect the ability of plants to absorb heavy metal(loid)s from the soil and the transfer of heavy metal(loid)s from plants' roots to their above-ground parts. The specific calculation formulas are listed as follows, in Equations (1)–(3).

$$BCF = C_{root}/C_{soil} \tag{1}$$

$$TF = C_{shoot}/C_{root} \quad (2)$$

$$BAF = C_{shoot}/C_{soil} \quad (3)$$

where C_{shoot} and C_{root} represent the concentrations of heavy metal(loid)s in the above-ground and below-ground parts of the plant, respectively (mg kg^{-1}), and C_{soil} represents the concentration of heavy metal(loid)s in the corresponding rhizosphere soil of the plant (mg kg^{-1}).

Table 2. Methods used to determine heavy metal(loid) contents in soil and plant samples.

Heavy Metal(loid)s	Soil Sample Pre-Treatment and Analysis Method	Plant Sample Pre-Treatment and Analysis Method
Cu	HJ 491-2019	GB 5009.268-2016
Cr	HJ 491-2019	GB 5009.268-2016
Ni	HJ 491-2019	GB 5009.268-2016
Zn	HJ 491-2019	GB 5009.268-2016
Pb	GB/T 17141-1997	GB 5009.268-2016
Cd	GB/T 17141-1997	GB 5009.268-2016
As	GB/T 22105.2-2008	GB 5009.268-2016
Sn	USEPA 6020B-2014	GB 5009.268-2016
Hg	GB/T 22105.1-2008	GB 5009.17-2021

3. Results and Discussion

3.1. Total Concentration of Heavy Metal(loid)s in the Soil

The concentrations of various heavy metal(loid)s in the rhizosphere soil of the investigated plants are illustrated in Figure 2, and their average concentrations and coefficient variations are listed in Table 3. As noted, the average concentrations of As, Cu, Zn, Sn, Pb, Cr, Ni, Cd, and Hg in the rhizosphere soil were 3.98×10^3 mg kg^{-1}, 2.83×10^3 mg kg^{-1}, 815 mg kg^{-1}, 176 mg kg^{-1}, 169 mg kg^{-1}, 72.2 mg kg^{-1}, 57.5 mg kg^{-1}, 5.69 mg kg^{-1}, and 0.200 mg kg^{-1}, respectively. Site 10 exhibited complex contamination patterns of heavy metal(loid)s, with notably elevated concentrations of As (2.56×10^4 mg kg^{-1}), Sn (372 mg kg^{-1}), Pb (397 mg kg^{-1}), Cr (208 mg kg^{-1}), and Ni (139 mg kg^{-1}), attributed to its proximity to the tailings. The average concentrations of As, Cu, Zn, and Cd exceeded their background values in other soils of the Yunnan Province by 316, 83.3, 10.5, and 49.1 times, while Pb, Cr, Ni, and Hg surpassed them by 4.46, 1.06, 1.59, and 5.02 times, respectively, highlighting the necessity for the remediation intervention. Beyond the elevated concentrations of heavy metal(loid)s, substantial variation exists among different sampling sites. The coefficients of variation across the sites all surpassed 45%, with As and Cr exhibiting the highest coefficients at 182% and 86.7%, respectively. This underscored significant spatial heterogeneity in heavy metal(loid) pollution in the soil, which can be associated with both anthropogenic and natural environmental factors. Specifically, some previous research on five Sn polymetallic deposits in Gejiu, including Malage, Songshujiao, Gaosong, Laochang, and Kafang, showed approximately 300 Mt of Sn ores at a mean grade of 1 wt% Sn, another 300 Mt of Cu ores averaging 2 wt% Cu, and 400 Mt of Pb-Zn ores with a mean grade of 7 wt% Pb/Zn [28] due to a protracted history of mining activities, including a Pb/Zn/Sn smelter, which resulted in the liberation and migration of the polymetallic elements [29–31]. Notably, As and Cr, unlike other heavy metal(loid) cations, generally exist in the form of oxygen-containing anions (AsO_4^{3-} and CrO_4^{2-}) in the soil [32,33], resulting in a strong affinity to soil mineral surfaces and, consequently, a lower mobilization capacity.

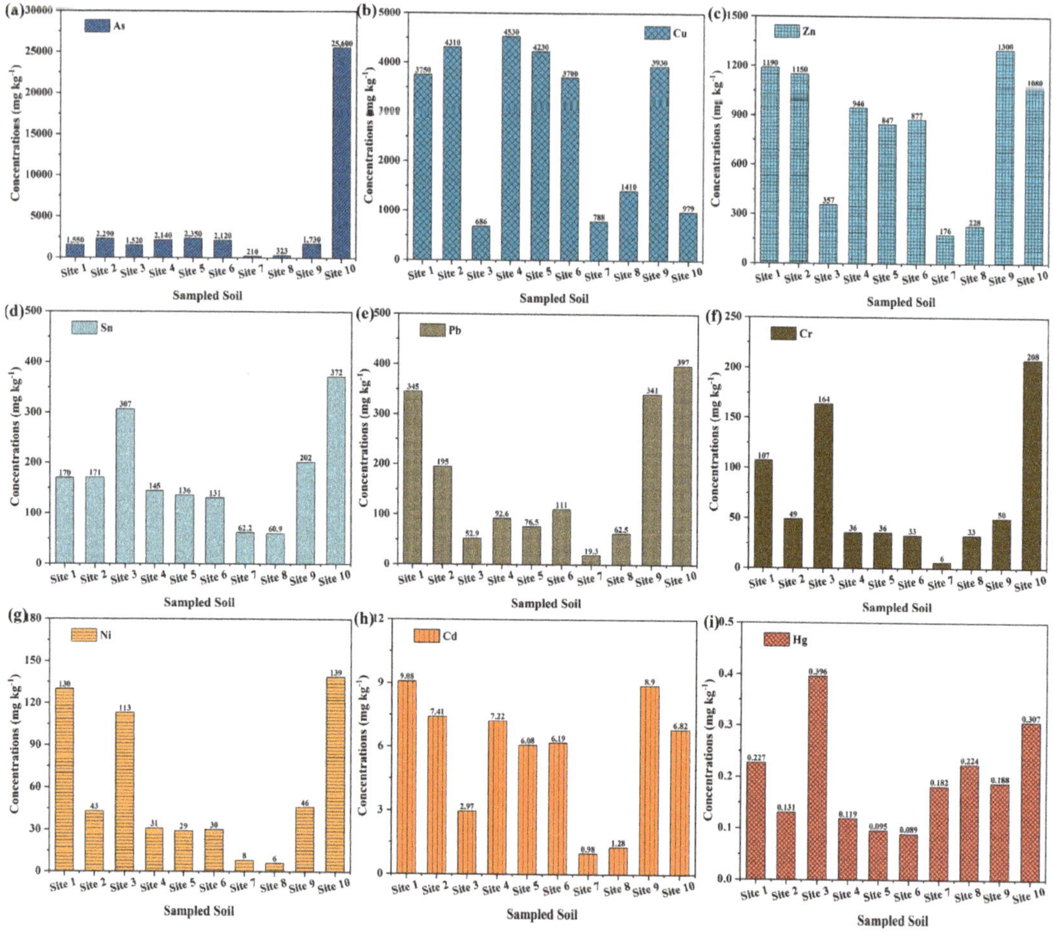

Figure 2. Distribution of different heavy metal(loid)s in the collected soil samples ((**a**) As; (**b**) Cu; (**c**) Zn; (**d**) Sn; (**e**) Pb; (**f**) Cr; (**g**) Ni; (**h**) Cd; and (**i**) Hg).

Table 3. Static concentrations of heavy metal(loid)s in the soil samples.

	As	Cu	Zn	Sn	Pb	Cr	Ni	Cd	Hg
Min. (mg kg^{-1})	2.56×10^3	686	176	60.9	19.3	6.00	6.00	0.980	0.089
Max. (mg kg^{-1})	210	4.53×10^3	1300	372	397	208	139	9.08	0.396
Average (mg kg^{-1})	3.98×10^3	2.83×10^3	815	176	169	72.2	57.5	5.69	0.200
Coefficient of variation	182%	54.8%	48.2%	53.1%	78.9%	86.7%	82.8%	49.0%	47.4%
Average contents in Yunnan Province, China [34]	11.0	34.0	77.7	/	37.9	68.1	36.2	0.120	0.0400
Soil environmental quality–risk control standard for soil contamination of agricultural land (GB 15618-2018) *	20.0	200	300	/	240	350	190	20.0	1.00

* Risk screening value, paddy field, pH > 7.5.

3.2. Concentration of Heavy Metal(loid)s in Plants

The concentrations of heavy metal(loid)s in both the above-ground and below-ground parts of the plants are depicted in Figure 3. Comparatively, the total concentrations of heavy metal(loid)s within the same plants from different sites exhibited minimal variation, consistently maintaining coefficients of variation below 10%. Regarding plant species, *P. capitata* exhibited the highest heavy metal(loid) uptake, with a total concentration of 689 mg kg^{-1}, followed by *B. lindleyana,* with 616 mg kg^{-1}. These findings align with some previous research, emphasizing the prevalence of these two plants in areas affected by mining activities, showcasing relatively high concentrations of various heavy metal(loid)s [35,36]. In general, *S. bimaculate*, *J. effuses*, and *L. formosa* shared a similar translocation behavior, with the concentration of all heavy metal(loid)s in below-ground parts higher than those in above-ground parts, while the others presented the opposite trend, possibly due to their well-developed root systems compared to plants, leading to the high biomass of the below-ground parts.

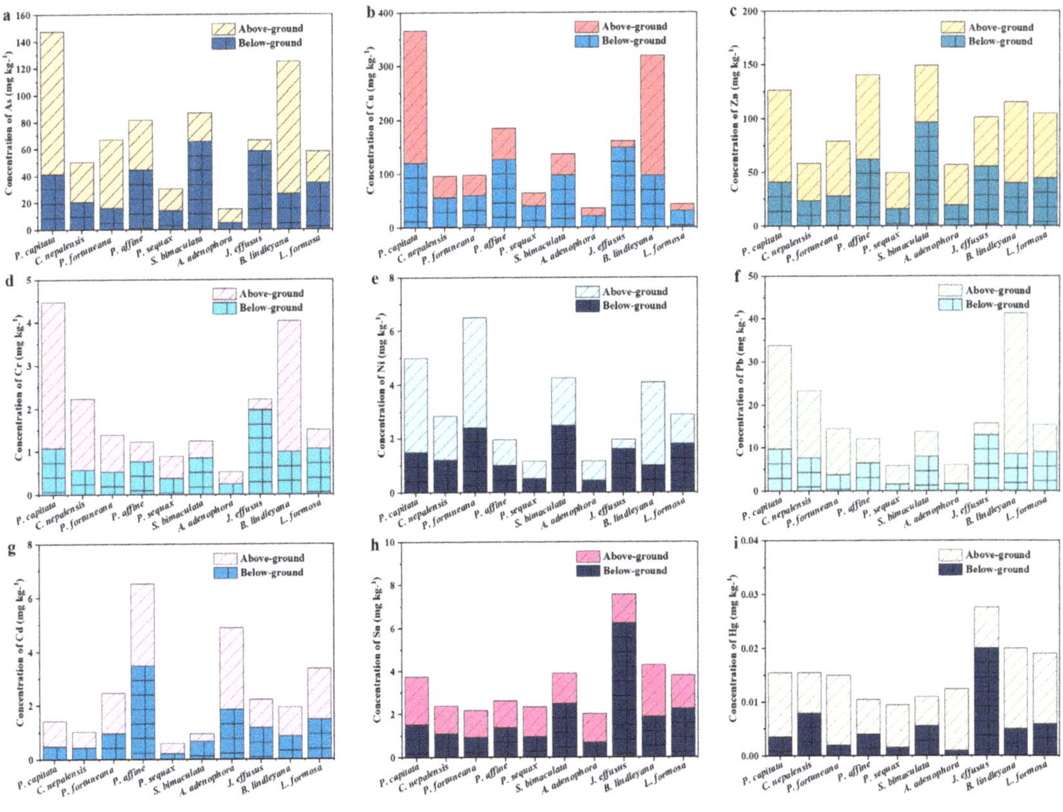

Figure 3. Concentration of heavy metal(loid)s in both above-ground and below-ground parts of the plants ((**a**) As; (**b**) Cu; (**c**) Zn; (**d**) Cr; (**e**) Ni; (**f**) Pb; (**g**) Cd; (**h**) Sn; (**i**) Hg).

From the elemental perspective, As and Cr display an analogous geochemical behaviors, inclined to form oxygen-based anions in soil, so they shared similar distributions in the plants except for *S. bimaculate*, *J. effuses*, and *L. formosa*. Other plants exhibited a great translocation ability of As and Cr as the concentration of the above-ground part significantly surpassed that of the below-ground part. Previous research indicated that, in alkaline oxidized soils (average pH = 8.28), As and Cr generally existed in the forms of As(V) and Cr(VI), respectively, and were readily absorbed by plant roots [37,38]. It is noteworthy that both As(V) and Cr(VI) can exhibit phytotoxic effects on plant growth, thereby impeding plant development [39–41]. To adapt to environments rich in these contaminants, plants transport As and Cr from their below-ground parts to their above-ground parts through compartmentalization to detoxify, resulting in relatively high concentrations in the above-ground tissues. It should be noted that these processes might involve physiological redox transformations of the elements. For example, in *Pteris vittata*, a plant species known for As hyperaccumulation, several mechanisms have been identified for its detoxification, including efficient uptake in roots via phosphate transporters *PvPht1;4*, reduction in roots via As(V) reductases, and translocation and metabolism from roots to fronds and storage into frond vacuoles via As(III) antiporters [42]. It has been reported that the reduced translocation of Cr(VI) to aerial plant parts occurs due to the conversion of Cr(VI) to Cr(III) inside the plants, along with the propensity of Cr(III) to bind to cell walls [38,43]. Compared to As, Cu, Zn, and Pb, plants exhibit significantly lower absorption and accumulation of Cd, primarily due to the relatively low total Cd content in the soil environment (with an average value of 5.69 mg kg^{-1}). Only *P. affine* demonstrates a relatively strong enrichment capability for Cd, with Cd contents in the above-ground and below-ground parts being comparable at 3.03 mg kg^{-1} and 3.49 mg kg^{-1}, respectively. In other plants, Cd tends to accumulate more in the above-ground parts. Recent research also reported the Cd phytoremediation potential of *P. affine* [44–46].

It is noteworthy that, despite the prolonged history of Sn mining in the area and the notably high average soil Sn content of 176 mg kg^{-1}, the ten predominant plant species displayed only moderate capabilities for absorbing and accumulating Sn, with an average concentration of 3.81 mg kg^{-1}. The highest recorded value was observed in *J. effusus*, reaching 7.56 mg kg^{-1} within 1.33 mg kg^{-1} in the above-ground part and 6.23 mg kg^{-1} in the below-ground part. This phenomenon may be attributed to the diminished bioavailability inherent in the specific type of Sn ores present. Previous research reported that, with the exception of oxidized Sn ores, the remaining four types of Sn ores were intricately linked to granitic intrusions, resulting in a relatively low bioavailability [28]. Additionally, only a few plants have been reported for Sn accumulation [47,48].

3.3. Bioconcentration and Translocation of Heavy Metal(loid)s in Plants

To explore the biological uptake of heavy metal(loid)s in the sampled plants, the bioconcentration factors (BCFs) of these plants were analyzed, as depicted in Figure 4. Notably, the relatively high concentration of heavy metal(loid)s in the rhizosphere soil did not consistently translate to higher concentrations in the plants. For instance, despite the rhizosphere soil at site 1 owning a total concentration of 7.25×10^3 mg kg^{-1}, the dominant plant *P. capitata* exhibited a total concentration of 689 mg kg^{-1}, with BCF values of 0.0298. This discrepancy may result from competitive adsorption and a lower bioavailability. For individual heavy metal(loid)s, despite the rhizosphere soil having average concentrations of As, Cu, Cr, and Sn of 3.98×10^3 mg kg^{-1}, 2.83×10^3 mg kg^{-1}, 68.1 mg kg^{-1}, and 176 mg kg^{-1}, respectively, the average BCF values for the dominant plants were lower than 0.05, indicating diverse phytoremediation mechanisms. Among the dominant plants, only *J. effusus* exhibited exceptional transferability of multiple heavy metal(loid)s, including As, Cu, Zn, Ni, Pb, and Cd, with BCF values of 0.181, 0.106, 0.244, 0.267, and 0.918, respectively. Klaudia et al. similarly observed that *J. effuses* played a pivotal role as a phytostabilizer for various heavy metals within the research site [14]. Additionally, Ladislas et al. utilized *J. effuses* for the removal of dissolved heavy metals, noting that Cd and Ni exhibited higher accumulation in roots compared to shoots [19].

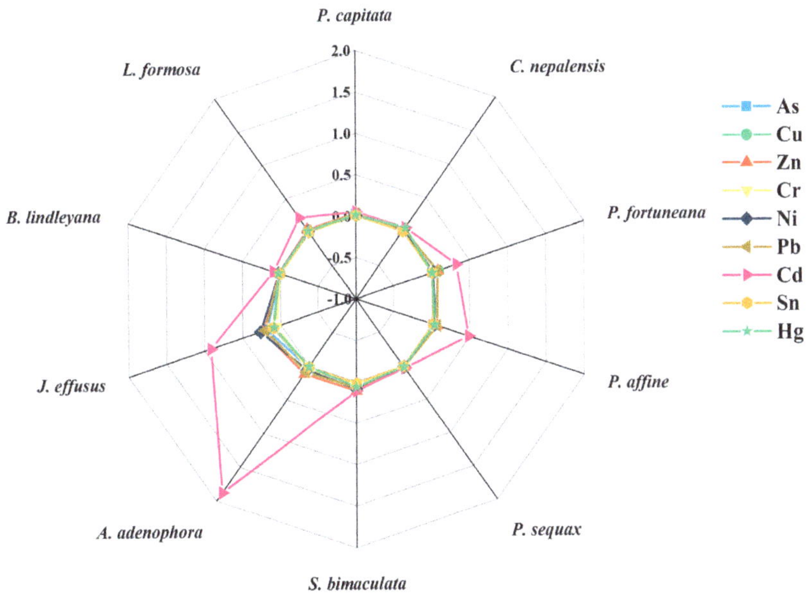

Figure 4. Bioconcentration factor (BCF) values of different heavy metal(loid)s in dominant plants.

The TF values elucidate the transfer and redistribution of heavy metal(loid)s among different plant parts. Plants with strong transferability prefer to absorb heavy metal(loid)s in the rhizosphere soil through their roots and transfer them to their shoots, protecting roots from toxicity. The average TF values decreased in the following order: Hg (3.59) > Pb (1.94) > As (1.59) > Zn (1.53) > Cr (1.48) > Ni (1.39) > Cd (1.30) > Sn (1.11) > Cu (0.84), as illustrated in Figure 5. Among these dominant plants, *P. capitata*, *C. nepalensis*, *P. fortuneana*, *P. sequax*, *A. adenophora*, and *B. lindleyana* exhibited strong transfer abilities for multiple heavy metal(loid)s, with average TF values exceeding 1, indicating their efficacy in bioextracting and biotranslocating heavy metal(loid)s away from the complex soil, to some extent. *P. capitata* displayed excellent biotransfer abilities for Zn, Cr, Cd, and Sn, with the highest TF values of 2.10, 3.17, 1.92, and 1.49, respectively. *B. lindleyana* demonstrated some greater biotransfer abilities for As (3.69), Cu (2.30), Ni (3.10), and Pb (3.85). Interestingly, only *P. capitata* and *B. lindleyana* exhibited TF values for Cu higher than 1, despite the relatively high concentrations of Cu in rhizosphere soils. Conversely, although the average concentration of Hg in rhizosphere soils was 0.196 mg kg^{-1}, significantly lower than Cu (2.83×10^3 mg kg^{-1}), the average TF values of Hg among the dominant plants were higher than those of Cu, indicating a higher bioavailability of Hg in the soils. *A. adenophora* exhibited the highest TF values for Hg of 11.5, despite the total concentration of Hg in the corresponding rhizosphere soil being 0.182 mg kg^{-1}. In addition to its capacity to phytoextract Hg, *A. adenophora* demonstrated significant potential for various heavy metal(loid)s, including As, Zn, Cr, Ni, Pb, Cd, and Sn, with IF values exceeding 1.0. This highlighted the versatility of *A. adenophora* as a potential candidate for a wide array of heavy metal removal applications. While *A. adenophora* is classified as an invasive species, it possesses attributes that render it suitable for ecological restoration, particularly in phytoremediation. Several invasive plant species have demonstrated tolerance to environmental stresses and have been identified as prospective candidates for the phytoremediation of contaminated sites. For instance, research has shown that invasive plants like *Typha latifolia*, *Ipomoea carnea*, and *Ricinus communis* have the ability to tolerate and even thrive in degraded environments, such as coal fly ash deposits, making them promising candidates for the restoration of these deposits [50]. Thus, *A. adenophora*, which possesses hyperaccumulation traits for heavy metal(loid)s, can be considered a potential phytoremediation candidate [51,52]. Nevertheless, it is crucial to implement effective monitoring and management strategies to prevent its spread beyond the intended remediation sites and mitigate any potential negative impacts on surrounding ecosystems. For other species, *S. bimaculata* and *J. effuses* displayed TF values of less than 1.0 for all heavy metal(loid)s, along with lower BCFs (<0.05), indicating their resistance and plant-stabilized barrier capabilities, consistent with total concentrations in their corresponding rhizosphere soils of 7.01×10^3 mg kg^{-1} (Site 6) and 2.12×10^3 mg kg^{-1} (Site 8). Although previous research reported a relatively high tolerance threshold for Zn in *S. bimaculata*, up to 56 mg L^{-1}, in this research, the TF and BCF values of Zn in *S. bimaculata* were 0.551 and 0.110, respectively, emphasizing the habitat-specific variations in heavy metal(loid) accumulation and tolerance among plants.

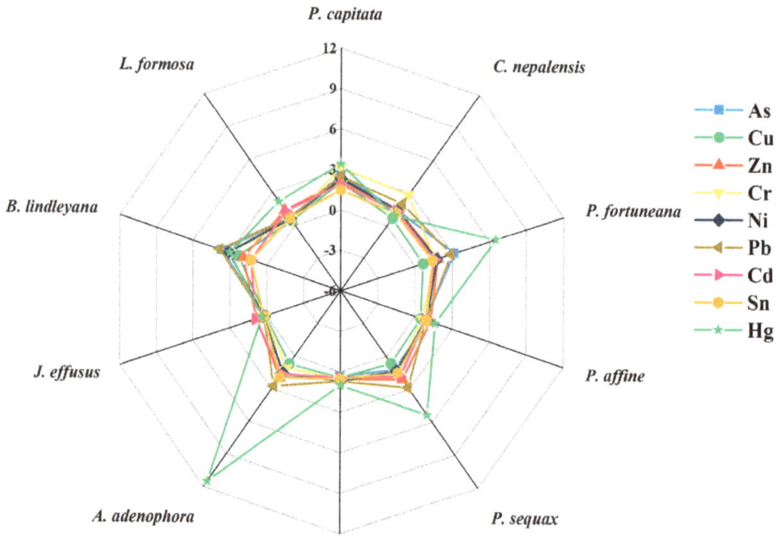

Figure 5. Transfer factor (TF) values of different heavy metal(loid)s in dominant plants.

3.4. Potential Candidates for Phytoremediation

Phytoremediation stands out as an economically advantageous approach when compared to alternative approaches like soil excavation, washing, or disposal. Utilizing dominant plants not only reduces the costs related to maintenance, such as mowing and replanting but also avoids soil disturbance, especially when potentially toxic metal(loid)s accumulate in easily harvestable plant parts, mainly the shoots. The suitability of the identified plants for phytoremediation in discharged mining areas is influenced by various factors. Previous research indicates that the efficiency of phytoextraction depends on two pivotal factors: the capacity of plants to accumulate metal(loid)s and their biomass production. The design considerations for phytoremediation systems involve the BAF, representing the ratio of metal(loid) content in the shoots to that in the soil, and plant productivity, measured by harvestable biomass. Preferably, metal(loid) accumulation in the shoots is more cost effective than in the roots, as the former allows for straightforward removal through pruning, whereas the latter necessitates uprooting the entire plant. Specific characteristics of the study sites, where shoots accumulate multiple heavy metal(loid)s from the soil, highlight the need for fast-growing plants that are easily harvestable and accumulate significant concentrations of heavy metal(loid)s in their harvestable tissues. Knowledge about plant species with BCFs and BAFs is crucial for selecting the most suitable plants for phytoremediation, potentially reducing operation and maintenance costs.

Figure 6 illustrated various BAF values of the heavy metal(loid)s in the dominant plants. Notably, despite the total Cd concentration in the rhizosphere soil at site 7 of only 0.98 mg kg^{-1}, *A. adenophora* exhibited the highest BAF value (3.11) for Cd, while the values ranged from 0.0420 to 0.829 for other species, demonstrating the highest accumulation in the shoot. It should be highlighted that the BCF of Cd in *A. adenophora* was also higher than 1 (1.66), indicating its potential as a phytoremediation candidate for Cd-contaminated soil. Previous research has emphasized the benefits of Cd accumulation against generalist pathogens, particularly under lower Cd levels, suggesting elemental defense as a key mechanism in the invasion success of *A. adenophora* in areas with prior industrial activity [51]. Moreover, *A. adenophora* has been widely identified as one of the most efficient hyperaccumulators of Cd, attributing its success to the metabolic pathways and enzymatically catalyzed reactions in hairy roots which contribute to the uptake, transformation, conjugation, and compartmentalization of Cd in vacuoles and/or cell walls, crucial detoxification sites in plants [53]. Therefore, *A. adenophora* emerges as the most representative species for phytoremediation, characterized by rapid growth, large leaves, and high Cd accumulation, making it suitable for phytoremediation purposes. The adaptability of *A. adenophora* to eroded and mining soils is evidenced by its rapid growth and substantial Cd accumulation in roots (BCF) and shoots (BAF).

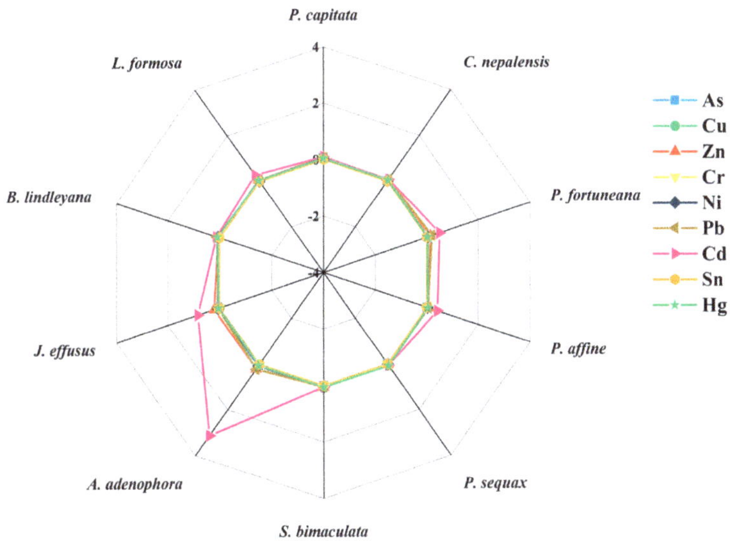

Figure 6. Bioaccumulation factor (BAF) values of different heavy metal(loid)s in dominant plants.

3.5. Prospects

In the studied area, the issue of As contamination has emerged as a significant concern due to its relatively high concentration. Notably, among the collected plant species, both *J. effusus* and *P. capitata* showed remarkable abilities for As accumulation. They exhibited the highest BCF and BAF values for As, indicating their superior tolerance to As contamination. While they may not strictly meet the criteria for hyperaccumulation [12], these species possess self-protection mechanisms against As toxicity in the soil, rendering them potential candidates for As accumulation in soil remediation and ecological restoration efforts within mining areas. Additionally, *A. adenophora* has emerged as a promising candidate for sites contaminated with multiple heavy metal(loid)s, demonstrating excellent accumulation capacities for various heavy metals (except Cu). The high TF values observed for all the elements suggest their potential in immobilizing heavy metal(loid)s from the soil and efficiently transporting them to the above-ground parts of the plant for further harvesting. However, the use of *A. adenophora* should be closely monitored and managed to prevent its spread beyond the intended remediation site and mitigate potential negative impacts on surrounding ecosystems. Interestingly, the soil in the area contained Cu concentrations second only to the concentrations of As. However, all the dominant plant species exhibited extremely low BCF and BAF values for Cu, indicating a limited bioavailability of Cu in the soil. Therefore, alternative physicochemical methods may need to be considered for remediating Cu-contaminated soil.

Furthermore, based on this investigation, a comprehensive analysis and validation of the correlation between heavy metal concentrations in plants within tailing dumpsites and their speciation in soils will be conducted. Although soil heavy metal risk control currently relies on total concentrations, the speciation of these metalloids is crucial as it dominates environmental risk. Thus, further investigation of the relationship between heavy metal forms and bioavailability is essential, providing a theoretical basis for relevant standards and policies [22].

4. Conclusions

This study, conducted in Gejiu city, Yunnan Province, focused on identifying potential candidates for phytoremediation in polymetallic-contaminated soils. The results highlighted the significant impact of historical mining activities on soil composition, particularly heavy metal(loid) contents in plants and rhizosphere soils. Different plants showed distinct adaptation strategies, including phytoextraction and phytostabilization. This study identified ten dominant plant species as promising candidates for phytoremediation due to their rapid growth, substantial biomass, adaptive advantage, successful colonization, and remediation capabilities. Among them, *P. capitata* and *J. effusus* exhibited promising potential for As phytoremediation, while *A. adenophora* showed efficiency in Cd phytoremediation. However, dominant plant species had a limited capacity for Cu accumulation, suggesting the need for alternative approaches in Cu-contaminated soils. Overall, this study provided valuable insights into the adaptive strategies employed by dominant plants in highly contaminated soils and identified specific plant candidates for effective phytoremediation strategies tailored to prevailing metal(loid) pollution.

Author Contributions: B.W.: sampling, investigation, formal analysis, and writing—original draft preparation; J.H.: experiments and investigation; X.W.: methodology, funding acquisition, and resources; X.N.: experiments and methodology; and F.Z.: supervision, conceptualization, formal analysis, visualization, and writing—review and editing. All authors have read and agreed to the published version of the manuscript.

Funding: This work was financially supported by the Key Technologies Research and Demonstration Project for Soil Remediation (No. 2018BC004), funded by the Yunnan Provincial Department of Science and Technology.

Data Availability Statement: The datasets generated and/or analyzed during the current study are available from the corresponding authors upon reasonable request.

Conflicts of Interest: Author Boxin Wang was employed by the company Shenzhen Shentai Environmental Technology Co., Ltd. The remaining authors declare that the research was conducted in the absence of any commercial or financial relationships that could be construed as a potential conflict of interest. The Shenzhen Shentai Environmental Technology Co., Ltd. had no role in the design of the study; in the collection, analyses, or interpretation of data; in the writing of the manuscript, or in the decision to publish the results.

References

1. Li, Z.; Ma, Z.; van der Kuijp, T.J.; Yuan, Z.; Huang, L. A review of soil heavy metal pollution from mines in China: Pollution and health risk assessment. *Sci. Total Environ.* **2014**, *468–469*, 843–853. [CrossRef] [PubMed]
2. Liu, L.; Li, W.; Song, W.; Guo, M. Remediation techniques for heavy metal-contaminated soils: Principles and applicability. *Sci. Total Environ.* **2018**, *633*, 206–219. [CrossRef] [PubMed]
3. Ali, H.; Khan, E.; Sajad, M.A. Phytoremediation of heavy metals—Concepts and applications. *Chemosphere* **2013**, *91*, 869–881. [CrossRef] [PubMed]

4. Álvarez-Mateos, P.; Alés-Álvarez, F.J.; García-Martín, J.F. Phytoremediation of highly contaminated mining soils by *Jatropha curcas* L. and production of catalytic carbons from the generated biomass. *J. Environ. Manag.* **2019**, *231*, 886–895. [CrossRef] [PubMed]
5. Prasad, M.N.V. Phytoremediation of Metal-Polluted Ecosystems: Hype for Commercialization. *Russ. J. Plant Physiol.* **2003**, *50*, 686–701. [CrossRef]
6. Mench, M.; Schwitzguebel, J.P.; Schroeder, P.; Bert, V.; Gawronski, S.; Gupta, S. Assessment of successful experiments and limitations of phytotechnologies: Contaminant uptake, detoxification and sequestration, and consequences for food safety. *Environ. Sci. Pollut. Res. Int.* **2009**, *16*, 876–900. [CrossRef] [PubMed]
7. Wang, J.; Aghajani Delavar, M. Techno-economic analysis of phytoremediation: A strategic rethinking. *Sci. Total Environ.* **2023**, *902*, 165949. [CrossRef] [PubMed]
8. Mendez, M.O.; Maier, R.M. Phytoremediation of mine tailings in temperate and arid environments. *Rev. Environ. Sci. Bio/Technol.* **2008**, *7*, 47–59. [CrossRef]
9. Yan, A.; Wang, Y.; Tan, S.N.; Mohd Yusof, M.L.; Ghosh, S.; Chen, Z. Phytoremediation: A Promising Approach for Revegetation of Heavy Metal-Polluted Land. *Front. Plant Sci.* **2020**, *11*, 359. [CrossRef]
10. Baker, A.J.M.; Brooks, R.R. Terrestrial higher plants which hyperaccumulate metal elements. A review of their distribution, ecology, and phytochemistry. *Biorecovery* **1989**, *1*, 81–126.
11. Mocek-Plociniak, A.; Mencel, J.; Zakrzewski, W.; Roszkowski, S. Phytoremediation as an Effective Remedy for Removing Trace Elements from Ecosystems. *Plants* **2023**, *12*, 1653. [CrossRef] [PubMed]
12. Li, J.-t.; Gurajala, H.K.; Wu, L.-h.; Van Der Ent, A.; Qiu, R.L.; Baker, A.J.M.; Tang, Y.-t.; Yang, X.-E.; Shu, W.-S. Hyperaccumulator Plants from China: A Synthesis of the Current State of Knowledge. *Environ. Sci. Technol.* **2018**, *52*, 11980–11994. [CrossRef] [PubMed]
13. Niu, X.; Jia, Y.; Wu, X.; Wang, S.; Hou, J.; Zhang, W. Phytoremediation potential of indigenous plants growing in soils affected by mine activities in Gejiu City, Yunnan Province. *Int. J. Phytoremediat.* **2022**, *25*, 880–888. [CrossRef] [PubMed]
14. Schachtschneider, K.; Chamier, J.; Somerset, V. Phytostabilization of metals by indigenous riparian vegetation. *Water SA* **2017**, *43*, 177–185. [CrossRef]
15. Zeidler, M.; Šipoš, J.; Banaš, M.; Černohorský, J. The successive trend of vegetation confirms the removal of non-indigenous woody species as an insufficient restoration action. *Biodivers. Conserv.* **2021**, *30*, 699–717. [CrossRef]
16. Albert, K.M. Role of revegetation in restoring fertility of degraded mined soils in Ghana: A review. *Int. J. Biodivers. Conserv.* **2015**, *7*, 57–80. [CrossRef]
17. Yang, X.e.; Long, X.; Ni, W.; Fu, C. *Sedum alfredii* H: A new Zn hyperaccumulating plant first found in China. *Chin. Sci. Bull.* **2002**, *47*, 1634–1637. [CrossRef]
18. Xu, W.; Du, Q.; Yan, S.; Cao, Y.; Liu, X.; Guan, D.X.; Ma, L.Q. Geographical distribution of As-hyperaccumulator *Pteris vittata* in China: Environmental factors and climate changes. *Sci. Total Environ.* **2022**, *803*, 149864. [CrossRef] [PubMed]
19. Visoottiviseth, P.; Francesconi, K.; Sridokchan, W. The potential of Thai indigenous plant species for the phytoremediation of arsenic contaminated land. *Environ. Pollut.* **2002**, *118*, 453–461. [CrossRef]
20. Oyuela Leguizamo, M.A.; Fernández Gómez, W.D.; Sarmiento, M.C.G. Native herbaceous plant species with potential use in phytoremediation of heavy metals, spotlight on wetlands—A review. *Chemosphere* **2017**, *168*, 1230–1247. [CrossRef]
21. Wu, B.; Peng, H.; Sheng, M.; Luo, H.; Wang, X.; Zhang, R.; Xu, F.; Xu, H. Evaluation of phytoremediation potential of native dominant plants and spatial distribution of heavy metals in abandoned mining area in Southwest China. *Ecotoxicol. Environ. Saf.* **2021**, *220*, 112368. [CrossRef] [PubMed]
22. Du, Y.; Tian, Z.; Zhao, Y.; Wang, X.; Ma, Z.; Yu, C. Exploring the accumulation capacity of dominant plants based on soil heavy metals forms and assessing heavy metals contamination characteristics near gold tailings ponds. *J. Environ. Manag.* **2024**, *351*, 119838. [CrossRef] [PubMed]
23. Bothe, H. Plants in Heavy Metal Soils. In *Detoxification of Heavy Metals*; Sherameti, I., Varma, A., Eds.; Springer: Berlin/Heidelberg, Germany, 2011; pp. 35–57.
24. Singh, S.; Karwadiya, J.; Srivastava, S.; Patra, P.K.; Venugopalan, V.P. Potential of indigenous plant species for phytoremediation of arsenic contaminated water and soil. *Ecol. Eng.* **2022**, *175*, 106476. [CrossRef]
25. Odoh, C.K.; Zabbey, N.; Sam, K.; Eze, C.N. Status, progress and challenges of phytoremediation—An African scenario. *J. Environ. Manag.* **2019**, *237*, 365–378. [CrossRef] [PubMed]
26. Mahar, A.; Wang, P.; Ali, A.; Awasthi, M.K.; Lahori, A.H.; Wang, Q.; Li, R.; Zhang, Z. Challenges and opportunities in the phytoremediation of heavy metals contaminated soils: A review. *Ecotoxicol. Environ. Saf.* **2016**, *126*, 111–121. [CrossRef] [PubMed]
27. Lee, S.-H.; Park, H.; Kim, J.-G. Current Status of and Challenges for Phytoremediation as a Sustainable Environmental Management Plan for Abandoned Mine Areas in Korea. *Sustainability* **2023**, *15*, 2761. [CrossRef]
28. Zhang, Y.H.; Zhou, J.X.; Tan, S.C.; Li, H.M.; Hao, S.; Jiang, Y.G.; He, X.H. Genesis of the oxidized Sn ores in the Gejiu district, Yunnan Province, SW China. *Ore Geol. Rev.* **2020**, *121*, 103474. [CrossRef]
29. Sun, S.; Dong, C.; Zhang, H.; Yang, H.; Zhang, N.; Bao, L. Heavy Metal Risk Assessment of Soils and Crops in the Intensive Mining Area of Gejiu, Yunnan Province. *J. Geosci. Environ. Prot.* **2022**, *10*, 128–139. [CrossRef]
30. Xiao, Y.; Liu, C.; Hu, N.; Wang, B.; Zheng, K.; Zhao, Z.; Li, T. Contributions of ectomycorrhizal fungi in a reclaimed poplar forest (*Populus yunnanensis*) in an abandoned metal mine tailings pond, southwest China. *J. Hazard. Mater.* **2023**, *448*, 130962. [CrossRef]

31. Li, C.; Dong, P.; Yan, J.; Gong, R.; Meng, Q.; Yao, J.; Yu, H.; Ma, Y.; Liu, B.; Xie, R. Analytical study on heavy metal output fluxes and source apportionment of a non-ferrous smelter in southwest China. *Environ. Pollut.* **2023**, *331*, 121867. [CrossRef]
32. Ascar, L.; Ahumada, I.; Richter, P. Influence of redox potential (Eh) on the availability of arsenic species in soils and soils amended with biosolid. *Chemosphere* **2008**, *72*, 1548–1552. [CrossRef]
33. Gil-Cardeza, M.L.; Ferri, A.; Cornejo, P.; Gomez, E. Distribution of chromium species in a Cr-polluted soil: Presence of Cr(III) in glomalin related protein fraction. *Sci. Total Environ.* **2014**, *493*, 828–833. [CrossRef] [PubMed]
34. Ma, P.; Zhang, Y.; Gao, L.; Guo, J.; Xu, L.; Ding, Y. Clustering and Principal Component Analysis for the Heavy Metal Contents of Soil in Yunnan Province. In Proceedings of the 2012 Second International Conference on Electric Technology and Civil Engineering, Bangalore, India, 3–4 August 2012; IEEE Computer Society: Washington, DC, USA, 2012; pp. 657–660.
35. Nedelkoska, T.V.; Doran, P.M. Characteristics of heavy metal uptake by plant species with potential for phytoremediation and phytomining. *Miner. Eng.* **2000**, *13*, 549–561. [CrossRef]
36. Zhang, Y.; Song, B.; Zhu, L.; Zhou, Z. Evaluation of the metal(loid)s phytoextraction potential of wild plants grown in three antimony mines in southern China. *Int. J. Phytoremediat.* **2020**, *23*, 781–790. [CrossRef] [PubMed]
37. Peng, H.; Yi, L.; Liu, C. Spatial distribution, chemical fractionation and risk assessment of Cr in soil from a typical industry smelting site in Hunan Province, China. *Environ. Geochem. Health* **2024**, *46*, 113. [CrossRef] [PubMed]
38. Shahid, M.; Shamshad, S.; Rafiq, M.; Khalid, S.; Bibi, I.; Niazi, N.K.; Dumat, D.C.; Rashid, M.I. Chromium speciation, bioavailability, uptake, toxicity and detoxification in soil-plant system: A review. *Chemosphere* **2017**, *178*, 513–533. [CrossRef] [PubMed]
39. Ran, M.; Lu, Y.; Ren, Y.; He, L.; Li, J. Efficient reduction of Cr(VI) by guava (*Psidium guajava*) leaf extract and its mitigation effect on Cr toxicity in rice seedlings. *J. Environ. Sci.* **2024**, *141*, 1–15. [CrossRef] [PubMed]
40. Bouhadi, M.; Lahmidi, A.; Am, A.; Elhajjouji, H.; Elkouali, M.; Talbi, M.; Fougrach, H. Study of the Toxicity and Translocation of Chromium (VI) in *Vicia faba* Plant. *Bull. Environ. Contam. Toxicol.* **2024**, *112*, 40. [CrossRef] [PubMed]
41. Farooq, M.A.; Islam, F.; Ali, B.; Najeeb, U.; Mao, B.; Gill, R.A.; Yan, G.; Siddique, K.H.M.; Zhou, W. Arsenic toxicity in plants: Cellular and molecular mechanisms of its transport and metabolism. *Environ. Exp. Bot.* **2016**, *132*, 42–52. [CrossRef]
42. Sun, D.; Zhang, X.; Yin, Z.; Feng, H.; Hu, C.; Guo, N.; Tang, Y.; Qiu, R.; Ma, L.Q.; Cao, Y. As-hyperaccumulator Pteris vittata and non-hyperaccumulator *Pteris ensiformis* under low As-exposure: Transcriptome analysis and implication for As hyperaccumulation. *J. Hazard. Mater.* **2023**, *458*, 132034. [CrossRef]
43. Han, F.X.; Su, Y.; Sridhar, B.B.M.; Monts, D.L. Distribution, transformation and bioavailability of trivalent and hexavalent chromium in contaminated soil. *Plant Soil.* **2004**, *265*, 243–252. [CrossRef]
44. Soubasakou, G.; Cavoura, O.; Damikouka, I. Phytoremediation of Cadmium-Contaminated Soils: A Review of New Cadmium Hyperaccumulators and Factors Affecting their Efficiency. *Bull. Environ. Contam. Toxicol.* **2022**, *109*, 783–787. [CrossRef] [PubMed]
45. Li, Q.; Liu, X.; Qin, Y. Cadmium phytoremediation potential of *Gnaphalium affine D. Don.* and physiological response in metal tolerance. *IOP Conf. Ser. Earth Environ. Sci.* **2020**, *546*, 042029. [CrossRef]
46. Padmavathiamma, P.K.; Li, L.Y. Phytoremediation Technology: Hyper-accumulation Metals in Plants. *Water Air Soil Pollut.* **2007**, *184*, 105–126. [CrossRef]
47. Liu, Y.; Xu, W.; Wang, Y.; Hao, W.; Zhou, Q.; Liu, J. Growth Responses and Accumulation Characteristics of Three Ornamental Plants to Sn Contamination in Soil. *Agriculture* **2021**, *11*, 205. [CrossRef]
48. Ashraf, M.A.; Maah, M.J.; Yusoff, I. Assessment of phytoextraction efficiency of naturally grown plant species at the former tin mining catchment. *Fresenius Environ. Bull.* **2012**, *21*, 523–533.
49. Ladislas, S.; Gérente, C.; Chazarenc, F.; Brisson, J.; Andrès, Y. Performances of Two Macrophytes Species in Floating Treatment Wetlands for Cadmium, Nickel, and Zinc Removal from Urban Stormwater Runoff. *Water Air Soil Pollut.* **2013**, *224*, 1408. [CrossRef]
50. Singh, S.; Saha, L.; Kumar, M.; Bauddh, K. Chapter 12—Phytoremediation potential of invasive species growing in mining dumpsite. In *Phytorestoration of Abandoned Mining and Oil Drilling Sites*; Bauddh, K., Korstad, J., Sharma, P., Eds.; Elsevier: Amsterdam, The Netherlands, 2021; pp. 287–305.
51. Dai, Z.C.; Cai, H.H.; Qi, S.S.; Li, J.; Zhai, D.L.; Wan, J.S.H.; Du, D.L. Cadmium hyperaccumulation as an inexpensive metal armor against disease in Crofton weed. *Environ. Pollut.* **2020**, *267*, 115649. [CrossRef]
52. Khan, I.U.; Qi, S.-S.; Gul, F.; Manan, S.; Rono, J.K.; Naz, M.; Shi, X.-N.; Zhang, H.; Dai, Z.-C.; Du, D.-L. A Green Approach Used for Heavy Metals 'Phytoremediation' Via Invasive Plant Species to Mitigate Environmental Pollution: A Review. *Plants* **2023**, *12*, 725. [CrossRef]
53. Agostini, E.; Talano, M.A.; González, P.S.; Oller, A.L.W.; Medina, M.I. Application of hairy roots for phytoremediation: What makes them an interesting tool for this purpose? *Appl. Microbiol. Biotechnol.* **2013**, *97*, 1017–1030. [CrossRef]

processes

Article

In Situ Remediation of Combined Ammonia and Nitrate Nitrogen Contamination Using Zero-Valent Iron-Enhanced Microorganisms in Acidic Groundwater: A Laboratory and Pilot-Scale Study

Junyi Chen [1], Yuchi Luo [2], Junda Zhang [3], Zexuan Lu [3], Yitong Han [2,*], Xiangxin Chen [1], Mingkui Li [1], Hanyun Fan [1] and Xipeng Wei [3,*]

[1] Guangzhou Runfang Environmental Protection Technology Co., Ltd., Guangzhou 510670, China
[2] Guangdong Provincial Academy of Environmental Science, Guangzhou 510045, China
[3] Guangdong Engineering Technology Research Center for Source Control of Combined Pollution in Mining Areas, Guangdong Provincial Key Laboratory of Chemical Pollution and Environmental Safety & MOE Key Laboratory of Theoretical Chemistry of Environment, South China Normal University, Guangzhou 510006, China
* Correspondence: hanyt1224@foxmail.com (Y.H.); weixipengyd@scnu.edu.cn (X.W.)

Abstract: In acidic groundwater, effectively removing both ammonia nitrogen (NH_4^+-N) and nitrate nitrogen (NO_3^--N) poses a challenge. This study focused on studying the removal of NH_4^+-N and NO_3^--N combined contaminations by zero-valent iron (ZVI) combined with microbial agents in both laboratory and field pilot-scale studies. Laboratory experiments showed that ZVI could reduce the denitrification stage from 15 days to 10 days by increasing solution pH and improving NO_3^--N reduction efficiency. In a field pilot test (at Qingyuan, Guangdong Province, China), high-pressure injection pumps were used to inject alkaline reagents to raise the pH to 7~8. Meanwhile, compressors were applied to aerate the groundwater to increase the dissolved oxygen (DO) concentration above 2 mg·L^{-1}. Subsequently, microbial agents of nitrobacteria were injected to initiate aerobic nitrification. As the DO level dropped below 2 mg·L^{-1}, agents of micro-ZVI and denitrifying bacteria were injected to stimulate autotrophic denitrification. Intermittent aeration was employed to modify the redox conditions in the groundwater to gradually eliminate NH_4^+-N and NO_3^--N. However, due to the effect of the low-permeability layers, adjustments in the frequency of remediation agent injection and aeration were necessary to achieve removal efficiencies exceeding 80% for both NH_4^+-N and NO_3^--N. This work aims to overcome the limitations of microbial remediation methods in the laboratory and the field and advance nitrogen pollution remediation technologies in groundwater.

Keywords: groundwater remediation; ammonia nitrogen; nitrate; nitration; denitrification; zero-valent iron

1. Introduction

Global groundwater pollution, including from pollutants such as petroleum, solvents, volatile organic compounds, pesticides, heavy metals, and nitrogen compounds, is a widespread and increasingly severe global issue [1–4]. Among various pollutants, the presence of ammonium nitrogen (NH_4^+-N) and nitrate nitrogen (NO_3^--N) in groundwater due to excessive application of agricultural fertilizers, agricultural drainage, urban and industrial wastewater, and surface water leakage have greatly impacted groundwater ecology and water safety [5–7]. NH_4^+-N is a highly oxygen-consuming substance that has toxic effects on humans and aquatic organisms [7]. The oxidation of 1 mg of NH_4^+-N to NO_3^--N requires the consumption of 4.57 mg of dissolved oxygen (DO); therefore, high concentrations of NH_4^+-N directly lead to severe oxygen depletion in groundwater, resulting in water-quality deterioration [8,9]. Furthermore, prolonged consumption of nitrogen-rich water increases the burden on the human kidneys and raises the potential risks of conditions such as methemoglobinemia, digestive tract cancer, bladder cancer, ovarian cancer, and diseases of the brain and nervous system [10–12]. The "Standard for Drinking Water Quality" [13] in China sets strict limits for NH_4^+-N and NO_3^--N contamination in drinking water at 0.5 mg/L and 10 mg/L, respectively. Therefore, the elimination of NH_4^+-N and NO_3^--N combined contamination in groundwater has long been a focal point for the government and the field of environmental remediation.

Microbial remediation technology has shown feasibility and potential in the practical application of removing NH_4^+-N and NO_3^--N contamination in groundwater [14,15]. For the removal of NH_4^+-N and NO_3^--N combined contamination in groundwater, a reliable approach involves the synergistic interaction between nitrifying bacteria-mediated nitrification and denitrification induced by denitrifying bacteria [16–18]. Nitrifying bacteria primarily oxidize NH_4^+-N to nitrite nitrogen (NO_2^--N), followed by the oxidation of NO_2^--N to NO_3^--N using the energy from the oxidation reaction to acquire energy and electrons for growth. Therefore, nitrifying bacteria do not require an external carbon source. However, denitrification is the process of reducing NO_3^--N to nitrogen gas (N_2) or NO_2^--N and typically occurs under oxygen-depleted conditions [14,19] The denitrification process is relatively slower and requires organic matter as an electron donor to proceed. Therefore, nitrification is usually more effective than denitrification in reducing NH_4^+-N and NO_3^--N combined contaminations in groundwater. The strategy to enhance denitrification efficiency has become the key issue in effectively removing the NH_4^+-N and NO_3^--N combined contamination [19,20]. Zero-valent iron (ZVI) is a commonly employed remediation material that finds extensive application in the remediation of groundwater and soil. ZVI demonstrates exceptional capabilities in reducing NO_3^--N [21]. For example, under alkaline conditions (pH range approximately 6–10), ZVI can reduce NO_3^--N to nitrite (NO_2^--N) (Equation (1)) or nitrogen gas (N_2) (Equation (1)), thereby enhancing denitrification [22,23].

$$NO_3^- + 3\,Fe^0 \rightarrow NO_2^- + 3\,Fe^{2+} + 2\,OH^- \quad (1)$$

$$NO_3^- + 10\,Fe^0 + 12\,H^+ \rightarrow 5/2\,N_2 + 10\,Fe^{2+} + 6\,H_2O \quad (2)$$

Additionally, acidic groundwater environment at the actual site is not conducive to the survival of denitrifying bacteria, while ZVI can react with acidic substances in groundwater, indirectly enhancing denitrification efficiency [23,24]. Therefore, through the synergistic effects of nitrifying bacteria, denitrifying bacteria, and ZVI, along with suitable site conditions and operational controls, it is feasible to enhance the removal efficiency for the co-pollutants of NH_4^+-N and NO_3^--N in groundwater. However, existing studies on ZVI-assisted microbial remediation have predominantly concentrated on single NH_4^+-N or NO_3^--N contamination, with limited studies on the concurrent elimination of NH_4^+-N and NO_3^--N combined pollution [6,25–27]. Specifically, ZVI-assisted microbial remediation studies have been primarily restricted to laboratory-scale studies, with few instances of field applications demonstrating synergistic improvement of denitrification by ZVI and microorganisms.

In the Qingyuan region of Guangdong, China, a typical site exhibits groundwater contamination with NH_4^+-N and NO_3^--N. This contamination has been worsened by the unconventional extraction of rare earth minerals, which involves the introduction of ammonium salt solutions into leach ponds, leaching heaps, or mountainous terrain, resulting in soil and groundwater acidification (with local soil pH levels ranging from 4.56 to 6.49). Consequently, this area presents a significantly severe case of NH_4^+-N and NO_3^--N combined contamination, making it an ideal candidate for our pilot-scale study.

In this study, laboratory experiments were conducted to simulate the groundwater contamination characteristics of the research area, validating the enhanced denitrification effects of ZVI and microbial agents. Subsequently, technologies based on ZVI and microbial agents were applied in on-site experiments to investigate the removal of NH_4^+-N and NO_3^--N combined contaminations. Additionally, the feasibility and operability of the remediation technology were assessed to analyze its practical application outcomes. This study is crucial for advancing the development of remediation technologies aimed at addressing the combined contamination of groundwater by NH_4^+-N and NO_3^--N.

2. Research Area and Methods

2.1. Materials

ZVI powder (>98%) was purchased from Gongyi City Meiqi Industry and Trade Co., Ltd., Zhengzhou, China. Sodium bicarbonate, Nessler's reagent, and KI, were purchased from the Aladdin Reagent Co., Shanghai, China. Amino sulfonic acid (analytically pure) was obtained from the Tianjin Damao Chemical Reagent Co., Tianjin, China.

2.2. Reaction Column Experiment in Laboratory

Laboratory experiments were conducted using column reactors to simulate the remediation process (Figure S1). The main component of the experimental setup is an organic glass column, 105 cm in height, with an inner diameter of 15 cm and an outer diameter of 17 cm. The described column is equipped with an outlet on one side at the top and an inlet on the opposite side at the bottom. Between the inlet and outlet, there are five distinct sampling ports. The inlet is connected to a water tank via a water pipe, with a peristaltic pump controlling the water flow rate. Each end of the glass column is filled with a 5 cm thick layer of coarse sand with a particle size ranging from 1 to 2 mm, which serves to filter, buffer, and protect the intermediate fillings. Adjacent to the coarse sand layer is a 20 cm thick layer of fine to medium sand with particle sizes ranging from 0.25 to 0.5 mm. The contaminated soil sample collected from the study area is placed next to the fine sand layer, simulating the aquifer zone and facilitating uniform water distribution. The middle 60 cm serves as the reactive fill zone, where the experimental requirements dictate the use of soil, sand, microbial agents and ZVI powder as the active fillings. The glass column is equipped with five sampling ports distributed in the middle section, with intervals of 15 cm between each port. The coarse sand and fine sand are thoroughly washed with distilled water, air-dried before use, and stored in a drying cabinet. The experimental water is synthetic groundwater with concentrations of NH_4^+-N and NO_3^--N set at 50 mg/L. The pH value of the newly prepared water ranges from 4.5 to 8.5. By varying the experimental conditions, such as different ratios of remediation materials, pollutant concentrations, solution pH, and DO, the impact of these factors on the removal efficiency for NH_4^+-N and NO_3^--N is investigated, gradually optimizing the remediation technology. The sampling frequency is once a day, and the removal efficiency for NH_4^+-N and NO_3^--N pollutants is calculated based on the concentration of pollutants at the final outlet.

To obtain nitrifying and denitrifying microbial communities, domesticated sludge from a sewage treatment plant in Qingyuan City was utilized. Representative sludge samples were collected from the secondary sedimentation tank of the sewage treatment plant. These sludge samples were exposed to different cultivation conditions, such as varying temperature, pH, and DO, etc., using culture media containing NH_4^+-N and NO_3^--N. This allowed the microbial communities to gradually adapt and grow in specific environments. Regular sampling (sampling once a day as needed) was conducted, and the concentrations of NH_4^+-N and NO_3^--N were measured to assess changes in the degradation capacity of the microbial communities. After a period of domestication, stability tests were performed on the domesticated sludge to confirm its high stability in degrading NH_4^+-N and NO_3^--N.

2.3. Hydrogeological and Climatic Characteristics of the Study Area

The pilot study site is located in Qingyuan City, Guangdong Province, China, at coordinates 113.712728° E and 24.103932° N (Figure 1). The site covers an area of approximately 27 acres, with a length of about 500 m from north to south and a width of about 300 m from east to west. The area is characterized by the accumulation of tailings resulting from illegal exploitation of rare earth minerals. It is situated in a low mountainous and hilly region, with generally higher western and northern topography compared to the central part. The low-lying areas exhibit relatively flat terrain. Groundwater in the area converges from the southern and western slopes of the pilot area towards the central valley and eventually discharges towards the southeast direction.

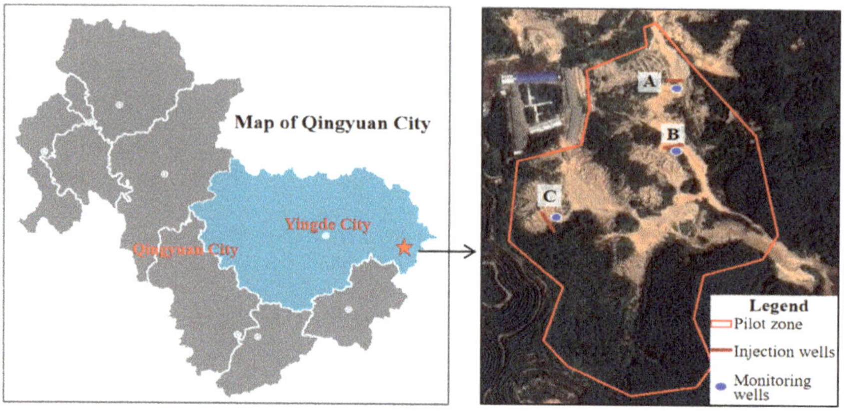

Figure 1. Location map of the pilot study area and research site. The stars indicate the locations within the study area, while points A, B, and C represent three representative research sites.

Based on hydrogeological surveys conducted in the pilot area, the predominant aquifer media consist of sandy loam and silty clay. The study area encompasses unconsolidated rock pore water and weathered rock fracture water. Unconsolidated rock pore water is mainly distributed in the Quaternary alluvial deposits, consisting of fine sand, medium sand, and coarse sand, etc. However, water permeability in clayey and silty clay layers is low, resulting in poor water-yielding potential, with individual well yields below 100 m^3/d. Annual fluctuations in water level for unconsolidated rock pore water range from 1 to 3 m, while those for weathered rock fracture water range from 2 to 5 m. The residual soil from the heap leaching of rare earth minerals and completely weathered granite form loose accumulations with a thickness of approximately 4 to 12 m. This type of groundwater primarily flows and discharges along the pores within the tailings, exhibiting good permeability and low water-yielding characteristics. In situ measurements indicate horizontal soil hydraulic conductivity (k_v) ranging from 4.6×10^{-6} to 3.6×10^{-5} cm/s and a vertical soil hydraulic conductivity (k_h) ranging from 4.1×10^{-6} to 2.5×10^{-5} cm/s. The measured soil porosity is approximately 30%. The calculated migration velocities of pollutants in the groundwater within the pilot area range from 0.01 to 0.12 m/d horizontally and from 0.01 to 0.10 m/d vertically, indicating a relatively slow and uneven distribution of pollutant migration within the groundwater.

The study area is located in the transitional zone from the South Subtropical Zone to the Central Subtropical Zone and experiences a subtropical monsoon climate. The annual average temperature ranges from 20.1 °C to 22.0 °C [28]. The groundwater dynamics in the surrounding area exhibit a seasonal pattern primarily influenced by rainfall. During the spring and summer seasons, precipitation is abundant, leading to increased groundwater recharge.

2.4. Groundwater Pollution Characteristics

The contamination at the site originated from the improper leaching process used in the past, which introduced leaching agents $(NH_4)_2SO_4$ and precipitants such as oxalic acid or NH_4HCO_3 into the mountain. The soil pH in the study area ranged from 4.56 to 6.49, indicating weak acidity. The monitoring results of the groundwater revealed that the

NH_4^+-N contamination concentrations in the research area ranged from 1.1 to 97.38 mg/L, with an average of 26.76 mg/L, while the NO_3^--N concentrations ranged from 0.21 to 51.35 mg/L, with an average of 21.2 mg/L. The initial average concentration of NO_2^--N was 0.06 mg/L, with a maximum concentration of 0.329 mg/L. The concentrations of NH_4^+-N, NO_3^-N, and NO_2^--N in the three pilot sites to be selected in this study are shown in Table 1.

Table 1. Nitrogen pollution concentration in different injection wells.

Sites	NH_4^+-N (mg/L)	NO_3^--N (mg/L)	NO_2^--N (mg/L)	Total Nitrogen (mg/L)
A	31.3	18.7	0.105	50.105
B	25.3	12.5	0.049	37.849
C	34.9	51.35	0.045	86.295

2.5. Analytical Methods

The DO and pH in the groundwater were monitored by a multi-parameter water quality analyzer (SX836, Shanghai Sanxin Instrument Factory, Shanghai, China). The Nessler's reagent spectrophotometric method was used to determine the NH_4^+-N content in water samples on a spectrophotometer (Metash UV-9000, Yourop Scientific Instruments (Shanghai) Co., Ltd., Shanghai, China), while the ultraviolet spectrophotometric method was employed to measure the NO_3^--N content [29]. The N-(1-Naphthyl) ethylenediamine spectrophotometric method was employed to determine the NO_2^--N content in water samples [30]. Groundwater samples were collected from source points or wellheads, ensuring that the collection process was free from the contamination of exogenous microbial communities. The total colony counts in the groundwater were determined using the plate count method or the spread plate method. Details of the detection methods for NH_4-N, NO_3-N, NO_2-N, and the total bacterial colonies have been listed in Text S1. The removal efficiencies (RE) for NH_4-N and/or NO_3-N were calculated by Equation (3):

$$RE\ (\%) = (C_0 - C_t)/C_0 \times 100\% \tag{3}$$

where C_0 represents the initial concentration of NH_4-N or NO_3-N before the injection of the ZVI and microbial reagents, and C_t refers to the concentration of residual NH_4-N or NO_3-N after incubation for t days.

3. Results and Discussion

3.1. The Column Experiment Results

In the column reactor without any additional bacteria or ZVI (control), the removal efficiencies for NH_4^+-N and NO_3^--N are less than 10% and 5%, respectively (Figure 2a). The minimal removal efficiency observed for NH_4^+-N could potentially be attributed to the presence of minute quantities of oxygen in the water circulating within the reactor, rather than to microbial degradation. Considering that the reaction column was filled with undisturbed soil samples sourced from the contaminated site, the results suggest a lack of a microbial community proficient in simultaneously removing NH_4^+-N and NO_3^--N under acidic conditions at the contaminated site [7,31]. This result also explains why the concentrations of NH_4^+-N and NO_3^--N in groundwater remain high despite decades of contamination at the pilot site.

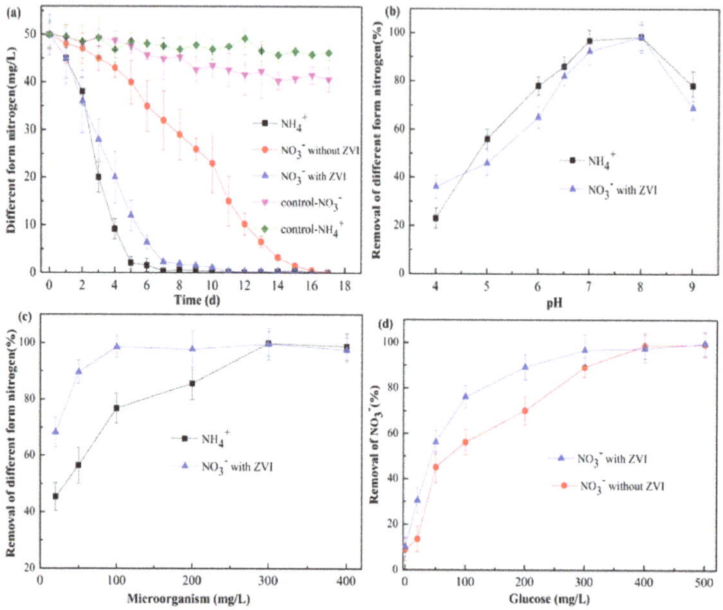

Figure 2. Influence of different factors on nitrification and denitrification efficiency. The effect of ZVI on the microbial degradation of NH_4^+-N and NO_3^--N, respectively (**a**), the effects of different pH (**b**), microbial mass concentration (**c**), and glucose concentration (**d**) on the removal of NH_4^+-N and NO_3^--N in a ZVI/microbial complex system (reaction conditions: $[ZVI]_0$ = 200 mg·L^{-1}, $[microorganism]_0$ = 20–400 mg·L^{-1}, $[glucose]_0$ = 0–500 mg·L^{-1}, and pH_{ini} = 4–9).

In the influent with an initial concentration of 50 mg/L of NH_4^+-N, the removal of NH_4^+-N by the column reactor with nitrobacteria reached 99.5% within 5 days (Figure 2a). However, it takes 15 days for denitrifying bacteria to remove 50 mg/L of NO_3^--N. Surprisingly, ZVI shortens the remediation period of NO_3^--N driven by denitrifying bacteria to about 10 days (Figure 2a). ZVI may increase denitrification via the following mechanisms: (i) ZVI is a strong reducing agent, and ZVI can provide electrons to denitrifying bacteria for the reduction of NO_3^--N to N_2 or NO_2^--N [6]. (ii) the iron oxide shells of ZVI, such as magnetite, produced by the reaction of ZVI with NO_3^--N, exhibit excellent adsorption capacity [32]. This property of ZVI allows it to adsorb NO_3^--N on its surface, facilitating the electron transfer process between the iron core of the ZVI and the NO_3^--N, thereby enhancing denitrification efficiency.

pH plays a significant role in affecting the redox potential of iron and nitrogen and the solubility of ZVI and iron minerals, as well as microbial activity, thereby exerting a considerable influence on the nitrification and denitrification processes [33]. Under pH conditions of 7–8, the removal efficiency for NH_4^+-N and NO_3^--N were the highest (both greater than 95%), but as the pH decreased to 4.0, the removal efficiency for NH_4^+-N and NO_3^--N significantly decreased from over 99% to around 40% or lower (Figure 2b). Generally, lower pH levels can inhibit microbial growth. Furthermore, under neutral and alkaline conditions, the redox potential of Fe(II)/Fe(III) is +200 mV, while that of NO_3^-/NO_2^- is +430 mV, facilitating the reaction of Fe^0/Fe(II) and reducing NO_3^-. However, under acidic conditions, the redox potential of Fe(II)/Fe(III) increases. For instance, the redox potential of Fe(II)/Fe(III) is +770 mV at pH 2, which may significantly inhibit the reduction of NO_3^--N [26]. Additionally, under acidic conditions, ZVI preferentially transfers electrons to H^+ to produce H_2 rather than reducing NO_3^--N, resulting in lower NO_3^--N reduction efficiency. Therefore, maintaining the pH at around 7–8 is more conducive to promoting the nitrification and denitrification processes. Therefore, under slightly acidic conditions (pH~4) at the contaminated site, the natural attenuation of NH_4^+-N and NO_3^--N co-pollution is very low. The issue of excessive groundwater acidity must be addressed first before achieving effective remediation of co-pollution at the actual contaminated site. Importantly, as the ZVI concentration increases from 0 to 500 mg/L, the solution pH gradually increases from acidic conditions (pH 4.5) to around 7.8 (Figure S2). This phenomenon provides strong evidence that ZVI can enhance nitrification and denitrification efficiency by neutralizing acidic substances.

The DO exerts distinct impacts on denitrification efficiency through nitrifying bacteria/denitrifying bacteria [34]. Nitrifying bacteria are strictly aerobic microorganisms that rely on ample DO to facilitate the oxidation of NH_4^+-N to NO_3^--N. As the DO concentration escalates from 0.1 to 4 mg/L, the removal efficiency for NH_4^+-N gradually rises from 11.2% to 100% (Figure S3). With elevated DO levels, nitrifying bacteria can effectively utilize oxygen for the oxidation of NH_4^+-N, thus enhancing denitrification efficiency. Conversely, denitrifying bacteria are typically facultative anaerobes, and as such, the removal efficiency for NO_3^--N decreases with rising DO levels.

The optimal concentrations of nitrifying bacteria and denitrifying bacteria for the removal of N-pollutants were 100 and 300 mg/L, respectively (Figure 2c). However, when the concentrations of bacteria exceed the optimal values, the removal efficiency for NH_4^+-N and NO_3^--N no longer increases. This result suggested that in addition to considering microbial concentrations, other factors such as nutrients, environmental conditions, and substrate concentrations should be comprehensively considered in the removal of N-pollutants [35,36].

The carbon source also affects the removal efficiency for NO_3^--N during the denitrification stage. As the glucose concentration increases from 0 to 300 mg/L, the removal efficiency for NO_3^--N gradually increases (Figure 2d). Glucose can serve as a carbon source to support growth and metabolism. In this study, the appropriate glucose concentration is 300–500 mg/L. As a result, the denitrification process requires organic compounds as electron donors, and the lack of carbon sources in the contaminated site may become a limiting factor for denitrification.

3.2. The Pilot Study for the Assessment of NH_4^+-N and NO_3^--N Combined Contaminations

Based on laboratory column test results, the in situ study involved injecting microbial agents in conjunction with micro-ZVI. The three chosen pilot test points were in situ acid-leached areas with high-concentration contamination (Figure 3a). Three injection wells (A, B, C) were planned to be perpendicular to groundwater, and a mixed arrangement of single and double rows was adopted. Monitoring wells MW1-MW3 were located downstream of the injection wells (Figure 3b). Tables 2 and 3 outline the information of injection wells and monitoring wells, respectively. Among them, the depth of the injection wells should ensure that the screen is 2 m below the groundwater level for proper material injection.

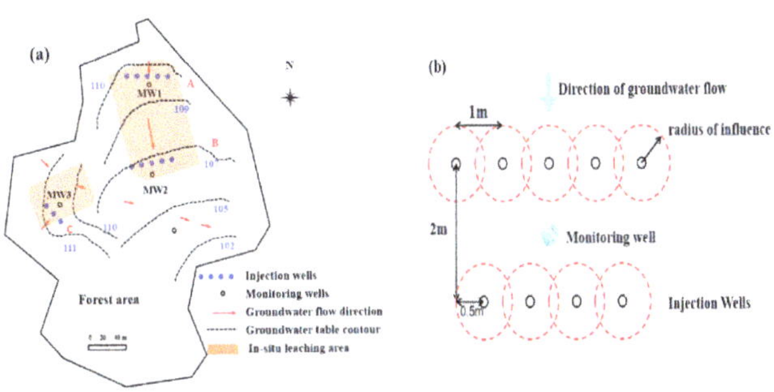

Figure 3. Pilot point position information: (**a**) map of the study site and (**b**) layout diagram of the injection well and monitoring well. The letters A, B, and C represent three representative research sites.

Table 2. In situ injection well parameters.

Injection Wells	Number of Rows	Number of Wells Per Row	Total Number of Injection Wells	Depth (m)	Actual Top Tube Length (m)	Length of Bobbin (m)
A	1	20	20	18	11	6.8
B	1	20	20	18	9	8.8
C	2	7	14	10.5	3.5	6.8

Table 3. List of groundwater monitoring well information.

Monitoring Well Number	Length of Upper Solid Pipe (m)	The Length of the Screen (m)	Settling Tube Length (m)
MW1	9	10	1
MW2	7	12.3	1
MW3	3	20	1

The metabolism of nitrifying and denitrifying bacteria occurs within a specific pH range (Figure 2). Alkali agents (i.e., sodium hydroxide and sodium bicarbonate) were firstly injected to maintain an appropriate pH condition. Following this, aeration was employed to raise the DO concentration above 2 mg·L^{-1}, creating an ideal aerobic environment for microbial nitrification reactions. After 5–7 days of nitrification, aeration ceased, and micro-ZVI was introduced to observe changes in NH_4^+-N and NO_3^--N concentrations in groundwater. Intermittent aeration was used to modify the groundwater's redox conditions, cycling between nitrification and denitrification reactions to progressively eliminate NH_4^+-N and NO_3^--N. Experience in the field suggested an injection pressure of 1.5 MPa, a flow rate of 300 L·h^{-1}, an impact radius of 1–2 m, aeration of each well every 5 days for 6 min, alkali agent injections every 18 days at 6 kg each time, and dosing of microbial agents and ZVI at 20 kg and 6 kg, respectively. For detailed dosing methods for nitrifying and denitrifying bacteria, consult Tables S1 and S2.

3.3. *Experimental Results of Pilot Site*

3.3.1. Monitoring Results of Total Iron Concentration in Groundwater

Generally, high concentrations of Fe^{2+} typically indicate higher ZVI activity and better mobility [37]. The concentrations of Fe^{2+} measured at different time intervals in the monitoring wells are shown in Table S3. The Fe^{2+} concentration steadily increased over the first 0~7 days post-ZVI injection, indicating that under high-pressure injection conditions, ZVI diffused into the downstream monitoring wells (~2 m downstream), although it is challenging to assess its exact quantity. Moreover, after 7 days, there was a slight decrease in Fe^{2+} concentration, suggesting that the reactivity of ZVI diminished after this period.

3.3.2. Monitoring Results of Injection Point A

The results from monitoring at injection point A, as illustrated in Figure 4, depict the variations in DO, pH, and different N concentrations at monitoring well MW1 over a 180-day remediation period post the introduction of remedial agents. Following the injection of alkali agents, the pH at MW1 spiked from 5.2 to 7.4 on the first day, stabilizing at around 7.0 thereafter (Figure 4a). Likewise, the DO levels rose swiftly to about 5.4 on the initial aeration day, maintaining at around 2 mg/L with continuous aeration. During the 1–26-day timeframe of nitrifying bacteria injection, the NH_4^+-N concentration plunged rapidly from 31.3 mg/L to below 5 mg/L. However, between 26 and 58 days, there was a slight uptick in both NH_4^+-N and NO_3^--N concentrations, notably peaking between 58 and 87 days. This re-rising concentration trend may be attributed to the runoff from the rainy season (Figure S4), which causes the release of N-pollutants from the aquifer soil. The high concentration of NH_4^+-N and NO_3^--N in the soil (Figure S5) indicated that the soil is highly likely to affect the distribution and movement of NH_4^+-N and NO_3^--N in groundwater through the adsorption, exchange, and degradation processes in the upper layer [14,38–40]. Similarly, points B and C (Figures 5 and 6) also experienced a rising trend in both NH_4^+-N and NO_3^--N concentrations between 58 and 87 days because of the runoff. Consequently, solely addressing NH_4^+-N and NO_3^--N in groundwater may not fully eradicate the pollution source, necessitating concurrent soil restoration to curb or diminish the release of pollutants into groundwater.

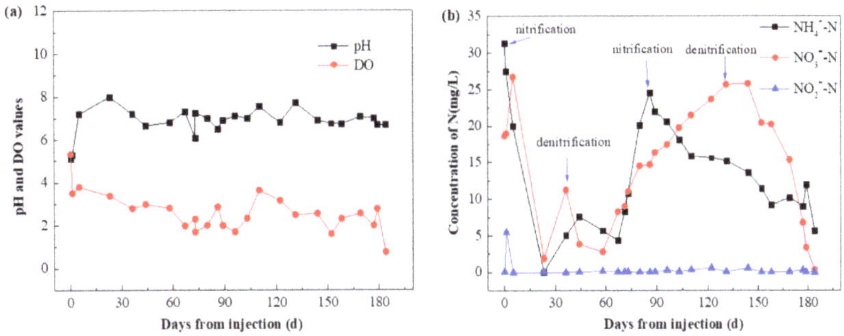

Figure 4. Variation of DO, pH (**a**) and different types of N concentration (**b**) during the point A remediation.

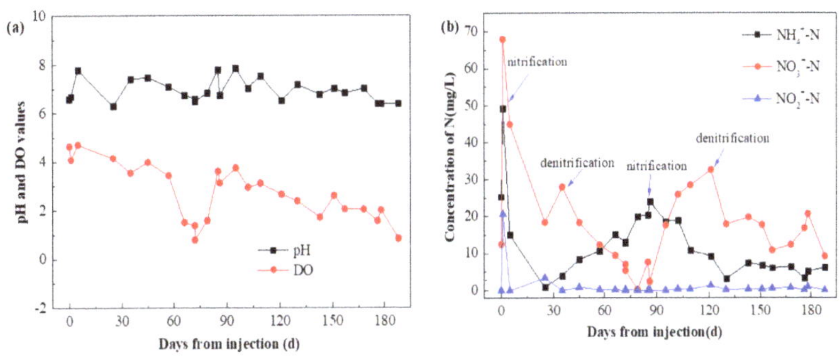

Figure 5. Variation of DO, pH (**a**), and different types of N concentration (**b**) during point B remediation.

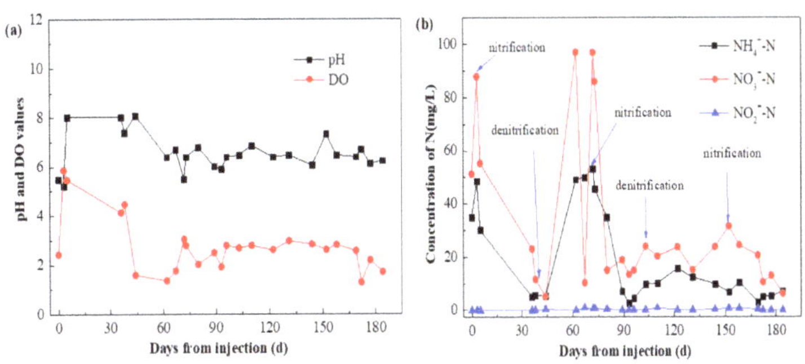

Figure 6. Variation of DO, pH (**a**), and different types of N concentration (**b**) during the point C remediation.

Upon escalating the frequency of nitrifying bacteria injections (without ZVI) between 88 and 124 days, NO_3^--N increased significantly with the decrease in NH_4^+-N, likely because (i) NO_3^--N is the oxidation product of NH_4^+-N, and the concentration of NO_3^--N would continue to increase when the concentration of NH_4^+-N decreased, and (ii) oxygen explosion is required during the NH_4^+-N remediation period, but high DO during NH_4^+-N remediation would inhibit the removal efficiency for NO_3^--N by denitrifying bacteria, resulting in the accumulation of NO_3^--N [41,42]. Following the injection of denitrification agents from 125 to 180 days, the NO_3^--N concentration gradually declined. Moreover, due to the depth of injection well A being greater than 10 m, the DO concentration in the groundwater lingered at a lower level, typically below 3 mg/L, even with continuous aeration, underscoring the significance of DO as a primary limiting factor in NH_4^+-N microbial remediation within deep groundwater. Through continual adjustments in aeration and the frequency of remedial agent injections, by day 180, the removal efficiency for NH_4^+-N and NO_3^--N had reached 82.50% and 98.80%, respectively. Notably, the NO_2^--N concentration remained below 1 mg/L for most of the study duration, except for a slight rise at the onset of the reaction.

3.3.3. Monitoring Results of Injection Point B

At injection point B (Figure 5), upon the introduction of alkali agents, the pH quickly increased from 6.6 to 7.8 on the first day, stabilizing at around 7.5 thereafter. Initially, during aeration, the DO concentration rose rapidly to approximately 4.7. However, reducing the aeration frequency led to a rapid decrease in DO concentration to below 2 mg/L. This decline can be attributed to the significant oxygen consumption during the nitrification process, where the removal of NH_4^+-N results in the depletion of oxygen [4]. Increasing the frequency of aeration would further elevate the DO concentration to 2 mg/L.

The NH_4^+-N concentration gradually declined from 25.3 mg/L to below 3 mg/L over days 1–30. Nevertheless, during the denitrification phase (31–60 days), the NH_4^+-N concentration increased back to 25 mg/L. Reinitiating the nitrification remediation between days 85 and 120, along with supplementing additional remediation fluid, led to a progressive decrease in NH_4^+-N concentration to below 5 mg/L. As a byproduct of nitrification, NO_3^--N exhibits an inverse relationship with the NH_4^+-N concentration. For instance, between 1 and 30 days, as the NH_4^+-N concentration decreased, the NO_3^--N concentration gradually rose to 45 mg/L. The concentration of NO_2^--N was also monitored during the remediation process, with the concentration remaining below 1 mg/L for the majority of the duration, except for a brief spike at the initial phase of the reaction. Through continuous adjustments in aeration and the frequency of remedial agent injections, by 180 days, the removal efficiency for NH_4^+-N and NO_3^--N had reached 87.5% and 86.4%, respectively, in comparison to their initial concentrations.

3.3.4. Monitoring Results at Injection Point C

The injection of alkali agents caused the pH to quickly rise from 5.2 to 7.4 on the first day, stabilizing at around 7.0 thereafter at injection point C (Figure 6). Subsequent aeration led to a rapid increase in the DO concentration from 2.4 to approximately 5.8 on the same day. During the nitrification phase, the NH_4^+-N concentration gradually decreased from 48.6 mg/L to below 5.6 mg/L within the first 30 days. However, reducing the aeration frequency resulted in a significant drop in the DO concentration to below 2 mg/L between 31 and 90 days.

Given the deeper groundwater level at monitoring well MW3, the DO concentration consistently remained low (<2 mg/L). During the denitrification phase, the NO_3^--N concentration at this well consistently stayed at a lower level (<20 mg/L), possibly due to conducive conditions for denitrifying bacteria in low DO concentrations facilitating the removal of NO_3^--N. Additionally, compared to the other monitoring points, this well displayed greater data variability, likely stemming from its distinct hydrogeological conditions. Site investigations identified the predominant soil type at injection point C as silty clay, characterized by a dense particle structure with limited permeability due to small gaps between the clay particles. These clay particles restrict the water flow, leading to lower permeability coefficients compared to sandy soil. Silty clay, often termed a "low-permeability layer" in groundwater systems, hinders the migration and diffusion of pollutants, slowing the spread of contaminants beyond the remediation area. This constraint may impede pollutant movement during remediation, prolonging both the duration and extent of remediation activities. By adjusting aeration and remedial agent injection frequency continuously, by the conclusion of the 180-day remediation period, the removal efficiency for NH_4^+-N and NO_3^--N had reached 94.50% and 92.90%, respectively. Meanwhile, NO_2^--N concentrations remained below 1.5 mg/L throughout the monitoring period.

3.3.5. Total Number of Colonies and NH_4^+-N Removal Efficiency

The total colony count serves as an indicator of the quantity and activity levels of microorganisms. Maintaining an optimal total colony count is crucial for preserving the diversity and equilibrium of microbial communities, ensuring the presence of microbes that adapt well to their surroundings and have degradation capabilities [43,44]. Figure 7a illustrates that prior to the introduction of bacterial agents, the total colony count at three locations hovered around 10^5 CFU/mL. The acidic condition of the groundwater before administering the remediation agent might inhibit microbial community growth, contributing to these initial readings. Post-injection, however, the count escalates significantly, surpassing 10^6 CFU/mL. It is well known that nitrifying bacteria ideally thrive in concentrations ranging from 10^5 to 10^6 CFU/mL, whereas denitrifying bacteria prefer a slightly higher concentration, typically between 10^6 and 10^7 CFU/mL [45]. However, the implementation in actual field conditions may require adjustments based on preliminary assessments and ongoing monitoring to optimize the effectiveness of the bioremediation process. It is essential to conduct trials to determine the most effective microbial concentration for a specific site and contamination scenario.

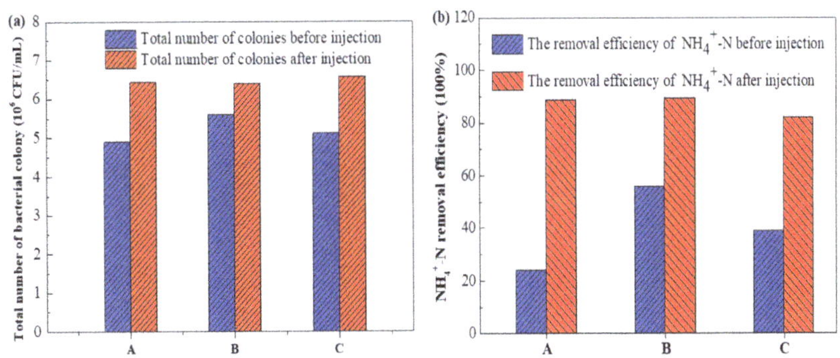

Figure 7. Total number of colonies (**a**) and removal rate of NH_4^+-N (**b**) before and after injection.

Additionally, a positive correlation exists between the total colony count and the removal of NH_4^+-N. Before the injection, the removal rates of NH_4^+-N at points A, B, and C were all below 60% (Figure 7b). Following the application of bacterial agents, these rates saw a substantial increase due to the growth of microbial colonies within the sediment at the bottom of the injection well, with efficiency reaching upwards of 80%.

3.3.6. Site Management and Operational Experience

Utilizing ZVI alongside microbial techniques to address NH_4^+-N and NO_3^--N pollutants at contaminated locations present significant challenges due to the complexity of the site's hydrogeology, which often results in outcomes that do not match the efficacy observed in controlled laboratory settings. Therefore, strengthening the management of on-site engineering implementation and dynamically adjusting remediation strategies are crucial for achieving improved remediation effects for the NH_4^+-N and NO_3^--N pollutants. Firstly, at locations where remediation effect has been effectively achieved, halting the injection of agents and shifting to regular monitoring can maintain effectiveness while being cost-efficient. Furthermore, it is necessary to adjust remediation parameters immediately based on the latest monitoring data. Particularly at locations where the removal efficiency for NH_4^+-N falls below 80%, intensifying aeration and adjusting the frequency of injecting solutions are essential measures to boost the nitrification process. In addition, the operational phase revealed issues like deformation, misalignment, and blockages in the pipework of injection wells (Figure S6). Addressing these problems involves critical maintenance actions such as excavating and replacing misaligned pipes, as well as applying aeration and pumping to clear blocked pipes of bio-silt. Based on our engineering operation experience, the above measures are of great significance for improving the removal efficiency for NH_4^+-N and NO_3^--N at the pilot site.

4. Conclusions

This study utilized both lab-based experiments and pilot field studies to prove that ZVI, when used alongside microbial agents, significantly improves the treatment of groundwater contaminated with high levels of NH_4^+-N and NO_3^--N combined contamination. The conclusions are as follows:

(1) The laboratory experiments showed that nitrifying bacteria could increase the removal efficiency for 50 mg/L of NH_4^+-N to over 99% within 5 days. DO concentration emerged as a crucial factor influencing nitrification efficiency, with higher DO levels significantly enhancing the nitrification process. The introduction of ZVI in the reaction columns reduced the denitrification-driven NO_3^--N remediation cycle from 15 days to approximately 10 days. This improvement was mainly attributed to ZVI elevating pH levels and facilitating the reduction of NO_3^--N, thereby promoting denitrification through microbial activity synergistically.

(2) The pilot-scale experiments demonstrated that the combination of ZVI with microbial agents substantially boosted the removal efficiency for NH_4^+-N and NO_3^--N combined contamination in groundwater. The injection of ZVI and microbial agents increased the removal efficiency for NH_4^+-N and NO_3^--N contaminants to above 80%. However, the presence of complex hydrogeological interactions, such as those between soil and groundwater and the existence of low-permeability layers, hindered the removal efficiency for N-pollutants. Adjusting the process parameters, including the remediation agent injection frequency and aeration frequency, is essential to meet the remediation requirements for NH_4^+-N and NO_3^--N combined contamination.

(3) In practical engineering applications, enhancing monitoring and remediation agent injection is critical. Locations with inadequate remediation outcomes require increased injection frequency. Addressing issues like sludge accumulation in well bottoms, voids around injection wells, and solid alkali precipitation necessitates improved maintenance of in situ injection wells. Strategies to enhance overall remediation efficiency involve supplementing filling materials, reinforcing injection wells, regular well maintenance, and optimizing alkali dosing levels.

Supplementary Materials: The following supporting information can be downloaded at https://www.mdpi.com/article/10.3390/pr12122768/s1, Text S1: Details of detection methods for NH_4-N, NO_3-N, NO_2-N, and the total bacterial colonies; Figure S1: Schematic diagram of reaction column in laboratory test; Figure S2: Effect of different ZVI mass concentrations on pH of aqueous solution; Figure S3: Effect of DO on removal efficiency for NH_4^+-N and NO_3^--N; Figure S4: The amount of rainfall in different remediation periods; Figure S5: The concentration of NH_4^+-N and NO_3^--N in soil at different points; Figure S6: Problems in injection wells and photos of work and results before and after rectification; Table S1: Parameters of in situ nitrification restorers; Table S2: Parameters of in situ denitrification restorers; Table S3: Geochemical parameters of groundwater in the pilot test area after ZVI injection.

Author Contributions: Conceptualization, J.C. and X.W.; formal analysis, Y.L., J.Z., Z.L., Y.H., X.C., M.L. and H.F.; investigation, J.C., J.Z., X.C., M.L. and H.F.; methodology, J.C., J.Z., Z.L. and

X.C.; supervision, X.W.; writing—original draft, J.C.; writing—review and editing, Y.H. and X.W. All authors have read and agreed to the published version of the manuscript.

Funding: This work was supported financially by the National Key Research and Development Program of China (No. 2022YFC3702300), the Guangdong Basic and Applied Basic Research Foundation (No. 2024A1515011091), the National Natural Science Foundation of China (No. 42207244), the Science and Technology innovation self-supporting projects by the Pearl River Water Resources Research Institute ([2022]YF005), and the Open Research Fund of the Key Laboratory of Water Security Guarantee in the Guangdong–Hong Kong–Marco Greater Bay Area of the Ministry of Water Resources (No. WSGBA-KJ202306).

Data Availability Statement: All data and materials supporting the findings of this study are available upon reasonable request.

Acknowledgments: We appreciate the assistance of the Analysis and Testing Center at the College of Ecological Environment, Guangdong University of Technology, and the Analysis and Testing Center at South China Normal University for their support with analytical testing. This work was financially supported by the central ecological environment fund for water pollution preventing and controlling project (XMHT-2022-SS-FW847) of China.

Conflicts of Interest: Authors Junyi Chen, Xiangxin Chen, Mingkui Li and Hanyun Fan were employed by the company Guangzhou Runfang Environmental Protection Technology Co., Ltd. The remaining authors declare that the research was conducted in the absence of any commercial or financial relationships that could be construed as a potential conflict of interest. The company had no role in the design of the study; in the collection, analyses, or interpretation of data; in the writing of the manuscript, or in the decision to publish the results.

References

1. Adimalla, N.; Li, P.; Qian, H. Evaluation of groundwater contamination for fluoride and nitrate in semi-arid region of Nirmal Province, South India: A special emphasis on human health risk assessment (HHRA). *Hum. Ecol. Risk Assess. Int. J.* **2019**, *25*, 1107–1124. [CrossRef]
2. Shen, S.; Ma, T.; Du, Y. Temporal variations in groundwater nitrogen under intensive groundwater/surface-water interaction. *Hydrogeol. J.* **2019**, *27*, 1753–1762. [CrossRef]
3. Tang, S.; Luo, Z.; Liao, J.; Liu, Z.; Xu, L.; Niu, J. Degradation and detoxification mechanisms of organophosphorus flame retardant tris(1,3-dichloro-2-propyl) phosphate (TDCPP) during electrochemical oxidation process. *Chin. Chem. Lett.* **2023**, *34*, 108090. [CrossRef]
4. Xiong, Y.; Du, Y.; Deng, Y.; Ma, T.; Li, D.; Sun, X.; Liu, G.; Wang, Y. Contrasting sources and fate of nitrogen compounds in different groundwater systems in the Central Yangtze River Basin. *Environ. Pollut.* **2021**, *290*, 118119. [CrossRef]
5. Yu, Q.J.; Cao, Q.; Connell, D.W. An overall risk probability-based method for quantification of synergistic and antagonistic effects in health risk assessment for mixtures: Theoretical concepts. *Environ. Sci. Pollut. Res. Int.* **2011**, *19*, 2627–2633. [CrossRef]
6. Zhang, Y.; Douglas, G.B.; Pu, L.; Zhao, Q.; Tang, Y.; Xu, W.; Luo, B.; Hong, W.; Cui, L.; Ye, Z. Zero-valent iron-facilitated reduction of nitrate: Chemical kinetics and reaction pathways. *Sci. Total Environ.* **2017**, *598*, 1140–1150. [CrossRef]
7. Zhang, Q.; Ren, F.; Li, F.; Chen, G.; Yang, G.; Wang, J.; Du, K.; Liu, S.; Li, Z. Ammonia nitrogen sources and pollution along soil profiles in an in-situ leaching rare earth ore. *Environ. Pollut.* **2020**, *267*, 15449. [CrossRef]
8. Deng, C.; Chen, Z.; Li, Y. Effective recovery of the nitritation process through hydrogen peroxide. *Environ. Sci. Pollut. Res.* **2024**, *31*, 28404–28417. [CrossRef]
9. Maharjan, A.K.; Mori, K.; Toyama, T. Nitrogen removal ability and characteristics of the laboratory-scale tidal flow constructed wetlands for treating ammonium-nitrogen contaminated groundwater. *Water* **2020**, *12*, 1326. [CrossRef]
10. Powlson, D.S.; Addiscott, T.M.; Benjamin, N.; Cassman, K.G.; Kok, T.M.; Grinsven, H.; L'hirondel, J.L.; Avery, A.A.; Kessel, C. When does nitrate become a risk for humans? *J. Environ. Qual.* **2008**, *37*, 291–295. [CrossRef]
11. Umezawa, Y.; Hosono, T.; Onodera, S.; Siringan, F.; Buapeng, S.; Delinom, R.; Yoshimizu, C.; Tayasu, I.; Nagata, T.; Taniguchi, M. Sources of nitrate and ammonium contamination in groundwater under developing Asian megacities. *Sci. Total Environ.* **2008**, *404*, 361–376. [CrossRef] [PubMed]
12. Yu, G.; Wang, J.; Liu, L. The analysis of groundwater nitrate pollution and health risk assessment in rural areas of Yantai, China. *BMC Public Health* **2020**, *20*, 43–52. [CrossRef] [PubMed]
13. *GB 5749-2022*; Standards for Drinking Water Quality. National Health Commission of the People's Republic of China: Beijing, China, 2022.
14. Li, S.; Zhang, Y.; Yin, S. Analysis of microbial community structure and degradation of ammonia nitrogen in groundwater in cold regions. *Environ. Sci. Pollut. Res.* **2020**, *27*, 44137–44147. [CrossRef] [PubMed]
15. Rajta, A.; Bhatia, R.; Setia, H.; Pathania, P. Role of heterotrophic aerobic denitrifying bacteria in nitrate removal from wastewater. *J. Appl. Microbiol.* **2020**, *128*, 1261–1278. [CrossRef]

16. Elisante, E.; Muzuka, A.N.N. Assessment of sources and transformation of nitrate in groundwater on the slopes of Mount Meru, Tanzania. *Environ. Earth Sci.* **2016**, *75*, 277. [CrossRef]
17. Guo, W.; Ying, X.; Zhao, N.; Yu, S.; Zhang, X.; Feng, H.; Zhang, Y.; Yu, H. Interspecies electron transfer between Geobacter and denitrifying bacteria for nitrogen removal in bioelectrochemical system. *Chem. Eng. J.* **2023**, *455*, 1385–8947. [CrossRef]
18. Xu, D.; Ling, H.; Li, Z.; Li, Y.; Chen, R.; Cai, S. Treatment of Ammonium-Nitrogen–Contaminated Groundwater by Tidal Flow Constructed Wetlands Using Different Substrates: Evaluation of Performance and Microbial Nitrogen Removal Pathways. *Water Air Soil Pollut.* **2022**, *233*, 159. [CrossRef]
19. Solomon, K.M.; Eldon, R.R.; Hullebusch, E.D.; Annachhatre, A.P. Nitrate removal from groundwater: A review of natural and engineered processes. *J. Water Supply Res. Technol.* **2018**, *67*, 885–902. [CrossRef]
20. Usher, K.M.; Kaksonen, A.H.; Cole, I.; Marney, D. Critical review: Microbially influenced corrosion of buried carbon steel pipes. *Int. Biodeterior. Biodegrad.* **2014**, *93*, 84–106. [CrossRef]
21. Lu, Q.; Jeen, S.-W.; Gui, L.; Gillham, R.W. Nitrate reduction and its effects on trichloroethylene degradation by granular iron. *Water Res.* **2017**, *112*, 48–57. [CrossRef]
22. Hou, J.; Wang, A.; Miao, L.; Wu, J.; Xing, B. The role of nitrate in simultaneous removal of nitrate and trichloroethylene by sulfidated zero-valent iron. *Sci. Total Environ.* **2022**, *829*, 154304. [CrossRef] [PubMed]
23. Liu, Y.; Wang, J. Reduction of nitrate by zero valent iron (ZVI)-based materials: A review. *Sci. Total Environ.* **2019**, *671*, 388–400. [CrossRef] [PubMed]
24. Wei, X.P.; Guo, Z.; Yin, H.; Yu, Y.; Dang, Z. Removal of heavy metal ions and polybrominated biphenyl ethers by sulfurized nanoscale zerovalent iron: Compound effects and removal mechanism. *J. Hazard. Mater.* **2021**, *414*, 125555. [CrossRef]
25. Song, W.; Gao, B.; Wang, H.; Xu, X.; Xue, M.; Zha, M.; Gong, B. The rapid adsorption-microbial reduction of perchlorate from aqueous solution by novel amine-crosslinked magnetic biopolymer resin. *Bioresour. Technol.* **2017**, *240*, 68–76. [CrossRef] [PubMed]
26. Zhang, Y.; Douglas, G.B.; Anna, H.; Cui, L.; Ye, Z. Microbial reduction of nitrate in the presence of zero-valent iron. *Sci. Total Environ.* **2019**, *646*, 1195–1203. [CrossRef]
27. Zhang, W.; Swaney, D.P.; Hong, B. Influence of rapid rural-urban population migration on riverine nitrogen pollution: Perspective from ammonia-nitrogen. *Environ. Sci. Pollut. Res.* **2017**, *24*, 27201–27214. [CrossRef] [PubMed]
28. Climate Bulletin of Qingyuan City of 2023. Available online: http://www.gdqy.gov.cn/jjqy/ljqy/jrfc/qyqh/content/post_1841096.html (accessed on 28 November 2024).
29. Lin, K.; Zhu, Y.; Zhang, Y.; Lin, H. Determination of ammonia nitrogen in natural waters: Recent advances and applications. *Trends Environ. Anal. Chem.* **2019**, *24*, e00073. [CrossRef]
30. Li, D.; Xu, X.; Li, Z.; Wang, T.; Wang, C. Detection methods of ammonia nitrogen in water: A review. *Trends Environ. Anal. Chem.* **2020**, *127*, 5890. [CrossRef]
31. Michael, O.R.; Stephen, R.B.; Morgan, P.; Smith, J.W.N.; Chrystina, D.B. Nitrate attenuation in groundwater: A review of biogeochemical controlling processes. *Water Res.* **2008**, *42*, 4215–4423. [CrossRef]
32. Liu, Y.; Wan, Y.; Ma, Z.; Dong, W.; Su, X.; Shen, X.; Yi, X.; Chen, Y. Effects of magnetite on microbially driven nitrate reduction processes in groundwater. *Sci. Total Environ.* **2023**, *855*, 15895. [CrossRef]
33. Xu, J.; Pu, Y.; Qi, W.; Yang, X.; Tang, Y.; Wan, P.; Adrian, F. Chemical removal of nitrate from water by aluminum-iron alloys. *Chemosphere* **2016**, *166*, 197–202. [CrossRef]
34. Kutvonen, H.; Rajala, P.; Carpén, L.; Malin, B. Nitrate and ammonia as nitrogen sources for deep subsurface microorganisms. *Front. Microbiol.* **2015**, *6*, 1079. [CrossRef] [PubMed]
35. Li, P.; Wang, Y.; Zuo, J.; Wang, R.; Zhao, J.; Du, Y. Nitrogen Removal and N_2O Accumulation during Hydrogenotrophic Denitrification: Influence of Environmental Factors and Microbial Community Characteristics. *Environ. Sci. Technol.* **2017**, *51*, 870–879. [CrossRef]
36. Niu, C.; Zhai, T.; Zhang, Q.; Wang, H.; Xiao, L. Research Advances in the Analysis of Nitrate Pollution Sources in a Freshwater Environment Using δ15 N-NO_3^- and δ18O-NO_3^-. *Int. J. Environ. Res. Public Health* **2021**, *18*, 11805. [CrossRef]
37. Wei, Y.T.; Wu, S.C.; Chou, C.M.; Che, C.H.; Tsai, S.M.; Lien, H.L. Influence of nanoscale zero-valent iron on geochemical properties of groundwater and vinyl chloride degradation: A field case study. *Water Res.* **2010**, *44*, 131–140. [CrossRef]
38. Li, D.; Zhou, Y.; Long, Q.; Li, R.; Lu, C. Ammonia nitrogen adsorption by different aquifer media: An experimental trial for nitrogen removal from groundwater. *Hum. Ecol. Risk Assess.* **2020**, *26*, 2434–2446. [CrossRef]
39. Liu, T.; Xia, X.; Liu, S.; Mou, X.; Qiu, Y. Acceleration of Denitrification in Turbid Rivers Due to Denitrification Occurring on Suspended Sediment in Oxic Waters. *Environ. Sci. Technol.* **2013**, *47*, 4053–4061. [CrossRef]
40. Yin, G.; Hou, L.; Liu, M.; Liu, Z.; Wayne, S.A. Novel Membrane Inlet Mass Spectrometer Method to Measure 15 NH_4^+ for Isotope-Enrichment Experiments in Aquatic Ecosystems. *Environ. Sci. Technol.* **2014**, *48*, 9555–9562. [CrossRef] [PubMed]
41. Wang, R.; Zheng, P.; Xing, Y.; Zhang, M.; Ghulam, A.; Zhao, A.; Li, W.; Wang, L. Anaerobic ferrous oxidation by heterotrophic denitrifying enriched culture. *J. Ind. Microbiol. Biotechnol.* **2014**, *41*, 803–809. [CrossRef]
42. Xu, D.; Li, Y.; Song, F.; Gu, T. Laboratory investigation of microbiologically influenced corrosion of C1018 carbon steel by nitrate reducing bacterium Bacillus licheniformis. *Corros. Sci.* **2013**, *77*, 385–390. [CrossRef]
43. Tong, K.; Zhang, Y.; Liu, G.; Ye, Z.; Paul, K. Treatment of heavy oil wastewater by a conventional activated sludge process coupled with an immobilized biological filter. *Int. Biodeterior. Biodegrad.* **2013**, *84*, 65–71. [CrossRef]

44. Wu, D.; Chen, G.; Zhang, X. Change in microbial community in landfill refuse contaminated with antibiotics facilitates denitrification more than the increase in ARG over long-term. *Sci. Rep.* **2017**, *7*, 41230. [CrossRef] [PubMed]
45. Goncalves da Silva, D. Nitrosomonas Eutropha D23—An In Vitro, Evaluation of Its Metabolic Phenotype and Evidence for Its Bio-Activity. Doctoral Thesis, University of Southampton, Southampton, UK, 2019; p. 272. Available online: http://eprints.soton.ac.uk/id/eprint/437367 (accessed on 28 November 2024).

processes

MDPI

Article

Performance and Bacteria Communities of a Full-Scale Constructed Wetland Treating the Secondary Effluent after Multi-Years' Operation

Siyuan Song [1,2], Baogui Wang [3], Tangwu Yang [1,4], Yumei Gu [1], Sheng Sheng [2], Dehua Zhao [1,*], Shuqing An [1,4] and Aimin Li [1,*]

1 State Key Laboratory of Pollution Control & Resource Reuse, School of Environment, School of Life Science, Nanjing University, Nanjing 210046, China
2 PowerChina Huadong Engineering Corporation Limited, Hangzhou 311122, China
3 Central Plains Environmental Protection Co., LCD, Zhengzhou 450000, China
4 Nanjing University Ecology Research Institute of Changshu (NJUecoRICH), Changshu 215500, China
* Correspondence: dhzhao@nju.edu.cn (D.Z.); liaimin@nju.edu.cn (A.L.)

Abstract: Constructed wetlands (CWs) had been widely used to treat the tailwater from sewage treatment plants. However, the enduring effectiveness of CWs was still unclear. Therefore, this study aimed to investigate the chemical oxygen demand (COD), ammonia nitrogen (NH_4^+-N), and total phosphorus (TP) removal efficiencies (RE) of the Hongze CW (HZ-CW) after multi-years' operation. The average COD RE were 7.6% and 15.14% in the 4th and 5th year, respectively. The NH_4^+-N RE was 78.33% and 46.04% in the 4th and 5th year, respectively, while the TP RE remained high at 66.86% and 64.68%. The high-throughput sequencing analysis revealed that the bacterial community of HZ-CW at the end of the 5th year exhibited a substantial abundance and diversity, and Proteobacteria and Bacteroidota were the dominant phyla with a relative abundance of 33.75–71.8% and 11.28–24.53% in different zones of HZ-CW. Ammonia oxidizing organisms (AOMs) presented much higher relative abundance (0.43–0.79%) in aerated pond (AP) and four free water surface flow CWs (FWS1–FWS4) than those of anammox bacteria, indicating the dominant role of nitrification in NH_4^+-N removal.

Keywords: full-scale constructed wetland; secondary effluent; bacterial community; multi-years' of operation

1. Introduction

Constructed wetlands (CWs) are a well-known and widely recognized solution, due to their good operation performance [1,2], cost-effectiveness [3], and ability to withstand impact loads [4–7]. This technology is capable of treating various types of wastewater, including agricultural wastewater, industrial effluent, polluted river water, and domestic sewage [8–11]. Currently, there is an increase in the application of CWs for the treatment of effluent from sewage treatment plants (STPs) driven by the increasingly stringent environmental protection measures in China [12,13]. In fact, in 2018, Jiangsu Province of China even proposed that "tail-water from STPs need to undergo ecological purification by CWs before being discharged into the surface water system".

The analysis of the operational effectiveness of tail-water CWs had observed a steady increase in conjunction with the rise in their application cases. It has been found that CWs exhibited high removal efficiencies for ammonia nitrogen (NH_4^+-N), chemical oxygen demand (COD) and total phosphorus (TP) [4,14]. Wu et al. conducted an extensive follow-up study for two and a half years on the tail-water from a wastewater treatment plant that was treated by a large-scale vertical-flow CW [4]. It was discovered that the average removal efficiencies for COD and NH_4^+-N by the CW were 53% and 72%, respectively. Additionally, the study highlighted that there was a significant positive correlation between inflow loads and treatment performance. Song et al. findings indicated that CWs exhibited effective removal efficiencies for NH_4^+-N, COD, and TP, with initial operation yielding percentages of 56.33%, 55.64% and 88.44%, respectively [3]. Zhu et al. have carried out a detailed 14-month follow-up investigation into the performance of a large-scale vertical-flow CWs treating wastewater treatment plant tail-water, taking into account seasonal variations [15]. The results of their study revealed that the comprehensive removal rates of COD, NH_4^+-N, and TP were 40.05%, 45.47%, and 62.55%, respectively. Notably, there was no significant seasonal difference in TP removal. An issue prevalent amongst these studies is the limited scope of their research phase, which largely focuses on the initial period of CWs establishment, and the lack of analysis on the long-term effectiveness of CWs after years of operation. However, it should be noted that the performance of CWs changes over time due to substrate blockage, saturation of adsorption capacity, decomposition and release of accumulated organic matter, and an increase in microbial diversity and abundance [16,17]. The tracking studies on the CWs' performance are few, which limits people's comprehensive understanding of them and is not conducive to their optimization and upgrading. Numerous studies have demonstrated that microbial processes serve as the primary pollutant removal pathway, prompting researchers to investigate microorganisms within CWs [15,18]. Zhu et al. discovered a considerable variety of microbial communities within the integrated vertical flow CW, with Proteobacteria, Planctomycetes, Bacteroidetes, and Acidobacteria emerging as the primary strains. Among these, the ratios of nitrifying bacteria, denitrifying bacteria, and anammox bacteria were calculated to fall within the range of 0.28–0.31%, 1.83–1.92%, and 0.34–0.40%, respectively; denitrification and anammox were identified as the core mechanisms for nitrogen removal [15]. Furthermore, Zhang et al. determined that anammox bacteria played a critical role in vertical flow CWs' nitrogen removal. Consequently, the functional properties of microbial communities in surface-flow CWs, including their specific microbial populations, require explicit exploration [18].

In this study, we focused on the performance and microbial communities of a full-scale surface-flow CW treating the secondary effluent after multi-years' operation. The objectives of this study were: (1) to assess the long-term removal efficiency of a large-scale surface-flow CW in removing pollutants such as COD, NH_4^+-N, and TP, as well as identifying seasonal variations; (2) to analyze the distribution of the bacterial communities in the CWs, with a particular focus on identifying functional microorganisms.

2. Materials and Methods

2.1. Site Description

The HZ-CW is located in the Hongze District of Jiangsu Province, East China, covering an expansive 55.58 × 10^4 m^2. This system was thoughtfully designed and comprises an aerated pond (AP), a facultative pond (FP), four free-water surface flow CW (FWS), and an ecological pond (EP) (Figure 1). The hydraulic retention times (HRT) of the system elements are 6.7 d, 4.1 d, 3.4 d, and 2.5 d for AP, FP, FWSs, and EP, respectively. The influent water for this system is taken from a sewage treatment plant and has a daily flow rate of 4 × 10^4 m^3. Emergent aquatic plants, specifically *Phragmites australis* (Cav.) Trin. Ex Steud., *Typha Orientalis* Presl, and *Thalia dealbata* Fraser, were deliberately selected for their extensive root systems and high biomass yields. For a more comprehensive understanding of the system's setup and management approach, reference Song and Liu's work [3].

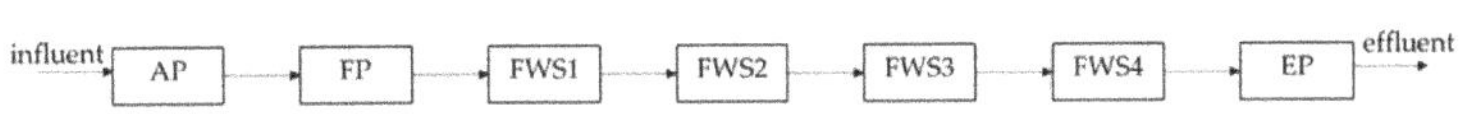

Figure 1. Schematic of HZ-CW system. AP, aerated pond; FP, facultative pond; FWS1–FWS4, four free water surface flow constructed wetlands follow the water flow; EP, ecological pond.

2.2. Water Quality Monitoring

The HZ-CW's inlet and outlet water quality were continuously monitored using an online detector from Nanjing Gangneng, China. The monitored parameters encompassed dissolved oxygen (DO), NH_4^+-N, COD, and TP. Hourly water samples were collected and analyzed while the average data for monitoring points in the HZ-CW were computed every 10 days.

2.3. Microorganism Sampling and Analysis

The bacterial communities in AP, FP, FW3s, and EP were analyzed at the end of the 5th year. Substrate samples of each processing zone were collected, and stored in the icebox during delivery to the laboratory. The substrate samples were collected from the surface layer (0~5 cm) at the bottom of each zone. Three sites were selected in each zone, and three samples were collected from each site and then the samples from the same site were combined for testing. The genomic DNA was extracted and further investigated according to Li et al. [19].

2.4. Data Analysis

The removal efficiency (RE) of pollutants was calculated using the percentage removal rate for each parameter as follows:

$$\text{RE (\%)} = \frac{(\text{InC} - \text{EffC})}{\text{InC}} \times 100 \quad (1)$$

where InC and EffC are the influent and effluent concentrations (mg L^{-1}), respectively. Mean (±s.d.) values were also calculated. Seasonal (spring: March–May; summer: June–August; autumn: September–November; winter: December–February) effects for contaminants removal were investigated using one-way analysis of variance (ANOVA). *t*-tests were used to compare parameters between the influent and the effluent. Nonparametric tests were used for non-normal distribution data. In addition, Tukey's *b*-test was used to determine the significance of differences among seasons, with two-sided $p < 0.05$ taken to indicate significance. All analyses were conducted using SPSS 22.0 (IBM Corp., Armonk, NY, USA).

3. Results and Discussion

3.1. The Performance of HZ-CW after Years Operation

3.1.1. The DO Variations

DO can influence the pollutant removal efficiencies of CWs [20–22]. Therefore, the influent and effluent DO concentrations of HZ-CW in the 4th, 5th, and 1st year ($n = 36$ for each year) were investigated and presented in Figure 2. As shown in Figure 2, the average influent concentration (InC) of DO in the 4th and 5th years were 3.08 and 1.82 mg L^{-1}, respectively ($n = 36$ for each year), both higher than that in the 1st year (the average DO was 1.4 mg L^{-1}). Conversely, the average DO EffC of 5.62 and 3.34 mg L^{-1} in the 4th and 5th years were lower than the average of 6.23 mg L^{-1} observed in the 1st year. The low DO EffC in the 5th year might be caused by the aging of solar aerators. Furthermore, the average DO EffC in summer and autumn of the 4th and 5th years was lower than those in spring and winter, which was partly associated with the low solubility and rapid consumption of DO in the high temperature.

3.1.2. The Performance for COD Removal

As shown in Figure 3a, the EffC of COD in the 4th and 5th years were 35.99 ± 8.65 mg L^{-1} and 24.09 ± 6.90 mg L^{-1}, both meeting the effluent requirement for COD (≤40 mg L^{-1}). *t*-tests revealed that InC and EffC of COD in both the 4th and 5th year presented a significant difference ($p < 0.01$). Compared to the 1st year (55.64%), the average COD RE declined in the 4th (7.60%) ($n = 36$) and 5th year (15.14%) ($n = 36$) (Figure 3c) and the average RE per season in the 4th and 5th year were lower than that in the 1st year (Figure 3b). One-way ANOVA showed that there was a significant difference in RE between the 4th~5th year and the 1st year ($p < 0.05$). Zhu et al. [15] and Zhang et al. [18] have shown that both full-scale integrated vertical-flow CWs and surface-flow CWs were able to achieve COD RE of 40.05% and 53%, respectively, when treating secondary effluent, which was comparable to the COD removal rate in 1st year of the study. These results demonstrated the effectiveness of CWs in COD removal, irrespective of the types of CWs, especially in the initial years of operation. However, the COD RE of the HZ-CW decreased significantly in the 4th and 5th years of operation. According to previous studies, the COD removal efficiency of CWs showed a significant positive correlation with the influent. In this study, the average COD InC in the 4th and 5th years were 43.93 ± 16.20 mg L^{-1} and 32.79 ± 17.05 mg L^{-1}, markedly lower than that in the 1st year (64.59 ± 27.57 mg L^{-1}), which was likely to be the dominant factor contributing to the COD RE decline in the 4th and 5th years. As shown in Figure 3a, the COD EffC reached lower than 40 mg L^{-1}, despite exceeding an influent COD concentration of 100 mg L^{-1} in the 5th year of operation, indicating HZ-CW still boasted strong COD removal capabilities. These results further support the notion that the decline of influent COD load was the primary driver for the decrease in COD RE.

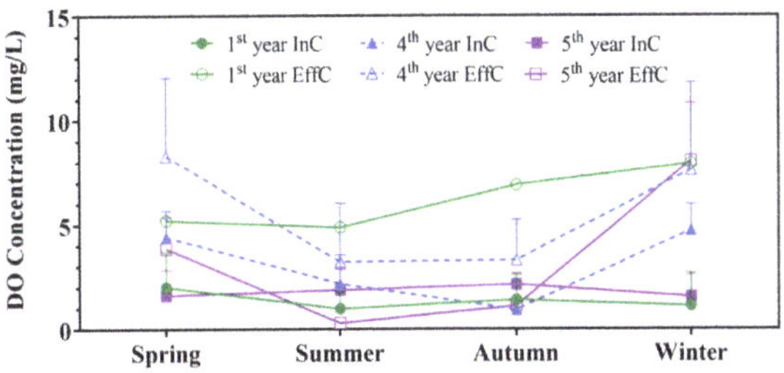

Figure 2. The influent and effluent DO concentrations of HZ-CW in the 4th, 5th and 1st year (n = 36 for each year), and the DO concentrations in different seasons (n = 9 for each season). Data are given as the mean ± s.d. InC, influent concentration; EffC, effluent concentration.

The One-way ANOVA also presented that the COD RE among seasons in the 1st year, 4th year, and 5th year showed no significant differences ($p > 0.05$) (Figure 3b). However, these findings differed from those studies by Zhu et al., who found that COD RE in winter was lower than in summer due to decreased microbial degradation capacity [15]. In fact, the removal of COD in CWs involves various mechanisms including precipitation, filtration, adsorption, and microbial degradation [23–25]. Temperature is believed to have a greater impact on microbial degradation, and a lesser effect on physico-chemical processes. Therefore, the small seasonal variation in COD removal in HZ-CW might be the result of the combined effects of various mechanisms.

3.1.3. The Performance for NH_4^+-N Removal

Nitrogen (N) is recognized as one of the key contributors to eutrophication in water; therefore, the removal performance of NH_4^+-N serves as a significant indicator for assessing the water purification effectiveness of the full-scale CWs [4,26,27]. As shown in Figure 4a, the average NH_4^+-N EffC in the 4th and 5th year were 0.84 ± 1.35 mg L^{-1} and 0.18 ± 0.23 mg L^{-1}, respectively, with almost all of the EffC of NH_4^+-N were lower than the effluent standard (Figure 4a). t-tests confirmed the significant difference between InC and EffC of NH_4^+-N ($p < 0.01$) during each year. However, the mean NH_4^+-N EffC in the 1st year was 4.40 ± 3.60 mg L^{-1}, failing to meet the effluent requirement, which might be due to the unstable establishment of the bacterial community and the relatively high InC of NH_4^+-N [15,28]. The average NH_4^+-N RE were 56.33%, 78.33%, and 46.04% in the 1st, 4th, and 5th year, respectively (Figure 4c), presenting a significant difference in RE between the 1st year and the 4th year ($p < 0.05$). According to the former studies, the NH_4^+-N RE in CWs had a positive correlation with the inflow load [4,15] in the 5th year, the average NH_4^+-N InC (1.63 ± 2.20 mg L^{-1}) was significantly lower than that in the 4th year (4.58 ± 4.31 mg L^{-1}), so the NH_4^+-N RE in the 5th year was significantly lower than that in the 4th year.

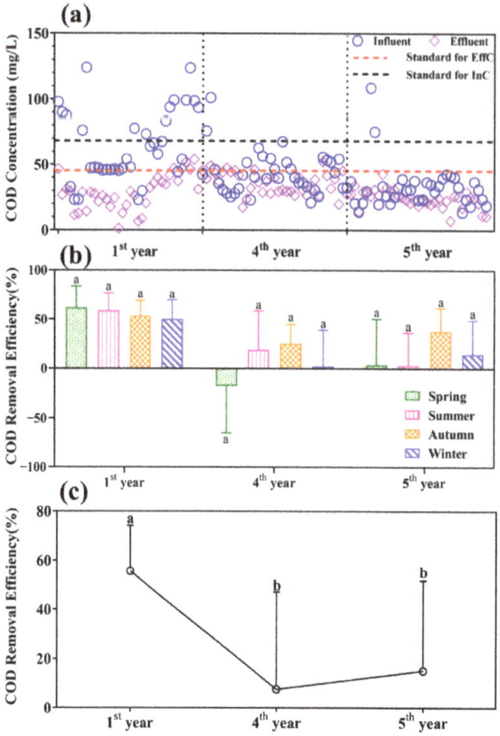

Figure 3. The performance of HZ-CW for COD removal during the study period. (**a**) The influent and effluent concentrations, (**b**) COD removal efficiencies (%) in different seasons (n = 9 for each season), (**c**) the mean COD removal efficiencies (%) in each year (the 1st year, 4th year, and 5th year, n = 36 for each year). Different letters indicate significant differences ($p < 0.05$) among different years. Data are given as the mean ± s.d. InC: influent concentration; EffC: effluent concentration.

This study also revealed a significant difference in NH_4^+-N RE between autumn and spring–winter in both the 4th and 5th years ($p < 0.05$) (Figure 4b), potentially due to the higher InC NH_4^+-N in the autumn of the 4th than that in the spring and winter. Similarly, the InC NH_4^+-N in the autumn of the 5th year was lower than that in the spring and winter of the same year. This result was inconsistent with former studies, which showed that a low temperature limited microbial activity and plant growth, thereby leading to a decrease in NH_4^+-N RE in CWs in winter and spring. Therefore, it is necessary to consider both temperature and NH_4^+-N InC when evaluating the removal effect of CWs on NH_4^+-N.

3.1.4. The Performance for TP Removal

As shown in Figure 5a, the average EffC of TP were 0.08, 0.67, and 0.08 mg L^{-1} in the 1st, 4th, and 5th year, respectively. Specifically, the mean EffC of TP in the 4th year failed to meet the effluent TP requirement of ≤0.4 mg L^{-1}, which was attributed to the elevated TP InC of 2.97 mg L^{-1}, approximately three times the design influent value, and higher than those recorded in the 1st (0.73 mg L^{-1}) and 5th year (0.30 mg L^{-1}). In the 1st and 5th years, there was no significant seasonal difference in TP RE ($p > 0.05$). Nevertheless, between summer and autumn–winter in the 4th year, there were seasonal variations ($p < 0.05$) (Figure 5b), potentially because of the high TP InC. Compared to the RE of NH_4^+-N and COD, the TP RE was relatively stable ($p > 0.05$), totalling 88.44%, 66.86%, and 64.68% in the 1st year, 4th year, and 5th year, respectively, which was higher than 60% (Figure 4c), as reported by Saggai, Ainouche [29].

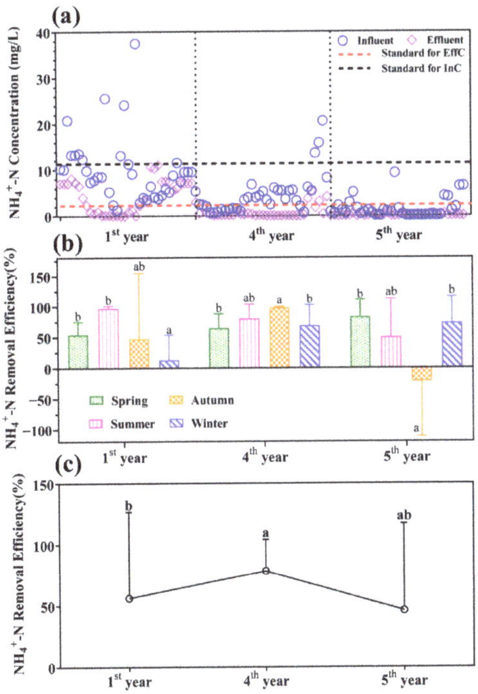

Figure 4. The performance of HZ-WC for NH_4^+-N removal during the study period. (**a**) The influent and effluent concentrations, (**b**) NH_4^+-N removal efficiencies (%) in different seasons during the study period (n = 9 for each season), (**c**) the mean NH_4^+-N removal efficiencies (%) in each year (the 1st year, 4th year, and 5th year, n = 36 for each year). Different letters indicate significant differences ($p < 0.05$) among different years. Data are given as the mean ± s.d. InC: influent concentration; EffC: effluent concentration.

Enhanced biological phosphorus removal (EBPR) processes have proven to be effective in removing TP from STPs through the accumulation of polyphosphate accumulating organisms (PAOs) [30]. However, the tail-water lacked organic matter and did not necessarily have alternating anaerobic/aerobic or anaerobic/anoxic conditions, so the role of EBPR in tail-water CWs was limited [15]. The substrate adsorption and filtration are the primary pathways for TP removal in these processes and are less affected by environmental factors, such as temperature. This explained the absence of seasonal variation in TP removal observed in the HZ-CW in the 1st and 5th years of operation. It is worth noting that clogging was reported as a factor affecting the RE of TP after years of operation [31], while this study found that the HZ-CW still exhibited effective TP removal, even after several years of operation.

3.2. The Bacterial Communities Diversity and Composition in HZ-CW

The bacterial communities in each zone of the HZ-CW at the end of the 5th year of operation were revealed by the high-throughput sequencing.

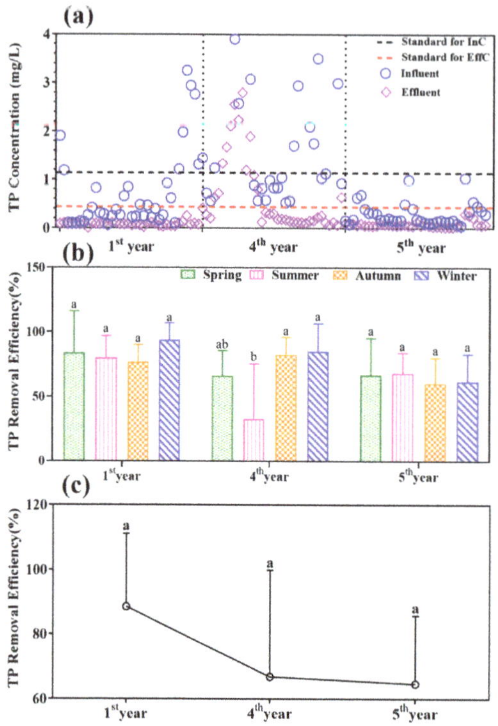

Figure 5. The performance of HZ-CW for TP removal during the study period. (**a**) The influent and effluent concentrations, (**b**) TP removal efficiencies (%) in different seasons ($n = 9$ for each season), (**c**) the mean TP removal efficiencies (%) in each year (the 1st year, 4th year, and 5th year, $n = 36$ for each year). Different letters indicate significant differences ($p < 0.05$) among different years. Data are given as the mean ± s.d. InC: influent concentration; EffC: effluent concentration.

3.2.1. The Richness and Diversity Analysis of the Bacterial Communities

Bacterial richness and diversity within each zone of the HZ-CW were listed in Table 1. The Coverage of the sequences were 0.969, 0.990, 0.966, 0.968, 0.967, 0.964, and 0.982 in AP, FP, FWS1, FWS2, FWS3, FWS4, and EP, respectively, which indicated that the majority of the OTUs have been captured [32]. ACE estimator showed that the community richness of FWS1, FWS2, FWS3, and FWS4 was similar to the ACE from 6223 to 6797, indicating their similar function. The community abundance of samples in the AP was slightly lower than that in the FWS1–4 with the ACE estimator of 5861, which might result from the relatively abundant matrix and the aerobic conditions [33]. The ACE estimators in FP and EP were 1997 and 3349, which might be caused by the greater depth of FP and EP (2.0 m and 1.5 m) and the non-aerated environment. Similar phenomena were observed for bacterial diversity, which FP and EP presented the lower diversity according to Shannon indices.

According to Wang et al., the Chao1 and Shannon index of bacterial communities in an STP were 1498 and 5.46 [30], respectively; both were significantly lower than those in the HZ-CW, except for those in FP. This result was attributed to the CWs' proximity to the natural environment and minimal artificial intervention. The high richness and diversity of microbial communities in the CWs indicate the greater adaptability in pollutant removal, and the CWs have greater potential for the elimination of emerging contaminants, such as antibiotics, estrogens, and personal care products [34,35].

Table 1. Comparison of phylotype coverage and diversity estimation of the 16S rRNA gene libraries at 97% similarity from the Miseq high-throughput analysis.

Samples	97% Similarity			
	OTUs	Coverage	Chao1	Shannon
AP	4371 ± 437 [ab]	0.969	5638 ± 137 [ab]	8.608 ± 1.41 [a]
FP	1464 ± 948 [c]	0.990	1883 ± 1014 [c]	5.486 ± 0.28 [b]
FWS1	4691 ± 1633 [ab]	0.966	6116 ± 2031 [ab]	8.894 ± 1.93 [a]
FWS2	4924 ± 1948 [a]	0.968	6099 ± 2384 [ab]	9.317 ± 2.40 [a]
FWS3	4661 ± 1831 [ab]	0.967	5989 ± 2205 [ab]	8.9785 ± 2.55 [a]
FWS4	5154 ± 761 [a]	0.964	3267 ± 2116 [bc]	9.331 ± 1.66 [a]
EP	2481 ± 1459 [bc]	0.982	6572 ± 2220 [a]	5.7825 ± 1.85 [b]

Coverage (Good's coverage), richness indices (chao1), diversity indices (Shannon) were calculated by the Mothur program. Different superscript lower case letters indicate statistically significant differences among HZ-CW elements ($p < 0.05$); AP, aerated pond; FP, facultative pond; FWS1–FWS4, four free water surface flow constructed wetlands follow the water flow; EP, ecological pond.

3.2.2. Bacterial Community Composition

The bacterial community structure in different samples was analyzed and shown in Figure 6. On the phylum level, Proteobacteria dominated in all sediment samples, and the relative abundance was 33.75%, 48.62%, 39.07%, 39.99%, 46.07%, 71.80%, and 36.74% in AP, FP, FWS1, FWS2, FWS3, FWS4, and EP, respectively. The result was similar to that reported by Hu et al. (30.38–52.28%) [23] and Li et al. (26.4–52.6%) [36]. The phylum Proteobacteria plays an integral role in carbon, nitrogen, and sulfur cycling within the ecosystem [37–39], with a significant prevalence observed in CWs, STPs, and soils [32,40]. The Bacillota was the subdominant phylum with a relative abundance of 23.36%, 24.53%, 14.04%, 18.83%, 18.79%, 11.28%, and 15.94%, respectively. Bacteroidota, Chloroflexota, and Actinomycetota were phyla with a relative abundance above 1.0% in at least one sample. These taxa exhibited a high level of prevalence in STPs. Actinomycetota was found to represent a significant proportion in FP, which was attributed to the anaerobic conditions situated at the bed of FP and the saprophytic tendency of Actinomycetota [41]. Acidobacteriota was also the top 10 phylum in the system with a relative abundance ranging from 0.4% to 3.1%, which could improve the degradation of plant residues. As shown in Figure 6a, FWS1, FWS2, FWS3, and FWS4 which were sequentially connected presented similar community composition on the phylum level.

On the genus level, 25 primary genera (relative abundance > 1.00% in at least one sample) were observed (Figure 6b). In AP, *Bacillus* was the dominant genus with a relative abundance of 15.30%, followed by *Rokubacteria*, *Hymenobacter*, *Phyllobacterium*, *unidentified-Deltaproteobacteria*, and *Sphingomonas*. In FP, the dominant genus was *Phyllobacterium*, followed by *Lactobacillus Delftia*, *unidentified-Cyanobacteria*, *Rhodococcus*, *Staphylococcus*, and *Megasphaera*. In FWS1, FWS2, FWS3, and FWS4, *Phyllobacterium* was the dominant genus with relative abundance of 9.33%, 6.39%, 8.04%, and 4.66%, respectively. For *Bacillus*, the relative abundances were 3.97%, 3.17%, 2%, and 11.84% in FWS1, FWS2, FWS3, and FWS4. Acinetobacter, Staphylococcus, and Clostridiales also presented high abundance. In EP, the relative abundance of Phyllobacterium was as high as 31.06%, similar to that in FP, which should result from a comparable ecosystem. The relative abundances of *Delftia*, *Acinetobacter*, *Bacillus*, *Staphylococcus*, as well as *Sphingomonas* in EP, were 11.93%, 8.82%, 4.49%, 2.76%, and 2.41%.

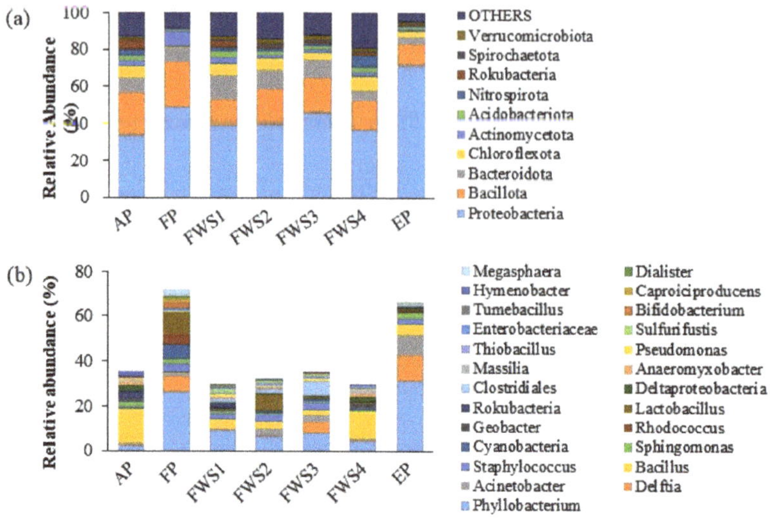

Figure 6. Relative abundance of bacterial community composition of each stage at the (**a**) phylum and (**b**) genus level by 16S rRNA MiSeq sequencing. The sum of the relative abundance of all the other phyla in the figure is grouped into "OTHERS". AP, aerated pond; FP, facultative pond; FWS1–FWS4, four free water surface flow constructed wetlands follow the water flow; EP, ecological pond.

3.2.3. Bacteria Related to Nitrogen and Phosphorus Removal

The bacteria related to nitrogen and phosphorus removal were listed in Table 2. The dominant ammonia oxidizers (AOB) were found to be *Unidentified_Nitrosomonadaceae*, with a relative abundance ranging from 0.171–0.593% in AP and FWS1–4, 0.026% in FP, and 0.036% in EP, respectively. *Nitrosomonas* was another AOB with a relative abundance below 0.013%. All samples obtained three genera of nitrite oxidizers (NOB), and the relative abundance were 0.25%, 0.026%, 0.18%, 0.25%, 0.19%, 0.19%, and 0.047% in AP, FP, FWS1, FWS2, FWS3, FWS4, and EP, respectively. Results showed that the relative abundance of AOB and NOB in AP and FWS1–4 was significantly higher than that in FP and EP, which was mainly caused by the lack of DO at the bottom of FP and EP. The relative abundance of nitrifying microorganisms was comparable to that of integrated subsurface-flow CW in Zhu's paper [15], which ensured the effectiveness of the CWs in removing NH_4^+-N. Three potential anammox bacteria were also found in sediment samples, *Candidatus_Anammoximicrobium*, *unidentified_Brocadiales*, and *Candidatus_Brocadia*, but their relative abundance in each region was less than 0.1%, which was significantly lower than other researches [15,18]. This was mainly because HZ-CW was a surface-flow CW, whereas those were subsurface-flow CWs. Therefore, based on the abundance of nitrifying bacteria and anammox bacteria, it could be inferred that nitrification, rather than anammox, was the main process of ammonia removal in surface-flow CWs.

Three putative genera of PAOs Thiothrix, Microlunatus, and Tessaracoccus were found in at least one sample, but the relative abundance was all below 0.01%. *Accumulibacter* and *Tetrasphaera*, the two typical phosphorus accumulating organisms (PAOs) in STPs, were not observed in all the samples, which might be caused by the absence of volatile fatty acids (VFA) or proteins and polysaccharides in the influent and the alternation of aerobic and anaerobic conditions. [30,42,43] The findings suggested that EBPR did not represent the principal mechanism for TP removal in the HZ-CW; instead, adsorption, filtration, and other processes appeared to be the driving factors. Consequently, TP removal remained unaffected by seasonal variances [44].

Table 2. The genera of nitrifying and polyphosphate accumulating bacteria in each stage of SHZ-CW.

	Genus	Relative Abundances (%)						
		AP	FP	FWS1	FWS2	FWS3	FWS4	EP
AOB	*unidentified_Nitrosomonadaceae*	0.178	0.026	0.287	0.171	0.222	0.593	0.036
	Nitrosomonas	0.013	0.011	0.007	0.007	0.008	0.005	0.002
NOB	*unidentified_Nitrospirae*	0.107	0.003	0.114	0.200	0.173	0.135	0.014
	unidentified_Nitrospiraceae	0.104	0.021	0.045	0.035	0.007	0.034	0.030
	Candidatus_Nitrotoga	0.042	0.002	0.022	0.013	0.010	0.019	0.003
Anammox	*Candidatus_Anammoximicrobium*	0.026	0.000	0.020	0.043	0.026	0.026	0.005
	unidentified_Brocadiales	0.000	0.000	0.000	0.001	0.002	0.018	0.000
PAOs	*Candidatus_Brocadia*	0.001	0.001	0.007	0.000	0.002	0.071	0.000
	Thiothrix	0.000	0.001	0.002	0.009	0.002	0.001	0.000
	Microlunatus	0.000	0.000	0.001	0.001	0.003	0.000	0.001

4. Conclusions

After several years of operation, the HZ-CW still exhibited a good contaminant removal efficiency, the average removal efficiencies of COD, NH_4^+-N, and TP were 7.6%, 78.33%, 66.86% and 15.14%, 64.68%, 46.04% in the 4th and 5th year, respectively. Fluctuation of the influent concentration was demonstrated as the main factor contributing to the inter- and intra-annual variation of COD, NH_4^+-N, and TP removal efficiency. The high-throughput sequencing analysis showed the high richness and diversity of the bacteria community in the HZ-CW, indicating the greater adaptability and potential for contaminants elimination. The relative abundance of ammonia-oxidizing organisms (AOMs) and anammox bacteria revealed that nitrification was the dominant NH_4^+-N removal pathway in the HZ-CW, and adsorption and filtration was the primary TP removal mechanism due to the fairly low abundance of PAOs. This study revealed that CWs were a promising choice for advanced treatment of tailwater even after multi-years' operation, which deepened the understanding of CW performance and could promote their wider application.

Author Contributions: Conceptualization, S.S. (Siyuan Song), D.Z. and S.A.; Data curation, S.S. (Siyuan Song) and Y.G.; Formal analysis, S.S. (Siyuan Song), D.Z. and B.W.; Resources, B.W.; Visualization, B.W.; Funding acquisition, S.A., A.L. and S.S. (Siyuan Song); Investigation, T.Y. and S.S. (Sheng Sheng); Writing—original draft, S.S. (Siyuan Song); Writing—review and editing, S.S. (Siyuan Song), B.W., D.Z. and A.L. All authors have read and agreed to the published version of the manuscript.

Funding: This research was funded by the Zhejiang Province Postdoctoral Research Funding Project (No.ZJ2021047); the Major Science and Technology Program for Water Pollution Control and Treatment (No.2017ZX07602002).

Data Availability Statement: Not applicable.

Conflicts of Interest: The authors declare no conflict of interest. The funders had no role in the design of this study; in the collection, analyses, or interpretation of data; in the writing of the manuscript; or in the decision to publish the results.

References

1. Vymazal, J. Constructed wetlands for wastewater treatment: Five decades of expericence. *Environ. Sci. Technol.* **2011**, *45*, 61–69. [CrossRef] [PubMed]
2. Vymazal, J.; Brezinova, T. Long term treatment performance of constructed wetlands for wastewater treatment in mountain areas: Four case studies from the Czech Republic. *Ecol. Eng.* **2014**, *71*, 578–583. [CrossRef]
3. Song, S.; Liu, B.; Zhang, W.; Wang, P.; Qiao, Y.; Zhao, D.; Yang, T.; An, S.; Leng, X. Performance of a large-scale wetland treatment system in treating tailwater from a sewage treatment plant. *Mar. Freshw. Res.* **2018**, *69*, 833–841. [CrossRef]
4. Wu, H.; Zhang, J.; Guo, W.; Liang, S.; Fan, J. Secondary effluent purification by a large-scale multi-stage surface-flow constructed wetland: A case study in northern China. *Bioresour. Technol.* **2018**, *249*, 1092–1096. [CrossRef] [PubMed]
5. Song, S.; Wang, P.; Liu, Y.; Zhao, D.; An, S. Effects of Oenanthe javanica on Nitrogen Removal in Free-Water Surface Constructed Wetlands under Low-Temperature Conditions. *Int. J. Environ. Res. Public Health* **2019**, *16*, 1420. [CrossRef]

6. Zhao, D.; Zhang, M.; Liu, Z.; Sheng, J.; An, S. Can cold-season macrophytes at the senescence stage improve nitrogen removal in integrated constructed wetland systems treating low carbon/nitrogen effluent? *Bioresour. Technol.* **2018**, *265*, 380–386. [CrossRef]
7. Gao, Y.; Zhang, W.; Gao, B.; Jia, W.; Miao, A.; Xiao, L.; Yang, L. Highly efficient removal of nitrogen and phosphorus in an electrolysis-integrated horizontal subsurface-flow constructed wetland amended with biochar. *Water Res.* **2018**, *139*, 301–310. [CrossRef]
8. Vymazal, J. Constructed wetlands for wastewater treatment. *Ecol. Eng.* **2005**, *25*, 475–477. [CrossRef]
9. Vymazal, J. Constructed wetlands for treatment of industrial wastewaters: A review. *Ecol. Eng.* **2014**, *73*, 724–751. [CrossRef]
10. Morvannou, A.; Choubert, J.M.; Vanclooster, M.; Molle, P. Modeling nitrogen removal in a vertical flow constructed wetland treating directly domestic wastewater. *Ecol. Eng.* **2014**, *70*, 379–386. [CrossRef]
11. Chen, S.K.; Jang, C.S.; Chou, C.Y. Assessment of spatiotemporal variations in river water quality for sustainable environmental and recreational management in the highly urbanized Danshui River basin. *Environ. Monit. Assess.* **2019**, *191*, 100. [CrossRef]
12. Liu, D.; Ge, Y.; Chang, J.; Peng, C.; Gu, B.; Chan, G.Y.S.; Wu, X. Constructed wetlands in China: Recent developments and future challenges. *Front. Ecol. Environ.* **2009**, *7*, 261–268. [CrossRef]
13. Zhang, D.Q.; Jinadasa, K.B.; Gersberg, R.M.; Liu, Y.; Ng, W.J.; Tan, S.K. Application of constructed wetlands for wastewater treatment in developing countries–a review of recent developments (2000–2013). *J. Environ. Manag.* **2014**, *141*, 116–131. [CrossRef]
14. Li, X.; Li, Y.; Lv, D.; Li, Y.; Wu, J. Nitrogen and phosphorus removal performance and bacterial communities in a multi-stage surface flow constructed wetland treating rural domestic sewage. *Sci. Total Environ.* **2020**, *709*, 136235. [CrossRef]
15. Zhu, T.; Gao, J.; Huang, Z.; Shang, N.; Gao, J.; Zhang, J.; Cai, M. Comparison of performance of two large-scale vertical-flow constructed wetlands treating wastewater treatment plant tail-water: Contaminants removal and associated microbial community. *J. Environ. Manag.* **2021**, *278*, 111564. [CrossRef]
16. Cooper, D.; Griffin, P.; Cooper, P. Factors affecting the longevity of sub-surface horizontal flow systems operating as tertiary treatment for sewage effluent. *Water Sci. Technol.* **2005**, *51*, 127–135. [CrossRef]
17. Nivala, J.; Rousseau, D.P.L. Reversing clogging in subsurface-flow constructed wetlands by hydrogen peroxide treatment: Two case studies. *Water Sci. Technol.* **2009**, *59*, 2037–2046. [CrossRef]
18. Zhang, M.; Zhao, D.; Chen, C.; Yang, J.; Lu, Q.; Zhang, N.; Leng, X.; An, S. The effect of re-startup strategies on the recovery of constructed wetlands after long-term resting operation. *Bioresour. Technol.* **2020**, *311*, 123583. [CrossRef]
19. Li, C.; Liu, Q.; Fan, J.; Peng, Y.; Du, R. Metagenomics-based interpretation of selective bioaugmentation promoting partial-denitrification coupling with anammox process reactivation in suspended sludge system. *Chem. Eng. J.* **2023**, *454*, 139977. [CrossRef]
20. Wang, X.; Tian, Y.; Zhao, X.; Peng, S.; Wu, Q.; Yan, L. Effects of aeration position on organics, nitrogen and phosphorus removal in combined oxidation pond-constructed wetland systems. *Bioresour. Technol.* **2015**, *198*, 7–15. [CrossRef]
21. Boog, J.; Nivala, J.; Aubron, T.; Wallace, S.; van Afferden, M.; Muller, R.A. Hydraulic characterization and optimization of total nitrogen removal in an aerated vertical subsurface flow treatment wetland. *Bioresour. Technol.* **2014**, *162*, 166–174. [CrossRef] [PubMed]
22. Dong, H.; Qiang, Z.; Li, T.; Jin, H.; Chen, W. Effect of artificial aeration on the performance of vertical-flow constructed wetland treating heavily polluted river water. *J. Environ. Sci.* **2012**, *24*, 596–601. [CrossRef] [PubMed]
23. Hu, N.; He, J.; Shi, W.; He, J.; Lv, B.; Liang, Y.; Huang, L. Ecological restoration for the Liangtan river by Rotating biological contactors combined with hybrid constructed wetlands. *J. Clean. Prod.* **2022**, *375*, 134189. [CrossRef]
24. Kadlec, R.H.; Wallace, S. *Treatment Wetlands*, 2nd ed.; CRC Press Taylor & Francis Group: Boca Raton, FL, USA, 2009.
25. Aiello, R.; Bagarello, V.; Barbagallo, S.; Iovino, M.; Marzo, A.; Toscano, A. Evaluation of clogging in full-scale subsurface flow constructed wetlands. *Ecol. Eng.* **2016**, *95*, 505–513. [CrossRef]
26. Du, L.; Trinh, X.; Chen, Q.; Wang, C.; Wang, H.; Xia, X.; Zhou, Q.; Xu, D.; Wu, Z. Enhancement of microbial nitrogen removal pathway by vegetation in Integrated Vertical-Flow Constructed Wetlands (IVCWs) for treating reclaimed water. *Bioresour. Technol.* **2018**, *249*, 644–651. [CrossRef] [PubMed]
27. Godos, I.; Vargas, V.A.; Blanco, S.; Gonzalez, M.C.; Soto, R.; Garcia-Encina, P.A.; Becares, E.; Munoz, R. A comparative evaluation of microalgae for the degradation of piggery wastewater under photosynthetic oxygenation. *Bioresour. Technol.* **2010**, *101*, 5150–5158. [CrossRef]
28. Wu, Y.; Han, R.; Yang, X.; Zhang, Y.; Zhang, R. Long-term performance of an integrated constructed wetland for advanced treatment of mixed wastewater. *Ecol. Eng.* **2017**, *99*, 91–98. [CrossRef]
29. Saggai, M.M.; Ainouche, A.; Nelson, M.; Cattin, F.; El Amrani, A. Long-term investigation of constructed wetland wastewater treatment and reuse: Selection of adapted plant species for metaremediation. *J. Environ. Manag.* **2017**, *201*, 120–128. [CrossRef]
30. Wang, B.; Jiao, E.; Guo, Y.; Zhang, L.; Meng, Q.; Zeng, W.; Peng, Y. Investigation of the polyphosphate-accumulating organism population in the full-scale simultaneous chemical phosphorus removal system. *Environ. Sci. Pollut. Res. Int.* **2020**, *27*, 37877–37886. [CrossRef]
31. Rousseau, D.P.L.; Horton, D.; Griffin, P.; Vanrolleghem, P.A.; Pauw, N.D. Impact of operational maintenance on the asset life of storm reed beds. *Water Sci. Technol.* **2005**, *51*, 243–250. [CrossRef]
32. Ansola, G.; Arroyo, P.; Saenz de Miera, L.E. Characterisation of the soil bacterial community structure and composition of natural and constructed wetlands. *Sci. Total Environ.* **2014**, *473–474*, 63–71. [CrossRef]

33. Wu, Y.; Han, R.; Yang, X.; Fang, X.; Chen, X.; Yang, D.; Zhang, R. Correlating microbial community with physicochemical indices and structures of a full-scale integrated constructed wetland system. *Appl. Microbiol. Biotechnol.* **2016**, *100*, 6917–6926. [CrossRef]
34. Zhang, Y.; You, X.; Huang, S.; Wang, M.; Dong, J. Knowledge Atlas on the Relationship between Water Management and Constructed Wetlands—A Bibliometric Analysis Based on CiteSpace. *Sustainability* **2022**, *14*, 8288. [CrossRef]
35. Zhang, Y.; Ji, Z.; Pei, Y. Nutrient removal and microbial community structure in an artificial-natural coupled wetland system. *Process Saf. Environ. Prot.* **2021**, *147*, 1160–1170. [CrossRef]
36. Li, D.; Ye, B.; Hou, Z.; Chu, Z.; Zheng, B. Long-term performance and microbial distribution of a filed-scale storing multi-pond constructed wetland with Ottelia acuminata for the treatment of non-point source pollution. *J. Clean. Prod.* **2020**, *262*, 121367. [CrossRef]
37. Li, H.; Liu, F.; Luo, P.; Chen, X.; Chen, J.; Huang, Z.; Peng, J.; Xiao, R.; Wu, J. Stimulation of optimized influent C:N ratios on nitrogen removal in surface flow constructed wetlands: Performance and microbial mechanisms. *Sci. Total Environ.* **2019**, *694*, 133575. [CrossRef]
38. Fumasoli, A.; Morgenroth, E.; Udert, K.M. Modeling the low pH limit of Nitrosomonas eutropha in high-strength nitrogen wastewaters. *Water Res.* **2015**, *83*, 161–170. [CrossRef]
39. Gao, J.; Zhu, T.; Liu, C.; Zhang, J.; Gao, J.; Zhang, J.; Cai, M.; Li, Y. Ammonium removal characteristics of heterotrophic nitrifying bacterium Pseudomonas stutzeri GEP-01 with potential for treatment of ammonium-rich wastewater. *Bioprocess Biosyst. Eng.* **2020**, *43*, 959–969. [CrossRef]
40. Chen, Y.; Wen, Y.; Tang, Z.; Huang, J.; Zhou, Q.; Vymazal, J. Effects of plant biomass on bacterial community structure in constructed wetlands used for tertiary wastewater treatment. *Ecol. Eng.* **2015**, *84*, 38–45. [CrossRef]
41. Li, D.; Chu, Z.; Zeng, Z.; Sima, M.; Huang, M.; Zheng, B. Effects of design parameters, microbial community and nitrogen removal on the field-scale multi-pond constructed wetlands. *Sci. Total Environ.* **2021**, *797*, 148989. [CrossRef]
42. Qiu, G.; Zuniga-Montanez, R.; Law, Y.; Thi, S.S.; Nguyen, T.Q.N.; Eganathan, K.; Liu, X.; Nielsen, P.H.; Williams, R.B.H.; Wuertz, S. Polyphosphate-accumulating organisms in full-scale tropical wastewater treatment plants use diverse carbon sources. *Water Res.* **2019**, *149*, 496–510. [CrossRef] [PubMed]
43. Fernando, E.Y.; McIlroy, S.J.; Nierychlo, M.; Herbst, F.A.; Petriglieri, F.; Schmid, M.C.; Wagner, M.; Nielsen, J.L.; Nielsen, P.H. Resolving the individual contribution of key microbial populations to enhanced biological phosphorus removal with Raman-FISH. *ISME J.* **2019**, *13*, 1933–1946. [CrossRef] [PubMed]
44. Tang, X.Q.; Huang, S.L.; Fciwem, M.S. Comparison of phosphorus removal between vertical subsurface flow constructed wetlands with different substrates. *Water Environ. J.* **2009**, *23*, 180–188. [CrossRef]

processes

MDPI

Article

Electrochemical Reduction Performance and Mechanism of 2,2′,4,4′-Tetrabromodiphenyl Ether (BDE-47) with Pd/Metal Foam Electrodes

Chenghao Liang [1], He Liu [2,*], Jiahao Liang [1], Xiaodong Du [1], Xueqin Tao [3] and Guining Lu [1,4,*]

1 School of Environment and Energy, South China University of Technology, Guangzhou 510006, China; liangch@foxmail.com (C.L.); liang_jiahao@mail.tsinghua.edu.cn (J.L.); dxdon@scut.edu.cn (X.D.)
2 South China Institute of Environmental Sciences, Guangzhou 510655, China
3 College of Resources and Environment, Zhongkai University of Agriculture and Engineering, Guangzhou 510225, China; taoxueqin@zhku.edu.cn
4 Key Laboratory of Pollution Control and Ecosystem Restoration in Industry Clusters, Ministry of Education, Guangzhou 510006, China
* Correspondence: liuhe@scies.org (H.L.); lutao@scut.edu.cn (G.L.)

Abstract: Polybrominated diphenyl ethers (PBDEs), a type of brominated flame retardant, are of global concern due to their environmental persistence, bioaccumulation, toxicity, and resistance to conventional remediation methods. In this study, the electrochemical reduction of 2,2′,4,4′-tetrabromodiphenyl ether (BDE-47) with Pd/Metal foam electrodes (Ni, Cu, and Ag) was investigated. The effect of Pd loadings was explored, and the results show that Pd loading enhances the debromination performance, with 15.16%Pd/Ni foam exhibiting the best efficiency, followed by 9.37%Pd/Cu and 10.26%Pd/Ag. The degradation mechanisms for Pd/Ni and Pd/Ag are primarily hydrogen atom transfer, while for Pd/Cu, electron transfer dominates. Among the reduction products, Pd/Ni foam shows the highest debromination capability. The impact of electrolytes, current intensity, and bromination degrees of PBDEs was evaluated for 15.16%Pd/Ni. The results reveal that the presence of electrolytes inhibits BDE-47 degradation; the degradation rate of BDE-47 increases with current density, peaks at 4 mA, and decreases as current rises; and 15.16%Pd/Ni foam can effectively degrade PBDEs with varying bromination levels. Additionally, cycling tests show a decrease in efficiency from 94.3% (first cycle) to 56.58% (fourth cycle), attributed to Pd loss and structural damage. The findings offer valuable insights for developing efficient, sustainable catalytic materials for the electrochemical degradation of PBDEs and other persistent organic pollutants.

Keywords: BDE-47; Pd/Metal foam electrodes; electrochemical reduction; degradation kinetics; debromination pathways; mechanism

1. Introduction

Polybrominated diphenyl ethers (PBDEs) have been extensively used as cost-effective and efficient flame retardants in a wide range of applications, including electronic devices (such as circuit boards and plastic casings), building materials, household items, furniture, and transportation vehicles [1]. However, with their widespread use over the past few decades, it has become evident that PBDEs possess harmful characteristics, such as environmental persistence, bioaccumulation, and endocrine-disrupting effects [2,3]. The mass production of fire-resistant products has resulted in significant amounts of PBDEs being released into the environment, leading to serious health and ecological concerns. As a result,

PBDEs have been globally banned from production and use to mitigate their adverse impacts [4]. PBDE pollution predominantly originates from the treatment of electronic waste, and its environmental impact is challenging to mitigate within a short timeframe [5–8]. Currently, most regions worldwide lack stringent regulations to control PBDE elimination during the electronic waste dismantling process. As a result, PBDE concentrations in the environment near dismantling sites continue to rise, particularly around non-standard dismantling facilities [9–11]. In some certain electronic waste dismantling areas such as Taizhou and Guiyu, the concentration of PBDEs in the soil ranges from approximately 400 to 3000 μg/kg [10,12–15]; however, in severely contaminated regions, the concentration can escalate significantly, reaching levels as high as 10 to 300 mg/kg [16,17]. Moreover, by September 2021, 87.9% of soil samples collected from Guiyu contained PBDEs at concentrations ranging from 0.345 to 401,000 ng/kg [18]. This highlights the persistent challenge of managing PBDE pollution in electronic waste recycling areas.

Currently, the primary environmental remediation technologies for PBDEs include reduction degradation, advanced oxidation, and biodegradation. The study on the degradation of PBDEs has focused on its reductive debromination. Although advanced oxidation and biodegradation methods are capable of completely removing PBDEs, their effectiveness is generally lower than that of reduction degradation methods [19–24], primarily due to the longer oxidative degradation time required for PBDEs. Shi et al. employed a concentrated sulfuric acid/potassium permanganate system to oxidatively degrade BDE-209; while the degradation rate of BDE-209 reached 90% in the first few minutes, achieving complete removal took at least three hours [25]. Liang et al. used a thermally activated persulfate system to degrade BDE-47; under conditions of PDS:BDE-47 = 1000:1 and 60 °C, 100% degradation could be achieved within three hours [26]. However, the high temperature and persulfate dosage required make this approach less economical for practical applications. In contrast, reduction methods such as ultraviolet, zero-valent metal reduction, and TiO_2 photocatalytic reduction can achieve 100% removal of PBDEs in just a few minutes [19,27–29], demonstrating that PBDEs are more efficiently degraded through reduction processes. However, a major limitation of the current reduction methods is the incomplete debromination of PBDEs, leading to the accumulation of lower-brominated PBDE intermediates. Nevertheless, with ongoing advancements in technology, researchers have found that foam metal/electrochemical catalytic systems exhibit excellent debromination effects on PBDEs while minimizing the buildup of intermediate products, and this technology holds significant promise as an efficient method for the removal of PBDEs [20,30,31].

Electrochemical remediation is an environmentally friendly and efficient technology for pollutant removal of both heavy metals and organic compounds, offering distinct advantages for treating pollutants in complex media like wastewater and soil [30,32]. Despite electrode damage, high maintenance costs, and energy consumption, the electrochemical method's operational simplicity, automation, and compact design make it an increasingly favored approach in environmental remediation [30,32,33]. In recent years, it has become a focal point of research and application due to its ability to generate in situ hydrogen atoms with strong reducing properties, enabling the effective degradation of a wide range of pollutants [34–36]. Currently, electrochemical reduction dehalogenation commonly employs catalysts such as Pd, Pt, Rh, and Ru. These metals are favored for their excellent catalytic properties, which facilitate efficient dehalogenation processes in the treatment of halogenated pollutants. Among them, Pd has excellent catalytic properties and is often used as a catalyst to produce active species in dehalogenation reactions [22,37,38]. Numerous studies have demonstrated that the incorporation of Pd significantly enhances catalyst performance and accelerates the degradation of PBDEs. Zhuang et al. initially synthesized Pd/Fe bimetallic materials for the degradation of BDE-21, demonstrating a degradation rate two orders of magnitude higher than that of Fe alone [39]. Subsequently, Wang et al. utilized Pd/TiO_2 under ultraviolet light to degrade BDE-47, also achieving complete degradation within approximately 10 min [40]. These studies emphasize that Pd has strong catalytic effects. In electrocatalytic hydrogenation reactions, Pd electrodes or Pd-modified electrodes notably increase the adsorption capacity of active hydrogen atoms at the electrode interface, which in turn boosts the degradation efficiency of pollutants [41–43]. Therefore, the Pd-modified electrode is likely to exhibit the capability to efficiently remove PBDEs while minimizing the generation and accumulation of lower brominated PBDEs.

Overall, this study aims to explore the electrochemical reduction and degradation removal performance of BDE-47 using Pd-modified foam metals with different electrochemical properties, including Ni, Cu, and Ag foam. Pd is chemically deposited onto these foam metals to create a variety of Pd-loaded electrode materials. BDE-47, a frequently detected and highly toxic environmental contaminant, is selected as a model compound for investigation. This study explores the degradation kinetics of BDE-47 involving Pd/Ni, Pd/Cu, and Pd/ Ag foam, and then analyzes the reduction products of BDE-47 to identify the dominant reaction mechanisms involved in its degradation with different foam materials. The electrode material with the most effective degradation performance is further examined to assess the impact of various environmental factors on its performance. Additionally, the cyclic stability and surface morphology changes of the material are investigated to better examine the foam material's long-term performance. The overall goal of this research is to develop a clean and efficient method for the removal of PBDEs based on electrochemical catalysis, providing new insights into the degradation of halogenated organic pollutants.

2. Materials and Methods

2.1. Chemicals

The authentic PBDE standards of 4-mono-PBDE (BDE-3), 4,4′-di-PBDE (BDE-15), 2,4,4′-tri-PBDE (BDE-28), 2,2′,4,4′-tetra-PBDE (BDE-47), 2,2′,4,4′,5-penta-PBDE (BDE-99), 2,2′,4,4′,6-penta-PBDE (BDE-100), 2,2′,4,4′,5,5′-hexa-PBDE (BDE-153), 2,2′,4,4′,5,6′-hexa-PBDE (BDE-154), and 2,2′,3,4,4′,5′,6-tetra-PBDE (BDE-183) were purchased from AccuStandard Inc., New Haven, CT, USA. The internal standard of 2,3,4,5-tetrachlorinated biphenyl (PCB-61) was purchased from ANPEL Laboratory Technologies Inc., Shanghai. The purities of all of the above authentic standards were greater than 99%. The Ni foam, Cu foam, and Ag foam were purchased from Yizhongtian New Material Co. (Kunshan, China). $PdNa_2Cl_4$ was purchased from Adamas Co., Ltd. (Shanghai, China). HPLC-grade solvents of methanol, acetonitrile, hexane, and acetone were purchased from Aladdin Bio-Chem Technology Co., Ltd. (Shanghai, China). The ultra-pure water was acquired from Millipore milli-Q system (Burlington, MA, USA).

2.2. Preparation of Metal Foam Electrodes

2.2.1. Pre-Treatment of Metal Foam Electrodes

The metal foam electrodes were first cut into dimensions of 20 × 50 mm (length × width). They were then immersed in acetone and sonicated for 60 min to remove all the possible organic contaminants from the surface. Next, the electrodes were washed with oxygen-free water and placed into an 80 g/L sulfuric acid solution. The electrodes were sonicated for 60 to 120 s to remove the oxide layer from the metal surface. After sonication, the electrodes were immediately rinsed with oxygen-free water. Finally, the metal foam electrodes were immersed in methanol and stored for future use.

2.2.2. Synthesis of Pd-Modified Metal Foam Electrodes

Based on the metal activity series, Pd^{4+} ions in the Na_2PdCl_4 solution were chemically reduced and spontaneously deposited onto the surface of the foam metal electrode, yielding the Pd-modified foam electrode (Pd/Metal foam) material required for this study. A total of three electrode materials, Pd/Ni foam, Pd/Cu foam, and Pd/Ag foam, were prepared by loading Pd particles onto Ni foam, Cu foam, and Ag foam by chemical deposition. Different volumes of 10 mM Na_2PdCl_4 solution were used to obtain the electrodes with different Pd particle loadings.

For synthesizing the Pd-modified metal foam electrodes, the pre-formulated 10 mM Na_2PdCl_4 was added to the pre-treated foam electrodes and adjusted to a total volume of 40 mL using oxygen-free water. Several pieces of foam metal with specific mass were placed into the 50 mL centrifuge tubes. Then, the tubes were vortexed and shaken at 2000 rpm for 1 h. Afterward, the foam materials were taken out, rinsed with oxygen-free water under slow flow, and vacuum freeze-dried for 24 h. Following the drying process, the samples were placed in drying dishes for later use. The theoretical Pd-loading ratios of Ni foam, Cu foam, and Ag foam electrodes are depicted in Table S1.

2.3. Degradation Procedures

The basic electrolyte for BDE-47 degradation experiments was prepared by mixing methanol and oxygen-free water in a 6:4 ratio, which contained 0.05 mol/L H_2SO_4 and 20 ppm of BDE-47. A cylindrical single-cell glass electrolyzer was used as the reaction vessel (Figure 1). At the beginning, 50 mL of the prepared electrolyte was added to the reactor, with rotor stirring set at 300 rpm. The anode was a platinum sheet electrode ($10 \times 10 \times 0.1$ mm) while the cathode consisted of a Pd/Metal foam electrode. The electrolysis was initiated by energizing the electrodes, and the reaction time was simultaneously recorded. Samples were taken at reaction intervals of 0, 5, 10, 20, 30, 45, 60, and 90 min. For each sample, 1 mL was transferred into a chromatographic vial for kinetic analysis using HPLC. An additional 1 mL was transferred into a 4 mL vial, and 1 mL of hexane extract (0.100 mg L^{-1} PCB-61 as the internal standard) was added. The mixture was shaken with a vortex shaker for 20 min to facilitate extraction. Finally, approximately 1 mL of the organic solvent containing possible degradation products was transferred to a new vial for GC/MS analysis.

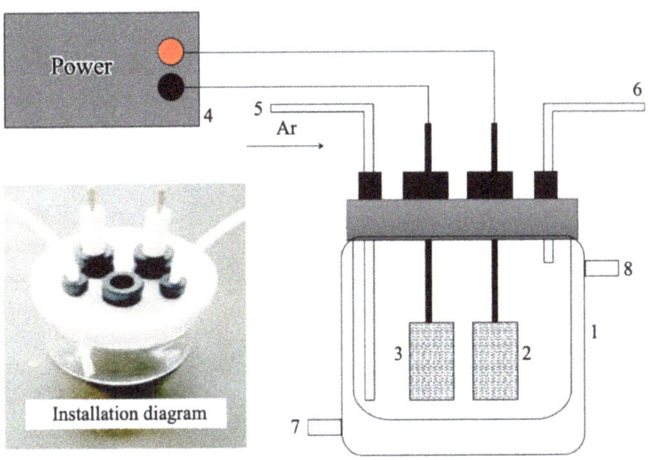

1. Electrolytic cell; 2. Anode; 3. Cathode; 4. Power supply or electrochemical workstation; 5. Ar inlet; 6. Ar outlet; 7. Water inlet; 8. Water outlet

Figure 1. Schematic diagram of the cylindrical single-cell glass electrolyzer.

2.4. Analytic Methods

High-Performance Liquid Chromatography (HPLC) was employed for the quantitative analysis of PBDEs. The analysis was carried out using a Phenomenex Luna C18 reverse-phase separation column (San Diego, CA, USA) with a flow rate of 1 mL/min. The mobile phase consisted of 15% water and 85% acetonitrile, and the column was maintained at a temperature of 40 °C.

Qualitative analysis of the debromination products of BDE-47 was performed with a Thermo Trace GC Ultra gas chromatography coupled to a DSQ II mass spectrometer detector (Thermo-Electron Corporation, Waltham, MA, USA). The mass spectrometer was equipped with an electron ionization (EI) source. An Agilent DB-5MS capillary column (30 m × 0.25 mm × 0.25 μm) (Santa Clara, CA, USA) was used for compound separation. The carrier gas of helium was set at a constant flow of 1 mL/min, and the temperature of the inlet was set at 280 °C with the splitless mode. The injection volume was set to 1 μL. The oven temperature was initiated at 120 °C for 1 min, then increased to 290 °C at a rate of 20 °C/min, then held for 2 min. All the samples were analyzed in both full scan mode and selective ion monitoring (SIM) mode. All the results were processed by Xcalibur software (Version 2.2.0).

2.5. Characterization of Pd/Metal Foam

The Zeiss Scanning Electron Microscopy (SEM) couped to Energy Dispersive X-ray Spectroscopy (EDS) (Oberkochen, Germany) were employed to visualize the microscopic surface properties of the Pd/Metal foam. Before SEM and EDS analysis, the samples were cut into appropriate sizes, vacuum-dried, and then characterized using SEM-EDS. The SEM was operated at an accelerating voltage of 10 kV, with a magnification range of 50–5000× and a working distance of 10 mm. For EDS analysis, the accelerating voltage was set to 5 kV, with a counting time of 100 s. The EDS measurements were conducted using a silicon drift detector to analyze the elemental composition of the materials.

3. Results and Discussion

3.1. Kinetics

In this study, we selected BDE-47 as a model compound, as it is the most toxic and frequently detected PBDE congeners in the environment. Five Pd/Ni foam electrodes with different Pd loading contents (2.12%, 4.17%, 8.07%, 15.16%, and 21.44%) were chosen for the degradation experiments. As illustrated in Figure 2a,b, increasing the Pd loading significantly enhances the catalytic degradation ability of the Ni foam electrode. With no Pd loaded, the Ni foam electrode achieved only a 9.5% degradation rate of BDE-47 over 90 min without current. In contrast, the Ni foam electrode achieved a 23.3% degradation rate of BDE-47 over 90 min with current. The involvement of Pd significantly enhanced the degradation of BDE-47, with 2.39%Pd/Ni foam exhibiting a 42.24% degradation rate within just 30 min. As the Pd loading increased, the degradation effect initially increased, peaking at 16.39%Pd/Ni foam, and then began to decrease at higher loadings. In addition, for Pd/Ni foam with Pd loading less than 16.39%, the reaction reached equilibrium in approximately 30 min. However, when the Pd loading was greater than 16.39%, the degradation rate slowed down, but the degradation rate still reached 90% within 90 min. The first-order kinetic fitting results revealed the BDE-47 degradation rates under different Pd loading amounts: 1.384, 1.964, 3.128, 5.644, and 2.291 h^{-1}, respectively. These results highlight the significant influence of Pd loading on the catalytic degradation performance of the Ni foam electrodes. It is noteworthy that during the degradation of BDE-47 with Pd/Ni foam, rapid degradation was observed at the beginning of the reaction, followed by a slower degradation equilibrium stage (Figure 3a–c). Further analysis of the equilibrium stage revealed that the degradation rate of BDE-47 during this stage was slow and comparable to that of Ni foam without Pd. This observation suggests that Pd may be consumed during the degradation process, leading to a subsequent equilibrium state in the reaction.

Although Pd/Cu bimetallic catalysts are extensively utilized in the catalytic degradation of halogen-containing organic compounds, their application in the form of Pd/Cu foam electrodes for the electrocatalytic degradation of halogenated organic pollutants remains limited [44]. Pd/Cu foam electrodes with Pd loadings of 1.25%, 2.48%, 4.86%, 7.16%, and 9.37% were prepared and evaluated. As illustrated in Figures 2c,d and 3d, unmodified Cu foam exhibited almost no degradation effect on BDE-47 without current. However, with current, the removal rate of BDE-47 reached 23.7% within 90 min, slightly higher than that of Ni foam (23.3%). When Pd was loaded onto Cu foam, the degradation efficiency of BDE-47 for 1.25%Pd/Cu, 2.48%Pd/Cu, and 4.86%Pd/Cu foams was similar to that of unmodified Cu foam. The reaction rates suggested that the addition of Pd slightly inhibited the degradation of BDE-47 at lower loading levels. However, as the Pd loading increased, the degradation efficiency improved progressively. When Pd was loaded onto Cu foam, the degradation efficiency of BDE-47 for 1.25%Pd/Cu, 2.48%Pd/Cu, and 4.86%Pd/Cu foams was similar to that of unmodified Cu foam. The reaction rates suggested that the addition of Pd slightly inhibited the degradation of BDE-47 at lower loading levels. Furthermore, the degradation behavior of BDE-47 using all Pd/Cu foam electrodes followed a first-order kinetic process, and unlike Pd/Ni foam, no equilibrium stage was observed during the reaction. This indicates consistent and sustained degradation activity throughout the reaction period.

Pd/Ag materials have been extensively studied in the field of electrocatalysis [45–47]. However, their degradation performance and mechanism for brominated organic compounds, such as PBDEs, have rarely been explored. Ag foam electrodes with Pd loadings of 1.36%, 2.70%, 5.30%, 7.83%, and 10.26% were prepared and tested for their degradation performance. As shown in Figures 2e,f and 3e, unmodified Ag foam exhibited no degradation effect on BDE-47 in the absence of current. When powered, the removal rate of BDE-47 by Ag foam reached 18.2% within 90 min, which is lower than the corresponding rates achieved by Pd/Ni foam and Pd/Cu foam under similar conditions. The addition of Pd onto Ag foam promoted the degradation of BDE-47, with the degradation efficiency positively correlated with the Pd loading amount. However, compared to Pd/Ni foam and Pd/Cu foam, the degradation rate of BDE-47 under 10.26%Pd/Ag foam was only 74.8% in 90 min, indicating that the enhancement provided by Pd on Ag foam was limited. This limitation may be attributed to the inherent physical and chemical properties of Ag, which are less conducive to catalytic performance compared to Ni or Cu. Similar to Pd/Cu foam, the degradation behavior of BDE-47 using Pd/Ag foam electrodes followed a first-order kinetic process, with no equilibrium stage observed throughout the reaction.

In summary, Pd loading was found to have the most significant effect on the catalytic performance of Ni foam, although the degradation rate of BDE-47 decreased over time due to the loss of Pd on the Ni foam surface. In contrast, Pd loading had a minimal effect on Cu foam at low levels but significantly improved the degradation performance of BDE-47 at higher loadings. On the other hand, Pd loading on Ag foam showed only limited improvement in degradation performance, even at high loadings, likely due to the inherent properties of Ag foam. These results highlight that the catalytic enhancement effect of Pd is material-dependent, with Ni foam being the most favorable substrate for BDE-47 degradation, followed by Cu foam, while Ag foam showed the least improvement.

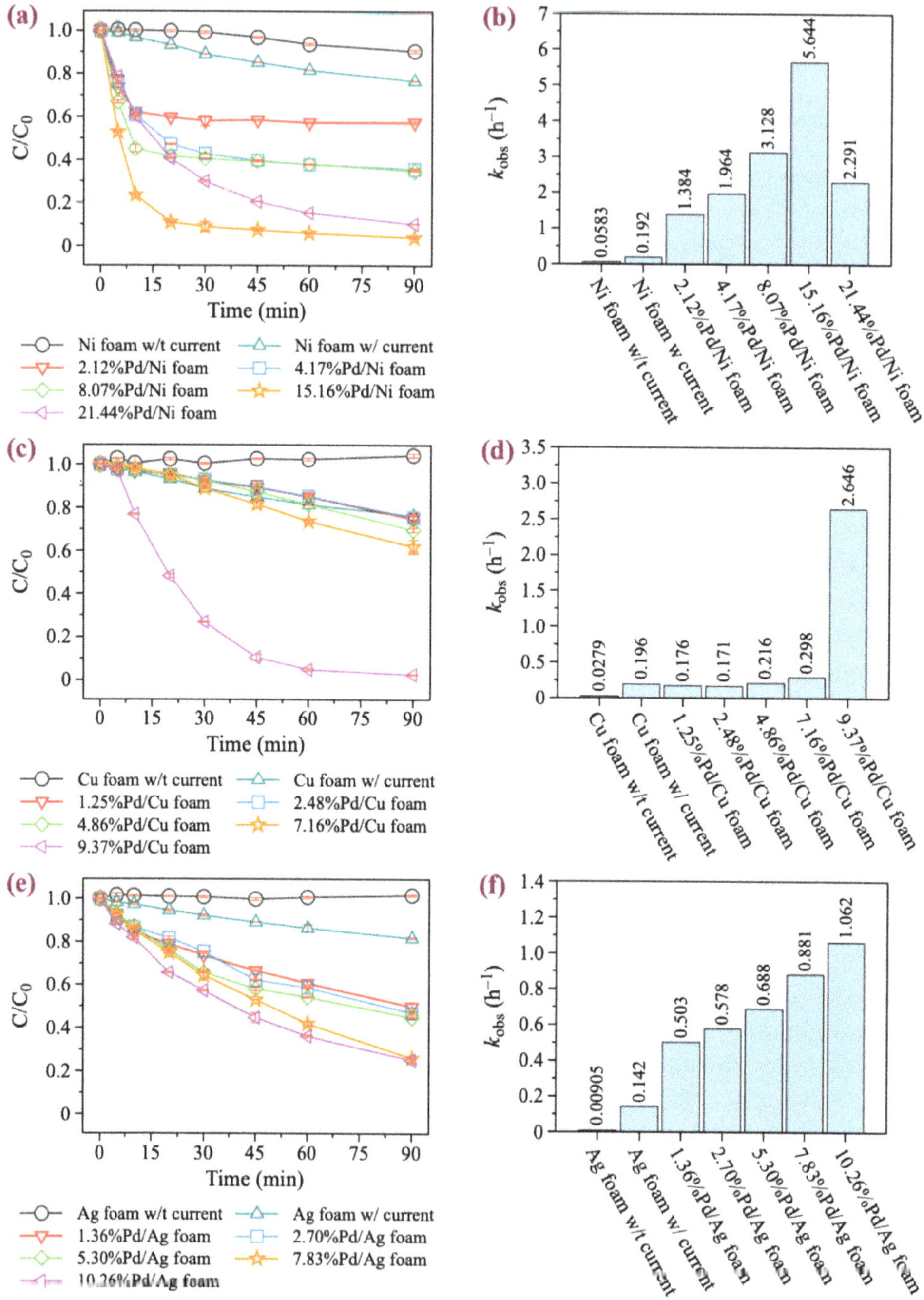

Figure 2. Degradation kinetics of BDE-47. (**a**,**c**,**e**) The degradation kinetics of BDE-47 with Pd/Ni, Pd/Cu, and Pd/Ag foam, respectively. (**b**,**d**,**f**) The kobs of BDE-47 with Pd/Ni, Pd/Cu, and Pd/Ag foam, respectively.

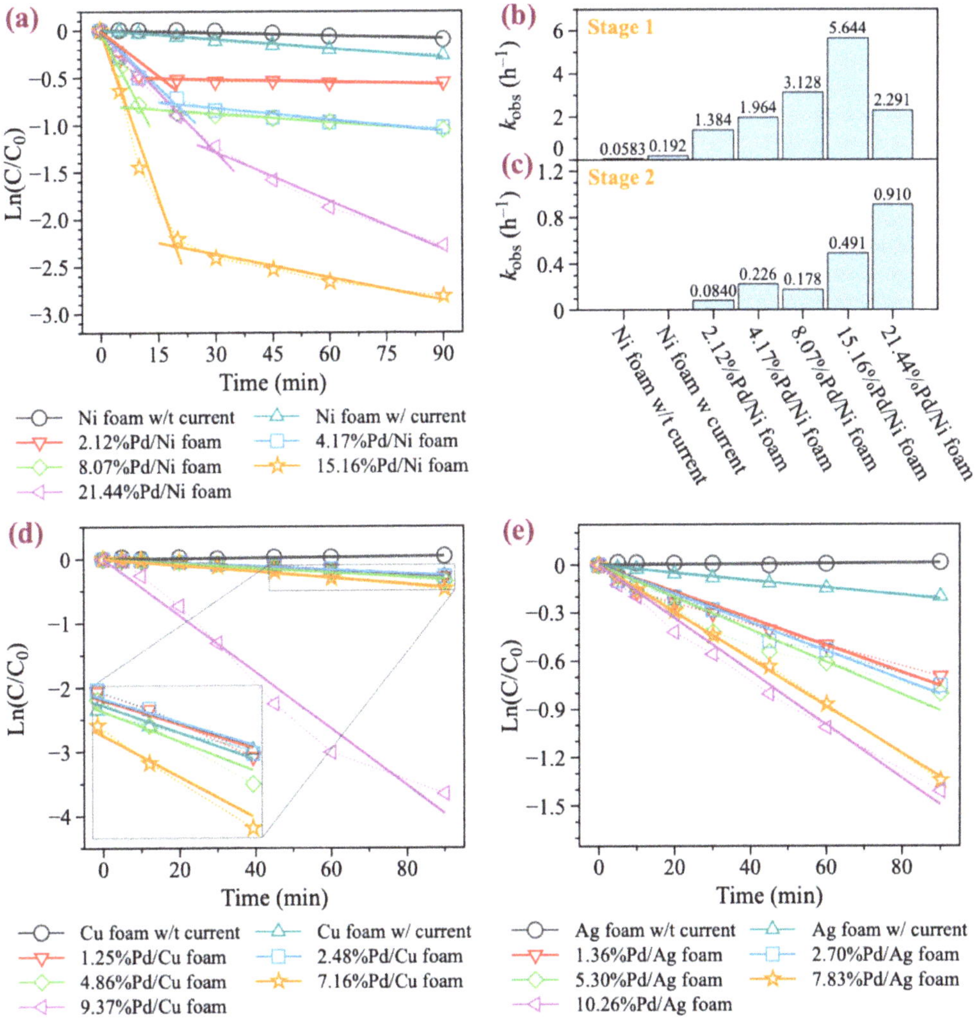

Figure 3. First-order kinetic fittings of BDE-47 with Pd/Ni, Pd/Cu, and Pd/Ag foam. (**a**) Fitted lines of BDE-47's degradation with Pd/Ni foam. (**b**,**c**) Fitted k_{obs} of BDE-47 with Pd/Ni foam at different degradation stages. (**d**) Fitted lines of BDE-47's degradation with Pd/Cu foam. (**e**) Fitted lines of BDE-47's degradation with Pd/Ag foam.

3.2. Degradation Products and Mechanisms

We further detected and analyzed the degradation products of BDE-47 under Pd/Ni, Pd/Cu, and Pd/Ag conditions to elucidate the degradation mechanism of BDE-47 on different electrode materials. The reductive debromination of polybrominated diphenyl ethers (PBDEs) primarily involves two mechanisms: the active hydrogen atom transfer mechanism and the electron transfer mechanism. By comparing the types and abundances of primary debromination products, the dominant degradation mechanism in the current system was inferred. BDE-47 has two primary debromination products: BDE-28, formed by the removal of an ortho-C/Br bond, and BDE-17, formed by the removal of a para-C/Br bond. The complete degradation pathway of BDE-47 is shown in Figure S1.

GC/MS analysis revealed various reduction products of BDE 47. Under Pd/Ni foam conditions, no significant debromination products were detected with low Pd loading (e.g., 2.12%Pd/Ni foam) (Figure S2). However, as the Pd loading increased, the abundance of products with different bromination levels gradually rose, with lower brominated PBDEs becoming the dominant products. This demonstrates that Pd/Ni foam exhibits an effective debromination capability for BDE-47. In particular, under 15.16%Pd/Ni foam, the abundance of debromination products was significant, with BDE-3 and diphenyl ether (DE) being the major products (Figure 4a). Nevertheless, analyzing the formation of BDE-28 and BDE-17 revealed that the abundance of BDE-17 consistently exceeded that of BDE-28, indicating that the degradation of BDE-47 under Pd/Ni foam is primarily governed by the active hydrogen atom transfer mechanism. Summarizing the product generation under Pd/Ni foam conditions, the reductive degradation pathway of BDE-47 proceeds as follows: BDE-47 → BDE-17 → BDE-8 → BDE-3 → DE.

For Pd/Cu foam, the amount of BDE-47 reduction products was found to be low under low Pd loading conditions (1.25%, 2.48%, 4.86%), with minimal difference in the abundances of BDE-17 and BDE-28 (Figure S3). However, as the Pd loading increased to 7.16%, the abundance of BDE-28 slightly exceeded that of BDE-17. At the highest Pd loading of 9.37% (Figure 4b), the abundance of BDE-28 significantly increased, with a notable rise in the amount of BDE-15. These observations strongly suggest that the reductive degradation of BDE-47 under Pd/Cu foam conditions is predominantly governed by the electron transfer mechanism. At lower Pd loading levels, however, neither the active hydrogen atom transfer mechanism nor the electron transfer mechanism contributed significantly to the degradation of BDE-47. This lack of dominance resulted in minimal differences in the abundances of BDE-28 and BDE-17, and the competition between these two mechanisms likely explains the lower degradation rates of BDE-47 observed under low Pd loading. Additionally, no formation of lower brominated PBDEs was detected in this system, indicating a limited debromination effect of Pd/Cu foam on BDE-47. In general, under Pd/Cu foam conditions, the reductive degradation pathway of BDE-47 can be described as BDE-47 → BDE-28 → BDE-15. This pathway, coupled with the absence of further debromination products, highlights the limited degradation capability of Pd/Cu foam, especially at lower Pd loadings.

For Pd/Ag foam, abundant reduction products of BDE-47 were observed under all Pd loading conditions (Figure S4). Further analysis revealed that the abundance of BDE-17 was consistently higher than that of BDE-28 (Figure 4c), suggesting that the degradation of BDE-47 under Pd/Ag foam conditions is predominantly governed by the active hydrogen atom transfer mechanism. Additionally, significant amounts of lower brominated PBDEs were detected under various Pd loading conditions, indicating that Pd/Ag foam has good debromination efficiency for BDE-47. However, due to its relatively slower degradation rate for BDE-47, its debromination effect is not as pronounced as that of Pd/Ni foam. It is worth noting that due to chromatographic limitations, we were unable to completely separate BDE-4 and BDE-7. However, based on the observed product distribution and degradation pathways, BDE-7 is considered to be the more likely product. In conclusion, the reductive degradation pathway of BDE-47 under Pd/Ag foam conditions can be summarized as BDE-47 → BDE-17 → BDE-7 → BDE-3 → DE. Although Pd/Ag foam shows good debromination performance, its overall efficiency is limited by its slower degradation rate compared to Pd/Ni foam.

Overall, Pd/Ni foam demonstrated the highest debromination efficiency and fastest degradation rate, driven by the active hydrogen atom transfer mechanism. Pd/Cu foam showed limited debromination but had an electron transfer mechanism that dominated at higher loadings, leading to efficient degradation. Pd/Ag foam achieved good debromination but was constrained by slower degradation rates, even though it produced abundant lower brominated PBDEs. The results highlight the role of Pd-loaded-foam electrodes in PBDEs' degradation performance and mechanism.

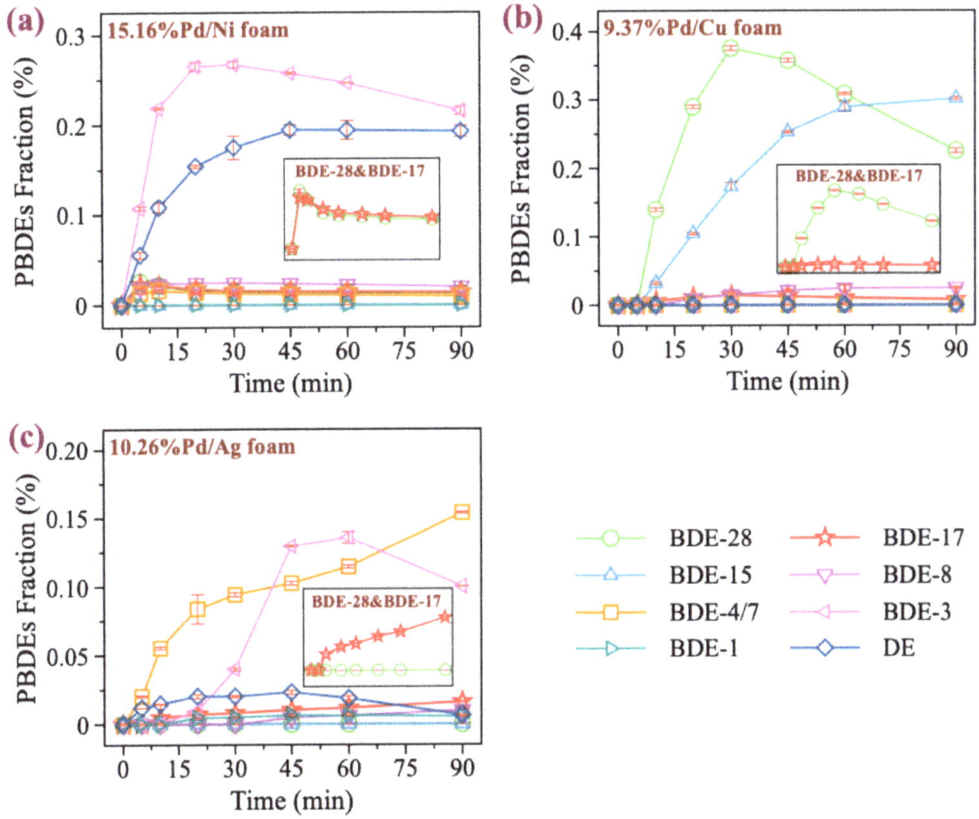

Figure 4. Production of the debromination products from BDE-47. (**a**–**c**) The products of BDE-47 in 15.16%Pd/Ni, 9.37%Pd/Cu, 10.26%Pd/Ag, respectively.

3.3. *Influencing Factors*

3.3.1. Electrolytes

The microscopic chemical properties of electrolytes play a crucial role in influencing electrochemical processes. So, we investigated the effects of four common electrolytes (NaCl, Na_2SO_4, KCl, and K_2SO_4) on the electrochemical reduction debromination of BDE-47 using the Pd/Ni foam electrode with 15.16%Pd loading. The electrolyte concentration was 0.05 mol/L. The experimental results showed that the debromination rates of BDE-47 were 71.03% for NaCl, 83.53% for Na_2SO_4, 74.36% for KCl, and 87.36% for K_2SO_4 (Figure 5). But compared to the previous results, it is obvious that all four electrolytes strongly inhibited the degradation of BDE-47 to varying degrees. For Cl salts (NaCl and KCl), the initial inhibition was attributed to the high concentration of Cl^- ions competing with BDE-47 for active hydrogen generated in the system, thereby reducing the availability of hydrogen for debromination. However, as the reaction progressed, Cl^- ions were gradually consumed, allowing for the degradation rate of BDE-47 to increase in the later stages. However, for SO_4^{2-} salts (Na_2SO_4 and K_2SO_4), an initial inhibition period was observed during the first 60 min of the reaction. This was likely due to the high concentration of SO_4^{2-}

ions temporarily inhibiting the activity of the Pd/Ni foam electrode. However, as the reaction continued, the Pd/Ni foam electrode became more active with continuous current, resulting in rapid degradation of BDE-47 after the inhibition period. Moreover, the presence of K^+ ions slightly improved the debromination performance compared to Na^+ ions under similar conditions. This is likely due to K^+ ions enhancing current conduction efficiency in the reaction system.

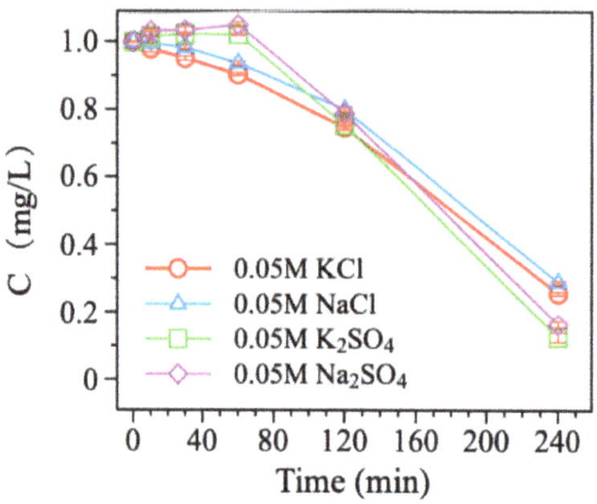

Figure 5. Degradation of BDE-47 in different electrolytes.

3.3.2. Current Intensity

The degradation performance of BDE-47 under different current intensities was explored using the 15.16%Pd/Ni foam electrode (Figure 6). The tested current intensities were 0, 0.5, 1.0, 2.0, 3.0, 4.0, 6.0, 8.0, and 10.0 mA. The results showed that the 15.16%Pd/Ni foam electrode could degrade BDE-47 without an applied current, but the degradation efficiency was relatively low, achieving only a 21.6% degradation rate after 90 min. As the current intensity increased, the degradation rate of BDE-47 improved, reaching the highest efficiency at 4.0 mA. These results demonstrate that the external electric field significantly enhances the catalytic degradation performance of the Pd/Ni foam electrode. However, at current intensities above 4.0 mA, the degradation efficiency began to decline, with a more pronounced inhibitory effect observed as the current intensity further increased. This decline in degradation performance at higher current intensities is likely due to the rapid loss of Pd on the surface of the Pd/Ni foam during the catalytic process, which hinders the continued degradation of BDE-47. These results suggest that the external electric field is critical for enhancing degradation, while excessive current intensity accelerates the depletion of Pd, thereby reducing the electrode's catalytic efficiency. Herein, we infer that optimizing the current intensity and controlling Pd loss on the surface of the Pd/Ni foam are essential for maintaining high degradation efficiency and prolonging the electrode's catalytic performance during the degradation of PBDEs.

3.3.3. PBDEs with Different Bromination Degrees

The number of aromatic rings, the benzene ring configuration, and the C/X bond position of aromatic bromides are the basic factors that affect the reaction [48,49]. The degradation performance of 15.16%Pd/Ni foam on PBDEs with varying degrees of bromine atom substitution was investigated, using nine representative PBDE monomers (BDE-183, BDE-154, BDE-153, BDE-100, BDE-99, BDE-47, BDE-28, BDE-15, and BDE-3) commonly found in commercial pentabromodiphenyl ether mixtures [50–52]. The results demonstrated that all nine PBDEs can be rapidly degraded within 90 min of reaction time, achieving degradation rates of over 80% (Figure 7). The degradation kinetics indicate that the catalytic degradation of PBDEs by Pd/Ni foam is not significantly affected by the degree of bromine atom substitution, suggesting that the number of bromine atoms in PBDEs does not play a limiting role in the reduction debromination process. This highlights the universality of the Pd/Ni foam electrode in effectively degrading PBDEs across a broad range of bromination levels.

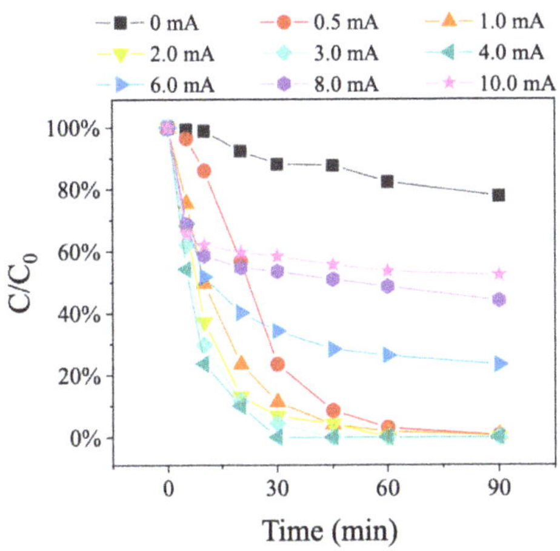

Figure 6. Degradation of BDE-47 in different current intensities.

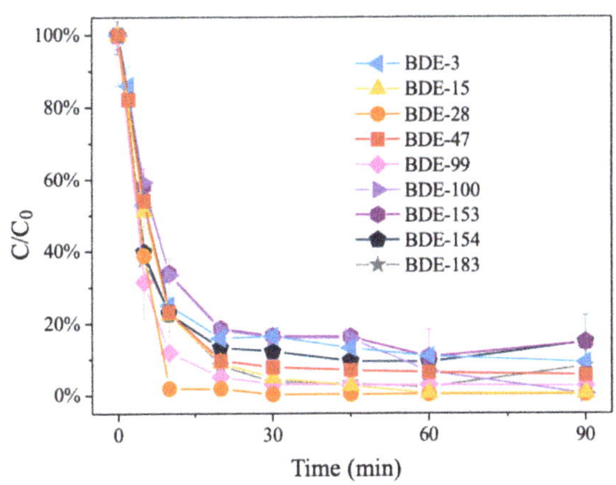

Figure 7. The degradation kinetics of PBDEs of different bromination levels with 15.16%Pd/Ni foam.

3.4. SEM Characterization

The 15.16%Pd/Ni foam electrode was selected for characterization using SEM and EDS to observe morphological changes before and after the 90 min reaction (Figure 8). The initial Ni foam exhibits a typical flower-like microstructure (Figure 8a). After Pd loading, the overall size of the "nickel flower" remains unchanged (Figure 8b); however, the edges of the flower become blunter and the "petals" appear to be much rougher. During the reaction, the surface morphology of the Pd/Ni foam undergoes significant changes. Initially (0 min), the structure maintains its flower-like shape, but as the reaction progresses, it first transforms into a "flower ball" (30 min) (Figure 8c) and ultimately into a granular structure (90 min) (Figure 8d). This morphological evolution is attributed to the electrochemical reaction occurring on the electrode surface, where continuous erosion and structural remodeling take place. The SEM images also further explain why the degradation rate of BDE-47 appears to be much slower as the reaction proceeds.

Figure 8. SEM of 15.16%Pd/Ni foam before and after a reaction. (**a**) The Ni foam without Pd loading. (**b**) The 15.16%Pd/Ni foam before reaction. (**c**) The 15.16%Pd/Ni foam after a 30 min reaction. (**d**) The 15.16%Pd/Ni foam after a 90 min reaction.

3.5. Electrode Cycling Performance

The stability of the electrode is crucial for its practical application. So, the 15.16%Pd/Ni foam electrode was further evaluated through cycling experiments to assess its performance. As shown in Figure 9, the degradation efficiency of BDE-47 decreased progressively over four consecutive cycles. Visible corrosion and slight damage to the electrode were observed, along with black particle shedding into the solution. In terms of kinetic performance, the system achieved optimal degradation in the first cycle, achieving a 94.3% removal rate of BDE-47 within 90 min. However, with subsequent cycles, the degradation efficiency declined to 64.26%, 58.74%, and 56.58% in the second, third, and fourth cycles, respectively. SEM analysis revealed significant surface corrosion and material shedding after the four cycles, with substantial Pd loss from the surface of the Pd/Ni foam (Figure 10). It is speculated that the reason for the corrosion and shedding of Pd/Ni foam may be the embrittlement and fracture of the material caused by the diffusion of hydrogen atoms in the metal phase. As Pd has a higher capacity for hydrogen storage compared to Ni, the repeated absorption and release of hydrogen atoms lead to the peeling and fragmentation of Pd. To enhance the long-term stability of Pd/Ni foam electrodes, it is essential to develop methods for securely anchoring Pd to the substrate and to minimize corrosion caused by overly acidic electrolytes when using Ni foam as the cathode.

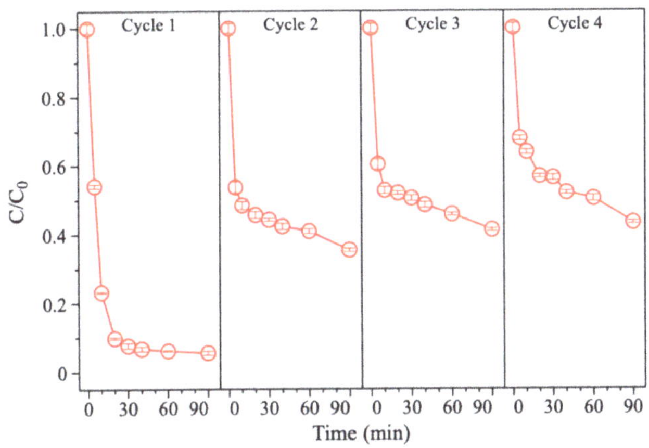

Figure 9. Performance of 15.16%Pd/Ni foam in degrading BDE-47 in four repetitive experiments.

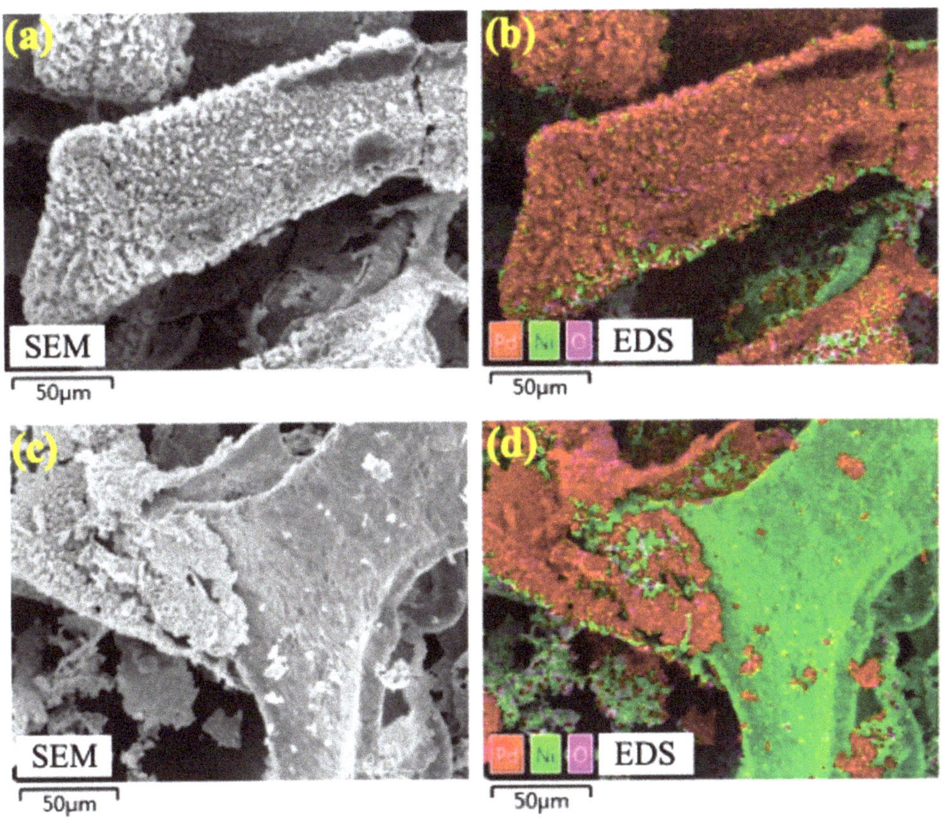

Figure 10. SEM and EDS of 15.16%Pd/Ni foam before and after four cycles of experiments. (**a**,**c**) The SEM of 15.16%Pd/Ni foam before and after four cycles of experiments. (**b**,**d**) The EDS of the distributions of 15.16%Pd/Ni foam before and after four cycles of experiments.

4. Conclusions

In conclusion, this study systematically investigated the electrochemical reduction debromination performance of polybrominated diphenyl ethers (PBDEs) using Pd/Ni, Pd/Cu, and Pd/Ag foams as catalytic electrodes. The results demonstrate that the loading of Pd significantly enhances the catalytic reduction debromination performance of the three electrode materials on BDE-47, with 15.16%Pd/Ni foam showing the most pronounced improvement, followed by 9.37%Pd/Cu and 10.26%Pd/Ag. An analysis of the primary reduction products of BDE-47 reveals that the degradation mechanism for Pd/Ni and Pd/Ag materials is predominantly governed by the active hydrogen atom transfer mechanism, while for the Pd/Cu material, the electron transfer mechanism plays a dominant role. Comparing the debromination efficiency of the reduction products, the Pd/Ni foam electrode exhibits the highest debromination capability. Subsequently, the 15.16%Pd/Ni foam, which demonstrated the best degradation performance, was selected to further investigate the effects of the electrolyte, current density, and the degree of bromination of PBDEs. The results indicate that the presence of an electrolyte significantly inhibits the degradation of BDE-47. The degradation rate of BDE-47 initially increases with the current density, reaches its peak efficiency at 4 mA, and then declines as the current density continues to rise. Moreover, 15.16%Pd/Ni foam exhibits effective degradation performance across PBDEs with varying degrees of bromination, underscoring its broad applicability and versatility. The cycling experiments further demonstrated the declining performance of the 15.16%Pd/Ni foam electrode, with degradation efficiency decreasing from 94.3% in the first cycle to 56.58% in the fourth cycle. The Pd loss and structural damage caused by hydrogen embrittlement and acidic corrosion were identified as key factors limiting the material's long-term stability. To enhance the practical application of Pd/Ni foam electrodes, future studies should focus on reducing Pd loss, improving the stability of the Ni material, minimizing hydrogen-induced embrittlement, and optimizing the reaction conditions to extend the electrode's lifespan. Overall, the findings provide valuable insights into the development of efficient and sustainable catalytic materials for the electrochemical degradation of PBDEs, while offering preliminary evidence of its potential adaptability to other structural-like persistent organic pollutants such as polybrominated biphenyls.

Supplementary Materials: The following supporting information can be downloaded at: https://www.mdpi.com/article/10.3390/pr13030853/s1, Table S1: Theoretical Pd loading ratios of Ni, Cu, and Ag foam electrodes; Figure S1: Debromination pathways of BDE-47; Figure S2: Generation of lower brominated products from BDE-47 with Pd/Ni foam of different Pd loadings; Figure S3: Generation of lower brominated products from BDE-47 with Pd/Cu foam of different Pd loadings; Figure S4: Generation of lower brominated products from BDE-47 with Pd/Ag foam of different Pd loadings.

Author Contributions: Writing—original draft preparation, C.L.; writing—review and editing, H.L., J.L., X.D. and G.L.; Investigation, C.L., H.L. and X.T. All authors have read and agreed to the published version of the manuscript.

Funding: This research was funded by the National Natural Science Foundation of China (Nos. 42077114 and 42277380) and the Local Innovation and Entrepreneurship Team Project of Guangdong Special Support Program (No. 2019BT02L218).

Data Availability Statement: The data presented in this study are available on request from the corresponding author due to the copyright of our experimental data.

Acknowledgments: The authors are grateful to the Ecological Restoration Research Group for their support in PBDEs' degradation experiments.

Conflicts of Interest: The authors declare no conflicts of interest.

References

1. Abbasi, G.; Li, L.; Breivik, K. Global Historical Stocks and Emissions of PBDEs. *Environ. Sci. Technol.* **2019**, *53*, 6330–6340. [CrossRef] [PubMed]
2. Anh, H.Q.; Nam, V.D.; Tri, T.M.; Ha, N.M.; Ngoc, N.T.; Mai, P.T.N.; Anh, D.H.; Minh, N.H.; Tuan, N.A.; Minh, T.B. Polybrominated Diphenyl Ethers in Plastic Products, Indoor Dust, Sediment and Fish from Informal e-Waste Recycling Sites in Vietnam: A Comprehensive Assessment of Contamination, Accumulation Pattern, Emissions, and Human Exposure. *Environ. Geochem. Health* **2017**, *39*, 935–954. [CrossRef] [PubMed]
3. Wang, R.; Cheng, H.; Bian, Z. Global Occurrence and Environmental Behavior of Novel Brominated Flame Retardants in Soils: Current Knowledge and Future Perspectives. *J. Hazard. Mater.* **2024**, *480*, 136298. [CrossRef]
4. Sharkey, M.; Harrad, S.; Abou-Elwafa Abdallah, M.; Drage, D.S.; Berresheim, H. Phasing-out of Legacy Brominated Flame Retardants: The UNEP Stockholm Convention and Other Legislative Action Worldwide. *Environ. Int.* **2020**, *144*, 106041. [CrossRef]
5. Labunska, I.; Harrad, S.; Wang, M.; Santillo, D.; Johnston, P. Human Dietary Exposure to PBDEs Around E-Waste Recycling Sites in Eastern China. *Environ. Sci. Technol.* **2014**, *48*, 5555–5564. [CrossRef]
6. Liang, J.; Wang, R.; Liu, H.; Xie, D.; Tao, X.; Zhou, J.; Yin, H.; Dang, Z.; Lu, G. Unintentional Formation of Mixed Chloro-Bromo Diphenyl Ethers (PBCDEs), Dibenzo-p-Dioxins and Dibenzofurans (PBCDD/Fs) from Pyrolysis of Polybrominated Diphenyl Ethers (PBDEs). *Chemosphere* **2022**, *308*, 136246. [CrossRef]
7. Liang, J.; Lu, G.; Wang, R.; Tang, T.; Huang, K.; Jiang, F.; Yu, W.; Tao, X.; Yin, H.; Dang, Z. The Formation Pathways of Polybrominated Dibenzo-p-Dioxins and Dibenzofurans (PBDD/Fs) from Pyrolysis of Polybrominated Diphenyl Ethers (PBDEs): Effects of Bromination Arrangement and Level. *J. Hazard. Mater.* **2020**, *399*, 123004. [CrossRef] [PubMed]
8. Li, T.-Y.; Zhou, J.-F.; Wu, C.-C.; Bao, L.-J.; Shi, L.; Zeng, E.Y. Characteristics of Polybrominated Diphenyl Ethers Released from Thermal Treatment and Open Burning of E-Waste. *Environ. Sci. Technol.* **2018**, *52*, 4650–4657. [CrossRef]
9. Guo, J.; Zhang, R.; Xu, Z. PBDEs Emission from Waste Printed Wiring Boards during Thermal Process. *Environ. Sci. Technol.* **2015**, *49*, 2716–2723. [CrossRef]
10. Gao, S.; Hong, J.; Yu, Z.; Wang, J.; Yang, G.; Sheng, G.; Fu, J. Polybrominated Diphenyl Ethers in Surface Soils from E-Waste Recycling Areas and Industrial Areas in South China: Concentration Levels, Congener Profile, and Inventory. *Environ. Toxicol. Chem.* **2011**, *30*, 2688–2696. [CrossRef]
11. Li, W.-L.; Ma, W.-L.; Jia, H.-L.; Hong, W.-J.; Moon, H.-B.; Nakata, H.; Minh, N.H.; Sinha, R.K.; Chi, K.H.; Kannan, K.; et al. Polybrominated Diphenyl Ethers (PBDEs) in Surface Soils across Five Asian Countries: Levels, Spatial Distribution, and Source Contribution. *Environ. Sci. Technol.* **2016**, *50*, 12779–12788. [CrossRef] [PubMed]
12. Liu, M.; Huang, B.; Bi, X.; Ren, Z.; Sheng, G.; Fu, J. Heavy Metals and Organic Compounds Contamination in Soil from an E-Waste Region in South China. *Environ. Sci. Process. Impacts* **2013**, *15*, 919. [CrossRef] [PubMed]
13. Tang, X.; Zeng, B.; Hashmi, M.Z.; Long, D.; Yu, B.; Ullah, N.; Shen, C.; Chen, Y. PBDEs and PCDD/Fs in Surface Soil Taken from the Taizhou e-Waste Recycling Area, China. *Chem. Ecol.* **2014**, *30*, 245–251. [CrossRef]
14. Zhang, S.; Xu, X.; Wu, Y.; Ge, J.; Li, W.; Huo, X. Polybrominated Diphenyl Ethers in Residential and Agricultural Soils from an Electronic Waste Polluted Region in South China: Distribution, Compositional Profile, and Sources. *Chemosphere* **2014**, *102*, 55–60. [CrossRef]
15. Huang, Y.; Zhang, D.; Yang, Y.; Zeng, X.; Ran, Y. Distribution and Partitioning of Polybrominated Diphenyl Ethers in Sediments from the Pearl River Delta and Guiyu, South China. *Environ. Pollut.* **2018**, *235*, 104–112. [CrossRef]
16. Alabi, O.A.; Bakare, A.A.; Xu, X.; Li, B.; Zhang, Y.; Huo, X. Comparative Evaluation of Environmental Contamination and DNA Damage Induced by Electronic-Waste in Nigeria and China. *Sci. Total Environ.* **2012**, *423*, 62–72. [CrossRef]
17. Labunska, I.; Harrad, S.; Santillo, D.; Johnston, P.; Brigden, K. Levels and Distribution of Polybrominated Diphenyl Ethers in Soil, Sediment and Dust Samples Collected from Various Electronic Waste Recycling Sites within Guiyu Town, Southern China. *Environ. Sci. Process. Impacts* **2013**, *15*, 503. [CrossRef]
18. Ling, S.; Zhou, S.; Tan, J.; Lu, C.; Fu, M.; Peng, C.; Zhang, W.; Hu, S.; Lin, K.; Zhou, B. Brominated Flame Retardants (BFRs) in Sediment from a Typical e-Waste Dismantling Region in Southern China: Occurrence, Spatial Distribution, Composition Profiles, and Ecological Risks. *Sci. Total Environ.* **2022**, *824*, 153813. [CrossRef]
19. Yu, K.; Gu, C.; Boyd, S.A.; Liu, C.; Sun, C.; Teppen, B.J.; Li, H. Rapid and Extensive Debromination of Decabromodiphenyl Ether by Smectite Clay-Templated Subnanoscale Zero-Valent Iron. *Environ. Sci. Technol.* **2012**, *46*, 8969–8975. [CrossRef]
20. Su, J.; Lu, N.; Zhao, J.; Yu, H.; Huang, H.; Dong, X.; Quan, X. Nano-Cubic Structured Titanium Nitride Particle Films as Cathodes for the Effective Electrocatalytic Debromination of BDE-47. *J. Hazard. Mater.* **2012**, *231–232*, 105–113. [CrossRef]
21. Wang, R.; Tang, T.; Lu, G.; Huang, K.; Yin, H.; Lin, Z.; Wu, F.; Dang, Z. Rapid Debromination of Polybrominated Diphenyl Ethers (PBDEs) by Zero Valent Metal and Bimetals: Mechanisms and Pathways Assisted by Density Function Theory Calculation. *Environ. Pollut.* **2018**, *240*, 745–753. [CrossRef] [PubMed]

22. Tang, T.; Lu, G.; Wang, R.; Chen, H.; Fang, Y.; Huang, K.; Zheng, J.; Zou, M.; Tao, X.; Yin, H.; et al. Debromination of Polybrominated Diphenyl Ethers (PBDEs) by Zero Valent Zinc: Mechanisms and Predicting Descriptors. *J. Hazard. Mater.* **2018**, *352*, 165–171. [CrossRef] [PubMed]
23. Wang, R.; Lu, G.; Lin, H.; Huang, K.; Tang, T.; Xue, X.; Yang, X.; Yin, H.; Dang, Z. Relative Roles of H-Atom Transfer and Electron Transfer in the Debromination of Polybrominated Diphenyl Ethers by Palladized Nanoscale Zerovalent Iron. *Environ. Pollut.* **2017**, *222*, 331–337. [CrossRef]
24. Lei, M.; Wang, N.; Zhu, L.; Tang, H. Peculiar and Rapid Photocatalytic Degradation of Tetrabromodiphenyl Ethers over Ag/TiO_2 Induced by Interaction between Silver Nanoparticles and Bromine Atoms in the Target. *Chemosphere* **2016**, *150*, 536–544. [CrossRef] [PubMed]
25. Shi, J.; Qu, R.; Feng, M.; Wang, X.; Wang, L.; Yang, S.; Wang, Z. Oxidative Degradation of Decabromodiphenyl Ether (BDE 209) by Potassium Permanganate: Reaction Pathways, Kinetics, and Mechanisms Assisted by Density Functional Theory Calculations. *Environ. Sci. Technol.* **2015**, *49*, 4209–4217. [CrossRef]
26. Liang, J.; Liu, H.; Zou, M.; Tao, X.; Zhou, J.; Dang, Z.; Lu, G. Degradation Efficiency and Mechanism of 2,2′,4,4′-Tetrabromodiphenyl Ether (BDE-47) by Thermally Activated Persulfate System. *Chemosphere* **2023**, *325*, 138396. [CrossRef] [PubMed]
27. Luo, S.; Yang, S.; Sun, C.; Gu, J.-D. Improved Debromination of Polybrominated Diphenyl Ethers by Bimetallic Iron–Silver Nanoparticles Coupled with Microwave Energy. *Sci. Total Environ.* **2012**, *429*, 300–308. [CrossRef]
28. Li, L.; Chang, W.; Wang, Y.; Ji, H.; Chen, C.; Ma, W.; Zhao, J. Rapid, Photocatalytic, and Deep Debromination of Polybrominated Diphenyl Ethers on Pd–TiO_2: Intermediates and Pathways. *Chem. Eur. J.* **2014**, *20*, 11163–11170. [CrossRef]
29. Du, X.; Li, H.; Liang, J.; Wang, R.; Huang, K.; Hayat, W.; Cai, L.; Tao, X.; Dang, Z.; Lu, G. Hydrogen-Donor-Controlled Polybrominated Dibenzofuran (PBDF) Formation from Polybrominated Diphenyl Ether (PBDE) Photolysis in Solutions: Competition Mechanisms of Radical-Based Cyclization and Hydrogen Abstraction Reactions. *Environ. Sci. Technol.* **2023**, *57*, 7777–7788. [CrossRef]
30. Konstantinov, A.; Bejan, D.; Bunce, N.J.; Chittim, B.; McCrindle, R.; Potter, D.; Tashiro, C. Electrolytic Debromination of PBDEs in DE-83TM Technical Decabromodiphenyl Ether. *Chemosphere* **2008**, *72*, 1159–1162. [CrossRef]
31. Peverly, A.A.; Pasciak, E.M.; Strawsine, L.M.; Wagoner, E.R.; Peters, D.G. Electrochemical Reduction of Decabromodiphenyl Ether at Carbon and Silver Cathodes in Dimethylformamide and Dimethyl Sulfoxide. *J. Electroanal. Chem.* **2013**, *704*, 227–232. [CrossRef]
32. Aguiar, T.; Baumann, L.; Albuquerque, A.; Teixeira, L.; De Souza Gil, E.; Scalize, P. Application of Electrocoagulation for the Removal of Transition Metals in Water. *Sustainability* **2023**, *15*, 1492. [CrossRef]
33. Cao, S.; Huang, H.; Shi, K.; Wei, L.; You, N.; Fan, X.; Yang, Z.; Zhang, W. Engineering Superhydrophilic/Superaerophobic Hierarchical Structures of Co-CH@NiFe-LDH/NF to Boost the Oxygen Evolution Reaction. *Chem. Eng. J.* **2021**, *422*, 130123. [CrossRef]
34. Qu, C.; Soomro, G.S.; Ren, N.; Liang, D.; Lu, S.; Xiang, Y.; Zhang, S. Enhanced Electro-Oxidation/Peroxone (In Situ) Process with a Ti-Based Nickel-Antimony Doped Tin Oxide Anode for Phenol Degradation. *J. Hazard. Mater.* **2020**, *384*, 121398. [CrossRef]
35. Xiong, M.; Chen, W.; Gu, S.; Zhang, D.; Ma, C.; Gu, H.; Meng, Y.; Jin, Y.; Xu, Z. Core-Shell Structured Fe/Ni@C Ternary Micro-Electrolysis Material for 4-Nitrochlorobenzen Removal: Performance Assessment and Mechanism Insight. *J. Clean. Prod.* **2022**, *372*, 133769. [CrossRef]
36. Tri, D.V.; Anh, N.T.; Luu, T.L.; Trippel, J.; Wagner, M. Electrochemical Degradation of Perfluorooctanoic Acid (PFOA) and Perfluorooctane Sulfonate (PFOS) in Water with Persulfate Catalyst Support. *Sep. Purif. Technol.* **2025**, *363*, 132076. [CrossRef]
37. Vanrenterghem, B.; Papaderakis, A.; Sotiropoulos, S.; Tsiplakides, D.; Balomenou, S.; Bals, S.; Breugelmans, T. The Reduction of Benzylbromide at Ag-Ni Deposits Prepared by Galvanic Replacement. *Electrochim. Acta* **2016**, *196*, 756–768. [CrossRef]
38. Zheng, Z.; Lu, G.; Wang, R.; Huang, K.; Tao, X.; Yang, Y.; Zou, M.; Xie, Y.; Yin, H.; Shi, Z.; et al. Effects of Surfactant on the Degradation of 2,2′,4,4′-Tetrabromodiphenyl Ether (BDE-47) by Nanoscale Ag/Fe Particles: Kinetics, Mechanisms and Intermediates. *Environ. Pollut.* **2019**, *245*, 780–788. [CrossRef]
39. Zhuang, Y.; Ahn, S.; Seyfferth, A.L.; Masue-Slowey, Y.; Fendorf, S.; Luthy, R.G. Dehalogenation of Polybrominated Diphenyl Ethers and Polychlorinated Biphenyl by Bimetallic, Impregnated, and Nanoscale Zerovalent Iron. *Environ. Sci. Technol.* **2011**, *45*, 4896–4903. [CrossRef] [PubMed]
40. Wang, R.; Tang, T.; Wei, Y.; Dang, D.; Huang, K.; Chen, X.; Yin, H.; Tao, X.; Lin, Z.; Dang, Z.; et al. Photocatalytic Debromination of Polybrominated Diphenyl Ethers (PBDEs) on Metal Doped TiO2 Nanocomposites: Mechanisms and Pathways. *Environ. Int.* **2019**, *127*, 5–12. [CrossRef] [PubMed]
41. Shen, Y.; Tong, Y.; Xu, J.; Wang, S.; Wang, J.; Zeng, T.; He, Z.; Yang, W.; Song, S. Ni-Based Layered Metal-Organic Frameworks with Palladium for Electrochemical Dechlorination. *Appl. Catal. B Environ.* **2020**, *264*, 118505. [CrossRef]

42. Li, J.; Luan, C.; Cui, Y.; Zhang, H.; Wang, L.; Wang, H.; Zhang, Z.; Zhao, B.; Zhang, H.; Zhang, X.; et al. Preparation and Characterization of Palladium/Polyaniline/Foamed Nickel Composite Electrode for Electrocatalytic Dechlorination. *Sep. Purif. Technol.* **2019**, *211*, 198–206. [CrossRef]
43. Sun, Z.; Wei, X.; Hu, X.; Wang, K.; Shen, H. Electrocatalytic Dechlorination of 2,4-Dichlorophenol in Aqueous Solution on Palladium Loaded Meshed Titanium Electrode Modified with Polymeric Pyrrole and Surfactant. *Colloids Surf. Physicochem. Eng. Asp.* **2012**, *414*, 314–319. [CrossRef]
44. Fan, J.; Du, H.; Zhao, Y.; Wang, Q.; Liu, Y.; Li, D.; Feng, J. Recent Progress on Rational Design of Bimetallic Pd Based Catalysts and Their Advanced Catalysis. *ACS Catal.* **2020**, *10*, 13560–13583. [CrossRef]
45. Bagger, A.; Ju, W.; Varela, A.S.; Strasser, P.; Rossmeisl, J. Electrochemical CO_2 Reduction: A Classification Problem. *ChemPhysChem* **2017**, *18*, 3266–3273. [CrossRef]
46. Wu, J.; Shan, S.; Luo, J.; Joseph, P.; Petkov, V.; Zhong, C.-J. PdCu Nanoalloy Electrocatalysts in Oxygen Reduction Reaction: Role of Composition and Phase State in Catalytic Synergy. *ACS Appl. Mater. Interfaces* **2015**, *7*, 25906–25913. [CrossRef]
47. Wu, Y.; Huo, X.; Zhang, W. Synergistic Pd/Cu Catalysis in Organic Synthesis. *Chem. Eur. J.* **2020**, *26*, 4895–4916. [CrossRef]
48. Neukermans, S.; Vorobjov, F.; Kenis, T.; De Wolf, R.; Hereijgers, J.; Breugelmans, T. Electrochemical Reduction of Halogenated Aromatic Compounds at Metal Cathodes in Acetonitrile. *Electrochim. Acta* **2020**, *332*, 135484. [CrossRef]
49. Guo, S.; Zhu, L.; Majima, T.; Lei, M.; Tang, H. Reductive Debromination of Polybrominated Diphenyl Ethers: Dependence on Br Number of the Br-Rich Phenyl Ring. *Environ. Sci. Technol.* **2019**, *53*, 4433–4439. [CrossRef]
50. Boucher, B.A.; Ennis, J.K.; Tsirlin, D.; Harris, S.A. A Global Database of Polybrominated Diphenyl Ether Flame Retardant Congeners in Foods and Supplements. *J. Food Compos. Anal.* **2018**, *69*, 171–188. [CrossRef]
51. Lee, H.J.; Kim, G.B. An Overview of Polybrominated Diphenyl Ethers (PBDEs) in the Marine Environment. *Ocean Sci. J.* **2015**, *50*, 119–142. [CrossRef]
52. Deng, C.; Chen, Y.; Li, J.; Li, Y.; Li, H. Environmental Pollution of Polybrominated Diphenyl Ethers from Industrial Plants in China: A Preliminary Investigation. *Environ. Sci. Pollut. Res.* **2016**, *23*, 7012–7021. [CrossRef]

processes

MDPI

Article

Utilizing Machine Learning Models with Molecular Fingerprints and Chemical Structures to Predict the Sulfate Radical Rate Constants of Water Contaminants

Ting Tang [1,2], Dehao Song [1], Jinfan Chen [1], Zhenguo Chen [3], Yufan Du [1], Zhi Dang [1,2,4] and Guining Lu [1,2,*]

1 School of Environment and Energy, South China University of Technology, Guangzhou 510006, China; tangt@scut.edu.cn (T.T.); gandalfrs@163.com (D.S.); chzdang@scut.edu.cn (Z.D.)
2 The Key Lab of Pollution Control and Ecosystem Restoration in Industry Clusters, Ministry of Education, South China University of Technology, Guangzhou 510006, China
3 SCNU (NAN'AN) Green and Low-Carbon Innovation Center, Guangdong Provincial Engineering Research Center of Intelligent Low-Carbon Pollution Prevention and Digital Technology, South China Normal University, Guangzhou 510006, China
4 Guangdong Provincial Key Laboratory of Solid Wastes Pollution Control and Recycling, South China University of Technology, Guangzhou 510006, China
* Correspondence: lutao@scut.edu.cn

Abstract: Sulfate radicals are increasingly recognized for their potent oxidative capabilities, making them highly effective in degrading persistent organic pollutants (POPs) in aqueous environments. These radicals excel in breaking down complex organic molecules that are resistant to traditional treatment methods, addressing the challenges posed by POPs known for their persistence, bioaccumulation, and potential health impacts. The complexity of predicting interactions between sulfate radicals and diverse organic contaminants is a notable challenge in advancing water treatment technologies. This study bridges this gap by employing a range of machine learning (ML) models, including random forest (DF), decision tree (DT), support vector machine (SVM), XGBoost (XGB), gradient boosting (GB), and Bayesian ridge regression (BR) models. Predicting performances were evaluated using R^2, RMSE, and MAE, with the residual plots presented. Performances varied in their ability to manage complex relationships and large datasets. The SVM model demonstrated the best predictive performance when utilizing the Morgan fingerprint as descriptors, achieving the highest R^2 and the lowest MAE value in the test set. The GB model displayed optimal performance when chemical descriptors were utilized as features. Boosting models generally exhibited superior performances when compared to single models. The most important ten features were presented via SHAP analysis. By analyzing the performance of these models, this research not only enhances our understanding of chemical reactions involving sulfate radicals, but also showcases the potential of machine learning in environmental chemistry, combining the strengths of ML with chemical kinetics in order to address the challenges of water treatment and contaminant analysis.

Keywords: machine learning models; advance oxidation; emerging contaminants; sulfate radicals; boosting models; SHAP analysis

1. Introduction

Sulfate radicals are known for their potent oxidative capabilities, making them an attractive agent for the degradation of persistent organic pollutants in aqueous environments [1]. These radicals are highly reactive, enabling them to break down complex organic molecules that are otherwise resistant to conventional treatment methods. This effectiveness is particularly crucial in addressing the challenge posed by persistent organic contaminants (POPs), which are known for their persistence, bioaccumulation, and potential adverse health effects. Studies have shown that sulfate radicals are more selective and show higher reactivity toward electron-rich pollutants; they can effectively degrade a wide range of these pollutants, including pharmaceuticals [2,3], personal care products [4], and industrial chemicals [5,6], often achieving higher degradation rates than other advanced oxidation processes. Furthermore, the generation of sulfate radicals can be achieved through various environmentally benign methods, notably through the homogenous activation of persulfate (PS) or peroxymonosulfate (PMS) using heat, ultraviolet (UV) light, base, or transition metals [7], thereby enhancing their appeal for sustainable water treatment technologies [8]. This aligns with the increasing global emphasis on green chemistry and sustainable solutions for environmental pollution.

Given the widespread use of $SO_4^{-}\bullet$ in the elimination of organic contaminates, obtaining the reaction rate constant is crucial for designing effective water treatment systems and for a comprehensive understanding of the underlying chemical mechanisms; however, considering the huge amount of chemicals is time consuming and requires extensive laboratory work to figure out the reaction kinetics of each chemical [9]. Neta et al. [10] measured the second-order rate constants (k) for the reaction of $SO_4^{-}\bullet$ with various substituted benzenes and benzoates, ranging from approximately 10^5 to 10^9 M^{-1} s^{-1}. Nguyen et al. [11] tested the second-order rate coefficients for the aqueous $SO_4^{-}\bullet$ with a series of organic aerosol compounds, such as 2-methyltetrol, 2-methyl-1,2,3-trihydroxy-4-sulfate, 2-methyl-1,2-dihydroxy-3-sulfate, and so on, which assisted the understanding the aerosol mass sinks. Therefore, predicting the reaction rate constants of organic contaminants toward sulfate radicals is essential, for it not only represents a significant leap in environmental chemistry, but is also a pragmatic step towards the optimization of water treatment processes. Accurate predictions enable us to prioritize and regulate the use of sulfate radicals based on the reactivity of different pollutants, thus streamlining the treatment process.

Quantitative structure−activity relationship (QSAR) models provide a valuable predictive framework, correlating chemical structure with reactive tendencies, and thus reducing the need for extensive empirical experimentation [12]. This methodology was applied to predict the toxicity [13] and activity [14,15] of chemicals. Sudhakaran et al. [16] developed QSAR models for ozonation and the •OH oxidation of organic micropollutants to predict $k_O{}^3$ and $k_{\bullet OH}$ based on multi-linear regression, resulting the $R^2 > 0.75$, and identified descriptors like double bond equivalence (DBE), ionization potential (IP), and electron affinity (EA) to affect k to different extents. Song et al. [3] built prediction models toward the kinetic constants of a suit of 15 PPCPs, incorporating model parameters such as the steady state of the radical concentration; moreover, the potential error induced by other radicals, such as $CO_3\bullet^-$, $Br\bullet^-$,were also taken into consideration, necessitating extensive laboratory work.

Traditional QSAR analysis employs predefined mathematical models and molecular descriptors, often requiring domain expertise for feature selection and generally offering clearer interpretability. However, it is less flexible and may struggle with complex, nonlinear data. The complexity of chemical reactions in environmental matrices often outstrips the capacity of traditional QSAR models to provide rapid and accurate predictions. This complexity is compounded by the vast number of potential contaminants and the variability of environmental conditions, which can alter reaction dynamics. Machine learning models emerge as a necessary evolution in this field, offering the ability to assimilate large datasets and uncover patterns.

To address these limitations and harness the complexity of chemical interactions, numerous machine learning (ML) models were developed and employed in the environmental field. These ML models were utilized across different scenarios to predict environmental problems, including forecasting regional daily to yearly $PM_{2.5}$ variations [17,18], predicting the change of various wastewater variables [19,20]. More specifically, Lu et al. developed multilayered neural network (NN) models to forecast the temperature-dependent site-specific rate constants of hydroxyl radical reactions with alkanes [21]; the results showed that the proposed NN models are robust in predicting the site-specific and overall rate constants. Not only used for prediction, ML models can also assist in classification, Cheng et al. [22] trained five different ML models to classify 3486 per- and polyfluoroalkyl substances (PFASs) from the OECD list, the multitask neural network, and graph-based models demonstrated the best performance. Generally, ensemble models like XGBoost (XGB) and gradient boosting (GB) models often perform well on a wide range of problems, but can be complex [23], while simple models like decision trees (DTs) are interpretable [24], but prone to overfitting. The selection of different machine learning models based on task demands is crucial to achieving robust prediction results.

Inspired by the recent successes in ML modeling in environmental chemistry, we investigate the efficacy of six distinct ML models—random forest, decision tree, support vector machine, XGB, GB, and Bayesian ridge regression—in the prediction of the reaction constants of organic contaminants with $SO_4^{-}\bullet$; the data were collected from peer-reviewed papers. These models spanning various categories including ensemble methods (RF, XGB, GB), predictive modeling (DT), discriminative models (SVM), and regularized regression (BR). Although there are already several studies on the prediction of the rate constant of organic contaminants mediated by sulfate radicals [25,26], this study compared the performances of six different machine learning models and two different features (Morgan-type fingerprints and chemical descriptors). Model performances were estimated by internal and external validation using R^2, RMSE, and MAE. Residual plots were presented to evaluate the ML performance. The SHAP values determined the top 10 important features. By applying these models to predict the reaction constants of organic contaminants with $SO_4^{-}\bullet$, our study aims to compare six machine learning models in order to identify the most effective approach for predicting the reaction constants of organic contaminants with $SO_4^{-}\bullet$. By employing two distinct feature types, we seek to evaluate which feature set provides more accurate and reliable predictions.

The novelty of our research lies in the comprehensive comparison of six distinct machine learning models. Furthermore, both Morgan fingerprints and chemical descriptors have been applied as predictive features. Through this combination of Ml models and chemical analysis, our research aims to establish the predictive modeling of reaction rates, offering insights that could revolutionize water treatment methodologies and our understanding of the fate of environmental pollutants.

2. Data Collection and Model Construction

2.1. Data Sets and MF

The reaction rate constants for the oxidative processes targeting a suit of organic contaminants, mediated by $SO_4^{-}\bullet$ radicals, were compiled from the peer-reviewed literatures [27–36], focusing on studies where the pH ranged from 5.5 to 7.5, and temperatures were within 20–25 °C. In certain studies, authors reported rate constants with k values. To normalize skewed distributions and reduce the impact of outliers, the data were transformed to the logarithmic scale when necessary, thereby enhancing the model accuracy and interpretability. The InChIKeys and SMILES codes for these chemicals were extracted via web crawling techniques from PubChem. This method ensured comprehensive and efficient data collection but was limited by the reliability and structure of the source websites. In instances where multiple reaction rate constants were reported for the same chemical, the average value was utilized. This approach, chosen to provide a representative rate constant for each chemical, potentially introduces a level of approximation, but is necessary to manage data variability. Following pre-cleaning, 724 chemicals, each accompanied by its corresponding experimental logk values, were used in model development. Using the RDKit program, these SMILES codes were converted into Morgan fingerprints, along with 209 chemical descriptors [37]. For some compounds lacking certain chemical descriptors, this resulted in 'nan' values in the dataset, potentially leading to errors in the model construction. To maintain data integrity and ensure model accuracy, these 'nan' values were systematically dropped from the dataset using the Python 'dropna' method. Following this data cleaning step, the dataset was reduced to 716 chemicals for analysis, applying chemical descriptors as features.

In the Morgan fingerprint analysis, a series of scans were conducted to determine the optimal length of the MF, with values including 512, 1024, 2048, 3072, 4096, and 8192. After selecting the fingerprint length yielding the optimal predictive performance, features were reduced through the following procedures: (1) dropping columns with a standard deviation greater than 0.1. This threshold was chosen as features with low variability (lower standard deviation) are less informative for models, and their removal can reduce noise and prevent overfitting. (2) Another method included calculating Pearson parameters and eliminating columns with Pearson correlations exceeding 0.96 [38]. This step was crucial to remove multicollinearity, as highly correlated features can distort the importance of variables and reduce the generalizability of the model. Following this methodology, we narrowed down the features to 168 chemical characteristics, deemed most relevant and robust for further model development. The entire database can be accessed via the file "SupInfoDataSet.xlsx", available in the Supporting Information section. This database comprises four sheets: "3072 MorganMFs" and "Chemical Descriptors", which, respectively detail the 3072-bit Morgan molecular fingerprints and chemical features of each chemical, generated using RDKit. The "Morgan MFs_test_train set" and "CDs_train_test set" sheets display the experimental and gradient boosting model-predicted $\log K_{SO_4^{-}\bullet}$ values, based on the 3072-bit Morgan MFs and chemical features, respectively.

2.2. Hyperparameter Optimization

A fivefold grid search was conducted for hyperparameters optimization [39]. This approach is advantageous as it allows for a comprehensive assessment of how different hyperparameter settings impact the model's performance, leading to the identification of an optimal set that enhances prediction accuracy. A range of values were considered, and the optimal hyperparameters were determined using the mean R^2 value on the validation set (Table 1). The hyperparameters across models, such as n_estimators (number of trees) and max_depth (tree depth) for ensemble methods like random forest and XGBoost, control model complexity and overfitting. In SVM, the Kernel determines the data transformation, C balances the classification accuracy and simplicity, and gamma affects the influence range of a single sample. For gradient boosting, learning_rate adjusts the contribution of each tree, enhancing performance tuning. In Bayesian ridge regression, alpha_1 and lambda_1 manage model complexity, while n_iter influences the convergence speed, collectively fine-tuning the balance between the model accuracy and generalization. The selected ranges for each hyperparameter were chosen based on their potential impact on model performance, with a focus on achieving an optimal balance between accuracy and generalizability. The final optimized hyperparameters for each model were also presented.

Table 1. Hyperparameters used for different models.

Models	Hyperparameters		
	Range	Molecular Fingerprints	Chemical Descriptors
RF	n_estimators:rage (1, 200, 20) max_depth: range (2, 40, 4)	n_estimators = 21 max_depth = 26	n_estimators = 40 max_depth = 20
DT	max_depth: range (1, 100, 10) min_samples_split: range (1, 10, 2) min_samples_leaf: range (1, 10, 2)	max_depth = 15 min_samples_split = 6 min_samples_leaf = 1	max_depth = 60 min_samples_split = 2 min_samples_leaf = 4
SVM	kernel: ['linear','rbf'] C:range (1, 10, 2) gamma: [0.005, 0.01, 0.05]	Kernel = rbf, C = 5 gamma = 0.05	Kernel = rbf, C = 4 gamma = 0.005
XGB	n_estimators: rage (10, 300, 50) min_child_weight: range (1, 5) max_depth: range (20, 100, 10)	n_estimators = 60 min_child_weight = 4 max_depth = 70	n_estimators = 200 min_child_weight = 4 max_depth = 20
GB	n_estimators:range (20, 100, 20) learning_rate': [0.01, 0.05, 0.1] max_depth':range (2, 10, 2)	n_estimators = 80 learning_rate = 0.1 max_depth = 8	n_estimators = 80 learning_rate = 0.1 max_depth = 6
BR	alpha_1: $[1 \times 10^{-6}, 1 \times 10^{-7}, 1 \times 10^{-8}]$ lambda_1: $[1 \times 10^{-4}, 5 \times 10^{-3}, 1 \times 10^{-3}]$ n_iter: [1, 30, 2]	alpha_1 = 1×10^{-8} lambda_1 = 0.005 n_iter = 20	alpha_1 = 1×10^{-7} lambda_1 = 0.005 n_iter = 1

2.3. ML Models and Validation

Six machine learning algorithms—random forest (RF), decision tree (DT), support vector machines (SVM), XGBoost (XGB), Bayesian ridge regression (BR), and gradient boosting (GB)—were utilized for predicting the reaction rate constant in the interactions between sulfate radicals and organic chemicals. These were selected for their diverse strengths and proven capabilities in various predictive modeling scenarios, ranging from capturing non-linear relationships to handling high-dimensional data. These predictive models were developed using scikit-learn packages [40]. Datasets were randomly split into an 80% training set and a 20% test set. Both internal and external validations were implemented to ascertain the reliability and robustness of the ML models, as well as their predictive capacity. Performance indices—correlation coefficient (R^2), root-mean-square deviation (RMSE), and mean absolute error (MAE)—were computed. The concurrent application of these metrics in both training and testing phases facilitated a comprehensive evaluation. Specifically, RMSE underscored the model's sensitivity to substantial errors, and MAE offered a direct estimation of average errors [40]. This dual approach steered the models towards a more balanced and resilient optimization.

$$R^2 = 1 - \frac{\sum_{i=1}^{n}\left(y_{exp} - y_{pred}\right)^2}{\sum_{i=1}^{n}\left(y_{exp} - \overline{y}_{exp}\right)^2} \quad (1)$$

$$RMSE = \sqrt{\frac{\sum_{i=1}^{n}\left(y_{exp} - y_{pred}\right)^2}{n}} \quad (2)$$

$$MAE = \frac{\sum_{i=1}^{n}\left|y_{exp} - y_{pred}\right|}{n} \quad (3)$$

2.4. Model Interpretation

A fundamental principle in developing QSAR/QSPR models is the mechanistic interpretation of the model, which necessitates predictions anchored in essential chemical interpretations [41]. In the realm of molecular descriptors, selected physicochemical properties are pivotal for optimal predictions. In this context, the SHapley Additive exPlanations (SHAP) method, an advanced interpretive technique, was employed to elucidate predictions. SHAP, renowned for elucidating machine learning model predictions by attributing feature importance values in the context of all possible feature combinations [42,43], was utilized to enhance our model's interpretability. By employing SHAP, we identified and visualized the top ten critical features to our predictions through violin plots, shedding light on the significant factors that drive the predictive performance. This approach not only underscores the importance of transparent model interpretation, but also aids researchers in pinpointing the essential features that govern the prediction of reaction rate constants, thereby enriching our model's explanatory power.

3. Results and Discussion

3.1. Effects of the MF Length

In this study, MFs are essential for converting molecular structures into fixed-length binary vectors, thereby enabling the quantitative assessment of molecular similarities and differences. This feature is vital for accurately and efficiently modeling molecular properties. To determine the optimal MF length that balances predictive performance with computational efficiency in the Morgan type, a preliminary study was undertaken. The R^2 of the test set was selected to evaluate the predicting performances; Figure 1 illustrates how the R^2 values were influenced by varying the length of MFs across different models. It was observed that longer MF lengths typically yield improved predictive performance for the tested models, except for the DT model. The variability in the performance of the DT model can be indicative of overfitting when provided with an increasingly complex feature space, a known limitation of decision trees. The selection of the optimal MF length was thus guided by a balance between maximizing predictive accuracy and maintaining computational efficiency. After evaluating the performance across various lengths, a 3072-bit length was chosen as the optimal size. This decision was based on its robust performance in enhancing R^2 values across models, without disproportionately increasing computational demands. While lengths beyond 3072 bits did show some incremental improvements in R^2, these were marginal and did not justify the significantly higher computational costs involved. Hence, 3072 bits was established as the optimal length for MFs in our study.

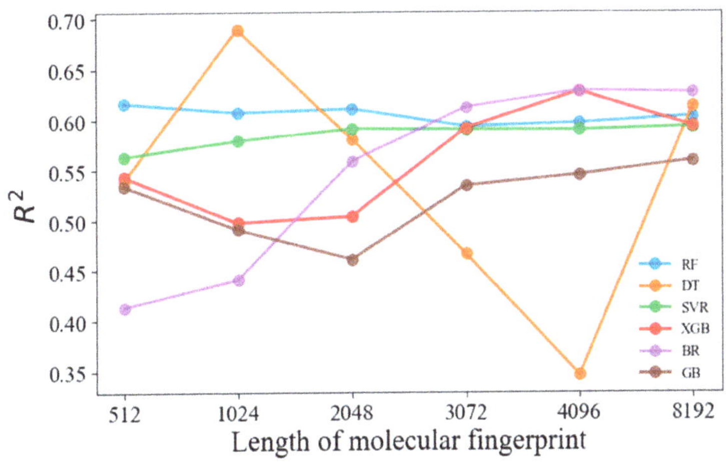

Figure 1. Effect of the Morgan fingerprint length on the R^2 values of selected machine learning models.

3.2. Internal and External Validation

The ML models generated R^2_{train} values in the range of 0.922~0.988 when using Morgan fingerprints (3072 bits) as features, suggesting good fitting of the training set (Table 2). To further assess the reliability of the developed models, external validation was performed using R^2_{test} as statistical criteria. The R^2_{test} was used to measure the correlation between experimental and predicted logk values; an R^2 value of 1 indicated that the regressions perfectly fit the data. In this study, the values ranged from 0.532~0.712. The SVM model has the lowest MAE and RMSE on both the training and test sets, which are crucial metrics for generalization, showing the smallest average error in the predictions. The results demonstrate that SVM methods emerge as the top-performing model for this particular task when using the R^2_{test} as the scoring factor. Contrastingly to the SVM model, the XGB and GB models yielded R^2_{test} values higher than 0.97 and R^2_{test} values higher than 0.60. Both have small MAE and RMSE numbers in their test sets, which demonstrated a robust predicting performance. BR also showed R^2_{train} values higher than 0.98 and R^2_{test} values around 0.597; however, when comparing with boosting models, we can see higher MAE and RMSE results on both the training and test sets. The results revealed that the ensemble and boosting methods (XGB, BR, GB) consistently outperformed the single models (DT) on both training and test sets [44]. Previous research also demonstrated that an ensemble is on generally more accurate than single base models [45].

Table 2. The performance parameters of machine learning models on training and test sets, using Morgan fingerprints.

Models	Morgan (3072 Bits)					
	Training Set			Test Set		
	R^2	MAE	RMSE	R^2	MAE	RMSE
RF	0.949	0.380	0.609	0.577	0.474	0.751
DT	0.922	0.416	0.633	0.532	0.500	0.790
SVM	0.974	0.340	0.547	0.712	0.379	0.620
XGB	0.988	0.338	0.600	0.641	0.404	0.691
BR	0.982	0.392	0.645	0.597	0.449	0.732
GB	0.972	0.348	0.583	0.620	0.429	0.712

Table 2 details the performance parameters of these models; our analysis revealed that ensemble and boosting methods consistently outperformed single models, aligning with the existing literature that suggests that ensemble approaches generally achieve higher accuracy than individual base models. The *p*-values for each pair of machine learning models, based on the R^2_{test} values derived from tenfold cross validation, were also calculated. The SVM model showed a significant difference with the rest of the models ($p < 0.05$), except for the gradient boost model ($p = 0.032$).

When choosing a model for tasks requiring high accuracy and reliability, like predicting reaction constants, the model should balance between fitting the training data and generalizing the unseen data. For example, the SVM model's robust performance underscores its suitability for complex predictive tasks, where minimizing errors is crucial. On the other hand, the high R^2_{train} and R^2_{test} scores of the XGB and GB models, combined with their low error metrics, highlight their potential in capturing the underlying patterns within the data without overfitting, making them reliable choices for predicting the reaction constants in varied chemical environments.

The predicted versus experimental logk values are presented in Figure 2; the diagonal dash represented the r = 1. The scatter being closer to this line indicated a better predicting performance; contrarily, the scatter far away from the diagonal indicated a bias between the predicted and experimental values. It is evident that the SVM, XGB, and GB models exhibit a high concentration of points near the diagonal, indicating a strong correlation between the predicted and experimental values. This suggests that these models have a higher predictive accuracy. However, there are some noticeable outliers, particularly in the DT model, where a few points significantly stray from the line of perfect correlation. These outliers indicate instances where the DT model failed to predict the logk values accurately, possibly due to the model's inability to capture complex patterns within the data.

The results calculated for models built on chemical descriptors are presented in Table 3. Considering the evaluated performance metrics, XGB and GB models exhibited the highest R^2 scores on both the training and test sets; this is consistent with the results displayed in Table 2. Specifically, the R^2 values exceeded 0.99 on the training set, and 0.8 for the test set, underscoring a superior predictive performance with chemical descriptors as features.

The GB model, in particular, showcased strong generalization capabilities across the training and test sets, displaying the lowest error metrics in the test set. Conversely, while the DT and SVM models perform well on the training set, a marked decrease in performance on the test set suggests overfitting. In contrast, the XGB model achieves the highest R^2 score on the training set and sustains this high level of performance on the test set, indicative of its excellent predictive ability with consistently low error rates. The BR model, however, exhibits the lowest performance on the training set and shows minimal improvement on the test set; this is indicative of potential underfitting or an inability to capture the dataset's complexity adequately. Statistical analyses for each pair of machine learning models based on the R^2_{test} values derived from tenfold cross validation revealed that the GB model was statistically significantly different from the other four models ($p < 0.05$), only presenting similarity to the RF model ($p = 0.648$).

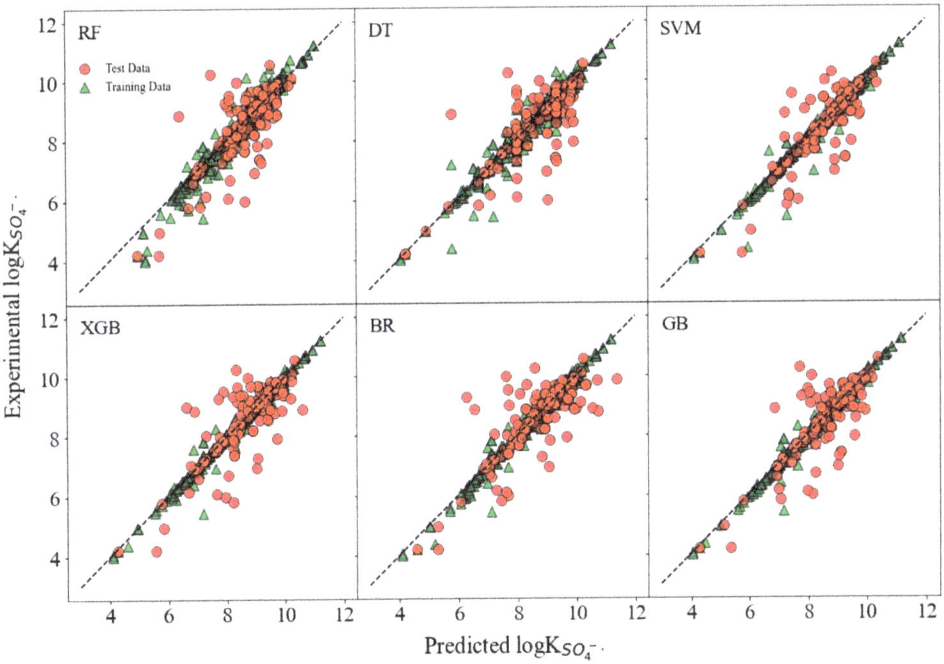

Figure 2. Plots of the predicted logk values from Morgan MFs versus the experimental values of both training and test sets.

Table 3. The performance parameters of machine learning models on training and test sets using chemical descriptors.

	Chemical Descriptors					
Models	**Training Set**			**Test Set**		
	R^2	**MAE**	**RMSE**	**R^2**	**MAE**	**RMSE**
RF	0.957	0.357	0.597	0.806	0.345	0.519
DT	0.955	0.429	0.759	0.724	0.366	0.618
SVM	0.984	0.570	0.849	0.540	0.496	0.799
XGB	0.994	0.346	0.651	0.788	0.309	0.542
BR	0.776	0.526	0.775	0.663	0.491	0.684
GB	0.992	0.329	0.594	0.844	0.275	0.465

In light of these results, the GB and XGB models emerge as the most dependable for deployment, characterized by high R^2 and low error metrics on unseen data, suggesting that they effectively capture the underlying data patterns without overfitting. Similar research from Li et al. [46] utilized different types of ML models to predict the net ecosystem carbon exchange. They also observed that the gradient boosting regression model was more accurate when compared to the other three models (support vector machine, stochastic gradient descent, and Bayesian ridge). Dineva et al. compared the ensemble boosting and bagging models for ground water potential prediction. The results revealed that bagging models like RF outperformed boosting models [47]. Such findings underscore the necessity of choosing the appropriate model based on the specific predictive task at hand, as no single approach fits all scenarios.

The logk values, as predicted from chemical descriptors verses experimental values, are presented in Figure 3. In XGB and GB models, the scatters showed high density proximate to the diagonal, indicating good predicting performances. The RF model shows a reasonable alignment with the diagonal and has a slightly more dispersed pattern, hinting at variability in its predictive accuracy. The DT model exhibits a more pronounced spread of points, which suggests a lesser degree of predictive accuracy when compared to the XGB and GB models. On the other hand, a notable pattern is observed in the SVM model plot where the predicted points align vertically along an axis. This alignment is a visual indicator of poor predictive performance, as the model predictions do not vary with the changes in the experimental logk values, resulting in a lack of correlation.

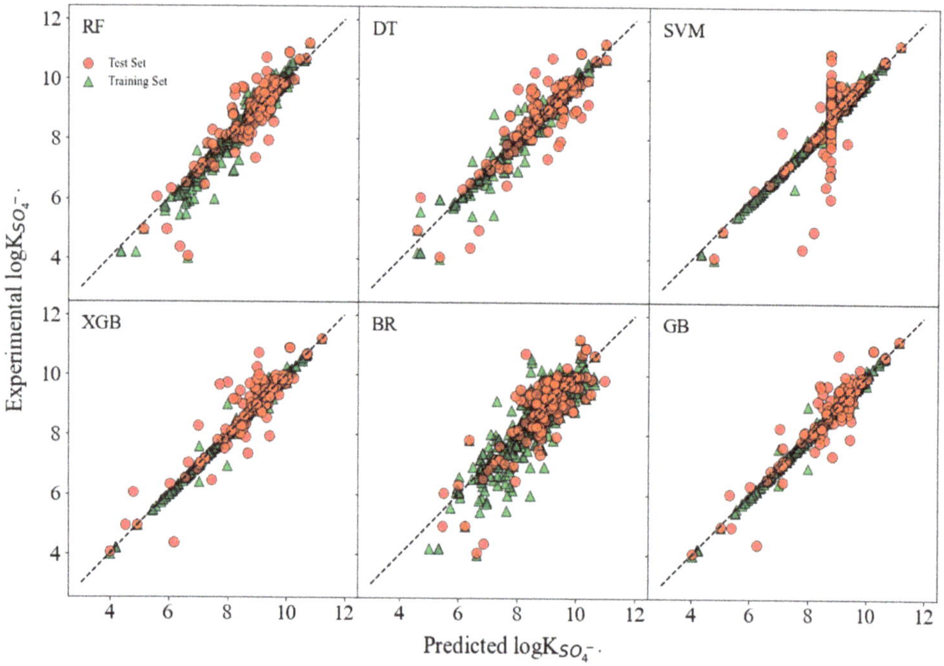

Figure 3. Plots of the predicted logk values from chemical descriptors versus experimental values of both training and test sets.

The differences between the observed and predicted values, known as residuals, are plotted against the predicted values in order to assess the performance of machine learning models [48]. Residual plots were created for gradient boosting models (Figure 4), which use Morgan fingerprints and chemical descriptors as features, respectively. The residuals in these models are randomly distributed across all predicted values, indicating effective prediction performance. When using MFs as features, there is a wider spread of residuals, especially in predictions ranging from eight to nine. Conversely, when featured as chemical descriptors, the residuals scatter evenly around the horizontal axis, demonstrating a strong fit without noticeable systematic errors. Most residuals lie within a −1 to 1 range, reflecting minor prediction errors, with only a few outliers. Both models show a concentration of residuals near zero, but the gradient boosting model, which utilizes chemical descriptors, exhibits a denser clustering around zero, thus signifying greater accuracy.

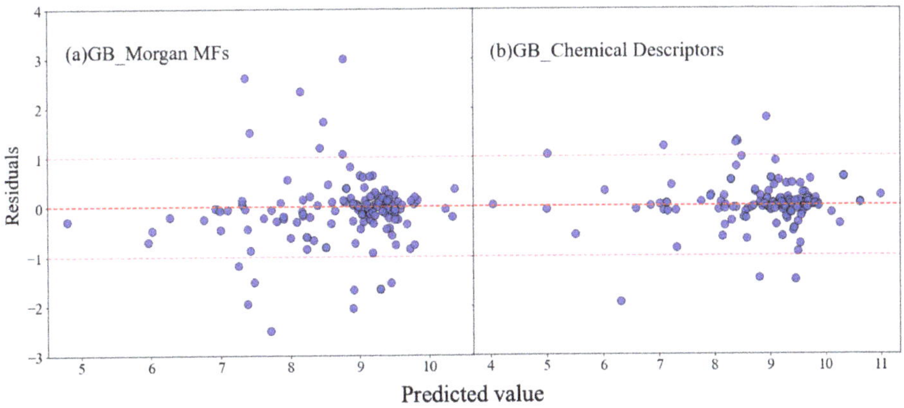

Figure 4. Residual plots of (**a**) the gradient boosting model established on MFs and (**b**) chemical descriptors, respectively. (The red dashed lines at residuals equals −1, 0 and 1 respectively represented reference lines to help evaluate the distribution of residuals in our model).

3.3. SHapley Additive exPlanations (SHAP) Analysis

SHAP analysis is a method developed to interpret machine learning models by quantifying the impact of each feature on the model's prediction. This approach provides a foundation for understanding the contribution of individual features to predictive outcomes, offering transparency in model decision-making processes. In this study, SHAP analysis was conducted, as presented in Figure 5. It illustrated a SHAP summary plot for GB models which use Morgan fingerprints and chemical descriptors, respectively, which quantify the contribution of individual features to the predictive outcomes. Each dot represents a SHAP value for a feature and an instance; the position on the x-axis indicates the impact of that value on the model's output. Features are ordered vertically by the sum of SHAP value magnitudes across all samples, which indicates the overall importance of the feature. The color gradient signifies the magnitude and direction of the feature's effect, with one end of the spectrum (blue) representing low feature values, and the other end (red) representing high values. A feature's value that pushes the prediction to the right of the mid line suggests an increased likelihood of the higher outcome predicted by the model, while a shift to the left indicates a lower outcome. In this study, Morgan fingerprints of 2897 bits and a chemical feature of SMR_VSA7 were the most impactful features effecting the prediction outcome. It offers insights into molecular volume and its potential impact on reactivity and interaction with biological targets. Chi3v was the second important chemical feature; it captures the complexity of molecular structures by quantifying the connectivity of atoms within a molecule, which can influence molecular stability and reactivity. This is followed by Balaban J, which provides a quantitative insight into the molecular complexity, which can affect molecular properties such as reactivity and biological activity [37]. SHAP analysis provides insights into the feature importance and their individual contributions to the model's predictive capabilities. The top ten impactful chemical descriptors and their description are presented in Table 4.

To summary, a 3072-bit length was identified as the optimal MF length, balancing the predictive accuracy with computational efficiency. Ensemble and boosting methods, such as XGB and GB, consistently outperformed single models, aligning with the literature on the efficacy of ensemble approaches in achieving higher accuracy. Additionally, SHAP analysis provided insights into the importance of individual features in the model predictions, highlighting the significance of specific molecular features like SMR_VSA7 and Chi3v on the outcome. Comprehensive analysis and significant findings were detailed in the conclusion section.

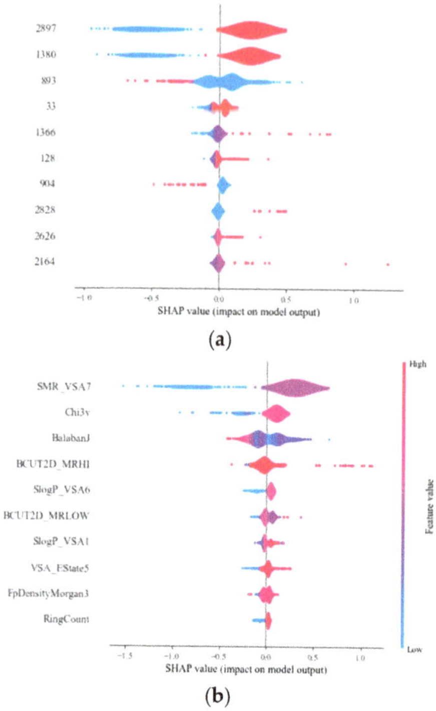

Figure 5. Importance of the top ten representative MF features (**a**) and chemical descriptors (**b**) and their impact on the model output for the Gradient Boosting model applied to logk.

Table 4. The top ten impactful features and their descriptions from SHAP analysis.

Feature Name	Description [37]
SMR_VSA7	MOE-type descriptors using MR contributions and surface area contributions
Chi3v	Third-order valence molecular connectivity index for topology analysis
Balaban J	A topological descriptor that quantifies the branching and complexity of cycles within a molecule's structure
BCUT2D_MRHI	Burden eigenvalues (BCUT) calculated from a 2D molecular graph, related to the highest eigenvalue of the molecular refractivity-weighted connectivity matrix
Slogp_VSA6	Log P partition coefficient, used to understand hydrophobic or lipophilic properties of the molecule
BCUT2D_MRLOW	The lowest eigenvalue of a molecule's Burden connectivity matrix
Slogp_VSA1	Volume descriptor for logP partition, smallest value bin
VSA_EState5	Distribution of electronic properties over the molecular surface
Fpdensitymorgan3	Density of features in a Morgan fingerprint with a radius of 3
RingCount	Total number of ring structures present in a molecule

4. Conclusions

This study employed six different machine learning models combined with Morgan molecular fingerprints and chemical features to develop prediction models for sulfate radicals-induced oxidation rate constants. The model performances were dependent on fingerprint length; for BR, XGB, and GB models, the R^2 value slightly decreased and then increased as the length was extended. For the SVM and RF models, which were not sensitive to the MF length, the R^2 remained relatively stable as the length of the MFs changed. For Morgan fingerprints, the SVM models demonstrated robust generalization capabilities, achieving the highest R^2 on the test set. For chemical descriptors, the GB and XGB models emerged as the most reliable for deployment, exhibiting high R^2 and low error metrics on the test data, indicating their effectiveness in capturing the underlying data patterns without overfitting. Models based on chemical features generally performed better than those based on Morgan fingerprints. SHAP analysis was conducted, and Morgan fingerprint bit 2897, along with the descriptor SMR_VSA7, were identified as the dominant features affecting the prediction performance. This study suggests the practicality of utilizing machine learning approaches to predict the reaction rate of organic contaminants in response to sulfate radicals, while also uncovering influential physical/chemical properties related to reaction rates. It also highlighted the importance of selecting appropriate machine learning models and features to achieve robust prediction performances.

Supplementary Materials: The following supporting information can be downloaded at: https://www.mdpi.com/article/10.3390/pr12020384/s1.

Author Contributions: Conceptualization, T.T. and G.L.; methodology, T.T.; software, T.T.; validation, D.S., J.C. and Z.C.; formal analysis, Y.D.; investigation, T.T.; resources, G.L.; data curation, T.T.; writing—original draft preparation, T.T.; writing—review and editing, T.T. and G.L.; visualization, D.S., J.C. and Z.C.; supervision, G.L. and Z.D.; project administration, T.T., G.L. and Z.D.; funding acquisition, T.T. All authors have read and agreed to the published version of the manuscript.

Funding: This research was supported by National Science Fund of China for Young Scholars [42107409], China postdoctoral Science Foundation [2021M701247] and GuangZhou Basic and Applied Basic Research Foundation [202201010223].

Data Availability Statement: The raw data supporting the conclusions of this article will be made available by the authors on request.

Conflicts of Interest: The authors declare no conflict of interest.

References

1. Giannakis, S.; Lin, K.-Y.A.; Ghanbari, F. A review of the recent advances on the treatment of industrial wastewaters by Sulfate Radical-based Advanced Oxidation Processes (SR-AOPs). *Chem. Eng. J.* **2021**, *406*, 127083. [CrossRef]
2. Hassani, A.; Scaria, J.; Ghanbari, F.; Nidheesh, P. Sulfate radicals-based advanced oxidation processes for the degradation of pharmaceuticals and personal care products: A review on relevant activation mechanisms, performance, and perspectives. *Environ. Res.* **2023**, *217*, 114789. [CrossRef] [PubMed]
3. Lian, L.; Yao, B.; Hou, S.; Fang, J.; Yan, S.; Song, W. Kinetic study of hydroxyl and sulfate radical-mediated oxidation of pharmaceuticals in wastewater effluents. *Environ. Sci. Technol.* **2017**, *51*, 2954–2962. [CrossRef] [PubMed]
4. Nfodzo, P.; Choi, H. Sulfate radicals destroy pharmaceuticals and personal care products. *Environ. Eng. Sci.* **2011**, *28*, 605–609. [CrossRef]
5. Li, W.; Orozco, R.; Camargos, N.; Liu, H. Mechanisms on the impacts of alkalinity, pH, and chloride on persulfate-based groundwater remediation. *Environ. Sci. Technol.* **2017**, *51*, 3948–3959. [CrossRef] [PubMed]
6. Yang, M.; Ren, X.; Hu, L.; Guo, W.; Zhan, J. Facet-controlled activation of persulfate by goethite for tetracycline degradation in aqueous solution. *Chem. Eng. J.* **2021**, *412*, 128628. [CrossRef]
7. Ji, Y.; Wang, L.; Jiang, M.; Lu, J.; Ferronato, C.; Chovelon, J.-M. The role of nitrite in sulfate radical-based degradation of phenolic compounds: An unexpected nitration process relevant to groundwater remediation by in-situ chemical oxidation (ISCO). *Water Res.* **2017**, *123*, 249–257. [CrossRef]
8. Lai, J.; Tang, T.; Du, X.; Wang, R.; Liang, J.; Song, D.; Dang, Z.; Lu, G. Oxidation of 1, 3-diphenylguanidine (DPG) by goethite activated persulfate: Mechanisms, products identification and reaction sites prediction. *Environ. Res.* **2023**, *232*, 116308. [CrossRef]
9. Pardue, H.L. Kinetic aspects of analytical chemistry. *Anal. Chim. Acta* **1989**, *216*, 69–107. [CrossRef]
10. Neta, P.; Madhavan, V.; Zemel, H.; Fessenden, R.W. Rate constants and mechanism of reaction of sulfate radical anion with aromatic compounds. *J. Am. Chem. Soc.* **1977**, *99*, 163–164. [CrossRef]
11. Tran, L.N.; Abellar, K.A.; Cope, J.D.; Nguyen, T.B. Second-Order Kinetic Rate Coefficients for the Aqueous-Phase Sulfate Radical ($SO4\bullet^-$) Oxidation of Some Atmospherically Relevant Organic Compounds. *J. Phys. Chem. A* **2022**, *126*, 6517–6525. [CrossRef] [PubMed]

12. Nirmalakhandan, N.N.; Speece, R.E. QSAR model for predicting Henry's constant. *Environ. Sci. Technol.* **1988**, *22*, 1349–1357. [CrossRef]
13. Agrawal, V.; Khadikar, P. QSAR prediction of toxicity of nitrobenzenes. *Bioorg. Med. Chem.* **2001**, *9*, 3035–3040. [CrossRef] [PubMed]
14. Du, Q.-S.; Huang, R.-B.; Chou, K.-C. Recent advances in QSAR and their applications in predicting the activities of chemical molecules, peptides and proteins for drug design. *Curr. Protein Pept. Sci.* **2008**, *9*, 248–259. [CrossRef]
15. Xiao, R.; Ye, T.; Wei, Z.; Luo, S.; Yang, Z.; Spinney, R. Quantitative Structure-Activity Relationship (QSAR) for the Oxidation of Trace Organic Contaminants by Sulfate Radical. *Environ. Sci. Technol.* **2015**, *49*, 13394–13402. [CrossRef]
16. Sudhakaran, S.; Amy, G.L. QSAR models for oxidation of organic micropollutants in water based on ozone and hydroxyl radical rate constants and their chemical classification. *Water Res.* **2013**, *47*, 1111–1122. [CrossRef] [PubMed]
17. Hu, X.; Belle, J.H.; Meng, X.; Wildani, A.; Waller, L.A.; Strickland, M.J.; Liu, Y. Estimating PM2. 5 concentrations in the conterminous United States using the random forest approach. *Environ. Sci. Technol.* **2017**, *51*, 6936–6944. [CrossRef]
18. Gupta, P.; Christopher, S.A. Particulate matter air quality assessment using integrated surface, satellite, and meteorological products: Multiple regression approach. *J. Geophys. Res. Atmos.* **2009**, *114*, D20205. [CrossRef]
19. Zhu, J.-J.; Anderson, P.R. Performance evaluation of the ISMLR package for predicting the next day's influent wastewater flowrate at Kirie WRP. *Water Sci. Technol.* **2019**, *80*, 695–706. [CrossRef]
20. Haimi, H.; Mulas, M.; Corona, F.; Vahala, R. Data-derived soft-sensors for biological wastewater treatment plants: An overview. *Environ. Model. Softw.* **2013**, *47*, 88–107. [CrossRef]
21. Lu, J.; Zhang, H.; Yu, J.; Shan, D.; Qi, J.; Chen, J.; Song, H.; Yang, M. Predicting Rate Constants of Hydroxyl Radical Reactions with Alkanes Using Machine Learning. *J. Chem. Inf. Model.* **2021**, *61*, 4259–4265. [CrossRef]
22. Cheng, W.; Ng, C.A. Using Machine Learning to Classify Bioactivity for 3486 Per- and Polyfluoroalkyl Substances (PFASs) from the OECD List. *Environ. Sci. Technol.* **2019**, *53*, 13970–13980. [CrossRef]
23. Kavzoglu, T.; Teke, A. Predictive Performances of ensemble machine learning algorithms in landslide susceptibility mapping using random forest, extreme gradient boosting (XGBoost) and natural gradient boosting (NGBoost). *Arab. J. Sci. Eng.* **2022**, *47*, 7367–7385. [CrossRef]
24. Yin, G.; Jameel Ibrahim Alazzawi, F.; Mironov, S.; Reegu, F.; El-Shafay, A.S.; Lutfor Rahman, M.; Su, C.-H.; Lu, Y.-Z.; Chinh Nguyen, H. Machine learning method for simulation of adsorption separation: Comparisons of model's performance in predicting equilibrium concentrations. *Arab. J. Chem.* **2022**, *15*, 103612. [CrossRef]
25. Ding Han, H.j. Prediction of Second-Order Rate Constants of Sulfate Radical with Aromatic Contaminants Using Quantitative Structure-Activity Relationship Mode. *Water* **2021**, *14*, 766. [CrossRef]
26. Sanches-Neto, F.O.; Dias-Silva, J.R.; Keng Queiroz Junior, L.H.; Carvalho-Silva, V.H. "pySiRC": Machine Learning Combined with Molecular Fingerprints to Predict the Reaction Rate Constant of the Radical-Based Oxidation Processes of Aqueous Organic Contaminants. *Environ. Sci. Technol.* **2021**, *55*, 12437–12448. [CrossRef]
27. Fabio Mercado, D.; Bracco, L.L.B.; Argues, A.; Gonzalez, M.C.; Caregnato, P. Reaction kinetics and mechanisms of organosilicon fungicide flusilazole with sulfate and hydroxyl radicals. *Chemosphere* **2018**, *190*, 327–336. [CrossRef] [PubMed]
28. Gabet, A.; Metivier, H.; Brauer, C.d.; Mailhot, G.; Brigante, M. Hydrogen peroxide and persulfate activation using UVA-UVB radiation: Degradation of estrogenic compounds and application in sewage treatment plant waters. *J. Hazard. Mater.* **2021**, *405*, 124693. [CrossRef] [PubMed]
29. Wang, Z.; Shao, Y.; Gao, N.; Lu, X.; An, N. Degradation of diethyl phthalate (DEP) by UV/persulfate: An experiment and simulation study of contributions by hydroxyl and sulfate radicals. *Chemosphere* **2018**, *193*, 602–610. [CrossRef] [PubMed]
30. Rickman, K.A.; Mezyk, S.P. Kinetics and mechanisms of sulfate radical oxidation of β-lactam antibiotics in water. *Chemosphere* **2010**, *81*, 359–365. [CrossRef] [PubMed]
31. Gupta, S.; Basant, N. Modeling the reactivity of ozone and sulphate radicals towards organic chemicals in water using machine learning approaches. *RSC Adv.* **2016**, *6*, 108448–108457. [CrossRef]
32. Real, F.J.; Acero, J.L.; Benitez, J.F.; Roldan, G.; Casas, F. Oxidation of the emerging contaminants amitriptyline hydrochloride, methyl salicylate and 2-phenoxyethanol by persulfate activated by UV irradiation. *J. Chem. Technol. Biotechnol.* **2016**, *91*, 1004–1011. [CrossRef]
33. Cvetnic, M.; Stankov, M.N.; Kovacic, M.; Ukic, S.; Bolanca, T.; Kusic, H.; Rasulev, B.; Dionysiou, D.D.; Loncaric-Bozic, A. Key structural features promoting radical driven degradation of emerging contaminants in water. *Environ. Int.* **2019**, *124*, 38–48. [CrossRef]
34. Cvetnic, M.; Tomic, A.; Sigurnjak, M.; Stankov, M.N.; Ukic, S.; Kusic, H.; Bolanca, T.; Bozic, A.L. Structural features of contaminants of emerging concern behind empirical parameters of mechanistic models describing their photooxidative degradation. *J. Water Process Eng.* **2020**, *33*, 101053. [CrossRef]
35. Shi, Y.; Yan, F.; Jia, Q.; Wang, Q. Normindex for predicting the rate constants of organic contaminants oxygenated with sulfate radical. *Environ. Sci. Pollut. Res.* **2020**, *27*, 974–982. [CrossRef]
36. Wojnárovits, L.; Takács, E. Rate constants of sulfate radical anion reactions with organic molecules: A review. *Chemosphere* **2019**, *220*, 1014–1032. [CrossRef] [PubMed]
37. RDKit: Open-Source Cheminformatics. Available online: http://www.rdkit.org (accessed on 10 February 2024).

38. Bouwmeester, R.; Martens, L.; Degroeve, S. Comprehensive and Empirical Evaluation of Machine Learning Algorithms for Small Molecule LC Retention Time Prediction. *Anal. Chem.* **2019**, *91*, 3694–3703. [CrossRef]
39. Pedregosa, F.; Varoquaux, G.; Gramfort, A.; Michel, V.; Thirion, B.; Grisel, O.; Blondel, M.; Prettenhofer, P.; Weiss, R.; Dubourg, V. Scikit-learn: Machine learning in Python. *J. Mach. Learn. Res.* **2011**, *12*, 2825–2830.
40. Hodson, T.O. Root-mean-square error (RMSE) or mean absolute error (MAE): When to use them or not. *Geosci. Model Dev.* **2022**, *15*, 5481–5487. [CrossRef]
41. OECD. *Guidance Document on the Validation of (Quantitative) Structure-Activity Relationship [(Q) SAR] Models*; Organisation for Economic Co-Operation and Development: Paris, France, 2014.
42. Lundberg, S.; Lee, S. A Unified Approach to Interpreting Model Predictions. *ArXiv* **2017**, arXiv:1705.07874.
43. Lundberg, S.M.; Erion, G.; Chen, H.; DeGrave, A.; Prutkin, J.M.; Nair, B.; Katz, R.; Himmelfarb, J.; Bansal, N.; Lee, S.-I. From local explanations to global understanding with explainable AI for trees. *Nat. Mach. Intell.* **2020**, *2*, 56–67. [CrossRef] [PubMed]
44. Bourel, M.; Cugliari, J.; Goude, Y.; Poggi, J.-M. Boosting diversity in regression ensembles. *Stat. Anal. Data Min. ASA Data Sci. J.* **2020**, 1–17. [CrossRef]
45. Odegua, R. An empirical study of ensemble techniques (bagging, boosting and stacking). In Proceedings of the Deep Learning IndabaX. 2019. Available online: https://www.researchgate.net/publication/338681864_An_Empirical_Study_of_Ensemble_Techniques_Bagging_Boosting_and_Stacking (accessed on 10 February 2024).
46. Cai, J.; Xu, K.; Zhu, Y.; Hu, F.; Li, L. Prediction and analysis of net ecosystem carbon exchange based on gradient boosting regression and random forest. *Appl. Energy* **2020**, *262*, 114566. [CrossRef]
47. Mosavi, A.; Sajedi Hosseini, F.; Choubin, B.; Goodarzi, M.; Dineva, A.A.; Rafiei Sardooi, E. Ensemble Boosting and Bagging Based Machine Learning Models for Groundwater Potential Prediction. *Water Resour. Manag.* **2021**, *35*, 23–37. [CrossRef]
48. Boldini, D.; Grisoni, F.; Kuhn, D.; Friedrich, L.; Sieber, S.A. Practical guidelines for the use of gradient boosting for molecular property prediction. *J. Cheminform.* **2023**, *15*, 73. [CrossRef]

processes

MDPI

Article

A Computational Framework for Design and Optimization of Risk-Based Soil and Groundwater Remediation Strategies

Xin Wang [1,2], Rong Li [2,*], Yong Tian [2], Bowei Zhang [2], Ying Zhao [3], Tingting Zhang [3] and Chongxuan Liu [2,*]

[1] School of Environment, Harbin Institute of Technology, Harbin 150090, China
[2] State Environmental Protection Key Laboratory of Integrated Surface Water-Groundwater Pollution Control, School of Environmental Science and Engineering, Southern University of Science and Technology, Shenzhen 518055, China
[3] Wisdri City Environment Protection Engineering Limited Company, Wuhan 430205, China
* Correspondence: lirong_sustech@yeah.net (R.L.); liucx@sustech.edu.cn (C.L.)

Highlights:

What are the main findings?

- A computational framework is developed for design of soil and groundwater remediation strategies.
- Machine-learning and process-based models are integrated to expedite computation in the framework.

What is the implication of the main finding?

- The applicability of the framework is successfully demonstrated at a field site contaminated with arsenic.

Abstract: Soil and groundwater systems have natural attenuation potential to degrade or detoxify contaminants due to biogeochemical processes. However, such potential is rarely incorporated into active remediation strategies, leading to over-remediation at many remediation sites. Here, we propose a framework for designing and searching optimal remediation strategies that fully consider the combined effects of active remediation strategies and natural attenuation potentials. The framework integrates machine-learning and process-based models for expediting the optimization process with its applicability demonstrated at a field site contaminated with arsenic (As). The process-based model was employed in the framework to simulate the evolution of As concentrations by integrating geochemical and biogeochemical processes in soil and groundwater systems under various scenarios of remedial activities. The simulation results of As concentration evolution, remedial activities, and associated remediation costs were used to train a machine learning model, random forest regression, with a goal to establish a relationship between the remediation inputs, outcomes, and associated cost. The relationship was then used to search for optimal (low cost) remedial strategies that meet remediation constraints. The strategy was successfully applied at the field site, and the framework provides an effective way to search for optimal remediation strategies at other remediation sites.

Keywords: remediation strategy; machine learning; soil and groundwater remediation; optimization; contaminated site

1. Introduction

Soil and groundwater contamination is still a major environmental problem in many countries. According to European Environment Agency, soil contaminants exist in almost 250,000 sites in Europe, where heavy metals and metalloids are the most common contaminants [1]. In China, a survey found that 16.1% of soil locations exceeded the regulatory control levels with the average contents of cadmium (Cd), arsenic (As), and mercury (Hg) exceeded the regulatory control levels by 7.0%, 2.7%, and 1.6%, respectively [2]. Various active remediation technologies have been developed and used to restore the contaminated soils in the contaminated sites. These technologies include soil replacement [3], stabilization/solidification [4], soil washing [5], and thermal remediation [6]. However, most of the active remediation technologies do not consider the natural attenuation potential in soil and groundwater environment, often leading to over-remediation and a significant cost burden [7,8]. Remediation technologies that can fully consider the combined effects of active remediation and natural attenuation are needed to significantly reduce remediation cost. Remediation cost is a major factor in assessing the feasibility of remediation strategies, especially in regions with the constraints of financial resources.

The geochemical and biogeochemical processes in soil and groundwater systems can degrade or attenuate contaminants. Contaminants can undergo dilution, dispersion, sorption, redox transformation, and degradation that may reduce their concentrations or toxicity to acceptable levels. Monitored natural attenuation (MNA) is a type of remediation technology that takes advantage of the geochemical and biogeochemical processes to treat contaminants in soil and groundwater systems with significant cost saving [9,10]. The challenge for applying MNA technologies is that it usually takes a long time [11,12]. For the contaminated lands with plans for near-future re-development, MNA will be difficult to meet the time constraint. Under such scenarios, active remediation technologies are often implemented without considering the natural attenuation potentials. The trade-off is the significant increase of remediation cost.

In this study, a numerical framework was developed that can fully consider the effects of active remediation technologies and natural attenuation in designing remediation strategies. The geochemical and biogeochemical processes were integrated using process-based models, which were used to simulate the effects of various remediation strategies with integrated active remediation technologies and natural attenuation on residual concentrations and distributions of contaminants in soil and groundwater systems. Extensive simulations were, however, required to find an optimal strategy that can minimize the cost of the remediation and meanwhile meet remediation requirements. To reduce the computational burden, selective simulations of remediation strategies were performed and the simulated results were used to train a machine learning (ML) model, random forest regression (RFR), with a goal to establish the relationships between the outcomes and costs of remediation strategies. The trained RFR was then used to search for an optimal remediation strategy within the constraints of the remediation requirements using a global optimization method. The approach was successfully demonstrated at a remediation site contaminated with As and can be applied at other contaminated sites for designing and optimizing remediation strategies.

2. Methods

2.1. Model Framework

The model framework for designing remediation strategies at a target site consists of a process-based simulation module, ML module, and optimization module (Figure 1). The process-based module was used to simulate reactive transport of contaminants under natural and remedial conditions to obtain various types of simulated data such as spatial and temporal changes of contaminant concentrations, contaminant migration paths and discharges to nearby sensitive locations such as river, and remediation costs for different remedial strategies. The obtained data were used to train an ML model in the ML module to establish the relationship between input and output variables in a target remediation site. In this study, the input variables (*xi* in Figure 1) are those related to hydrological and biogeochemical boundary and initial conditions, as well as those related to remedial activities, such as a set of remediation strategies to be adopted at the site, the locations and extents for each adopted technology, the locations and amounts of injecting reagents, etc. The output variables (*Yi* in Figure 1) are the simulated outcomes obtained from the process-based model, such as residual contaminant concentrations and their spatiotemporal distributions in groundwater and soil, contaminant flux and its spatiotemporal distributions, etc. These variables were used in both the process-based and ML modules. The trained ML model was then used by the optimization module to find a suite of optimal strategies that meet the remediation requirements for site decision makers to select a final strategy to be implemented. In the framework, the remediation requirements are often the time constraints and the regulatory control levels. For example, for a site contaminated with As, the remediation requirement for groundwater concentration is <10 ug/L.

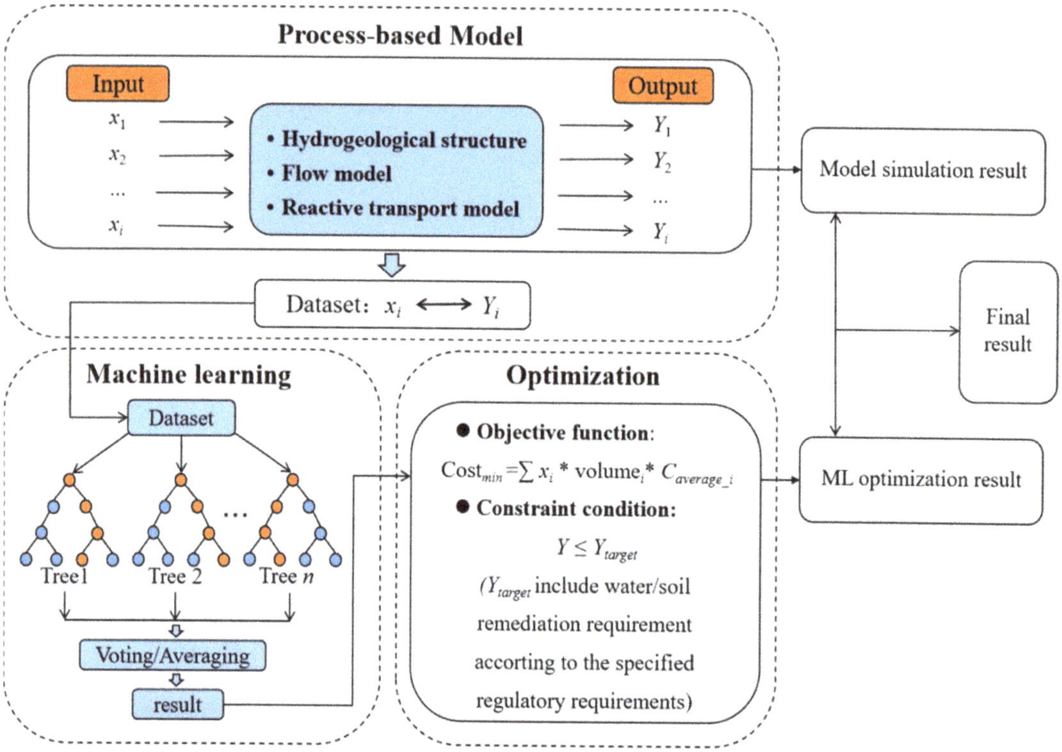

Figure 1. The framework for designing and optimizing remediation strategies for a contaminated site.

2.1.1. Process-Based Simulation Module

PFLOTRAN, an open-source, massively parallel simulator of subsurface flow and reactive transport [13,14] was used in this study to simulate contaminant transport under both natural and remedial conditions. The simulator solves Richards equation for subsurface flow [15].

$$\frac{\partial}{\partial t}(ns\rho) + \nabla \cdot (\rho q) = \Omega_{\omega} \tag{1}$$

$$q = -\frac{kk_r(s)}{u}(\nabla P - W_w \rho g z) \tag{2}$$

where n is the porosity (−), s is the saturation degree (m^3/m^3), ρ is the water density ($kmol/m^3$), q is the Darcy velocity (m/s), k is the intrinsic permeability (m^2), k_r is the relative permeability (−), μ is the viscosity (Pa·s), P is the pressure (Pa), W_w is the molecular weight of water (kg/kmol), g is the gravity (m/s^2), and z is the vertical component of the position vector (m). Water density and viscosity are functions of temperature and pressure. The source/sink term Ω_ω (kmol/m^3·s) has the form:

$$\Omega_\omega = \frac{qM}{W_\omega}\delta(r - r_{ss}) \quad (3)$$

where qM denotes a mass rate in kg/m^3/s, and r_{ss} denotes the location of the source/sink.

The solute reactive transport in the model was simulated using the following equations [13]:

$$\frac{\partial}{\partial t}\left(n\sum_{\alpha} sC_j^{\alpha}\right) + \nabla \cdot \sum_{\alpha} Q_j^{\alpha} = -\sum_{m} v_{jm}R_m \quad (4)$$

for the jth primary species, and

$$\frac{\partial \theta_m}{\partial t} = \overline{V_m}R_m \quad (5)$$

for the mth mineral. In Equations (3) and (4), C_j^{α} and Q_j^{α} denote the total concentration and flux, respectively. The R_m is the reaction rate, v_{jm} is the reaction stoichiometric coefficients, θ_m is the mineral volume fraction, and V_m is the molar volume.

2.1.2. ML Module

Random forest is one of the ensemble ML methods that has a strong generalization performance and nonlinear mapping capability with a high accuracy to overcome the problem of overfitting and instability during training and testing [16,17]. Random forest is now widely used in various environmental fields, such as soil quality assessment [18], surface water pollution [19], lake trophic status [20], groundwater pollution [17], and air quality prediction [21]. There are two types of random forest methods: random forest regression (RFR) for regression objectives, and random forest classification (RFC) for classification objectives [22]. In this study, RFR was selected to establish the relationship between remediation strategy, outcomes, and cost. RFR is an extension of regression trees that generates many regression trees (typically hundreds or several thousand) to learn and make predictions independently, and aggregates the predictions by averaging the predictions obtained from multiple regression trees to get the final predictions [23]. The overall data were randomly split into learning and testing data, in which 70% of data was for training and 30% for testing. The predictive performance of the RFR model was assessed using the coefficient of determination (R^2) and Nash–Sutcliffe efficiency (NSE), both of which denote the model error relative to the total variation in the response variable. The higher NSE and R^2, the higher the predictive accuracy. The equations of R^2 and NSE are given as follows [24]:

$$R^2 = \frac{\left(\Sigma(O_i - \overline{O})(E_i - \overline{E})\right)^2}{\Sigma(O_i - O)^2\,\Sigma(E_i - \overline{E})^2} \quad (6)$$

$$NSE = 1 - \frac{\Sigma_{i=1}^{n}(O_i - E_i)^2}{\Sigma_{i=1}^{n}(O_i - \overline{O})^2} \quad (7)$$

2.1.3. Optimization Module

After obtaining the ML-based relationship between the remediation strategy, outcomes, and cost, the optimization algorithm called "the Shuffled Complex Evolution method developed at the University of Arizona algorithm (SCE-UA)" was used to find an optimal remediation strategy with a minimum remediation cost. SCE-UA algorithm [25,26] is one of the widely used global optimization methods that has a strong nonlinear mapping capability with a good optimization effect and high stability.

The SCE-UA algorithm starts by selecting complexes P_{comp} and each complex containing m, which results in a sample size of $s = P_{comp} \times m$ points. Then, these points are sorted in an order of increasing function value and stored in an array D. The competitive complex evolution (CCE) algorithm is used to evolve each complex separately, and then shuffled and reassigned to new complexes to enable information sharing [27]. The process of evolution and shuffling are repeated until the convergence criteria are satisfied [25].

2.2. Demonstration of the Framework at a Remediation Site

2.2.1. Remediation Site

The remediation site is located in Guangzhou city, Guangdong Province, China, spreading over an area of approximately 124,000 m^2 (Figure 2). It is in the southern subtropical monsoon climate area, with an average annual temperature of 23 °C, and nearby Shabei River. The average annual precipitation at the site is about 1700 mm, of which 80% occurs in the rainy season from April to September. Shabei River flows at the western boundary of the site, which belongs to the tidal section of the Pearl River Estuary in southern China. The altitude at the site is high in the northeast and low in the southwest. The soil at the site was contaminated with As at several locations (Figure 2), and will be re-developed to establish a research campus after remediation.

2.2.2. Model Domain and Properties

Hydrogeological structure of the model domain for the site was constructed by the high-precision digital elevation model (DEM) from the BIGEMAP software [28] and drilling data from site investigation report, from which the model domain was mesh-discretized. After discretization, the model domain was divided into four layers based on the kriging interpolation using the drilling data and hydrogeological parameters (Table S1). A total number of 69,069 grid cells were discretized with each horizontal grid cell size in the range of 5–30 m. There were 21 vertical layers, divided into two parts. Specifically, the depth of each vertical layer within 8 m underground was 0.5 m, and the depth of remaining vertical layer was 2 m.

In this study, Shabei river was treated as the time-varying specified head boundary at the west side of the model domain. Every four-hour tidal water level taken from National Marine Data and Information Service [29] was used as hydraulic head in the river. The top boundary (ground surface) was specified as the infiltration boundary with the precipitation data taken from literature [30]. No flow boundary was assumed for the bottom of the model domain since the underlying formation is the aquitard. The remaining boundary of the model domain was placed on the watershed divides as the no flux boundary. Since the division of the model domain was based on a watershed approach, the model boundary was larger than site boundary. To determine the initial flow conditions, the model was first burn-in for 20 years to reach a dynamic steady state. Arsenic was the main contaminant in the study site.

The measured aqueous chemical compositions, including As concentrations in Shabei river, were used as the chemical boundary conditions, and the measured aqueous chemical compositions in groundwater were used as initial chemical conditions for the groundwater in the model domain. The distribution of soil As in the study site was obtained from the site investigation report, and spatially interpolated using three-dimensional scatter interpolation software from MATLAB R2019b (Figure 2e). The initial distribution of As concentration in groundwater was assumed to be in equilibrium with solid phase (Figure 2d). The initial distributions of DOC, DO, Fe (II), CO_2, and pH are shown in Figure S1. Various geochemical and biogeochemical reactions that affect As reactive transport were considered in the reactive transport model [31,32]. These reactions are described in supporting information (Text S1). The reaction parameters and boundary conditions as discussed above were summarized in Tables S2 and S3.

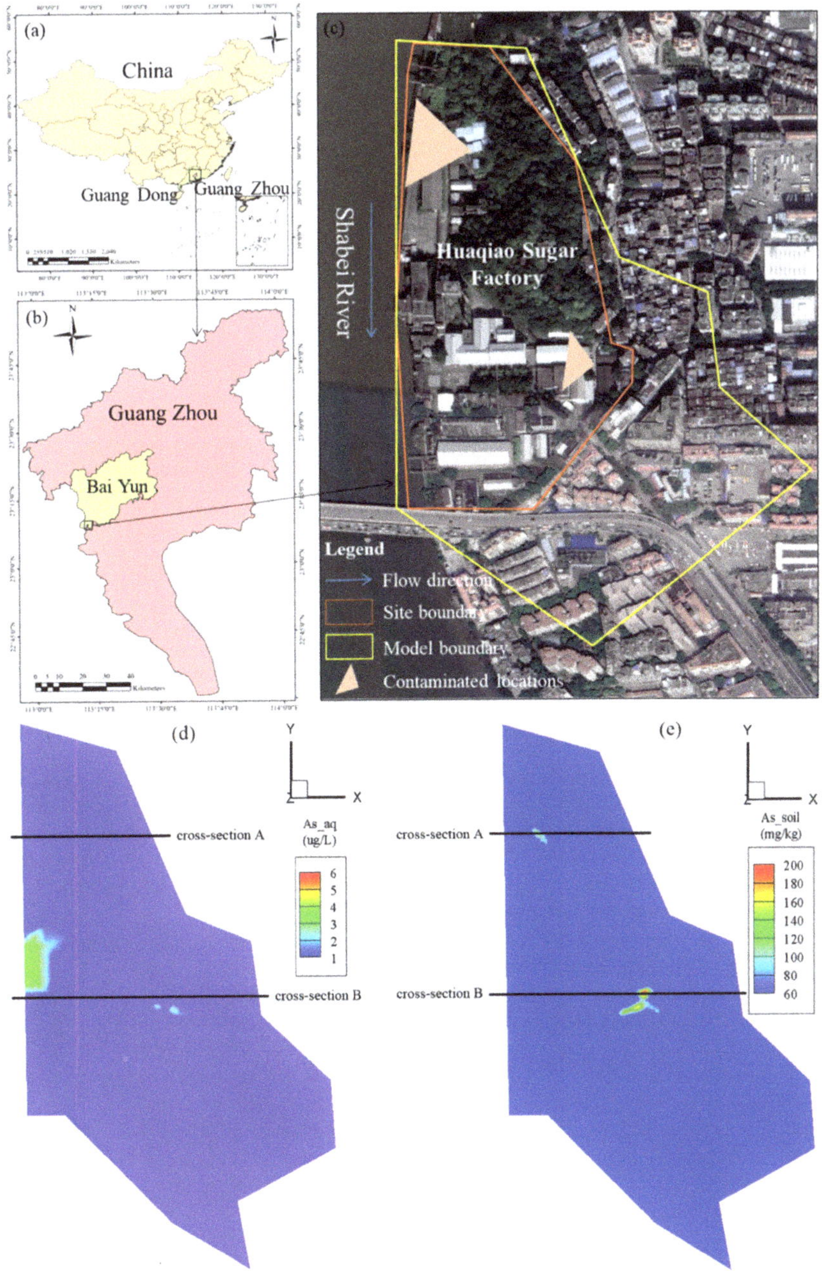

Figure 2. The location of the study site including site and modeling boundaries, and contaminated locations (**a**–**c**). The initial distributions of arsenic concentrations in groundwater (**d**) and soil (**e**). Here, As_aq and As_soil represented groundwater arsenic concentration, soil arsenic concentration, respectively. The cross-sections A and B were the cross-section of the contaminated locations, respectively.

2.2.3. Implementation of the Model Framework

For the targeted remediation site, numerical simulations under different scenarios of remedial activities found that the partial excavation and replacement of the contaminated soils with clean soils are required to meet the constraint of time for the re-development of the site. However, the amount and locations of the contaminated soils to be excavated can be optimized with the residual contaminated soil left for natural attenuation. At this site, full excavation of the contaminated soil would result in immediate clean-up of the site, but with a significant cost burden. On the other hand, too much contaminated soil left for natural attenuation might result in the long-term risk of groundwater contamination and contaminant discharge to the nearby river beyond the regulatory limits. Therefore, the goal is to find a remediation strategy with a lowest cost that the site can meet regulatory requirements for the re-development of the site. In this study, the amount of the contaminated soil to be excavated was used to calculate the cost for the corresponding remediation strategy. The simulated risk effects of the residual contaminated soils on groundwater quality and contaminant discharge to the nearby river were treated as the remediation outcomes, which were used to establish a relationship between the remediation strategy, outcomes, and cost, which was then used to find an optimal remediation strategy. In this study, the risk standard for As was divided into two levels from the local government (Table 1): level I, the As concentration in groundwater does not exceed 10 ug/L [33]; level II, the As concentration in groundwater is less than 10 ug/L. Meanwhile, the As content in the soil is less than 60 mg/kg [34].

Table 1. Optimal contaminant limits in this study.

Standard	Contaminant	Unit	Value	Reference
Level I	Arsenic	ug/L	10	[33]
Level II	Arsenic	mg/kg	60	[34]

For those complicated cases with multiple contaminants requiring multiple remediation technologies, extensive simulations will be required to find an optimal remediation strategy. To reduce the computational cost, one can select some representative scenarios for simulations, and the results can then be used to train the ML model to establish the relationship of remediation strategy, outcomes, and cost, which can then be used to find an optimal strategy using the global optimization module.

In the ML step, hyperparameters are important metrics whose values are set before starting the learning process. Here, a grid search optimization method is used to adjust the hyperparameters used in RFR to improve the model prediction accuracy [35,36]. In this study, the RFR model under water and soil remediation requirements were created using the Python-based, open-source code Scikit-learn. Table S4 provides the values of the hyperparameter in this study. The parameters used in the optimization step (SCE-UA algorithm) include the number of complexes (P_{comp}), the maximum number of function evaluations allowed during optimization (n_{max}), maximum number of evolution loops before convergency (k_{stop}), and the percentage change allowed in k_{stop} loops before convergency (P_{cento}). The specific values of the parameters in the SCE algorithm for this study are provided in Table S5.

3. Results and Discussion

3.1. Results of Natural Attenuation Simulated from the Process-Based Model

The natural attenuation scenario requires the modeling and evaluation of contaminant degradation rates and pathways and the prediction of contaminant concentration at down gradient receptor points, especially when a plume is still expanding/migrating. Figure 3 shows the snapshots of the As concentration distributions, including the contaminated areas, simulated using the process-based model under the natural attenuation condition for the entire site. The initial As concentrations in soil at two locations along cross-sections

A and B exceeded the As screening level value (60 mg/kg) in soil with the highest As concentration of 208 mg/kg in cross-section B. All the initial As concentrations in groundwater were below the groundwater standard of 10 ug/L. The result is consistent with the field survey report on the Former Site of Guangzhou Huaqiao Sugar Factory [37]. However, simulation results suggested that the As concentration in groundwater would gradually increase with time as a result of As release from soil when groundwater redox condition changes and the reductive dissolution of As-ferrihydrite in the soil. After 10 years, the highest As concentration in groundwater at cross-section B would exceed 10 ug/L, suggesting that there is a risk of groundwater contamination if the contaminated soil is not treated. On the other hand, the As concentration in the contaminated soil will gradually decrease, but it takes long time to reach below 60 mg/kg. Processes that can affect the natural attenuation of As include adsorption and desorption, as well as site sediment heterogeneity, affecting the flow velocity. The calculated flux of As from the contaminated site to the Shabei river is small and would not increase the As concentration in the river as a result of dilution (Figure 3e). The groundwater mobility in this study site is poor, leading to a very slow migration process of As which takes a long time to achieve the remediation goals. Therefore, the longer attenuation and monitoring time were selected for this specific site. However, this is a site-specific issue, depending on the characteristics of the site and types of contaminants. The primary objective of site modeling is to demonstrate that natural processes of contaminant degradation will reduce contaminant concentrations below regulatory standards or risk-based levels in an appropriate time frame before potential exposure pathways are completed. If the groundwater mobility of the contaminated site is large and the degradation rate of contaminants is fast, such a long simulation time may not be required. In addition, sampling and sample analysis must be conducted during the process to confirm that clean-up is proceeding at rates consistent with cleanup objectives. A long-term monitoring program is underway at the site that will monitor groundwater As concentration. The results can be used to validate approach in future.

These results indicated that active remedial actions, such as the partial excavation and the replacement of the contaminated soil with clean soil, or the establishment of a barrier to isolate the source zone to prevent As dispersion in the soil, are required. The barrier approach would, however, affect groundwater concentration in the isolated area. Therefore, the choice in the source zone is to partially excavate the contaminated soil and that partially left for natural attenuation with the goal to minimize the excavation amount without leading to groundwater contamination.

3.2. Results of Excavation Strategies from the Process-Based Model

Figures 4–6 show the results of three scenarios with partial excavation of the contaminated soil and partial left for natural attenuation simulated using the process-based model. In scenario 1, 3142 m^3 of the highly contaminated soil is removed. The residual contaminated soil still has As concentration as high as 205 mg/kg that can gradually release As to groundwater (Figure 4). After 10 years, the As concentration in groundwater at cross-section B would exceed 10 ug/L, posing a risk to groundwater contamination. When the volume of the excavated soil is increased in scenario 2 to 2.8 times that in scenario 1, the highest As concentration in soils decreases to 135 mg/kg. The residual contaminated soil would still release As to groundwater (Figure 5). However, the As concentration in groundwater is no longer above 10 ug/L. In scenario 3, the contaminated soil is completely excavated (168,077 m^3), and both As concentrations in groundwater and soil meet the regulatory requirements after the excavation (Figure 6). Other scenarios with variable volumes of the excavated soil are also simulated (results not shown), generating a set of data linking the remediation cost (volume of excavated soil) and outcomes (aqueous As concentrations, As concentration in residual soil, contaminated volumes of water and soil, and flux discharge to river). These results were used to find an optimal remediation strategy, as described in the next section.

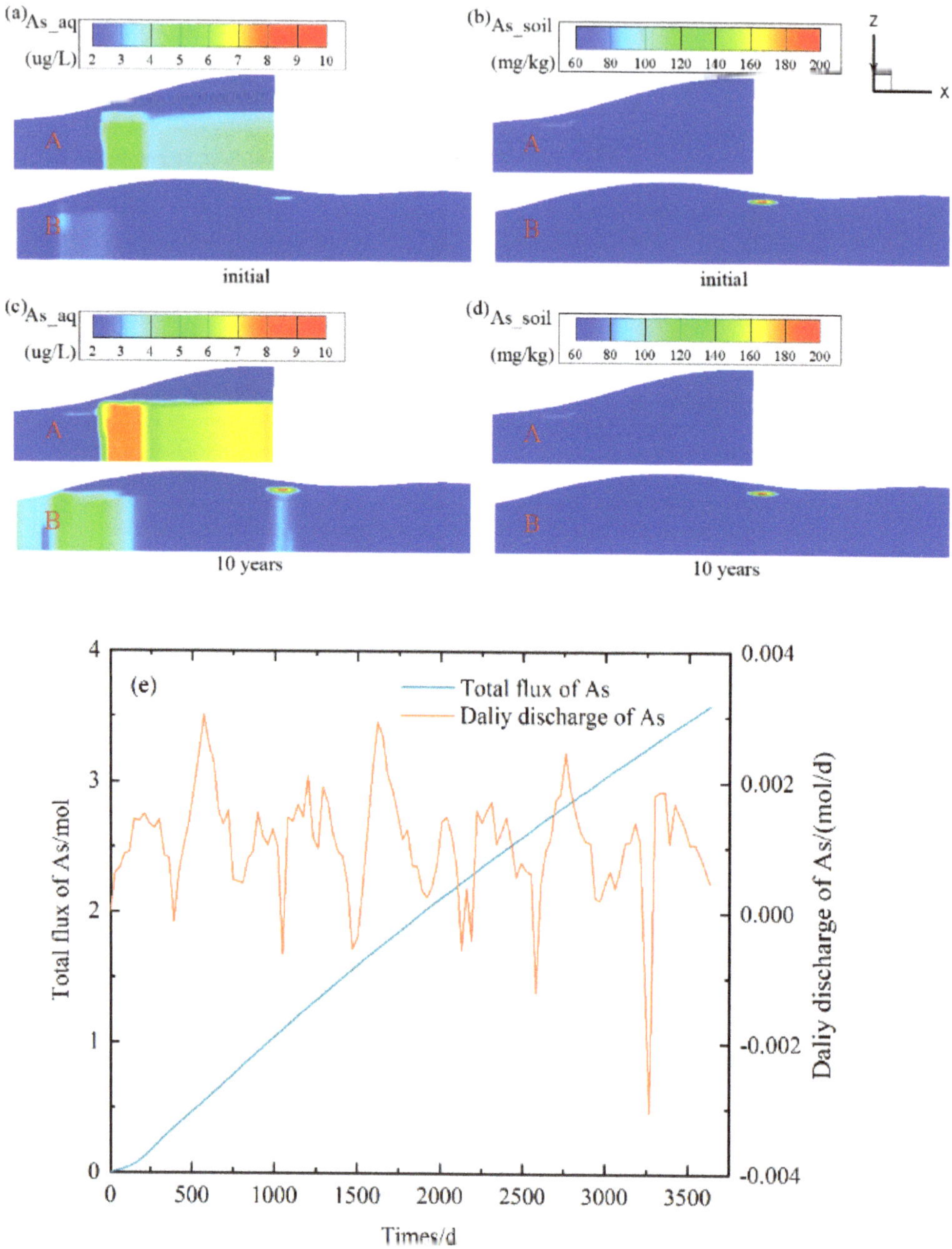

Figure 3. Initial and simulated (after 10 years) distributions of groundwater arsenic concentrations (**a**,**c**) and soil arsenic concentrations (**b**,**d**) along cross-sections A and B without any soil treatments. Here, As_aq and As_soil represented groundwater arsenic concentration, soil arsenic concentration, respectively. The locations of cross-sections A and B were provided in Figure 2. Plot (**e**) is the total flux and daily discharge of As from the site to the river.

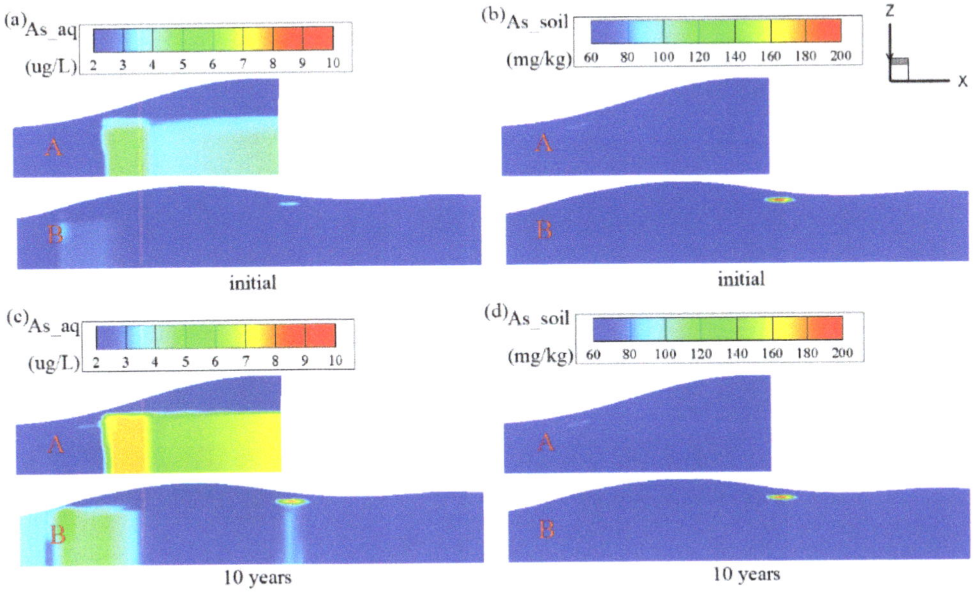

Figure 4. Initial distributions of As concentration in groundwater (**a**) and in soil (**b**), The simulated distributions of As concentrations in groundwater (**c**) and in soil (**d**) after 10 years of the partial excavation of highly contaminated soil (scenario 1). Here, As_aq and As_soil represented groundwater arsenic concentration, soil arsenic concentration, respectively.

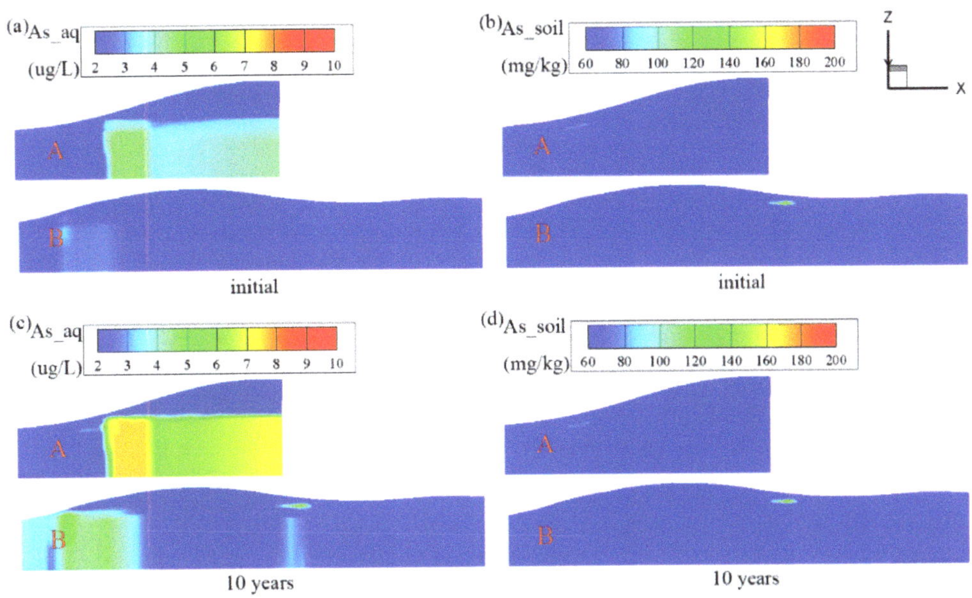

Figure 5. Initial distributions of As concentration in groundwater (**a**) and in soil (**b**), The simulated distributions of As concentrations in groundwater (**c**) and in soil (**d**) after 10 years of the partial excavation of highly contaminated soil (scenario 2). Here, As_aq and As_soil represented groundwater arsenic concentration, soil arsenic concentration, respectively.

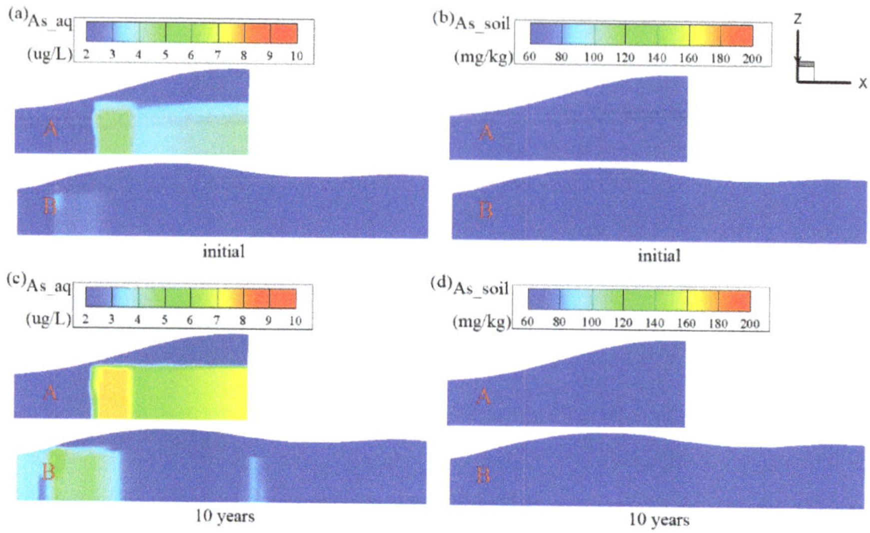

Figure 6. Initial distributions of As concentration in groundwater (**a**) and in soil (**b**), The simulated distributions of As concentrations in groundwater (**c**) and in soil (**d**) after 10 years of the complete excavation of highly contaminated soil (scenario 3). Here, As_aq and As_soil represented groundwater arsenic concentration, soil arsenic concentration, respectively.

3.3. Results of Optimal Remediation Strategy from ML and Optimization Modules

The simulated results from the process-based models (Figure 7a,c) indicated that the volumes of the contaminated groundwater and the residual contaminated soil decrease with the increasing volume of excavated soil. However, for the same volume of the excavated soil, there are multiple of remediation outcomes because there are multiple placement selections for the same volume of the excavated soil at the site. For each specific volume of the excavated soil, there should be one strategy that leads to a minimal volume of contaminated groundwater and residual contaminated soil. For the same volume of the excavated soil, the excavation strategy leading to the mineral volume of contaminated groundwater is not necessarily corresponding to the strategy leading to the minimal volume of the residual contaminated soil. The optimization strategy for this case is then to find the minimal volumes of contaminated groundwater and the residual contaminated soil as a function of the excavated volume of the contaminated soil. Such a function can be obtained from the simulation outcomes using the process-based model and be used by decision makers to select an optimal remediation strategy. However, extensive simulations would be required to generate enough data for each excavated volume of the contaminated soil.

Instead of the extensive simulations using the process-based model, the ML method, RFR, was used in this study to establish the relationship between the excavated volume of the contaminated soil and the minimal volumes of contaminated groundwater and residual contaminated soil. To provide enough data for training the ML model, 341 excavation scenarios were simulated using the process-based model with the data partially shown in (Figure 7a,c). Specifically, the excavation locations, including vertical and horizontal distributions of the excavated soil used in the process-based model as the inputs, the volumes of the contaminated groundwater with As concentration exceeding 10 ug/L, and the volume of the residual contaminated soil with concentration larger than 60 mg/kg were used as the outputs to train the RFR model. The trained model has, respectively, R^2 and *NSE* values of 0.9998 and 0.9995 for contaminated groundwater, and 0.9637 and 0.9965 for the residual contaminated soil (Table S6, Figures S2 and S3), indicating the model was well trained.

After training the RFR model, it was then used in the SCE-UA optimization algorithm to find a set of optimal remediation strategies as a function of excavated soil, i.e., minimal volumes of contaminated groundwater and residual contaminated soil as a function of the excavated soil (Figure 7b,d) as well as corresponding horizontal and vertical locations of the excavated soil. Such a function can be used by decision makers to select an optimal remediation strategy based on the cost and remediation requirements. For example, the minimum amount of the contaminated soil that needs to be excavated is 11,068 m^3 in order to avoid groundwater contamination with a corresponding excavation location provided in Figure 8. Note that for this relatively simple case, the optimal function can also be obtained from extensive simulations from the process-based model. However, for this case, it took 18 h for each simulation case using the process-based model, while the time cost for a RFR model training and validation was only about 10 s, indicating the magnitude of the reduction of computation burden.

The remediation strategy, as shown in Figure 8, was successfully implemented at the target remediation site. Compared with the complete excavation of the contaminated soils (168,077 m^3), the adopted strategy only excavated 6.6% of the total contaminated soil, leading to a significant saving of the remediation cost.

The model framework proposed in this study is also applicable to other heavy metals. In this study, As was the only metallic contaminant in this site while other heavy metal (such as Cd and Hg) were not observed. However, this is a site-specific issue requiring a large amount of site survey data to construct simulation model (e.g., drilling data, contaminant data, tidal water level, DEM, etc.). If Cd and Hg contamination is to be studied, it is necessary to re-locate the contaminated site and collect site survey data to build a new model.

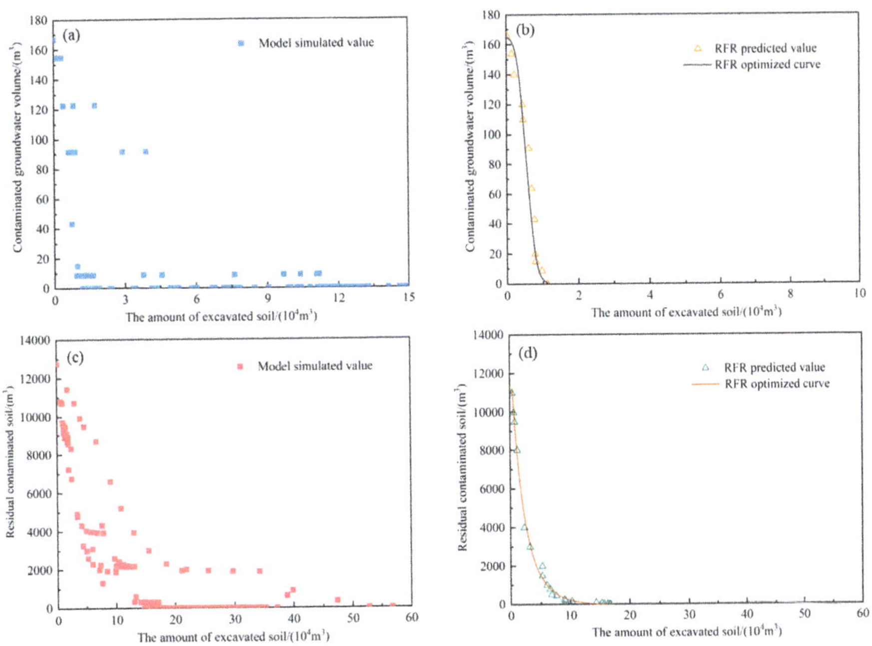

Figure 7. The simulated results of the volume of the contaminated groundwater (**a**) and the volume of residual contaminated soil (**c**) as a function of volume the excavated contaminated soil. Symbols are the results from the process-based model. Plot (**b**) and plot (**d**) are the optimized profile from the RFR model between the volume of excavated contaminated soil and the minimal volumes of contaminated groundwater (**b**) and the residual contaminated soil (**d**).

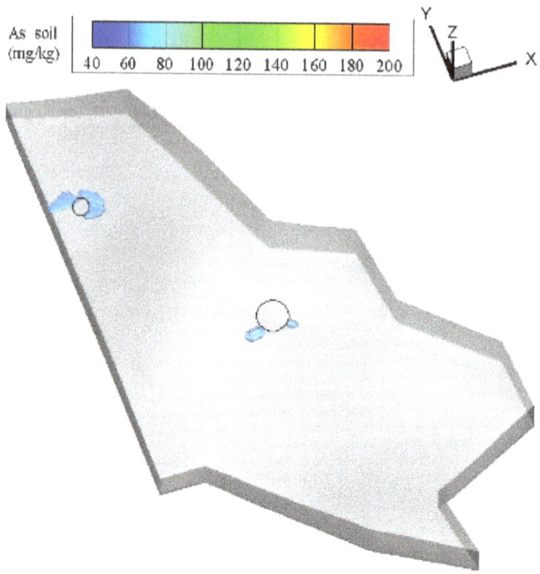

Figure 8. The excavation locations corresponding to the minimum amount of the contaminated soil to be excavated to prevent groundwater contamination as described in the text. The As_soil was soil arsenic concentration. The blue blocks are the residual contaminated soil exceeded 60 mg/kg. The black circles are the excavation locations.

4. Conclusions

A computational framework was developed in this study for searching optimal soil and groundwater remediation strategies. The approach relies on the process-based model that can simulate the integrated effects of various remediation approaches on the remediation outcomes. Extensive simulations are, however, required if only the process-based model is used to search for optimal remediation strategies, especially for the sites when multiple remediation technologies are required to restore soil and groundwater systems. The framework uses an ML approach in coupling with an optimization algorithm to expedite the searching process and to reduce the computation cost. The framework is especially useful for searching risk-based remediation approaches that can take the advantage of the natural attenuation potential in soil and groundwater. In such a case, only some of the contaminant sources need to be treated, while others can be left for natural attenuation without generating the risks of groundwater contamination or contaminant discharge to sensitive locations. The framework can provide various alternative remediation strategies for decision makers to select.

The framework was successfully applied at a field site contaminated with As. For this specific site, the framework provided various simulation scenarios of remediation outcomes, from which only the combination of partial excavation of the contaminated soil and partial maintenance for natural attenuation is a viable choice of remediation strategy based on regulatory and site re-development constraints. The optimization problem became to search for excavation volumes and locations that would generate the minimal volume of contaminated groundwater and residual contaminated soil after remediation. Under the constraint that the residual contaminated soil would not cause groundwater contamination and risk to the nearby river, an optimal excavation strategy, including the volume and location, was found. The result was successfully demonstrated at the field site. This site is relatively simple, however, because the field engineering work only involves the excavation of the contaminated soil and replacement with clean soil. The framework can also be applied to other sites with complicated contamination history and multiple contaminants requiring the combination of various active remediation technologies.

Supplementary Materials: The following supporting information can be downloaded at: https://www.mdpi.com/article/10.3390/pr10122572/s1, Figure S1: The initial distribution of pH (a), CO_2 (b), DO (c), Fe(II) (d) and DOC (e); Figure S2: The performance of RFR model under groundwater remediation requirement; Figure S3: The performance of RFR model under soil remediation requirement; Table S1: Hydrogeological parameters used in this study; Table S2: Reaction parameters used in this study; Table S3: Boundary conditions of chemical species in this study; Table S4: Optimized RFR hyperparameter values; Table S5: The specific parameters of SCE algorithm in this study; Table S6: R^2 and *NSE* of the RFR model under different remediation requirement; Text S1: The geochemical and biogeochemical reactions that affect As reactive transport in the model.

Author Contributions: Conceptualization, methodology, investigation, software, formal analysis, visualization, writing—original draft, X.W.; Conceptualization, supervision, writing—review & editing, R.L.; Software, writing—review & editing, Y.T.; Investigation, validation, B.Z.; Investigation, validation, Y.Z.; Investigation, validation, T.Z.; Conceptualization, supervision, writing—review & editing, funding acquisition, C.L. All authors have read and agreed to the published version of the manuscript.

Funding: This work was supported by National key research and development program (project No.: 2019YFC1803903), and by the National Natural Science Foundation of China (No. 41907166, 41830861), and the Program for Guangdong Introducing Innovative and Entrepreneurial Teams (2017ZT07Z479).

Institutional Review Board Statement: Not applicable.

Informed Consent Statement: Not applicable.

Data Availability Statement: The data that support the findings of this study are available from the author (11849583@mail.sustech.edu.cn) on reasonable request.

Acknowledgments: This research is supported by Center for Computational Science and Engineering of Southern University of Science and Technology. In addition, the authors also thank the helpful discussion from Yajie Wu, Qiaomei Feng, Shiwen Hu and Kun Gao.

Conflicts of Interest: The authors declare no conflict of interest.

Abbreviations

As	Arsenic
Cd	Cadmium
Hg	Mercury
BNS	Bulletin on Natural Survey
MNA	Monitored natural attenuation
RFR	Random forest regression
RFC	Random forest classification
ML	Machine learning
R^2	The coefficient of determination
NSE	Nash-Sutcliffe Efficiency
SCE-UA	The Shuffled Complex Evolution method developed at the University of Arizona algorithm
CCE	The competitive complex evolution algorithm
DEM	Digital Elevation Model
P_{comp} v	The number of complexes
n_{max}	The maximum number of function evaluations allowed during optimization
k_{stop}	Maximum number of evolution loops before convergency
P_{cento}	The percentage change allowed in kstop loops before convergency
n	Porosity
s, m^3/m^3	Saturation degree

ρ, kmol/m^3	Water density
q, m/s	Darcy velocity
k, m^2	Intrinsic permeability
k_r	Relative permeability
μ, Pa s	Viscosity
P, Pa	Pressure
W_w, kg/kmol	Molecular weight of water
g, m/s^2	Gravity
z, m	Vertical component of the position vector
Ω_ω, kmol/m^3·s	Source/sink term
qM, kg/m^3/s	Mass rate
r_{ss}	The location of the source/sink
C_j^α	Total concentration
Q_j^α	Flux
R_m	Reaction rate
v_{jm}	Reaction stoichiometric coefficients
θ_m	Mineral volume fraction
V_m	Molar volume

References

1. European Environment Agency. Available online: https://www.eea.europa.eu/ (accessed on 1 January 2022).
2. BNS. Bulletin on Natural Survey of Soil Contamination in 2014. 2014. Available online: http://www.gov.cn/xinwen/2014-04/17/content_2661765.htm (accessed on 24 April 2022).
3. Douay, F.; Roussel, H.; Pruvot, C.; Loriette, A.; Fourrier, H. Assessment of a remediation technique using the replacement of contaminated soils in kitchen gardens nearby a former lead smelter in Northern France. *Sci. Total Environ.* **2008**, *401*, 29–38. [CrossRef] [PubMed]
4. Moon, D.H.; Grubb, D.G.; Reilly, T.L. Stabilization/solidification of selenium-impacted soils using Portland cement and cement kiln dust. *J. Hazard. Mater.* **2009**, *168*, 944–951. [CrossRef] [PubMed]
5. Hu, W.; Niu, Y.L.; Zhu, H.; Dong, K.; Wang, D.Q.; Liu, F. Remediation of zinc-contaminated soils by using the two-step washing with citric acid and water-soluble chitosan. *Chemosphere* **2021**, *282*, 131092. [CrossRef] [PubMed]
6. Wei, Y.M.; Wang, F.; Liu, X.; Fu, P.R.; Yao, R.X.; Ren, T.T.; Shi, D.Z.; Li, Y.Y. Thermal remediation of cyanide-contaminated soils: Process optimization and mechanistic study. *Chemosphere* **2020**, *239*, 124707. [CrossRef]
7. Nejad, Z.D.; Jung, M.C.; Kim, K.H. Remediation of soils contaminated with heavy metals with an emphasis on immobilization technology. *Environ. Geochem. Health* **2018**, *40*, 927–953. [CrossRef]
8. Gong, Y.Y.; Zhao, D.Y.; Wang, Q.L. An overview of field-scale studies on remediation of soil contaminated with heavy metals and metalloids: Technical progress over the last decade. *Water Res.* **2018**, *147*, 440–460. [CrossRef]
9. EPA. Use of Monitored Natural Attenuation at Superfund, RCRA Corrective Action, and Underground Storage Tank Sites. OSWER. April 21. OSWER Directive No. 9200.4-17P. 1999. Available online: https://semspub.epa.gov/work/HQ/159152.pdf (accessed on 24 January 2022).
10. Sarkar, D.; Ferguson, M.; Datta, R.; Birnbaum, S. Bioremediation of petroleum hydrocarbons in contaminated soils: Comparison of biosolids addition, carbon supplementation, and monitored natural attenuation. *Environ. Pollut.* **2005**, *136*, 187–195. [CrossRef]
11. Rugner, H.; Finkel, M.; Kaschl, A.; Bittens, M. Application of monitored natural attenuation in contaminated land management—A review and recommended approach for Europe. *Environ. Sci. Policy* **2006**, *9*, 568–576. [CrossRef]
12. Khan, F.I.; Husain, T. Risk-based monitored natural attenuation—A case study. *J. Hazard. Mater.* **2001**, *85*, 243–272. [CrossRef]
13. Mills, R.T.; Lu, C.; Lichtner, P.C.; Hammond, G.E. Simulating subsurface flow and transport on ultrascale computers using PFLOTRAN. *J. Phys. Conf. Ser.* **2007**, *78*, 012051. [CrossRef]
14. Wu, R.J.; Chen, X.Y.; Hammond, G.; Bisht, G.; Song, X.H.; Huang, M.Y.; Niu, G.Y.; Ferre, T. Coupling surface flow with high-performance subsurface reactive flow and transport code PFLOTRAN. *Environ. Modell. Softw.* **2021**, *137*, 104959. [CrossRef]
15. Lichtner, P.C.; Hammond, G.E.; Lu, C.; Karra, S.; Bisht, G.; Andre, B.; Mills, R.; Kumar, J. *PFLOTRAN User Manual: A Massively Parallel Reactive Flow and Transport Model for Describing Surface and Subsurface Processes. Los Alamos National Laboratory (LANL)*; U.S. Department of Energy: Washington, DC, USA, 2015. [CrossRef]
16. Breiman, L. Random forests. *Mach. Learn.* **2001**, *45*, 5–32. [CrossRef]
17. He, S.; Wu, J.H.; Wang, D.; He, X.D. Predictive modeling of groundwater nitrate pollution and evaluating its main impact factors using random forest. *Chemosphere* **2022**, *290*, 133388. [CrossRef]
18. Dai, L.J.; Ge, J.S.; Wang, L.Q.; Zhang, Q.; Liang, T.; Bolan, N.; Lischeid, G.; Rinklebe, J. Influence of soil properties, topography, and land cover on soil organic carbon and total nitrogen concentration: A case study in Qinghai-Tibet plateau based on random forest regression and structural equation modeling. *Sci. Total Environ.* **2022**, *821*, 153440. [CrossRef]

19. Harrison, J.W.; Lucius, M.A.; Farrell, J.L.; Eichler, L.W.; Relyea, R.A. Prediction of stream nitrogen and phosphorus concentrations from high-frequency sensors using Random Forests Regression. *Sci. Total Environ.* **2021**, *763*, 143005. [CrossRef]
20. Li, B.; Yang, G.S.; Wan, R.R.; Hormann, G.; Huang, J.C.; Fohrer, N.; Zhang, L. Combining multivariate statistical techniques and random forests model to assess and diagnose the trophic status of Poyang Lake in China. *Ecol. Indic.* **2017**, *83*, 74–83. [CrossRef]
21. Guo, B.; Zhang, D.M.; Pei, L.; Su, Y.; Wang, X.X.; Bian, Y.; Zhang, D.H.; Yao, W.Q.; Zhou, Z.X.; Guo, L.Y. Estimating PM2.5 concentrations via random forest method using satellite, auxiliary, and ground-level station dataset at multiple temporal scales across China in 2017. *Sci Total Environ* **2021**, *778*, 146288. [CrossRef] [PubMed]
22. Breiman, L.; Friedman, J.H.; Olshen, R.A.; Stone, C.J. *Classification and Regression Trees*, 1st ed.; Routledge: New York, NY, USA, 1984. [CrossRef]
23. Rahmati, O.; Choubin, B.; Fathabadi, A.; Coulon, F.; Soltani, E.; Shahabi, H.; Mollaefar, E.; Tiefenbacher, J.; Cipullo, S.; Bin Ahmad, B.; et al. Predicting uncertainty of machine learning models for modelling nitrate pollution of groundwater using quantile regression and UNEEC methods. *Sci. Total Environ.* **2019**, *688*, 855–866. [CrossRef]
24. Wang, X.; Li, R.; Tian, Y.; Liu, C.X. Watershed-scale water environmental capacity estimation assisted by machine learning. *J. Hydrol.* **2021**, *597*, 126310. [CrossRef]
25. Duan, Q.Y.; Sorooshian, S.; Gupta, V.K. Optimal Use of the Sce-Ua Global Optimization Method for Calibrating Watershed Models. *J. Hydrol.* **1994**, *158*, 265–284. [CrossRef]
26. Zhang, Y.Y.; Shao, Q.X. Uncertainty and its propagation estimation for an integrated water system model: An experiment from water quantity to quality simulations. *J. Hydrol.* **2018**, *565*, 623–635. [CrossRef]
27. Tan, J.W.; Duan, Q.Y.; Gong, W.; Di, Z.H. Differences in parameter estimates derived from various methods for the ORYZA (v3) Model. *J. Integr. Agr.* **2022**, *21*, 375–388.
28. Chengdu Bigemap Data Processing Ltd. Bigemap. Available online: http://www.bigemap.com/ (accessed on 24 April 2022).
29. China Oceanic Information Network. Available online: http://www.nmdis.org.cn/ (accessed on 1 April 2022).
30. He, J.; Yang, K.; Tang, W.J.; Lu, H.; Qin, J.; Chen, Y.Y.; Li, X. The first high-resolution meteorological forcing dataset for land process studies over China. *Sci. Data* **2020**, *7*, 25. [CrossRef] [PubMed]
31. Huang, K.; Liu, Y.Y.; Yang, C.; Duan, Y.H.; Yang, X.F.; Liu, C.X. Identification of Hydrobiogeochemical Processes Controlling Seasonal Variations in Arsenic Concentrations Within a Riverbank Aquifer at Jianghan Plain, China. *Water Resour. Res.* **2018**, *54*, 4294–4308. [CrossRef]
32. Yang, C.; Zhang, Y.K.; Liu, Y.; Yang, X.; Liu, C. Model-Based Analysis of the Effects of Dam-Induced River Water and Groundwater Interactions on Hydro-Biogeochemical Transformation of Redox Sensitive Contaminants in a Hyporheic Zone. *Water Resour. Res.* **2018**, *54*, 5973–5985. [CrossRef]
33. Environmental Protection Agency, Inorganic Contaminant Accumulation in Potable Water Distribution Systems, Office of Groundwater and Drinking Water, USA. 2006. Available online: https://www.epa.gov/dwreginfo/inorganic-contaminant-accumulation-potable-water-distribution-systems (accessed on 24 April 2022).
34. Ministry of Ecology and Environment of the People's Republic of China. *Soil Environmental Quality-Risk Control Standard for Soil Contamination of Development Land (GB36600-2018)*; China Environment Publishing Group: Beijing, China, 2018; Volume 6. (In Chinese)
35. Huan, J.; Li, H.; Li, M.B.; Chen, B. Prediction of dissolved oxygen in aquaculture based on gradient boosting decision tree and long short-term memory network: A study of Chang Zhou fishery demonstration base, China. *Comput. Electron. Agr.* **2020**, *175*, 105530. [CrossRef]
36. Pedregosa, F.; Varoquaux, G.; Gramfort, A.; Michel, V.; Thirion, B.; Grisel, O.; Blondel, M.; Prettenhofer, P.; Weiss, R.; Dubourg, V.; et al. Scikit-learn: Machine Learning in Python. *J. Mach. Learn. Res.* **2011**, *12*, 2825–2830.
37. Guangzhou CAOMUFAN Ecological REsearch Co., Ltd. Caomufan. Available online: https://www.caomufan.com/ (accessed on 1 April 2022).

processes

MDPI

Article

Screening and Optimization of Soil Remediation Strategies Assisted by Machine Learning

Bowei Zhang [1], Xin Wang [2] and Chongxuan Liu [1,*]

[1] State Environmental Protection Key Laboratory of Integrated Surface Water-Groundwater Pollution Control, School of Environmental Science and Engineering, Southern University of Science and Technology, Shenzhen 518055, China; 12131080@mail.sustech.edu.cn

[2] Shenzhen Urban Public Safety and Technology Institute, Shenzhen 518046, China; 11849583@mail.sustech.edu.cn

* Correspondence: liu188@hotmail.com

Abstract: A numerical approach assisted by machine learning was developed for screening and optimizing soil remediation strategies. The approach includes a reactive transport model for simulating the remediation cost and effect of applicable remediation technologies and their combinations for a target site. The simulated results were used to establish a relationship between the cost and effect using a machine learning method. The relationship was then used by an optimization method to provide optimal remediation strategies under various constraints and requirements for the target site. The approach was evaluated for a site contaminated with both arsenic and polycyclic aromatic hydrocarbons at a former shipbuilding factory in Guangzhou City, China. An optimal strategy was obtained and successfully implemented at the site, which included the partial excavation of the contaminated soils and natural attenuation of the residual contaminated soils. The advantage of the approach is that it can fully consider the natural attenuation capacity in designing remediation strategies to reduce remediation costs and can provide cost-effective remediation strategies under variable constraints for policymakers. The approach is general and can be applied for screening and optimizing remediation strategies at other remediation sites.

Keywords: reactive transport model; machine learning; soil remediation; natural attenuation; co-contaminated site; optimal strategies

1. Introduction

Heavy metals and hazardous organic chemicals are common contaminants in soils [1–5]. Various approaches have been proposed to remediate soil contaminants at industrial sites, including monitored natural attenuation (MNA), in-situ stabilization, in-situ thermal desorption, excavation (EX), ex-situ treatments, etc. [6–12]. Low-intensity remediation technologies such as MNA have cost-effective advantages over high-intensity remediation technologies such as the EX and ex-situ treatments because they can use the natural attenuation capacity in remediating contaminants [13,14]. A challenge with low-intensity remediation technologies is that they often require long durations for natural processes to attenuate contaminants [15–17]. This will pose a problem for a site targeted for re-development that often has a remediation duration constraint. High-intensity technologies such as EX can significantly shorten remediation durations to meet the site re-development targets [18–20]. However, the trade off is the significant increase in remediation cost, especially when the volume of contaminated soils is large [19,21]. It would be ideal to combine the high- and low-intensity remediation technologies, using the high-intensity technologies to remediate soils containing the contaminants with a high risk of exposure, and low-intensity technologies to treat low risk contaminants. The challenge is how to design and optimize such a remediation strategy.

The design and optimization of the mixed low- and high-intensity remediation technologies often require the prediction of the fate and transport of contaminants during and after remediation [22–24]. Reactive transport models (RTMs) have been used to simulate the fate and transport of contaminants under both natural and remedial conditions that can be used to assist the design and optimization of the remediation strategies [25,26]. The computational cost is, however, the constraint for applying RTMs in designing and optimizing remediation technologies, especially for a large site with a complex mixture of different contaminants [27–29]. Recent progress in combining RTMs with machine learning (ML) has been demonstrated to be effective and can significantly reduce computation costs [17]. The application of such an RTM-ML approach in designing and optimizing remediation strategies is, however, still limited, especially for the sites contaminated with both heavy metals and organic contaminants.

This paper reports a study using the RTM-ML approach to design and optimize a remediation strategy that combines low- and high-intensity technologies to remediate a site contaminated with both heavy metals (arsenic, As) and organic contaminants (polycyclic aromatic hydrocarbons, PAHs). A biogeochemical model was first established to describe As and PAH reactions and their interactions in soils, which were then incorporated into the RTM. The RTM was then used to simulate various remediation scenarios using single and multiple combined remediation technologies. The simulated results were used to train an ML model to establish the cost-and-effect relationship for different combinations of remediation technologies for the site. The relationship was then used by an optimization algorithm to find optimal remediation strategies combining low- and high-intensity technologies under various requirements and constraints at the site. An optimal strategy was then selected that was successfully implemented at the site. This study demonstrated a complete procedure that can be used to design and optimize soil remediation strategies that combine low- and high-intensity technologies.

2. Materials and Methods

2.1. Study Site

A former industrial site was selected as a target remediation site that was located in the Liwan District, Guangdong Province, China, which was the former site of the Guangzhou Shipyard International Co. (Guangzhou, Guangdong Province, China) (Figure 1). The stratigraphy of the site, from top to bottom, consists of the following layers: miscellaneous fill soil, silty clay, muddy soil, fully weathered mudstone, and strongly weathered mudstone (Figure S1).

Figure 1. Location of (**a**) the study site, site boundaries (red line in (**b**)), and model boundaries (blue line in (**c**)).

To investigate the soil contamination, the soil samples were taken to the laboratory for analysis. The soil samples were dried, sieved, and passed through a 2 mm mesh and were then stored for measurements. For As measurement, the soil was first digested with HCl and HNO_3, and its concentrations were then measured using atomic fluorescence spectrometry (AFS, AFS-9130, Titan Instruments, China) [30]. For PAH concentration measurement, the soil was extracted using the Soxhlet extraction method, and the PAHs in the extracts were measured using gas chromatography-mass spectrometry (GC-MS, QP2010

Shimadzu, Kyoto, Japan) [31]. The measurement quality control for As was conducted using standard reference soil (GSS-5), with a determination accuracy of 97.5–100.3%. The PAHs measurement quality control was conducted using the soil spiked with standard solutions that had recovery ratios of 79–137% for benzo[a]pyrene (Bap) and 69–82% for dibenz(a,h)anthracene (Dba).

The soil at the site contained As with a concentration range of 1.6 to 210.2 mg/kg, two PAH compounds (Bap and Dba) with Bap concentrations ranging from 0.001 to 4.31 mg/kg, and a Dba concentration from 0.001 to 0.75 mg/kg. Their concentrations in groundwater were below regulatory standards: As: <10 μg/L and PAHs: <10 ng/L. Soil samples were also taken to analyze microbial communities at the site by DNA extraction, 16S rRNA sequencing, and subsequent high-throughput sequencing analysis. The detailed sampling procedure and results are provided in Supporting Materials (Text S1).

2.2. Model Framework

The framework for screening and optimizing remediation strategies for a contaminated site comprises three key components: RTM, ML, and optimization (Figure 2). The RTM is used to simulate the fate and transport of contaminants under various remedial conditions. The results include remediation effects (Figure 2, Y_i), such as the residual soil contaminant volumes, groundwater contaminant concentrations, and contaminant discharges to nearby rivers, and remediation parameters (Figure 2, x_i), such as the locations and volumes for the implementation of different remediation technologies, the types or concentrations of remediation reagents, etc., that are related to the cost of a remediation strategy. The results are used to train an ML model with a goal to establish the relationships between remediation parameters and remediation effects. The relationships are then used by an optimization algorithm to identify optimal remediation strategies under various constraints of site-specific remediation requirements. The framework is general and can be applied to any contaminated site.

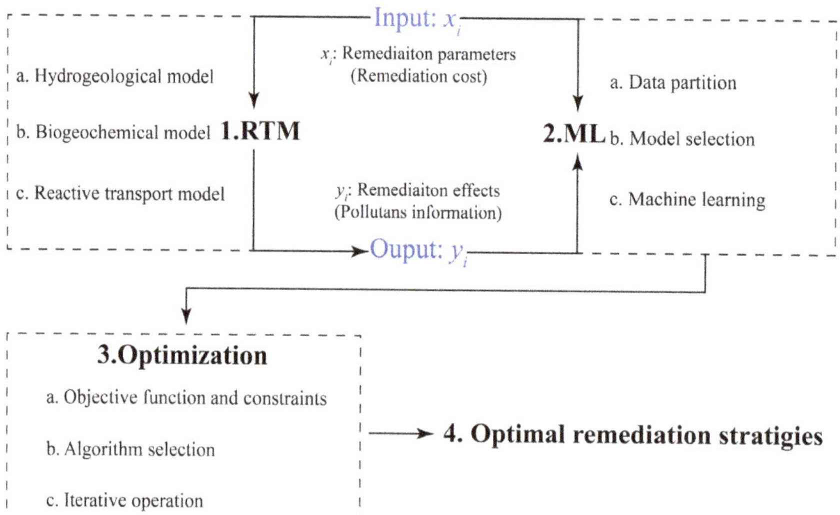

Figure 2. A model framework for screening and optimizing remediation strategies.

2.2.1. Reactive Transport Model

A reactive transport simulation software, PFLOTRAN (v5.0, USA) [32], was used as an RTM in this study. In the software, the groundwater flow is simulated using the Richards' equation as follows:

$$\frac{\partial}{\partial t}\left(\varphi \sum_{\alpha} s\rho\right) + \nabla \cdot \sum_{\alpha} (\rho \mathbf{q}) = Q_{\omega} \tag{1}$$

$$\mathbf{q} = -\frac{kk_r}{\mu} \nabla (P - W_{\omega}\rho g z) \tag{2}$$

where φ is the porosity [-], s is the saturation degree [m^3/m^3], ρ is the water density [$kmol/m^3$], $\mathbf{q}$ is the Darcy velocity [m/s], and Q_{ω} represents a source/sink of water [$kmol/m^3 \cdot s$]. The Darcy flux, $\mathbf{q}$, is calculated following Darcy's Law, where k is the intrinsic permeability [m^2], k_r is the relative permeability [-], μ is the viscosity [Pa·s], P is the pressure [Pa], W_{ω} is the molecular weight of water [kg/kmol], g is the gravity [m/s^2], and z is the vertical component of the position vector [m].

The reactive transport of dissolved contaminants such as organic carbon (DOC), As, Bap, and Dba can be described as follows:

$$\frac{\partial}{\partial t}\left(\varphi \sum sC_j\right) + \nabla \cdot \sum \sigma \Omega_j = -\sum_{m} v_{jm} R_m \tag{3}$$

where C_j is the total concentration of component j [mol/L], Ω_j is the total flux of component j [mol/(L s)], v_{jm} is the stoichiometric coefficient of chemical component j in the mth reaction, and R_m is the mth reaction rate [mol/(L s)].

Various geochemical and biogeochemical reactions controlling the transformation of contaminants can be incorporated into the RTM. In this study, reactions related to As, Bap, and Dba were considered (Figure S2 and Table S3). Because the degradation pathways for the PAHs and As sorption and desorption are often related to iron biogeochemistry [33,34], iron redox reactions were also considered in this study. At the site, the iron(III) content in the soil, as determined by XRF, ranged from 1.6% to 5.5%.

A Monod-type model was used to describe the bioreduction of ferrihydrite, which was often treated as a surrogate for the sorbents of metal and organic contaminants in subsurface systems [35–38]:

$$CH_2O + 7H^+ + 4Fe(OH)_3 \xrightarrow{k_1} 4Fe^{2+} + HCO_3^- + 10H_2O \tag{4}$$

$$R_1 = k_1 C_{Fe^{3+}} \frac{C_{DOC}}{K_{DOC,1} + C_{DOC}} \tag{5}$$

where k_2 is the oxidation rate constant [L/(mol·h)], $C_{Fe}{}^{2+}$ is the ferrous concentration [mol/L], C_{O_2} is the DO concentration [mol/L].

The aerobic microbial respiration was also considered in this study with the following rate expression [39,40]:

$$Fe^{2+} + 0.25O_2 + 2.5H_2O \xrightarrow{k_2} Fe(OH)_3 + 2H^+ \tag{6}$$

$$R_2 = k_2 C_{Fe^{2+}} C_{O_2} \tag{7}$$

where k_2 is the oxidation rate constant [L/(mol·h)], $C_{Fe}{}^{2+}$ is the ferrous concentration [mol/L], and C_{O_2} is the DO concentration [mol/L].

The aerobic microbial respiration was also considered in this study with the following rate expression [39,40]:

$$CH_2O + O_2 \xrightarrow{k_3} H_2O + CO_2 \tag{8}$$

$$R_3 = k_3 X \frac{C_{DOC}}{C_{DOC} + K_{DOC,3}} \frac{C_{O_2}}{C_{O_2} + K_{O_2,3}} \tag{9}$$

where k_3 is the respiration rate constant [1/s], X is the biomass concentration [mol/L], C_{O_2} is the DO concentration [mol/L], C_{DOC} is the DOC concentration [mol/L], and K_{DOC} and K_{O_2} are, respectively, the half-saturation constants with respect to DOC and DO [mol/L].

The rate of DOC production from the transformation of particulate organic carbon (POC) is described below [39,40]:

$$POC \overset{\alpha_{DOC}}{\rightarrow} DOC \tag{10}$$

$$R_4 = \alpha_{DOC}(K_{d,DOC} C_{DOC} - C_{POC}) \tag{11}$$

where C_{POC} is the POC concentration [mol/m^3], $K_{d,DOC}$ is the distribution coefficient between DOC and POC [L/g], and α_{DOC} is the mass transfer coefficient [1/h].

For PAHs, both aerobic and anaerobic biodegradation processes were considered in this study [25,41]. The aerobic reactions and rates are described as follows:

$$C_{20}H_{12} + 23O_2 \overset{k_5}{\rightarrow} 6H_2O + 20CO_2 \tag{12}$$

$$C_{22}H_{14} + 25.5O_2 \overset{k_5}{\rightarrow} 7H_2O + 22CO_2 \tag{13}$$

$$R_5 = -k_5 X \frac{C_{PAHs}}{C_{PAHs} + K_{PAH,5}} \frac{C_{O_2}}{C_{O_2} + K_{O_2,5}} \tag{14}$$

where k_5 is the aerobic biodegradation rate constant [1/s], C_{PAHs} is the PAH concentration [mol/L], $K_{PAH,5}$ is the half-saturation constant with respect to PAHs [mol/L], respectively, and other variables are as described before.

The anaerobic reactions and rates are described as follows:

$$C_{20}H_{12} + 92Fe(OH)_3 + 184H^+ \overset{k_6}{\rightarrow} 236H_2O + 20CO_2 + 92Fe^{2+} \tag{15}$$

$$C_{22}H_{14} + 102Fe(OH)_3 + 204H^+ \overset{k_6}{\rightarrow} 262H_2O + 22CO_2 + 102Fe^{2+} \tag{16}$$

$$R_6 = -k_6 X \frac{C_{PAHs}}{C_{PAHs} + K_{PAH,6}} \frac{C_{Fe^{3+}}}{C_{Fe^{3+}} + K_{Fe^{3+}}} \frac{I_{O_2}}{I_{O_2} + C_{O_2}} \tag{17}$$

where k_6 is the anaerobic biodegradation rates [1/s], $K_{PAH,6}$ and $K_{Fe}{}^{3+}$ are the half-saturation constants with respect to PAHs and ferrihydrite [mol/m^3], respectively, I_{O_2} is the inhibition constant during anaerobic biodegradation [mol/L], and other variables are described as before.

The co-metabolism of multi-substrate PAH biodegradation was also considered in this study using the following expression [42]:

$$R_i' = R_i \frac{C_i}{K_{si} + \sum_{j=1}^{n} \frac{K_{si}}{K_{sj}} C_j} \tag{18}$$

where K_{si} and K_{sj} are the half rate constants for the degradation of PAH compounds i and j [mol/L], and C_i and C_j are the concentrations of PAH compound i and j [mol/L], respectively.

Both As and PAHs can sorb to subsurface materials [43,44]. Their competitive sorption was derived by assuming that their sorption follows the Langmuir sorption isotherm as follows:

$$[\mathrm{As}] = \frac{K_{\mathrm{As}} * C_{\mathrm{As}} - \frac{K_{\mathrm{PAHs}} C_{\mathrm{PAHs}}}{1+K_{\mathrm{PAHs}} C_{\mathrm{PAHs}}}}{1 + K_{\mathrm{As}} * C_{\mathrm{As}} - \frac{K_{\mathrm{PAHs}} C_{\mathrm{PAHs}}}{1+K_{\mathrm{PAHs}} C_{\mathrm{PAHs}}}} * S_t \tag{19}$$

where K_{As} and K_{PAHs} are the sorption affinity constants for As and PAHs [L/mol], respectively, C_{As} and C_{PAHs} are the As and PAH concentrations in the aqueous phase [mol/L], and S_t is the total sorption site density [mol/m^3]. The values of these parameters can be found in Table S2.

For the RTM implementation, a remediation site has to be discretized. For the target site, the high-precision data from a digital elevation map (DEM) and site characterization were used in the discretization. An unstructured triangular grid was generated by COMSOL Multiphysics software (v6.0, USA) with a denser grid in the contaminated areas. Vertically, the bedrock lies approximately 15 m below the ground surface, while contaminants are mainly located at depths of 0–3 m. Consequently, variable vertical discretization intervals were used in this study with an interval of 0.5 m, from 0 to 3 m depth, 1 m, from 3 to 5 m depth, and 5 m, from 5 to 20 m depth, resulting in a total number of the numerical grids of 86,076. The model domain covered the entire study area with the following boundary conditions: tidal water level in the nearby Zhujiang River for the eastern boundary (Figure S3b); ground surface with local precipitation infiltration at the top boundary (Figure S3a); no-flux condition at the bottom boundary where is an impermeable layer; and no-flux condition at other boundaries because they were placed at the watershed edge [45]. The measured chemical compositions in the nearby Zhujiang River water samples were used as the chemical concentration conditions at the eastern boundary. The initial concentration distributions of soil contaminants (such as As, Bap, and Dba) were derived from the site characterization and were interpolated by MATLAB R2019b. The initial As and PAH concentrations in groundwater were estimated by assuming their equilibrium with the concentrations in the soil.

There were 7 contaminated zones to be remediated at the site (Figure 3a). For each remediation zone, the effects of individual remediation technologies were simulated using the RTM. The candidate technologies that are acceptable to local agencies include MNA, enhanced natural attenuation (ENA), in-situ stabilization, and EX. The simulation results (Figures S5–S8) indicated that only the EX treatment could meet the site remediation requirements, i.e., the groundwater concentrations of all the contaminants must be lower than drinking water standards (As <10 μg/L, PAHs <10 ng/L [46–48]) and the concentrations of residual soil contaminants must be below the standards of industrial land development (As <60 mg/kg, PAHs <0.55 mg/kg [49,50]). However, the volume for the EX treatment can be optimized by treating heavily contaminated soils using the EX treatment, the slightly contaminated soils with enhanced natural attenuation (ENA), and the residual contaminated soils treated with MNA. Consequently, for the subsequent simulations, the EX treatment was placed in the center of radius R1, which was the place with the highest concentrations of contaminants. The ENA was from radius R1 to R2, and the MNA for the remaining contaminated area, where the concentrations of contaminants were relatively low in each zone (Figure 3b). For each zone, R1 and R2 and the corresponding depth D1 and D2 from the ground surface were the variables to be optimized with a goal to reduce the total remediation cost. In this study, the main cost was associated with the volume for the EX and treatment under the constraints that soil and groundwater quality values meet the regulatory standards after remediation. The values for R1, R2, D1, and D2 were assumed to be independent and could be different for different remediation zones (Zone 1–7). This leads to 4 variables for each zone, yielding 28 variables to be optimized for the site with 7 remediation zones.

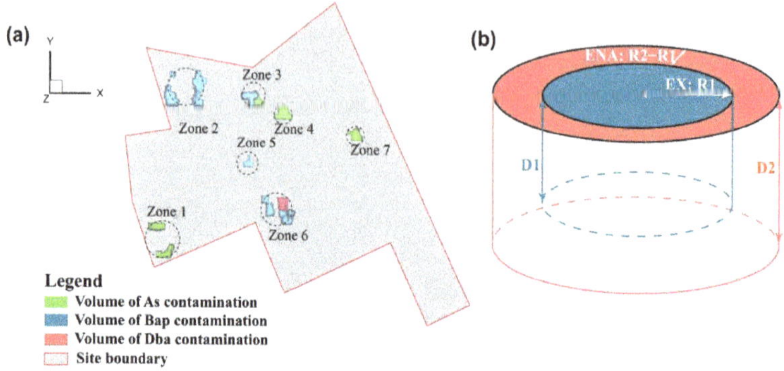

Figure 3. Illustration of (**a**) 7 remediation zones and (**b**) remediation variables for each remediation zone where the center region with a radius R1 and depth D1 is to be treated with EX, the region from R1 to R2 with a depth of D2 to be treated with ENA, and the remaining contaminated soil outside R2 to be treated with MNA in each zone.

2.2.2. ML Model

In this study, the RTM was first used to simulate remediation effects with different combinations of the remediation technologies for the 7 remediation zones. The simulated results were used to train an ML model to establish the relationship between the inputs (variables R1, R2, D1, and D2 for 7 zones) and outputs (maximum concentrations of the contaminants in groundwater and the residual contaminated soil volumes for MNA). The reliability of the ML-based relationship was evaluated during the training process using variables R^2 and RMSE. The data required for training a reliable ML-based relationship was continuously generated by the RTM until the ML model achieved a reliable accuracy. For the current case, 291 cases were simulated for training the ML model, satisfying the ratio requirement of sample size to feature size (280 cases at least) [51]. Four different ML models were evaluated in this study: ridge regression, support vector machine (SVR), random forest, and extreme gradient boosting (XGBoost). The five-fold cross-validation was used in this study to obtain a set of optimal hyperparameters for constructing the ML model, where the training and test datasets were divided at a 3:1 ratio. The model with the highest precision was chosen.

2.2.3. Optimization of Remediation Strategies

With the establishment of the ML-based relationship between the remediation input and output variables, an optimization algorithm developed at the University of Arizona (SCE-UA [52]) was used to search for optimal strategies and associated parameters by minimizing the total remediation cost (e.g., EX treatment volume) under different regulatory constraints and site requirements. The SCE-UA algorithm was chosen due to its excellent performance and efficiency in global search [53–55] with the detailed optimization procedure described elsewhere [56].

3. Results and Discussion

3.1. Simulated Groundwater Flow and Reactive Transport of Contaminants under Natural Conditions

Groundwater flow fluctuated daily and seasonally, as revealed by the fluctuation of groundwater levels at the monitoring wells at the site (Figures 4 and 5). The fluctuation of the groundwater levels was mainly attributed to the fluctuation of nearby river water levels as a result of tidal and seasonal precipitation events (Figure S3). The groundwater levels simulated by the RTM matched well with the observed monitoring wells with a Nash coefficient of 0.83 (Figure 4a), indicating a good agreement between the simulated and observed data [57]. The groundwater flow directions at the site changed frequently near the riverside (e.g., GW4). This change diminished with an increasing distance from the river (e.g., GW1). The frequent reversals of groundwater flow directions would increase the exchange of river water with groundwater, promoting the biogeochemical transformation of soil contaminants [58–60].

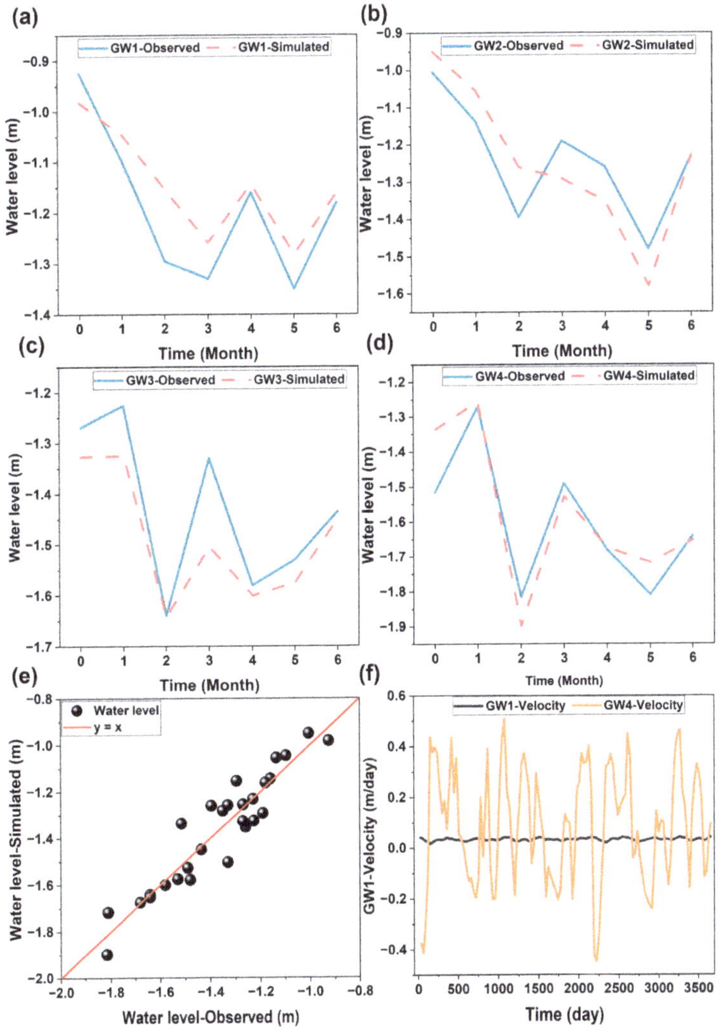

Figure 4. Observed and simulated groundwater levels at (**a**–**d**) individual monitoring wells, (**e**) all observed and simulated data, and the simulated groundwater velocity at monitoring wells (**f**) GW1 and GW4.

The effect of MNA was simulated to predict the fate and reactive transport of the contaminants under natural conditions. The results showed that the As concentration in groundwater gradually increased with time near the contaminant sources, reaching 13.2 μg/L after 10 years (Figure 6a and Figure S5) as a result of the slow reduction of sorbent ferrihydrite. Meanwhile, the As concentration in the soil slowly decreased as it was released to groundwater. The simulated result also showed that PAH oxidation enhanced the As release under anaerobic conditions, a phenomenon consistent with the literature reports [61,62]. The biogeochemical processes related to iron reduction and PAH biodegradation considered in this study were feasible at the site where the main microbial compositions were *Proteobacteria* (42–65%) and *Firmicutes* (23–34%) at the phylum level and *Bacillus, Hydrogenophaga,* and *Acinetobacter* at the genus level (Table 1 and Figure S2). These microorganisms can reduce iron oxides and degrade PAHs.

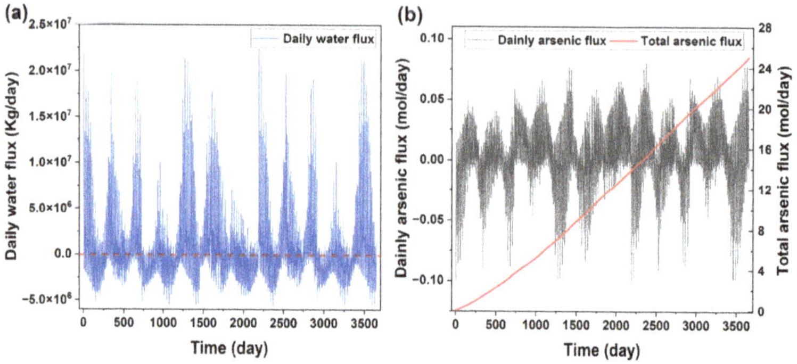

Figure 5. The flux of water (plot (**a**)) and As (plot (**b**)) from the site to the river.

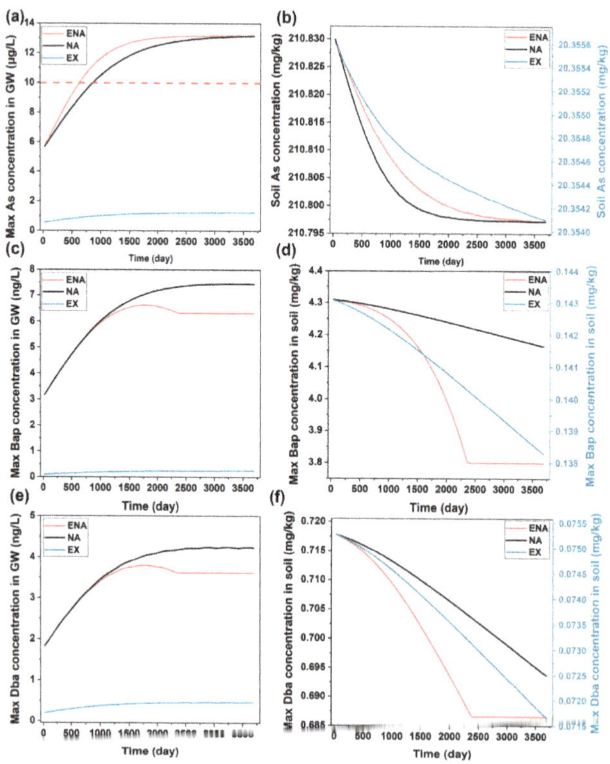

Figure 6. The concentration changes in groundwater ((**a**): As, (**c**): Bap, (**e**): Dba) and soil ((**b**): As, (**d**): Bap, (**f**): Dba) under conditions of MNA, ENA, and EX.

Table 1. Microorganisms and their functions at the target site.

Genus	Relative Abundance	Related Microbial Function	Reference
Bacillus	13–30%	Iron bioreduction, PAHs aerobic and anaerobic biodegradation	[63–65]
Hydrogenophaga	5–32%	PAHs aerobic and anaerobic biodegradation	[63,66]
Acinetobacter	2–30%	PAHs aerobic biodegradation	[67]

The As flux from the site to the nearby Zhujiang River was small (Figure 5b) and did not significantly increase the As concentration in the river as it was diluted by the high water flux in the river (Figure 5a). The result indicated that the MNA could remediate the contaminant in the soils at the site without affecting the nearby river water quality. However, it could not meet the time constraint for the site that was targeted for subsequent redevelopment after remediation within a short time duration. It may also cause a problem of groundwater contamination during the long duration of the MNA. Consequently, the MNA alone was not enough to remediate the site.

The simulated concentrations of both Bap and Dba in groundwater were lower than the regulatory level (10 ng/L, Figure 6a), while their concentrations in the soil were significantly over the regulatory level (Figure 6b) during the simulation period. In the simulation, the microbial degradation of PAHs was considered. However, their solubility was low, leading to the low biodegradability of PAHs in the soil [68,69]. The results of As and PAHs both indicated that active remediation of the contaminated soil was needed for the targeted site.

3.2. Simulated Effects of Active Remediation Technologies

The effects of two active remediation technologies, ENA and EX, were individually simulated to evaluate the changes in contaminant concentrations in groundwater and soil. ENA was implemented with the addition of organic material (lactate in this study) as the electron donor and carbon source for microorganism growth [70,71]. The implementation of the ENA resulted in a faster As release rate from the soil than MNA due to the enhanced ferrihydrite reduction by microorganisms with the addition of organic matter (Figure 6a,b). The faster As release led to the earlier arrival of As concentrations in groundwater to above regulatory level, indicating that the ENA would also potentially cause groundwater contamination. The microbial degradation of Bap and Dba was also enhanced by the addition of organic matter. However, the rate decreased significantly when the added organic matter was diminished with time, indicating that it needs to be replenished continuously to maintain the degradation [72].

The EX is a simple and time-effective remediation strategy for contaminant removal [18,73,74]. The concentrations of As, Bap, and Dba after EX can both meet groundwater and soil remediation goals (Figures 6 and S8). However, the cost for the EX was much higher than the MNA and ENA as it requires the excavation of a large amount of the contaminated soil and subsequent treatment. Considering the efficiency and cost of the MNA, ENA, and EX, it would be ideal to combine them with the goal of minimizing the remediation cost under the constraints of regulatory and local requirements. The remediation requirements at the site were various, including the discharge of contaminants from the site to the nearby river that would not affect river water quality, no groundwater contamination was allowed during and after the remediation, contaminant concentrations in the soil must meet regulatory criteria, and the remediation duration must meet site-redevelopment schedule. Given these requirements, a remediation strategy, as described in Figure 3, was proposed, which separated each remediation zone into three spatial regions defined by parameters R1, R2, D1, and D2. In order to determine the values of these parameters that would give the minimal remediation cost under the constraints of the remediation requirements, RTM was first used to simulate the reactive transport of contaminants with different combinations of these parameters. The data generated from the RTM simulations were then used to establish the relationships linking these parameter values with corresponding remediation costs, which were then used to find an optimal remediation strategy, as described in the next section.

3.3. Remediation Strategies and Optimization

The RTM-simulated results showed that different remediation strategies, as generated by adjusting remediation parameters (R1, R2, D1, and D2) for each remediation zone, would lead to different residual volumes of the contaminated soil for subsequent MNA (Figure 7a) and different groundwater concentrations (Figure 7b) as a function of the total volume of the contaminated soil treated with EX and ENA. A larger excavated volume generally, as expected, led to lower groundwater concentrations and residual volumes of the contaminated soil. The result also showed that for a specific EX volume of the contaminated soil, groundwater concentrations and residual volumes of the contaminated soil varied. This is because the different placements of the EX zones within the contaminated areas would lead to different remediation results. The placement of the EX at a highly contaminated location within each zone led to lower groundwater concentrations and residual volumes of the contaminated soil despite the same amount of the excavated contaminated soil. These results indicated that extensive RTM simulations are required to find an optimal combination of the remediation parameters (R1, R2, D1, and D2), which would yield the lowest cost of remediation.

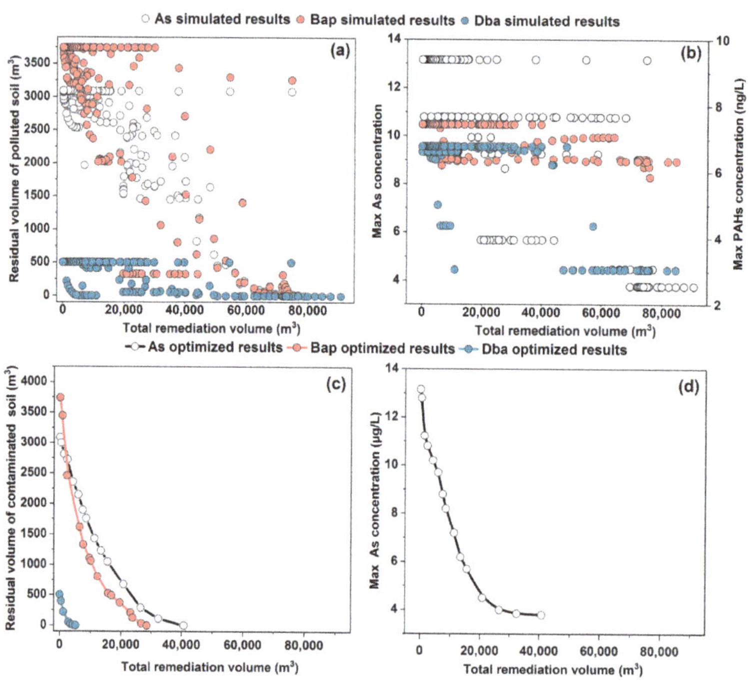

Figure 7. Relationship between the residual volume of the contaminated soil vs. the total remediation volume of the contaminated soil (excavated volume) (**a**) and maximum concentrations of the contaminants in groundwater as a function of total remediation volume of the contaminated soil (**b**). The optimized residual volume of the contaminated soil as a function of total remediation volume from the ML-SCE-UA results (**c**) and the maximum As concentration in groundwater as a function of remediation volume (**d**).

In this study, instead of performing extensive RTM simulations, the results from a select set of simulations, as shown in Figure 7, were used to train an ML model to establish a relationship linking the remediation parameters (R1, R2, D1, and D2) with the simulated results of groundwater concentrations and residual volumes of the contaminants at the site. When properly trained, such an ML-based relationship could replace the RTM to generate simulation results from other combinations of remediation parameters (R1, R2, D1, and D2). In this study, various ML methods were compared to evaluate their performances in fitting the RTM-simulated results, as shown in Figures S9 and S10, and it was found that the XGBoost method had the best performance.

The ML-based relationship from the XGBoost method was then used by an optimization algorithm (SCE-UA) to search for an optimal combination of remediation parameters (R1, R2, D1, and D2) that would give the lowest cost of the remediation and meet the site remediation requirements (Figure 7c,d). During the optimization process, the remediation parameters (R1, R2, D1, and D2) were optimized for each contaminant (As, Bap, and Dba). The optimization results indicated that to meet the groundwater quality standard for As (<10 μg/L) at the site, 5768 m^3 of the contaminated soil with the high concentration has to be excavated, while the other soil can be left for the MNA. Since the simulated groundwater concentrations of Bap and Dba are always less than the groundwater quality standards, the EX treatment of 5768 m^3 of the contaminated soil with the MNA for the residual contaminated soil can meet all groundwater standards for both As and PAH contaminants. The residual contaminated soil would not pose a health risk if kept at the site despite the fact that the concentrations of the contaminants in the residual contaminated soil still exceeded the soil quality standards. If the site needs to meet both the groundwater standards and the soil standards, then the excavated volume would be 40,755 m^3 for the As-contaminated soil, 21,922 m^3 for the Bap-contaminated soil, and 5149 m^3 for the Dba-contaminated soil. Because of the partial overlap of the soils contaminated with these contaminants (Figure 3), the total excavation volume would be less than their addition. The optimization results also showed that adding ENA with a volume of 6910 m^3 would reduce the excavation volume (Table S3). The other residual contaminated soils with their contaminant concentrations above the background levels were left for MNA. The result from the optimization process indicated that the combination of EX and MNA was the most optimum method to treat the contaminated soil. EX is a time-efficient but costly technology for soil remediation, while MNA is a low-cost technology but requires a long duration of remediation action. For targeting a site for re-development, a long duration of remediation is not acceptable. The combination of the EX and MNA can have advantages and minimize the shortcomings of each technology. Here, the EX was optimized to treat the contaminated soils with high concentrations of contaminants, while the areas containing low concentrations of contaminants were left for MNA. Because the areas with low concentrations of contaminants have no risk of health risk exposure, they can be left at the site for MNA without a need for active remediation. Consequently, the combined EX and MNA can reduce the amount of the contaminated soil for EX treatment, and meanwhile, it can significantly reduce the time duration of active remediation.

Under the actual field conditions, the selection of remediation strategies should be considered under different constraints suitable to the specific sites, such as economic, environmental, and social impacts, besides the selection of remediation technologies to meet both regulatory groundwater and soil standards [75]. In this context, our framework can readily consider the different constraints during the optimization step that can be tailored to a specific site. At this site, a remediation strategy that combines EX and MNA was eventually selected to meet both groundwater and soil standards because of the current regulatory requirements for the site redevelopment. From the view of remediation cost, treating 5768 m^3 of contaminated soil with EX, followed by MNA for the residual contaminated soil, was cost-effective and could reduce the cost by 91.5%, assuming that the cost for MNA is negligible compared to EX. As a validation of the approach, the simulated results of aqueous As concentrations were compared with the measured concentrations at the monitoring wells (Figure S11). The simulated results generally matched well with the measured results, indicating that the approach was effective and could be used for screening and optimizing soil remediation strategies.

A similar framework to the one reported previously [17] was used in this study, which consists of RTM, ML, and optimization algorithms. However, the previous study focused more on the development of the framework itself with a demonstration site having a relatively simple contamination history and biogeochemistry. The current study focused on solving a real and complex problem using the framework on a site contaminated with both organic and inorganic contaminants, featuring multiple contamination zones as shown in Table 2. This significantly increases the complexity of the biogeochemistry and optimization, providing a realistic demonstration of the framework's application and an optimal remediation strategy for the site. Additionally, the simulated results presented here were compared with the measured data monitored in this study, validating the effectiveness of the framework.

Table 2. The comparison between the current and previous study.

	Current Study	Previous Study
Contaminants	As, Bap and Dba	As
Contaminated zones	7	2
ML model	XGBoost	Random forest
Parameters to be optimized	28	4
Optimal remediation strategies	EX, ENA and MNA	EX and MNA
Result and validation	Simulated and validated	Simulated only

4. Prospect and Limitation

The framework used in this study was general and can be adapted to other contaminated sites by incorporating site-specific data such as hydrology, geochemistry, and biogeochemistry, as well as the site requirements and constraints for remediation. Being module-based, the framework allows any advancements in each module to be readily updated. For example, the ML part is rapidly advancing, becoming more intelligent in handling high-dimensional and large-scale datasets [76–78]. Such an advance can be readily incorporated into the ML module in the framework. Similarly, the optimization method is also fast evolving with progress, such as swarm intelligence algorithms, hybrid algorithms, and nature-inspired algorithms that are computationally efficient and fast converging with a strong global search capability [79–81]. Such advances can also be incorporated into the framework by updating the optimization module. One major limitation of the application of the framework is that extensive site characterization of hydrology and biogeochemistry properties is required to obtain a reliable set of input data for the RTM, which is the foundation of the framework. The high-precision prediction of remediation effects from different potential remediation approaches relies on the accuracy of simulated RTM results, which will ultimately determine the selection of an optimal remediation strategy during the optimization step.

5. Conclusions

This study developed a numerical approach that can be used for screening and optimizing remediation strategies to clean up contaminants in soil and groundwater systems. The approach consisted of three components: (1) RTM simulations of the fate and transport of contaminants under natural and remedial conditions for preliminary screening of potential remediation technologies; (2) using the RTM simulation data to train an ML model for establishing the relationships linking remediation parameters, such as sizes and locations, for the implementation of specific remediation technologies with remediation effects, such as residual concentrations in groundwater and soil, their subsequent changes after remediation, and the costs associated with the implementation of remediation technologies, such as the excavation volume of the contaminated soil; and (3) using the relationship and global optimization algorithm to find optimal remediation strategies. Using a field site that was contaminated with As and PAHs as an example, this study demonstrated the application of the approach in finding optimal remediation strategies. The optimal approach selected from this study under the constraint of local regulations and site requirements was successfully implemented.

While a site co-contaminated with As and PAHs was demonstrated, the developed approach is general for screening and optimizing remediation strategies at any other sites. One important adaptation in applying the approach to other sites is the biogeochemical part in the RTM, including biogeochemical reactions and parameters, which may be specific to the target site. The ML part of the approach is to reduce computational cost because extensive RTM simulations are required to generate enough data for establishing the relationships between remediation parameters and remediation effects. Such relationships can be established with a limited number of RTM simulations when ML is used. The ML part can be ignored if the remediation site is relatively small and biogeochemical reactions are relatively simple to describe the fate and transport of contaminants. For a large remediation site or a site contaminated with multiple contaminants with complex geochemical behaviors, the ML part is recommended to be included.

Supplementary Materials: The following supporting information can be downloaded at: https://www.mdpi.com/article/10.3390/pr12061157/s1, Text S1: Methods for DNA extraction and microbial structure analysis; Figure S1: Geological stratigraphy of the borehole at the site; Figure S2: Relative abundance of microbial organisms on phylum level (a), genus level (b) and net analysis among microbial communities at genus level (c); Figure S3: Seasonal precipitation (a) and river stage fluctuations (b) of Zhujiang River; Figure S4: The locations of the cross-sections for As (black dotted line) and PAHs (Bap and Dba for blue dotted line) in demonstrating the simulation results shown in Figures S3–S6; Figure S5: The concentration changes of As (a,c,g), Bap (b,e,h), and Dba (c,f,i) in groundwater during 10-year simulation under MNA conditions; Figure S6: The concentration changes of As (a,c,g), Bap (b,e,h), and Dba (c,f,i) in groundwater during 10-year simulation under ENA conditions; Figure S7: The concentration changes of As (a,c,g), Bap (b,e,h), and Dba (c,f,i) in groundwater during 10-year simulation under in-situ stabilization conditions; Figure S8: The concentration changes of As (a,c,g), Bap (b,e,h), and Dba (c,f,i) in groundwater during 10-year simulation under EX conditions; Figure S9: The performance of various ML models in fitting the RTM data, which was evaluated by index of R^2 (a) and root mean square error (RMSE) (b); Figure S10: The relationship between the groundwater concentration and residual volume of the contaminated soil ((a), As; (b), Bap; (c), Dba) obtained from RTM and ML (XGBoost); Figure S11: The measured and simulated concentrations (As, Bap, and Dba) at three monitoring wells: (a) GW1, (b) GW2, (c) GW4, and (d) in all wells combined; Table S1: Parameters for the biogeochemical reactions used in this study; Table S2: Hydrogeological parameters used in this study; Table S3: The optimal remediation parameters under different remediation requirements. Refs. [40,41,82–89] are cited in Supplementary Materials.

Author Contributions: Conceptualization, Methodology, Software, Formal analysis, Investigation, Writing—original draft, Visualization, B.Z.; Methodology, Software, Validation, Investigation, X.W.; Conceptualization, Writing—review & editing, Supervision, Funding acquisition, C.L. All authors have read and agreed to the published version of the manuscript.

Funding: This work was funded by the National Key Research and Development Program of China (project No.: 2019YFC1803903), Program for Guangdong Introducing Innovative and Entrepreneurial Teams (2017ZT07Z479) and by High level of special funds (G03050K001).

Data Availability Statement: All data and materials supporting the findings of this study are available upon reasonable request.

Acknowledgments: This work was also supported by the Center for Computational Science and Engineering at Southern University of Science and Technology.

Conflicts of Interest: The authors declare no conflicts of interest.

References

1. Li, T.K.; Liu, Y.; Lin, S.J.; Liu, Y.Z.; Xie, Y.F. Soil Pollution Management in China: A Brief Introduction. *Sustainability* **2019**, *11*, 556. [CrossRef]
2. Orecchio, S. Assessment of polycyclic aromatic hydrocarbons (PAHs) in soil of a Natural Reserve (Isola delle Femmine) (Italy) located in front of a plant for the production of cement. *J. Hazard. Mater.* **2010**, *173*, 358–368. [CrossRef] [PubMed]
3. Wang, D.G.; Alaee, M.; Byer, J.; Liu, Y.J.; Tian, C.G. Fugacity approach to evaluate the sediment-water diffusion of polycyclic aromatic hydrocarbons. *J. Environ. Monit.* **2011**, *13*, 1589–1596. [CrossRef] [PubMed]
4. Zhang, H.B.; Luo, Y.M.; Wong, M.H.; Zhao, Q.G.; Zhang, G.L. Distributions and concentrations of PAHs in Hong Kong soils. *Environ. Pollut.* **2006**, *141*, 107–114. [CrossRef] [PubMed]
5. Zhao, F.J.; Ma, Y.B.; Zhu, Y.G.; Tang, Z.; McGrath, S.P. Soil Contamination in China: Current Status and Mitigation Strategies. *Environ. Sci. Technol.* **2015**, *49*, 750–759. [CrossRef] [PubMed]
6. Han, L.J.; Li, J.S.; Xue, Q.; Che, Z.; Zhou, Y.Y.; Poon, C.S. Bacterial-induced mineralization (BIM) for soil solidification and heavy metal stabilization: A critical review. *Sci. Total Environ.* **2020**, *746*, 140967. [CrossRef] [PubMed]
7. Hussain, A.; Rehman, F.; Rafeeq, H.; Waqas, M.; Asghar, A.; Afsheen, N.; Rahdar, A.; Bilal, M.; Iqbal, H.M.N. In-situ, Ex-situ, and nano-remediation strategies to treat polluted soil, water, and air-A review. *Chemosphere* **2022**, *289*, 133252. [CrossRef]
8. Shah, V.; Daverey, A. Phytoremediation: A multidisciplinary approach to clean up heavy metal contaminated soil. *Environ. Technol. Innov.* **2020**, *18*, 100774. [CrossRef]
9. Senila, M.; Cadar, O.; Senila, L.; Angyus, B.S. Simulated Bioavailability of Heavy Metals (Cd, Cr, Cu, Pb, Zn) in Contaminated Soil Amended with Natural Zeolite Using Diffusive Gradients in Thin-Films (DGT) Technique. *Agriculture* **2022**, *12*, 321. [CrossRef]
10. Sarkar, D.; Ferguson, M.; Datta, R.; Birnbaum, S. Bioremediation of petroleum hydrocarbons in contaminated soils: Comparison of biosolids addition, carbon supplementation, and monitored natural attenuation. *Environ. Pollut.* **2005**, *136*, 187–195. [CrossRef]
11. Falciglia, P.P.; Giustra, M.G.; Vagliasindi, F.G.A. Low-temperature thermal desorption of diesel polluted soil: Influence of temperature and soil texture on contaminant removal kinetics. *J. Hazard. Mater.* **2011**, *185*, 392–400. [CrossRef] [PubMed]
12. Zhang, B.; Guo, Y.; Huo, J.; Xie, H.; Xu, C.; Liang, S. Combining chemical oxidation and bioremediation for petroleum polluted soil remediation by BC-nZVI activated persulfate. *Chem. Eng. J.* **2020**, *382*, 123055. [CrossRef]
13. Schwartz, M.O.; Kgomanyane, J. Modelling natural attenuation of heavy-metal groundwater contamination in the Selebi-Phikwe mining area, Botswana. *Environ. Geol.* **2008**, *54*, 819–830. [CrossRef]
14. Zhao, H.R.; Xia, B.C.; Qin, J.Q.; Zhang, J.Y. Hydrogeochemical and mineralogical characteristics related to heavy metal attenuation in a stream polluted by acid mine drainage: A case study in Dabaoshan Mine, China. *J. Environ. Sci.* **2012**, *24*, 979–989. [CrossRef] [PubMed]
15. Chen, C.; Zhang, X.M.; Chen, J.A.; Chen, F.Y.; Li, J.H.; Chen, Y.C.; Hou, H.B.; Shi, F. Assessment of site contaminated soil remediation based on an input output life cycle assessment. *J. Clean. Prod.* **2020**, *263*, 121422. [CrossRef]
16. Gong, Y.Y.; Zhao, D.Y.; Wang, Q.L. An overview of field-scale studies on remediation of soil contaminated with heavy metals and metalloids: Technical progress over the last decade. *Water Res.* **2018**, *147*, 440–460. [CrossRef] [PubMed]
17. Wang, X.; Li, R.; Tian, Y.; Zhang, B.W.; Zhao, Y.; Zhang, T.T.; Liu, C.X. A Computational Framework for Design and Optimization of Risk-Based Soil and Groundwater Remediation Strategies. *Processes* **2022**, *10*, 2572. [CrossRef]
18. Liu, L.W.; Li, W.; Song, W.P.; Guo, M.X. Remediation techniques for heavy metal-contaminated soils: Principles and applicability. *Sci. Total Environ.* **2018**, *633*, 206–219. [CrossRef]
19. Pavel, L.V.; Gavrilescu, M. Overview of Ex Situ Decontamination Techniques for Soil Cleanup. *Environ. Eng. Manag. J.* **2008**, *7*, 815–834. [CrossRef]
20. Rajendran, S.; Priya, T.A.K.; Khoo, K.S.; Hoang, T.K.A.; Ng, H.S.; Munawaroh, H.S.H.; Karaman, C.; Orooji, Y.; Show, P.L. A critical review on various remediation approaches for heavy metal contaminants removal from contaminated soils. *Chemosphere* **2022**, *287*, 132369. [CrossRef]
21. Rahman, S.; Rahman, I.M.M.; Hasegawa, H. Management of arsenic-contaminated excavated soils: A review. *J. Environ. Manag.* **2023**, *346*, 118943. [CrossRef] [PubMed]
22. Locatelli, L.; Binning, P.J.; Sanchez-Vila, X.; Sondergaard, G.L.; Rosenberg, L.; Bjerg, P.L. A simple contaminant fate and transport modelling tool for management and risk assessment of groundwater pollution from contaminated sites. *J. Contam. Hydrol.* **2019**, *221*, 35–49. [CrossRef]
23. Masi, M.; Ceccarini, A.; Iannelli, R. Multispecies reactive transport modelling of electrokinetic remediation of harbour sediments. *J. Hazard. Mater.* **2017**, *326*, 187–196. [CrossRef]
24. Qin, X.S.; Huang, G.H.; He, L. Simulation and optimization technologies for petroleum waste management and remediation process control. *J. Environ. Manag.* **2009**, *90*, 54–76. [CrossRef]
25. Lu, G.P.; Clement, T.P.; Zheng, C.M.; Wiedemeier, T.H. Natural attenuation of BTEX compounds: Model development and field-scale application. *Groundwater* **1999**, *37*, 707–717. [CrossRef] [PubMed]
26. Olson, M.R.; Sale, T.C. Implications of soil mixing for NAPL source zone remediation: Column studies and modeling of field-scale systems. *J. Contam. Hydrol.* **2015**, *177*, 206–219. [CrossRef] [PubMed]
27. Hansen, S.K.; Pandey, S.; Karra, S.; Vesselinov, V.V. CHROTRAN 1.0: A mathematical and computational model for in situ heavy metal remediation in heterogeneous aquifers. *Geosci. Model. Dev.* **2017**, *10*, 4525–4538. [CrossRef]

28. Verardo, E.; Atteia, O.; Rouvreau, L.; Siade, A.; Prommer, H. Identifying remedial solutions through optimal bioremediation design under real-world field conditions. *J. Contam. Hydrol.* **2021**, *237*, 103751. [CrossRef]
29. Yadav, B.; Ch, S.; Mathur, S.; Adamowski, J. Estimation of in-situ bioremediation system cost using a hybrid Extreme Learning Machine (ELM)-particle swarm optimization approach. *J. Hydrol.* **2016**, *543*, 373–385. [CrossRef]
30. Tao, C.; Wang, Y.; Cui, W.; Zou, B.; Zou, Z.; Tu, Y. A transferable spectroscopic diagnosis model for predicting arsenic contamination in soil. *Sci. Total Environ.* **2019**, *669*, 964–972. [CrossRef]
31. Pongpiachan, S.; Hattayanone, M.; Tipmanee, D.; Suttinun, O.; Khumsup, C.; Kittikoon, I.; Hirunyatrakul, P. Chemical characterization of polycyclic aromatic hydrocarbons (PAHs) in 2013 Rayong oil spill-affected coastal areas of Thailand. *Environ. Pollut.* **2018**, *233*, 992–1002. [CrossRef]
32. Hammond, G.E.; Lichtner, P.C.; Mills, R.T. Evaluating the performance of parallel subsurface simulators: An illustrative example with PFLOTRAN. *Water Resour. Res.* **2014**, *50*, 208–228. [CrossRef] [PubMed]
33. Herbel, M.; Fendorf, S. Biogeochemical processes controlling the speciation and transport of arsenic within iron coated sands. *Chem. Geol.* **2006**, *228*, 16–32. [CrossRef]
34. Zhao, Z.W.; Meng, Y.; Yuan, Q.K.; Wang, Y.H.; Lin, L.M.; Liu, W.B.; Luan, F.B. Microbial mobilization of arsenic from iron-bearing clay mineral through iron, arsenate, and simultaneous iron-arsenate reduction pathways. *Sci. Total Environ.* **2021**, *763*, 144613. [CrossRef] [PubMed]
35. Campos-Pereira, H.; Kleja, D.B.; Sjöstedt, C.; Ahrens, L.; Klysubun, W.; Gustafsson, J.P. The Adsorption of Per- and Polyfluoroalkyl Substances (PFASs) onto Ferrihydrite Is Governed by Surface Charge. *Environ. Sci. Technol.* **2020**, *54*, 15722–15730. [CrossRef]
36. Meng, S.; Wang, H.L.; Liu, H.; Yang, C.H.; Wei, Y.; Hou, D.L. Evaluation of the ability of ferrihydrite to bind heavy metal ions: Based on formation environment, adsorption reversibility and ageing. *Appl. Geochem.* **2014**, *45*, 114–119. [CrossRef]
37. Qian, G.R.; Chen, W.; Lim, T.T.; Chui, P. In-situ stabilization of Pb, Zn, Cu, Cd and Ni in the multi-contaminated sediments with ferrihydrite and apatite composite additives. *J. Hazard. Mater.* **2009**, *170*, 1093–1100. [CrossRef]
38. Zhang, P.; Liu, A.J.; Huang, P.; Min, L.J.; Sun, H.W. Sorption and molecular fractionation of biochar-derived dissolved organic matter on ferrihydrite. *J. Hazard. Mater.* **2020**, *392*, 122260. [CrossRef] [PubMed]
39. Gu, C.H.; Hornberger, G.M.; Mills, A.L.; Herman, J.S.; Flewelling, S.A. Nitrate reduction in streambed sediments: Effects of flow and biogeochemical kinetics. *Water Resour. Res.* **2007**, *43*. [CrossRef]
40. Liu, Y.Y.; Xu, F.; Liu, C.X. Coupled Hydro-Biogeochemical Processes Controlling Cr Reductive Immobilization in Columbia River Hyporheic Zone. *Environ. Sci. Technol.* **2017**, *51*, 1508–1517. [CrossRef]
41. Geng, X.L.; Boufadel, M.C.; Lee, K.; Abrams, S.; Suidan, M. Biodegradation of subsurface oil in a tidally influenced sand beach: Impact of hydraulics and interaction with pore water chemistry. *Water Resour. Res.* **2015**, *51*, 3193–3218. [CrossRef]
42. Guha, S.; Peters, C.A.; Jaffe, P.R. Multisubstrate biodegradation kinetics of naphthalene, phenanthrene, and pyrene mixtures. *Biotechnol. Bioeng.* **1999**, *65*, 491–499. [CrossRef]
43. Raven, K.P.; Jain, A.; Loeppert, R.H. Arsenite and arsenate adsorption on ferrihydrite: Kinetics, equilibrium, and adsorption envelopes. *Environ. Sci. Technol.* **1998**, *32*, 344–349. [CrossRef]
44. Tunega, D.; Zaoui, A. Adsorption of polycyclic aromatic hydrocarbons on FeOOH polymorphs: A theoretical study. *Surf. Sci.* **2021**, *706*, 121795. [CrossRef]
45. Reggiani, P.; Rientjes, T.H.M. Flux parameterization in the representative elementary watershed approach: Application to a natural basin. *Water Resour. Res.* **2005**, *41*. [CrossRef]
46. Chakraborti, D.; Rahman, M.M.; Das, B.; Murrill, M.; Dey, S.; Mukherjee, S.C.; Dhar, R.K.; Biswas, B.K.; Chowdhury, U.K.; Roy, S.; et al. Status of groundwater arsenic contamination in Bangladesh: A 14-year study report. *Water Res.* **2010**, *44*, 5789–5802. [CrossRef]
47. Podgorski, J.; Berg, M. Global threat of arsenic in groundwater. *Science* **2020**, *368*, 845–850. [CrossRef] [PubMed]
48. Zhang, Y.; Zhang, L.F.; Huang, Z.P.; Li, Y.N.; Li, J.F.; Wu, N.; He, J.H.; Zhang, Z.Z.; Liu, Y.Q.; Niu, Z.G. Pollution of polycyclic aromatic hydrocarbons (PAHs) in drinking water of China: Composition, distribution and influencing factors. *Ecotoxicol. Environ. Safe* **2019**, *177*, 108–116. [CrossRef]
49. Xie, K.T.; Xie, N.E.; Liao, Z.Y.; Luo, X.S.; Peng, W.J.; Yuan, Y. Bioaccessibility of arsenic, lead, and cadmium in contaminated mining/smelting soils: Assessment, modeling, and application for soil environment criteria derivation. *J. Hazard. Mater.* **2023**, *443*, 130321. [CrossRef]
50. Zhao, Y.Y.; Duan, F.A.; Cui, Z.J.; Hong, J.L.; Ni, S.Q. Insights into the vertical distribution of the microbiota in steel plant soils with potentially toxic elements and PAHs contamination after 60 years operation: Abundance, structure, co-occurrence network and functionality. *Sci. Total Environ.* **2021**, *786*, 147338. [CrossRef]
51. Zhu, J.-J.; Yang, M.; Ren, Z.J. Machine Learning in Environmental Research: Common Pitfalls and Best Practices. *Environ. Sci. Technol.* **2023**, *57*, 17671–17689. [CrossRef] [PubMed]
52. Duan, Q.Y.; Sorooshian, S.; Gupta, V.K. Optimal Use of the Sce-Ua Global Optimization Method for Calibrating Watershed Models. *J. Hydrol.* **1994**, *158*, 265–284. [CrossRef]
53. Gopala, S.P.; Kawamura, A.; Amaguchi, H.; Takasaki, T.; Azhikodan, G. A bootstrap approach for the parameter uncertainty of an urban-specific rainfall-runoff model. *J. Hydrol.* **2019**, *579*, 124195. [CrossRef]
54. Green, C.H.; van Griensven, A. Autocalibration in hydrologic modeling: Using SWAT2005 in small-scale watersheds. *Environ. Model. Softw.* **2008**, *23*, 422–434. [CrossRef]

55. Qi, W.; Zhang, C.; Fu, G.T.; Zhou, H.C. Quantifying dynamic sensitivity of optimization algorithm parameters to improve hydrological model calibration. *J. Hydrol.* **2016**, *533*, 213–223. [CrossRef]
56. Duan, Q.Y.; Ajami, N.K.; Gao, X.G.; Sorooshian, S. Multi-model ensemble hydrologic prediction using Bayesian model averaging. *Adv. Water Resour.* **2007**, *30*, 1371–1386. [CrossRef]
57. Moriasi, D.N.; Gitau, M.W.; Pai, N.; Daggupati, P. Hydrologic and Water Quality Models: Performance Measures and Evaluation Criteria. *Trans. ASABE* **2015**, *58*, 1763–1785. [CrossRef]
58. Harvey, J.W.; Bohlke, J.K.; Voytek, M.A.; Scott, D.; Tobias, C.R. Hyporheic zone denitrification: Controls on effective reaction depth and contribution to whole-stream mass balance. *Water Resour. Res.* **2013**, *49*, 6298–6316. [CrossRef]
59. Shuai, P.; Chen, X.Y.; Song, X.H.; Hammond, G.E.; Zachara, J.; Royer, P.; Ren, H.Y.; Perkins, W.A.; Richmond, M.C.; Huang, M.Y. Dam Operations and Subsurface Hydrogeology Control Dynamics of Hydrologic Exchange Flows in a Regulated River Reach. *Water Resour. Res.* **2019**, *55*, 2593–2612. [CrossRef]
60. Son, K.; Fang, Y.L.; Gomez-Velez, J.D.; Byun, K.; Chen, X.Y. Combined Effects of Stream Hydrology and Land Use on Basin-Scale Hyporheic Zone Denitrification in the Columbia River Basin. *Water Resour. Res.* **2022**, *58*, e2021WR031131. [CrossRef]
61. Meckenstock, R.U.; Boll, M.; Mouttaki, H.; Koelschbach, J.S.; Tarouco, P.C.; Weyrauch, P.; Dong, X.Y.; Himmelberg, A.M. Anaerobic Degradation of Benzene and Polycyclic Aromatic Hydrocarbons. *J. Microbiol. Biotechnol.* **2016**, *26*, 92–118. [CrossRef] [PubMed]
62. Shen, X.F.; Dong, W.H.; Wan, Y.Y.; Feng, K.J.; Liu, Y.; Wei, Y.J. Influencing mechanisms of siderite and magnetite, on naphthalene biodegradation: Insights from degradability and mineral surface structure. *J. Environ. Manag.* **2021**, *299*, 113648. [CrossRef] [PubMed]
63. Chen, C.; Zhang, Z.; Xu, P.; Hu, H.; Tang, H. Anaerobic biodegradation of polycyclic aromatic hydrocarbons. *Environ. Res.* **2023**, *223*, 115472. [CrossRef]
64. Wang, N.; Xue, X.M.; Juhasz, A.L.; Chang, Z.Z.; Li, H.B. Biochar increases arsenic release from an anaerobic paddy soil due to enhanced microbial reduction of iron and arsenic. *Environ. Pollut.* **2017**, *220*, 514–522. [CrossRef] [PubMed]
65. Zada, S.; Zhou, H.; Xie, J.; Hu, Z.; Ali, S.; Sajjad, W.; Wang, H. Bacterial degradation of pyrene: Biochemical reactions and mechanisms. *Int. Biodeterior. Biodegrad.* **2021**, *162*, 105233. [CrossRef]
66. Yan, Z.; Zhang, Y.; Wu, H.; Yang, M.; Zhang, H.; Hao, Z.; Jiang, H. Isolation and characterization of a bacterial strain: *Hydrogenophaga* sp. PYR1 for anaerobic pyrene and benzo[a]pyrene biodegradation. *RSC Adv.* **2017**, *7*, 46690–46698. [CrossRef]
67. Brzeszcz, J.; Kaszycki, P. Aerobic bacteria degrading both n-alkanes and aromatic hydrocarbons: An undervalued strategy for metabolic diversity and flexibility. *Biodegradation* **2018**, *29*, 359–407. [CrossRef] [PubMed]
68. Baltrons, O.; Lopez-Mesas, M.; Vilaseca, M.; Gutierrez-Bouzan, C.; Le Derf, F.; Portet-Koltalo, F.; Palet, C. Influence of a mixture of metals on PAHs biodegradation processes in soils. *Sci. Total Environ.* **2018**, *628–629*, 150–158. [CrossRef]
69. Eggen, T.; Majcherczyk, A. Removal of polycyclic aromatic hydrocarbons (PAH) in contaminated soil by white rot fungus Pleurotus ostreatus. *Int. Biodeterior. Biodegrad.* **1998**, *41*, 111–117. [CrossRef]
70. Blazquez-Palli, N.; Rosell, M.; Varias, J.; Bosch, M.; Soler, A.; Vicent, T.; Marco-Urrea, E. Multi-method assessment of the intrinsic biodegradation potential of an aquifer contaminated with chlorinated ethenes at an industrial area in Barcelona (Spain). *Environ. Pollut.* **2019**, *244*, 165–173. [CrossRef]
71. Schneidewind, U.; Haest, P.J.; Atashgahi, S.; Maphosa, F.; Hamonts, K.; Maesen, M.; Calderer, M.; Seuntjens, P.; Smidt, H.; Springael, D.; et al. Kinetics of dechlorination by Dehalococcoides mccartyi using different carbon sources. *J. Contam. Hydrol.* **2014**, *157*, 25–36. [CrossRef]
72. Geng, X.L.; Pan, Z.; Boufadel, M.C.; Ozgokmen, T.; Lee, K.; Zhao, L. Simulation of oil bioremediation in a tidally influenced beach: Spatiotemporal evolution of nutrient and dissolved oxygen. *J. Geophys. Res.-Oceans* **2016**, *121*, 2385–2404. [CrossRef]
73. Khan, S.; Naushad, M.; Lima, E.C.; Zhang, S.X.; Shaheen, S.M.; Rinklebe, J. Global soil pollution by toxic elements: Current status and future perspectives on the risk assessment and remediation strategies-A review. *J. Hazard. Mater.* **2021**, *417*, 126039. [CrossRef]
74. Tomei, M.C.; Daugulis, A.J. Ex Situ Bioremediation of Contaminated Soils: An Overview of Conventional and Innovative Technologies. *Crit. Rev. Environ. Sci. Technol.* **2013**, *43*, 2107–2139. [CrossRef]
75. Naseri-Rad, M.; Berndtsson, R.; Persson, K.M.; Nakagawa, K. INSIDE: An efficient guide for sustainable remediation practice in addressing contaminated soil and groundwater. *Sci. Total Environ.* **2020**, *740*, 139879. [CrossRef]
76. Pyo, J.; Hong, S.M.; Kwon, Y.S.; Kim, M.S.; Cho, K.H. Estimation of heavy metals using deep neural network with visible and infrared spectroscopy of soil. *Sci. Total Environ.* **2020**, *741*, 140162. [CrossRef] [PubMed]
77. Janga, J.K.; Reddy, K.R.; Raviteja, K.V.N.S. Integrating artificial intelligence, machine learning, and deep learning approaches into remediation of contaminated sites: A review. *Chemosphere* **2023**, *345*, 140476. [CrossRef] [PubMed]
78. Yan, H.; Zheng, Q.; Zeng, L. Conditional generative adversarial networks for groundwater contamination characterization and source identification. *J. Hydrol.* **2024**, *632*, 130900. [CrossRef]
79. Zhang, Y.; Chen, H.; Lv, M.; Hou, Z.; Wang, Y. Swarm intelligence machine-learning-assisted progressive global optimization of DNAPL-contaminated aquifer remediation strategy. *Water Supply* **2023**, *23*, 1250–1266. [CrossRef]
80. Pham, Q.B.; Tran, D.A.; Ha, N.T.; Islam, A.R.M.T.; Salam, R. Random forest and nature-inspired algorithms for mapping groundwater nitrate concentration in a coastal multi-layer aquifer system. *J. Clean. Prod.* **2022**, *343*, 130900. [CrossRef]

81. Yaseen, Z.M.; Melini Wan Mohtar, W.H.; Homod, R.Z.; Alawi, O.A.; Abba, S.I.; Oudah, A.Y.; Togun, H.; Goliatt, L.; Ul Hassan Kazmi, S.S.; Tao, H. Heavy metals prediction in coastal marine sediments using hybridized machine learning models with metaheuristic optimization algorithm. *Chemosphere* **2024**, *352*, 141329. [CrossRef] [PubMed]
82. Liu, H.; Hu, Z.; Zhou, M.; Zhang, H.; Li, Z.; Zhang, H.; Hu, J.; Yao, X.; Lou, L.; Xi, C.; et al. Airborne microorganisms exacerbate the formation of atmospheric ammonium and sulfate. *Environ. Pollut.* **2020**, *263*, 114293. [CrossRef] [PubMed]
83. Lu, H.; Li, Y.; Shan, X.; Abbas, G.; Zeng, Z.; Kang, D.; Wang, Y.; Zheng, P.; Zhang, M. A holistic analysis of ANAMMOX process in response to salinity: From adaptation to collapse. *Sep. Purif. Technol.* **2019**, *215*, 342–350. [CrossRef]
84. Yang, C.; Zhang, Y.K.; Liu, Y.Y.; Yang, X.F.; Liu, C.X. Model-Based Analysis of the Effects of Dam-Induced River Water and Groundwater Interactions on Hydro-Biogeochemical Transformation of Redox Sensitive Contaminants in a Hyporheic Zone. *Water Resour. Res.* **2018**, *54*, 5973–5985. [CrossRef]
85. Huang, K.; Liu, Y.Y.; Yang, C.; Duan, Y.H.; Yang, X.F.; Liu, C.X. Identification of Hydrobiogeochemical Processes Controlling Seasonal Variations in Arsenic Concentrations Within a Riverbank Aquifer at Jianghan Plain, China. *Water Resour. Res.* **2018**, *54*, 4294–4308. [CrossRef]
86. Zarnetske, J.P.; Haggerty, R.; Wondzell, S.M.; Bokil, V.A.; González-Pinzón, R. Coupled transport and reaction kinetics control the nitrate source-sink function of hyporheic zones. *Water Resour. Res.* **2012**, *48*. [CrossRef]
87. Urbain, V.; Mobarry, B.; de Silva, V.; Stahl, D.A.; Rittmann, B.E.; Manem, J. Integration of performance, molecular biology and modeling to describe the activated sludge process. *Water Sci. Technol.* **1998**, *37*, 223–229. [CrossRef]
88. Herold, M.; Greskowiak, J.; Ptak, T.; Prommer, H. Modelling of an enhanced PAH attenuation experiment and associated biogeochemical changes at a former gasworks site in southern Germany. *J. Contam. Hydrol.* **2011**, *119*, 99–112. [CrossRef]
89. Brauner, J.S.; Widdowson, M.A. Sequential electron acceptor model for evaluation of bioremediation of petroleum hydrocarbon contaminants in groundwater. *Ann. N. Y. Acad. Sci.* **1997**, *829*, 263–279. [CrossRef]

processes

MDPI

Review

Lead (Pb) Contamination in Soil and Plants at Military Shooting Ranges and Its Mitigation Strategies: A Comprehensive Review

Zafer Alasmary

Soil Science Department, College of Food and Agricultural Sciences, King Saud University, P.O. Box 2460, Riyadh 11451, Saudi Arabia; zalasmari@ksu.edu.sa; Tel.: +966-50540-6503

Abstract: Heavy metals, especially lead (Pb), is the major cause of pollution in the military shooting range soils. Bullets, which are primarily made of Pb, are a substantial source of this pollution. On speciation, this Pb is distributed into its different metal forms. Different physicochemical properties of the soil like pH, moisture content, cation exchange capacity (CEC), and organic matter play a very crucial role in the distribution, transformation, and bioavailability of the Pb. The concentration of Pb found in different shooting ranges is examined. Moreover, bullet weathering and the availability of contaminants in the soil are influenced by the physicochemical properties of the soil. For the management of firing range pollution, a variety of strategies have been investigated, including soil washing, phytoremediation, and chemical stabilization. This review focus on the pollution status of different shooting ranges, the impact of the physicochemical properties of soil on the distribution, speciation, and transformation of Pb, and different mitigation strategies to control Pb pollution in military shooting ranges.

Keywords: shooting ranges; Pb; soil; physicochemical properties; Pb speciation; mitigation strategies

1. Introduction

Heavy metals such as Pb, especially from shooting ranges, regularly contaminate the soil [1–3]. The average shot has 97% of Pb, 2% of antimony, 0.5% of arsenic, and sporadically, nickel [3,4]. The use of Pb in explosives has a long history because of its unique qualities. Compared with less dense materials, its high density gives projectiles more momentum and range. In addition, due to a low melting point of 327.4 °C, it is easy to mold into bullets and shoot. Its low cost, malleability, and corrosion resistance help to reduce gun barrel abrasion [5]. According to [6] bullets, which are primarily made of Pb, are a substantial source of this pollution. Pb emerges as the primary contaminant, with concentrations reaching up to 150 g kg^{-1} in the soils from military shooting ranges [2,3,7]. Due to the annual deposit of 60,000 tons of Pb present on shooting range grounds in the US, the United States Environmental Protection Agency has designated Pb residues in such soils as hazardous materials [8]. In particular, in Sweden, shotgun ammunition uses 500–600 tons of Pb yearly. In the USA, ammunition, shots, and bullets included approximately 75,100 and 73,900 tons of Pb, respectively, in 2011 and 2012 [9]. When fired, bullets and shots build up in shooting range soils where they can undergo a variety of changes, including oxidation and hydration, under the right circumstances [10]. Pb pollution concerns are highlighted by the startling 58,300 tons of Pb that are dumped into the soil annually in the

USA alone [11]. The increased Pb values reported by many countries, including Sweden, England, Denmark, and Finland range from 10,000 mg/kg to 100,000 mg/kg [1].

Moreover, according to [1], Pb concentrations in the shooting range soils range from 11 to 1000 mg kg^{-1} and can even surpass 20,000 mg kg^{-1}. The current levels greatly surpass the regulatory limit of 300 mg Pb kg^{-1} soil set by the European Council Directive (1986) and beyond the overall screening requirement of 400 mg Pb kg^{-1} soil set by the United States. The number of military and nonmilitary firing ranges in the United States is estimated to be between 16,000 and 18,000, while the annual environmental release of Pb ammunition in Europe is estimated to be in the "tens of thousands" [11,12]. Meanwhile, it has been demonstrated that in shooting range soils, the solubility of secondary minerals that form a weathering crust around the corroding bullets limits the activity of Pb^{2+} in solution [3]. Some examples of these compounds are $PbCO_3$ (cerussite), $Pb_3(CO_3)_2(OH)_2$ (hydrocerussite), PbO (massicot or litharge), and to a lesser extent, $PbSO_4$ (anglesite), which can be found in soil, depending on the condition [13]. Typically, anglesite (Pb sulfate, $PbSO_4$), cerussite (Pb carbonate, $PbCO_3$), and minium (Pb (2^+, 4^+) oxide, Pb_3O_4) are produced by oxidizing galena near the surface with other minerals [7]. Pb sulfide, often known as galena (PbS), is a frequently found main component in sulfide ore deposits. Weathering induces the gradual oxidation of galena by air oxygen, resulting in the formation of either sulfate (anglesite) or carbonate (cerussite) through the chemical reaction [8] as described below:

$$PbS + 2O_2 \rightarrow PbSO_4 \text{ (Pb sulfate: anglesite)}$$

$$PbS + CO_2 + H_2O + 2O_2 \rightarrow PbCO_3 + H_2SO_4$$

Previous studies have shown that cerussite is usually formed at a pH higher than 6, and anglesite is formed at a pH lower than 6 [14,15]. Upon the dissolution of Pb from primary minerals, it has a tendency to either mix with carbonate or sulfate ions to produce insoluble Pb carbonate or Pb sulfate, or alternatively, it can be absorbed by ferric hydroxide [10]. Furthermore, soil organic matter, clay minerals, and manganese or iron oxides can significantly bind aqueous Pb^{2+} through adsorption mechanisms [11]. The use of Pb-containing ammunition on shooting ranges significantly increases the danger of Pb pollution in nearby agricultural areas [14]. Because of its insolubility and resistance to microbial degradation, Pb has a prolonged residence time in soils, which contributes to its persistence over an extended period of time. Pb in soil has a half-life of 740 to 5900 years. Pb background levels in soils typically vary between 10 and 30 mg/kg [16]. Meanwhile, Pb concentrations in soils have been observed to reach as high as 1104 mg/kg because of shooting practices [17]. It is possible that there are 1108 spent Pb bullets per hectare in the soil at shooting ranges. Pb concentrations in shooting range soils frequently exceed legal limits and pose a serious risk to human health [18]. The intake of poisoned plant products or exposure to Pb through contaminated soils can cause serious health problems and even death [18,19].

Elevated Pb levels can disrupt ecosystem functionality by adversely affecting essential soil organisms like microbes and nematodes, leading to impaired processes such as organic matter decomposition and nutrient cycling [20]. These metalloids of Pb undergo transformation from bullet fragments into more mobile forms and accumulate in a range of organisms, including earthworms, plants, and birds [21]. The extensive soil contamination observed across different shooting ranges underscores a significant ecological risk [22,23]. To assess the environmental impact of accumulated pollutants, bioassays, particularly short-term ones like seed germination and root elongation tests, are employed to evaluate their effects [23]. The guidelines of the Organization for Economic Cooperation and Development (OECD) recommend lettuce (*Lactuca sativa* L.) as a suitable test plant for chemical testing. Previous seed germination tests using *L. sativa* have indicated that elevated metalloid contamination in shooting range soil contributes to phytotoxicity [24]. Considerable environmental and health problems are associated with the use of Pb in munitions resulting in soil pollution at shooting ranges. Therefore, it is essential to comprehend how Pb behaves in soils, how it is absorbed by plants, and how it could affect the food chain to adopt efficient regulatory measures and lessen the negative effects of Pb pollution. This review focuses on synthesizing the current knowledge of the fate and ultimate risks presented by Pb associated with military training activities. Our focus will be on understanding and identifying trends in the literature regarding Pb metal toxicity in shooting ranges. We will primarily focus on its abiotic interactions with soil and its biotic transformation by plant uptake. This review is based on Pb derived from small-arms range bullets as the most likely source of contamination from military training. Objectives of the review paper are as follows:

- To provide a comprehensive assessment of the potential risk of Pb on the vegetation growing in shooting ranges
- To examine the effect of Pb on the physiochemical properties of soil

2. Accumulation and Distribution of Pb in Shooting Range Soils

Shooting ranges have gained notoriety as significant sources of Pb pollution in the environment, ranking second only to the battery industry in Pb emissions [25]. Recent research has intensified the focus on shooting ranges as Pb pollution culprits, particularly due to the escalating use of small-arms ammunition for shooting practices [2]. Once Pb from bullets and shots enters the soil, it undergoes various chemical and physical transformations that enhance its mobility and bioavailability, potentially leading to elevated Pb concentrations in shooting range soils [8]. Studies have consistently shown that shooting ranges accumulate Pb at concentrations significantly exceeding natural levels, which typically range from 10 to 30 mg/kg [26]. Notably, Pb pollution from ammunition is a global issue, with documented extreme Pb levels in shooting range soils worldwide [25]. For instance, levels as high as 97,600 mg kg^{-1} have been recorded in the US, 29,200 mg kg^{-1} in Japan, 66,961 mg kg^{-1} in Canada, 38,386 mg/kg in Botwana, 206,600 mg/kg in New Zealand, and 300,000 mg/kg in the Netherlands [2,27]. Recent years have shown an alarming increase in the buildup of Pb in shooting range soils, with some ranges registering concentration as high as 300,000 mg/kg, and this Pb accumulation is alarming [28]. Soil properties, shooting frequency, and the age of shooting ranges play significant roles in influencing Pb concentration levels, making comprehensive assessment and remediation strategies imperative [29]. The longer that spent bullet and shoot remain in shooting range soil, the higher the potential for the oxidation process to transform them into more soluble mineral forms, further exacerbating Pb mobility and bioavailability [25]. Moreover, it has been observed that rifle shooting ranges have a tendency to accumulate more Pb than pistol shooting ranges, mostly because rifle ammunition contains more Pb [3]. Table 1 shows the concentration of Pb in various shooting ranges of the world.

Table 1. Pb contamination in the soils of shooting areas across the different countries.

Location and Year of Study	Number of Studied Shooting Ranges	Concentration of Pb Found in Soil (mg/kg)	References
Switzerland (2010)	2	621 ± 12.35	[29]
Norway (2010)	1	22,000 ± 31.47	[30]
USA (2011)	3	20,350 ± 21.62	[31]
Canada (2011)	4	24,100 ± 78.54	[32]
South Korea (2012)	1	4625 ± 19.03	[33]

Table 1. *Cont.*

Location and Year of Study	Number of Studied Shooting Ranges	Concentration of Pb Found in Soil (mg/kg)	References
Australia (2012)	4	6403 ± 00	[2]
Finland (2012)	2	30,300 ± 113.48	[34]
South Korea (2013)	1	11,800 ± 52.11	[3]
Norway (2013)	5	30,000 ± 163.90	[35]
China (2014)	1	2563 ± 41.53	[36]
Netherland (2014)	1	2158 ± 55.10	[37]
Australia (2015)	3	30,600 ± 134.88	[38]
Finland (2015)	2	23,000 ± 105.49	[18]
China (2015)	3	6019 ± 45.18	[39]
Nigeria (2016)	1	17,500 ± 59.74	[40]
Australia (2016)	4	2145 ± 21.92	[41]
Spain (2016)	1	724 ± 12.47	[42]
Pakistan	1	1331 ± 19.04	[43]
South Korea (2017)	1	3436 ± 49.13	[44]
Norway (2017)	7	2700 ± 25.97	[32]
Nigeria (2017)	1	26,933 ± 00	[45]
Spain (2017)	1	710 ± 19.16	[46]
Switzerland (2017)	2	620 ± 10.41	[47]
Botswana (2017)	1	38,300 ± 158.51	[26]
Lithuania (2018)	1	654 ± 45.12	[48]
Alaska (2019)	1	2800 ± 27.49	[49]
Poland (2019)	1	3800 ± 14.03	[50]

Additionally, Pb accumulation and dispersion in shooting range soils are significantly influenced by soil characteristics, such as pH, organic matter (OM), CEC, soil type and moisture, etc. [51]. Meanwhile, it has also been suggested that the frequency of shooting activities and the age of the shooting range are key factors affecting Pb concentration in soils [26].

3. Factor Influencing Pb Dispersal

The speciation of Pb (distribution of its different metal forms) within shooting range soils is influenced by various factors, including soil properties such as pH, CEC and chemical processes like association–dissociation, oxidation–reduction reactions, precipitation–dissolution, sorption–desorption. These mechanisms collectively determine the mobility and bioavailability of Pb in soil; for example, soil pH affects Pb bioavailability, with lower pH leading to increased dissolution and mobility of Pb. CEC plays a crucial role, with higher CEC resulting in greater binding and retention of Pb in the soil. Additionally, chemical processes like sorption–desorption and precipitation–dissolution affect how Pb interacts with soil particles and compounds. Understanding these complex interactions is essential for managing Pb contamination in shooting range soil and assessing its environmental impact [52]. Moreover, both pH and EC are crucial factors which affect Pb speciation, but generally, pH has a dominant role in this speciation due to its link with wide-ranging effects on dissolution–precipitation reactions, adsorption–desorption processes, and bioaccessibility across different digestive environments [49,52].

4. Effects of Soil Physicochemical Properties in Accumulation of Pb

The physical and chemical properties of soil play a crucial role in influencing the distribution, mobility, solubility, bioavailability, bioaccessibility, and fate of Pb in shooting range soils. Notably, higher soil moisture levels, acidic soil pH, and increased soil organic matter create conducive environments for the weathering and conversion of metallic Pb projectiles into more reactive secondary Pb compounds. These factors significantly affect the behavior of Pb in shooting range environments [8].

4.1. Effect of Soil pH

Soil pH has a significant impact on solubility, mobility, and bioavailability of Pb in soils near shooting ranges, as supported by numerous studies [5]. According to [53] the presence of acidic soils in these environments creates a favorable environment for a number of processes linked to the corrosion, weathering, and alteration of Pb bullets and shoots. According to [54] a lower pH, which dissolves the crust of secondary minerals that forms on the surface of metallic Pb, speeds up the conversion of metallic Pb shoots and bullets into secondary minerals and exposes new metallic Pb surfaces to additional weathering [55]. Additionally, the acidic nature of soil having high moisture content can cause the dissolution of Pb minerals found on the surface of bullets, enabling the leaching of Pb into subsoils. Shooting ranges have acidic soil that range in pH from 5.4 to 6.4; for instance, it was found that Pb migrated from the surface soil to subsoils at a rate of 7.5% to 46%, compared to only 6% to 18% in a shooting range of alkaline pH 9.3. Additionally, acidic sandy soils at the surface show a lower affinity for heavy metals like Pb and frequently have unsaturated sorption sites for Pb due to the limited buildup of Pb on the surface of soil [55,56]. It is significant to remember that when hydrogen ions are consumed in the presence of oxygen, the corrosion of metallic Pb bullets and shoots within acidic soils can lead to an elevation in the pH of the nearby area [11].

In conclusion, the mobility and bioavailability of Pb in shooting range soils are significantly influenced by soil pH; while alkaline circumstances seem to increase Pb stability and retention in these environments, acidic ones tend to increase Pb mobility.

4.2. Effect of Soil CEC

A key factor in determining a soil's binding capacity is its soil CEC, which measures the number of exchangeable cations that a specific soil mass can adsorb. Pb has a lower mobility and bioavailability in soils with higher CEC due to their greater propensity to bind with Pb. Shooting ranges with high CEC have the potential to collect substantial Pb concentrations, despite the fact that this relationship is not universally applicable [8]. Cao et al. [57] compared two shooting ranges in Florida. A high CEC of 42.3 cmolkg^{-1} was found to be correlated with a total Pb concentration of 48,300 mg kg^{-1} at the MPR shooting range. While the CEC of 8.52 cmolkg^{-1} with a total Pb concentration of 12,710 mg kg^{-1} was found at the TRR shooting range in Florida. Additionally, a study by [58] showed that a rifle shooting range with a CEC of 10.75 cmol/kg had a total Pb concentration of 38,386 mg/kg, whereas a comparable range with a CEC of 8.00 cmol/kg showed a total Pb concentration of 25,193 mg/kg [59]. In investigations addressing Pb contamination at various shooting ranges, CEC values were found to vary from 1.05 to 50.1 meq 100 g^{-1}. The total Pb values varied accordingly, ranging from 5040 to 60,600 mg/kg [10]. It is crucial to stress that Pb mobility, bioavailability, and density are not solely controlled by soil CEC. The weathering and transformation of Pb are also influenced by organic matter and soil pH [2]. An Australian study that found variability in Pb contamination levels and CEC across various shooting ranges serves as an illustration of this intricacy.

4.3. Effect of Soil Moisture

Pb's weathering, mobility, transformation, and fate are significantly influenced by soil moisture, a fundamental physical feature of soil. The circumstances for Pb shots and bullets to corrode and erode are caused by soil moisture. Reactive gasses like oxygen and carbon dioxide can diffuse to the surface of the Pb, and moisture helps to generate a thin watery layer on the shoots' surface. As a result of this diffusion process, the weathering compounds of Pb such as Pb oxides, carbamates, phosphates, and sulfates are produced, which eventually leads to their solubilization and dissolution [26]. Liu et al. [53] conducted a study to determine the impact of moisture present in the soil, on weathering, and the transformation rate of metallic Pb powder. The research showed that during the course of seven days, the metallic Pb in X-ray diffraction (XRD) spectra vanished and was replaced by hydrocerussite, cerussite ($PbCO_3$), and massicot (PbO). The same results were seen in the field experiment where all metallic Pb was converted into hydrocerussite, cerussite ($PbCO_3$), and massicot (PbO). This shows that abraded Pb in shooting range soils experiences a rapid change into secondary Pb compounds. However, in contrast to the Pb powder used in the above experiment, a typical shoot may take many years to go through this transformation into secondary Pb minerals. The amount of moisture in the soil has a significant effect on how Pb shots turn into more reactive substances like cerussite. The surface of Pb shots develops a coating of cerussite when soil moisture is present, and this layer combines with water in watery circumstances to form hydrocerussite, as shown by the equation below:

$$3PbCO_3(s) + H_2O(l) \rightarrow 4Pb_3(CO_3)_2(OH)_2(s) + CO_2(g) \text{ (hydrocerussite)}$$

Additionally, the cerussite and hydrocerussite layers on the Pb shots and bullets dissolve and transform into other Pb compounds when soil moisture and other favorable soil conditions, such as soil pH, are present. This exposes the Pb shot surface to further weathering and transformation. Pure water does not erode the metallic Pb shoot or bullet, but it acts as a medium for the diffusion of oxidizing gasses to metal surface and causes the creation of secondary Pb compounds [60].

4.4. *Effect of Soil Texture*

The distribution of soil particles, like clay, silt, and sand within the mineral composition of soil is referred to as soil texture, and it is crucial in determining how Pb changes and ends up in shooting range soils [51]. Due to their large pores, sandy soils have higher infiltration rates and less water retention than clay soils, which result in varied moisture-holding capabilities, with clay soils having the highest capacity and sandy soils having the lowest. Additionally, compared to sandy and silt soils, clay soils often have higher organic matter (OM) contents [61,62]. The type and texture of the soil significantly influence the rate of the weathering of Pb bullet and shoot soil predominantly composed of sand in berms, and bullets incubated in sandy soils exhibited lower total Pb concentrations compared to clay soils due to slower weathering rates. By forming a thin layer on the surfaces of Pb bullets and shoots, allowing reactive gasses (CO_2 and O_2) to penetrate, corrosion is sped up and Pb weathering is aided. Pb bullets are further chemically weathered by soils with lower pH and more OM, which causes metallic Pb to change into ionic Pb [63].

In a study conducted by [64] the total Pb concentration present in sand berms was significantly lower than that in the soil berms as a result of slower weathering rates, which were attributed to a lower moisture retention capacity and a lower organic matter content, both of which are essential elements in the transformatiokon of metallic Pb into ionic Pb [65]. Pb shotgun pellets undergo speciation during periods of 6 to 25 years, with Pb shots or bullets in shooting range surroundings generating a protective layer of Pb compounds [49]. Reduced particle size, on the other hand, may lead to enhanced weathering rates of Pb powder, inhibiting the creation of a protective coating and promoting total conversion into secondary Pb minerals [66].

5. Environmental Impacts of Pb

Significant environmental and health issues have been highlighted by the accumulation of Pb in shooting ranges, which is a naturally occurring heavy metal that makes about 90–99% of the core composition of small-weapons projectiles [25]. Over the past 50 years, extensive research has been performed on Pb mobility and its density in shooting range soils [32]. The concentration of 23 and 80 g/kg of Pb can be found in these soils [32]. Pb is very pH-dependent when it comes to solubility, being most soluble at high pH values of about 10.5 [20]. Despite elevated Pb concentrations in the soils of shooting ranges, surface and pore waters in the vicinity generally maintain low Pb levels. This low Pb level is maintained due to circumneutral pH conditions, the formation of less soluble Pb compounds, and abundant surfaces for Pb adsorption [65,67]. Pb exposure carries significant health hazards, especially for young infants who are more susceptible to Pb absorption. Behavioral problems, difficulties with learning, kidney failure, hearing and vision impairment, stunted growth, brain damage, and hearing loss are health problems linked to Pb exposure [25]. Pb exposure in adults can lead to memory loss, high blood pressure, joint pain, nervous system problems, kidney and intestinal disorders, and pregnancy complications [68]. The effects of Pb in small-arms ammunition on groundwater and surface water are generally regarded as insignificant, with surface water contamination primarily happening during periods of heavy precipitation [69]. Several environmental parameters, such as soil characteristics, rainwater percolation, and co-contaminants, have an impact on Pb mobility [1]. Pb mobility is influenced by the bullet's composition, with older ammunition showing slower rates of corrosion and dissolution than newer Pb alloy ammunition. Pb species are mainly found in carbonate and oxide complexes, such as hydrocerussite, cerussite, litharge, and massicot, in shooting range soil. Based on the mineralogical circumstances, Pb can also attach to sulfur and phosphorus [70].

6. Mechanisms and Effects of Pb Toxicity in Plants

The accumulation of Pb in plants and crops close to shooting range is a serious issue since it could enter the food chain. High levels of Pb have been identified in a number of plant tissues, including Vetiver grass tissue (1390–1450 ppm/kg) and plant leaves (up to 70 mg/kg). Depending on the plant type and the manner of Pb absorption, plant roots can absorb Pb concentrations ranging from 1347.1 to 3825.8 mg kg^{-1} (dry weight) [20]. The concentration of Pb in the soil has a direct impact on the pH, electrical conductivity, and mineral composition of plants [71]. According to research, plants close to shooting ranges frequently exceed the limits advised by groups like the WHO and FAO [51]. Increased soil Pb levels can have a negative impact on soil quality and agricultural production. Even minute levels of Pb can interfere with a variety of biological processes, reducing plant development and agricultural yields. Reactive oxygen species (ROS) such as hydrogen peroxide (H_2O_2) can be produced as a result of Pb exposure. ROS can cause oxidative stress, enzyme inactivation, and lipid peroxidation, all of which lower agricultural productivity [72]. Malondialdehyde (MDA), a byproduct of lipid peroxidation that causes oxidative stress in plants under high-stress conditions, can also be produced by Pb exposure [73]. After contamination, effective remediation is necessary to return soil Pb levels to normal. Critical plant activities such as cell division, chlorophyll production, transpiration, root elongation, seedling growth, and seed germination are all negatively impacted by Pb [50]. Through interactions with metabolic enzymes and the displacement of necessary ions, Pb poisoning alters the permeability of cell membranes [73]. Additionally, Pb exposure causes an overproduction of ROS, which inhibits ATP synthesis, damages DNA, and exacerbates lipid peroxidation. To combat these effects and prevent against oxidative damage, plants have evolved defensive mechanisms, such as enzyme systems and non-enzymatic antioxidant systems [74]. Through passive ion exchange mechanisms, plants make it easier for Pb to be taken up from the soil. Then this Pb moved from roots to shoots. Different plant tissues accumulate Pb differently, with roots often having higher quantities than stems and shoots. Pb is largely transported from roots to shoots via the root apoplast; Pb absorption is influenced by parameters such as root surface area and mycorrhization. Strong Pb binding in lignified root tissues restricts Pb transport to plant tissues farther from the roots [50]. In some instances, Pb deposition on leaf surfaces can cause the total Pb concentration in plant leaves to be relatively high, which can affect the plants' capacity to absorb and accumulate Pb. Differential Pb accumulation in plant organs has effects on soil remediation techniques and raises the possibility of Pb re-entry into the food chain when herbivores consume polluted plant matter [20].

Plant Tolerance Against Pb Toxicity

Although it has no nutritional value and is not necessary for plant growth, Pb is naturally found in some plant species at concentrations that range between 2.1 and 2.5 mg/kg (dry weight) [20]. However, it has been determined that soil Pb values between 100 and 500 mg/kg are harmful to plants. Similarly, Pb concentrations in plant tissues between 30 and 300 mg/kg are regarded as hazardous and can have negative impacts on plant growth, photosynthesis, dry weight, and root development [75]. Significant growth loss in pine trees under active shooting range circumstances has been noted and is attributed to damaged roots and mycorrhizal fungi. After the shelling stopped, pine trees showed signs of recovery and even grew taller than the trees at the control site. Pb accumulation in plant roots can impede root growth by preventing root tip cell division [18]. Various plant species growing in Pb-polluted soil from shooting ranges have shown suppression of root elongation. Microtubule damage from high Pb concentration in plant roots can cause blockages in prometaphase cells. Pb exposure in plants can cause chlorosis and stunted growth, as well as other harmful effects on the physiology of the plant [50]. The physiological functions of hormones, electron transport, membrane structure, and water absorption are all inhibited by pb toxicity in plants. Because abscisic acid concentrations rise with elevated Pb levels, stomata close. Pb can interfere with the activities and tertiary structure of enzymes, especially those with thiol groups (-SHs) or carboxyl groups (-COOHs). By additionally blocking carboxyl groups (-COOHs) in enzymes, Pb ions in plant tissues can further decrease enzyme function [76]. Reduced chlorophyll synthesis, interference with the uptake of vital nutrients like magnesium and iron, decreased photosynthesis rates due to chloroplast and chlorophyll degradation, and Pb-induced stomatal closure, which restricts the amount of carbon dioxide (CO_2) available, are all effects of Pb toxicity in plant leaves [76]. Pb has the ability to impair respiratory procedures and lessen the formation of adenosine 5′-triphosphate (ATP). Additionally, it may prevent cations and nitrate ions from entering plant roots, which would interfere with nutrient intake and nitrogen metabolism [25]. Pb exposure results in elevated reactive oxygen species (ROS) levels in plant tissues, which leads to oxidative stress and damages cell membranes and causes lipid peroxidation [76]. Despite the toxicity of Pb, certain plants have different strategies for tolerating high Pb accumulation. These plants have the option of using the "accumulator" or "excluder" approaches. By precipitating Pb into less harmful forms or tying it to the carboxylate groups (COOHs) of uronic acid, excluder plants keep low, non-toxic Pb concentrations in their tissues, preventing roots from absorbing it. In order to reduce Pb toxicity, accumulator plants aggressively concentrate Pb in their tissues at a wide range of soil concentrations. These resilient plants are essential for phytoremediation, which helps to regulate and control Pb pollution in shooting ranges [6]. Depending on their particular physiological adaptations, they can either accumulate or exclude Pb. For example, *Equisetum arvense* displays excluder characteristics by maintaining low Pb concentrations in plant tissues even at high soil Pb levels. Other plants, such as *Lactuca sativa*, tolerate Pb toxicity due to high organic matter content that complexes Pb and reduces its toxicity. These plants play a vital role in mitigating Pb pollution in shooting ranges through phytoremediation efforts [77].

7. Mitigation Strategies for Pb-Impacted Soil in Military Shooting Ranges

7.1. Stabilization

In order to reduce Pb's solubility, bioavailability, and mobility, stabilization is a frequently used remediation strategy for Pb-contaminated soil [78]. Due to its affordability, environmental friendliness, adaptability, simplicity, and effectiveness, this technology has attracted a lot of attention in real-world applications [79]. As stabilizing agents, a variety of substances, such as fly ash, cement, blast furnace slag, phosphate rock, red mud, and steel are frequently used [80]. The operation of cement-based stabilization through precipitation and physical and chemical fixing mechanisms has drawn particular interest for Pb-contaminated soils [81]. It is renowned for its high energy usage and carbon footprint [82]. As potential substitutes for immobilizing Pb-contaminated soil, phosphate-based materials like phosphoric acid and phosphate rock have demonstrated their potential. These have the capacity to transform Pb into stable forms like pyromorphite [$Pb_5(PO_4)_3(OH, Cl, F)$] [82]. Phosphate rock encourages the development of phosphate minerals in Pb-contaminated soil, greatly lowering Pb's water solubility, phytoavailability, and bioaccessibility [83]. Phosphoric acid can produce molecules that resemble chloropyromorphite, which can further reduce the solubility and bioavailability of Pb in highly contaminated soil when paired with KCl as a source of chloride ions. When utilizing phosphate-based chemicals to increase the solubility of phosphate and Pb-related phases, an acidic environment is frequently necessary [20].

According to research, Pb solubility in soil can be significantly decreased by simultaneously applying compounds that are both acidic and phosphate based. According to Li et al. [77] phosphate rock activated with oxalic acid (OA) demonstrated better Pb conversion from non-residual to residual fractions, hence increasing its stabilization efficiency. Additionally, according to XRD analysis, a phosphate- binder made of monopotassium phosphate, reactive magnesia, and OA-activated phosphate rock effectively reduced Pb leachability from polluted soil by forming stable Pb-related precipitates like fluoropyromorphite ($Pb_5(PO_4)_3F$) [84].

7.2. Phytoremediation

Researchers have focused on practical and long-lasting solutions in their search for environmentally friendly methods to reduce Pb pollution in shooting range soils. A highly effective and environmentally beneficial method for cleaning up severely damaged soils is phytoremediation [68]. In order to handle a variety of environmental contaminants, like trace metals, heavy metals, organic chemicals, and radioactive substances, phytoremediation strategically employs modified green plants. The purpose of this strategy is to make it easier to remove, immobilize, contain, and stabilize these contaminants inside the soil matrix [85]. The goal of phytoremediation is to improve soil by enhancing the uptake and transport of contaminants through plants through a variety of plant-driven mechanisms, including chemical, biological, and physical processes. Phytostabilization, phytoextraction, phytovolatilization, and rhizofiltration are important phytoremediation methods [68,85]. Pb is immobilized in the soil by plants through a process known as phytostabilization [2]. This is accomplished by Pb being absorbed and precipitated in the root matrix, which effectively reduces its mobility within the soil matrix. Phytoextraction is the process in which plant roots actively absorb the Pb from the soil, and then translocate it into the above-ground biomass [68]. Pb is taken up by plants and then released as secondary Pb products by transpiring in the atmosphere, and the process is known as phytovolatilization. This process is particularly prominent in plants that are actively growing because they release Pb through transpiration through their leaves after absorbing it along with water [68].

Recent research has demonstrated how phytoremediation may successfully treat shooting range soils. For instance, Dinake et al. [15] found that Pisum sativum had extraordinary phytoextraction capabilities, attaining an impressive 96.23% Pb removal efficiency from shooting range soil contaminated with more than 1331 mg/kg of Pb. *Agrostis capillaris* displayed efficient phytoremediation in a research study by Tariq et al. [42] on a Spanish shooting range by absorbing 1107 mg/kg of Pb through its roots and translocating roughly 135 mg/kg into its shoots. Grass pea plants (*Lathyrus sativus* L.) are also used for phytoremediation. Similar findings were made in the United Kingdom by Sneddon et al. [86] who discovered that *Lolium perenne*, a perennial ryegrass, absorbed 38 mg/kg of Pb into its shoots when it was growing in soil contaminated with Pb at concentrations ranging from 43.89 to 159.98 mg/kg of Pb. Compared to other soil remediation methods, including chemical and physical amendments, phytoremediation has clear advantages. By preserving soil structures and habitats for living things, it minimizes the environmental impact compared to other soil removal techniques. It is also economical, as it does not introduce foreign chemicals into the ecosystem [20]. Additionally, plants utilized in phytoremediation assist in reducing the effects of rainfall runoff and soil erosion by stabilizing the soil through their root systems. The release of chemicals into the soil by plant roots served as nutrition for bacteria in the rhizosphere emphasizes the symbiotic relationship between plants and soil microbial communities. In comparison to areas farther from plant roots, this area has a higher microbial density [87]. The United States Environmental Protection Agency (USEPA) promotes phytoremediation as a practical strategy for reducing and managing Pb pollution in shooting ranges due to its wide range of advantages [68]. Further advantages and disadvantages are mentioned in Table 2.

Table 2. A brief comparison of the Pb mitigation strategies at shooting ranges.

Mitigation Strategy	Descriptions	Advantage	Disadvantage	Reference
Chemical stabilization	This process involves the addition of stabilizing agents, such as phosphates, silicates, or carbonates, to the contaminated soil. These agents react with the Pb ions to form stable, insoluble compounds which are less mobile and less bioavailable. Generally, precipitation, adsorption, and complexation mechanisms are involved in this process.	• Cost effective • In Situ treatment • Rapid treatment • Long-term stability • Cost effective	• Soil specific • Can deteriorate soil quality • Potential for rebound	[88]
Phytoremediation	Phytoremediation involves the use of specific plants known as hyperaccumulators, which have the ability to absorb and accumulate heavy metals like Pb from the soil through their roots. The Pb is then translocated to the above-ground parts of the plant, where it can be harvested and removed. Overall, phytostabilization, phytoextraction, and rhizodegradation are the common mechanisms which are used in phytoremediation.	• Environmentally friendly • Cost effective • Long-term sustainability	• Limited to certain contaminants • Time consuming • Plant specific	[89]

Table 2. *Cont.*

Mitigation Strategy	Descriptions	Advantage	Disadvantage	Reference
Microbial bioremediation	Microorganisms, such as bacteria and fungi, are utilized to degrade, transform, or immobilize environmental pollutants, including heavy metals like Pb. Biosorption, bioaccumulation, bioprecipitation, and biovolatilization are the common mechanisms which are used in microbial bioremediation.	• Cost effective • Environmentally friendly • On-site treatment • Low environmental impact	• Limited effectiveness • Site specific • Time consuming • Limited to biodegradable contaminants	[90]
Soil washing	It is an ex situ remediation technique used to remove contaminants, such as heavy metals, from soil. Overall, soil excavation, physical separation, chemical extraction, separation, and washwater treatment are the general steps which are followed in this practice.	• Highly effective • Reduce the volume of contaminated soil • Adjustable according to soil type	• Energy consumption • Soil disruption • Residual contamination	[91]

7.3. Microbial Bioremediation

In combating Pb toxicity, bioremediation using bacterial community plays an important role. In addressing Pb toxicity, bioaugmentation stands out as a widely studied technique. Bioaugmentation involves the deliberate introduction of specific microbial communities to augment the native microbial populations at contaminated sites, thereby enhancing the degradation of contaminants or toxic substance [92]. Native microbes present at contaminated sites exist in limited quantities and are often insufficient for effective contamination reduction. Bioaugmentation addresses this limitation by introducing a substantial number of targeted microbes to facilitate comprehensive biodegradation. Microbial methods can effectively remediate soil contaminated with heavy metals. For instance, the bacterium *Rhodobacter sphaeroides* contributes to altering the speciation of Pb, converting it into inert forms with reduced bioavailability. Achal et al. explored the mechanisms behind Pb absorption by the bacterium *Leclercia adecarboxylata*, highlighting the role of extracellular polymeric substances (EPS) produced by these bacteria in binding with Pb, preventing its entry into the cell [93]. *Kocuria flava* employs calcite formation via urease, a prominent enzyme, to chelate Pb effectively, reducing its availability [94]. Bacteria, including *Kocuria flava*, are vital in Pb bioremediation due to their ability to sequester Pb, modify its oxidation state, or induce precipitation [95]. *Pseudomonas aeruginosa* has ability to reduce Pb concentrations in soil, emphasizing the potential of bacteria in Pb-contaminated soil remediation [96]. Microalgae like *Phormidium ambiguum* and *Scenedesmus quadricauda* (Chlorophyta) have been utilized for Pb removal from freshwater [97]. Fungi offer an alternative approach to mitigate Pb stress, employing distinct detoxification mechanisms compared to other organisms [98]. These mechanisms encompass intracellular processes involving polyphosphates, sulfur compounds, peptides, organic acids, and intracellular compartmentalization, as well as extracellular mechanisms like precipitation, chelation, and cell wall binding, which play a pivotal role in metal detoxification [99].

7.4. Chemical-Assisted Soil Washing

Chemical-assisted soil washing is a remediation that effectively removes particular contaminants from soil by using a variety of chemical agents, such as acids, bases, salts, chelators, and surfactants [94]. Chelators have shown significant potential in removing harmful metals from polluted soil during the washing process, among various chemical agents. Chelators can remove target pollutants from the soil surface by forming stable and soluble metal–chelator complexes [95]. Due to the durability of metal–EDTA complexes, ethylenediaminetetraacetic acid (EDTA) has traditionally been a preferred chelator, especially for remediating Pb-contaminated soils [57,98]. However, issues have been raised regarding EDTA's poor biodegradability and prolonged persistence in soils, which could result in secondary pollution and adverse effects on the ecosystem [57,96]. In consequence of these issues, biodegradable chelators are becoming more popular as EDTA substitutes for the remediation of metal-contaminated soil [88,89,97]. It is crucial, but unexplored, to compare the efficacy of biodegradable chelators to EDTA for remediating Pb-contaminated shooting range soil. Chelator-assisted soil washing successfully removes metal pollutants, but it may interfere with the binding of residual metals with the soil components, thus increasing their mobility and leaching [73,99].

Combining chemical immobilization with chelator-assisted washing offers a workable solution to these problems. Through the effective extraction of labile Pb from contaminated soil, the solubility of the remaining Pb in the washed-soil residue is subsequently decreased. This approach has been successful with biodegradable chelators like EDDS, GLDA, and HIDS, which offer a more sustainable option to persistent chelators like EDTA. Following chelator-assisted washing, residual Pb is only slightly mobile after a two-step post-treatment involving $FeCl_3$ and a CaO solution. A potential remediation solution for Pb-contaminated shooting range soil is provided by this combined approach [29].

8. Planned Future Use of Shooting Range Site

The planned future use of a shooting range site plays a critical role in shaping the Pb mitigation strategy for several reasons. This future use impacts risk assessment by setting different cleanup standards; for example, sites intended for residential or recreational purposes need more stringent decontamination compared to those for industrial use. The objectives of remediation are directly influenced by what the land will become, with residential areas requiring a deeper clean than commercial spaces [67].

The economic aspect of remediation is also tied to future use, where more intensive cleanup might be justified for high-value land. Regulatory compliance is another factor; understanding the end use helps in adhering to specific environmental laws [95]. Moreover, community acceptance can vary; for instance, methods like phytoremediation might be preferable for future parks but less so for industrial zones [33].

The sustainability of the mitigation method chosen must align with the intended use, such as using permanent vegetative caps for nature reserves but not for sites where construction is planned. Future use also dictates the need for ongoing monitoring post-remediation to ensure safety. Some remediation techniques could restrict future development options, like soil replacement, which might limit underground construction. The environmental impact of the site after cleanup is considered, with priority given to ecological restoration for conservation areas [90]. Lastly, funding for remediation is often directed toward sites with the most significant potential impact or health risks, based on their planned future use [90]. In summary, by considering the future use of a shooting range site in the planning of Pb mitigation, it can be stated that cleanup efforts are both effective and sustainable, tailored to the site's ultimate purpose, thereby optimizing resource use and minimizing environmental risks in the long term.

9. Conclusions and Future Perspectives

In summary, the presence of heavy metals, particularly Pb, in military shooting range soils represents a significant and concerning source of environmental pollution. Pb-based bullets are a primary contributor to this contamination. Through speciation analysis, it has become evident that Pb can exist in various chemical forms within these soils, and specific physicochemical soil properties, including pH, moisture content, CEC, and organic matter, heavily influence its distribution, transformation, and bioavailability. The strategies discussed include Pb recovery and recycling initiatives, soil stabilization techniques, routine soil monitoring to assess contamination levels, THE proper disposal of Pb-contaminated materials, and adopting Pb-free ammunition. It is crucial to acknowledge that addressing the pollution status of military shooting ranges is imperative, as the impact of soil physicochemical properties on Pb distribution, speciation, and transformation underscores the need for site-specific solutions. Current mitigation strategies involve containment methods like bullet traps and backstops to collect spent Pb, alongside substitution with non-Pb alternatives like steel or bismuth, though these have met resistance due to cost and performance issues. Legislative efforts and educational programs are also incrementally promoting the reduction in Pb use. Looking forward, innovative approaches are anticipated to include fully biodegradable or encapsulated ammunition to prevent Pb leaching, and advanced remediation techniques like bioremediation and nanotechnology to clean up existing contamination. The development of smart systems for the real-time monitoring of Pb levels and policy enhancements toward creating Pb-free zones in ecologically sensitive areas are also on the horizon.

However, challenges persist, including economic barriers to adopting more expensive non-Pb ammunition and cultural resistance from traditional shooting enthusiasts. Opportunities for change arise from technological advancements that could lower costs, increased public health awareness, and collaborative efforts among scientists, policymakers, and the shooting community. Future research should focus on long-term efficacy studies of new mitigation technologies and understanding the ecological impacts of alternative materials. The integration of AI and IoT for smarter environmental management, along with comprehensive predictive models for Pb dispersion, could revolutionize how shooting ranges operate, ensuring they align with both ecological and public health standards. Ultimately, the goal is to mitigate the legacy of Pb pollution, fostering environments that are safer for both human communities and wildlife.

Funding: This research received no external funding.

Data Availability Statement: Not applicable.

Acknowledgments: The author would like to express his heartfelt gratitude to the Deanship of Scientific Research at King Saud University for its generous support and encouragement of scientific research and researchers. He also greatly appreciates the editors' and anonymous reviewers' insightful comments and recommendations.

Conflicts of Interest: The author declares that he has no personal relationships or competing financial interests that could have appeared to influence the work.

References

1. Chrastný, V.; Komárek, M.; Hájek, T. Lead contamination of an agricultural soil in the vicinity of a shooting range. *Environ. Monit. Assess.* **2010**, *162*, 37–46. [CrossRef] [PubMed]
2. Smith, C. Areas of Major Concentration in the Use and Traffic of Small Arms. In *Small Arms Control*; Routledge: London, UK, 2019; pp. 83–126.
3. Vantelon, D.; Lanzirotti, A.; Scheinost, A.C.; Kretzschmar, R. Spatial distribution and speciation of lead around corroding bullets in a shooting range soil studied by micro-X-ray fluorescence and absorption spectroscopy. *Environ. Sci. Technol.* **2005**, *39*, 4808–4815. [CrossRef] [PubMed]
4. Sanderson, P.; Naidu, R.; Bolan, N.; Bowman, M. Critical review on chemical stabilization of metal contaminants in shooting range soils. *J. Hazard. Toxic Radioact. Waste* **2012**, *16*, 258–272. [CrossRef]

5. Ma, L.Q.; Hardison, D.W.; Harris, W.G.; Cao, X.; Zhou, Q. Effects of soil property and soil amendment on weathering of abraded metallic Pb in shooting ranges. *Water Air Soil Pollut.* **2007**, *178*, 297–307. [CrossRef]
6. Schindler, M.; Weatherhead, K.; Mantha, H. The release of incidental nanoparticles during the weathering of gunshot residue in soils of a shooting range in Ontario, Canada. *Can. Mineral.* **2021**, *59*, 69–89. [CrossRef]
7. Rapp, G.; Rapp, G. Metals and related minerals and ores. In *Archaeomineralogy*; Springer: Berlin/Heidelberg, Germany, 2009; pp. 143–182. [CrossRef]
8. Lottermoser, B.; Lottermoser, B.G. Sulfidic mine wastes. In *Mine Wastes: Characterization, Treatment and Environmental Impacts*; Springer: Berlin/Heidelberg, Germany, 2010; pp. 43–117. [CrossRef]
9. Elizondo-Álvarez, M.A.; Uribe-Salas, A.; Nava-Alonso, F. Flotation studies of galena (PbS), cerussite (PbCO3) and anglesite (PbSO4) with hydroxamic acids as collectors. *Miner. Eng.* **2020**, *155*, 106456. [CrossRef]
10. Liu, Y.; Molinari, S.; Dalconi, M.C.; Valentini, L.; Ricci, G.; Carrer, C.; Ferrari, G.; Artioli, G. The leaching behaviors of lead, zinc, and sulfate in pyrite ash contaminated soil: Mineralogical assessments and environmental implications. *J. Environ. Chem. Eng.* **2023**, *11*, 109687. [CrossRef]
11. Ajala, M.A.; Abdulkareem, A.S.; Kovo, A.S.; Tijani, J.O.; Ajala, E.O. Synthesis of Ag2O-TiO2-Kaolinite Clay Nanocomposite for Efficient Removal of Mn2+, Fe3+, Cu2+, and Pb2+ and Pathogens in Mining Wastewater. *Water Air Soil Pollut.* **2024**, *235*, 42. [CrossRef]
12. Mana, S.C.A.; Fatt, N.T.; Ashraf, M.A. The fate and transport of arsenic species in the aquatic ecosystem: A case study on Bestari Jaya, Peninsular Malaysia. *Environ. Sci. Pollut. Res.* **2017**, *24*, 22799–22807. [CrossRef] [PubMed]
13. Sharma, P.; Kumar, S. Bioremediation of heavy metals from industrial effluents by endophytes and their metabolic activity: Recent advances. *Bioresour. Technol.* **2021**, *339*, 125589. [CrossRef]
14. Redwan, M.; Rammlmair, D.; Berkh, K. Secondary minerals in a calcareous environment: An example from Um Gheig Pb/Zn mine site, Eastern Desert, Egypt. *Environ. Earth Sci.* **2021**, *80*, 274–302. [CrossRef]
15. Dinake, P.; Mokgosi, S.M.; Kelebemang, R.; Kereeditse, T.T.; Motswetla, O. Pollution risk from Pb towards vegetation growing in and around shooting ranges—A review. *Environ. Pollut. Bioavailab.* **2021**, *33*, 88–103. [CrossRef]
16. Yin, D.; Wang, X.; Chen, C.; Peng, B.; Tan, C.; Li, H. Varying effect of biochar on Cd, Pb and As mobility in a multi-metal contaminated paddy soil. *Chemosphere* **2016**, *152*, 196–206. [CrossRef] [PubMed]
17. Ahmad, M.; Lee, S.S.; Moon, D.H.; Yang, J.E.; Ok, Y.S. A review of environmental contamination and remediation strategies for heavy metals at shooting range soils. In *Environmental Protection Strategies for Sustainable Development*; Springer: Dordrecht, The Netherlands, 2012; pp. 437–451. [CrossRef]
18. Zwolak, A.; Sarzyńska, M.; Szpyrka, E.; Stawarczyk, K. Sources of soil pollution by heavy metals and their accumulation in vegetables: A review. *Water Air Soil Pollut.* **2019**, *230*, 1–9. [CrossRef]
19. Naeem, N.; Khalid, N.; Sarfraz, W.; Ejaz, U.; Yousaf, A.; Rizvi, Z.F.; Ikram, S. Assessment of lead and cadmium pollution in soil and wild plants at different functional areas of Sialkot. *Bull. Environ. Contam. Toxicol.* **2021**, *107*, 336–342. [CrossRef]
20. Bandara, T.; Vithanage, M. Phytoremediation of shooting range soils. In *Phytoremediation: Management of Environmental Contaminants, Volume 3*; Springer: Cham, Switzerland, 2016; pp. 469–488. [CrossRef]
21. Khan, A.Z.; Khan, S.; Muhammad, S.; Baig, S.A.; Khan, A.; Nasir, M.J.; Azhar, M.; Naz, A. Lead contamination in shooting range soils and its phytoremediation in Pakistan: A greenhouse experiment. *Arab. J. Geosci.* **2021**, *14*, 1–7. [CrossRef]
22. Bai, J.; Zhao, X. Ecological and human health risks of heavy metals in shooting range soils: A meta assessment from China. *Toxics* **2020**, *8*, 32. [CrossRef]
23. Česynaitė, J.; Praspaliauskas, M.; Pedišius, N.; Sujetovienė, G. Biological assessment of contaminated shooting range soil using earthworm biomarkers. *Ecotoxicology* **2021**, *30*, 2024–2035. [CrossRef]
24. Barker, A.J.; Clausen, J.L.; Douglas, T.A.; Bednar, A.J.; Griggs, C.S.; Martin, W.A. Environmental impact of metals resulting from military training activities: A review. *Chemosphere* **2021**, *265*, 129–138. [CrossRef]
25. Sehube, N.; Kelebemang, R.; Totolo, O.; Laetsang, M.; Kamwi, O.; Dinake, P. Lead pollution of shooting range soils. *S. Afr. J. Chem.* **2017**, *70*, 21–28. [CrossRef]
26. Ni, S.; Rahman, S.; Kasai, S.; Yoshioka, S.; Wong, K.H.; Mashio, A.S.; Hasegawa, H. Remediation of lead-contaminated shooting range soil: Biodegradable chelator-assisted washing and subsequent post-treatment using $FeCl_3$ and CaO. *Environ. Technol. Innov.* **2023**, *31*, 103–111. [CrossRef]
27. Dien, N.T.; Hirai, Y.; Koshiba, J.; Sakai, S.-i. Factors affecting multiple persistent organic pollutant concentrations in the air above Japan: A panel data analysis. *Chemosphere* **2021**, *277*, 130–141. [CrossRef] [PubMed]
28. Conesa, H.M.; Wieser, M.; Gasser, M.; Hockmann, K.; Evangelou, M.W.H.; Studer, B.; Schulin, R. Effects of three amendments on extractability and fractionation of Pb, Cu, Ni and Sb in two shooting range soils. *J. Hazard. Mater.* **2010**, *181*, 845–850. [CrossRef]

29. Heier, L.S.; Meland, S.; Ljønes, M.; Salbu, B.; Strømseng, A.E. Short-term temporal variations in speciation of Pb, Cu, Zn and Sb in a shooting range runoff stream. *Sci. Total Environ.* **2010**, *408*, 2409–2417. [CrossRef] [PubMed]
30. Fayiga, A.O.; Saha, U.; Ma, L.Q. Chemical and physical characterization of lead in three shooting range soils in Florida. *Chem. Speciat. Bioavailab.* **2011**, *23*, 163–169. [CrossRef]
31. Laporte-Saumure, M.; Martel, R.; Mercier, G. Evaluation of physicochemical methods for treatment of Cu, Pb, Sb, and Zn in Canadian small arm firing ranges backstop soils. *Water Air Soil Pollut.* **2010**, *213*, 171–189. [CrossRef]
32. Selonen, S.; Liiri, M.; Strömmer, R.; Setälä, H. The fate of lead at abandoned and active shooting ranges in a boreal pine forest. *Environ. Toxicol. Chem.* **2012**, *31*, 2771–2779. [CrossRef] [PubMed]
33. Amorim, A.; Black, D.; Borkhetaria, P.; Gambill, M.; Gray, M.; Jin, H.; Pergola, S.; Qiang, K.; Rosenberger, M.; Smires, A. Life Cycle Assessment of Firearms and Ammunition. 2019. Available online: https://digitalcollections.drew.edu/CLA/SpatialDataCenter/Spring2020/LCA_Firearms.pdf (accessed on 22 January 2025).
34. Okkenhaug, G.; Amstätter, K.; Lassen Bue, H.; Cornelissen, G.; Breedveld, G.D.; Henriksen, T.; Mulder, J. Antimony (Sb) contaminated shooting range soil: Sb mobility and immobilization by soil amendments. *Environ. Sci. Technol.* **2013**, *47*, 6431–6439. [CrossRef] [PubMed]
35. Liu, Y.; Fang, Z.; Xie, C.; Li, J. Analysis of existing speciation and evaluation of heavy metals pollution of soil in a shooting range. *Nat. Environ. Pollut. Technol.* **2014**, *13*, 449–558.
36. Luo, W.; Verweij, R.A.; van Gestel, C.A.M. Contribution of soil properties of shooting fields to lead biovailability and toxicity to *Enchytraeus crypticus*. *Soil Biol. Biochem.* **2014**, *76*, 235–241. [CrossRef]
37. Sanderson, P.; Naidu, R.; Bolan, N. Ecotoxicity of chemically stabilised metal (loid) s in shooting range soils. *Ecotoxicol. Environ. Saf.* **2014**, *100*, 201–208. [CrossRef] [PubMed]
38. Li, Y.; Zhu, Y.; Zhao, S.; Liu, X. The weathering and transformation process of lead in China's shooting ranges. *Environ. Sci. Process. Impacts* **2015**, *17*, 1620–1633. [CrossRef] [PubMed]
39. Etim, E.U. Distribution of soil-bound lead arising from rainfall-runoff events at impact berm of a military shooting range. *J. Environ. Prot.* **2016**, *7*, 623–634. [CrossRef]
40. Sanderson, P.; Naidu, R.; Bolan, N. The effect of environmental conditions and soil physicochemistry on phosphate stabilisation of Pb in shooting range soils. *J. Environ. Manag.* **2016**, *170*, 123–130. [CrossRef] [PubMed]
41. Rodríguez-Seijo, A.; Lago-Vila, M.; Andrade, M.L.; Vega, F.A. Pb pollution in soils from a trap shooting range and the phytoremediation ability of *Agrostis capillaris* L. *Environ. Sci. Pollut. Res.* **2016**, *23*, 1312–1323. [CrossRef]
42. Tariq, S.R.; Ashraf, A. Comparative evaluation of phytoremediation of metal contaminated soil of firing range by four different plant species. *Arab. J. Chem.* **2016**, *9*, 806–814. [CrossRef]
43. Islam, M.N.; Park, J.-H. Immobilization and reduction of bioavailability of lead in shooting range soil through hydrothermal treatment. *J. Environ. Manag.* **2017**, *191*, 172–178. [CrossRef] [PubMed]
44. Etim, E.U. Lead removal from contaminated shooting range soil using acetic acid potassium chloride washing solutions and electrochemical reduction. *J. Health Pollut.* **2017**, *7*, 22–31. [CrossRef]
45. Rodríguez-Seijo, A.; Cachada, A.; Gavina, A.; Duarte, A.C.; Vega, F.A.; Andrade, M.L.; Pereira, R. Lead and PAHs contamination of an old shooting range: A case study with a holistic approach. *Sci. Total Environ.* **2017**, *575*, 367–377. [CrossRef] [PubMed]
46. Tandy, S.; Meier, N.; Schulin, R. Use of soil amendments to immobilize antimony and lead in moderately contaminated shooting range soils. *J. Hazard. Mater.* **2017**, *324*, 617–625. [CrossRef]
47. Sujetovienė, G.; Česynaitė, J. Assessment of toxicity to earthworm Eisenia fetida of lead contaminated shooting range soils with different properties. *Bull. Environ. Contam. Toxicol.* **2019**, *103*, 559–564. [CrossRef] [PubMed]
48. Barker, A.J.; Mayhew, L.E.; Douglas, T.A.; Ilgen, A.G.; Trainor, T.P. Lead and antimony speciation associated with the weathering of bullets in a historic shooting range in Alaska. *Chem. Geol.* **2020**, *553*, 119–127. [CrossRef]
49. Lago-Vila, M.; Rodríguez-Seijo, A.; Vega, F.A.; Arenas-Lago, D. Phytotoxicity assays with hydroxyapatite nanoparticles lead the way to recover firing range soils. *Sci. Total Environ.* **2019**, *690*, 1151–1161. [CrossRef]
50. Kelebemang, R.; Dinake, P.; Sehube, N.; Daniel, B.; Totolo, O.; Laetsang, M. Speciation and mobility of lead in shooting range soils. *Chem. Speciat. Bioavailab.* **2017**, *29*, 143–152. [CrossRef]
51. Mariussen, E.; Johnsen, I.V.; Strømseng, A.E. Distribution and mobility of lead (Pb), copper (Cu), zinc (Zn), and antimony (Sb) from ammunition residues on shooting ranges for small arms located on mires. *Environ. Sci. Pollut. Res.* **2017**, *24*, 10182–10196. [CrossRef]
52. Chen, M.; Daroub, S.H.; Ma, L.Q.; Harris, W.G.; Cao, X. Characterization of lead in soils of a rifle/pistol shooting range in central Florida, USA. *Soil Sediment Contam.* **2002**, *11*, 1–17. [CrossRef]
53. Liu, R.; Gress, J.; Gao, J.; Ma, L.Q. Impacts of two best management practices on Pb weathering and leachability in shooting range soils. *Environ. Monit. Assess.* **2013**, *185*, 6477–6484. [CrossRef]

54. Aslam, M.; Aslam, A.; Sheraz, M.; Ali, B.; Ulhassan, Z.; Najeeb, U.; Zhou, W.; Gill, R.A. Lead toxicity in cereals: Mechanistic insight into toxicity, mode of action, and management. *Front. Plant Sci.* **2021**, *11*, 587–595. [CrossRef] [PubMed]
55. Smith, P.; Calvin, K.; Nkem, J.; Campbell, D.; Cherubini, F.; Grassi, G.; Korotkov, V.; Le Hoang, A.; Lwasa, S.; McElwee, P. Which practices co-deliver food security, climate change mitigation and adaptation, and combat land degradation and desertification? *Glob. Change Biol.* **2020**, *26*, 1532–1575. [CrossRef] [PubMed]
56. Aziz, M.A.; Wattoo, F.M.; Khan, F.; Hassan, Z.; Mahmood, I.; Anwar, A.; Karim, M.F.; Akram, M.T.; Manzoor, R.; Khan, K.S. Biochar and polyhalite fertilizers improve soil's biochemical characteristics and sunflower (*Helianthus annuus* L.) yield. *Agronomy* **2023**, *13*, 483. [CrossRef]
57. Cao, X.; Ma, L.Q.; Chen, M.; Hardison, D.W., Jr.; Harris, W.G. Lead transformation and distribution in the soils of shooting ranges in Florida, USA. *Sci. Total Environ.* **2003**, *307*, 179–189. [CrossRef] [PubMed]
58. Zhu, Y.; Che, R.; Tu, B.; Miao, J.; Lu, X.; Li, J.; Zhu, Y.; Wang, F. Contamination and remediation of contaminated firing ranges—An overview. *Front. Environ. Sci.* **2024**, *12*, 135–147. [CrossRef]
59. Sanderson, P.; Naidu, R.; Bolan, N. Effectiveness of chemical amendments for stabilisation of lead and antimony in risk-based land management of soils of shooting ranges. *Environ. Sci. Pollut. Res.* **2015**, *22*, 8942–8956. [CrossRef]
60. Klik, B.; Gusiatin, Z.M.; Kulikowska, D. A holistic approach to remediation of soil contaminated with Cu, Pb and Zn with sewage sludge-derived washing agents and synthetic chelator. *J. Clean. Prod.* **2021**, *311*, 127–134. [CrossRef]
61. Liu, Z.; Xu, Z.; Xu, L.; Buyong, F.; Chay, T.C.; Li, Z.; Cai, Y.; Hu, B.; Zhu, Y.; Wang, X. Modified biochar: Synthesis and mechanism for removal of environmental heavy metals. *Carbon Res.* **2022**, *1*, 8–17. [CrossRef]
62. Aziz, M.A.; Khan, K.S.; Khalid, R.; Shabaan, M.; Alghamdi, A.G.; Alasmary, Z.; Majrashi, M.A. Integrated application of biochar and chemical fertilizers improves wheat (*Triticum aestivum*) productivity by enhancing soil microbial activities. *Plant Soil* **2024**, *502*, 433–448. [CrossRef]
63. Yin, C.-Y.; Mahmud, H.B.; Shaaban, M.G. Stabilization/solidification of lead-contaminated soil using cement and rice husk ash. *J. Hazard. Mater.* **2006**, *137*, 1758–1764. [CrossRef] [PubMed]
64. Larson, S.L.; Malone, P.G.; Weiss, C.A.; Martin, W.A.; Trest, C.; Fabian, G.; Warminsky, M.F.; Mackie, D.; Tasca, J.J.; Wildey, J. Amended ballistic sand studies to provide low maintenance lead containment at active small arms firing range systems. *US Army Corps Eng. Eng. Res. Dev. Cent.* **2007**, *34*, 58–67.
65. Mariussen, E.; Johnsen, I.V.; Strømseng, A.E. Selective adsorption of lead, copper and antimony in runoff water from a small arms shooting range with a combination of charcoal and iron hydroxide. *J. Environ. Manag.* **2015**, *150*, 281–287. [CrossRef] [PubMed]
66. Nnaji, N.D.; Onyeaka, H.; Miri, T.; Ugwa, C. Bioaccumulation for heavy metal removal: A review. *SN Appl. Sci.* **2023**, *5*, 125–134. [CrossRef]
67. Migliorini, M.; Pigino, G.; Bianchi, N.; Bernini, F.; Leonzio, C. The effects of heavy metal contamination on the soil arthropod community of a shooting range. *Environ. Pollut.* **2004**, *129*, 331–340. [CrossRef] [PubMed]
68. Hoaghia, M.-A.; Cadar, O.; Moisa, C.; Roman, C.; Kovacs, E. Heavy metals and health risk assessment in vegetables grown in the vicinity of a former non-metallic facility located in Romania. *Environ. Sci. Pollut. Res.* **2022**, *29*, 40079–40093. [CrossRef]
69. Grammenou, A.; Petropoulos, S.A.; Thalassinos, G.; Rinklebe, J.; Shaheen, S.M.; Antoniadis, V. Biostimulants in the soil–plant interface: Agro-environmental implications—A review. *Earth Syst. Environ.* **2023**, *7*, 583–600. [CrossRef]
70. Mishra, R.; Datta, S.P.; Annapurna, K.; Meena, M.C.; Dwivedi, B.S.; Golui, D.; Bandyopadhyay, K. Enhancing the effectiveness of zinc, cadmium, and lead phytoextraction in polluted soils by using amendments and microorganisms. *Environ. Sci. Pollut. Res.* **2019**, *26*, 17224–17235. [CrossRef]
71. Verma, S.; Dubey, R.S. Lead toxicity induces lipid peroxidation and alters the activities of antioxidant enzymes in growing rice plants. *Plant Sci.* **2003**, *164*, 645–655. [CrossRef]
72. Kastori, R.; Petrović, M.; Petrović, N. Effect of excess lead, cadmium, copper, and zinc on water relations in sunflower. *J. Plant Nutr.* **1992**, *15*, 2427–2439. [CrossRef]
73. Chandwani, S.; Kayasth, R.; Naik, H.; Amaresan, N. Current status and future prospect of managing lead (Pb) stress through microbes for sustainable agriculture. *Environ. Monit. Assess.* **2023**, *195*, 479–485. [CrossRef]
74. Evangelou, M.W.H.; Hockmann, K.; Pokharel, R.; Jakob, A.; Schulin, R. Accumulation of Sb, Pb, Cu, Zn and Cd by various plants species on two different relocated military shooting range soils. *J. Environ. Manag.* **2012**, *108*, 102–107. [CrossRef]
75. Gjorgieva Ackova, D. Heavy metals and their general toxicity on plants. *Plant Sci. Today* **2018**, *5*, 15–19. [CrossRef]
76. Scoriza, R.N.; Correia, M.E.F. Establishment of leguminous trees in the soil of a shooting range. *Floresta E Ambiente* **2019**, *26*, 2017–2025. [CrossRef]
77. Li, J.-s.; Wang, Q.; Chen, Z.; Xue, Q.; Chen, X.; Mu, Y.; Poon, C.S. Immobilization of high-Pb contaminated soil by oxalic acid activated incinerated sewage sludge ash. *Environ. Pollut.* **2021**, *284*, 117–120. [CrossRef] [PubMed]

78. Du, Y.-J.; Wei, M.-L.; Reddy, K.R.; Jin, F.; Wu, H.-L.; Liu, Z.-B. New phosphate-based binder for stabilization of soils contaminated with heavy metals: Leaching, strength and microstructure characterization. *J. Environ. Manag.* **2014**, *146*, 179–188. [CrossRef]
79. Rahman, S.; Jii, N.; Ni, S.; Harada, Y.; Mashio, A.S.; Begum, Z.A.; Rahman, I.M.M.; Hasegawa, H. Biodegradable chelator-assisted washing and stabilization of arsenic-contaminated excavated soils. *Water Air Soil Pollut.* **2022**, *233*, 213–221. [CrossRef]
80. Wang, Q.; Li, J.-s.; Tang, P.; Fang, L.; Poon, C.S. Sustainable reclamation of phosphorus from incinerated sewage sludge ash as value-added struvite by chemical extraction, purification and crystallization. *J. Clean. Prod.* **2018**, *181*, 717–725. [CrossRef]
81. Su, X.; Zhu, J.; Fu, Q.; Zuo, J.; Liu, Y.; Hu, H. Immobilization of lead in anthropogenic contaminated soils using phosphates with/without oxalic acid. *J. Environ. Sci.* **2015**, *28*, 64–73. [CrossRef] [PubMed]
82. Zhang, Y.; Wang, X.; Ji, H. Stabilization process and potential of agro-industrial waste on Pb-Contaminated soil around Pb–Zn mining. *Environ. Pollut.* **2020**, *260*, 114–122. [CrossRef] [PubMed]
83. Mishra, S.; Bharagava, R.N.; More, N.; Yadav, A.; Zainith, S.; Mani, S.; Chowdhary, P. Heavy metal contamination: An alarming threat to environment and human health. In *Environmental Biotechnology: For Sustainable Future*; Springer: Singapore, 2019; pp. 103–125. [CrossRef]
84. Tangahu, B.V.; Sheikh Abdullah, S.R.; Basri, H.; Idris, M.; Anuar, N.; Mukhlisin, M. A review on heavy metals (As, Pb, and Hg) uptake by plants through phytoremediation. *Int. J. Chem. Eng.* **2011**, *2011*, 939–946. [CrossRef]
85. Nedjimi, B. Phytoremediation: A sustainable environmental technology for heavy metals decontamination. *SN Appl. Sci.* **2021**, *3*, 286–294. [CrossRef]
86. Sneddon, J.; Clemente, R.; Riby, P.; Lepp, N.W. Source-pathway-receptor investigation of the fate of trace elements derived from shotgun pellets discharged in terrestrial ecosystems managed for game shooting. *Environ. Pollut.* **2009**, *157*, 2663–2669. [CrossRef]
87. Teng, Z.; Shao, W.; Zhang, K.; Huo, Y.; Zhu, J.; Li, M. Pb biosorption by Leclercia adecarboxylata: Protective and immobilized mechanisms of extracellular polymeric substances. *Chem. Eng. J.* **2019**, *375*, 122–133. [CrossRef]
88. Cui, W.; Li, X.; Duan, W.; Xie, M.; Dong, X. Heavy metal stabilization remediation in polluted soils with stabilizing materials: A review. *Environ. Geochem. Health* **2023**, *45*, 4127–4163. [CrossRef] [PubMed]
89. Yaashikaa, P.R.; Kumar, P.S.; Jeevanantham, S.; Saravanan, R. A review on bioremediation approach for heavy metal detoxification and accumulation in plants. *Environ. Pollut.* **2022**, *301*, 119–126. [CrossRef]
90. Rahman, Z.; Singh, V.P. Bioremediation of toxic heavy metals (THMs) contaminated sites: Concepts, applications and challenges. *Environ. Sci. Pollut. Res.* **2020**, *27*, 27563–27581. [CrossRef]
91. Koul, B.; Taak, P.; Koul, B.; Taak, P. Ex situ soil remediation strategies. In *Biotechnological Strategies for Effective Remediation of Polluted Soils*; Springer: Singapore, 2018; pp. 39–57. [CrossRef]
92. Schmidt, H.-P.; Kammann, C.; Niggli, C.; Evangelou, M.W.H.; Mackie, K.A.; Abiven, S. Biochar and biochar-compost as soil amendments to a vineyard soil: Influences on plant growth, nutrient uptake, plant health and grape quality. *Agric. Ecosyst. Environ.* **2014**, *191*, 117–123. [CrossRef]
93. Achal, V.; Pan, X.; Zhang, D.; Fu, Q. Bioremediation of Pb-contaminated soil based on microbially induced calcite precipitation. *J. Microbiol. Biotechnol.* **2012**, *22*, 244–247. [CrossRef]
94. Kang, C.-H.; Kwon, Y.-J.; So, J.-S. Bioremediation of heavy metals by using bacterial mixtures. *Ecol. Eng.* **2016**, *89*, 64–69. [CrossRef]
95. Kalita, D.; Joshi, S.R. Study on bioremediation of Lead by exopolysaccharide producing metallophilic bacterium isolated from extreme habitat. *Biotechnol. Rep.* **2017**, *16*, 48–57. [CrossRef]
96. Saleem, M.; Asghar, H.N.; Ahmad, W.; Akram, M.A.; Saleem, M.U.; Khan, M.Y.; Naveed, M.; Zahir, Z.A. Prospects of bacterial-assisted remediation of metal-contaminated soils. In *Agro-Environmental Sustainability: Volume 2: Managing Environmental Pollution*; Springer: Cham, Switzerland, 2017; pp. 41–58. [CrossRef]
97. Shanab, S.; Essa, A.; Shalaby, E. Bioremoval capacity of three heavy metals by some microalgae species (Egyptian Isolates). *Plant Signal. Behav.* **2012**, *7*, 392–399. [CrossRef] [PubMed]
98. Hassan, S.E.; Hijri, M.; St-Arnaud, M. Effect of arbuscular mycorrhizal fungi on trace metal uptake by sunflower plants grown on cadmium contaminated soil. *New Biotechnol.* **2013**, *30*, 780–787. [CrossRef] [PubMed]
99. Bellion, M.; Courbot, M.; Jacob, C.; Blaudez, D.; Chalot, M. Extracellular and cellular mechanisms sustaining metal tolerance in ectomycorrhizal fungi. *FEMS Microbiol. Lett.* **2006**, *254*, 173–181. [CrossRef] [PubMed]

processes

MDPI

Review

Microbial Degradation of Soil Organic Pollutants: Mechanisms, Challenges, and Advances in Forest Ecosystem Management

Pengfei Liu [1,2], Shizhi Wen [1,*], Shanshan Zhu [1], Xi Hu [1] and Yamin Wang [2]

1 College of Life and Environmental Sciences, Central South University of Forestry and Technology, Changsha 410004, China; pengfei524@126.com (P.L.); 2006zhushanshan@163.com (S.Z.); huxi@csuft.edu.cn (X.H.)
2 College of Elementary Education, Changsha Normal University, Changsha 410199, China; liam.ho.edu@gmail.com
* Correspondence: wenshizhi@163.com

Abstract: With industrialization and widespread chemical use, soil organic pollutants have become a major environmental issue. Forest ecosystems, among the most important on Earth, have unique potential for controlling and remediating soil pollution. This article explores the mechanisms of microbial community degradation of organic pollutants, their adaptability across forest ecological conditions, and the effects of environmental factors on degradation efficiency. For example, acidic pH (pH < 5.5) favors PAH degradation, near-neutral pH (6.0–7.5) enhances pharmaceutical and PPCP degradation, and alkaline conditions (pH > 7.5) facilitate petroleum hydrocarbon, VOC, and PPCP breakdown. Optimal microbial degradation occurs with humidity levels between 60% and 80%, and SOM content of 2–5%. This review analyzes advancements in microbial degradation technologies for forest ecosystem soil pollution treatment, including genetic engineering, composting, bioaugmentation, and bio-stimulation techniques, and their integration with phytoremediation. The review also addresses the challenges of real-world implementation, such as maintaining microbial diversity, managing pollutant complexity, adapting to environmental changes, and highlighting future research opportunities. The next decade will focus on synthetic biology, omics technologies, microbial-electrochemical systems, community dynamics, eco-engineering, and plant-microbe synergy to develop efficient, sustainable bioremediation strategies.

Keywords: forest ecosystem; soil pollution; microbial degradation; organic pollutants; bioremediation efficiency; forest ecosystem health

1. Introduction

Soil pollution poses a serious threat to global ecological safety and human health [1]. In forest ecosystems, human activities introduce a variety of organic pollutants into forest soil, disrupting the ecological balance [2]. In particular, industrial areas and regions with frequent human activity often exceed safety standards for pollutants in forest soil, jeopardizing the health and stability of these ecosystems. These pollutants not only impact microbial diversity and activity but may also infiltrate groundwater and enter the food chain, causing widespread harm [3,4].

Microbial degradation technology offers a promising approach to address soil pollution [5–8]. Microorganisms, with their unique metabolic capabilities, can efficiently decompose and transform harmful contaminants. This study focuses on the application of microbial degradation in forest ecosystems, exploring its potential and environmental adaptability to provide a scientific foundation for soil pollution control [9]. Bioremediation, composting, bioaugmentation, bioventing, and phytoremediation are effective soil remediation techniques, each with specific applications and limitations.

This research addresses the increasing problem of organic matter pollution in forest soils, threatening ecosystem stability and biodiversity. The scope of this study focuses on the microbial degradation of organic pollutants in forest soils, including persistent organic pollutants (POPs), petroleum hydrocarbons, pharmaceuticals, and volatile organic compounds (VOCs). It explores the mechanisms and environmental adaptability of microbial communities, the impact of soil and forest ecosystem factors on degradation efficiency, and the practical challenges and opportunities in applying microbial remediation techniques in real-world forest environments. Specifically, this study will do the following: (i) identify and analyze common organic pollutants in forest soils, such as persistent organic pollutants (POPs), petroleum hydrocarbons, pharmaceuticals, and volatile organic compounds (VOCs); (ii) examine the mechanisms and environmental adaptability of microbial communities in degrading these pollutants under various forest ecological conditions; (iii) evaluate the influence of soil properties, forest cover, and environmental factors (temperature, moisture) on microbial degradation efficiency; and (iv) address research voids related to the integration of synthetic biology, omics technologies, microbial-electrochemical systems, community dynamics, eco-engineering approaches, and plant-microbe synergy to enhance the efficiency and sustainability of microbial degradation and bioremediation strategies for soil organic pollutants.

2. Soil Organic Pollutants in Forest Ecosystems

2.1. *Hazards of Organic Pollutants in Forest Soil*

Forest soils are increasingly affected by various organic pollutants, each presenting unique challenges (Figure 1). Persistent organic pollutants (POPs), such as DDT and PCBs, are found in regions like the Amazon Rainforest and the Arctic [10,11], primarily due to historical pesticide use and atmospheric transport. Petroleum hydrocarbons, including PAHs and alkanes, severely impact areas such as the Canadian oil sands and the Niger Delta due to extraction and spills [12]. Pharmaceuticals and personal care products (PPCPs) contaminate river basins like the Yangtze and Danube, primarily from agricultural runoff and wastewater discharge [13]. Volatile organic compounds (VOCs), including benzene and toluene, are prominent in urban and industrial areas such as Beijing and Los Angeles [14,15], driven by vehicular and industrial emissions.

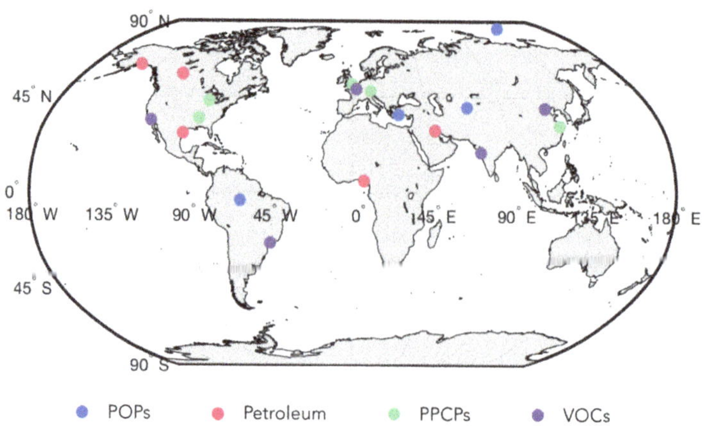

Figure 1. Global situation regarding organic pollutants in forest soils [10,12–14,16–22].

Organic pollutants in forest soils pose significant threats to environmental health and ecosystem stability. As shown in Figure 2, POPs, such as pesticides and industrial chemicals, persist in the environment, disrupting microbial communities, inhibiting plant growth, and causing bioaccumulation in wildlife, with severe human health risks in areas of intensive agriculture or industry [23–25]. Petroleum hydrocarbons degrade soil structure, reduce aeration, and impair microbial functions, often originating from oil spills, industrial sites, and transportation routes [26,27], while PPCPs and VOCs, introduced through wastewater discharge, improper disposal, and industrial emissions, disrupt microbial and plant health and cause endocrine disruption in wildlife (details are illustrated in Supplementary Materials Table S1). Collectively, these pollutants underscore the urgent need for effective management and remediation strategies to safeguard soil health and ecosystem stability [28].

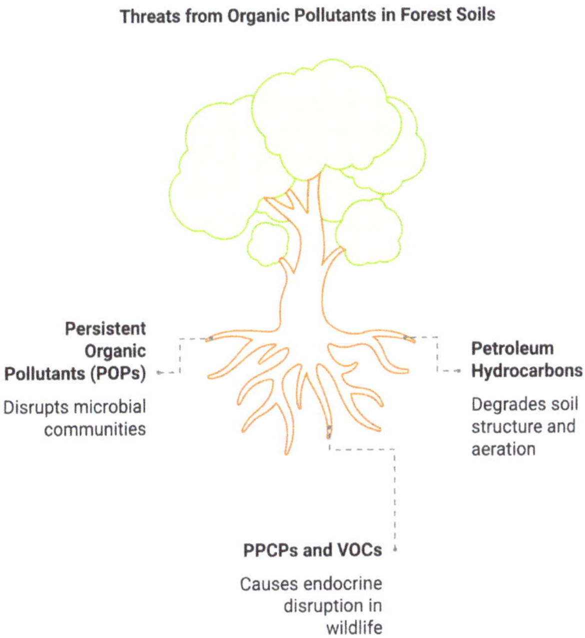

Figure 2. The hazards of organic pollutants in forest soil.

2.2. Sources of Organic Matter Contamination in Forest Soils

Organic matter contamination in forest soils originates primarily from four sources: agricultural runoff, industrial discharges, wastewater discharge, and oil spills or leaks, as illustrated in Figure 3. Agricultural runoff is the largest contributor, accounting for 35% of total contamination, primarily due to the presence of pesticides and fertilizers [29]. Industrial discharges contribute 25%, with predominant pollutants including PCBs, dioxins, and heavy metals. Wastewater discharge accounts for 20%, introducing pharmaceuticals and personal care products as key contaminants. Finally, oil spills and leaks make up 15%, involving petroleum hydrocarbons and heavy metals.

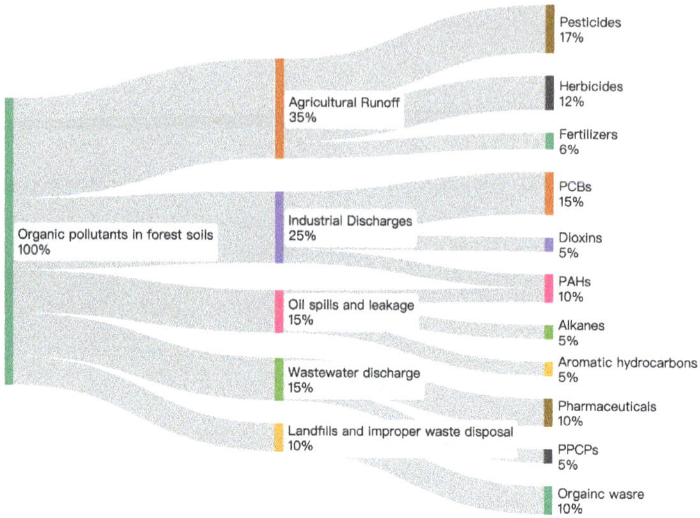

Figure 3. Sources of soil pollution [30–36].

2.3. Microbial Treatment

Microbial treatments for organic matter contamination in forest soils offer various strategies to address pollution through natural processes [37]. Bioremediation, composting, bioaugmentation, bioventing, and phytoremediation are effective soil remediation techniques, each with specific applications and limitations.

As shown in Figure 4, bioremediation is a highly versatile and cost-effective approach that utilizes microbial species, such as *Pseudomonas* and *Phanerochaete chrysosporium*, to degrade contaminants. However, its effectiveness depends on careful optimization of environmental conditions and the selection of appropriate microbial strains. Additionally, bioremediation may result in incomplete degradation products and often requires sufficient time to achieve desired outcomes [38–40]. Composting is an effective method for enhancing soil fertility and reducing waste volume, relying on key microorganisms such as bacteria and fungi to facilitate degradation. However, its success requires careful management to ensure complete degradation and to mitigate potential greenhouse gas emissions, such as methane, which can occur if the process is not properly controlled [41–44]. Bioaugmentation involves the introduction of specific microorganisms to enhance the degradation of particular pollutants, providing tailored solutions for targeted remediation. However, this approach may face challenges related to microbial integration into native ecosystems and compatibility with environmental conditions, highlighting the need for improved strategies to enhance microbial adaptation and minimize ecological disruption [45–48]. Bioventing is a remediation technique that enhances oxygen availability to stimulate microbial activity, making it particularly effective for treating contaminants under moderately aerobic conditions. However, it is ineffective for pollutants that degrade anaerobically and requires precise environmental control to ensure optimal performance [49–53]. Phytoremediation, particularly when combined with microbial assistance, utilizes plants to remove contaminants and stabilize soil, making it a cost-efficient approach for addressing surface-level contamination. However, its effectiveness relies on the careful selection of appropriate plant-microbe combinations, and it carries the potential risk of transferring pollutants into the food chain, which must be managed to ensure environmental and ecological safety [54–59].

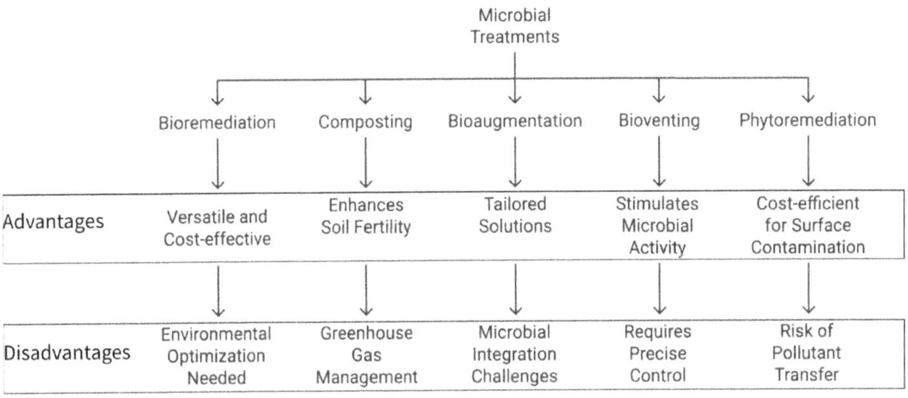

Figure 4. Advantages and disadvantages of microbial treatments for organic matter contamination in forest soils.

Each method has its advantages and limitations, making the choice of treatment dependent on the specific contamination scenario and site conditions. To gain a better understanding of the mechanisms by which these methods remove organic pollutants from soil, it is essential to investigate how the microorganisms utilized in these processes degrade soil organic pollutants (Section 3).

3. Mechanism of Microbial Degradation of Soil Organic Pollutants

3.1. Complex Processes and Key Species

Microorganisms, existing in diverse populations across different soil textures and depths in forest ecosystems, play a crucial role as decomposers in the degradation of organic pollutants. Their enzymatic activity is central to this process, with enzymes acting as biological catalysts that transform complex organic pollutants into simpler, less toxic compounds [60]. Each enzyme targets specific bonds or functional groups within pollutant molecules, enabling precise and efficient breakdown [61]. The effectiveness of this degradation largely depends on the chemical nature of the pollutants, which range from petroleum hydrocarbons to synthetic chemicals such as pesticides and pharmaceuticals. Each enzyme targets a specific type of bond or functional group within the pollutant molecules (Table 1).

Table 1. Types of enzymatic reactions in microbial degradation of soil organic pollutants.

Types of Enzymatic Reactions	Enzyme	Mechanism	References
Hydrolysis	Hydrolases	Hydrolysis involves the addition of water to break chemical bonds in complex organic molecules, resulting in simpler, often more polar and less toxic products.	[62,63]
Oxidation-Reduction (Redox) Reactions	Oxidases, Oxygenases, Dehydrogenases Reductases	Oxidation: Enzymes introduce oxygen into organic molecules or remove electrons, leading to the breakdown of the pollutants. Reduction: Enzymes add electrons to pollutants, often removing halogens or nitro groups, resulting in less toxic or more biodegradable compounds.	[61,64–66]

Table 1. *Cont.*

Types of Enzymatic Reactions	Enzyme	Mechanism	References
Dehalogenases	Dehalogenases	Dehalogenation involves the removal of halogen atoms (e.g., chlorine, bromine) from organic compounds. This process is crucial for detoxifying halogenated pollutants, which are often persistent and toxic.	[67,68]
Ring Cleavage	Dioxygenases Peroxidases	Ring cleavage involves the breaking of aromatic rings in complex pollutants, such as PAHs and phenolic compounds. This step is critical for the complete mineralization of these compounds.	[69,70]
Hydroxylation	Hydroxylases Monooxygenases	Hydroxylation introduces hydroxyl groups (-OH) into organic molecules, increasing their solubility and making them more susceptible to further degradation.	[71–73]

Certain microorganisms have evolved specialized mechanisms, including co-metabolism, to degrade specific pollutants like PAHs and chlorinated compounds. Additionally, microbial consortia, where multiple species collaborate, enhance the efficiency of degradation through synergistic interactions (Table 2). Microbial degradation of soil organic pollutants is a complex process that involves the action of various microbial species and enzymes. For instance, the degradation of PAHs is primarily carried out by Ascomycetes (e.g., *Fusarium oxysporum*), Basidiomycetes (e.g., *Laccaria bicolor*), and Actinobacteria (e.g., *Streptomyces griseus*) [74–77]. These microorganisms utilize oxygenases to cleave the aromatic rings in PAHs, making them more susceptible to further degradation. For PPCPs, Pseudomonas putida and Bacillus subtilis are key species involved in the oxidation and hydroxylation of complex organic compounds. Similarly, the dehydrogenases and peroxidases produced by these microbes play critical roles in the reduction and oxidation of pollutants, facilitating their breakdown [78–80]. The degradation of VOCs, such as benzene and toluene, is often carried out by *Pseudomonas putida* and *Rhodococcus* spp. [81], which use monooxygenases to oxidize these compounds. In the case of petroleum hydrocarbons, *Alcanivorax borkumensis* and *Rhodococcus* sp. [82–84] utilize specialized hydroxylases to degrade long-chain alkanes and aromatic hydrocarbons. These enzymes break down pollutants by adding oxygen atoms or removing electrons, transforming complex compounds into simpler, less toxic molecules.

Halogen-containing organic pollutants, such as polychlorinated biphenyls (PCBs), dichlorodiphenyltrichloroethane (DDT), and trichloroethylene (TCE), are persistent and toxic due to strong carbon–halogen bonds (e.g., C-Cl). Microbial degradation involves reductive dechlorination, oxidative dehalogenation, and hydrolytic processes. *Dehalococcoides* spp. (bacteria) and *Archaeoglobus fulgidus* (archaea) perform reductive dechlorination, removing chlorine atoms from PCBs and TCE under anaerobic conditions and converting them into less toxic intermediates like ethene or biphenyls, with efficiencies up to 70% in 20 days [85,86]. *Pseudomonas putida* and *Rhodococcus* spp. use oxygenases for oxidative dehalogenation of DDT, achieving 60–80% degradation in aerobic forest soils within 14 days [87]. Challenges include low bioavailability, toxicity to microbes, and slow kinetics, requiring bioaugmentation or biostimulation for enhancement [88].

Microplastics, increasingly prevalent in forest soils due to agricultural plastics and atmospheric deposition, pose a novel challenge. Microbial degradation involves bacteria (*Pseudomonas* spp., *Rhodococcus* spp.) and fungi (*Aspergillus niger*, *Penicillium chrysogenum*) that produce enzymes like hydrolases, oxidases, and esterases to break down polyethylene (PE), polypropylene (PP), and polystyrene (PS) [89,90]. Kumar et al. [91] reported that *Rhodococcus ruber* degrades PE by 15% over 60 days in forest soils, forming smaller fragments via oxidative cleavage. Fungi like *Aspergillus niger* achieve 25% PS degradation through hydrolysis, but efficiencies are low due to microplastics' recalcitrance and limited enzyme specificity [92]. Environmental factors (e.g., pH, temperature) and microbial consortia enhance degradation, yet microplastics reduce soil microbial diversity and activity, necessitating advanced strategies like bioelectrochemical systems [93].

Table 2. Some types of organic pollutants and their microbial degradation mechanism(s).

Type of Pollutant	Microbial Species	Mechanism of Degradation	Ref.
Pesticides			
DDT	Bacteria: *Sphingomonas* spp. Fungi: *Phanerochaete chrysosporium*	Reduction and Hydrolysis: Reduction of DDT to DDE, hydrolysis to DDA (dichlorodiphenyldichloroethane)	[94]
Aldrin	Bacteria: *Burkholderia cepacia* Fungi: *Aspergillus niger*	Oxidative Degradation: Oxidation of Aldrin to Dieldrin, further degradation to less toxic forms	[95]
Endrin	Bacteria: *Pseudomonas putida* Fungi: *Trichoderma harzianum*	Dechlorination: Dechlorination and hydrolysis of Endrin to less toxic products	[78]
Chlordane	Bacteria: *Mycobacterium* spp. Fungi: *Cunninghamella echinulata*	Oxidative and Hydrolytic Degradation: Oxidation of Chlordane to less toxic metabolites	[96]
Lindane	Bacteria: *Rhodococcus* spp. Fungi: *White-rot fungi*	Ring Cleavage: Ring cleavage and mineralization of Lindane to non-toxic products	[82]
Industrial Chemicals			
PCBs	Bacteria: *Burkholderia* spp. Fungi: *Phanerochaete chrysosporium*	Dechlorination: Microbial dechlorination of PCBs to less chlorinated and less toxic forms	[97]
Dioxins	Bacteria: *Dechloromonas* spp. Fungi: *Cunninghamella elegans*	Reductive Dechlorination: Reduction of chlorine atoms from dioxins to less toxic forms	[98]
Furans	Bacteria: *Pseudomonas* spp. Fungi: *Lentinus edodes*	Oxidative Degradation: Oxidation of furans to less harmful products	[99]
Alkanes			
n-Hexane	Bacteria: *Pseudomonas putida* Fungi: *Phanerochaete chrysosporium*	Oxidative Biodegradation: Conversion of n-Hexane to less harmful products through hydroxylation	[100]
n-Heptane	Bacteria: *Mycobacterium* spp. Fungi: *Aspergillus niger*	Hydroxylation: Oxidation of n-Heptane to heptane-1-ol, followed by further oxidation	[101]
n-Octane	Bacteria: *Rhodococcus* spp. Fungi: *White-rot fungi*	Terminal Oxidation: Terminal oxidation of n-Octane to fatty acids and further degradation	[102]
Aromatic Hydrocarbons			

Table 2. *Cont.*

Type of Pollutant	Microbial Species	Mechanism of Degradation	Ref.
Benzene	Bacteria: *Pseudomonas putida* Fungi: *Phanerochaete chrysosporium*	Ring Cleavage: Conversion of benzene to catechol and further breakdown via ring cleavage	[103]
Toluene	Bacteria: *Pseudomonas putida* Fungi: *Aspergillus niger*	Monooxygenation: Oxidation of toluene to toluene-4-monooxygenase, further oxidized to benzoic acid	[79]
Ethylbenzene	Bacteria: *Pseudomonas* spp. Fungi: *White-rot fungi*	Oxidative Degradation: Oxidation of ethylbenzene to ethylbenzene-1,2-diol, then to catechol	[80]
Xylenes	Bacteria: *Pseudomonas putida* Fungi: *Phanerochaete chrysosporium*	Oxidative Degradation: Oxidation of xylenes to methylbenzoic acids and further breakdown	[104]
PAHs			
Naphthalene	Bacteria: *Pseudomonas putida* Fungi: *Phanerochaete chrysosporium*	Ring Cleavage: Conversion of naphthalene to catechol and further breakdown	[74]
Anthracene	Bacteria: *Mycobacterium* spp. Fungi: *Aspergillus niger*	Ring Cleavage and Oxidation: Conversion to anthraquinone and further breakdown	[75]
Phenanthrene	Bacteria: *Sphingomonas* spp. Fungi: *White-rot fungi*	Ring Cleavage and Oxidation: Conversion to phenanthrene-2,3-diol and further breakdown	[76]
Benzo[a]pyrene	Bacteria: *Mycobacterium* spp. Fungi: *Phanerochaete chrysosporium*	Ring Cleavage and Oxidation: Conversion to less toxic metabolites	[77]
Antibiotics			
Ciprofloxacin	Bacteria: *Pseudomonas* spp. Fungi: *Aspergillus niger*	Oxidative Degradation: Conversion to less toxic derivatives	[105]
Tetracycline	Bacteria: *Bacillus* spp. Fungi: *Trichoderma harzianum*	Hydrolysis and Oxidation: Hydrolysis to inactive forms and oxidative cleavage	[106]
Hormones			
Estrogens	Bacteria: *Comamonas testosteroni* Fungi: *Phanerochaete chrysosporium*	Hydroxylation and Oxidation: Conversion to less active metabolites	[107]
Progesterone	Bacteria: *Sphingomonas* spp. Fungi: *White-rot fungi*	Oxidative Degradation: Conversion to less active forms through hydroxylation	[108]
Sunscreens			
Oxybenzone	Bacteria: *Pseudomonas putida* Fungi: *Aspergillus niger*	Oxidative Degradation: Conversion to less toxic metabolites	[109]
Octocrylene	Bacteria: *Pseudomonas* spp. Fungi: *Penicillium chrysogenum*	Hydrolysis and Oxidation: Conversion to less harmful products through hydrolysis	[110]
Synthetic Fragrances			
Phthalates	Bacteria: *Burkholderia* spp. Fungi: *Aspergillus niger*	Hydrolysis: Hydrolysis of phthalates to phthalic acid and further degradation	[111]
Musk Compounds	Bacteria: *Sphingomonas* spp. Fungi: *White-rot fungi*	Oxidative Degradation: Oxidation of musk compounds to less toxic metabolites	[112]
Solvents			

Table 2. *Cont.*

Type of Pollutant	Microbial Species	Mechanism of Degradation	Ref.
Acetone	Bacteria: *Pseudomonas* spp. Fungi: *Aspergillus niger*	Oxidative Degradation: Conversion to acetic acid and further oxidation	[81]
Ethanol	Bacteria: *Zymomonas mobilis* Fungi: *Saccharomyces cerevisiae*	Fermentation: Conversion to acetaldehyde and acetic acid via fermentation	[113]
Methanol	Bacteria: *Methylobacterium* spp. Fungi: *Aspergillus niger*	Oxidative Degradation: Conversion to formaldehyde and further oxidation	[114]
Industrial Emissions			
Formaldehyde	Bacteria: *Methylobacterium* spp. Fungi: *Phanerochaete chrysosporium*	Oxidative Degradation: Conversion to formic acid and further breakdown	[115]
Styrene	Bacteria: *Pseudomonas putida* Fungi: *Aspergillus niger*	Oxidative Degradation: Conversion to styrene oxide and further breakdown	[116]
Trichloroethylene	Bacteria: *Dehalococcoides* spp. Fungi: *White-rot fungi*	Reductive Dechlorination: Conversion to less toxic forms via dechlorination	[117]
BTEX			
Benzene	Bacteria: *Pseudomonas putida* Fungi: *Phanerochaete chrysosporium*	Ring Cleavage: Conversion to catechol and further breakdown	[118]
Toluene	Bacteria: *Pseudomonas putida* Fungi: *Aspergillus niger*	Monooxygenation: Conversion to benzoic acid	[119]
Xylenes	Bacteria: *Pseudomonas putida* Fungi: *Phanerochaete chrysosporium*	Oxidative Degradation: Conversion to methylbenzoic acids	[119]

Microbial enzymes are essential for the biodegradation of organic pollutants in soils. Oxygenases, such as dioxygenases and monooxygenases, are involved in the breakdown of aromatic hydrocarbons and are crucial for the ring cleavage of pollutants like PAHs and aromatic hydrocarbons. Dehydrogenases play a vital role in the reduction of pollutants, facilitating the breakdown of compounds like VOCs and PPCPs [75]. These enzymes are responsible for the oxidation or reduction of pollutants, enhancing their bioavailability and subsequent microbial degradation. Peroxidases, commonly found in fungi like *Phanerochaete chrysosporium*, assist in the oxidative degradation of pollutants like chlorinated hydrocarbons and pesticides, breaking down the complex molecules into simpler, non-toxic compounds [119].

Future research on the microbial degradation of soil organic pollutants should focus on unraveling the molecular mechanisms underlying enzymatic pathways and microbial community interactions, particularly in response to diverse environmental conditions and complex pollutant mixtures. Emphasis should be placed on integrating multi-omics technologies, machine learning models, and advanced gene editing tools to enhance microbial efficiency and adaptability. Moreover, addressing the degradation of emerging contaminants, the formation and toxicity of secondary metabolites, and the synergistic effects of microbial consortia remains critical.

3.2. Microbial Groups and Their Roles in Soil Degradation

Microorganisms in forest ecosystems play a crucial role in degrading organic pollutants through diverse enzymatic pathways, as previously described. In addition to the key degraders like *Pseudomonas*, *Rhodococcus*, and *Phanerochaete chrysosporium*, soil-dwelling counterparts such as sulfate-reducing bacteria (SRB), sulfur-oxidizing bacteria (SOB), iron-reducing bacteria (IRB), and iron-oxidizing bacteria (IOB), along with their archaeal equivalents, significantly influence soil properties and pollutant degradation.

These microbial groups modify soil redox potential, pH, and structure, thereby affecting the bioavailability and degradation of organic pollutants like PAHs, petroleum hydrocarbons, and chlorinated compounds.

3.2.1. SRB and Archaea

SRB, such as *Desulfovibrio* spp. (bacteria) and *Archaeoglobus fulgidus* (archaea), reduce sulfate to sulfide under anaerobic conditions, lowering redox potential (Eh < −200 mV) and creating reducing environments conducive to degrading chlorinated hydrocarbons (e.g., TCE) via reductive dechlorination. However, sulfide production can increase soil acidity (pH < 5.5), potentially inhibiting other microbial activities, as observed in waterlogged forest soils [120]. SRB also contribute to PAH degradation under sulfate-reducing conditions, albeit at lower efficiencies (e.g., 35% pyrene removal) [77].

3.2.2. SOB and Archaea

SOB, such as *Thiobacillus* spp. (bacteria) and *Sulfolobus* spp. (archaea), oxidize sulfur compounds (e.g., sulfide to sulfate) under aerobic or microaerobic conditions, increasing soil pH and redox potential (Eh > 300 mV). This enhances aerobic degradation of hydrocarbons (e.g., alkanes) by increasing oxygen availability, but excessive sulfide oxidation can lead to soil alkalinity (pH > 7.5), reducing microbial diversity and pollutant bioavailability, as reported in coniferous forest soils [121].

3.2.3. IRB and Archaea

IRB, such as *Shewanella* spp. (bacteria) and *Methanosarcina* spp. (archaea), reduce ferric iron (Fe^{3+}) to ferrous iron (Fe^{2+}) under anaerobic conditions, lowering redox potential and facilitating pollutant degradation (e.g., PAHs, petroleum hydrocarbons) via electron transfer. This process stabilizes soil structure but can increase acidity, impacting microbial activity, as seen in boreal forest soils [120]. IRB enhance bioelectrochemical degradation in SMFCs, improving PAH removal rates [122].

3.2.4. IOB and Archaea

IOB, such as *Leptothrix* spp. (bacteria) and *Ferroplasma* spp. (archaea), oxidize ferrous iron to ferric iron under aerobic conditions, raising redox potential and supporting oxidative degradation of pollutants like VOCs and PAHs. However, iron oxide accumulation can compact soil, reducing permeability and microbial access to pollutants, as observed in tropical rainforest soils [123].

Bacteria and archaea differ in their metabolic versatility and environmental resilience. Bacteria like *Desulfovibrio* and *Thiobacillus* are more abundant and exhibit rapid growth, dominating soil degradation under fluctuating conditions. Archaea, such as *Archaeoglobus* and *Ferroplasma*, thrive in extreme environments (e.g., low pH, high salinity), offering unique degradation pathways but with slower growth rates, making them less dominant in forest soils [123–125].

4. Factors Influencing Microbial Degradation Efficiency

The efficiency of microbial degradation in forest ecosystems is influenced by a myriad of factors, both environmental and biological [126–129]. Understanding these factors is crucial for enhancing the effectiveness of microbial remediation strategies.

4.1. Forest Cover Types

The type of forest cover influences the microbial degradation process through its effect on soil conditions and the microbial community structure. Different forest cover types have different levels of pH, temperature, and humidity, and the differences in vegetation and soil will lead to different types of microorganisms (Table 3). Deciduous forests, for example, typically have a higher organic matter content due to leaf litter, fostering a rich microbial community capable of degrading a wide array of pollutants [130]. Coniferous forests, on the other hand, tend to have more acidic soils, which can limit microbial activity and thus degradation rates. The diversity of plant species in mixed forests can support a more diverse microbial community, potentially enhancing the degradation of a broader spectrum of pollutants [131]. The interaction between forest cover type and microbial degradation is a complex interplay of soil chemistry, microbial ecology, and pollutant characteristics, illustrating the need for ecosystem-specific approaches to bioremediation [132].

Table 3. Characteristics of different forest cover types and microorganisms in them.

Forest Type	Characteristics	Microbial Differences	Ref.
Evergreen Coniferous Forest	pH: Slightly acidic to neutral Temperature: Cool to temperate Humidity: High year-round	High fungal diversity, especially ectomycorrhizal fungi Lower bacterial diversity compared to broadleaf forests	[133,134]
Deciduous Broadleaf Forest	pH: Slightly acidic to neutral Temperature: Warm summers, cold winters Humidity: Seasonal variation, high in summer, low in winter	Diverse microbial community, especially during leaf litter decomposition Abundance of decomposer fungi and bacteria during fall	[135,136]
Mixed Forest (Coniferous and Broadleaf)	pH: Varies with soil type and tree species Temperature: Variable, depending on location Humidity: Variable, depending on location and season	Microbial diversity reflects a mix of coniferous and broadleaf forests High fungal diversity, including mycorrhizal fungi	[137,138]
Tropical Rainforest	pH: Slightly acidic Temperature: Hot and humid year-round Humidity: Very high	Extremely diverse microbial community High abundance of fungi and bacteria due to rich organic matter	[139,140]
Dry Forest/Savanna	pH: Slightly acidic to neutral Temperature: Hot during the day, cooler at night Humidity: Low to moderate, seasonal, rains,	Adapted to drought conditions, lower microbial diversity Fungi and bacteria tolerant to dry conditions	[141,142]
Boreal Forest	pH: Slightly acidic Temperature: Cold year-round Humidity: High, with significant snowfall	Low microbial diversity due to cold temperatures Adapted to cold conditions, including psychrophilic microorganisms	[143,144]

4.2. Soil Properties

Soil properties, including pH, temperature, moisture content, and organic matter content, are critical determinants of microbial degradation efficiency.

4.2.1. pH

In forest ecosystems, soil pH plays a critical role in the effectiveness of microbial remediation technologies (Figure 5). It not only directly influences the survival and activity of microorganisms but also affects the chemical forms, bioavailability, and degradation potential of pollutants [145]. The pH value of soil has a profound effect on microbial diversity and abundance.

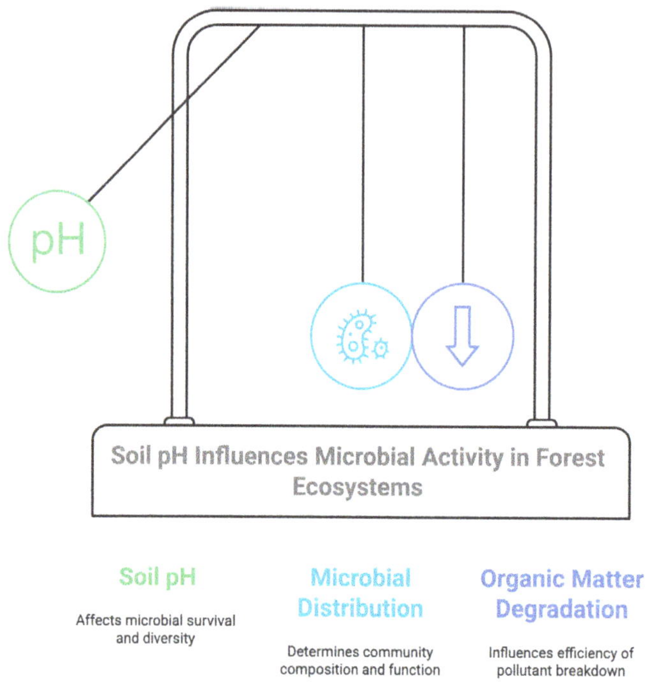

Figure 5. The effect of pH on organic matter degradation.

In forest ecosystems, soil pH highly influences microbial distribution, which impacts microbial community composition and its role in organic matter degradation. In acidic soils (pH < 5.5), predominantly hosted microorganisms include *Ascomycetes* (e.g., *Fusarium oxysporum*), *Basidiomycetes* (e.g., *Laccaria bicolor*), and *Acidobacteria* (e.g., *Acidobacterium capsulatum*), which are adapted to low-pH environments but may exhibit reduced enzymatic activity [146,147]. In neutral soils (pH 6.0–7.5), a broader range of microbes thrive, including Actinobacteria like *Streptomyces griseus*, Firmicutes such as *Bacillus subtilis*, and Proteobacteria like *Pseudomonas fluorescens* [148]. These microbes perform highly efficient organic matter decomposition under balanced pH conditions. In alkaline soils (pH > 7.5), the dominant microbes are *Alcaligenes* (e.g., *Alcaligenes faecalis*), *Nitrobacter* (e.g., *Nitrobacter winogradskyi*), and *Clostridia* (e.g., *Clostridium acetobutylicum*), though their degradation efficiency may decline due to limited microbial diversity and nutrient availability [149,150]. The microbial degradation of soil organic pollutants—such as polycyclic aromatic hydrocarbons (PAHs), pharmaceuticals and personal care products (PPCPs), volatile organic compounds (VOCs), and petroleum hydrocarbons—is a complex process driven by these species and their specialized enzymes. For instance, Ascomycetes, Basidiomycetes, and Actinobacteria degrade PAHs using oxygenases to cleave aromatic rings, while Pseudomonas putida and Bacillus subtilis oxidize PPCPs via dehydrogenases and hydroxylases. Similarly, VOCs like benzene and toluene are broken down by Pseudomonas putida and *Rhodococcus* spp. using monooxygenases, and petroleum hydrocarbons are degraded by *Alcanivorax borkumensis* and *Rhodococcus* sp. through hydroxylases. These enzymes—oxygenases, dehydrogenases, and peroxidases (e.g., from fungi like *Phanerochaete chrysosporium*)—facilitate the oxidation, reduction, and ring cleavage of pollutants, transforming complex compounds into simpler, less toxic molecules and enhancing their bioavailability for further microbial breakdown.

Various microorganisms in forest soils exhibit specialized capabilities for degrading a range of organic pollutants (Table 4). Ascomycetes and Basidiomycetes are effective in breaking down polycyclic aromatic hydrocarbons (PAHs) and pesticides through oxidative processes involving enzymes such as laccases and peroxidases [151]. Actinobacteria are involved in the hydrolysis and oxidation of aromatic compounds, although their degradation rates are generally lower compared to other groups [152]. They excel in breaking down petroleum hydrocarbons and pharmaceuticals via oxidative cleavage and hydroxylation. *Firmicutes* perform anaerobic degradation and fermentation, effectively treating petroleum hydrocarbons and pharmaceuticals. Proteobacteria and *Pseudomonas* are prominent in degrading a broad spectrum of organic pollutants, including VOCs and PPCPs, through oxidative and metabolic processes [153]. Bacteroidetes and *Alcaligenes* are significant for their hydrolytic and oxidative degradation capabilities [152]. *Nitrobacter* and *Clostridia* contribute to the degradation of nitrogen-containing organic compounds and complex organic molecules, respectively [154]. These microorganisms employ diverse metabolic pathways to break down hazardous organic pollutants, making them essential for bioremediation in polluted forest soils. As shown in Table 4, in acidic soils (pH < 5.5), microorganisms such as Ascomycetes, Basidiomycetes, and Acidobacteria are primarily involved in the degradation of PAHs, phenols, pesticides, and other organic compounds through oxidative degradation processes using enzymes like laccases, peroxidases, and hydrolytic enzymes. In neutral soils (pH 6.0–7.5), microorganisms like Actinobacteria, Firmicutes, Proteobacteria, and Bacteroidetes play a significant role in degrading petroleum hydrocarbons, PPCPs, VOCs, and pharmaceuticals, employing oxidative cleavage, hydroxylation, fermentation, and hydrolytic processes. In alkaline soils (pH > 7.5), organisms such as Alcaligenes, Nitrobacter, Clostridia, and Bacillus help degrade petroleum hydrocarbons, PPCPs, and other organic compounds through oxygenase-mediated oxidation, nitrification, and anaerobic fermentation, while Pseudomonas and Arthrobacter engage in the oxidative degradation of aromatic hydrocarbons. The microorganisms contribute to the breakdown of complex molecules via enzyme-mediated processes, enhancing the efficiency of bioremediation in diverse soil conditions.

Table 4. Degradation capabilities of various microorganisms for organic pollutants in forest soil.

Microorganism	Types of Organic Matter	Specific Organic Matter	Degradation Principle	Ref.
Acidic Soils (pH < 5.5)				
Ascomycetes	PAHs Phenols	Naphthalene Phenol	Oxidative degradation via enzymes such as laccases and peroxidases.	[155]
Basidiomycetes	PAHs Pesticides	Pyrene DDT	Ligninolytic enzymes (laccases, peroxidases) oxidize aromatic compounds.	[156]
Acidobacteria	PAHs Pesticides	Fluoranthene Aldrin	Hydrolysis and oxidation of aromatic rings and side chains.	[157]

Table 4. *Cont.*

Microorganism	Types of Organic Matter	Specific Organic Matter	Degradation Principle	Ref.
Neutral Soils (pH 6.0–7.5)				
Actinobacteria	Petroleum Hydrocarbons PPCPs	Toluene Caffeine	Oxidative cleavage of aromatic rings, hydroxylation, and mineralization.	[158]
Firmicutes	Petroleum Hydrocarbons PPCPs	Benzene Diclofenac	Anaerobic degradation, hydrolysis, and fermentation of complex compounds.	[159]
Proteobacteria	Petroleum Hydrocarbons PPCPs VOCs	Xylene Estrone	Oxidative degradation, cometabolism, and mineralization of organic pollutants.	[160]
Bacteroidetes	Pharmaceuticals PPCPs	Erythromycin Bisphenol A	Hydrolytic and oxidative processes, breaking down complex organic molecules.	[161]
Alkaline Soils (pH > 7.5)				
Alcaligenes	Petroleum Hydrocarbons VOCs	Octane Trichloroethylene	Oxygenase-mediated oxidation and degradation of hydrocarbons.	[162]
Nitrobacter	PPCPs Organic Nitrogen Compounds	Ammonium Nitrobenzene	Nitrification and oxidation of nitro-containing compounds.	[163]
Clostridia	Petroleum Hydrocarbons PPCPs	Butyrate, Hexachlorobenzene	Anaerobic fermentation and degradation of complex organic compounds.	[164]
Bacillus	Petroleum Hydrocarbons PPCPs VOCs	Dodecane, Paracetamol	Hydrocarbon oxidation, enzyme-mediated degradation of organic pollutants.	[165]
Arthrobacter	PAHs Petroleum Hydrocarbons	Anthracene Cyclohexane	Oxidative degradation of aromatic hydrocarbons and heterocyclic compounds.	[166]
Pseudomonas	PAHs Petroleum Hydrocarbons, PPCPs VOCs	Phenanthrene Naphthalene Atrazine Toluene	Enzyme-mediated oxidation, cometabolic degradation, and mineralization.	[158]

4.2.2. Temperature

Over the past decade, the global average temperature has shown a clear upward trend due to climate change [167]. According to National Aeronautics and Space Administration (NASA) and National Oceanic and Atmospheric Administration (NOAA) data, the global average temperature increased by approximately 0.9 °C from 2013 to 2023. This rise is consistent with long-term warming trends attributed to anthropogenic climate change. Over the last decade, temperature changes in major forest regions worldwide have generally shown an upward trend, consistent with global warming patterns (as shown in Figure 6).

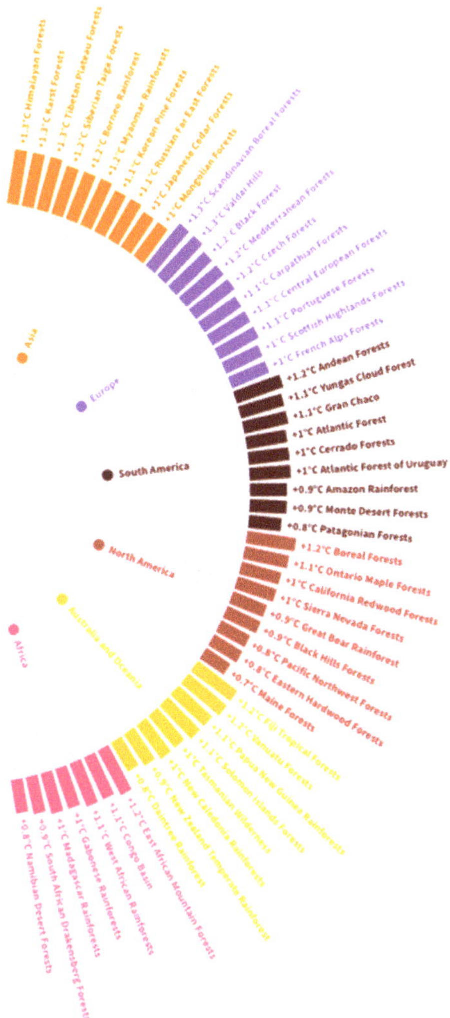

Figure 6. Temperature changes in major forest regions globally over the last decade [168–173].

Temperature significantly influences the types, diversity, and richness of microorganisms in forest soils (Figure 7) [174,175]. In cold environments (below 5 °C), psychrophiles like *Psychrobacter* dominate, but microbial diversity and richness are generally low due to fewer species' adaptation to extreme cold. In moderate temperatures (15–25 °C), mesophilic microorganisms such as Bacillus and Actinobacteria thrive, leading to the highest levels of microbial diversity and richness [175,176]. However, in high-temperature environments (above 30 °C), thermophiles like Thermococcus become prevalent, with a noticeable decrease in microbial diversity and richness due to the selective pressure of heat [177]. These patterns highlight the adaptability of microbial communities to varying temperature conditions and their differing efficiencies in pollutant degradation, with mesophiles being most effective under moderate conditions, thermophiles thriving in extreme heat, and psychrophiles struggling in cold climates, emphasizing the broader implications of climate change on microbial ecosystems and their environmental functions.

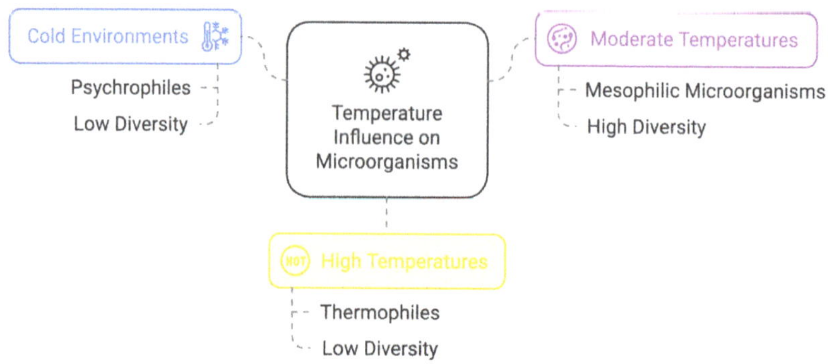

Figure 7. Effect of temperature on the types and diversity of microorganisms in forest soils [168–173].

Moreover, temperature affects microbial degradation processes by modulating degradation rates, enzyme activity, and microbial metabolic activity [178–180]. Understanding these effects is crucial for optimizing bioremediation strategies and managing soil health in varying temperature conditions.

4.2.3. Oxygen Levels

Under aerobic conditions, where oxygen is readily available, microorganisms such as Pseudomonas putida, *Rhodococcus* spp., and Bacillus subtilis utilize oxygen as a terminal electron acceptor in oxidative degradation pathways. Oxygen serves as a key substrate for oxygenases (e.g., monooxygenases and dioxygenases), which catalyze the cleavage of aromatic rings and oxidation of aliphatic hydrocarbons, such as petroleum hydrocarbons (e.g., alkanes, PAHs) and VOCs (e.g., benzene, toluene) [181]. These enzymes introduce oxygen atoms into pollutant molecules, increasing their solubility and susceptibility to further enzymatic breakdown, resulting in faster and more complete mineralization into carbon dioxide, water, and biomass.

Aerobic degradation is typically more efficient due to higher energy yields from oxidative phosphorylation, enabling rapid microbial growth and pollutant removal. For instance, a study by Wang et al. [122] demonstrated that *Pseudomonas putida* achieved a 90% degradation rate of toluene in contaminated forest soil under aerobic conditions within 7 days, leveraging oxygen-dependent monooxygenases to convert toluene into benzoic acid and subsequent intermediates. This efficiency is particularly pronounced for hydrophobic pollutants like PAHs, where oxygen enhances bioavailability by reducing adsorption to soil organic matter (SOM). In anaerobic conditions, where oxygen is absent, facultative and obligate anaerobes such as Clostridia, *Desulforamulus aquiferis*, and *Dehalococcoides* spp. rely on alternative electron acceptors (e.g., nitrate, sulfate, or ferric iron) for pollutant degradation [77]. These microorganisms employ fermentation, reductive dechlorination, and other anaerobic pathways, which are generally slower and less efficient than aerobic processes due to lower energy yields. For example, anaerobic degradation of VOCs like trichloroethylene (TCE) involves reductive dechlorination by *Dehalococcoides* spp., transforming TCE into less toxic ethene, but this process can take weeks to months, with efficiencies often below 50% compared to aerobic degradation [182].

The slower rate stems from the complexity of anaerobic metabolic pathways, such as the need for syntrophic interactions among microbial consortia to share electrons and degrade recalcitrant pollutants. Zhang et al. [77] reported that *Desulforamulus aquiferis* degraded pyrene and benzo[a]pyrene in anaerobic forest soil with a removal efficiency of only 35% over 30 days, highlighting the limitations of oxygen scarcity. However, anaerobic conditions are advantageous in waterlogged or deep soil layers where oxygen diffusion is limited, enabling the degradation of specific pollutants like chlorinated hydrocarbons and certain PAHs that resist aerobic breakdown.

4.2.4. Redox Potential

Redox potential is a critical environmental factor influencing microbial degradation of organic pollutants in forest soils, directly determining metabolic pathways and efficiency. Under high redox potential (>300 mV) aerobic conditions, microorganisms such as *Pseudomonas putida* and *Rhodococcus* spp. use oxygen as an electron acceptor, rapidly and efficiently degrading PAHs, alkanes, and VOCs via oxygenases; for instance, Johnsen et al. [183] reported removal rates of 80%, 70%, and 60% for naphthalene, phenanthrene, and pyrene, respectively, within 14 days. In low redox potential (<−200 mV) anaerobic conditions, microorganisms like *Clostridia* and *Dehalococcoides* spp. rely on nitrate, sulfate, or iron for fermentation and reductive dechlorination, degrading chlorinated hydrocarbons (e.g., tetrachloroethylene, PCE) and some PAHs, but with lower efficiency [184]. Microbes adapt to redox gradients through metabolic flexibility (e.g., facultative anaerobes switching pathways), with community synergies (e.g., *Pseudomonas* and *Clostridia*) achieving 60% phenanthrene degradation. Case studies demonstrate that manipulating redox potential via bioelectrochemical systems (e.g., SMFCs) can significantly boost degradation efficiency; for example, Wang et al. [185] increased phenanthrene degradation from 25% to 65% by adjusting potential. However, challenges such as soil heterogeneity, microbial stability, and electrode degradation limit large-scale applications. Future research should integrate omics technologies, synthetic biology, and microbial-electrochemical systems to optimize redox conditions, developing sustainable remediation strategies for polluted forest soils.

4.2.5. Humidness

Humidness in forest ecosystem has a profound influence on microbial degradation [186]. Water availability directly affects the survival, growth, and activity of microorganisms [187]. While adequate moisture enhances microbial activity, excessive water can cause soil hypoxia, impairing the metabolism of aerobic microorganisms. Conversely, insufficient moisture reduces microbial activity and slows degradation rates. Therefore, regulating and maintaining optimal soil moisture conditions is critical for optimizing microbial pollutant treatment in forest ecosystems [188].

Under optimal moisture levels (60–80% of soil moisture), microbial communities thrive, enabling efficient degradation of pollutants such as PAHs and petroleum hydrocarbons through active enzymatic processes [189]. In dry soils, microbial activity is suppressed, with drought-tolerant taxa like *Actinobacteria* dominating but performing less efficiently [190]. In waterlogged soils, anaerobic conditions favor facultative and obligate anaerobes like *Clostridia*, which can degrade VOCs, though these processes are generally slower than aerobic degradation [191]. In dry soils, microbial communities shift towards drought-resistant taxa like Firmicutes, while in moist soils, the diversity increases, incorporating phyla such as Proteobacteria, Bacteroidetes, and Acidobacteria. This enhanced diversity in moist conditions supports a more comprehensive degradation process for pollutants like petroleum hydrocarbons and VOCs [192]. However, excessive moisture can lead to the proliferation of anaerobic microorganisms, such as *Methanogens*, which degrade specific pollutants but at lower overall efficiencies due to oxygen limitations [193].

Moisture levels also influence pollutant bioavailability in soil. In dry conditions, hydrophobic pollutants adsorb more tightly to soil particles, reducing accessibility for microbial degradation. In contrast, optimal moisture levels increase pollutant solubility, transport, and desorption from soil particles, thereby enhancing bioavailability and microbial degradation efficiency [194]. These conditions regulate enzyme production and activity, with enzymes such as oxygenases and hydrolases functioning most effectively under moderate moisture conditions, accelerating the breakdown of organic pollutants [195]. The optimal humidity ranges for biodegradation of various pollutants are illustrated in Figure 8.

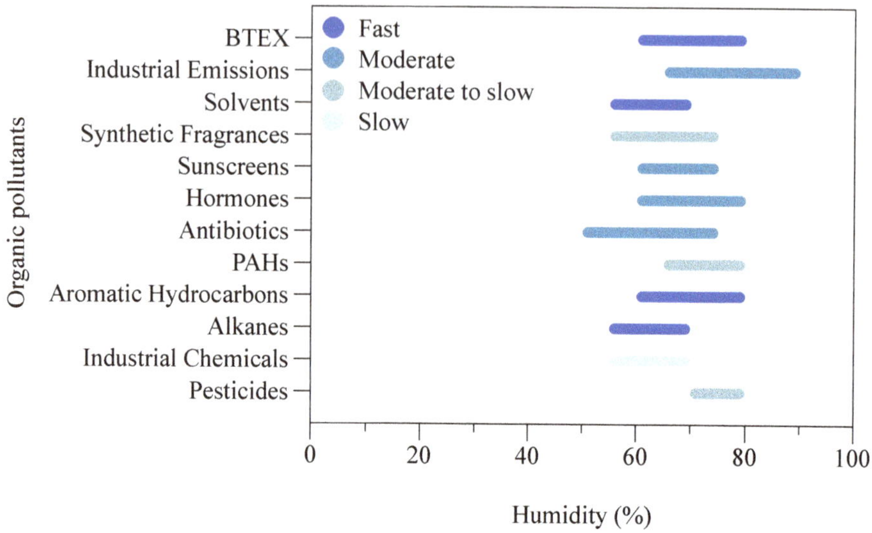

Figure 8. Effects of humidity on the microbial degradation of various organic pollutants in forest soils.

4.2.6. Soil Organic Matter

Soil organic matter (SOM) plays a crucial role in shaping microbial communities and enhancing the degradation of organic pollutants in forest soils. It acts as a nutrient reservoir, providing essential carbon and energy sources that support microbial growth and enzyme synthesis [196]. This facilitates the breakdown of complex pollutants such as pesticides and aromatic hydrocarbons by microorganisms like Pseudomonas putida, which rely on SOM for metabolic activity [197]. Additionally, SOM improves pollutant bioavailability by forming micelle-like structures with hydrophobic compounds, thereby increasing their accessibility to microbial degradation.

SOM also adsorbs pollutants, creating stable complexes that allow for slow but sustained biodegradation over time. This dual role of enhancing immediate bioavailability while enabling long-term stabilization is critical in the degradation of persistent pollutants like PCBs and PAHs by species such as *Sphingomonas* and *Burkholderia* [198,199]. Furthermore, SOM supports microbial colonization and biofilm formation, providing a protective environment that enhances microbial resilience and cooperative degradation efforts.

Beyond improving bioavailability and colonization, SOM buffers the toxicity of certain pollutants by adsorbing harmful compounds, creating conditions conducive to microbial growth [200,201]. It also promotes microbial diversity by offering a range of carbon sources, enabling different species, including bacteria, fungi, and actinomycetes, to collaborate in degrading various pollutants [202–204]. Overall, SOM enhances the efficiency and sustainability of bioremediation processes in forest soils (Table 5).

Table 5. Effects of soil organic matter on microbial degradation of organic pollutants in forest soils.

Pollutant Group	Impact of SOM	Optimal SOM Content (%)	Degradation Rate Impact	Ref.
Pesticides	SOM increases bioavailability by adsorbing and releasing pollutants.	3–5%	Enhanced (70–85%)	[205–207]
Industrial Chemicals	High SOM can limit degradation by sequestering pollutants.	2–4%	Moderate (60–75%)	[205,208,209]
Alkanes	SOM aids in emulsification, enhancing microbial uptake.	1–3%	Fast (75–90%)	[210–212]
Aromatic Hydrocarbons	SOM enhances solubility and bioavailability.	4–6%	High (80–95%)	[213,214]
PAHs	SOM adsorbs PAHs, but microbial enzymes can release them.	3–5%	Moderate (65–85%)	[214–216]
Antibiotics	SOM can reduce toxicity and improve microbial tolerance.	4–6%	Moderate (60–80%)	[217,218]
Hormones	SOM enhances stability, promoting slow release.	2–4%	Moderate (65–75%)	[219,220]
Sunscreens	SOM helps in dispersing hydrophobic molecules.	3–5%	Moderate (60–75%)	[221–223]
Synthetic Fragrances	SOM reduces volatilization, aiding microbial degradation.	3–5%	Moderate (70–80%)	[224]
Solvents	Low SOM minimizes competition with naturally occurring compounds.	1–2%	High (80–90%)	[225–228]
Industrial Emissions	SOM can trap volatile compounds but also facilitate microbial uptake.	3–5%	Moderate (65–80%)	[229,230]
BTEX	SOM improves solubility and bioavailability.	4–6%	High (85–95%)	[231–233]

4.3. Types of Pollutant

The biodegradability of soil pollutants in forest ecosystems is heavily influenced by their chemical nature. Factors such as molecular structure, functional groups, solubility, stability, and bioavailability determine how easily microorganisms can access and degrade these compounds [234]. Understanding these relationships helps in predicting the persistence of pollutants and developing effective bioremediation strategies.

The intricate relationship between the chemical nature of pollutants and their biodegradability underscores the complexity of soil pollution management in forest ecosystems. Organic pollutants, encompassing a wide array of compounds from hydrocarbons to pesticides, present varying challenges for microbial degradation. The structure of these molecules plays a pivotal role in determining their susceptibility to breakdown processes [235]. Simple organic compounds, characterized by fewer carbon rings and a lack of complex branching, are typically more amenable to microbial attack. This is largely due to the relative ease with which microorganisms can access and metabolize these simpler structures, utilizing them as sources of carbon and energy.

In contrast, complex organic pollutants such as PAHs and chlorinated compounds, notorious for their persistence in the environment, pose significant challenges for microbial degradation [236]. The stability of PAHs, conferred by their multiple aromatic rings, and the resistance of chlorinated compounds, due to the presence of chlorine atoms, make these pollutants less readily degradable. These characteristics impede the enzymatic activity of microorganisms, often requiring specific enzymes capable of breaking these strong chemical bonds. Furthermore, the presence of functional groups, including hydroxyl and carboxyl groups, can influence the degradation process. Hydroxyl groups increase solubility through hydrogen bonding with water molecules and serve as reactive sites for oxidation and substitution reactions. Carboxyl groups, being both polar and ionizable, further enhance solubility, especially at higher pH levels where they form negatively charged carboxylate ions. These groups are also highly reactive, participating in esterification, redox processes, and interactions with other environmental components. Collectively, these functional groups influence the mobility, transformation, and potential ecological impact of pollutants in aquatic and terrestrial systems [237–239]. These groups may alter the solubility and reactivity of the pollutants, thereby affecting their availability to microbes and the efficiency of degradation enzymes [240].

Besides, compounds containing halogens (e.g., chlorine, bromine) or nitro groups (e.g., nitrobenzene, TNT) are generally more resistant to biodegradation [68,241,242]. The carbon-halogen bond is relatively stable, and halogenated compounds can be toxic to microorganisms. Examples include chlorinated solvents (e.g., trichloroethylene) and pesticides (e.g., DDT). The nitro group is electron-withdrawing, making the compound more stable and less reactive.

Stable compounds that resist chemical breakdown are generally more persistent in the environment and less biodegradable, such as microplastics. The effects of microplastics on soil microorganisms in forest ecosystems are reflected in many aspects, such as reducing the number and activity of microorganisms, changing the structure of microbial community, and indirectly affecting microorganisms through changing the physical and chemical properties of soil [243]. The presence of microplastics will reduce the number and activity of soil microorganisms and interfere with their normal physiological functions, which may lead to a decrease in the ability of microorganisms to decompose organic matter, which in turn affects the nutrient cycling of soil and the overall function of forest ecosystems. In addition, microplastics may also change the community structure of soil microorganisms, some microbial species sensitive to microplastics may decrease, and the microbial species tolerant to or able to use microplastics may increase [244]. Such changes in community structure not only affect soil biodiversity but also pose potential threats to the stability and functioning of forest ecosystems. In addition to their direct effects on microorganisms, microplastics may also indirectly affect microorganisms by altering the physical and chemical properties of soil. For example, microplastics may reduce soil permeability, water retention, and fertility, thereby altering the living environment and activity of microorganisms. These impacts may pose potential threats to the health and stability of forest ecosystems [245].

Water solubility also plays an important part in the process of biodegradation of organic pollutants. Hydrophilic (water-soluble) compounds are generally more accessible to microorganisms and therefore more readily biodegradable [246]. Examples include many simple alcohols and acids. Hydrophobic (water-insoluble) compounds, such as high-molecular-weight PAHs and hydrocarbons, tend to adsorb strongly to soil organic matter and are less accessible to microorganisms. This reduces their biodegradability.

The differential biodegradability of pollutants highlights the necessity for tailored remediation strategies [247]. The diverse nature of soil contaminants demands a multifaceted approach to bioremediation, one that takes into account the specific characteristics of each pollutant and the capabilities of the microbial community. Enhancing microbial diversity within the soil is crucial for this purpose, as a varied microbial population is more likely to contain species or strains capable of degrading a wider range of pollutants. Strategies such as bioaugmentation, where specific microbial degraders are introduced to the soil, and biostimulation, which involves modifying environmental conditions to favor the growth of degrading microorganisms, can be employed to enhance the degradation of challenging pollutants [248].

Future research in microbial degradation in forest ecosystems will likely focus on harnessing microbial consortia and synthetic biology for targeted pollutant removal, exploring mycorrhizal-microbial symbioses for enhanced bioremediation, and discovering novel enzymes to degrade persistent pollutants. Additionally, investigating how climate change impacts microbial communities and their degradation efficiency could provide insights for adaptive remediation strategies. Integrating omics technologies with ecological modeling will be key in predicting microbial responses to environmental stressors, paving the way for more effective, sustainable pollution management.

5. Technological Advances and Methodological Approaches

5.1. Genetic Engineering of Microorganisms

Genetic engineering stands at the forefront of enhancing microbial degradation capabilities [249]. By modifying the genetic makeup of microorganisms, scientists have been able to create strains with enhanced abilities to degrade specific pollutants, such as recalcitrant hydrocarbons and complex pesticides. This involves the introduction of genes encoding for particular enzymes that can break down pollutants into less harmful substances [250]. For instance, the engineering of bacterial strains to express enzymes capable of degrading polychlorinated biphenyls (PCBs) has shown promise in remediation efforts [251]. This approach not only increases the efficiency of degradation but also expands the range of pollutants that can be addressed through microbial processes. Practical examples of using genetically engineered microorganisms to degrade soil organic pollutants are shown in Table 6.

Table 6. Practical examples of using genetically engineered microorganisms to degrade soil organic pollutants.

Pollutant Category	Microbial Species	Genetic Engineering Technology	Pollutant Name	Degradation Time	Degradation End Product	Ref.
Persistent Organic Pollutants (POPs)	*Bacillus subtilis*	Expression of dehalogenase genes	DDT	7 days	Non-toxic products	[252]
	Pseudomonas putida	Introduction of linA gene	Lindane	10 days	Degradation products (less toxic)	[253]
	Rhodococcus erythropolis	Engineering of dioxygenase genes	PCBs	14 days	Non-chlorinated compounds	[254]
	Mycobacterium vanbaalenii	Genetic modification for biphenyl dioxygenase	PCBs	12 days	Degradation intermediates	[255]
	Cupriavidus necator	Introduction of organophosphorus hydrolase gene	Parathion	9 days	Non-toxic products	[256]
Petroleum Hydrocarbons	*Alcanivorax borkumensis*	Alkane hydroxylase genes	Alkanes (crude oil)	15 days	Carbon dioxide and water	[83]
	Pseudomonas aeruginosa	Modification of n-alkane-degrading enzymes	Long-chain alkanes	8 days	Alkanes and fatty acids	[257]
	Bacillus licheniformis	PAH dioxygenase genes	PAHs	11 days	Less toxic degradation products	[258]
	Rhodococcus sp.	Enhanced catabolic pathways	Petroleum hydrocarbons	10 days	Less complex hydrocarbons	[259]
	Burkholderia cepacia	PAH-degrading plasmids	High molecular weight PAHs	12 days	Degradation intermediates	[260]

Table 6. *Cont.*

Pollutant Category	Microbial Species	Genetic Engineering Technology	Pollutant Name	Degradation Time	Degradation End Product	Ref.
Pharmaceuticals and Personal Care Products (PPCPs)	*Pseudomonas putida*	Tetracycline-degrading genes	Tetracycline	6 days	Degradation products	[261]
	Escherichia coli	Esterase genes for steroid degradation	Estrone, Estradiol	7 days	Non-toxic products	[262]
	Bacillus thuringiensis	Sulfonamide-degrading enzymes	Sulfonamide antibiotics	8 days	Degradation products	[263]
	Rhizobium leguminosarum	Advanced biotransformation pathways	Various PPCPs	9 days	Less harmful products	[264]
	Acinetobacter baumannii	Synthetic estrogen-degrading pathways	Synthetic estrogens	10 days	Degradation products	[265]
Volatile Organic Compounds (VOCs)	*Pseudomonas putida*	Toluene-degrading operon	Toluene	5 days	Benzene and other derivatives	[266]
	Burkholderia fungorum	Xylenes-degrading pathways	Xylenes	7 days	Degradation products	[267]
	Rhodococcus sp.	VOC-degrading plasmids	Benzene, Ethylbenzene	6 days	Less harmful products	[268]
	Comamonas testosteroni	Halogenated VOC-degrading pathways	Trichloroethylene	8 days	Degradation products	[269]
	Mycobacterium smegmatis	Specialized VOC degradation pathways	Various VOCs	9 days	Degradation products	[270]

Genetic engineering enhances microbial degradation of soil organic pollutants by expanding catabolic pathways, improving enzyme efficiency, increasing resilience to harsh environmental conditions, enabling co-metabolic processes, and accelerating remediation time. However, potential risks include ecological disruptions, horizontal gene transfer, uncontrolled persistence of engineered microbes, and the generation of toxic byproducts. To mitigate these risks and maximize benefits, future efforts should prioritize enhancing field adaptability, systematically monitoring ecological impacts, and developing robust containment strategies for the safe and effective application of engineered microorganisms in environmental remediation.

5.2. Next-Generation Microbial Agents

Future research is poised to focus on developing next-generation microbial agents with enhanced degradation capabilities. This involves harnessing genetic engineering and synthetic biology to create microorganisms specifically tailored to degrade complex and recalcitrant pollutants under a variety of environmental conditions [271]. By integrating computational biology and machine learning algorithms, researchers can design microbial strains with optimized metabolic pathways for the efficient breakdown of pollutants, potentially revolutionizing the field of bioremediation [272].

The next generation of biological agents has shown great application potential in the degradation of soil organic matter (Table 7). These biologics are able to decompose organic pollutants efficiently and environmentally in soil, such as pesticide residues and petroleum hydrocarbons, by utilizing the metabolic capacity and enzyme systems of specific microorganisms. They can not only effectively reduce environmental pollution but also restore soil health and improve soil fertility and biodiversity. The development and application of a new generation of biologics has brought innovative solutions to the field of soil remediation and environmental protection, contributing to sustainable soil use and management.

Table 7. Next-generation biological agents for degrading soil organic pollutants, their target pollutants, mechanisms, and references.

Biological Agent	Target Pollutants	Mechanism	Ref.
Mixotrophic Cyanobacteria and Microalgae	Organic pollutants, hydrocarbons	Mixotrophic pathways enhance degradation efficiency and carbon sequestration	[273]
Earthworms	Pesticides and organic pollutants	Gut enzymes and microbial stimulation in processed soil	[274]
Microbial Consortia	PAHs, petroleum hydrocarbons, pharmaceuticals	Synergistic microbial interactions for efficient degradation	[275]
Photocatalytic Materials with Microbes (ICPB)	Refractory organic pollutants (e.g., VOCs)	Combines photocatalysis and biodegradation	[276]
Genetically Modified Microorganisms	PCBs, PAHs, xenobiotic organic compounds	Enhanced gene transfer for pollutant degradation pathways	[277]
Rhizosphere Microbial Synergies	Hydrophobic and persistent organic pollutants	Enhanced degradation through plant-microbe interactions in the rhizosphere	[278]

Next-generation biological agents provide substantial benefits for the degradation of soil organic pollutants by enhancing efficiency, adapting to harsh conditions, targeting specific contaminants, and fostering synergistic interactions, while maintaining cost-effectiveness and sustainability. Potential risks include ecological disruptions, horizontal gene transfer, uncontrolled persistence, and the generation of toxic byproducts. To fully realize their potential and minimize risks, future efforts should prioritize optimized agent design, robust containment strategies, and comprehensive ecological risk assessments.

5.3. Nanotechnologies

The integration of nanotechnology is the integration of cutting-edge research results in nanotechnology with technologies and methods in other fields to achieve more efficient and accurate innovative applications. This interdisciplinary integration has not only promoted the development of nanomaterials and nanodevices but also brought revolutionary changes to the fields of environmental remediation, biomedicine, and energy science and technology [279,280].

Recent research on nanotechnology for microbial degradation of organic pollutants in forest ecosystems shows remarkable progress (Table 8). The application of nanotechnology can effectively improve the efficiency and rate of microbial degradation of organic pollutants. As a catalyst or carrier, nanomaterials can enhance the enzyme activity of microorganisms and promote the biodegradation process of organic pollutants. In addition, nanotechnology can also improve the growth conditions of microorganisms, enhancing their adaptability and survival rate in polluted environments to ensure the stability and sustainability of microbial degradation processes. These research results provide a new idea and method for the bioremediation of organic pollutants in forest ecosystems and have broad applications prospects [281].

Nanotechnologies enhance microbial degradation of soil organic pollutants by increasing pollutant bioavailability, accelerating degradation rates, enhancing microbial activity, and enabling the breakdown of persistent contaminants such as PAHs and VOCs. Additionally, they integrate seamlessly with existing remediation strategies, including phytoremediation and bioaugmentation. However, potential risks include ecotoxic effects on soil microbial communities, long-term environmental persistence of nanoparticles, facilitation of horizontal gene transfer, production of secondary pollutants, and unresolved regulatory challenges. To fully leverage their benefits while minimizing risks, future efforts must focus on designing environmentally safe nanomaterials, evaluating their long term ecological impacts, and developing comprehensive regulatory frameworks for their application in soil remediation.

Table 8. Examples of nanotechnologies used to enhance microbial degradation of soil organic pollutants, target pollutants, and mechanisms.

Nanotechnology	Target Pollutants	Mechanism	Ref.
Nanobiosurfactants	Organic pollutants, pesticides, and herbicides	Enhances pollutant bioavailability for microbial degradation	[282]
Carbon Nanotubes (CNTs)	Persistent organic pollutants (POPs), heavy metals	Facilitates adsorption and transport of pollutants, improving microbial accessibility	[283]
Zero-Valent Iron Nanoparticles (nZVI)	Petroleum hydrocarbons, VOCs	Catalyzes degradation and immobilization of contaminants	[284]
Nanophytoremediation	Organic pollutants, heavy metals	Assists plant-microbe interactions to detoxify pollutants	[285]
Nanobiosorbents	Agrochemicals and organic pollutants	Enhances microbial bioremediation by adsorbing and concentrating pollutants	[282]
Green Nanomaterials (e.g., nano-chitosan)	Pesticides and industrial pollutants	Improves pollutant adsorption while minimizing toxicity to microbes	[286]
Metallic Nanoparticles	Organic dyes, antibiotics, and pesticides	Facilitates enzymatic degradation and pollutant breakdown	[287]
Modified nZVI Nanoparticles	Hydrocarbons, heavy metals	Reduced toxicity to microbes, promoting enhanced microbial activity	[283]

5.4. Bio-Electrochemical Technologies

Bioelectrochemical technologies enhance the degradation of soil organic pollutants through the intricate coupling of microbial metabolism and electrochemical reactions [288]. The core mechanism hinges on extracellular electron transfer (EET), where electroactive microbes transfer electrons to electrodes via direct pathways (nanowires, membrane proteins) or indirect mediators (e.g., flavins) or accept electrons from electrodes, driving pollutant oxidation or reduction [289]. For instance, in anaerobic settings, electrodes substitute oxygen as electron acceptors, facilitating the breakdown of PAHs or alkanes; in aerobic conditions, electrodes supply electrons to boost oxidase activity, accelerating phenol degradation [122]. Recent soil microbial fuel cell (SMFC) breakthroughs include Mohanakrishna et al.'s 46% petroleum hydrocarbon removal with carbon electrodes, underscoring the pivotal role of electrode-surface biofilms [290]; Zhou et al.'s 2.7-fold PAH degradation increase (22.5% removal of five- and six-ring PAHs) by mixing soil with sand (2:1) to enhance mass transfer, highlighting substrate permeability's impact on EET [291]; and Wang et al.'s 70% chlorinated hydrocarbon removal in dual-chamber SMFCs, combining electroosmosis with microbial synergy [292]. Research shows that biochar enhances electrode conductivity and microbial adhesion, while surfactants improve pollutant solubility, yet challenges persist at the electrode–soil interface, including electron transfer efficiency, microbial community stability, and electrode degradation over time. Large-scale deployment is hampered by mass transfer resistance, soil heterogeneity, and microbial dependence on electrodes. BES offers low energy consumption, sustainability, and energy recovery potential but is limited by prolonged treatment durations, efficiency variability, and engineering complexities.

Briefly, genetic engineering of microorganisms offers significant potential for improving organic matter degradation in soils by enhancing the metabolism of recalcitrant compounds like PAHs and PCBs and increasing stress tolerance. However, challenges remain due to competition with native microbes, horizontal gene transfer, and regulatory barriers. Next-generation microbial agents, such as consortia, rhizosphere-targeted microbes, and mixotrophic algae, enhance pollutant degradation through synergistic microbial-plant interactions but face difficulties in ensuring consistent performance. Nanotechnologies, including zero-valent iron nanoparticles (nZVI) and nanobiosurfactants, boost bioavailability and degradation rates, especially for persistent pollutants, but pose risks like toxicity to native microbes and high costs. Future research should focus on developing non-toxic, biodegradable nanomaterials, improving microbial survivability, and ensuring ecological containment. Combining genetic engineering, microbial consortia, and nanotechnologies could leverage their respective strengths, such as targeting specific pollutants, enhancing pollutant bioavailability, and achieving faster degradation rates. Additionally, large-scale studies in diverse environments are essential to address the gap between laboratory results and real-world applications. Effective cost management and scalability are critical to ensure practical deployment on a wider scale.

Bioremediation methods like genetic engineering, bioaugmentation, and biostimulation each offer unique benefits for degrading organic pollutants in forest soils but face distinct challenges. Genetic engineering enhances microbial efficiency by introducing genes (e.g., in *Pseudomonas* or *Rhodococcus*) to target recalcitrant pollutants like PCBs and PAHs, achieving higher degradation rates, though it risks biosafety, regulatory issues, and horizontal gene transfer. Bioaugmentation adds specialized microbes (e.g., for hydrocarbons) to boost degradation but struggles with microbial survival and integration into native communities. Biostimulation stimulates native microbes with nutrients or environmental adjustments (e.g., pH, oxygen) to enhance hydrocarbon degradation cost-effectively, yet its efficacy depends on the natural microbial capacity and precise condition control, limiting scalability.

Future research should emphasize ecological safety, regulatory clarity, and real-time monitoring tools. Minimizing unintended consequences, such as the spread of engineered genes or nanoparticle toxicity, requires comprehensive risk assessments and containment strategies. Collaboration between scientists, policymakers, and industries is crucial to develop clear guidelines for these technologies. Additionally, advancements in biosensors and analytical tools will enable real-time evaluation of remediation efficiency, safety, and ecological impacts, ensuring sustainable and effective soil restoration.

Despite the promising potential of bioremediation, several challenges remain that hinder its widespread implementation. One significant obstacle is the rivalry among microbial communities. Microbial communities in contaminated environments often compete for resources, and this competition can hinder the efficiency of pollutant degradation. For example, when introducing engineered or enhanced strains, they may struggle to outcompete native microbes, leading to reduced degradation efficiency. Additionally, there is the possibility of harmful byproduct formation during degradation. Some microorganisms may generate intermediate metabolites that are toxic or recalcitrant, complicating the bioremediation process and potentially leading to secondary pollution. Lastly, cost-effectiveness remains a major concern, particularly for large-scale applications. While genetic engineering and bioaugmentation can enhance degradation rates, the high costs associated with producing GMOs and scaling these methods in real-world environments must be addressed for bioremediation to become more widely adopted.

Future research should focus on leveraging innovative methods such as metagenomics and synthetic biology to overcome existing bioremediation challenges. Metagenomics, for instance, allows for the comprehensive analysis of microbial communities in polluted environments without the need for cultivation. By identifying the functional genes involved in pollutant degradation, metagenomics can help uncover new microbial species and enzymes that are effective in breaking down complex pollutants, thus improving bioremediation efficiency. Additionally, synthetic biology holds great promise in designing customized microorganisms that are more efficient in degrading specific pollutants. Through gene editing techniques like CRISPR/Cas9 [293], engineered microbes can be optimized to withstand harsh environmental conditions and improve their long-term stability in the field. Moreover, synthetic biology can facilitate the production of biodegradable materials and bio-based catalysts, reducing the environmental impact and cost of large-scale remediation efforts.

Several recent studies have begun to address these challenges using innovative approaches. For example, metagenomics was used in a study by Liu et al. [294], where researchers analyzed the microbial communities in an oil-contaminated site and identified new genes responsible for hydrocarbon degradation. This discovery led to the development of novel bioremediation strategies tailored to specific pollutants. Similarly, synthetic biology has been applied in the field of heavy metal remediation, where engineered bacteria, such as *Pseudomonas fluorescens*, have been designed to absorb and detoxify arsenic and cadmium. In a study by Sharma et al. [295], genetically modified strains demonstrated significantly higher degradation rates than their wild counterparts, highlighting the potential of synthetic biology to address large-scale contamination issues.

6. Conclusions

Addressing soil pollution in forest ecosystems is a critical environmental challenge, and microbial degradation has emerged as a promising solution. This process leverages the natural metabolic capabilities of microorganisms to transform harmful organic pollutants into less toxic substances, thereby safeguarding ecosystem health and promoting soil functionality. The study emphasizes the importance of understanding microbial community dynamics and optimizing environmental conditions to enhance biodegradation processes. This approach aligns with sustainable forest management practices and reinforces the ecological balance necessary for long-term ecosystem stability resilience.

Advances in genetic engineering, metagenomics, and biostimulators have significantly improved the potential for microbial remediation, providing innovative tools to enhance degradation efficiency. By integrating these technologies, scientists have deepened their understanding of microbial ecology and developed novel strategies for addressing complex pollutants. However, challenges remain, including maintaining microbial diversity and adapting to pollutant complexity. Overcoming these hurdles will require interdisciplinary collaboration, combining insights from microbiology, ecology, and environmental engineering to develop robust, practical solutions. Future research must focus on refining microbial techniques and ensuring their ecological sustainability, contributing to effective and comprehensive soil pollution management.

The next decade of research on microbial degradation of soil organic pollutants is expected to focus on several innovative areas. First, synthetic biology will drive the engineering of microbial strains and consortiums with enhanced pollutant-degrading capabilities through genome modifications and optimized metabolic pathways. In parallel, omics technologies such as metagenomics and transcriptomics will allow for a deeper understanding of microbial communities and functional genes involved in pollutant degradation, leading to the discovery of novel degrading microbes. The integration of microbial-electrochemical systems will gain traction, with bio-electrochemical technologies enhancing pollutant breakdown through improved electron transfer. Research will also explore microbial community dynamics and adaptation, focusing on microbial resilience and the evolution of degradation pathways under varying environmental stressors. Moreover, eco-engineering approaches combining microbes with soil amendments will aim to improve both pollutant degradation and soil restoration. Lastly, the synergy between microbes and plants in phytoremediation will be further developed, with research focusing on microbial support for plant health in degrading complex pollutants in the rhizosphere. These advancements hold promise for creating more efficient, sustainable, and multifaceted bioremediation strategies in contaminated soils.

A pivotal development in this landscape is the rise of next-generation microbial agents, which are set to redefine bioremediation strategies. These agents—genetically engineered microorganisms tailored for heightened specificity, efficiency, and environmental adaptability—leverage cutting-edge synthetic biology tools, such as CRISPR-Cas9, and AI-driven design to optimize their pollutant-degrading capabilities. These advancements position next-generation microbial agents as superior alternatives to traditional bioremediation methods, particularly for tackling persistent pollutants like PAHs and emerging contaminants such as microplastics and PPCPs.

In terms of their proportion in future bioremediation strategies, next-generation microbial agents are expected to assume an increasingly significant role, potentially comprising 30–50% of applied approaches as the technology matures and regulatory frameworks evolve. This projection is supported by their demonstrated superiority in degradation efficiency and specificity, as evidenced by laboratory successes and preliminary field trials. Moreover, their integration with emerging technologies—such as nanotechnology and plant-microbe synergies—will further amplify their effectiveness and sustainability. However, their widespread adoption hinges on overcoming key challenges, including ecological risks, long-term stability, and cost considerations, which will require extensive field validation and risk assessments. As these barriers are addressed, next-generation microbial agents are poised to become a cornerstone of efficient, sustainable, and multifaceted bioremediation strategies, particularly in managing contaminated forest soils and addressing the complexities of modern pollutants.

Supplementary Materials: The following supporting information can be downloaded at https://www.mdpi.com/article/10.3390/pr13030916/s1: Table S1: The hazards of organic pollutants in forest soil.

Author Contributions: P.L.: conceptualization, formal analysis, investigation, data curation, writing—original draft preparation; S.W.: software, visualization, investigation; S.Z.: writing—review and editing, supervision, resources; X.H.: formal analysis, validation, supervision, writing—review and editing, resources; Y.W.: software, visualization, investigation. All authors have read and agreed to the published version of the manuscript.

Funding: This research was funded by the Scientific Research Program of Hunan Provincial Department of Education, (Nos. 23B1134 and 21A0154), and the Natural Science Foundation of Hunan Province (No. 2024JJ5646).

Conflicts of Interest: The authors declare no conflict of interest.

References

1. Bharat Raj, S.; Lal, R.; Blum, W.H.; Valentin, C.; Stewart, B.A. Soil Pollution and Contamination. In *Methods for Assessment of Soil Degradation*; CRC Press eBooks; Taylor & Francis Group: Abington, UK, 2020. [CrossRef]
2. Zhikun, C.; Muhammad, I.; Guanghua, J.; Weixi, W.; Biao, H.; Yingmei, L.; Yanxia, Z.; Yizhe, Y.; Qingtao, L.; Zhao, Z.; et al. Toxic elements pollution risk as affected by various input sources in soils of greenhouses, kiwifruit orchards, cereal fields, and forest/grassland. *Environ. Pollut.* **2023**, *338*, 122639. [CrossRef]
3. Su, C.T.; Zou, S.Z.; Tang, J.S.; Liang, B. Influence of different ecosystems on the soil moisture and karst effect in Luota, West Hunan Province of China. *Carbonates Evaporites* **2018**, *33*, 187–193. [CrossRef]
4. Vanguelova, E.I.; Benham, S.; Pitman, R.; Moffat, A.J.; Broadmeadow, M.; Nisbet, T.; Durrant, D.; Barsoum, N.; Wilkinson, M.; Bochereau, F.; et al. Chemical fluxes in time through forest ecosystems in the UK—Soil response to pollution recovery. *Environ. Pollut.* **2010**, *158*, 1857–1869. [CrossRef] [PubMed]
5. Emmanuel, A.; Rebeca, B. Microbial Remediation of Heavy Metals Contamination in Soil. Zenodo (CERN European Organization for Nuclear Research). 2022. Available online: https://zenodo.org/records/7081192 (accessed on 1 November 2021).

6. Janick, F.A.; James, W.; Stanislav, M.; Michael, A.C. Soil and Land Pollution. In *Environmental and Pollution Science*, 3rd ed.; Elsevier: Amsterdam, The Netherlands, 2019. [CrossRef]
7. María, J.G.-G.; Houshang, A.; Izzie, A.; Vikki, A.; Alexandra, M.; Brett, R.; Sky, H.; Prosser, J.A.; Jacqui, H. Phytoremediation of microbial contamination in soil by New Zealand native plants. *Appl. Soil Ecol.* **2021**, *167*, 104040. [CrossRef]
8. Niego, A.G.T.; Rapior, S.; Thongklang, N.; Raspé, O.; Hyde, K.D.; Mortimer, P. Reviewing the contributions of macrofungi to forest ecosystem processes and services. *Fungal Biol. Rev.* **2023**, *44*, 100294. [CrossRef]
9. Martina, S.G.; Colin, D.C.; Killham, K.; James, I.P.; Lesley Anne, G. Bacterial diversity promotes community stability and functional resilience after perturbation. *Environ. Microbiol.* **2005**, *7*, 301–313. [CrossRef]
10. Corsolini, S.; Sarà, G. The trophic transfer of persistent pollutants (HCB, DDTs, PCBs) within polar marine food webs. *Chemosphere* **2017**, *177*, 189–199. [CrossRef]
11. Talreja, N.; Hegde, C.; Kumar, E.M.; Chavali, M. Emerging Environmental Contaminants: Sources, Consequences and Future Challenges. In *Green Technologies for Industrial Contaminants*; John Wiley & Sons: Hoboken, NJ, USA, 2025; pp. 119–149.
12. Sonibare, O.; Alimi, H.; Jarvie, D.; Ehinola, O. Origin and occurrence of crude oil in the Niger delta, Nigeria. *J. Pet. Sci. Eng.* **2008**, *61*, 99–107. [CrossRef]
13. Zhang, D.; Liu, W.; Yin, C.; She, L.; Ren, J.; Xu, Q.; Wang, S.; Peng, Y. Occurrence of contaminants of emerging concern in surface and waste water from the Yangtze River chemical contiguous zone, China: Distribution, sources and ecological risk assessment. *Sci. Total Environ.* **2024**, *949*, 175151. [CrossRef]
14. Gu, S.; Guenther, A.; Faiola, C. Effects of anthropogenic and biogenic volatile organic compounds on Los Angeles air quality. *Environ. Sci. Technol.* **2021**, *55*, 12191–12201. [CrossRef]
15. Okweye, P.; Garner, K.; McCullers, K.; Hutchinson, M.; Sheeley, N. The Presence of Contaminants of Emerging Concern (CECs) and Volatile Organic Compounds (VOCs) in Northern Alabama Aquatic Ecosystems. *J. Environ. Sci. Eng. B* **2021**, *10*, 77–89.
16. Clarke, N.; Fuksová, K.; Gryndler, M.; Lachmanová, Z.; Liste, H.-H.; Rohlenová, J.; Schroll, R.; Schröder, P.; Matucha, M. The formation and fate of chlorinated organic substances in temperate and boreal forest soils. *Environ. Sci. Pollut. Res.* **2009**, *16*, 127–143. [CrossRef] [PubMed]
17. Wu, Q.; Cao, Y.; Yu, T.; Yang, J.; Fan, S.; Feng, C.; Liu, Z.; Huang, C. A Scientometric Analysis and Visualization of Forest Soil Contamination Research from Global Perspectives. *Forests* **2024**, *15*, 1068. [CrossRef]
18. Aichner, B.; Bussian, B.; Lehnik-Habrink, P.; Hein, S. Levels and spatial distribution of persistent organic pollutants in the environment: A case study of German forest soils. *Environ. Sci. Technol.* **2013**, *47*, 12703–12714. [CrossRef] [PubMed]
19. Štrbac, S.; Kašanin-Grubin, M.; Stojić, N.; Pezo, L.; Lončar, B.; Tognetti, R.; Pucarević, M. Persistent organic pollutants in soil samples from mountain beech forests across Europe. *Plant Soil* **2024**, *495*, 313–339. [CrossRef]
20. Belis, C.; Offenthaler, I.; Uhl, M.; Nurmi-Legat, J.; Bassan, R.; Jakobi, G.; Kirchner, M.; Knoth, W.; Kräuchi, N.; Levy, W. A comparison of Alpine emissions to forest soil and spruce needle loads for persistent organic pollutants (POPs). *Environ. Pollut.* **2009**, *157*, 3185–3191. [CrossRef]
21. Gong, P.; Xu, H.; Wang, C.; Chen, Y.; Guo, L.; Wang, X. Persistent organic pollutant cycling in forests. *Nat. Rev. Earth Environ.* **2021**, *2*, 182–197. [CrossRef]
22. Nam, J.J.; Gustafsson, O.; Kurt-Karakus, P.; Breivik, K.; Steinnes, E.; Jones, K.C. Relationships between organic matter, black carbon and persistent organic pollutants in European background soils: Implications for sources and environmental fate. *Environ. Pollut.* **2008**, *156*, 809–817. [CrossRef]
23. Valentin, L.; Nousiainen, A.; Mikkonen, A. Introduction to organic contaminants in soil: Concepts and risks. In *Emerging Organic Contaminants in Sludges: Analysis, Fate and Biological Treatment*; Springer: Berlin/Heidelberg, Germany, 2013; pp. 1–29.
24. Gaur, N.; Narasimhulu, K.; PydiSetty, Y. Recent advances in the bio-remediation of persistent organic pollutants and its effect on environment. *J. Clean. Prod.* **2018**, *198*, 1602–1631. [CrossRef]
25. Jones, K.C.; De Voogt, P. Persistent organic pollutants (POPs): State of the science. *Environ. Pollut.* **1999**, *100*, 209–221. [CrossRef]
26. Kuppusamy, S.; Maddela, N.R.; Megharaj, M.; Venkateswarlu, K.; Kuppusamy, S.; Maddela, N.R.; Megharaj, M.; Venkateswarlu, K. Ecological impacts of total petroleum hydrocarbons. In *Total Petroleum Hydrocarbons: Environmental Fate, Toxicity, and Remediation*; Springer: Berlin/Heidelberg, Germany, 2020; pp. 95–138.
27. Pinedo, J.; Ibáñez, R.; Lijzen, J.; Irabien, A. Assessment of soil pollution based on total petroleum hydrocarbons and individual oil substances. *J. Environ. Manag.* **2013**, *130*, 72–79. [CrossRef]
28. Sharma, I.; Khare, N.; Singh, K.; Dahiya, V.S. Environmental Applications and Emerging Pollutants: Monitoring and Remediation Techniques. In *Biotechnology for Environmental Sustainability*; Springer: Berlin/Heidelberg, Germany, 2025; pp. 637–660.
29. Zhu, X.; Liu, S.; Gao, X.; Gu, Y.; Yu, Y.; Li, M.; Chen, X.; Fan, M.; Jia, Y.; Tian, L. Typical emerging contaminants in sewage treatment plant effluent, and related watersheds in the Pearl River Basin: Ecological risks and source identification. *J. Hazard. Mater.* **2024**, *476*, 135046. [CrossRef] [PubMed]
30. Havugimana, E.; Bhople, B.S.; Kumar, A.; Byiringiro, E.; Mugabo, J.P.; Kumar, A. Soil pollution–major sources and types of soil pollutants. *Environ. Sci. Eng.* **2017**, *11*, 53–86.

31. Duarte, R.M.; Matos, J.T.; Senesi, N. Organic pollutants in soils. In *Soil Pollution*; Elsevier: Amsterdam, The Netherlands, 2018; pp. 103–126.
32. Zwolak, A.; Sarzyńska, M.; Szpyrka, E.; Stawarczyk, K. Sources of soil pollution by heavy metals and their accumulation in vegetables: A review. *Water Air Soil Pollut.* **2019**, *230*, 164. [CrossRef]
33. Alloway, B. Soil pollution and land contamination. In *Pollution, Causes, Effects and Control*; Royal Society of Chemistry: London, UK, 1996; Volume 3, pp. 352–377.
34. Fischer, J.R.; Zapata, F.; Dubelman, S.; Mueller, G.M.; Jensen, P.D.; Levine, S.L. Characterizing a novel and sensitive method to measure dsRNA in soil. *Chemosphere* **2016**, *161*, 319–324. [CrossRef]
35. Knox, A.; Gamerdinger, A.; Adriano, D.; Kolka, R.; Kaplan, D. Sources and practices contributing to soil contamination. *Bioremediation Contam. Soils* **1999**, *37*, 53–87.
36. Cachada, A.; Pato, P.; Rocha-Santos, T.; da Silva, E.F.; Duarte, A. Levels, sources and potential human health risks of organic pollutants in urban soils. *Sci. Total Environ.* **2012**, *430*, 184–192. [CrossRef]
37. Cui, X.; Cao, X.; Xue, W.; Xu, L.; Cui, Z.; Zhao, R.; Ni, S.-Q. Integrative effects of microbial inoculation and amendments on improved crop safety in industrial soils co-contaminated with organic and inorganic pollutants. *Sci. Total Environ.* **2023**, *873*, 162202. [CrossRef]
38. Reid, A.S.; Jerald, L.S. Phytoremediation, Bioaugmentation, and the Plant Microbiome. *Environ. Sci. Technol.* **2022**, *56*, 16602–16610. [CrossRef]
39. Sanjana, M.; Prajna, R.; Katti, U.S.; Kavitha, R.V. Bioremediation—The recent drift towards a sustainable environment. *Environ. Sci. Adv.* **2024**, *3*, 1097–1110. [CrossRef]
40. Muhammad, M.; Batool, S.; Hivare, V.; Li, W.-J.; Waheed, A.; Sinha, D. Chapter 1—Bioremediation techniques—Classification, principles, advantages, limitations, and prospects. In *Microbiome-Assisted Bioremediation*; Parray, J.A., Li, W.-J., Eds.; Academic Press: Cambridge, MA, USA, 2024; pp. 1–23. [CrossRef]
41. Sun, Y.; Shaheen, S.M.; Ali, E.F.; Abdelrahman, H.; Sarkar, B.; Song, H.; Rinklebe, J.; Ren, X.; Zhang, Z.; Wang, Q. Enhancing microplastics biodegradation during composting using livestock manure biochar. *Environ. Pollut.* **2022**, *306*, 119339. [CrossRef] [PubMed]
42. Zhao, K.; Yin, X.; Wang, N.; Chen, N.; Jiang, Y.; Deng, L.; Xiao, W.; Zhou, K.; He, Y.; Zhao, X.; et al. Optimizing the management of aerobic composting for antibiotic resistance genes elimination: A review of future strategy for livestock manure resource utilization. *J. Environ. Manag.* **2024**, *370*, 122766. [CrossRef]
43. Lahori, A.H.; Tunio, M.; Ahmed, S.R.; Mierzwa-Hersztek, M.; Vambol, V.; Afzal, A.; Kausar, A.; Vambol, S.; Umar, A.; Muhammad, A. Role of pressmud compost for reducing toxic metals availability and improving plant growth in polluted soil: Challenges and recommendations. *Sci. Total Environ.* **2024**, *951*, 175493. [CrossRef]
44. Lin, C.; Cheruiyot, N.K.; Bui, X.T.; Ngo, H.H. Composting and its application in bioremediation of organic contaminants. *Bioengineered* **2022**, *13*, 1073–1089. [CrossRef] [PubMed]
45. López, A.M.Q.; dos Santos Silva, A.L. Biostimulation and Bioaugmentation. In *Genomics Approach to Bioremediation*; John Wiley & Sons: Hoboken, NJ, USA, 2023; pp. 53–68. [CrossRef]
46. Kumar, R.; Kaushal, S.; Verma, N.; Kumar, P.; Thakur, N.; Kumar, A.; Kumar, S.; Umar, A.; Almas, T.; Pal, K.; et al. Nano bioaugmentation for textile dye remediation: A sustainable approach for health and environment management. *J. Mol. Liq.* **2024**, *415*, 126254. [CrossRef]
47. Kansour, M.K.; Al-Mailem, D.M. Bioremediation of two oil-contaminated Kuwaiti hyper-saline soils by cross bioaugmentation and the role of indigenous halophilic/halotolerant hydrocarbonoclastic bacteria. *Environ. Technol. Innov.* **2023**, *32*, 103259. [CrossRef]
48. Shan, X.; Guo, H.; Ma, F.; Shan, Z. Enhanced treatment of synthetic wastewater by bioaugmentation with a constructed consortium. *Chemosphere* **2023**, *338*, 139520. [CrossRef]
49. Nivetha, N.; Srivarshine, B.; Sowmya, B.; Rajendiran, M.; Saravanan, P.; Rajeshkannan, R.; Rajasimman, M.; Pham, T.H.T.; Shanmugam, V.; Dragoi, E.-N. A comprehensive review on bio-stimulation and bio-enhancement towards remediation of heavy metals degeneration. *Chemosphere* **2023**, *312*, 137099. [CrossRef]
50. Milano, G.; Colosio, A.; Minotta Quebradas, M.J.; Pratobevera, A.; Daffara, V.; Saccomanno, M.F. Biologic augmentation of rotator cuff repair with microfragmented autologous subacromial bursal tissue enveloped in a patch of compressed autologous long head of biceps tendon tissue: The *Bio-Ravioli* technique. *JSES Int.* **2024**, *8*, 1010–1015. [CrossRef]
51. Eyupoglu, S. 8—Characterization techniques for bio-fillers/bio-plasticizers. In *Sustainable Fillers/Plasticizers for Polymer Composites*; Suyambulingam, I., Divakaran, D., Rangappa, S.M., Siengchin, S., Eds.; Elsevier Science Ltd: Amsterdam, The Netherlands, 2025; pp. 185–210. [CrossRef]
52. Samaei, M.R.; Mortazavi, S.B.; Bakhshi, B.; Jafari, A.J.; Shamsedini, N.; Mehrazmay, H.; Ansarizadeh, M. Investigating the effects of combined bio-enhancement and bio-stimulation on the cleaning of hexadecane-contaminated soils. *J. Environ. Chem. Eng.* **2022**, *10*, 106914. [CrossRef]

53. Vijay Pradhap Singh, M.; Ravi Shankar, K. Next-generation hybrid technologies for the treatment of pharmaceutical industry effluents. *J. Environ. Manag.* **2024**, *353*, 120197. [CrossRef]
54. Min, T.; Luo, T.; He, H.; Qin, J.; Wang, Y.; Cheng, L.; Ru, S.; Li, J. Dissolved organic matter–assisted phytoremediation potential of cotton for Cd-contaminated soil: A relationship between dosage and phytoremediation efficiency. *Environ. Sci. Pollut. Res.* **2022**, *29*, 84640–84650. [CrossRef]
55. Liu, N.; Zhao, J.; Du, J.; Hou, C.; Zhou, X.; Chen, J.; Zhang, Y. Non-phytoremediation and phytoremediation technologies of integrated remediation for water and soil heavy metal pollution: A comprehensive review. *Sci. Total Environ.* **2024**, *948*, 174237. [CrossRef]
56. Kaur, H.; Kumar, A.; Bindra, S.; Sharma, A. Phytoremediation: An emerging green technology for dissipation of PAHs from soil. *J. Geochem. Explor.* **2024**, *259*, 107426. [CrossRef]
57. Yin, F.; Li, J.; Wang, Y.; Yang, Z. Biodegradable chelating agents for enhancing phytoremediation: Mechanisms, market feasibility, and future studies. *Ecotoxicol. Environ. Saf.* **2024**, *272*, 116113. [CrossRef]
58. Montreemuk, J.; Stewart, T.N.; Prapagdee, B. Bacterial-assisted phytoremediation of heavy metals: Concepts, current knowledge, and future directions. *Environ. Technol. Innov.* **2024**, *33*, 103488. [CrossRef]
59. Gerhardt, K.E.; Huang, X.-D.; Glick, B.R.; Greenberg, B.M. Phytoremediation and rhizoremediation of organic soil contaminants: Potential and challenges. *Plant Sci.* **2009**, *176*, 20–30. [CrossRef]
60. Xiaodan, S.; Yuqian, Y.; Jiahui, L.; Andrey, S.; Smirnov, P.V.; Yakov, K. Organic Mulching Increases Microbial Activity in Urban Forest Soil. *Forests* **2022**, *13*, 1352. [CrossRef]
61. Saravanan, A.; Kumar, P.S.; Dai-Viet, N.V.; Jeevanantham, S.; Karishma, S.; Yaashikaa, P.R. A review on catalytic-enzyme degradation of toxic environmental pollutants: Microbial enzymes. *J. Hazard. Mater.* **2021**, *419*, 126451. [CrossRef]
62. Kanwar, P.; Yadav, N.; Srivastava, S. Microbial Biodiversity and Bioremediation: A Systematic, Biological and Metabolic Engineering Tool. In *Microbiology-2.0 Update for a Sustainable Future*; Gupta, J., Verma, A., Eds.; Springer Nature: Singapore, 2024; pp. 77–93.
63. Harirchi, S.; Rafieyan, S.; Nojoumi, S.A.; Etemadifar, Z. Microbial Biodegradation and Metagenomics in Remediation of Environmental Pollutants: Enzymes and Mechanisms. In *Omics Insights in Environmental Bioremediation*; Kumar, V., Thakur, I.S., Eds.; Springer Nature: Singapore, 2022; pp. 487–514.
64. Karigar, C.S.; Rao, S.S. Role of microbial enzymes in the bioremediation of pollutants: A review. *Enzym. Res.* **2011**, *2011*, 805187. [CrossRef]
65. Aysha, A.; Muhammad, R.; Ashraf, S.S. Oxidoreductases for the remediation of organic pollutants in water—A critical review. *Crit. Rev. Biotechnol.* **2018**, *38*, 971–988. [CrossRef]
66. Dave, S.; Das, J. Chapter 13—Role of microbial enzymes for biodegradation and bioremediation of environmental pollutants: Challenges and future prospects. In *Bioremediation for Environmental Sustainability*; Saxena, G., Kumar, V., Shah, M.P., Eds.; Elsevier: Amsterdam, The Netherlands, 2021; pp. 325–346. [CrossRef]
67. Pankaj Kumar, G.; Manvi, G. Bioremediation of Organic Pollutants in Soil–Water System: A Review. *Biotech* **2023**, *12*, 36. [CrossRef]
68. Salishcheva, O.; Burlachenko, A.; Tarasova, Y.; Moldagulova, N.; Yustratov, V. Biodegradation of organic compounds in wastewater. *BIO Web Conf.* **2023**, *64*, 01003. [CrossRef]
69. Marcela Alejandra, S.; Ana Silvia, T.; Mónica Lucrecia, B.; Mariana, G.; César Nicolás, P.; María Isabel, F.; Laura, L.; Laura Lidia, V. Evaluation of bioremediation strategies for treating recalcitrant halo-organic pollutants in soil environments. *Ecotoxicol. Environ. Saf.* **2020**, *202*, 110929. [CrossRef]
70. Chakraborty, J.; Das, S. Molecular perspectives and recent advances in microbial remediation of persistent organic pollutants. *Environ. Sci. Pollut. Res.* **2016**, *23*, 16883–16903. [CrossRef]
71. Lin, S.; Wei, J.; Yang, B.; Zhang, M.; Zhuo, R. Bioremediation of organic pollutants by white rot fungal cytochrome P450: The role and mechanism of CYP450 in biodegradation. *Chemosphere* **2022**, *301*, 134776. [CrossRef]
72. Pieper, D.H.; Martins dos Santos, V.t.A.P.; Golyshin, P.N. Genomic and mechanistic insights into the biodegradation of organic pollutants. *Curr. Opin. Biotechnol.* **2004**, *15*, 215–224. [CrossRef]
73. Lü, H.; Wei, J.-L.; Tang, G.-X.; Chen, Y.-S.; Huang, Y.-H.; Hu, R.; Mo, C.-H.; Zhao, H.-M.; Xiang, L.; Li, Y.-W.; et al. Microbial consortium degrading of organic pollutants: Source, degradation efficiency, pathway, mechanism and application. *J. Clean. Prod.* **2024**, *451*, 141913. [CrossRef]
74. Zehnle, H.; Otersen, C.; Benito Merino, D.; Wegener, G. Potential for the anaerobic oxidation of benzene and naphthalene in thermophilic microorganisms from the Guaymas Basin. *Front. Microbiol.* **2023**, *14*, 1279865. [CrossRef]
75. Garcés Mejía, A.C.; Pino, N.J.; Peñuela, G.A. Effect of Secondary Metabolites Present in Brassica nigra Root Exudates on Anthracene and Phenanthrene Degradation by Rhizosphere Microorganism. *Environ. Eng. Sci.* **2017**, *35*, 203–209. [CrossRef]

76. Dong, J.; Kang, Y.; Kuang, S.; Ma, H.; Li, M.; Xiao, J.; Wang, Y.; Guo, Z.; Wu, H. Combined biological effects of polystyrene microplastics and phenanthrene on Tubifex tubifex and microorganisms in wetland sediment. *Chem. Eng. J.* **2023**, *462*, 142260. [CrossRef]
77. Zhang, Z.; Sun, J.; Gong, X.; Wang, C.; Wang, H. Anaerobic biodegradation of pyrene and benzo[a]pyrene by a new sulfate-reducing *Desulforamulus aquiferis* strain DSA. *J. Hazard. Mater.* **2023**, *459*, 132053. [CrossRef] [PubMed]
78. Wójcik, A.; Perczyk, P.; Wydro, P.; Broniatowski, M. Incorporation of cyclodiene pesticides and their polar metabolites to model membranes of soil bacteria. *J. Mol. Liq.* **2020**, *298*, 112019. [CrossRef]
79. Su, F.; Wang, F.; Zhang, C.; Lu, T.; Zhang, S.; Zhang, R.; Qi, X.; Liu, P. Ameliorating substance accessibility for microorganisms to amplify toluene degradation and power generation of microbial fuel cell by using activated carbon anode. *J. Clean. Prod.* **2022**, *377*, 134481. [CrossRef]
80. Toabaita, M.; Vangnai, A.S.; Thiravetyan, P. Removal of Ethylbenzene from Contaminated Air by Zamioculcas Zamiifolia and Microorganisms Associated on Z. Zamiifolia Leaves. *Water Air Soil Pollut.* **2016**, *227*, 115. [CrossRef]
81. Kita, Y.; Amao, Y. Acetoacetate Production from CO_2 and Acetone with Acetone Carboxylase from Photosynthetic Bacteria Rhodobacter Capsulatus. *Catal. Surv. Asia* **2023**, *27*, 67–74. [CrossRef]
82. Shanthi, T.R.; Vasanthy, M.; Mohamed Hatha, A.A. Bioremediation of Lindane-Contaminated Soil and Water Ecosystems: A Review. In *Organic Pollutants: Toxicity and Solutions*; Vasanthy, M., Sivasankar, V., Sunitha, T.G., Eds.; Springer International Publishing: Cham, Switzlerand, 2022; pp. 199–227.
83. Kadri, T.; Magdouli, S.; Rouissi, T.; Brar, S.K. Ex-situ biodegradation of petroleum hydrocarbons using Alcanivorax borkumensis enzymes. *Biochem. Eng. J.* **2018**, *132*, 279–287. [CrossRef]
84. Deng, S.; Wang, B.; Zhang, H.; Qu, R.; Sun, S.; You, Q.; She, Y.; Zhang, F. Degradation and enhanced oil recovery potential of *Alcanivorax borkumensis* through production of bio-enzyme and bio-surfactant. *Bioresour. Technol.* **2024**, *400*, 130690. [CrossRef]
85. Cheng, J.; Yuan, J.; Li, S.; Yang, X.; Lu, Z.; Xu, J.; He, Y. Promoted reductive removal of chlorinated organic pollutants co-occurring with facilitated methanogenesis in anaerobic environment: A systematic review and meta-analysis. *Crit. Rev. Environ. Sci. Technol.* **2022**, *52*, 2582–2609. [CrossRef]
86. Khomyakova, M.; Slobodkin, A. Transformation of methoxylated aromatic compounds by anaerobic microorganisms. *Microbiology* **2023**, *92*, 97–118. [CrossRef]
87. Jaiswal, S.; Singh, D.K.; Shukla, P. Degradation effectiveness of hexachlorohexane (γ-HCH) by bacterial isolate *Bacillus cereus* SJPS-2, its gene annotation for bioremediation and comparison with *Pseudomonas putida* KT2440. *Environ. Pollut.* **2023**, *318*, 120867. [CrossRef] [PubMed]
88. Su, X.; Xie, M.; Han, Z.; Xiao, Y.; Wang, R.; Shen, C.; Hashmi, M.Z.; Sun, F. Resuscitation-promoting factor accelerates enrichment of highly active tetrachloroethene/polychlorinated biphenyl-dechlorinating cultures. *Appl. Environ. Microbiol.* **2023**, *89*, e0195122. [CrossRef] [PubMed]
89. Tao, H.; Zhou, L.; Yu, D.; Chen, Y.; Luo, Y.; Lin, T. Effects of polystyrene microplastics on the metabolic level of *Pseudomonas aeruginosa*. *Sci. Total Environ.* **2024**, *922*, 171335. [CrossRef]
90. Safdar, A.; Ismail, F.; Iftikhar, H.; Majid Khokhar, A.; Javed, A.; Imran, M.; Safdar, B. Determination of Biodegradation Potential of *Aspergillus niger*, *Candida albicans*, and *Acremonium sclerotigenum* on Polyethylene, Polyethylene Terephthalate, and Polystyrene Microplastics. *Int. J. Microbiol.* **2024**, *2024*, 7682762. [CrossRef]
91. Kumar, A.; Mishra, S.; Pandey, R.; Yu, Z.G.; Kumar, M.; Khoo, K.S.; Thakur, T.K.; Show, P.L. Microplastics in terrestrial ecosystems: Un-ignorable impacts on soil characterises, nutrient storage and its cycling. *TrAC Trends Anal. Chem.* **2023**, *158*, 116869. [CrossRef]
92. Zhang, X.; Li, Y.; Ouyang, D.; Lei, J.; Tan, Q.; Xie, L.; Li, Z.; Liu, T.; Xiao, Y.; Farooq, T.H. Systematical review of interactions between microplastics and microorganisms in the soil environment. *J. Hazard. Mater.* **2021**, *418*, 126288. [CrossRef] [PubMed]
93. Wang, H.; Zhou, Q. Bioelectrochemical systems—A potentially effective technology for mitigating microplastic contamination in wastewater. *J. Clean. Prod.* **2024**, *450*, 141931. [CrossRef]
94. Lew, S.; Lew, M.; Biedunkiewicz, A.; Szarek, J. Impact of Pesticide Contamination on Aquatic Microorganism Populations in the Littoral Zone. *Arch. Environ. Contam. Toxicol.* **2013**, *64*, 399–409. [CrossRef]
95. Bose, S.; Kumar, P.S.; Vo, D.-V.N.; Rajamohan, N.; Saravanan, R. Microbial degradation of recalcitrant pesticides: A review. *Environ. Chem. Lett.* **2021**, *19*, 3209–3228. [CrossRef]
96. Zhu, G.; Du, R.; Du, D.; Qian, J.; Ye, M. Keystone taxa shared between earthworm gut and soil indigenous microbial communities collaboratively resist chlordane stress. *Environ. Pollut.* **2021**, *283*, 117095. [CrossRef]
97. Zeldes, B.M.; Keller, M.W.; Loder, A.J.; Straub, C.T.; Adams, M.W.W.; Kelly, R.M. Extremely thermophilic microorganisms as metabolic engineering platforms for production of fuels and industrial chemicals. *Front. Microbiol.* **2015**, *6*, 1209. [CrossRef]
98. Schlüter, M.; Hentzel, T.; Suarez, C.; Koch, M.; Lorenz, W.G.; Böhm, L.; Düring, R.-A.; Koinig, K.A.; Bunge, M. Synthesis of novel palladium(0) nanocatalysts by microorganisms from heavy-metal-influenced high-alpine sites for dehalogenation of polychlorinated dioxins. *Chemosphere* **2014**, *117*, 462–470. [CrossRef]

99. Zanellati, A.; Spina, F.; Bonaterra, M.; Dinuccio, E.; Varese, G.C.; Scarpeci, T.E. Screening and evaluation of phenols and furans degrading fungi for the biological pretreatment of lignocellulosic biomass. *Int. Biodeterior. Biodegrad.* **2021**, *161*, 105246. [CrossRef]
100. Wang, Z.; Hu, L.; He, J.; Zhou, G.; Chen, Z.; Wang, Z.; Chen, J.; Hayat, K.; Hrynsphan, D.; Tatsiana, S. Mechanisms of N, N-dimethylacetamide-facilitated n-hexane removal in a rotating drum biofilter packed with bamboo charcoal-polyurethane composite. *Bioresour. Technol.* **2023**, *372*, 128600. [CrossRef]
101. Joshi, N.; Keshav, A.; Poonia, A.K. Reactive separation of gallic acid using tributyl phosphate dissolved in 2-octanone, laurylalcohol and heptane. *Chem. Data Collect.* **2020**, *25*, 100325. [CrossRef]
102. Zhou, Z.; Zhang, C.-J.; Liu, P.-F.; Fu, L.; Laso-Pérez, R.; Yang, L.; Bai, L.-P.; Li, J.; Yang, M.; Lin, J.-Z.; et al. Non-syntrophic methanogenic hydrocarbon degradation by an archaeal species. *Nature* **2022**, *601*, 257–262. [CrossRef]
103. Panwar, R.; Mathur, J. Remediation of polycyclic aromatic hydrocarbon-contaminated soils using microbes and nanoparticles: A review. *Pedosphere* **2023**, *33*, 93–104. [CrossRef]
104. Wang, Y.; Yuan, D.; Sun, L.; Xu, S.; Wan, S. Conductive and hydrophobically modified sodium alginate hydrogels for enhanced gaseous para-xylene removal by functional bacteria. *Chem. Eng. J.* **2023**, *461*, 142005. [CrossRef]
105. Xu, Q.; Shi, Y.; Ke, L.; Qian, L.; Zhou, X.; Shao, X. Ciprofloxacin enhances cadmium toxicity to earthworm Eisenia fetida by altering the gut microorganism composition. *Environ. Pollut.* **2023**, *333*, 122106. [CrossRef] [PubMed]
106. Wang, Q.; Li, X.; Yang, Q.; Chen, Y.; Du, B. Evolution of microbial community and drug resistance during enrichment of tetracycline-degrading bacteria. *Ecotoxicol. Environ. Saf.* **2019**, *171*, 746–752. [CrossRef]
107. Mantegazza, G.; Dalla Via, A.; Licata, A.; Duncan, R.; Gardana, C.; Gargari, G.; Alamprese, C.; Arioli, S.; Taverniti, V.; Karp, M.; et al. Use of kefir-derived lactic acid bacteria for the preparation of a fermented soy drink with increased estrogenic activity. *Food Res. Int.* **2023**, *164*, 112322. [CrossRef] [PubMed]
108. Liu, S.-S.; Chen, J.; Zhang, J.-N.; Liu, Y.-S.; Hu, L.-X.; Chen, X.-W.; Liu, S.; Xu, X.-R.; Ying, G.-G. Microbial transformation of progesterone and dydrogesterone by bacteria from swine wastewater: Degradation kinetics and products identification. *Sci. Total Environ.* **2020**, *701*, 134930. [CrossRef]
109. Chen, S.; Xie, J.; Wen, Z. Removal of pharmaceutical and personal care products (PPCPs) from waterbody using a revolving algal biofilm (RAB) reactor. *J. Hazard. Mater.* **2021**, *406*, 124284. [CrossRef]
110. Suleiman, M.; Schröder, C.; Kuhn, M.; Simon, A.; Stahl, A.; Frerichs, H.; Antranikian, G. Microbial biofilm formation and degradation of octocrylene, a UV absorber found in sunscreen. *Commun. Biol.* **2019**, *2*, 430. [CrossRef]
111. Chen, W.; Wang, Z.; Xu, W.; Tian, R.; Zeng, J. Dibutyl phthalate contamination accelerates the uptake and metabolism of sugars by microbes in black soil. *Environ. Pollut.* **2020**, *262*, 114332. [CrossRef]
112. Kruis, A.J.; Bohnenkamp, A.C.; Patinios, C.; van Nuland, Y.M.; Levisson, M.; Mars, A.E.; van den Berg, C.; Kengen, S.W.M.; Weusthuis, R.A. Microbial production of short and medium chain esters: Enzymes, pathways, and applications. *Biotechnol. Adv.* **2019**, *37*, 107407. [CrossRef]
113. Senne de Oliveira Lino, F.; Bajic, D.; Vila, J.C.C.; Sánchez, A.; Sommer, M.O.A. Complex yeast–bacteria interactions affect the yield of industrial ethanol fermentation. *Nat. Commun.* **2021**, *12*, 1498. [CrossRef] [PubMed]
114. Jeffrey, L.C.; Maher, D.T.; Chiri, E.; Leung, P.M.; Nauer, P.A.; Arndt, S.K.; Tait, D.R.; Greening, C.; Johnston, S.G. Bark-dwelling methanotrophic bacteria decrease methane emissions from trees. *Nat. Commun.* **2021**, *12*, 2127. [CrossRef] [PubMed]
115. Dai, M.; Li, F.; Zhang, J.; Shi, Q.; Wu, Y.; Kong, Q. Treatment of formaldehyde-containing wastewater and power generation by constructed wetland–microbial fuel cells enhanced by formaldehyde-degrading bacteria. *J. Water Process Eng.* **2024**, *59*, 104984. [CrossRef]
116. Oveisi, F.; Fallah, N.; Nasernejad, B. Biodegradation of synthetic wastewater containing styrene in microbial fuel cell: Effect of adaptation of microbial community. *Fuel* **2021**, *305*, 121382. [CrossRef]
117. Ma, J.; Xie, M.; Zhao, N.; Wang, Y.; Lin, Q.; Zhu, Y.; Chao, Y.; Ni, Z.; Qiu, R. Enhanced trichloroethylene biodegradation: The mechanism and influencing factors of combining microorganism and carbon-iron materials. *Sci. Total Environ.* **2023**, *878*, 162720. [CrossRef] [PubMed]
118. Roig, M.G.; Pedraz, M.A.; Sánchez, J.M. Application of immobilized microorganisms in the biodegradation of alkyl sulphate and linear alkyl benzene sulphonate surfactants. *Int. Biodeterior. Biodegrad.* **1996**, *37*, 258–259. [CrossRef]
119. Verkholiak, N.S.; Peretyatko, T.B. Destruction of toluene and xylene by sulfate-reducing bacteria. *Ecol. Noospherol.* **2019**, *30*, 95–100. [CrossRef] [PubMed]
120. Muyzer, G.; Stams, A.J. The ecology and biotechnology of sulphate-reducing bacteria. *Nat. Rev. Microbiol.* **2008**, *6*, 441–454. [CrossRef]
121. Ranadev, P.; Revanna, A.; Bagyaraj, D.J.; Shinde, A.H. Sulfur oxidizing bacteria in agro ecosystem and its role in plant productivity—A review. *J. Appl. Microbiol.* **2023**, *134*, lxad161. [CrossRef]
122. Wang, H.; Xing, L.; Zhang, H.; Gui, C.; Jin, S.; Lin, H.; Li, Q.; Cheng, C. Key factors to enhance soil remediation by bioelectrochemical systems (BESs): A review. *Chem. Eng. J.* **2021**, *419*, 129600. [CrossRef]

123. Emerson, D.; Moyer, C. Isolation and characterization of novel iron-oxidizing bacteria that grow at circumneutral pH. *Appl. Environ. Microbiol.* **1997**, *63*, 4784–4792. [CrossRef]
124. Golyshina, O.V.; Timmis, K.N. *Ferroplasma* and relatives, recently discovered cell wall-lacking archaea making a living in extremely acid, heavy metal-rich environments. *Environ. Microbiol.* **2005**, *7*, 1277–1288. [CrossRef]
125. Huber, H.; Stetter, K.O. Archaeoglobus. In *Bergey's Manual of Systematics of Archaea and Bacteria*; John Wiley & Sons: Hoboken, NJ, USA, 2015; pp. 1–5.
126. Liying, J.; Timothy, P.; Buenfeld, N.R.; Stephen, R.S. A critical review of the physiological, ecological, physical and chemical factors influencing the microbial degradation of concrete by fungi. *Build. Environ.* **2022**, *214*, 108925. [CrossRef]
127. Ruyan, Z.; Xiaolu, Y.; Qifeng, T.; Xinggang, W.; Chang, L.; Chengliang, L.; Feng, L. Soil microbial community structure, metabolic potentials and influencing factors in a subtropical mountain forest ecosystem of China. *Environ. Pollut. Bioavailab.* **2020**, *32*, 69–78. [CrossRef]
128. Yao, Z.; Lan, C.; Jie-Ying, W.; Yi, L.; Jun, W.; Yaoxin, G.; Chengjie, R.; Halbert, B.; Huihui, S.; Fazhu, Z. Differences and influencing factors of microbial carbon use efficiency in forest rhizosphere soils at different altitudes in Taibai Mountain, China. *Chin. J. Plant Ecol.* **2023**, *47*, 275–288. [CrossRef]
129. Jadhav, S.S.; Sharma, S.B.; Sibi, G. Microbial Degradation of Petroleum Hydrocarbons and Factors Influencing the Degradation Process. *Bioprocess Eng.* **2019**, *3*, 6–11. [CrossRef]
130. Hugo, S.; Martin, L.; Farrukh, J. An Analysis of Fast Learning Methods for Classifying Forest Cover Types. *Appl. Artif. Intell.* **2020**, *34*, 691–709. [CrossRef]
131. Boyka, M. Microbial diversity and enzymatic activity of soils in coniferous forest ecosystems. *Bulg. J. Soil Sci. Agro-Chem. Ecol.* **2020**, *54*, 43–54.
132. Teresa, F.; Ana Cristina, G. Editorial: Forest species and stands regeneration. *Front. For. Glob. Change* **2023**, *6*, 1208267. [CrossRef]
133. Su, F.; Xu, S.; Sayer, E.J.; Chen, W.; Du, Y.; Lu, X. Distinct storage mechanisms of soil organic carbon in coniferous forest and evergreen broadleaf forest in tropical China. *J. Environ. Manag.* **2021**, *295*, 113142. [CrossRef]
134. Waring, R.H.; Franklin, J.F.J.S. Evergreen Coniferous Forests of the Pacific Northwest: Massive long-lived conifers dominating these forests are adapted to a winter-wet, summer-dry environment. *Science* **1979**, *204*, 1380–1386. [CrossRef]
135. Yi-Zhen, S.; Senlin, W.; Yushan, L.; Yun, C.; Zuyuan, H.; Jing, W.; Fengqin, L.; Zuoqiang, Y. Importance of Bark Physicochemical Properties in an Epiphytic Bryophyte Community within a Temperate Deciduous Broadleaf Forest. *Diversity* **2023**, *15*, 688. [CrossRef]
136. Johansson, M.; Gyllin, M.; Witzell, J.; Küller, M.J.U.F.; Greening, U. Does biological quality matter? Direct and reflected appraisal of biodiversity in temperate deciduous broad-leaf forest. *Urban For. Urban Green.* **2014**, *13*, 28–37. [CrossRef]
137. Renhui, M.; Jun, M.; Yinzhan, L.; Yanchun, L.; Zhongling, Y.; Ming, G. Variability of Aboveground Litter Inputs Alters Soil Carbon and Nitrogen in a Coniferous–Broadleaf Mixed Forest of Central China. *Forests* **2019**, *10*, 188. [CrossRef]
138. Knoke, T.; Ammer, C.; Stimm, B.; Mosandl, R. Admixing broadleaved to coniferous tree species: A review on yield, ecological stability and economics. *Eur. J. For. Res.* **2008**, *127*, 89–101. [CrossRef]
139. Andressa Monteiro, V.; Júlia Brandão, G.; Jéssica, A.M.; Erika, B.; Kabir, G.P.; Siu Mui, T.; Brendan, J.M.B. Soil microbes under threat in the Amazon Rainforest. *Trends Ecol. Evol.* **2023**, *38*, 693–696. [CrossRef]
140. Francielle Roberta Dias de, L.; Polyana, P.; Ediu, C.S.; Isabela, C.F.V.; Jakeline Rosa de, O.; Cláudia Carvalhinho, W.; Alberto Vasconcellos, I.; David, C.W.; Nilton, C.; Bruno Teixeira, R.; et al. Geochemistry signatures of mercury in soils of the Amazon rainforest biome. *Environ. Res.* **2022**, *215*, 114147. [CrossRef]
141. Koné, A.W.; Kassi, S.-P.A.Y.; Koffi, B.Y.; Masse, D.; Maïga, A.A.; Tondoh, J.E.; Kisaka, O.M.; Touré, G.-P.T. *Chromolaena odorata* (L.) K&R (*Asteraceae*) invasion effects on soil microbial biomass and activities in a forest-savanna mosaic. *CATENA* **2021**, *207*, 105619. [CrossRef]
142. Eduarda Martiniano de Oliveira, S.; Marcela de Castro Nunes Santos, T.; Hanster, S.; Eduardo Eiji, M.; Fausto Weimar Acerbi, J.; José Roberto Soares, S. Carbon-diversity hotspots and their owners in Brazilian southeastern Savanna, Atlantic Forest and Semi-Arid Woodland domains. *For. Ecol. Manag.* **2019**, *452*, 117575. [CrossRef]
143. Liao, Y.-C.-Z.; Pu, H.-X.; Jiao, Z.-W.; Palviainen, M.; Zhou, X.; Heinonsalo, J.; Berninger, F.; Pumpanen, J.; Köster, K.; Sun, H. Enhancing boreal forest resilience: A four-year impact of biochar on soil quality and fungal communities. *Microbiol. Res.* **2024**, *283*, 127696. [CrossRef]
144. Roman, D.; Colin, M.; Rebecca, E.H.; Amy, M.W.; Wong, R.K.; David Livingston, C.; Madeline, G.Z.; Patrick, F.S. Arctic sea ice retreat fuels boreal forest advance. *Science* **2024**, *383*, 877–884. [CrossRef]
145. Naji, M.B.; Nicholas, B.; Jonathan, R.L. Microbial degradation of isosaccharinic acid at high pH. *ISME J.* **2014**, *9*, 310–320. [CrossRef]
146. Brulé, C.; Frey-Klett, P.; Pierrat, J.C.; Courrier, S.; Gérard, F.; Lemoine, M.C.; Rousselet, J.L.; Sommer, G.; Garbaye, J. Survival in the soil of the ectomycorrhizal fungus *Laccaria bicolor* and the effects of a mycorrhiza helper *Pseudomonas fluorescens*. *Soil Biol. Biochem.* **2001**, *33*, 1683–1694. [CrossRef]

147. Duo, Y.; Xiaolong, C.; Jingbo, H.; Zhao, F.J.; Peng, L. Soil properties and microbial functional attributes drive the response of soil multifunctionality to long-term fertilization management. *Appl. Soil Ecol.* **2023**, *192*, 105095. [CrossRef]
148. Vaithyanathan, V.K.; Cabana, H.; Vaidyanathan, V.K. Remediation of trace organic contaminants from biosolids: Influence of various pre-treatment strategies prior to *Bacillus subtilis* aerobic digestion. *Chem. Eng. J.* **2021**, *419*, 129966. [CrossRef]
149. Deni, J.; Penninckx, M.J. Influence of long-term diesel fuel pollution on nitrite-oxidising activity and population size of *nitrobacter* spp. in soil. *Microbiol. Res.* **2004**, *159*, 323–329. [CrossRef]
150. List, C.; Hosseini, Z.; Lederballe Meibom, K.; Hatzimanikatis, V.; Bernier-Latmani, R. Impact of iron reduction on the metabolism of *Clostridium acetobutylicum*. *Environ. Microbiol.* **2019**, *21*, 3548–3563. [CrossRef] [PubMed]
151. Lakhani, S.; Acharya, D.; Sakariya, R.; Sharma, D.; Patel, P.; Shah, M.; Prajapati, M. A comprehensive study of bioremediation for pharmaceutical wastewater treatment. *Clean. Chem. Eng.* **2022**, *4*, 100073. [CrossRef]
152. Aparicio, J.D.; Raimondo, E.E.; Saez, J.M.; Costa-Gutierrez, S.B.; Álvarez, A.; Benimeli, C.S.; Polti, M.A. The current approach to soil remediation: A review of physicochemical and biological technologies, and the potential of their strategic combination. *J. Environ. Chem. Eng.* **2022**, *10*, 107141. [CrossRef]
153. Kumar, V.; Agrawal, S.; Bhat, S.A.; Américo-Pinheiro, J.H.P.; Shahi, S.K.; Kumar, S. Environmental impact, health hazards, and plant-microbes synergism in remediation of emerging contaminants. *Clean. Chem. Eng.* **2022**, *2*, 100030. [CrossRef]
154. Liu, J.; Zhang, H.; Zhang, P.; Wu, Y.; Gou, X.; Song, Y.; Tian, Z.; Zeng, G. Two-stage anoxic/oxic combined membrane bioreactor system for landfill leachate treatment: Pollutant removal performances and microbial community. *Bioresour. Technol.* **2017**, *243*, 738–746. [CrossRef]
155. Jasu, A.; Lahiri, D.; Nag, M.; Ray, R.R. Chapter 17—Fungi in bioremediation of soil organic pollutants. In *Fungi Bio-Prospects in Sustainable Agriculture, Environment and Nano-Technology*; Sharma, V.K., Shah, M.P., Parmar, S., Kumar, A., Eds.; Academic Press: Cambridge, MA, USA, 2021; pp. 381–405. [CrossRef]
156. Baldrian, P. Wood-inhabiting ligninolytic basidiomycetes in soils: Ecology and constraints for applicability in bioremediation. *Fungal Ecol.* **2008**, *1*, 4–12. [CrossRef]
157. Lladó, S.; Žifčáková, L.; Větrovský, T.; Eichlerová, I.; Baldrian, P. Functional screening of abundant bacteria from acidic forest soil indicates the metabolic potential of Acidobacteria subdivision 1 for polysaccharide decomposition. *Biol. Fertil. Soils* **2016**, *52*, 251–260. [CrossRef]
158. Berger, T.; Poyntner, C.; Margesin, R. Culturable bacteria from an Alpine coniferous forest site: Biodegradation potential of organic polymers and pollutants. *Folia Microbiol.* **2021**, *66*, 87–98. [CrossRef]
159. Ling, L.; Fu, Y.; Jeewani, P.H.; Tang, C.; Pan, S.; Reid, B.J.; Gunina, A.; Li, Y.; Li, Y.; Cai, Y.; et al. Organic matter chemistry and bacterial community structure regulate decomposition processes in post-fire forest soils. *Soil Biol. Biochem.* **2021**, *160*, 108311. [CrossRef]
160. Jiang, L.; Zhang, D.; Song, M.; Guan, G.; Sun, Y.; Li, J.; Cheng, X.; Luo, C.; Zhang, G. The positive role of root decomposition on the bioremediation of organic pollutants contaminated soil: A case study using PCB-9 as a model compound. *Soil Biol. Biochem.* **2022**, *171*, 108726. [CrossRef]
161. Krohn, C.; Jin, J.; Wood, J.L.; Hayden, H.L.; Kitching, M.; Ryan, J.; Fabijański, P.; Franks, A.E.; Tang, C. Highly decomposed organic carbon mediates the assembly of soil communities with traits for the biodegradation of chlorinated pollutants. *J. Hazard. Mater.* **2021**, *404*, 124077. [CrossRef] [PubMed]
162. Uttarotai, T.; Sutheeworapong, S.; Crombie, A.T.; Murrell, J.C.; Mhuantong, W.; Noirungsee, N.; Wangkarn, S.; Bovonsombut, S.; McGenity, T.J.; Chitov, T. Genome characterisation of an isoprene-degrading Alcaligenes sp. isolated from a tropical restored forest. *Biology* **2022**, *11*, 519. [CrossRef]
163. Laffite, A.; Florio, A.; Andrianarisoa, K.S.; Creuze des Chatelliers, C.; Schloter-Hai, B.; Ndaw, S.M.; Periot, C.; Schloter, M.; Zeller, B.; Poly, F.; et al. Biological inhibition of soil nitrification by forest tree species affects populations. *Environ. Microbiol.* **2020**, *22*, 1141–1153. [CrossRef]
164. Zenteno-Rojas, A.; Martínez-Romero, E.; Castañeda-Valbuena, D.; Rincón-Molina, C.I.; Ruíz-Valdiviezo, V.M.; Meza-Gordillo, R.; Villalobos-Maldonado, J.J.; Vences-Guzmán, M.Á.; Rincón-Rosales, R. Structure and diversity of native bacterial communities in soils contaminated with polychlorinated biphenyls. *AMB Express* **2020**, *10*, 124. [CrossRef]
165. Sarwan, J. Role of isolates of Bacillus species for biodegradation of multiple contaminants. *J. Sustain. Environ. Manag.* **2022**, *1*, 292–298. [CrossRef]
166. Sato, Y.; Tsukamoto, T.; Sato, M. Rapid degradation of carbaryl by two novel strains of *Arthrobacter* spp. isolated from forest soil. *J. For. Res.* **1999**, *4*, 275–280. [CrossRef]
167. Young, P.C. Data-based mechanistic modelling and forecasting globally averaged surface temperature. *Int. J. Forecast.* **2018**, *34*, 314–335. [CrossRef]
168. Malhi, Y.; Aragão, L.E.O.C.; Galbraith, D.; Huntingford, C.; Fisher, R.; Zelazowski, P.; Sitch, S.; McSweeney, C.; Meir, P. Exploring the likelihood and mechanism of a climate-change-induced dieback of the Amazon rainforest. *Proc. Natl. Acad. Sci. USA* **2009**, *106*, 20610–20615. [CrossRef]

169. Silva de Miranda, P.L.; Dexter, K.G.; Swaine, M.D.; de Oliveira-Filho, A.T.; Hardy, O.J.; Fayolle, A. Dissecting the difference in tree species richness between Africa and South America. *Proc. Natl. Acad. Sci. USA* **2022**, *119*, e2112336119. [CrossRef] [PubMed]
170. Crabbe, R.A.; Dash, J.; Rodriguez-Galiano, V.F.; Janous, D.; Pavelka, M.; Marek, M.V. Extreme warm temperatures alter forest phenology and productivity in Europe. *Sci. Total Environ.* **2016**, *563–564*, 486–495. [CrossRef] [PubMed]
171. Nowacki, G.J.; Abrams, M.D. Is climate an important driver of post-European vegetation change in the Eastern United States? *Glob. Change Biol.* **2015**, *21*, 314–334. [CrossRef] [PubMed]
172. Naveendrakumar, G.; Vithanage, M.; Kwon, H.-H.; Chandrasekara, S.S.K.; Iqbal, M.C.M.; Pathmarajah, S.; Fernando, W.C.D.K.; Obeysekera, J. South Asian perspective on temperature and rainfall extremes: A review. *Atmos. Res.* **2019**, *225*, 110–120. [CrossRef]
173. Zhao, P.; Zhu, Y.; Zhang, R. An Asian–Pacific teleconnection in summer tropospheric temperature and associated Asian climate variability. *Clim. Dyn.* **2007**, *29*, 293–303. [CrossRef]
174. Wang, S.; Chen, X.; Li, D.; Wu, J. Effects of soil organism interactions and temperature on carbon use efficiency in three different forest soils. *Soil Ecol. Lett.* **2021**, *3*, 156–166. [CrossRef]
175. Tang, Z.; Sun, X.; Luo, Z.; He, N.; Sun, O.J. Effects of temperature, soil substrate, and microbial community on carbon mineralization across three climatically contrasting forest sites. *Ecol. Evol.* **2018**, *8*, 879–891. [CrossRef]
176. Mitra, D.; Mondal, R.; Khoshru, B.; Senapati, A.; Radha, T.K.; Mahakur, B.; Uniyal, N.; Myo, E.M.; Boutaj, H.; Sierra, B.E.G.; et al. Actinobacteria-enhanced plant growth, nutrient acquisition, and crop protection: Advances in soil, plant, and microbial multifactorial interactions. *Pedosphere* **2022**, *32*, 149–170. [CrossRef]
177. Arbab, S.; Ullah, H.; Khan, M.I.U.; Khattak, M.N.K.; Zhang, J.; Li, K.; Hassan, I.U. Diversity and distribution of thermophilic microorganisms and their applications in biotechnology. *J. Basic Microbiol.* **2022**, *62*, 95–108. [CrossRef]
178. Carter, D.O.; Yellowlees, D.; Tibbett, M. Temperature affects microbial decomposition of cadavers (*Rattus rattus*) in contrasting soils. *Appl. Soil Ecol.* **2008**, *40*, 129–137. [CrossRef]
179. Varjani, S.J.; Upasani, V.N. A new look on factors affecting microbial degradation of petroleum hydrocarbon pollutants. *Int. Biodeterior. Biodegrad.* **2017**, *120*, 71–83. [CrossRef]
180. Liu, H.; Wu, Y.; Ai, Z.; Zhang, J.; Zhang, C.; Xue, S.; Liu, G. Effects of the interaction between temperature and revegetation on the microbial degradation of soil dissolved organic matter (DOM)—A DOM incubation experiment. *Geoderma* **2019**, *337*, 812–824. [CrossRef]
181. Mallick, S.; Chakraborty, J.; Dutta, T.K. Role of oxygenases in guiding diverse metabolic pathways in the bacterial degradation of low-molecular-weight polycyclic aromatic hydrocarbons: A review. *Crit. Rev. Microbiol.* **2011**, *37*, 64–90. [CrossRef]
182. Mirza, B.S.; Sorensen, D.L.; Dupont, R.R.; McLean, J.E. Dehalococcoides abundance and alternate electron acceptor effects on large, flow-through trichloroethene dechlorinating columns. *Appl. Microbiol. Biotechnol.* **2016**, *100*, 2367–2379. [CrossRef]
183. Johnsen, A.R.; Wick, L.Y.; Harms, H. Principles of microbial PAH-degradation in soil. *Environ. Pollut.* **2005**, *133*, 71–84. [CrossRef]
184. Friis, A.K.; Heimann, A.C.; Jakobsen, R.; Albrechtsen, H.-J.; Cox, E.; Bjerg, P.L. Temperature dependence of anaerobic TCE-dechlorination in a highly enriched *Dehalococcoides*-containing culture. *Water Res.* **2007**, *41*, 355–364. [CrossRef]
185. Hong, X.; Qin, J.; Chen, R.; Yuan, L.; Zha, J.; Huang, C.; Li, N.; Ji, X.; Wang, Z. Phenanthrene-induced apoptosis and its underlying mechanism. *Environ. Sci. Technol.* **2017**, *51*, 14397–14405. [CrossRef] [PubMed]
186. Ayumi, K.; Naoki, M.; Akira, O. Response of microbial respiration from fine root litter decomposition to root water content in a temperate broad-leaved forest. *Plant Root* **2013**, *7*, 77–82. [CrossRef]
187. Zhu, M.; Zheng, J.; Xie, J.; Zhao, D.; Qiao, Z.-W.; Huang, D.; Luo, H.-B. Effects of environmental factors on the microbial community changes during medium-high temperature Daqu manufacturing. *Food Res. Int.* **2022**, *153*, 110955. [CrossRef]
188. Aytac, Z.; Xu, J.; Raman Pillai, S.K.; Eitzer, B.D.; Xu, T.; Vaze, N.; Ng, K.W.; White, J.C.; Chan-Park, M.B.; Luo, Y. Enzyme-and relative humidity-responsive antimicrobial fibers for active food packaging. *ACS Appl. Mater. Interfaces* **2021**, *13*, 50298–50308. [CrossRef]
189. Premnath, N.; Mohanrasu, K.; Rao, R.G.R.; Dinesh, G.; Prakash, G.S.; Ananthi, V.; Ponnuchamy, K.; Muthusamy, G.; Arun, A. A crucial review on polycyclic aromatic Hydrocarbons-Environmental occurrence and strategies for microbial degradation. *Chemosphere* **2021**, *280*, 130608. [CrossRef] [PubMed]
190. Behera, S.; Das, S. Potential and prospects of *Actinobacteria* in the bioremediation of environmental pollutants: Cellular mechanisms and genetic regulations. *Microbiol. Res.* **2023**, *273*, 127399. [CrossRef] [PubMed]
191. Lv, L.; Sun, L.; Yuan, C.; Han, Y.; Huang, Z. The combined enhancement of RL, nZVI and AQDS on the microbial anaerobic-aerobic degradation of PAHs in soil. *Chemosphere* **2022**, *307*, 135609. [CrossRef] [PubMed]
192. Brown, R.W.; Bull, I.D.; Journeaux, T.; Chadwick, D.R.; Jones, D.L. Volatile organic compounds (VOCs) allow sensitive differentiation of biological soil quality. *Soil Biol. Biochem.* **2021**, *156*, 108187. [CrossRef]
193. Sengupta, K.; Pal, S. A review on microbial diversity and genetic markers involved in methanogenic degradation of hydrocarbons: Futuristic prospects of biofuel recovery from contaminated regions. *Environ. Sci. Pollut. Res.* **2021**, *28*, 40288–40307. [CrossRef]

194. Wang, J.; Bao, H.; Pan, G.; Zhang, H.; Li, J.; Li, J.; Cai, J.; Wu, F. Combined application of rhamnolipid and agricultural wastes enhances PAHs degradation via increasing their bioavailability and changing microbial community in contaminated soil. *J. Environ. Manag.* **2021**, *294*, 112998. [CrossRef] [PubMed]
195. Cheng, M.; Chen, D.; Parales, R.E.; Jiang, J. Oxygenases as powerful weapons in the microbial degradation of pesticides. *Annu. Rev. Microbiol.* **2022**, *76*, 325–348. [CrossRef]
196. Tiessen, H.; Cuevas, E.; Chacon, P. The role of soil organic matter in sustaining soil fertility. *Nature* **1994**, *371*, 783–785. [CrossRef]
197. Filonov, A.E.; Puntus, I.F.; Karpov, A.V.; Kosheleva, I.A.; Kashparov, K.I.; Slepenkin, A.V.; Boronin, A.M. Efficiency of naphthalene biodegradation by Pseudomonas putida G7 in soil. *J. Chem. Technol. Biotechnol. Int. Res. Process Environ. Clean Technol.* **2004**, *79*, 562–569. [CrossRef]
198. Quero, G.M.; Cassin, D.; Botter, M.; Perini, L.; Luna, G.M. Patterns of benthic bacterial diversity in coastal areas contaminated by heavy metals, polycyclic aromatic hydrocarbons (PAHs) and polychlorinated biphenyls (PCBs). *Front. Microbiol.* **2015**, *6*, 1053. [CrossRef] [PubMed]
199. Calogero, R.; Arcadi, E.; Fabiano, F.; Rizzo, C.; Romeo, T.; Greco, S. PCB bioremediation potential of thermophilic strains from shallow hydrothermal vent (Vulcano Island). *J. Water Process Eng.* **2024**, *61*, 105330. [CrossRef]
200. Sun, Y.; Xiong, X.; He, M.; Xu, Z.; Hou, D.; Zhang, W.; Ok, Y.S.; Rinklebe, J.; Wang, L.; Tsang, D.C. Roles of biochar-derived dissolved organic matter in soil amendment and environmental remediation: A critical review. *Chem. Eng. J.* **2021**, *424*, 130387. [CrossRef]
201. Reijonen, I.; Metzler, M.; Hartikainen, H. Impact of soil pH and organic matter on the chemical bioavailability of vanadium species: The underlying basis for risk assessment. *Environ. Pollut.* **2016**, *210*, 371–379. [CrossRef] [PubMed]
202. Chen, L.; Zhou, S.; Zhang, Q.; Zou, M.; Yin, Q.; Qiu, Y.; Qin, W. Effect of organic material addition on active soil organic carbon and microbial diversity: A meta-analysis. *Soil Tillage Res.* **2024**, *241*, 106128. [CrossRef]
203. Esmaeilzadeh-Salestani, K.; Bahram, M.; Seraj, R.G.M.; Gohar, D.; Tohidfar, M.; Eremeev, V.; Talgre, L.; Khaleghdoust, B.; Mirmajlessi, S.M.; Luik, A. Cropping systems with higher organic carbon promote soil microbial diversity. *Agric. Ecosyst. Environ.* **2021**, *319*, 107521. [CrossRef]
204. Maron, P.-A.; Sarr, A.; Kaisermann, A.; Lévêque, J.; Mathieu, O.; Guigue, J.; Karimi, B.; Bernard, L.; Dequiedt, S.; Terrat, S. High microbial diversity promotes soil ecosystem functioning. *Appl. Environ. Microbiol.* **2018**, *84*, e02738-17. [CrossRef] [PubMed]
205. Bollag, J.-M.; Myers, C.J.; Minard, R.D. Biological and chemical interactions of pesticides with soil organic matter. *Sci. Total Environ.* **1992**, *123*, 205–217. [CrossRef]
206. Spark, K.; Swift, R. Effect of soil composition and dissolved organic matter on pesticide sorption. *Sci. Total Environ.* **2002**, *298*, 147–161. [CrossRef]
207. Farenhorst, A. Importance of soil organic matter fractions in soil-landscape and regional assessments of pesticide sorption and leaching in soil. *Soil Sci. Soc. Am. J.* **2006**, *70*, 1005–1012. [CrossRef]
208. Mikutta, R.; Kleber, M.; Torn, M.S.; Jahn, R. Stabilization of soil organic matter: Association with minerals or chemical recalcitrance? *Biogeochemistry* **2006**, *77*, 25–56. [CrossRef]
209. Kögel-Knabner, I. The macromolecular organic composition of plant and microbial residues as inputs to soil organic matter. *Soil Biol. Biochem.* **2002**, *34*, 139–162. [CrossRef]
210. Mead, R.; Xu, Y.; Chong, J.; Jaffé, R. Sediment and soil organic matter source assessment as revealed by the molecular distribution and carbon isotopic composition of n-alkanes. *Org. Geochem.* **2005**, *36*, 363–370. [CrossRef]
211. Wiesenberg, G.L.; Schwarzbauer, J.; Schmidt, M.W.; Schwark, L. Source and turnover of organic matter in agricultural soils derived from n-alkane/n-carboxylic acid compositions and C-isotope signatures. *Org. Geochem.* **2004**, *35*, 1371–1393. [CrossRef]
212. Endo, S.; Grathwohl, P.; Haderlein, S.B.; Schmidt, T.C. Characterization of sorbent properties of soil organic matter and carbonaceous geosorbents using n-alkanes and cycloalkanes as molecular probes. *Environ. Sci. Technol.* **2009**, *43*, 393–400. [CrossRef]
213. Ukalska-Jaruga, A.; Smreczak, B.; Klimkowicz-Pawlas, A. Soil organic matter composition as a factor affecting the accumulation of polycyclic aromatic hydrocarbons. *J. Soils Sediments* **2019**, *19*, 1890–1900. [CrossRef]
214. Yang, Y.; Zhang, N.; Xue, M.; Tao, S. Impact of soil organic matter on the distribution of polycyclic aromatic hydrocarbons (PAHs) in soils. *Environ. Pollut.* **2010**, *158*, 2170–2174. [CrossRef]
215. Yang, Y.; Zhang, N.; Xue, M.; Lu, S.; Tao, S. Effects of soil organic matter on the development of the microbial polycyclic aromatic hydrocarbons (PAHs) degradation potentials. *Environ. Pollut.* **2011**, *159*, 591–595. [CrossRef] [PubMed]
216. Yang, Y.; Tao, S.; Zhang, N.; Zhang, D.; Li, X. The effect of soil organic matter on fate of polycyclic aromatic hydrocarbons in soil: A microcosm study. *Environ. Pollut.* **2010**, *158*, 1768–1774. [CrossRef]
217. Zhang, Y.-L.; Lin, S.-S.; Dai, C.-M.; Shi, L.; Zhou, X.-F. Sorption–desorption and transport of trimethoprim and sulfonamide antibiotics in agricultural soil: Effect of soil type, dissolved organic matter, and pH. *Environ. Sci. Pollut. Res.* **2014**, *21*, 5827–5835. [CrossRef]

218. Chen, K.-L.; Liu, L.-C.; Chen, W.-R. Adsorption of sulfamethoxazole and sulfapyridine antibiotics in high organic content soils. *Environ. Pollut.* **2017**, *231*, 1163–1171. [CrossRef]
219. Nardi, S.; Pizzeghello, D.; Ertani, A. Hormone-like activity of the soil organic matter. *Appl. Soil Ecol.* **2018**, *123*, 517–520. [CrossRef]
220. Scaglia, B.; Pognani, M.; Adani, F. Evaluation of hormone-like activity of the dissolved organic matter fraction (DOM) of compost and digestate. *Sci. Total Environ.* **2015**, *514*, 314–321. [CrossRef] [PubMed]
221. Misic, C.; Harriague, A.C.; Trielli, F. Organic matter recycling in a beach environment influenced by sunscreen products and increased inorganic nutrient supply (Sturla, Ligurian Sea, NW Mediterranean). *Sci. Total Environ.* **2011**, *409*, 1689–1696. [CrossRef] [PubMed]
222. Zhou, L.; Ji, Y.; Zeng, C.; Zhang, Y.; Wang, Z.; Yang, X. Aquatic photodegradation of sunscreen agent p-aminobenzoic acid in the presence of dissolved organic matter. *Water Res.* **2013**, *47*, 153–162. [CrossRef] [PubMed]
223. Sánchez-Quiles, D.; Tovar-Sánchez, A. Are sunscreens a new environmental risk associated with coastal tourism? *Environ. Int.* **2015**, *83*, 158–170. [CrossRef]
224. Rocha, F.; Ratola, N.; Homem, V. *Fragrances in the Environment: Properties, Applications, and Emissions*; Springer: Cham, Switzerland, 2023.
225. Tfaily, M.M.; Chu, R.K.; Tolić, N.; Roscioli, K.M.; Anderton, C.R.; Paša-Tolić, L.; Robinson, E.W.; Hess, N.J. Advanced Solvent Based Methods for Molecular Characterization of Soil Organic Matter by High-Resolution Mass Spectrometry. *Anal. Chem.* **2015**, *87*, 5206–5215. [CrossRef]
226. Gentile, L.; Floudas, D.; Olsson, U.; Persson, P.; Tunlid, A. Fungal decomposition and transformation of molecular and colloidal fractions of dissolved organic matter extracted from boreal forest soil. *Soil Biol. Biochem.* **2024**, *195*, 109473. [CrossRef]
227. Ahangar, A.G.; Smernik, R.J.; Kookana, R.S.; Chittleborough, D.J. The effect of solvent-conditioning on soil organic matter sorption affinity for diuron and phenanthrene. *Chemosphere* **2009**, *76*, 1062–1066. [CrossRef]
228. Ran, Y.; Sun, K.; Ma, X.; Wang, G.; Grathwohl, P.; Zeng, E.Y. Effect of condensed organic matter on solvent extraction and aqueous leaching of polycyclic aromatic hydrocarbons in soils and sediments. *Environ. Pollut.* **2007**, *148*, 529–538. [CrossRef]
229. Lehmann, J.; Kleber, M. The contentious nature of soil organic matter. *Nature* **2015**, *528*, 60–68. [CrossRef]
230. Lal, R. Challenges and opportunities in soil organic matter research. *Eur. J. Soil Sci.* **2009**, *60*, 158–169. [CrossRef]
231. Li, H.; Gu, X.; Song, J.; Hui, K.; Chen, G.; Tan, W.; Wang, H.; Jiang, Y.; Yuan, Y. Effects of Soil-groundwater Environmental Factors on BTEX Transport and Transformation: A Review. *J. Environ. Chem. Eng.* **2024**, *12*, 113697. [CrossRef]
232. Ali, M.; Song, X.; Wang, Q.; Zhang, Z.; Che, J.; Chen, X.; Tang, Z.; Liu, X. Mechanisms of biostimulant-enhanced biodegradation of PAHs and BTEX mixed contaminants in soil by native microbial consortium. *Environ. Pollut.* **2023**, *318*, 120831. [CrossRef]
233. Ahmed, N.; Ok, Y.S.; Jeon, B.-H.; Kim, J.R.; Chae, K.-J.; Oh, S.-E. Assessment of benzene, toluene, ethyl-benzene, and xylene (BTEX) toxicity in soil using sulfur-oxidizing bacterial (SOB) bioassay. *Chemosphere* **2019**, *220*, 651–657. [CrossRef] [PubMed]
234. Shahi, A.; Aydin, S.; Ince, B.; Ince, O. Evaluation of microbial population and functional genes during the bioremediation of petroleum-contaminated soil as an effective monitoring approach. *Ecotoxicol. Environ. Saf.* **2016**, *125*, 153–160. [CrossRef]
235. Qiang, F.; Yi-Zhen, S.; Senlin, W.; Fengqin, L.; Guohang, T.; Yun, C.; Zuoqiang, Y.; Ye, Y. Soil Microbial Distribution Depends on Different Types of Landscape Vegetation in Temperate Urban Forest Ecosystems. *Front. Ecol. Evol.* **2022**, *10*, 858254. [CrossRef]
236. Prasenjit, G.; Soumyo, M. Environmental contamination by heterocyclic Polynuclear aromatic hydrocarbons and their microbial degradation. *Bioresour. Technol.* **2021**, *341*, 125860. [CrossRef]
237. Aboul-Kassim, T.A.; Simoneit, B.R. Interaction mechanisms between organic pollutants and solid phase systems. In *Pollutant-Solid Phase Interactions Mechanisms, Chemistry and Modeling*; Springer: Berlin/Heidelberg, Germany, 2001; pp. 107–167.
238. Hayat, M.T.; Xu, J.; Ding, N.; Mahmood, T. Dynamic behavior of persistent organic pollutants in soil and their interaction with organic matter. In *Molecular Environmental Soil Science at the Interfaces in the Earth's Critical Zone*; Springer: Berlin/Heidelberg, Germany, 2010; pp. 217–222.
239. Chen, Y.; Li, H.; Yin, Y.; Shan, S.; Huang, T.; Tang, H. Effect of microplastics on the adherence of coexisting background organic contaminants to natural organic matter in water. *Sci. Total Environ.* **2023**, *905*, 167175. [CrossRef]
240. Thiloka, K.; Paul, D.P.; Julia, H.; Gregory, D. Greener extraction of polycyclic aromatic hydrocarbons from soil and sediment using eucalyptus oil. *Environ. Chem. Lett.* **2022**, *20*, 2757–2764. [CrossRef]
241. Chaudhry, G.R.; Chapalamadugu, S. Biodegradation of halogenated organic compounds. *Microbiol. Rev.* **1991**, *55*, 59–79. [CrossRef]
242. Tratnyek, P.G.; Edwards, E.; Carpenter, L.; Blossom, S. Environmental occurrence, fate, effects, and remediation of halogenated (semi)volatile organic compounds. *Environ. Sci. Process. Impacts* **2020**, *22*, 465–471. [CrossRef] [PubMed]
243. Zhai, Y.; Bai, J.; Chang, P.; Liu, Z.; Wang, Y.; Liu, G.; Cui, B.; Peijnenburg, W.; Vijver, M.G. Microplastics in terrestrial ecosystem: Exploring the menace to the soil-plant-microbe interactions. *TrAC Trends Anal. Chem.* **2024**, *174*, 117667. [CrossRef]
244. Othman, A.R.; Hasan, H.A.; Muhamad, M.H.; Ismail, N.I.; Abdullah, S.R.S. Microbial degradation of microplastics by enzymatic processes: A review. *Environ. Chem. Lett.* **2021**, *19*, 3057–3073. [CrossRef]

245. Ren, X.; Yin, S.; Wang, L.; Tang, J. Microplastics in plant-microbes-soil system: A review on recent studies. *Sci. Total Environ.* **2022**, *816*, 151523. [CrossRef]
246. Ioannis, A.; Ioannis, P.; Berkant, K.; Dimitrios, K. Microplastics as carriers of hydrophilic pollutants in an aqueous environment. *J. Mol. Liq.* **2022**, *350*, 118182. [CrossRef]
247. Dziurzynski, M.; Gorecki, A.; Pawlowska, J.; Istel, L.; Decewicz, P.; Golec, P.; Styczynski, M.; Poszytek, K.; Rokowska, A.; Gorniak, D.; et al. Revealing the diversity of bacteria and fungi in the active layer of permafrost at Spitsbergen island (Arctic)—Combining classical microbiology and metabarcoding for ecological and bioprospecting exploration. *Sci. Total Environ.* **2023**, *856*, 159072. [CrossRef]
248. Coban, O.; De Deyn, G.B.; van der Ploeg, M. Soil microbiota as game-changers in restoration of degraded lands. *Science* **2022**, *375*, 990. [CrossRef]
249. Gavin, J.K.; Jennifer, A.D. CRISPR-Cas guides the future of genetic engineering. *Science* **2018**, *361*, 866–869. [CrossRef]
250. Gaurav, P.; Deviram, G.; Urvashi, A.; Prasuna, R.G.; Thangavel, M.; Arivalagan, P. Biological approaches practised using genetically engineered microbes for a sustainable environment: A review. *J. Hazard. Mater.* **2021**, *405*, 124631. [CrossRef]
251. Koichi, F. Microbial Degradation of Polychlorinated Biphenyls (PCBs). In *Biodegradation and Detoxification of Environmental Pollutants*; CRC Press eBooks; Taylor & Francis Group: Abington, UK, 2018. [CrossRef]
252. Gopinath, K.P.; Rajagopal, M.; Krishnan, A.; Sreerama, S.K. A review on recent trends in nanomaterials and nanocomposites for environmental applications. *Curr. Anal. Chem.* **2021**, *17*, 202–243. [CrossRef]
253. Margesin, R.; Schinner, F. Biodegradation of organic pollutants at low temperatures. In *Biotechnological Applications of Cold-Adapted Organisms*; Margesin, R., Schinner, F., Eds.; Springer: Berlin/Heidelberg, Germany, 1999; pp. 271–289. [CrossRef]
254. Zambrano-Pinto, M.V.; Tinizaray-Castillo, R.; Riera, M.A.; Maddela, N.R.; Luque, R.; Díaz, J.M.R. Microplastics as vectors of other contaminants: Analytical determination techniques and remediation methods. *Sci. Total Environ.* **2024**, *908*, 168244. [CrossRef]
255. Qutob, M.; Rafatullah, M.; Muhammad, S.A.; Alosaimi, A.M.; Alorfi, H.S.; Hussein, M.A. A Review of Pyrene Bioremediation Using Mycobacterium Strains in a Different Matrix. *Fermentation* **2022**, *8*, 260. [CrossRef]
256. Manzano, M.; Morán, A.C.; Tesser, B.; González, B. Role of eukaryotic microbiota in soil survival and catabolic performance of the 2,4-D herbicide degrading bacteria Cupriavidus necator JMP134. *Antonie Van Leeuwenhoek* **2007**, *91*, 115–126. [CrossRef] [PubMed]
257. Varjani, S.J.; Upasani, V.N. Biodegradation of petroleum hydrocarbons by oleophilic strain of Pseudomonas aeruginosa NCIM 5514. *Bioresour. Technol.* **2016**, *222*, 195–201. [CrossRef] [PubMed]
258. Eskandari, S.; Hoodaji, M.; Tahmourespour, A.; Abdollahi, A.; Baghi, T.; Eslamian, S.; Ostad-Ali-Askari, K. Bioremediation of polycyclic aromatic hydrocarbons by Bacillus Licheniformis ATHE9 and Bacillus Mojavensis ATHE13 as newly strains isolated from oil-contaminated soil. *J. Geogr. Environ. Earth Sci. Int.* **2017**, *11*, 1–11. [CrossRef]
259. Pi, Y.; Chen, B.; Bao, M.; Fan, F.; Cai, Q.; Ze, L.; Zhang, B. Microbial degradation of four crude oil by biosurfactant producing strain Rhodococcus sp. *Bioresour. Technol.* **2017**, *232*, 263–269. [CrossRef]
260. Kim, T.J.; Lee, E.Y.; Kim, Y.J.; Cho, K.-S.; Ryu, H.W. Degradation of polyaromatic hydrocarbons by Burkholderia cepacia 2A-12. *World J. Microbiol. Biotechnol.* **2003**, *19*, 411–417. [CrossRef]
261. Kagle, J.; Porter, A.W.; Murdoch, R.W.; Rivera-Cancel, G.; Hay, A.G. Chapter 3 Biodegradation of Pharmaceutical and Personal Care Products. In *Advances in Applied Microbiology*; Academic Press: Cambridge, MA, USA, 2009; Volume 67, pp. 65–108.
262. Zheng, Z.; He, J.; Dong, C.; Lo, I.M.C. Photoelectrochemical sewage treatment by sulfite activation over an optimized $BiVO_4$ photoanode to simultaneously promote PPCPs degradation, H_2 evolution and *E. coli* disinfection. *Chem. Eng. J.* **2021**, *419*, 129418. [CrossRef]
263. Górny, D.; Guzik, U.; Hupert-Kocurek, K.; Wojcieszyńska, D. Naproxen ecotoxicity and biodegradation by *Bacillus thuringiensis* B1(2015b) strain. *Ecotoxicol. Environ. Saf.* **2019**, *167*, 505–512. [CrossRef]
264. Holt, L.M.; Laursen, A.E.; McCarthy, L.H.; Bostan, I.V.; Spongberg, A.L. Effects of land application of municipal biosolids on nitrogen-fixing bacteria in agricultural soil. *Biol. Fertil. Soils* **2010**, *46*, 407–413. [CrossRef]
265. Chaturvedi, P.; Shukla, P.; Giri, B.S.; Chowdhary, P.; Chandra, R.; Gupta, P.; Pandey, A. Prevalence and hazardous impact of pharmaceutical and personal care products and antibiotics in environment: A review on emerging contaminants. *Environ. Res.* **2021**, *194*, 110664. [CrossRef]
266. Villaverde, S.; Fernández-Polanco, F. Spatial distribution of respiratory activity in **Pseudomonas** putida 54G biofilms degrading volatile organic compounds (VOC). *Appl. Microbiol. Biotechnol.* **1999**, *51*, 382–387. [CrossRef]
267. Andreolli, M.; Lampis, S.; Poli, M.; Gullner, G.; Biró, B.; Vallini, G. Endophytic Burkholderia fungorum DBT1 can improve phytoremediation efficiency of polycyclic aromatic hydrocarbons. *Chemosphere* **2013**, *92*, 688–694. [CrossRef]
268. Jeong, E.; Hirai, M.; Shoda, M. Removal of *o*-xylene using biofilter inoculated with *Rhodococcus* sp. *BTO62*. *J. Hazard. Mater.* **2008**, *152*, 140–147. [CrossRef]
269. Duc, H.D. Degradation of chlorotoluenes by Comamonas testosterone KT5. *Appl. Biol. Chem.* **2017**, *60*, 457–465. [CrossRef]

270. McNerney, R.; Mallard, K.; Okolo, P.I.; Turner, C. Production of volatile organic compounds by mycobacteria. *FEMS Microbiol. Lett.* **2012**, *328*, 150–156. [CrossRef]
271. Monika, J.; Shilpi, S.; Satyawati, S. Development of next-generation formulation against *Fusarium oxysporum* and unraveling bioactive antifungal metabolites of biocontrol agents. *Sci. Rep.* **2021**, *11*, 22895. [CrossRef]
272. David, T.J. Setting the standards for machine learning in biology. *Nat. Rev. Mol. Cell Biol.* **2019**, *20*, 659–660. [CrossRef]
273. Subashchandrabose, S.R.; Ramakrishnan, B.; Megharaj, M.; Venkateswarlu, K.; Naidu, R. Mixotrophic cyanobacteria and microalgae as distinctive biological agents for organic pollutant degradation. *Environ. Int.* **2013**, *51*, 59–72. [CrossRef]
274. Mishra, C.S.K.; Samal, S.; Samal, R.R. Evaluating earthworms as candidates for remediating pesticide contaminated agricultural soil: A review. *Front. Environ. Sci.* **2022**, *10*, 924480. [CrossRef]
275. Zhang, T.; Zhang, H. Microbial Consortia Are Needed to Degrade Soil Pollutants. *Microorganisms* **2022**, *10*, 261. [CrossRef] [PubMed]
276. Zuo, W.; Zhang, L.; Zhang, Z.; Tang, S.; Sun, Y.; Huang, H.; Yu, Y. Degradation of organic pollutants by intimately coupling photocatalytic materials with microbes: A review. *Crit. Rev. Biotechnol.* **2021**, *41*, 273–299. [CrossRef] [PubMed]
277. Top, E.M.; Springael, D.; Boon, N. Catabolic mobile genetic elements and their potential use in bioaugmentation of polluted soils and waters. *FEMS Microbiol. Ecol.* **2002**, *42*, 199–208. [CrossRef] [PubMed]
278. Chaudhry, Q.; Blom-Zandstra, M.; Gupta, S.K.; Joner, E. Utilising the Synergy between Plants and Rhizosphere Microorganisms to Enhance Breakdown of Organic Pollutants in the Environment (15 pp). *Environ. Sci. Pollut. Res.* **2005**, *12*, 34–48. [CrossRef]
279. Antonio, V. Nanotechnology, bionanotechnology and microbial cell factories. *Microb. Cell Factories* **2010**, *9*, 53. [CrossRef]
280. Günther, J.; Patrick van, R.; Barbara Santos de, M.; Alexander, B. Ferritin: A Versatile Building Block for Bionanotechnology. *Chem. Rev.* **2015**, *115*, 1653–1701. [CrossRef]
281. Karpenko, A.A.; Odintsov, V.S.; Aleksandra, I. Micro-nano-sized polytetrafluoroethylene (teflon) particles as a model of plastic pollution detection in living organisms. *Environ. Sci. Pollut. Res.* **2021**, *29*, 11281–11290. [CrossRef]
282. Dhanapal, A.R.; Thiruvengadam, M.; Vairavanathan, J.; Venkidasamy, B.; Easwaran, M.; Ghorbanpour, M. Nanotechnology Approaches for the Remediation of Agricultural Polluted Soils. *ACS Omega* **2024**, *9*, 13522–13533. [CrossRef]
283. Alazaiza, M.Y.D.; Albahnasawi, A.; Ali, G.A.M.; Bashir, M.J.K.; Copty, N.K.; Amr, S.S.A.; Abushammala, M.F.M.; Al Maskari, T. Recent Advances of Nanoremediation Technologies for Soil and Groundwater Remediation: A Review. *Water* **2021**, *13*, 2186. [CrossRef]
284. Yadav, N.; Garg, V.K.; Chhillar, A.K.; Rana, J.S. Detection and remediation of pollutants to maintain ecosustainability employing nanotechnology: A review. *Chemosphere* **2021**, *280*, 130792. [CrossRef]
285. Arshad, I.; Noor, A.; Rashid, H.; Hussan, M.U.; Waqas, M.; Fatima, N.; Anwer, A.; Ijaz, V.; Qayyum, A.; Arshad, F.; et al. A comprehensive review on role of nanotechnology in the soil pollutants remediation. *Agric. Sci. J.* **2022**, *4*, 17–38. [CrossRef]
286. Nakum, J.; Bhattacharya, D. Various Green Nanomaterials Used for Wastewater and Soil Treatment: A Mini-Review. *Front. Environ. Sci.* **2022**, *9*, 724814. [CrossRef]
287. Hemalatha, I.; Harika, D.; Karnena, M.K. Sustainable Nano-Bioremediation Approaches for the Treatment of Polluted Soils. *Nat. Environ. Pollut. Technol.* **2022**, *21*, 1818–1826. [CrossRef]
288. He, Y.; Zhou, Q.; Mo, F.; Li, T.; Liu, J. Bioelectrochemical degradation of petroleum hydrocarbons: A critical review and future perspectives. *Environ. Pollut.* **2022**, *306*, 119344. [CrossRef]
289. Wang, X.; Cai, Z.; Zhou, Q.; Zhang, Z.; Chen, C. Bioelectrochemical stimulation of petroleum hydrocarbon degradation in saline soil using U-tube microbial fuel cells. *Biotechnol. Bioeng.* **2012**, *109*, 426–433. [CrossRef] [PubMed]
290. Mohanakrishna, G.; Venkata Mohan, S.; Sarma, P.N. Bio-electrochemical treatment of distillery wastewater in microbial fuel cell facilitating decolorization and desalination along with power generation. *J. Hazard. Mater.* **2010**, *177*, 487–494. [CrossRef]
291. Zhou, Y.; Zou, Q.; Fan, M.; Xu, Y.; Chen, Y. Highly efficient anaerobic co-degradation of complex persistent polycyclic aromatic hydrocarbons by a bioelectrochemical system. *J. Hazard. Mater.* **2020**, *381*, 120945. [CrossRef] [PubMed]
292. Chen, F.; Liang, B.; Li, Z.-L.; Yang, J.-Q.; Huang, C.; Lyu, M.; Yuan, Y.; Nan, J.; Wang, A.-J. Bioelectrochemical assisted dechlorination of tetrachloroethylene and 1,2-dichloroethane by acclimation of anaerobic sludge. *Chemosphere* **2019**, *227*, 514–521. [CrossRef]
293. Doudna, J.A.; Charpentier, E. The new frontier of genome engineering with CRISPR-Cas9. *Science* **2014**, *346*, 1258096. [CrossRef]
294. Liu, Q.; He, W.; Zhang, W.; Wang, L.; Tang, J. Metagenomic analysis reveals the microbial response to petroleum contamination in oilfield soils. *Sci. Total Environ.* **2024**, *912*, 168972. [CrossRef]
295. Sharma, P.; Sirohi, R.; Tong, Y.W.; Kim, S.H.; Pandey, A. Metal and metal(loids) removal efficiency using genetically engineered microbes: Applications and challenges. *J. Hazard. Mater.* **2021**, *416*, 125855. [CrossRef] [PubMed]

processes

MDPI

Review

Trends and Hotspots in Soil Minerals' Impacts on Carbon Stability Research: A Bibliometric Analysis Based on Web of Science

Xiaoyu Meng [1,2], Bing Xia [2,*], Wenjing Gao [2,3], Wei Chen [2], Qianjia He [2], Jiazhong Qian [1], Zhixiang Chen [1,2], Hongfeng Chen [2], Xiaoyu Zhang [4,*] and Rongrong Ying [4]

1 School of Resource and Environmental Engineering, Hefei University of Technology, Hefei 230009, China; xyu@mail.hfut.edu.cn (X.M.); qianjiazhong@hfut.edu.cn (J.Q.); 2022170852@mail.hfut.edu.cn (Z.C.)
2 Anhui Academy of Eco-Environmental Science Research, Hefei 230071, China; wjgao@mail.ustc.edu.cn (W.G.); chenwei2014@pku.org.cn (W.C.); heqianjia@mail.ustc.edu.cn (Q.H.); ahhkyjdzx@aaesr.cn (H.C.)
3 School of Earth and Space Sciences, University of Science and Technology of China, Hefei 230026, China
4 Nanjing Institute of Environmental Sciences, Ministry of Ecology and Environment, Nanjing 210042, China; yrr@nies.org
* Correspondence: xiabing@nies.org (B.X.); zhangxiaoyu@nies.org (X.Z.)

Abstract: The association with minerals is the most critical factor influencing the stability of organic carbon in soil. It is essential to gain an in-depth understanding of the research progress and future development trends regarding the impact of soil minerals on organic carbon stability both domestically and internationally and to identify current key issues; a total of 1834 research papers from the Web of Science Core Collection database were selected as the data source. These papers were examined using CiteSpace, HistCite, VOSviewer, and Origin 9.1 tools. The analysis categorized and visualized data based on countries, institutions, journals, disciplines, and keywords, utilizing indices like the number of articles and total/average citation frequency. The results indicate that the number of publications on the study of soil minerals and their impact on organic carbon stability has been increasing from 2013 to 2023. China and the United States have significantly led in the number of publications in this field. However, research collaborations among countries also exhibit significant regional characteristics. The Chinese Academy of Sciences (CAS) has the most publications. Yet, its average frequency of local citations is only 0.81 per paper, significantly lower than the top ten average of 2.23 per paper. The journal with the highest number of articles in this field is *Nature Communications*, with 248 articles. The main research hotspots focus on aspects such as the adsorption of organic carbon by minerals, catalytic transformation, and redox reactions. Future research should build on this foundation to focus more on detailed mechanisms, particularly the long-term effects of different environmental factors and time scales on soil carbon stability.

Keywords: soil minerals; organic carbon stability; bibliometrics; visualization; prospect

1. Introduction

As a dynamic global carbon reservoir, soil contains more organic carbon than vegetation and the atmosphere [1]. The estimated global stock of soil organic carbon (SOC) ranges from 1500 to 1600 Pg (Pg = 10^{15} g) at a soil depth of 1 m, and from 2376 to 2456 Pg at a depth of 2 m [2] (Figure 1). The release or activation of even a small fraction of SOC can significantly impact atmospheric greenhouse gases and global climate. Approximately 14% of the atmospheric CO_2 increase in the last decade has been reported to be attributed to SOC instability [3]. However, the combination of soil minerals with organic matter can effectively prevent the decomposition of organic matter. Soil minerals are a crucial component of soil carbon stocks, accounting for more than 60% of the total [4]. Particulate organic carbon (POC) primarily originates from partially decomposed plant residues and represents a relatively unstable organic carbon component. In contrast, mineral-associated organic carbon (MAOC) is mainly formed from recalcitrant microbial metabolites and residues, characterized by slow turnover and excellent stability [5]. Therefore, studying the stability of SOC from the perspective of soil minerals is of great significance for assessing global carbon stocks and the carbon cycle [6].

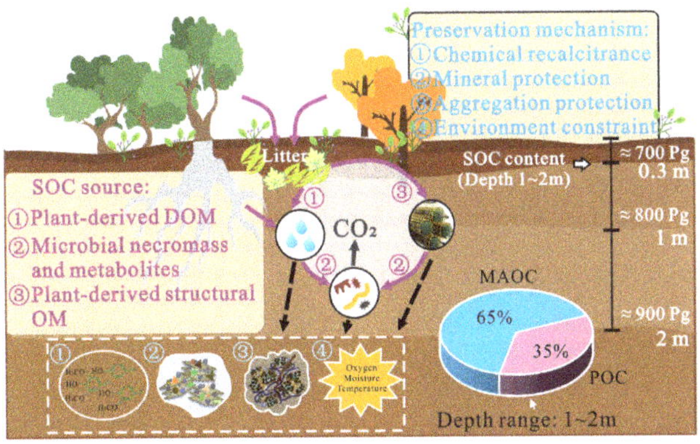

Figure 1. Global organic carbon stocks (modified from Feng et al. [7]). DOM: dissolved organic matter, OM: organic matter.

Earlier studies have indicated that carbon derived from above-ground plant residues is the soil's primary carbon source and regulates the rate of SOC accumulation [8]. Although it was previously believed that single molecular structures constituted "stable" SOC [9], advances in isotope tracing techniques have revealed that the so-called "stable" carbon pool contains a significant amount of SOC with unstable molecular structures, such as sugars and proteins [10]. Krull et al. [11] summarized SOC stability into two main mechanisms: biochemical resistance and physical protective effects. Later, Lützow et al. [12] further categorized three mechanisms: selective protection, spatial segregation of biological and SOC, and interaction with soil mineral surfaces and metal ions [13–16]. Among these, the interaction of SOC with soil minerals is widely recognized as the most stable mechanism [17]. Studies have confirmed that interactions with minerals can alter the morphology and properties of SOC through redox, polymerization, and catalytic reactions, directly affecting its stability and bioavailability [18–21]. Additionally, the association complexes formed through adsorption, complexation, and co-precipitation between SOC and minerals can protect SOC from being degraded [22–25]. The catalytic polymerization of SOC on mineral surfaces is considered a key step in preserving and accumulating SOC in soils [26,27]. Moreover, SOC can change the structure of minerals through chelation, solubilization, co-precipitation, and redox reactions, significantly affecting their combination with SOC, which is also crucial for the fate of SOC [28–30]. Despite numerous studies on how soil minerals affect SOC stability, there is a gap in the literature that analyzes and predicts future research directions or technological innovations in this field from the perspectives of changes in research trends and hot topics.

Using visual analysis tools such as CiteSpace [31], HistCite [32], and VOSviewer [33], along with bibliometric methods, this paper macroscopically quantifies the nature and direction of both domestic and international research on the impacts of soil minerals on the stability of SOC over the past decade. By clustering, analyzing, and visualizing the retrieved literature, we reveal and describe the development history and structural relationships of scientific studies on the interactions between soil minerals and organic carbon from 2013 to 2023. It is important to note that this review adopts a bibliometric analysis approach, which differs from traditional narrative reviews. Unlike narrative reviews that focus on qualitative synthesis, bibliometric analysis relies on quantitative methods to systematically analyze large datasets of scientific literature [34]. As such, this paper includes detailed Section 2 to ensure transparency and reproducibility of the analysis. This study outlines the historical evolution and emerging trends in global research on the impact of soil minerals on carbon stability, identifies key gaps in current knowledge, and proposes potential directions for future research.

2. Materials and Methods

2.1. Data Collation

For this study, the Web of Science (WOS) Core Collection database under Clarivate Analytics was used as the source for literature searches, covering the period from 1 January 2013 to 30 October 2023. The search was executed on 1 December 2023, using the following subject search terms: TS = (carbon or soil organic matter or mineral-bound organic carbon) and (soil mineral or iron or aluminum or cadmium or barium or copper) and (microbial residue C or plant residue C) and (stabilizat or affect). The selected types of literature were research articles and reviews. A total of 1834 articles were retrieved and imported into HistCite, including contributions from 13,848 authors, 307 journals, and 96,307 references.

2.2. Statistical Method

This paper employed various analytical tools (Figure 2), including the built-in analysis tool of the WOS database, CiteSpace literature visualization [31], HistCite citation analysis software [32], VOSviewer [33]. In the application of HistCite [32], the key parameters are the Total Local Citation Score (TLCS) and the Total Global Citation Score (TGCS). CiteSpace and VOSviewer were used to extract important noun phrases from the titles, abstracts, and keywords of the documents for co-word, co-occurrence, and emergent term analysis. This is to identify the research focus and trends regarding the role of soil minerals in organic carbon stability [31]. The analysis in CiteSpace covered the time span from 2013 to 2023, with a time node of one year and the discipline category as the node type. The node strength was set to Cosine by default, the threshold was TOP [25], and Pathfinder and Pruning Sliced Networks in the network pruning function area were selected for conducting the map analysis. VOSviewer was used to analyze countries and keywords, with the data type selected from WOS and the counting method set to Full counting. The keyword parameter threshold was set to 54 to generate the final map. Data analyzed by HistCite, VOSviewer, and CiteSpace were all from the articles downloaded in text format from the WOS database. Please note that this study is based on 1834 documents from the WoS database, potentially omitting relevant studies from other sources (e.g., Scopus, Google Scholar). Future research could integrate multiple databases and apply text mining and NLP to further explore topic evolution and key technological advances.

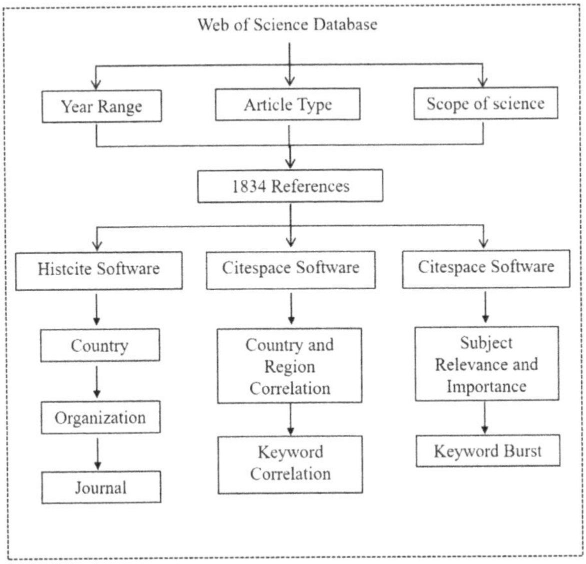

Figure 2. Data analysis flowchart.

3. Results and Discussion

3.1. Trends in the Number of Publications

Changes in the number of publications can effectively indicate research intensity in a particular field over a given period, providing great significance for analyzing the development dynamics during that period and predicting future trends [35]. The global atmospheric CO_2 concentration surpassed 400 mg/L for the first time in 2013 [36], after which the number of publications concerning the topic of this study showed an approximately linear increase (Figure 3). According to Raza et al. [2], 64% of the 46,080 SOC-related publications were published within the last decade (2014–2023). As shown in Figure 3, publications on the influences of minerals on SOC stability have steadily increased over the past decade. The year 2022 saw the highest number of publications, reaching 232, and by 2023, the cumulative number of publications had reached 1834. Publications in this field accounted for 4–8% of SOC-themed publications in the past decade. Several global initiatives have been launched with discussions on increasing SOC inventories as a potential solution for climate change mitigation, including the "4 pour 1000" initiative (https://4p1000.org, accessed on 7 December 2024), RECSOIL (https://www.fao.org/global-soil-partnership/areas-of-work/recarbonization-of-global-soils, accessed on 7 December 2024), the Koronivia Joint Work on Agriculture (https://www.fao.org/koronivia, accessed on 7 December 2024), the REDD+project (https://www.fao.org/redd/overview/en/, accessed on 7 December 2024), and the European Soil Strategy. These initiatives have effectively promoted research on the role of minerals in stabilizing organic carbon, making the interaction between minerals and SOC an increasingly popular topic in global soil carbon research. Moreover, the growing number of academic achievements in this field indicates a deepening understanding among researchers regarding the effects of minerals on organic carbon stability.

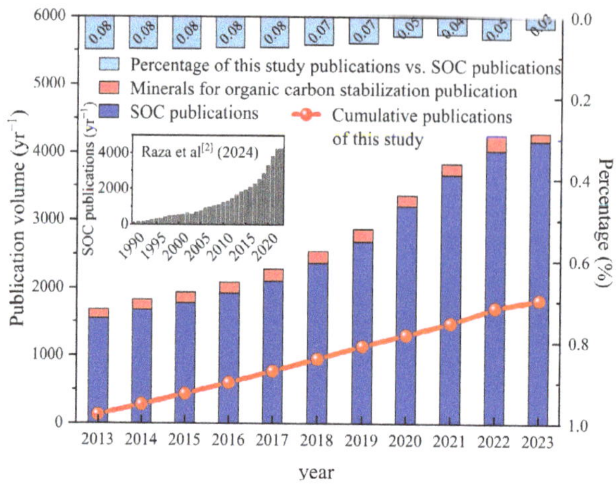

Figure 3. Annual number of publications and cumulative number of publications during 2013–2023 [2].

3.2. Top 10 Countries/Regions in Terms of Total Number of Publications

The number of publications (records) in a particular field of research can indicate the level of interest in this research field among different countries and reflect current global trends in advanced research. Additionally, the number and frequency with which a region's or country's articles are indexed and cited by the Science Citation Index (SCI) reveal the overall scientific research capacity and influence of that country or region in the field [37]. In the WOS Core Collection database, research papers on the impacts of soil minerals on organic carbon stability were retrieved from 108 countries/regions. These countries/regions include the United States, China, Germany, the United Kingdom, and Japan, with a primary concentration in Europe, East Asia, and North America (Figure 4). As shown in Table 1, from 2013 to 2023, the United States published 846 articles, achieving the highest total citation frequency (1385 times), thereby demonstrating its dominant position in this research field. China ranks second globally in terms of the number of publications, with a total of 673 articles, which is 428 more than Germany, the third-ranked country. This indicates that the United States and China are leading significantly in the number of publications in this field. This is consistent with the findings of Raza et al. in their global soil carbon research [2]. The results of scientific papers, policy direction, and financial support reflect the dominance of these two countries in this field. China, a populous country with a developed agricultural economy, emphasized green agriculture and sustainable development in its 14th Five-Year Plan. This initiative aims to reduce global carbon emissions [38]. The United States, as a developed country, boasts numerous high-level research institutions and researchers, holding significant academic influence globally [38]. In addition, the publication volume by country indicates that researchers with a high focus on this field are primarily concentrated in developed countries. This is mainly due to these countries' strong research capabilities, abundant natural resources, and a high demand for addressing global issues such as climate change. Research institutions and universities in these countries have advanced experimental facilities and stable funding support (as illustrated in Table 2). At the same time, their diverse geographical and ecological environments provide varied conditions for studying the interactions between soil minerals and SOC. Among these countries, the total local citation score per research paper (TLCSPR) in China (0.71) is lower than the world's top ten average of 1.01. This suggests that Chinese research papers in this field are less influential. Canada and the United States have the highest total local citation score per research paper, with values of

1.84 and 1.64, respectively. Furthermore, China's total global citation score per research paper (TGCSPR) is 231.82, ranking ninth among the top ten countries. This also reflects the gap between China and the United States.

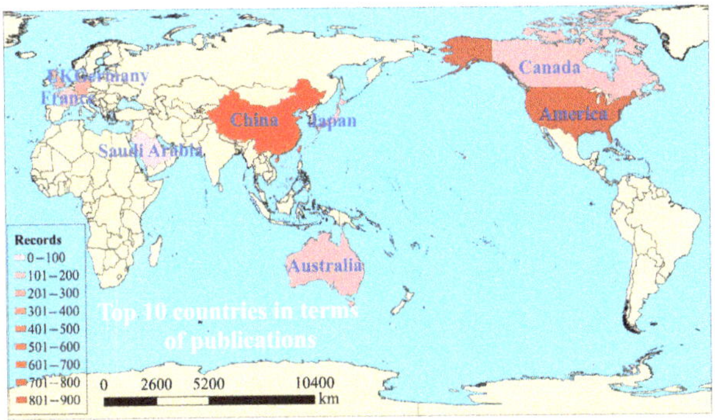

Figure 4. Distribution of top 10 countries in terms of number of publications.

Table 1. Top 10 countries in research on the impacts of soil minerals on organic carbon stability from 2013 to 2023.

Country	Records	Total Local Citation Score (TLCS)	Total Local Citation Score per Research Paper (TLCSPR)	Total Global Citation Score (TGCS)	Total Global Citation Score per Research Paper (TGCSPR)
USA	846	1385	1.64	291,602	344.68
China	673	479	0.71	156,013	231.82
Germany	245	324	1.32	72,253	294.91
UK	228	164	0.72	59,700	261.84
Japan	129	158	1.22	43,746	339.12
Australia	124	50	0.40	33,501	270.17
France	121	74	0.61	28,870	238.60
Canada	103	190	1.84	35,800	347.57
Switzerland	103	147	1.43	30,626	297.34
Saudi Arabia	91	19	0.21	14,463	158.93

Table 2. Top 10 institutions researching the impacts of soil minerals on organic carbon stability from 2013 to 2023.

Institution	Records	Total Local Citation Score (TLCS)	Total Local Citation Score per Research Paper (TLCSPR)	Total Global Citation Score (TGCS)	Total Global Citation Score per Research Paper (TGCSPR)
Chinese Academy of Sciences	176	142	0.81	49,833	283.14
Stanford University	78	211	2.71	29,181	374.12
University of California, Berkeley	76	319	4.20	29,542	388.71
University of Chinese Academy of Sciences	62	47	0.76	16,700	269.35
Massachusetts Institute of Technology	51	51	1.00	21,763	426.73
California Institute of Technology	49	123	2.51	13,714	279.88
Lawrence Berkeley National Laboratory	46	161	3.50	13,904	302.26
Columbia University	45	167	3.71	20,739	460.87
Harvard University	43	69	1.60	22,383	520.53
Tsinghua University	40	61	1.53	12,259	306.48

In this study, the VOSviewer visual analysis tool (version 1.6.20) was employed to analyze the collaborative relationships in the field of soil minerals' impacts on organic carbon stability research between China (including Taiwan) and countries worldwide. Nodes represent keywords. The node size represents the number of occurrences of keywords. The threshold value of keyword parameters is set to 54; that is, keywords appear at least 54 times. Different colors represent different clusters. The size of the circles in the figure indicates the level of activity and the number of articles published by a country or region. In contrast, the distance of connecting lines between countries or regions represents the closeness of their cooperation, with shorter distances indicating closer collaboration. As shown in Figure 5, in this field, countries do not work in isolation—those with higher publication volumes tend to collaborate more closely, reflecting a shared commitment to soil pollution remediation. Figure 5 further illustrates the differences in regional cooperation, showing that China and the United States lead the research in this field. China collaborates more closely with countries in the Asia-Pacific region, whereas the United States has stronger cooperation with Western countries. International collaboration has shown distinct regional characteristics, and future international cooperation and knowledge sharing in this field still need to be further enhanced.

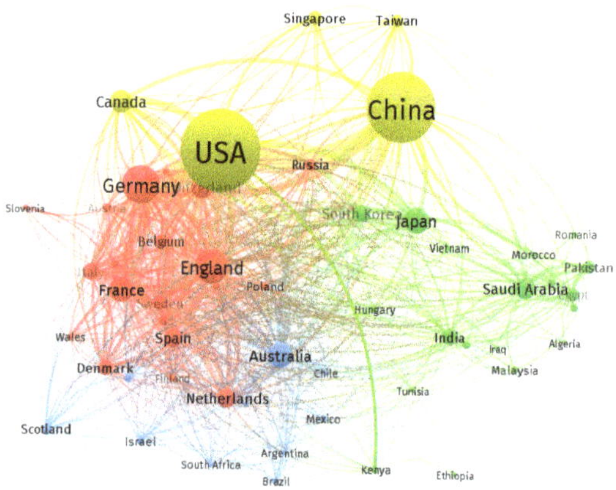

Figure 5. Cooperation between China (including Taiwan) and other countries in the field of research on the impacts of soil minerals on organic carbon stability from 2013 to 2023.

3.3. *Key Research Institutions*

According to the results from the citation analysis software HistCite, research papers on the impacts of soil minerals on organic carbon stability involve a total of 1266 research institutions. The top 10 research institutions in terms of the number of publications were analyzed. As shown in Table 2, research institutions from China and the United States dominate the top 10 list in terms of the records (number of publications). Three Chinese institutions—Chinese Academy of Sciences (CAS), University of Chinese Academy of Sciences, and Tsinghua University—are included in the top 10 list, with CAS leading significantly by publishing a total of 176 articles, indicating its high academic activity in this field. The rest are all research institutions from the United States. The ranking of institutional publication volume explains the differences in publication volume between countries. In terms of TGCS, CAS ranks first with 49,833 citations, followed by the University of California, Berkeley, with 29,542 citations, both demonstrating strong academic influence. However, considering the indicator of TLCSPR, CAS, despite having a higher number of publication records, has only 0.81 citations per paper, significantly lower than the average of the top 10 institutions (2.23 citations per paper). This suggests that CAS, as a representative of China's research institutions, still lags behind European and American research organizations in terms of the quality of academic papers.

3.4. Main Publication Journals

An in-depth analysis of journals within a particular field of study allows for the identification of core SCI source journals in that field. Table 3 lists the top 10 journals with the most articles published in the field of soil minerals–organic carbon stability. In 2020, the journal with the highest impact factor was *Nature*, at 64.8, followed by *Science* and *Nature Communications*, with impact factors of 56.9 and 16.6, respectively. *Scientific Reports* has the lowest impact factor among these journals, at 4.9. *Nature Communications* has the highest number of publications in this field. The top three journals in terms of TGCS are *Nature Communications* (77,630), *Nature* (63,463), and *Science* (44,760), indicating that these journals are highly influential in this field.

Table 3. Top 10 journals with the most publications in the field of soil minerals' impacts on organic carbon stability research from 2013 to 2023.

Journals	Records	Total Local Citation Score (TLCS)	Total Global Citation Score (TGCS)	Important Factor (5 Years)	Important Factor (2020)	Country
Nature Communications	248	190	77,630	17	16.6	UK
Nature	146	214	63,463	60.9	64.8	UK
Science	112	223	44,760	54.5	56.9	USA
Proceedings of The National Academy of Sciences of The United States of America	71	194	22,750	12	11.1	USA
Journal of The American Chemical Society	69	237	24,154	15.1	15	USA
Scientific Reports	56	70	9966	4.9	4.6	UK
Angewandte Chemie-International Edition	47	115	14,114	15.3	16.6	Germany
Science Advances	31	113	6027	15.4	13.6	USA
Frontiers in Plant Science	29	83	3458	6.8	5.6	Switzerland
Advanced Science	27	61	4261	16.7	15.1	USA

3.5. Discipline Co-Occurrence Analysis

As can be seen from the co-occurrence analysis mapping of the primary disciplines (Figure 6), among the top ten disciplines, aside from Geoscience and Environmental Engineering, the other eight disciplines have larger red outer circles. This reveals that there has been a parallel development across these disciplines and a rapid growth rate in multidisciplinary fields in recent years. Additionally, the dense connecting lines between disciplines indicate that research on the impact of soil minerals on organic carbon stability exhibits characteristics of interdisciplinary integration and close interrelation.

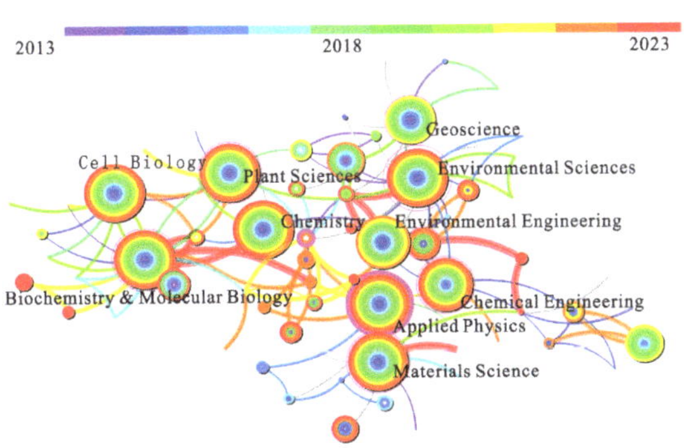

Figure 6. Analysis of main disciplines involved in research on the impacts of soil minerals on organic carbon stability from 2013 to 2023.

Figure 7 illustrates the frequency and centrality of various disciplines related to the impacts of soil minerals on organic carbon stability. The parameter set in CiteSpace is centrality, which can reflect the importance of a node within the system; stronger centrality indicates greater importance. Despite Chemistry having a high number of publications, its centrality is only 0.09, indicating that the impacts of soil minerals on organic carbon stability are not a primary research focus within this discipline. In contrast, Environmental Sciences has a centrality of 0.63, significantly higher than that of other disciplines. This suggests that Environmental Sciences serves as a crucial hub for interdisciplinary integration and connection. However, the cumulative number of publications in this discipline over the past ten years is only 136, indicating substantial potential for further research on the impacts of soil minerals on organic carbon stability. The low frequency and high centrality of environmental science in this field suggest that its research potential needs to be further explored. Research on organic carbon stability within environmental science has received widespread attention, such as the source of plant litter, occlusion within aggregates, incorporation in organo-mineral complexes, and location within the soil profile. Wetlands under different hydrological conditions [39] and the impacts of various natural and anthropogenic disturbances [40] have been studied. However, the regulatory mechanisms of environmental factors on minerals and SOC interactions and the effects of minerals on organic carbon stability under climate change, land use changes, and pollutant interference still require further research.

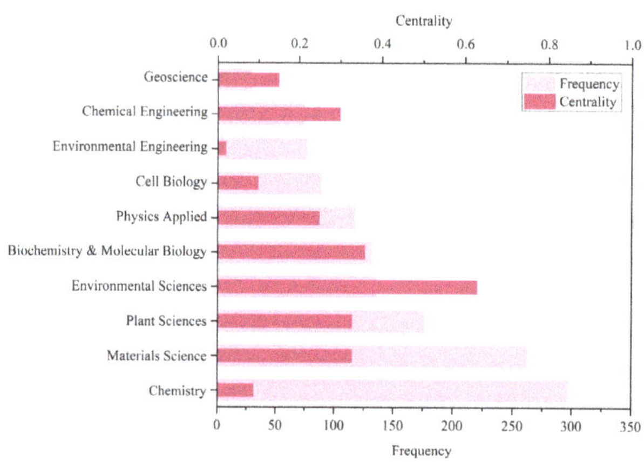

Figure 7. Centrality analysis of disciplines engaged in research on the impacts of soil minerals on organic carbon stability from 2013 to 2023.

3.6. Analysis of Research Hotspots and Frontiers at Home and Abroad

3.6.1. Research Hotspots at Home and Abroad

The keyword co-occurrence network model (Figure 8) was constructed using VOSviewer, a literature visualization and analysis software. By analyzing the keywords, researchers can gain insights into current research hotspots and future trends in the field [41]. In VOSviewer, keywords with a frequency of occurrence greater than six were visualized and analyzed. Among these, the keyword with the highest frequency of occurrence was "iron", appearing 563 times.

In the co-occurrence network model, the size of each circle represents the importance of each keyword; larger circles indicate higher frequency and greater significance within the research direction. The different colors of the circles represent different clusters to which the keywords belong. From Figure 8, it is evident that the keyword co-occurrence network is divided into six clusters. The six clusters are further summarized into two major categories:

"direct effects of minerals on organic carbon stability" and "indirect effects of inputs (quantity, quality) and degradation of vegetation production on organic carbon stability".

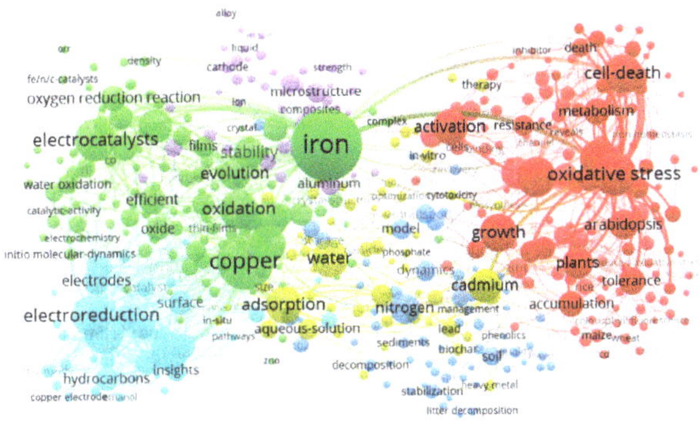

Figure 8. Distribution of research hotspots on the impacts of soil minerals on organic carbon stability from 2013 to 2023 in China.

The core terms within the green cluster are "iron", "catalyst", and "efficient". This cluster primarily focuses on the mineral-mediated catalytic conversion of SOC. Research has proved that minerals, acting as catalysts, can increase the reaction rate without altering the overall standard Gibbs free energy change of the reaction [42]. This effect can be attributed to three main aspects. First, minerals increase the concentration of SOC and oxidants, accelerating the reaction rate [43,44]. Secondly, minerals provide vacant electron orbits to accept bonded electron pairs from SOC or act as proton donors/acceptors to catalyze the reaction [45,46]. Thirdly, transition-metal-based mineral oxides or doped minerals can promote catalytic functions through cyclic valence changes and electron transfer [47–49]. In particular, redox-active iron minerals play crucial roles in many biogeochemical pathways [12,14]. As shown in Figure 7, there has been a growing interest in recent years in studying the effects of iron minerals on the stability of SOC. Clay minerals with different properties exhibit varying capacities to protect SOC. While laboratory studies indicate that the mass of clay minerals plays a crucial role in SOC protection, further mechanistic studies are still needed to explore this aspect [50]. In the future, efforts should be made to have a more comprehensive understanding of the function of minerals and the contributions of soil minerals to the transformation of SOC.

The core vocabularies within the light blue cluster include "electroreduction", "electrodes", "hydrocarbons", and "surfaces". This cluster focuses on the interactions between SOC and redox-active minerals. The direct electron transfer reactions between SOC and minerals and the indirect processes for the formation of reactive oxygen species (ROS) or microorganisms are essential interactions in the soil environment. First of all, although SOC encompasses a diverse range of surface functions from oxidation to reduction, it typically acts as an electron donor in redox reactions with soil minerals, resulting in the reduction of mineral metals. Additionally, redox reactions between soil minerals and SOC can generate ROS. These ROS, including hydrogen peroxide, superoxide, and hydroxyl radicals, are crucial in the process of SOC oxidation. They influence the molecular size, the presence of oxygen-containing functional groups, and the degree of SOC mineralization [51–53], depending on the specific reactants and conditions [42].

The core words and expressions within the yellow cluster are "adsorption", "aqueous-solution", and "cadmium". This cluster focuses on exploring the mechanisms of mineral adsorption of SOC. In natural soil systems, metal ion adsorption typically occurs through reactions with organic compounds, metal (oxygen) hydroxides, and clay mineral surfaces [54]. According to the research findings, entropy-driven hydrophobic repulsion, van der Waals forces, or hydrogen bonding may be the primary mechanism of adsorption under specific conditions [55,56]. For example, regarding electrostatic attraction, electrostatic bonding between soil mineral surfaces and organic molecules can occur through cation exchange. This reaction occurs when positively charged organic molecules, such as protonated amines, replace inorganic cations on the exchange complex [57]. The adsorption process generally begins with the formation of a monolayer on the mineral surface, which then evolves into a multilayer molecular structure driven by complex biological and chemical processes [58]. The mineral adsorption of SOC often determines the stability of SOC in the soil environment. This is why the adsorption mechanism predominates in the research [59]. Additionally, different metals exhibit varying affinities for different mineral species. For instance, cadmium, a heavy metal, is toxic to soil biota and is one of the most common metal pollutants in the soils of industrial, agricultural, and even urban areas [60,61]. Consequently, the adsorption effect of heavy metals on SOC has been extensively studied in recent years. This is due to the severe pollution of global water resources by heavy metals [62]. At the same time, adsorption offers advantages over other technologies, such as simplicity in operation, small footprint, stable effects, low cost, and great regeneration potential [63,64]. While various laboratory studies suggest that the quality of the clay minerals plays an important role in SOC protection, this has not been investigated by more mechanistic studies. Some existing literature indicates that the type of reactive mineral surfaces controls the mode of organic matter (OM) adsorption and assumes that bonding mechanisms control the turnover period [50].

For the purple cluster, the core terms include "microstructures", "cathodes", "crystals", "ions", and "composites". This group focuses on the interactions between mineral ions and SOC, mainly in the form of co-precipitation. Co-precipitation is a key process in the association of SOC with soil minerals, akin to adsorption. However, the mixtures formed through co-precipitation play a more critical role in the cycling of carbon and mineral elements than those formed through adsorption [65]. Negatively charged mineral surfaces repel organic anions, but multivalent cations in the exchange complex facilitate their binding. In neutral and alkaline soils, Ca^{2+} and Mg^{2+} are the dominant cations, while in acidic soils, hydroxy polymers of Fe^{3+} and Al^{3+} predominate [50]. These positively charged ions are adsorbed onto negatively charged mineral surfaces and promote the adsorption of negatively charged long-chain organic molecules through cation bridging [12,66,67]. Unlike Fe^{3+} and Al^{3+} ions, Ca^{2+} ions do not form strong coordination complexes with organic ions. However, cation bridging interactions are weaker than ligand-exchange-mediated adsorption in OM binding [68,69]. Generally, the organic carbon (OC) adsorption sequence on clay minerals follows the order of oxides–hydroxides > 2:1 clay minerals > 1:1 clay minerals, a trend controlled by the SSA and CEC of the respective minerals.

The core terms for the red and blue clusters are "plants", "oxidative stress", "cell-death", "activation", "nitrogen", "dynamics", "litter decomposition", and "sedimentation", respectively. These clusters focus on the indirect effects of vegetation growth and degradation on the stabilization of SOC. It is commonly believed that most of the humic materials in soil are closely associated with the surfaces of colloidal minerals. Evidence for this comes from the following findings: organic materials associated with fine particles and clay [70–72] are older and have longer turnover times than those associated with sand particles [73,74]. According to Xu Jiahui et al. [74], the chemical properties of vegetation litter, such as lignin content, are major factors determining its degradation rate. Among abiotic regulatory factors, reduced soil moisture and changes in soil nitrogen and phosphorus content due to warming may promote plant root growth [75], thereby increasing carbon inputs. However, reduced moisture also decreases SOC chemical stability [76], exacerbating its loss. On the other hand, under normal environmental conditions, plants maintain balance by regulating the production and removal of ROS. This balance can be disrupted by various abiotic stress factors, such as heavy metals [77]. Consequently, ROS levels can be used to monitor the extent of oxidative stress in plants.

A deeper review of the research hotspots reveals that existing studies primarily focus on several key areas, including the catalytic transformation of SOC mediated by minerals, redox reactions between SOC and active minerals, co-precipitation of mineral ions and SOC, the mechanisms of mineral adsorption on SOC, and the indirect effects of plant decomposition and formation on the stabilization of SOC. These directions highlight the central role of minerals in soil carbon stability research, providing clear focal points and a systematic framework for research in this field. Clay minerals with different properties exhibit varying capacities to protect SOC, making it difficult to determine the contribution of minerals in various mechanisms. Although several laboratory studies indicate that the quality of clay minerals plays an important role in SOC protection, this has yet to be explored through more mechanistic research [50]. Some existing literature suggests that the type of reactive mineral surface controls the adsorption pattern of OM and hypothesizes that bonding mechanisms control the turnover cycle. While this hypothesis holds true for in vitro studies, where residence time is related to desorption potential, long-term studies do not examine the correlation between specific bonding mechanisms and carbon turnover time. Furthermore, the existing knowledge based on time series mostly suggests a correlation between OC turnover time and the abundance of certain mineral phases. However, such studies cannot exclude the potential influence of factors such as soil aggregation, anoxia, and changes in microbial ecology on C turnover [50]. Therefore, this focus also implies that other potential mechanisms or environmental variables, such as the synergistic effects between microbes and minerals [78], the dynamic behavior of amorphous minerals [79], and the minerals and SOC interactions under climate change [80], may be relatively overlooked in current research trends. To address this, we suggest that future research could combine high-resolution characterization techniques (such as synchrotron X-ray absorption spectroscopy, nuclear magnetic resonance, etc.) and long-term incubation experiments to reveal further how different minerals influence the stability mechanisms of SOC.

3.6.2. Analysis of Research Frontiers

Emergent keywords are terms that appear suddenly and are frequently cited, used for analyzing research frontiers and hotspots over different periods [81]. Specifically, "begin" and "end" denote the start and end of a keyword's emergence period, respectively, while "strength" signifies the intensity of its emergence; a higher number indicates a more influential keyword [82]. Using the keyword burstiness analysis in CiteSpace software (version 6.3.R1), the detection model in the Control Panel was configured as follows: $f(x) = \alpha e^{(-\alpha x)}$, with $\alpha 1 / \alpha 0 = 2.0$, The Number of States set to 2, $\gamma[0,1] = 1.0$, and Minimum Duration = 2. The top ten most frequently cited keywords were calculated based on this configuration. The keyword emergence analysis of literature on the impact of soil minerals on organic carbon stability is summarized in Figure 9. Among these, keywords such as "Electroreduction", "Hydrocarbon", "Lipid peroxidation", "Cell death", and "Protein" have shown a higher intensity of emergence after 2013, with values of 8.07, 7.83, 5.49, 5.22, and 4.47, respectively. These values reveal the current research hotspots in this field to a certain extent.

As mentioned above, different scholars hold varying views on the stabilization mechanisms of SOC in the soil. Nonetheless, the binding of SOC to minerals is recognized as a primary stabilization mechanism [12,14,16]. This binding involves several aspects, including ligand exchange, multivalent cation bond bridges, complexation, and van der Waals forces [12,13]. The main minerals contributing to OM stabilization in soil environments include phyllosilicates (layered aluminum silicates, commonly referred to as clay minerals), metal oxides, hydroxides, and hydroxy-oxides (such as hematite, goethite, and pyroxenes), as well as short-range ordered aluminosilicates (such as opal, imogolite) [83]. Iron oxide minerals are favored over other mineral species for their larger specific surface area, more substantial adsorption capabilities, and higher preservation potential [12,14]. Therefore, they are regarded as some of the soil minerals that play a key role in the stabilization of SOC [13,84], and their content directly affects the soil's ability to adsorb SOC [16]. Additionally, numerous studies have demonstrated that Fe (III) oxides promote SOC degradation through direct or catalytic redox reactions [85–88]. Recent advancements in isotope tracing techniques have revealed that the carbon pool contains significant amounts of SOC with unstable molecular structures. These less stable organic carbons are more susceptible to degradation during redox processes with soil minerals, forming organic molecules with smaller sizes and richer O-functional groups [89,90]. Moreover, reactive-mineral-mediated processes can promote the mineralization of SOC through redox reactions, rather than protecting the soil environment [51,91]. For instance, data from Swedish forest soil surveys indicate that exchangeable Mn concentrations are negatively correlated with SOC content due to the oxidative decomposition of recalcitrant organic carbon by Mn peroxidase, which obscures conservation and production effects [87]. Electroreduction and iron oxide highlight the critical role of redox reactions on mineral surfaces in the adsorption, transformation, and protection of SOC. Hydrocarbon and protein emphasize the selectivity and diversity of interactions between different organic carbon molecules and minerals, influencing their stability. Lipid peroxidation and cell death further show how biological factors and their metabolic products regulate the minerals and organic carbon system through environmental chemical processes. These associations suggest that research should focus on the interfacial reactions between minerals and SOC, the coupling of biological and chemical processes, and the differential effects of various molecular types on stability to gain a more comprehensive understanding of the dynamic mechanisms of soil carbon stability.

Figure 9. Keyword emergence from 2013 to 2023 (Red represents the start and end years).

As analyzed above, all these properties contribute to the high citation frequency of these keywords in the analysis of the impacts of soil minerals on organic carbon stability. In other words, these keywords have represented the frontiers and hotspots in the field of SOC stability research, particularly the interaction mechanisms between SOC and soil minerals, in the past ten years.

4. Conclusions

This paper presents a systematic bibliometric analysis of the research on the impact of soil minerals on SOC stability included in the WOS database from 2013 to 2023. Overall, research in this field shows an increasing trend, with a total of 1834 publications. The research focuses on aspects such as the adsorption of SOC by minerals, catalytic transformation, and redox reactions. It is important to emphasize that with advancements in data analysis technology, future research could utilize methods such as knowledge graphs and deep learning to build more intelligent models for identifying and predicting research hotspots, thereby enhancing the depth and accuracy of the analysis. The specific conclusions are as follows:

(1) Global Research Trends and Collaboration: Research on the stability and interactions of soil minerals and SOC has continuously grown worldwide, particularly in Asia, Europe, and North America. While China and the United States lead in publication volume, research collaboration among countries shows distinct regional characteristics. Specifically, China collaborates more closely with countries in the Asia-Pacific region, while the United States primarily cooperates with European and North American countries. Nevertheless, there remains significant potential for further enhancement of international cooperation and knowledge sharing in this field.

(2) Major Research Institutions and Journals: The Chinese Academy of Sciences leads global research in this field, followed closely by the University of California, Berkeley. Regarding journal publications, *Nature Communications*, *Nature*, and *Science* are the most frequently cited journals in this area, indicating that research in this field has reached the international forefront.

(3) Co-occurrence Analysis: Environmental science is highly central to research on the impacts of soil minerals on organic carbon stability, serving as a key intersection between disciplines. However, over the past decade, research publications in this field have been relatively few, with only 136 papers, suggesting that environmental science's potential has not been fully realized. Future research should strengthen interdisciplinary collaboration between environmental science, chemistry, cell biology, and other fields to promote a deeper exploration of the mechanisms underlying the impacts of soil minerals on organic carbon stability.

(4) Research Themes and Future Directions: Current research primarily focuses on the mineral-mediated catalytic transformation of SOC, redox reactions, co-precipitation mechanisms, and adsorption mechanisms. However, some research gaps remain, such as the multi-factor interactions between minerals and SOC, the role of amorphous minerals, and the impact of environmental changes (e.g., climate change) on carbon stability. Future research should emphasize these complex interaction mechanisms, particularly the long-term effects of different environmental factors and time scales on soil carbon stability.

Author Contributions: Conceptualization, X.M. and B.X.; methodology, X.M., W.G. and W.C.; software, X.M., Q.H. and J.Q.; validation, X.M. and Z.C.; formal analysis, X.M.; investigation, H.C. and X.Z.; resources, X.M. and R.Y.; data curation, X.M.; writing—original draft preparation, X.M.; writing—review and editing, X.M., B.X. and X.Z.; visualization, X.M. and W.G.; supervision, W.C. and J.Q; funding acquisition, B.X. and X.Z.; All authors have read and agreed to the published version of the manuscript.

Funding: This research was funded by the Preparation and Application of Highly Efficient Iron-carbon Based Catalysts for Urban Legacy Chlorinated Organic Matter Contaminated Sites (No. SRHB-2023–07 2023hb0016) and the Anhui Province Soil Environmental Background Value Development Project (No. ZXBZ2021–02).

Data Availability Statement: The data presented in this study are available on request from the corresponding author.

Conflicts of Interest: The authors declare no conflicts of interest.

References

1. Lehmann, J.; Kleber, M. The contentious nature of soil organic matter. *Nature* **2015**, *528*, 60–68. [CrossRef] [PubMed]
2. Raza, S.; Irshad, A.; Margenot, A.; Zamanian, K.; Li, N.; Ullah, S.; Mehmood, K.; Ajmal Khan, M.; Siddique, N.; Zhou, J.; et al. Inorganic carbon is overlooked in global soil carbon research: A bibliometric analysis. *Geoderma* **2024**, *443*, 116831. [CrossRef] [PubMed]
3. Friedlingstein, I.; Jones, M.W.; Sullivan, O.M.; Andrew, R.M.; Hauck, J.; Peters, G.P.; Peters, W.; Pongratz, J.; Sitch, S.; Le Quéré, C.; et al. Global Carbon Budget 2019. *Earth Syst. Sci. Data* **2019**, *11*, 1783–1838. [CrossRef]
4. Xu, Z.; Tsang, D.C.W. Mineral-mediated stability of organic carbon in soil and relevant interaction mechanisms. *Eco-Environ. Health* **2024**, *3*, 59–76. [CrossRef] [PubMed]
5. Sokol, N.W.; Sanderman, J.; Bradford, M.A. Pathways of mineral-associated soil organic matter formation Integrating the role of plant carbon source, chemistry, and point of entry. *Glob. Chang. Biol.* **2018**, *25*, 12–24. [CrossRef]
6. Li, Q.; Wang, L.; Fu, Y.; Lin, D.; Hou, M.; Li, X.; Hu, D.; Wang, Z. Transformation of soil organic matter subjected to environmental disturbance and preservation of organic matter bound to soil minerals: A review. *J. Soils Sediment.* **2022**, *23*, 1485–1500. [CrossRef]
7. Feng, X.; Dai, G.; Liu, T.; Jia, J.; Zhu, E.; Liu, C.; Zhao, Y.; Wang, Y.; Kang, E.; Xiao, J.; et al. Understanding the mechanisms and potential pathways of soil carbon sequestration from the biogeochemistry perspective. *Sci. China Earth Sci.* **2024**, *67*, 3386–3396. [CrossRef]
8. Jenkinson, D.S. Studies on decomposition of plant material in soil. V. The effects of plant cover and the soil type on the loss of carbon from 14C-labelled rye grass decomposition under field conditions. *J. Soil Sci.* **1977**, *28*, 424–434. [CrossRef]
9. Melillo, J.M.; Aber, J.D.; Muratore, J.F. Nitrogen and Lignin Control of Hardwood Leaf Litter Decomposition Dynamics. *Ecology* **1982**, *63*, 621–626. [CrossRef]
10. Kiem, R.; Kögel-Knabner, I. Contribution of lignin and polysaccharides to the refractory carbon pool in C-depleted arable soils. *Soil Biol. Biochem.* **2003**, *35*, 101–118. [CrossRef]
11. Krull, E.S.; Baldock, J.A.; Skjemstad, J.O. Importance of mechanisms and processes of the stabilisation of soil organic matter for modelling carbon turnover. *Funct. Plant Biol. FPB* **2003**, *30*, 207–222. [CrossRef] [PubMed]
12. Lützow, M.V.; Kögel-Knabner, I.; Ekschmitt, K.; Matzner, E.; Guggenberger, G.; Marschner, B.; Flessa, H. Stabilization of organic matter in temperate soils: Mechanisms and their relevance under different soil conditions—A review. *Eur. J. Soil Sci.* **2006**, *57*, 426–445. [CrossRef]
13. Wang, L.; Ying, R.R.; Shi, J.Q.; Long, T.; Lin, Y.S. Advancement in study on adsorption of organinc matter on soil mineral and its mechanism. *Acta Pedol. Sin.* **2017**, *54*, 805–818.
14. Throckmorton, H.M.; Bird, J.A.; Monte, N.; Doane, T.; Firestone, M.K.; Horwath, W.R. The soil matrix increases microbial C stabilization in temperate and tropical forest soils. *Biogeochemistry* **2015**, *122*, 35–45. [CrossRef]
15. Six, J.; Conant, R.T.; Paul, E.A.; Paustian, K. Stabilization mechanisms of soil organic matter: Implications for C-saturation of soils. *Plant Soil* **2002**, *241*, 155–176. [CrossRef]
16. Mcnally, S.R.; Beare, M.H.; Curtin, D.; Meenken, E.D.; Kelliher, F.M.; Calvelo Pereira, R.; Shen, Q.; Baldock, J. Soil carbon sequestration potential of permanent pasture and continuous cropping soils in New Zealand. *Glob. Chang. Biol.* **2017**, *23*, 4544–4555. [CrossRef]
17. Schrumpf, M.; Kaiser, K.; Guggenberger, G.; Persson, T.; Kögel-Knabner, I.; Schulze, E.D. Storage and stability of organic carbon in soils as related to depth, occlusion within aggregates, and attachment to minerals. *Biogeosciences* **2013**, *10*, 1675–1691. [CrossRef]
18. Li, H.; Reinhart, B.; Moller, S.; Herndon, E. Effects of C/Mn Ratios on the Sorption and Oxidative Degradation of Small Organic Molecules on Mn-Oxides. *Environ. Sci. Technol.* **2022**, *57*, 741–750. [CrossRef]
19. Hui, L.; Fernanda, S.; Kristen, B.; Elizabeth, H. A Critical Review on the Multiple Roles of Manganese in Stabilizing and Destabilizing Soil Organic Matter. *Environ. Sci. Technol.* **2021**, *55*, 12136–12152. [CrossRef]
20. Sunda, W.G.; Kieber, D.J. Oxidation of humic substances by manganese oxides yields low-molecular-weight organic substrates. *Nature* **1994**, *367*, 62–64. [CrossRef]
21. Fang, Q.; Lu, A.; Hong, H.; Kuzyakov, Y.; Algeo, T.J.; Zhao, L.; Olshansky, Y.; Moravec, B.; Barrientes, D.M.; Chorover, J. Mineral weathering is linked to microbial priming in the critical zone. *Nat. Commun.* **2023**, *14*, 345. [CrossRef] [PubMed]
22. Karen, I.; Graham, P.; Elisa, L.; Caroline, P.; Neil, G.; Thomas, W.; Christian, M.; Leon, B.; Jesus, O.; Nina, F.; et al. Towards a mechanistic understanding of carbon stabilization in manganese oxides. *Nat. Commun.* **2015**, *6*, 7620. [CrossRef]
23. Kleber, M.; Sollins, P.; Sutton, R. A Conceptual Model of Organo-Mineral Interactions in Soils: Self-Assembly of Organic Molecular Fragments into Zonal Structures on Mineral Surfaces. *Biogeochemistry* **2007**, *85*, 9–24. [CrossRef]
24. Faust, J.C.; Tessin, A.; Fisher, B.J.; Zindorf, M.; Papadaki, S.; Hendry, K.R.; Doyle, K.A.; März, C. Millennial scale persistence of organic carbon bound to iron in Arctic marine sediments. *Nat. Commun.* **2021**, *12*, 275. [CrossRef]
25. Lehmann, J.; Hansel, C.M.; Kaiser, C.; Kleber, M.; Maher, K.; Manzoni, S.; Nunan, N.; Reichstein, M.; Schimel, J.P.; Torn, M.S.; et al. Persistence of soil organic carbon caused by functional complexity. *Nat. Geosci.* **2020**, *13*, 529–534. [CrossRef]

26. Xiao, K.; Zhao, Y.; Liang, C.; Zhao, M.; Moore, O.W.; Otero-Fariña, A.; Zhu, Y.; Johnson, K.; Peacock, C.L. Introducing the soil mineral carbon pump. *Nat. Rev. Earth Environ.* **2023**, *4*, 135–136. [CrossRef]
27. Moore, O.W.; Curti, L.; Woulds, C.; Bradley, J.A.; Babakhani, P.; Mills, B.J.W.; Homoky, W.B.; Xiao, K.; Bray, A.W.; Fisher, B.J.; et al. Long-term organic carbon preservation enhanced by iron and manganese. *Nature* **2023**, *621*, 312–317. [CrossRef]
28. Masiello, C.A.; Chadwick, O.A.; Southon, J.; Torn, M.S.; Harden, J.W. Weathering controls on mechanisms of carbon storage in grassland soils. *Glob. Biogeochem. Cycles* **2004**, *18*, GB4023. [CrossRef]
29. Franks, M.; Duncan, E.; King, K.; Vázquez-Ortega, A. Role of Fe- and Mn-(oxy)hydroxides on carbon and nutrient dynamics agricultural soils: A chemical sequential extraction approach. *Chem. Geol.* **2021**, *561*, 120035. [CrossRef]
30. Yu, G.; Xiao, J.; Hu, S.; Polizzotto, M.L.; Zhao, F.; McGrath, S.P.; Li, H.; Ran, W.; Shen, Q. Mineral Availability as a Key Regulator of Soil Carbon Storage. *Environ. Sci. Technol.* **2017**, *51*, 4960–4969. [CrossRef] [PubMed]
31. Synnestvedt, M.C.; Chen, C.; Holmes, J.H. CiteSpace II: Visualization and knowledge discovery in bibliographic databases. AMIA. Annual Symposium proceedings/AMIA Symposium. *AMIA Symp.* **2005**, *2005*, 724.
32. Garfield, E.; Paris, S.W.; Stock, W.G. HistCite: A software tool for informetric analysis of citation linkage. *NFD Inf. -Wiss. Und Prax.* **2006**, *57*, 391–400.
33. Eck, N.J.P.V.; Waltman, L.R. VOSviewer: A Computer Program for Bibliometric Mapping. *ERIM Rep. Ser. Res. Manag.* **2009**, *84*, 523–538. [CrossRef]
34. Donthu, N.; Kumar, S.; Mukherjee, D.; Pandey, N.; Lim, W.M. How to conduct a bibliometric analysis: An overview and guidelines. *J. Bus. Res.* **2021**, *133*, 285–296. [CrossRef]
35. Wang, Y.H.; Li, C.L.; Wang, H.; Liang, B.M.; Lv, S.J. Research status and hotspots of green infrastructure in the field of ecological environment. *Acta Ecol. Sin.* **2022**, *42*, 2510–2521.
36. Kodaira, H. Responding to climate change and expectations for research. *Paddy Water Environ.* **2014**, *12*, 211–212. [CrossRef]
37. Li, Y.; Li, J.; Xie, S. Bibliometric analysis: Global research trends in biogenic volatile organic compounds during 1991–2014. *Environ. Earth Sci.* **2017**, *76*, 1–13. [CrossRef]
38. Hu, J.; Dong, J.; Xu, D.; Yang, Q.; Liang, J.; Li, N.; Wang, H. Trends in Global Agricultural Carbon Emission Research: A Bibliometric Analysis. *Agronomy* **2024**, *14*, 2617. [CrossRef]
39. Zhao, H.; Jia, J.; Zhao, Q.; Wang, J.; Gao, Y.; Huang, Y.; Chen, G. Soil organic carbon stabilization and associated mineral protection in typical coastal wetlands under different hydrologic conditions. *Front. Mar. Sci.* **2022**, *9*, 1031561. [CrossRef]
40. Yang, J.; Li, A.; Yang, Y.; Li, G.; Zhang, F. Soil organic carbon stability under natural and anthropogenic-induced perturbations. *Earth-Sci. Rev.* **2020**, *205*, 103199. [CrossRef]
41. Zhang, Y.T.; Xiao, H.B.; Nie, X.D.; Li, Z.W.; Deng, C.X.; Zhou, M. Evolution of research on soil erosion at home and abroad in the past 30 years—Based on bibliometric analysis. *Acta Pedol. Sin.* **2020**, *57*, 797–810.
42. Kleber, M.; Bourg, I.C.; Coward, E.K.; Hansel, C.M.; Myneni, S.C.B.; Nunan, N. Dynamic interactions at the mineral–organic matter interface. *Nat. Rev. Earth Environ.* **2021**, *2*, 402–421. [CrossRef]
43. Chacon, S.S.; García-Jaramillo, M.; Liu, S.Y.; Ahmed, M.; Kleber, M. Differential capacity of kaolinite and birnessite to protect surface associated proteins against thermal degradation. *Soil Biol. Biochem.* **2018**, *119*, 101–109. [CrossRef]
44. Aldersley, M.F.; Joshi, P.C.; Price, J.D.; Ferris, J.P. The role of montmorillonite in its catalysis of RNA synthesis. *Appl. Clay Sci.* **2011**, *54*, 1–14. [CrossRef]
45. Laszlo, P. Chemical Reactions on Clays. *Science* **1987**, *235*, 1473–1477. [CrossRef]
46. Wu, D.; Huang, S.; Zhang, X.; Ren, H.; Jin, X.; Gu, C. Iron Minerals Mediated Interfacial Hydrolysis of Chloramphenicol Antibiotic under Limited Moisture Conditions. *Environ. Sci. Technol.* **2021**, *55*, 9569–9578. [CrossRef]
47. Xia, Q.; Chen, J.; Dong, H. Effects of Organic Ligands on the Antibacterial Activity of Reduced Iron-Containing Clay Minerals: Higher Extracellular Hydroxyl Radical Production Yet Lower Bactericidal Activity. *Environ. Sci. Technol.* **2023**, *57*, 6888–6897. [CrossRef]
48. Xu, Z.; Yu, Y.; Xu, X.; Tsang, D.C.W.; Yao, C.; Fan, J.; Zhao, L.; Qiu, H.; Cao, X. Direct and Indirect Electron Transfer Routes of Chromium(VI) Reduction with Different Crystalline Ferric Oxyhydroxides in the Presence of Pyrogenic Carbon. *Environ. Sci. Technol.* **2022**, *56*, 1724–1735. [CrossRef]
49. Liu, X.; Dong, H.; Zeng, Q.; Yang, X.; Zhang, D. Synergistic Effects of Reduced Nontronite and Organic Ligands on Cr(VI) Reduction. *Environ. Sci. Technol.* **2019**, *53*, 13732–13741. [CrossRef]
50. Singh, M.; Sarkar, B.; Sarkar, S.; Churchman, J.; Bolan, N.; Mandal, S.; Menon, M.; Purakayastha, T.J.; Beerling, D.J. Chapter Two —Stabilization of Soil Organic Carbon as Influenced by Clay Mineralogy. In *Advances in Agronomy*; Sparks, D.L., Ed.; Academic Press: Cambridge, MA, USA, 2018; Volume 148, pp. 33–84; ISBN 0065-2113.
51. Chen, N.; Fu, Q.; Wu, T.; Cui, P.; Fang, G.; Liu, C.; Chen, C.; Liu, G.; Wang, W.; Wang, D.; et al. Active Iron Phases Regulate the Abiotic Transformation of Organic Carbon during Redox Fluctuation Cycles of Paddy Soil. *Environ. Sci. Technol.* **2021**, *55*, 14281–14293. [CrossRef]

52. Trusiak, A.; Treibergs, L.A.; Kling, G.W.; Cory, R.M. The role of iron and reactive oxygen species in the production of CO_2 in arctic soil waters. *Geochim. Cosmochim. Acta* **2018**, *224*, 80–95. [CrossRef]
53. Lv, Y.; Liu, J.; Zhu, R.; Zhu, J.; Chen, Q.; Liang, X.; He, H. Photoreductive Dissolution of Iron (Hydr)oxides and Its Geochemical Significance. *ACS Earth Space Chem.* **2022**, *6*, 811–829. [CrossRef]
54. Sipos, P.; Tóth, A.; Kis, V.K.; Balázs, R.; Kovács, I.; Németh, T. Partition of Cd, Cu, Pb and Zn among mineral particles during their sorption in soils. *J. Soils Sediment.* **2019**, *19*, 1775–1787. [CrossRef]
55. Lu, Q.; Wang, J.; Wang, Z.; Xie, L.; Liu, Q.; Zeng, H. Molecular Insights into the Interaction Mechanism Underlying the Aggregation of Humic Acid and Its Adsorption on Clay Minerals. *Environ. Sci. Technol.* **2023**, *57*, 24. [CrossRef]
56. Willemsen, J.A.R.; Bourg, I.C. Molecular dynamics simulation of the adsorption of per- and polyfluoroalkyl substances (PFASs) on smectite clay. *J. Colloid Interface Sci.* **2021**, *585*, 337–346. [CrossRef]
57. Wang, X.; Lee, C. Adsorption and desorption of aliphatic amines, amino acids and acetate by clay minerals and marine sediments. *Mar. Chem.* **1993**, *44*, 1–23. [CrossRef]
58. Huang, X.; Li, Y.; Guggenberger, G.; Kuzyakov, Y.; Liu, B.-F.; Wu, J. Direct evidence for thickening nanoscale organic films at soil biogeochemical interfaces and its relevance to organic matter preservation. *Environ. Sci. Nano* **2020**, *7*, 2747–2758. [CrossRef]
59. Chen, M.D.; Cui, X.Y. Mechanisms and influencing factors of soil organic carbon sequestration by minerals. *Chin. J. Eco-Agric.* **2022**, *30*, 175–183. [CrossRef]
60. Soylak, M.; Akkaya, Y.; Elçi, L. Monitoring trace metal levels in Yozgat-Turkey: Determinations of some metal ions in roadside soils. *Trace Elem. Electrolytes* **2001**, *18*, 176–180.
61. Turkoglu, O.; Saraçoğlu, S.; Soylak, M. Trace metal levels in soil samples from crossroads in Kayseri-Ankara motorway. *Trace Elem. Electrolytes* **2003**, *20*, 225–229.
62. Qiu, L.; Suo, C.; Zhang, N.; Yuan, R.; Chen, H.; Zhou, B. Adsorption of heavy metals by activated carbon: Effect of natural organic matter and regeneration methods of the adsorbent. *Desalination Water Treat.* **2022**, *252*, 148–166. [CrossRef]
63. Gupta, V.K.; Ali, I. Removal of lead and chromium from wastewater using bagasse fly ash—A sugar industry waste. *J. Colloid Interface Sci.* **2004**, *271*, 321–328. [CrossRef] [PubMed]
64. Renu; Agarwal, M.; Singh, K. Heavy metal removal from wastewater using various adsorbents: A review. *J. Water Reuse Desalination* **2016**, *7*, 387–419. [CrossRef]
65. Mikutta, R.; Lorenz, D.; Guggenberger, G.; Haumaier, L.; Freund, A. Properties and reactivity of Fe-organic matter associations formed by coprecipitation versus adsorption: Clues from arsenate batch adsorption. *Geochim. Et Cosmochim. Acta* **2014**, *144*, 258–276. [CrossRef]
66. Sposito, G.; Skipper, N.T.; Sutton, R.; Park, S.; Soper, A.K.; Greathouse, J.A. Surface geochemistry of the clay minerals. *Proc. Natl. Acad. Sci. USA* **1999**, *96*, 3358–3364. [CrossRef]
67. Yang, J.; Liu, J.; Hu, Y.; Rumpel, C.; Bolan, N.; Sparks, D. Molecular-level understanding of malic acid retention mechanisms in ternary kaolinite-Fe(III)-malic acid systems: The importance of Fe speciation. *Chem. Geol.* **2017**, *464*, 69–75. [CrossRef]
68. Benke, M.B.; Mermut, A.R.; Shariatmadari, H. Retention of dissolved organic carbon from vinasse by a tropical soil, kaolinite, and Fe-oxides. *Geoderma* **1999**, *91*, 47–63. [CrossRef]
69. Kaiser, K.; Zech, W. Sorption of dissolved organic nitrogen by acid subsoil horizons and individual mineral phases. *Eur. J. Soil Sci.* **2000**, *51*, 403–411. [CrossRef]
70. Eusterhues, K.; Rumpel, C.; Kleber, M.; Kögel-Knabner, I. Stabilisation of soil organic matter by interactions with minerals as revealed by mineral dissolution and oxidative degradation. *Org. Geochem.* **2003**, *34*, 1591–1600. [CrossRef]
71. Quideau, S.A.; Chadwick, O.A.; Trumbore, S.E.; Johnson-Maynard, J.L.; Graham, R.C.; Anderson, M.A. Vegetation control on soil organic matter dynamics. *Org. Geochem.* **2001**, *32*, 247–252. [CrossRef]
72. Scharpenseel, H.; Becker-Heidmann, P. Shifts in 14C Patterns of Soil Profiles Due to Bomb Carbon, Including Effects of Morphogenetic and Turbation Processes. *Radiocarbon* **1989**, *31*, 627–636. [CrossRef]
73. Balesdent, J. The significance of organic separates to carbon dynamics and its modelling in some cultivated soils. *Eur. J. Soil Sci.* **1996**, *47*, 485–493. [CrossRef]
74. Xu, J.; Sun, Y.; Gao, L.; Cui, X. A review of the factors influencing soil organic carbon stability. *Chin. J. Eco-Agric.* **2018**, *26*, 222–230. [CrossRef]
75. Keuper, F.; Dorrepaal, E.; van Bodegom, P.M.; van Logtestijn, R.; Venhuizen, G.; van Hal, J.; Aerts, R. Experimentally increased nutrient availability at the permafrost thaw front selectively enhances biomass production of deep-rooting subarctic peatland species. *Glob. Change Biol.* **2017**, *23*, 4257–4266. [CrossRef]
76. Wang, H.; Wang, J.; Yang, Y.J.; Chen, L.; Liu, S.R. Effects of throughfall reduction on soil organic carbon concentration and chemical compositions in the Castanopsis hystrix plantation. *J. Beijing For. Univ.* **2022**, *44*, 102–111. [CrossRef]
77. Mathew, S.; Tiwari, D.K.; Tripathi, D. Interaction of carbon nanotubes with plant system: A review. *Carbon Lett.* **2020**, *31*, 167–176. [CrossRef]

78. Li, G.L.; Zhou, C.H.; Fiore, S.; Yu, W.H. Interactions between microorganisms and clay minerals: New insights and broader applications. *Appl. Clay Sci.* **2019**, *177*, 91–113. [CrossRef]
79. Konrad-Schmolke, M.; Halama, R.; Wirth, R.; Thomen, A.; Klitscher, N.; Morales, L.; Schreiber, A.; Wilke, F.D.H. Mineral dissolution and reprecipitation mediated by an amorphous phase. *Nat. Commun.* **2018**, *9*, 1637. [CrossRef]
80. Wang, M.; Zhang, S.; Guo, X.; Xiao, L.; Yang, Y.; Luo, Y.; Mishra, U.; Luo, Z. Responses of soil organic carbon to climate extremes under warming across global biomes. *Nat. Clim. Chang.* **2024**, *14*, 98–105. [CrossRef]
81. Zhang, T.T.; He, D.; Feng, S.; Chen, K. Status and trends of global protected area management—Based on bibliometric analysis. *For. Econ.* **2021**, *43*, 35–59. [CrossRef]
82. Li, J.; Cai, B.Q.; Xu, L. Research on cyberspace security in China based on CiteSpace. *J. Fujian Univ. Technol.* **2018**, *16*, 174–178. [CrossRef]
83. Spears, D. Introduction to Clay Minerals. Chemistry, Origins, Uses and Environmental Significance. *Clay Miner.-Clay Min.* **1993**, *28*, 161–162. [CrossRef]
84. Pinheiro, É.F.M.; de Campos, D.V.B.; de Carvalho Balieiro, F.; Anjos, L.H.C.D.; Pereira, M.G. Tillage systems effects on soil carbon stock and physical fractions of soil organic matter. *Agric. Syst.* **2015**, *132*, 35–39. [CrossRef]
85. Lelde, K.; Gry, L.; Anders, T.; Per, P. Oxidation of a Dimethoxyhydroquinone by Ferrihydrite and Goethite Nanoparticles: Iron Reduction versus Surface Catalysis. *Environ. Sci. Technol.* **2017**, *51*, 9053–9061. [CrossRef]
86. Studenroth, S.; Huber, S.G.; Kotte, K.; Schöler, H.F. Natural abiotic formation of oxalic acid in soils: Results from aromatic model compounds and soil samples. *Environ. Sci. Technol.* **2013**, *47*, 1323–1329. [CrossRef]
87. Stendahl, J.; Berg, B.; Lindahl, B.D. Manganese availability is negatively associated with carbon storage in northern coniferous forest humus layers. *Sci. Rep.* **2017**, *7*, 15487. [CrossRef]
88. Sun, Z.; Qian, X.; Shaaban, M.; Wu, L.; Hu, J.; Hu, R. Effects of iron(III) reduction on organic carbon decomposition in two paddy soils under flooding conditions. *Environ. Sci. Pollut. Res.* **2019**, *26*, 12481–12490. [CrossRef]
89. Gu, Y.; Lensu, A.; Ki, S.; Ojala, A.; Talo, A. Iron and pH Regulating the Photochemical Mineralization of Dissolved Organic Carbon. *ACS Omega* **2017**, *2*, 1905–1914. [CrossRef]
90. Page, S.E.; Logan, J.R.; Cory, R.M.; McNeill, K. Evidence for dissolved organic matter as the primary source and sink of photochemically produced hydroxyl radical in arctic surface waters. *Environ. Sci. Process. Impacts* **2014**, *16*, 807–822. [CrossRef]
91. Chen, C.; Hall, S.J.; Coward, E.; Thompson, A. Iron-mediated organic matter decomposition in humid soils can counteract protection. *Nat. Commun.* **2020**, *11*, 2255. [CrossRef]

MDPI AG
Grosspeteranlage 5
4052 Basel
Switzerland
Tel.: +41 61 683 77 34

Processes Editorial Office
E-mail: processes@mdpi.com
www.mdpi.com/journal/processes

www.ingramcontent.com/pod-product-compliance
Lightning Source LLC
Chambersburg PA
CBHW061544220926
51850CB00005B/248